313

S-193 p.302

SPACE-BASED
RADAR
HANDBOOK

The Artech House Radar Library

David K. Barton, *Series Editor*

Modern Radar System Analysis by David K. Barton

Introduction to Electronic Warfare by D. Curtis Schleher

High Resolution Radar by Donald R. Wehner

RGCALC: Radar Range Detection Software and User's Manual by John E. Fielding and Gary D. Reynolds

Principles and Applications of Millimeter-Wave Radar, Charles E. Brown and Nicholas C. Currie, eds.

Mulitple-Target Tracking with Radar Applications by S. Blackman

Solid-State Radar Transmitters by Edward D. Ostroff, *et al.*

Logarithmic Amplification by Richard Smith Hughes

Radar Propagation at Low Altitudes by M.L. Meeks

Radar Cross Section by Eugene F. Knott, *et al.*

Radar Anti-Jamming Techniques by M.V. Maksimov, *et al.*

Radar System Design and Analsis by S.A. Hovanessian

Monopulse Principles and Techniques by Samuel M. Sherman

High Resolution Radar Imaging by Dean L. Mensa

Radar Detection by J.V. DiFranco and W.L. Rubin

Handbook of Radar Measurement by D.K. Barton and H.R. Ward

Radar Technology, Eli Brookner, ed.

The Scattering of Electromagnetic Waves from Rough Surfaces by Petr Beckmann and Andre Spizzichino

Radar Range-Performance Analysis by Lamont V. Blake

Interference Suppression Techniques for Microwave Antennas and Transmitters by Ernest R. Freeman

Radar Reflectivity of Land Sea by Maurice W. Long

Aspects of Modern Radar, by Eli Brookner, *et al.*

Analog Automatic Control Loops in Radar and EW by R.S. Hughes

Electronic Homing Systems by M.V. Maksimov and G.I. Gorgonov

Principles of Modern Radar Systems by Michel H. Carpentier

Secondary Surveillance Radar by Michael C. Stevens

Multifunction Array Radar by Dale R. Billetter

Radar Reflectivity Measurement: Techniques and Applications, Nicholas C. Currie, ed.

Space-Based Radar Handbook, Leopold J. Cantafio, ed.

SPACE-BASED RADAR HANDBOOK

Leopold J. Cantafio
editor

ARTECH HOUSE

Cantafio, Leopold J.   89-308 CIP
   Space-Based Radar Handbook.

   p.   cm.
   Bibliography: p.
   Includes index.
   1. Space based radar.   I. Cantafio, Leopold J.
TL3285.S65   1989   621.3848—dc19
ISBN   0-89006-281-1:

Copyright © 1989
ARTECH HOUSE, INC.
685 Canton Street
Norwood, MA 02062

All rights reserved. Printed and bound in the United States of America. No part of this book may be reproduced or utilized in any form or by any means, electronic or mechanical, including photocopying, recording, or by any information storage and retrieval system, without permission in writing from the publisher.

International Standard Book Number: 0-89006-281-1
Library of Congress Catalog Card Number: 89-308

10   9   8   7   6   5   4   3   2   1

CONTENTS

Preface		xiii
Chapter 1	SPACE-BASED RADAR SYSTEMS by L.J. Cantafio	1
1.1	Introduction	1
1.2	SBR Types	2
	1.2.1 Type I SBR	2
	1.2.2 Type II SBR	2
	1.2.3 Type III SBR	3
1.3	Systems Considerations	3
	1.3.1 Orbit Selection	3
	1.3.2 Rationale	6
	1.3.3 Radiation Environment Effects	10
1.4	SBR System Descriptions	10
	1.4.1 STS Rendezvous Radar	10
	1.4.2 SEASAT-A System	14
	1.4.3 Shuttle Imaging Radar	15
	1.4.4 GEOS-3	15
	1.4.5 USSR Cosmos 1500 Side-Looking Radar	19
	1.4.6 USSR Synthetic Aperture Radar Polyus-V	21
	1.4.7 Apollo Lunar Sounder Radar	23
1.5	SBR Systems of the Future	26
	1.5.1 Future Rendezvous Radar Missions	26
	1.5.2 SBR in Future Remote Sensing Missions	28
	1.5.3 SBR Concept for Global Air Traffic Surveillance	32
	1.5.4 Military SBR Systems	33
	1.5.5 Military SBR Applications	35
	1.5.6 SBR for the Proposed United Nations International Satellite Monitoring Agency	36
1.6	SBR Issues	38
	1.6.1 SBR System Costs	39

	1.6.2	Survivability and Vulnerability	41
	1.6.3	Nuclear Prime Power	41
	1.6.4	SBR System Calibration	42
	1.6.5.	Clutter and Interference	42
	1.6.6	Launcher Vehicle Capabilities	43
References			44

Chapter 2   ORBITAL CONSIDERATIONS FOR SPACE-BASED RADAR by V.A. Chobotov   47
2.1   Kepler's Laws   47
2.2   Orbit Equations   47
2.3   Time of Flight   50
2.4   Coordinate Systems   51
    2.4.1   ECI Reference Frame   51
    2.4.2   Latitude-Longitude Coordinate Frame   52
    2.4.3   Azimuth-Elevation Coordinate Frame   53
2.5   Orbital Elements   53
    2.5.1   Summary of Orbit Equations   56
2.6   Gravitational Perturbations   58
    2.6.1   Earth's Oblateness Effects   58
2.7   Orbital Systems   62
    2.7.1   Launch Window Considerations   63
    2.7.2   Ground Trace Considerations   66
    2.7.3   Highly Eccentric, Critically Inclined $Q = 2$ Orbits (Molniya)   68
2.8   Satellite-Based Radar Systems   71
    2.8.1   Global Coverage   71
    2.8.2   Continuous Global Coverage—Polar Constellations   72
    2.8.3   Inclined Walker Constellations   76
References   81

Chapter 3   IONOSPHERIC ENVIRONMENT AND EFFECTS ON SPACE-BASED RADAR DETECTION by D.L. Knepp and J.T. Reinking   83
3.1   Introduction   83
    3.1.1   Ionization Irregularity Description   85
3.2   Received Signal Description   90
    3.2.1   Received Signal First-Order Statistics   91
    3.2.2   Target Statistics   91
    3.2.3   Propagation Channel Statistics   92
    3.2.4   Propagation Channel Coherence   94
3.3   Radar System Characteristics   96
3.4   Probability of Detection   96

		3.4.1	Receiver Model	96

- 3.4.1 Receiver Model — 96
- 3.4.2 Nonfluctuating Target — 98
- 3.4.3 Single-Burst Detection — 99
- 3.4.4 M out of N Detection — 100
- 3.4.5 Independent Bursts — 100
- 3.4.6 Constant Propagation Channel — 101
- 3.4.7 Noncoherent Integration—Constant Propagation Channel — 101
- 3.4.8 Numerical Methods — 102
- 3.4.9 General Correlation of Burst Returns — 102
- 3.5 Results — 104
  - 3.5.1 Type I Fading — 109
  - 3.5.2 General Burst Coherence — 109
  - 3.5.3 M Other than 1 — 112
  - 3.5.4 Noncoherent Integration — 113
- 3.6 Conclusions — 115
- References — 118

Chapter 4 SAR IN SPACE—THE THEORY, DESIGN, ENGINEERING AND APPLICATION OF A SPACE-BASED SAR SYSTEM by S.W. McCandless — 121

- 4.1 Space-Based SAR Design Principles — 122
  - 4.1.1 Basic Principles of Aperture Synthesis — 122
  - 4.1.2 Ambiguity Relationships — 127
  - 4.1.3 The Radar Equations Interpreted for SAR — 132
- 4.2 End-to-End System Description — 135
  - 4.2.1 System Design and Technology Considerations — 136
  - 4.2.2 System Implementations Including ISAR — 139
  - 4.2.3 Image Processing — 143
- 4.3 An Assessment of Pacing Technologies — 149
  - 4.3.1 Nonlimiting Technologies — 149
  - 4.3.2 Limiting Technologies — 152
- References — 165

Chapter 5 BISTATIC RADAR IN SPACE by P. Hartl and H.M. Braun — 167

- 5.1 Comparison of Bistatic and Monostatic Systems — 167
- 5.2 Bistatic Systems — 167
  - 5.2.1 Low Earth Orbit (LEO) Systems — 167
  - 5.2.2 Systems with Geostationary (GEO) Transmitter — 172
  - 5.2.3 Parasitic Radar Systems — 173
- 5.3 Performance Considerations — 176
  - 5.3.1 Bistatic Geometry — 176
  - 5.3.2 Spatial Performance — 180
  - 5.3.3 Radiometric Performance — 186

5.4	Examples of Bistatic Radars	189
	5.4.1 Bistatic Parasitic Radar (BIPAR)	189
	5.4.2 Bistatic Synthetic Aperture Radar (BISAR)	191
	5.4.3 BISAR with a Geostationary Transmitter	193
5.5	Summary	194
References		195

Chapter 6   RENDEZVOUS RADAR by J.W. Locke and L.J. Cantafio   197

6.1	Rendezvous Radar Missions	197
6.2	Space Shuttle Rendezvous Radar	200
6.3	Future Rendezvous Radar Missions	204
6.4	OMV Rendezvous Radar	206
6.5	Earth Clutter Effects	211
	6.5.1 Radar Parameters	211
	6.5.2 Radar-Earth Geometry	214
	6.5.3 Clutter Area	216
	6.5.4 Clutter-Target Range Ratio	220
	6.5.5 Clutter Frequency Spread	221
	6.5.6 Clutter Reflectivity Coefficient	223
	6.5.7 Summary of Earth Clutter Effects	224
6.6	Background Noise Sources	225
References		226

Chapter 7   RADAR ALTIMETERS FOR SPACE VEHICLES
by T.J. Lund   229

7.1	Principles of Space-Based Radar Altimeters	229
	7.1.1 Properties of Waveforms for Radar Altimeters	230
	7.1.2 Range Equation for Radar Altimeters	233
7.2	Description and Performance of Radar Altimeter Types	237
	7.2.1 Short-Pulse Radar Altimeter	237
	7.2.2 Pulse Compression and High Resolution Radar Altimeters	242
	7.2.3 Linearly Frequency Modulated Radar Altimeter	245
	7.2.4 Phase-Shift Keyed Radar Altimeter	252
7.3	Description of Radar Altimeters Used in Space	258
	7.3.1 Radar Altimeter for the Saturn I Launch Vehicle	258
	7.3.2 Radar Altimeter for the Surveyor Lunar Lander	260
	7.3.3 Radar Altimeter for the Apollo Lunar Module	262
	7.3.4 Radar Altimeter for Skylab S-193	264
	7.3.5 Radar Altimeter for the Viking Mars Lander	265
	7.3.6 Radar Altimeter for the GEOS-C Spacecraft	268
	7.3.7 Radar Altimeter for the SEASAT-A Ocean Dynamics Satellite	272

		7.3.8	Radar Altimeter for the Pioneer Venus Orbiter	274
		7.3.9	Radar Altimeter for the GEOSAT Satellite	276
	7.4	Radar Altimeters of the Future		277
References				279

Chapter 8 SCATTEROMETERS AND OTHER MODEST-RESOLUTION SYSTEMS by R.K. Moore — 281

- 8.1 Introduction — 281
- 8.2 Fundamentals of Measurement — 282
  - 8.2.1 Resolution Techniques — 282
  - 8.2.2 Amplitude Measurement (Scatterometry) — 289
- 8.3 Ocean-Surface Wind-Vector Measurement — 292
- 8.4 Scatterometer Systems — 299
  - 8.4.1 Pencil-Beam Systems — 299
  - 8.4.2 Fan-Beam Systems — 305
- 8.5 Real-Aperture Imaging Radars in Space — 309
- 8.6 Measuring Winds Aloft from Space — 312
- 8.7 Summary — 315
- References — 316

Chapter 9 THERMAL CONTROL FOR SPACE-BASED RADAR by L.M. Herold and M.S. Busby — 319

- 9.1 Introduction — 319
- 9.2 Thermal Design Requirements — 320
  - 9.2.1 Radar Equipment Requirements — 320
  - 9.2.2 Mission Parameters — 325
  - 9.2.3 Heat Dissipation of the Electrical Power System — 334
- 9.3 Trades to Minimize Thermal Management System Weight — 338
- 9.4 Thermal Control Components — 340
  - 9.4.1 Thermal Management Systems — 341
  - 9.4.2 Thermal Storage — 347
- References — 348

Chapter 10 RADAR CROSS SECTION (RCS) OF SATELLITES AND OTHER SPACE-BASED TARGETS by J.W. Curtis — 349

- 10.1 Introduction — 349
- 10.2 Space Targets — 354
- 10.3 Airborne Targets — 362
- 10.4 Other Targets — 366
- 10.5 RCS Measurement — 370
- 10.6 Summary — 371
- References — 371

Chapter 11 SBR CLUTTER AND INTERFERENCE by G.A. Andrews and K. Gerlach — 373

11.1	Introduction	373
11.2	Characterization of SBR Clutter	377
	11.2.1 Amplitude of Clutter	377
	11.2.2 Clutter Spectral Distributions	392
11.3	Unique Features of SBR Clutter	395
	11.3.1 Antenna Pattern Considerations	395
	11.3.2 Range Resolution and Ambiguities	402
	11.3.3 Platform Motion Effects	405
11.4	Clutter-Rejection Techniques	413
	11.4.1 Radar Doppler Processors	413
	11.4.2 Motion Compensation Techniques	425
	11.4.3 Doppler Beam-Sharpening	441
11.5	Space-Based Interference Considerations	442
11.6	Interference/Rejection Techniques	447
11.7	Main and Auxiliary Antenna Considerations	451
	11.7.1 Main Antenna	451
	11.7.2 Auxiliary Antennas	452
	11.7.3 Bandwidth-Aperture Dispersion	459
	11.7.4 Phase Center Matching	463
	11.7.5 Other Issues	464
1..8	Sidelobe Cancellers	465
	11.8.1 SLC Algorithm	465
	11.8.2 Implementation	467
	11.8.3 Limitations	468
11.9	Interactions of Clutter and Interference Processing	472
References		475

Chapter 12	SPACE ANTENNA TECHNOLOGY by L.J. Cantafio	481
12.1	Requirements	481
	12.1.1 Requirements for Type I SBR Antennas	482
	12.1.2 Requirements for Type II SBR Antennas	483
	12.1.3 Requirements for Type III SBR Antennas	484
	12.1.4 Lens Antenna Requirements	489
	12.1.5 Other SBR Antenna Requirements	494
12.2	Selected Space-Based Antenna Designs and Concepts	498
	12.2.1 US Space Deployable Antennas	499
	12.2.2 USSR Space Deployable Antennas	521
12.3	Testing Space Antennas	524
12.4	Future of Space Antenna Technology	524
References		526

Chapter 14 ON-BOARD RADAR SIGNAL PROCESSORS by E.E. Schwartzlander, Jr. ... 531

14.1	Introduction	531
14.2	Generic Signal Processors	532
	14.2.1 Computer Architectures	533
	14.2.2 Array Processors	537
14.3	VLSI Implementation of a Generic Signal Processor	539
	14.3.1 VLSI Components	539
	14.3.2 Processor Implementation	544
14.4	Special-Purpose Signal Processor Development	548
	14.4.1 Processor Development Approach	549
	14.4.2 FFT Processor Implementation	549
	14.4.3 The Delay Commutator Circuit	551
	14.4.4 Arithmetic Realization	554
14.5	Digital Radar Beamformer Case Study	556
	14.5.1 The Chip Development Process	556
	14.5.2 Digital Beam Forming	557
	14.5.3 Data Rates	559
	14.5.4 Algorithm Selection	560
	14.5.5 VLSI Implementation	563
	14.5.6 The Payoff	564
References		565
Chapter 15	PRIME POWER SYSTEMS IN SPACE by J.E. Boretz	567
15.1	Space Power System Classifications	567
	15.1.1 Introduction	567
	15.1.2 System Selection	567
	15.1.3 Space Power Configurations	568
15.2	Solar Array-Battery Systems (S/A-B)	569
	15.2.1 Solar Array Performance	569
	15.2.2 Battery System Characteristics	572
15.3	Organic Rankine Cycle (System No. 1)	579
15.4	Closed Brayton Cycle (System No. 2)	585
15.5	Free Piston Stirling Engine Power Systems (System No. 3)	593
15.6	Supercritical Cycle Power Systems (System No. 4)	602
15.7	Potassium Rankine Cycle (System No. 5)	605
15.8	Alkali Metal Thermoelectric Conversion (AMTEC) System (System No. 6)	610
15.9	Thermoelectric Conversion Cycles (Systems Nos. 7, 8, 9)	611
15.10	Thermionic Conversion Systems (System No. 10)	616
	15.10.1 Candidate Thermionic Conversion Systems	621
References		636

Chapter 16	SPACE-BASED RADAR STRUCTURES by E. Kovalcik	637
16.1	General Requirements for SBR Structures	637
16.2	Areal Structural Concepts	640
	16.2.1 Solid-Surface Construction	640
	16.2.2 Mesh Surface Construction	641
	16.2.3 Single-Piece Solid-Surface Reflector Concepts	641
	16.2.4 Deployable Solid-Surface Reflector Concepts	643
	16.2.5 Deployable Mesh Concepts	652
16.3	Linear Deployable Structure Concepts	659
	16.3.1 Lattice Mast Booms	661
	16.3.2 Folding Articulated Square Truss Mast (FASTMAST)	663
	16.3.3 Metallic Strip Booms	665
	16.3.4 Folding Beam Concepts	666
16.4	Erectable Structural Concepts	666
16.5	Structural Design Considerations for SBR Antennas	667
16.6	Material Selection for SBR Structures	669
References		674
Index		677

PREFACE

We could state that God created radar when he endowed the bat and porpoise with the ability to transmit and receive signals that ultrasonically detect the presence of objects at long range. Much later in time, L. Spallanzani performed a study of bats and reported his findings in 1793. Still later, in 1922, Guglielmo Marconi spoke to the IRE and urged the use of short waves for radio detection. He also described tests in which the detections of objects several miles away were obtained. In autumn of that same year, Taylor and Young at NRL detected wooden ships by using continuous wave transmission at a wavelength of five meters. Then, in 1925, Breit and Tuve used pulsed radar to measure the height of the ionosphere. Subsequently, many engineers and scientists developed sophisticated radars for ground, airborne, and shipborne platforms. The newest platform for radar became a satellite in space when during the mid-1960s the Gemini spacecraft used a radar to complete a rendezvous with another satellite. Since then radars in space have been used for altimetry, ocean observation, remote sensing, mapping, and navigation. The only limitations to the use of radar in space will be cost and the creativity of the space-based radar (SBR) system engineer.

I hope that this SBR handbook will be a useful tool in reducing the technology to practice, for, as Leonardo da Vinci said, "Practice following theory, since practice without theory is like a ship without a rudder, withering in water and not knowing where it is going." This book is the first to treat the general subject of SBR, but it will not be the last. Each of the contributing authors has been urged by their editor to produce a book on each of the specialized areas treated.

The editor acknowledges each of the following: the management of TRW and, in particular, Dr. L.A. Hromas, manager of the System Technology Laboratory for providing the environment in which this book could be prepared; (2) Ms. Emogene Jackson for typing the drafts of Chapter 1 and other parts of this book; my wife, Norma, for her patience, inspiration, and understanding during the many hours required in the preparation of the manuscript; and all of the contributing authors for their excellent work.

Leopold J. Cantafio
Palos Verdes Peninsula, California
January 1989

Apologia

The original plan for this book called for 16 chapters, including one devoted to transmitting-receiving modules. However, due to circumstances beyond the control of the Editor and the Publisher, the proposed Chapter 13, titled T/R Modules, will not appear in this volume. For this omission, we wish to offer our sincere apologies.

Chapter 1
SPACE-BASED RADAR SYSTEMS
Leopold J. Cantafio
TRW

1.1 INTRODUCTION

Significant developments have been made in *space-based radar* (SBR) systems and technology since the first satellite with a radar payload was launched (the Gemini radar). A new rendezvous radar was developed for the space shuttle and has become operational. The unmanned *orbital maneuvering vehicle* (OMV) will use a new low-cost rendezvous radar that is expected to be operational during the early 1990s. *Synthetic aperture radar* (SAR) types of SBR have been used by the US and USSR for earth and planetary exploration. Altimeters have been used on many satellites. The technology of SBR subsystems has been developed in the areas of antennas, transmitters, receivers, solid-state transmit-receive or transceiver (T/R) modules, signal processors, and prime power. This chapter will serve as an introduction to our SBR handbook and will review SBR systems with the intention of providing a description that is not too sketchy to be substantive. Consequently, selected systems and technology will be discussed. Maximum use is made of references to subsequent chapters of this book containing illustrations and tabulated data that can be useful to professionals who want to obtain rapid answers to SBR problems. We will describe several SBR systems for rendezvous and earth and planetary exploration missions. We will discuss systems considerations, such as the space environment, orbit selection, rationale for SBR, and critical issues. Many topics must be omitted, such as electronic countermeasures and sensitive military system developments. This chapter and this book should be considered as a status report on the new frontier for radar systems.

1.2 SBR TYPES

There are three types of radar that have been and can be based in space. Space-based radars that are typical of Type I are the small, short-range rendezvous radars such as those used on the Shuttle, Apollo, and Gemini programs [1–4]. Type II SBR includes the earth and planetary resources radars used for mapping, scatterometers, altimeters, and subsurface probing [5–9]. Side-looking SAR techniques are typical of the mapping radars such as those used on the SEASAT satellite (June 1978) and the Space Shuttle (November 1981) with the Shuttle Imaging Radar-A (SIR-A.) The Type III SBR includes the large phased array surveillance radar proposed for multimission defense, air traffic control, and disarmament functions [10–14].

1.2.1 Type I SBR

The *rendezvous radar* is the tracking sensor for a guidance system. The Gemini and Apollo programs demonstrated the first operational experience with the rendezvous maneuver. The successful performance of the rendezvous radars in these programs effectively opened the door to many possible missions that may be performed in space. The Ku-band Integrated Radar and Communications Subsystem (IRACS), designed for the space shuttle's orbiter vehicle, demonstrated rendezvous, satellite retrieval, and station-keeping missions. The maiden voyage for this radar was aboard Challenger STS-7 on June 22, 1983 [15]. In the foreseeable future, unmanned vehicles such as the Orbital Manuevering Vehicle (OMV) will be conducting the majority of the rendezvous missions. This US vehicle is under development for NASA [16] by TRW. A new rendezvous radar is being built for the OMV. A complete treatment of the rendezvous radars, IRACS, and OMV is given in Chapter 6 of this text [17].

1.2.2 Type II SBR

Remote sensing of the earth from space began in 1960 with the launch of the first Television and Infrared Observation Satellite (TIROS) for weather. Remote sensing of the earth from space by radar began in 1975 with the launch of the GEOS-C by NASA [18–20], and continued with the SEASAT in 1978, the SIR-A on the shuttle in 1981, and the SIR-B in 1984 on the shuttle STS-17 [21]. The requirements and design details for the TYPE II SBR are discussed in Chapters 4, 7, and 8 of this book [22–24] (see also Meneghini and Kozu [25]). Selected details of the SEASAT, GEOS, and Apollo Lunar Sounder radars are given in Section 1.4. Type II SBRs measure many parameters, including altitude, reflectivity (both surface and subsurface), terrain features, ocean wave height, surface winds over water, ocean dynamics, sea ice, rainfall rate, and cloud height.

1.2.3 Type III SBR

This type of SBR represents the most significant challenge to the radar systems engineer. The mission is the surveillance of large areas and the detection or tracking of many targets. Before the design of a Type III SBR can begin, the requirements for the surveillance radar systems must be specified. These requirements should include, but would not be limited to [26], (a) target radar cross section (RCS) model, (b) target velocity and acceleration (maximum), (c) number of targets, (d) probability of detection, (e) probability of false alarm and false alarm time, (f) track accuracy, (g) minimum target spacing, (h) designation error, (i) warning time, (j) length of detection fence, (k) revisit time, (l) clutter model, and (m) weather model. With these requirements as a minimum input to the design study, the orbit selection may begin and parameter trade-offs can be made. The influence of the space environment, interference, and clutter must be considered. The shuttle or Space Transportation System (STS) will be the launch vehicle for any future SBR, and therefore the capabilities of the STS ought to be examined. The rationale for large surveillance radars in space should also be considered.

Target characteristics and requirements for its coverage, track data rate and revisit rate are important parameters. The radar subclutter visibility capability, antenna size, scan rate, and grazing angle limitations also determine the orbit selected for the SBR. The space environment can determine the orbit selected if the natural radiation lifetime dosage that the SBR electronics receives is too large. Finally, there is the requirement to use the lowest possible number of satellites to minimize the total system cost.

1.3 SYSTEMS CONSIDERATIONS

1.3.1 Orbit Selection

There are many factors that contribute to the selection of the orbit to be used for each type of SBR and particularly a large surveillance SBR. The orbital parameters of period, altitude, and velocity are the first consideration. The velocity for a satellite in a circular orbit around the earth is given by [27]:

$$V_c = \sqrt{\frac{\mu}{r}} \quad (1.1)$$

where r is the distance of the satellite from the center of the earth, and μ is the product of the universal gravitational constant and the mass of the earth. The period of an earth satellite is given by [27]:

$$T = \frac{2\pi\mu}{\sqrt{V_a^3 V_p^3}} \qquad (1.2)$$

where V_a is the velocity of the satellite at apogee and V_p is the velocity of the satellite at perigee. For a circular orbit, $V_a = V_p$, and the period of a circular orbiting satellite is

$$T_c = \frac{2\pi\mu}{V_c^3} \qquad (1.3)$$

Table 1.1 shows selected calculations of circular orbit velocity and the period when the radius of the earth is 20.903 $(10)^6$ ft (6.371×10^6 m), μ is 1.4069 $(10)^{16}$ ft^3 s^2 (3.9839×10^{14} m^3/s^2), and one nautical mile 6076.1 ft (1852 m).

Table 1.1
Selected Orbital Parameters

Altitude		Velocity		Period
(nmi)	(km)	(ft/s)	(m/s)	(min)
99	183	25,587	7799	88
414	767	24,520	7474	100
912	1689	23,074	7034	120
2262	4189	20,157	6144	180
5612	10,393	15,999	4876	360
19,369	35,871	10,079	3072	1440

Many studies concerning the design of satellite constellations for optimal coverage have been reported [28–33]. Luders and Ginsberg [28] describe an analytical solution to the problem of achieving continuous coverage of latitudinally bounded zones of the globe. Emara and Leondes [29] solved the problem of simultaneous observations by at least four satellites using a constellation of the minimum number of satellites. Ballard [30] extended earlier work by Walker [31] and analyzed rosette constellations that provided the largest possible great-circle range between an observer anywhere on the earth's surface and the nearest sub-satellite point. Single, double, triple, and quadruple visibility were provided by various constellations. Beste [32] designed satellite constellations that provided single and triple continuous coverage by the minimum number of satellites. All of these studies determined coverage for satellites with sensors that observe only angles around the nadir. Electro-optical sensors and mapping radars typically provide this coverage. However, these studies do not have results for SBR surveillance sensors that must detect targets in clutter. These sensors typically have a "nadir hole" extending 20 to 30° from the nadir in which the signal-to-clutter ratio (SCR) is too large for reliable detection. This is shown in Figure 1.1 for a 50° maximum grazing angle and a 3° minimum grazing angle. The minimum grazing angle is a limit set by the atmospheric attenuation allocated in the SBR loss budget and the refraction angle error. To illustrate the different results that can be obtained, consider a requirement to provide continuous coverage of the earth from an orbital altitude of 10,371 km (5600 nmi). For a single sensor on each satellite with no grazing angle limitations, a constellation of six satellites can provide the required continuous coverage from polar orbits. The satellites would be equally distributed in two orbital planes, using the study results given by Harney [33]. However, if the sensor in the SBR were limited to grazing angles between 3 and 60°, the required coverage could be provided by a constellation of 10 satellites. This constellation consists of one satellite in each of 10 equally spaced orbital planes at an inclination of 49.4°, resembling the Walker 10/10/8 constellation [31]. If the grazing angles extend between 3 and 70°, a 14-satellite constellation in a Walker 14/14/12 configuration provides a continuous, global, twofold coverage. The inclination angle of each orbital plane is 49.4°. Figure 1.2 shows a "snapshot" of the constellation of satellites. A complete discussion of the orbital considerations for SBR is given in Chapter 2 [34].

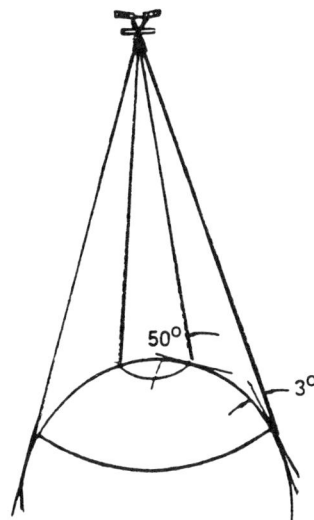

Figure 1.1 Illustration of SBR coverage and NADIR hole.

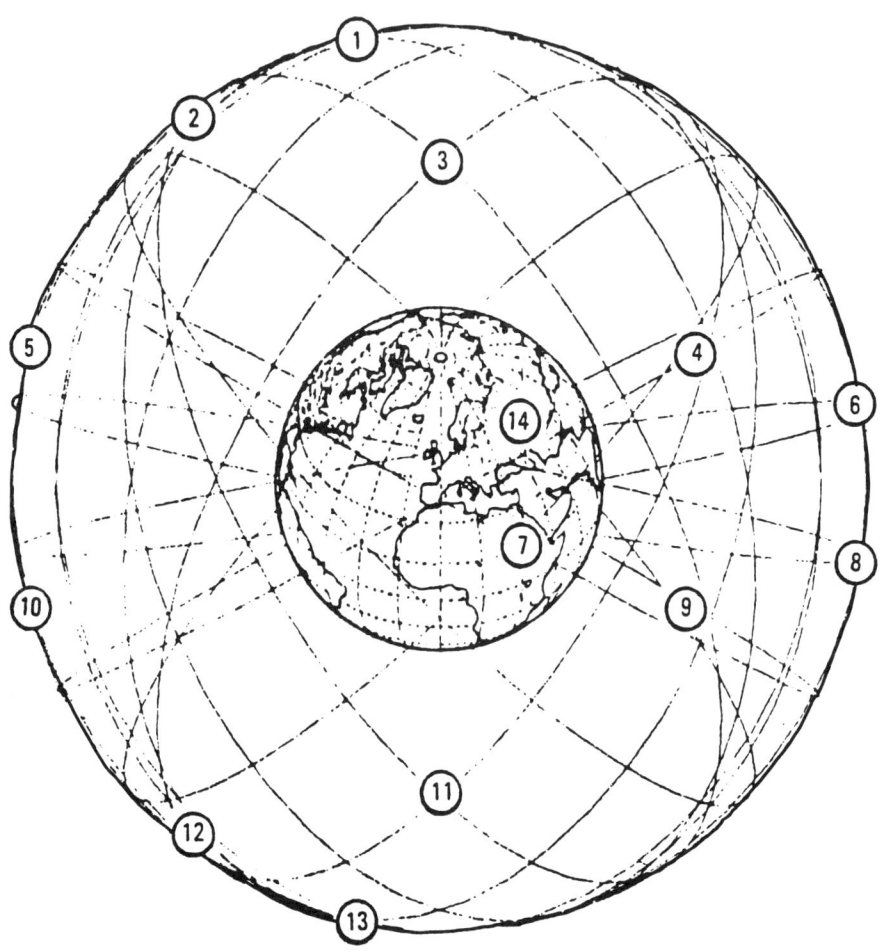

Figure 1.2 Snapshot of satellite constellation [12].

1.3.2 Rationale

When we consider sensors that are required for missions such as space, ocean, air, and missile defense systems, this suggests the use of SBRs. The advantages of such radars deployed in space compared with ground-based radars are described below.

(a) Coverage in both space and time is limited only by the orbit selected and the number of satellites. Large scale continuous observation can be obtained as shown in Figure 1.3 and 1.4 [33]. In Figure 1.3, the required numbers of vehicles

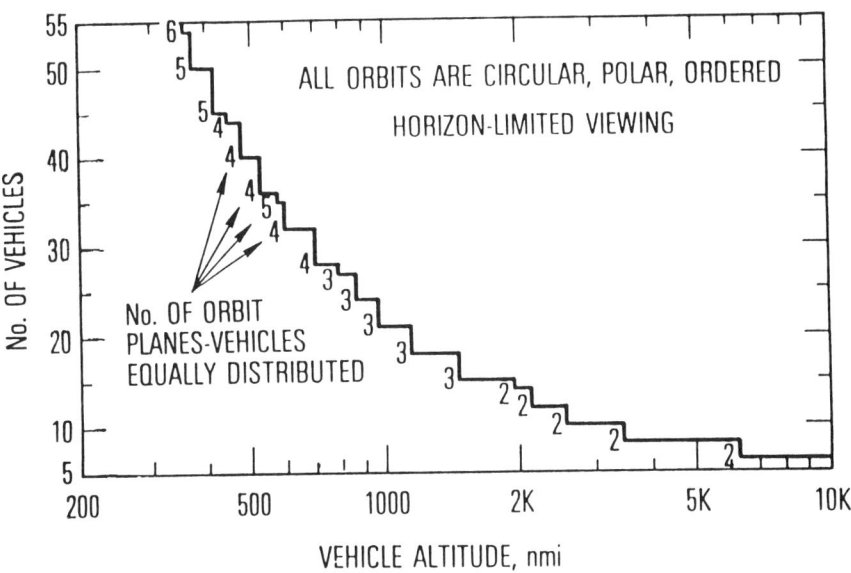

Figure 1.3 Global coverage by polar orbits [33].

are shown as well as the number of orbital planes in which they are distributed to provide continuous coverage of the entire earth's surface from circular polar orbits. We can see that six vehicles in two orbital planes may be used for vehicle altitudes greater than about 6000 nmi. There is no nadir hole in the satellite coverage. Figure 1.4 is an illustration of the special case of equatorial orbits and the number of vehicles required for continuous coverage. This situation is limited to the use of wide swaths that extend up to the specific latitudes indicated. Thus, four vehicles can cover a 60° swath when the vehicles are at altitudes greater than about 6000 nmi. Temporal coverage is illustrated in Figure 1.5, which shows the maximum time for viewing ground objects from a space vehicle if the objects are tracked [33]. Therefore, a ground object can be observed for more than 7000 s when the orbital altitude is 6000 nmi.

(b) The SBR can perform multiple missions when it uses an electronic scanning antenna, as shown in Figure 1.6. For example, a system of radar satellites can (1) search a fence formed completely around the continental United States (CONUS) to detect bombers at a distance from the coast [36, 37]; (2) search a fence over the poles to detect intercontinental ballistic missiles (ICBMs) before they can be detected by the Ballistic Missile Early Warning System (BMEWS); (3) monitor potential sites for space launches from any foreign country; (4) perform surveillance of ocean areas; (5) search a sea-launched ballistic missile (SLBM)

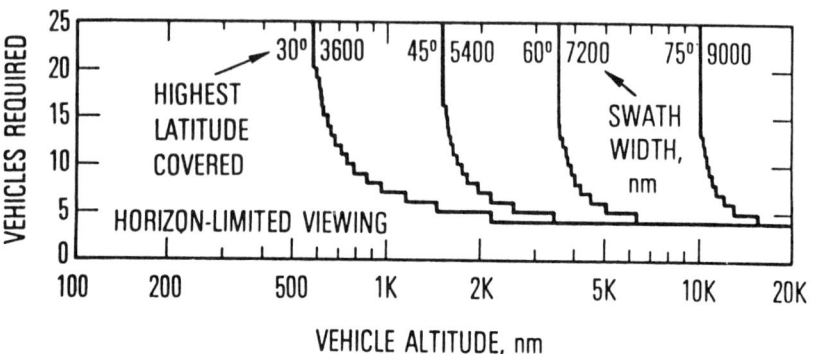

Figure 1.4 Zonal coverage by equatorial orbits [33].

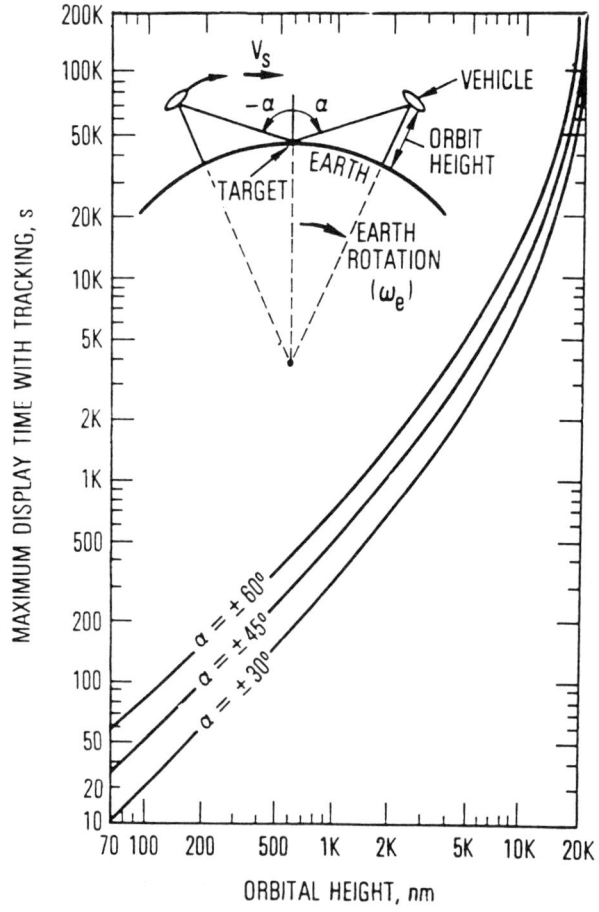

Figure 1.5 Maximum time for viewing ground objects [33].

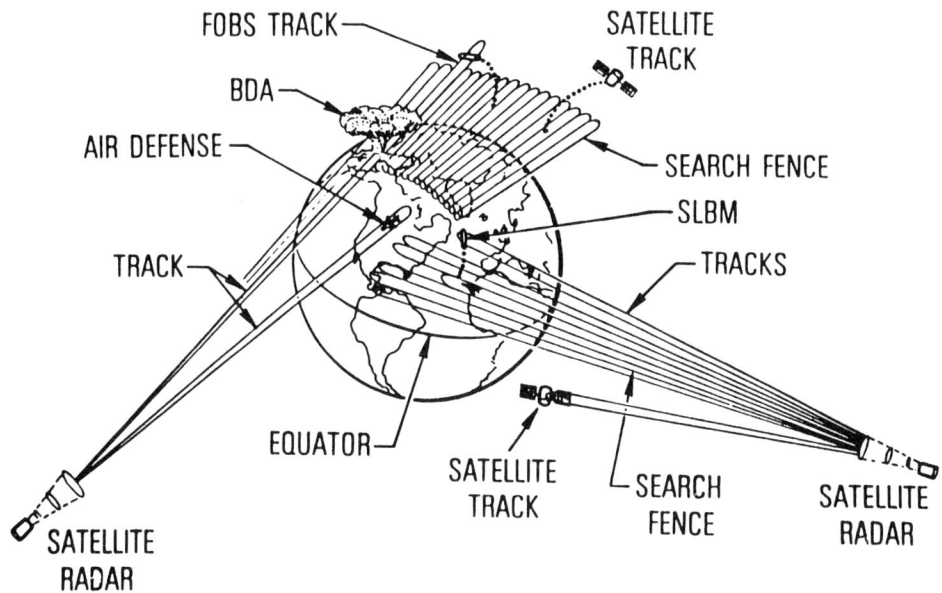

Figure 1.6 Multimission SBR system [26].

detection fence; and (6) detect objects in space that appear to be threats to US synchronous satellites. The number of missions is limited only by the weight and prime power available, but even these limitations can be relaxed when the Space Shuttle is the planned launch vehicle. Therefore, the only real limitations are technology and cost.

(c) Atmospheric propagation problems can be minimized by proper selection of operating frequencies and geometry.

(d) No overseas stations are required if data are read out to CONUS via relay satellites. Hence, the SBR system allows the US to be politically independent, and the loss of tracking stations in a foreign country can have no effect on US system capabilities.

The factors that affect the pace of development of large radar systems in space are given in the two paragraphs to follow.

(a) The technologies of large antenna structures, large phased arrays, large weights, and large prime power systems in space are considered to be in their early stages. A radar clutter model is still to be determined, and *electronic countercountermeasures* (ECCM) techniques are expected to offer technological challenges to the SBR systems engineer.

(b) The funds that can be reasonably spent on a space-based multimission radar system are an important factor to be determined. Even with the use of the

Space Shuttle to reduce the cost per pound of payload into orbit, large investments are expected for the SBR system.

1.3.3 Radiation Environment Effects

Space-based radars can encounter particle radiation in space that may be due to both natural phenomena and nuclear detonations. The satellite must be designed to operate for a reasonable lifetime in the natural space environment, which is a function of orbital altitude. When operating in the midaltitude orbits, exposure to the earth's Van Allen belts will be predictable, and the effect on the radar electronics will be a function of the inherent hardness level of the components and the shielding used. Figure 1.7 shows the trapped radiation data [35], which describes the proton and electron flux that has been measured as a function of altitude. Figure 1.8 [38, 39] shows the total five-year dose in rad (Si) that satellites in orbits between 350 and 6500 nmi altitudes will experience as a function of the aluminum shielding used. Current technology in integrated circuit hardening should produce a total dose hardness of about $5(10)^5$ rad (Si) for devices suitable for the SBR T/R modules. This hardness level is adequate for SBR deployments in many of the candidate orbits with a mission life in the natural environment of several years. A hardness of $5(10)^6$ rad (Si) which may be achievable is required for five-year mission life. Survival of a saturated nuclear environment typical of a high altitude nuclear burst, requires a hardness of 1 to $5(10)^7$ rad (Si), depending on the specific orbit. The development and consistent fabrication of devices that are this hard is relatively uncertain.

1.4 SBR SYSTEM DESCRIPTIONS

Type I and Type II space-based radars have been deployed by the US and the USSR. This section describes some of these SBR systems including the shuttle's rendezvous radar, the SEASAT-A SAR, the Shuttle Imaging Radar (SIR), the GEOS-C altimeter, the Cosmos-1500 side-looking radar, the Polyus-V SAR, and the Apollo Lunar Sounder Radar.

1.4.1 STS Rendezvous Radar [1, 17]

The Integrated Radar and Communications Subsystem (IRACS) was developed by Hughes Aircraft Company for use on the Space Transportation Systems. The IRACS is a coherent, range-gated, Ku-band, pulsed doppler radar that searches for, acquires, and tracks other orbiting objects and provides the spatial measurement data needed to perform rapid and efficient rendezvous with those objects. Table 1.2 gives the characteristics of the IRACS radar.

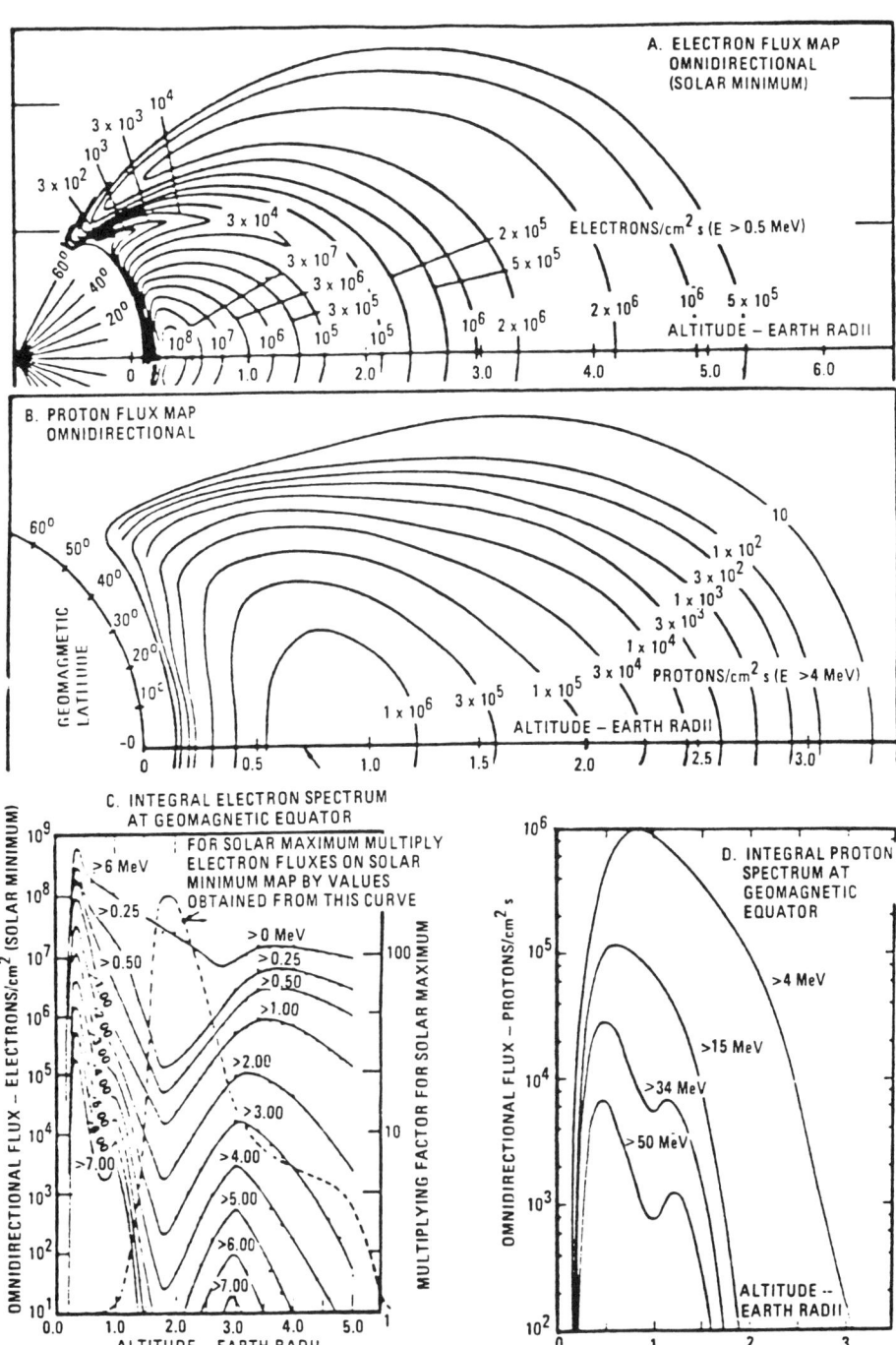

Figure 1.7 Trapped Radiation Data [35].

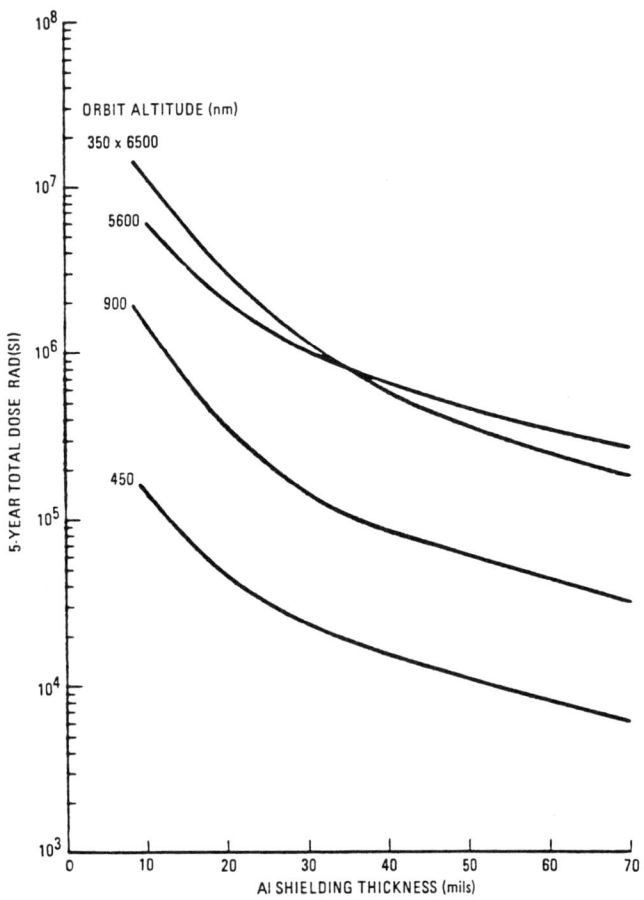

Figure 1.8 Total dose *versus* shielding thickness for five-year mission [38, 39].

Table 1.2
IRACS Radar Characteristics [1, 17]

Detection Performance	
P_d:	99%
Target:	1 m^2; Swerling I
Range:	12 nmi
False-Alarm Rate:	1/h
Search Scan:	± 30° cone

Table 1.2 cont'd.

Track Performance	
Angle Accuracy (3Σ):	8 mrad
Angle Rate (3Σ):	0.14 mrad/s
Range Accuracy (3Σ):	80 ft; $R < 1.3$ nmi
	1% of R; $1.3 < R < 4.9$ nmi
	300 ft; $4.9 < R < 12$ nmi
Range Rate Accuracy (3Σ):	1 ft/s; $R < 10$ nmi
System Parameters	
RF:	13.75–14.02 GHz
PRF:	0.3, 3, 7 kHz
Pulsewidth:	0.122, 4.15, 8.3, 16.6, 33.2, 66.4 μs
System Noise Temperature:	≈1585 K
Antenna	
Type:	Parabola; prime-focus feed
Diameter:	36 in.
Depth:	12.5 in. overall
Gain:	38.4 dB at 13.8 GHz
Beamwidth:	1.68°
Polarization:	Linear
Transmitter	
Type:	TWT
Peak Power:	50 W
Gain:	≈44 dB
Receiver	
Type:	Single-channel monopulse
Noise Figure:	<5 dB; GaAs FET LNA
System-Physical	
Deployed Assembly Weight:	135 lbs
Radar Processor Weight:	31 lbs
Electronics Volume:	≈3 ft^3
Prime Power:	460 W

The IRACS performs both radar and communication functions for the STS. In the pulsed doppler radar mode, IRACS performs the rendezvous function as described. In the communication mode, IRACS searches for, acquires, and tracks the relay satellites of the Tracking and Data Relay Satellite System (TDRSS) to provide two-way communication between the Space Shuttle and ground tracking stations. A detailed description of the IRACS radar and its performance is given in Chapter 6 [17].

1.4.2 SEASAT-A System

The SEASAT-A program was managed for the NASA Office of Applications by the California Institute of Technology Jet Propulsion Laboratory (JPL). The mission for SEASAT-A was to demonstrate that measurements of the ocean dynamics are feasible. The measurements included topography, surface winds, gravity waves, surface temperature, extent and age of sea ice, ocean features, and salinity. Precision of the GEOID measurement was specified as ± 10 cm [40].

The primary and secondary objectives of the SAR experiment on SEASAT-A included:

Primary: (a) to obtain radar imagery of ocean wave patterns in deep oceans; (b) to obtain ocean wave patterns and water-land interaction data in coastal regions; and (c) to obtain radar imagery of sea and fresh water ice and snow cover.

Secondary: (a) to obtain radar imagery of land surfaces; (b) to obtain data for mapping of the earth's surface; (c) to obtain data for estimates of land and sea surface roughness, ice type, differentiation of surface materials, vegetation, and land forms; (d) to obtain data for monitoring changes in the environment; (e) to obtain a demonstration of all weather, day-night measurement capability; and (f) to obtain data useful for designing future high-resolution spaceborne radar systems.

The SEASAT-A satellite was launched at 6:12 PM (PST) on June 26, 1978. The orbital altitude was 783 km at apogee and 778 km at perigee. The retrograde polar orbit had an inclination angle of 108° and a period of 100.5 minutes. Three radar and two radiometer sensors were carried on the spacecraft. The coherent SAR operated at 1.275 GHz. The radar altimeter operated in the 12–14 GHz band and covered a 1.6 km swath directly below the spacecraft. The wind scatterometer operated at 14.599 GHz and covered two swaths, each 400 km wide and offset on either side of the spacecraft. Four antennas were used to measure wind speed in the range from 4 to 28 m/s. The microwave radiometer had five frequency channels at 6.6, 10.6, 18, 21, and 37.6 GHz. A swath 1000 km wide, centered at the nadir, was covered. The visible and IR radiometer covered a single swath that was 1800 km wide, symmetrical about the nadir.

SEASAT-A collected data until October 9, 1978, when a short circuit developed at the slip rings between the solar array and the power distribution bus. Figure 8.16 in Chapter 8 [24] shows an example of the radar imagery collected and reduced in the ground station.

The SEASAT-A synthetic aperture radar was a focused SAR consisting of five subsystems: (1) spacecraft radar antenna, (2) spacecraft radar sensor, (3) spacecraft-to-ground data link, (4) ground data recorder and formatter, and (5) ground data processor. The antenna was a microstrip array of eight panels [41] that were fed by a corporate-feed network and operated at 1275 MHz. Details of the SEASAT-A antenna are discussed in Chapter 12 [42]. The solid-state radar transmitter generated a nominal peak power of 800 W with a linear frequency modulation (LFM) that is derived from a stable local oscillator (STALO). Figure 1.9 shows a simplified block diagram of the SEASAT-A radar transmitter. The antenna illuminated a 100 km swath width at the surface of the earth with an antenna elevation beamwidth of 6° oriented at an angle of 20° with respect to the nadir. Upon reception of the reflection signal by the receiver in the radar sensor, the return signal is amplified by a sensitivity time-controlled RF amplifier.

This signal and a fraction of the radar STALO were then combined and transmitted to a ground station by an analog data link. At the ground station, the data line demodulator recovers the radar sensor STALO and radar return signal. The synchronously demodulated video signal recovered was then converted into digital form by the radar data recorder and formatter subsystem. Upon conversion, the signal was buffered and recorded by a high density magnetic tape recorder. The radar data processor subsequently converted the digital recorded data into a two-dimensional map of the radar cross section of the area observed by the antenna. The SAR system generated a 25-m resolution radar map in elevation (across track) by time-gated compressed radar return signals and in azimuth (along track) by focusing the coherent radar returns during the data processing interval in the earth-based signal processor. Total SAR on-orbit weight was 223 kg; required radar prime power was 624 W. Table 1.3 shows a summary of the SEASAT SAR parameters and performance. Additional details of the SEASAT SAR are given in Chapter 4 [22].

1.4.3 Shuttle Imaging Radar

The technology developed for the SEASAT-A SAR formed the basis for the shuttle imaging radar series, SIR-A and SIR-B. Minor differences in the antenna are discussed in Chapter 12 [42]. The L-band radar transmitter was used with slight bandwidth changes so that resolution was 40 m on SIR-A and 20 m on SIR-B. Swath width was 50 km for both radars. Orbital altitudes were 240 and 220 km, respectively, so that radar range and incidence angles were different. Details of the shuttle imaging radar are given in Chapter 4 [22].

1.4.4 GEOS-3 [18–20]

The Geodynamics Experimental Ocean Satellite (GEOS-3) was a remote sensing satellite that contained five instruments in the experiment package. These were (1) an SBR altimeter, (2) two C-band transponders, (3) an S-band transponder, (4) laser retroreflectors, and (5) a radio doppler system. The purpose of the GEOS-3 satellite was to perform experiments in support of the application of geodetic satellite techniques to geoscience investigations such as earth physics and

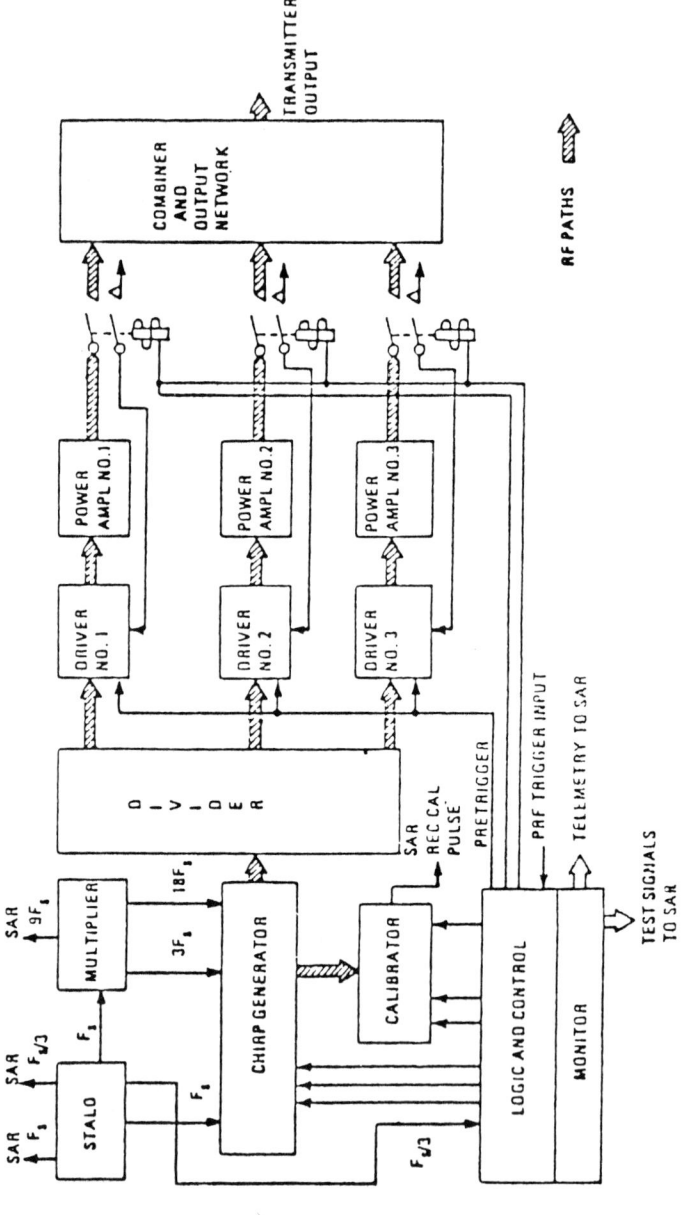

Figure 1.9 SEASAT radar transmitter.

Table 1.3
SEASAT Synthetic Aperture Radar

Antenna	
Type:	Planar phased array (10.74m × 2.16 m)
Beamwidth:	1.1 AZ, 6 EL (1 dB points)
Look Angle:	20° depression, 90° with respect to velocity vector
Gain:	34.7 dB
Polarization:	Horizontal
Weight:	113 kg
Transmitter	
Type:	Solid-state transistor
Efficiency:	38%
RF Carrier:	1275 MHz
Peak Power:	800W (nom), 1125W (max)
Pulse Length:	33.8 μs
PRF:	1463, 1540, 1645 pulses/s
Duty Cycle:	0.056 (max)
Average Power:	44.5W (nom), 62.6W (max)
Waveform:	Pulse, LFM, 19-MHz bandwidth
Receiver	
Noise Temperature:	550 K
Bandwidth:	22 MHz
System Input Noise	−127.42 dBW
AGC Time Constant:	5 s
STC Gain Variation:	9 dB
STALO Stability:	3×10^{-10} in 5 ms
Recorder:	25 kbs digital
System Weight:	110 kg (excluding antenna)
Total Prime Power:	624W (max)
Resolution:	25 m
Swath Width:	100 km
Swath Length:	2000 km per pass
Swath Orientation:	Right side of orbit path
Signal-to-Noise Ratio:	9 dB (nom)

oceanography. The SBR altimeter mission objective on the GEOS-3 satellite was to perform an in-orbit experiment: (a) to determine the feasibility and utility of a spaceborne radar altimeter to map the topography of the ocean surface with an absolute accuracy of ±5 m, and with a relative accuracy of 1 to 2 m; (b) to determine the feasibility of measuring wave height; (c) to determine the feasibility of measuring the deflection of the vertical at sea; and (d) to contribute to the technology leading to a future operational altimeter satellite system with a 10 cm measurement capability.

The GEOS-C satellite (designation changed to GEOS-3 after successful orbit) was launched on April 9, 1975. The nominal orbit parameters were mean altitude of 843 km, inclination angle of 115°, eccentricity of 0.000, period of 101.8 minutes. The GEOS-3 spacecraft was an eight-sided aluminum shell topped by a truncated pyramid. The satellite width was 132 cm (53 in), height was 81 cm (32 in), and weight was 340 kg (750 lbs).

The GEOS-C radar altimeter was a precision SBR altimeter (made by the same contractor that built the Skylab altimeter), developed [65] primarily to measure ocean surface topography and sea state. A complex multimode radar system, it had two distinct radar-gathering modes (global and intensive modes) and two corresponding self-test and calibration modes for use in on-orbit functional test and instrument calibration. The key performance features were its capability to provide (1) precise satellite-to-ocean surface height measurements (precision of 50 cm in the global mode and 20 cm in the intensive mode at an output rate of one per second) for use in mapping the shape of the ocean surface, and (2) data that can be processed to estimate peak-to-trough ocean wave height (wave heights in the range of 2 to 10 m can be estimated to an accuracy of 25%). Several key areas of technology are included in the design: (a) high frequency logic circuitry with a 160 MHz clock and four-phase division for 1.56 ns resolution; (b) wideband (100 MHz) LFM pulse compression system with a compression ratio of 100:1 and a compressed pulsewidth of 12.5 ns; (c) high-speed sample-and-hold (S/H) circuitry for accurate sampling of wideband (50 MHz) noisy video return signals; and (d) design and packaging of high-voltage (12 kV) power supplies for space application.

The instrument weighs 68 kg (150 lbs) and occupies a volume of 0.119 m^3 (4.2 ft^3), including the antenna, which is a 0.6 m (24 in) diameter parabolic dish. The instrument is packaged in two basic sections, an RF section and attached electronics section, which are both mounted to a center cylindrical disk baseplate with a diameter of 0.65 m (26 in). The major subsystems contained in the RF section are (1) the IM transmitter (chirp generator, up-converter, 1 W driver TWT and high-voltage power supply, 2 kW output TWT and high-voltage power supply), (2) the GM transmitter (magnetron and high-voltage power supply), (3) the RF switch assembly (RF switches, waveguide runs, calibrated attenuation path, and T/R switch), and (4) the receiver front end (down-converter and preamplifier). The major subsystems contained in the attached electronics section are (a) the IF receiver (IF amplifiers, filters, pulse compressor, detectors), (b) the signal processor (AGC, acquisition, and tracking functions implemented with analog and digital circuitry on multilayer board assemblies), (c) the frequency synthesizer, (d) the mode control circuitry, (e) the calibrate-and-test circuitry, and (f) the low-voltage power supply. The nominal power required for operation was 71 W for the global mode and 126 W for the intensive mode (16 waveform samplers). Additional details on the GEOS-C altimeter are given in Chapter 7 [23].

1.4.5 USSR Cosmos 1500 Side-Looking Radar

The USSR launched the Cosmos 1500 oceanographic satellite on September 28, 1983 into a nominal 650 km polar orbit [43]. The satellite was the first of a series intended to provide continuous world ocean observations for civil and military missions. The sensors provide *side-looking radar* (SLR), radiometric, and visual coverage of oceans and ice zones for land- and sea-based users through an operational distribution network [43]. Table 1.4 gives the characteristics of the Cosmos 1500 system, and Table 1.5 provides a summary of the parameters and performance of the real-beam SLR. The radar operates at a frequency of 9500 MHz with a magnetron transmitter that has a peak power output of 100 kW. The antenna is a slotted waveguide that is 11 m in length and 4 cm in height. Cosmos 1500 has demonstrated many significant capabilities: (a) routine automatic picture transmission of SLR images of earth; (b) mapping of inhomogeneities of Antarctic and Greenland ice cover, which were previously not detected; (c) radar images of polar regions of multiyear and first-year ice zones; (d) mapping of elongated zones of ice cover continuity disturbances; (e) tracking of sea ice drift by using a series of radar images of the same water area; (f) detection of oil slicks, wind fields, and currents; (g) guidance of ships trapped in arctic ice during October and November 1983.

Table 1.4
COSMOS 1500 System Characteristics [43]

Orbit
   Apogee   676 km
   Perigee   649 km
   Period   97.7 min
   Inclination   82.6°
   Launch Date   28 September 1983
Sensors
   Type: Radar (SLR), optical, radiometer
   Operation: Simultaneous radar and optical at 4 Hz scanning frequency
   Coverage: Radar swath width 460 km
                  Optical swath width 1930 km
                  Radar grazing angles 22° to 52°
   Wavelengths: Radar 3.15 cm
                      Optical 0.5–0.6 μm
                                   0.6–0.7 μm
                                   0.7–0.8 μm
                                   0.8–1.1 μm
                  Radiometer 0.8 cm
                                   1.35 cm
                                   8.5 cm

Table 1.4 cont'd.

Resolution: Radar: 1.5–2 km
              Optical: 1.5 km
Processing
   On Board: Geometric distortion removal creates linear image structure recording of 6.5 minutes of data
   Ground stations at Moscow, Novosibirsk, Khabarovsk, and ships at sea ($\approx$500)
Data Link
   466 MHz and 137 MHz
   7 Channels of Data

Table 1.5
COSMOS 1500 SBR Parameters and Performance [43]

Type:	Real-Beam Side-Looking Radar
Frequency; Wavelength	9500 MHz; 3.15 cm
Antenna	
Type:	Slotted waveguide
Size:	11.085 m × 40 mm
No. of Slots:	480
Illumination:	Cosine on a pedestal
Beamwidth:	0.20° × 42°
Gain:	35 dB
Sidelobes:	−22 dB to −25 dB
Waveguide:	Copper 23 × 10 mm cross section
Polarization:	Vertical
Swing Angle:	35° from nadir
Noise Temperature:	300 K
Transmitter	
Type:	Magnetron
Power:	100 kW peak, 30 W ave.
Pulsewidth:	3 μs
PRF:	100 pulses/s
Loss:	1.7 dB
Receiver	
Type:	Superheterodyne
Noise Power:	−140 dBW
Loss:	1.7 dB
Pulses Integrated:	8 Noncoherent
LNA Noise Temperature:	150–200 K
LNA Gain:	15 dB
Dynamic Range:	30 dB
IF:	30 MHz ± 0.1 MHz
Input Power	400 W
Range:	700 km (min), 986 km (max)
SNR:	0 dB on $\sigma^0 = -20$ dB

The orbit of Cosmos 1500 allows complete earth coverage each 1.41 days for the optical sensors, and each 5.9 days for the radar sensor. Subequent launches of the Cosmos 1500 type of satellite have occurred in September 1984, July 1985, and September 1987 by Cosmos 1602, Cosmos 1766, and Cosmos 1869. The SLR on Cosmos 1869 was not able to function because its five-segment, twelve-meter long antenna failed to deploy to its full length [44]. The Soviets ultimately intend to have three of the Cosmos 1500 type of spacecraft operating in orbit simultaneously by 1990 to provide good coverage for ice monitoring and ocean surface observations.

1.4.6 USSR Synthetic Aperture Radar Polyus-V

During June 1983, the USSR launched the Venera-15 and Venera-16 spacecraft. Each spacecraft had a weight of 4000 kg, and included a synthetic aperture radar and altimeter as part of its payload [45]. The mission performed by the Venera was to conduct a radar survey of the surface of Venus. The footprint for each radar antenna is shown in Figure 1.10, as seen from an average orbital altitude of 1200 km. The SAR antenna orientation was 10° from the nadir. An elliptical orbit with a period of 24 hours was utilized by the Soviets. At pericenter, the orbital altitude was approximately 1000 km, and it was approximately 60,000 km at apocenter. The pericenter was near 60° N latitude, and the inclination angle to the equator was chosen to ensure that the spacecraft passed near the north pole of the planet. The radar collected data from altitudes in the range from 700 to 2000 km when the orbital velocity was 9000 m/s. Swath width was 120–150 km. For a coherent signal processing time of 4 ms, the unfocused 8 cm wavelength SAR provides a map resolution of 1354 m of the surface of Venus.

The USSR designated the radar as the Polyus-V SAR. It has two antennas: a 6 m × 1.4 m antenna and a 1 m diameter altimeter antenna. The 1 m paraboloid is oriented along the subsatellite point and the SAR antenna has its electrical axis inclined 10° from the vertical. A simplified block diagram of the radar is shown in Figure 1.11. The same transmitter is used by both radars, but different waveforms are generated by the 8 cm wavelength transmitter. Table 1.6 gives a summary of the waveforms transmitted by the Polyus-V radar, the parameters, and its performance in both modes. The length of the mapping swath per session is 8000 km. At the end of the mapping session for each orbit, the data are relayed to earth. The average data transmitted is 75 Mb. The data transmission rate is about 10^5 b/s. A 64 m diameter antenna near Moscow and the Deep Space Communications Center at Yvpatoriya receive the Polyus-V radar data.

Processing of the data is performed at the computing centers of OKB MEI and the Institute of Radio Engineering and Electronics of the USSR Academy of Sciences. The radar data concerning the structure of a portion of the northern hemisphere of Venus covered about $1.15 (10)^8$ km^2 in area. The spacecraft have provided the first images of the surface with photographic quality. Figure 1.12

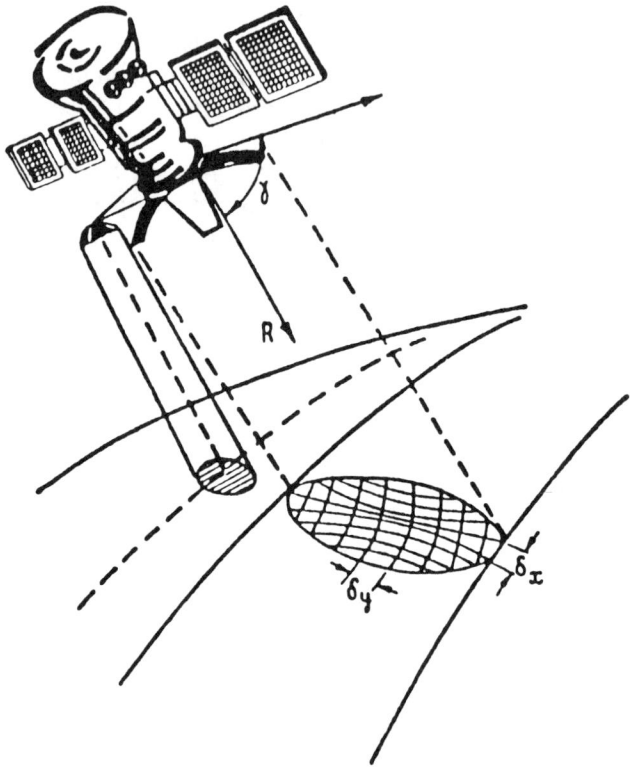

Figure 1.10 Polyus-Venera SAR footprint.

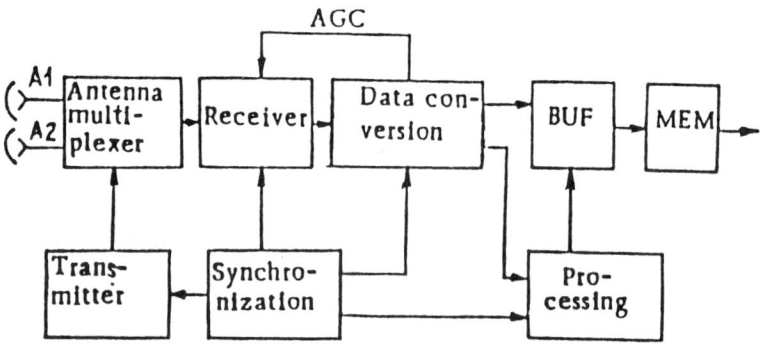

Figure 1.11 Polyus-Venera radar block diagram.

Table 1.6
USSR Polyus-V Radar Parameters and Performance

Functions:	SAR and altimeter
Wavelength:	8 cm
Effective Radiated Power (incl. microwave losses):	8 W
Type of Signal Transmitted:	Phase-coded packets
SAR Synthesis Time:	3.8 ms
Packet Duration:	15 ms
Phase-Coded PRF (SAR Mode):	5 kHz
Phase Code:	31-element (altimeter); 127-element (SAR) M-sequence
Element Duration:	1.54 µs
Size of Side-Looking Antenna:	6 m × 1.4 m
Antenna Orientation:	10° from subsatellite point
Diameter of Altimeter Antenna:	1 m
SAR Antenna Beamwidth:	0.93° × 4°
Mean Surface Resolution:	1.3 km × 1.3 km (earth-based processing)
Width of Mapping Swath:	130 km
Length of Mapping Swath:	8000 km (per session)
Size of Averaged Surface for Altitude Measurement:	8 km × 50 km
rms Error in Altitude:	50 m
Processor:	4-bit digital I and Q
Memory:	Buffer and tape recorder

shows a poor quality reproduction of the mapping and simultaneous profile data collected by the altimeter on January 17, 1984 by Venera-16. The arrow corresponds to Patera Cleopatra and also shows the depth of the crater Maxwell Montes.

1.4.7 Apollo Lunar Sounder Radar

The Apollo Lunar Sounder Experiment (ALSE) was assigned by NASA the Scientific Experiment Designation S-209 [46]. Its primary objective was the detection and location of subsurface discontinuities ("sounding"). Secondary objectives included the generation of a lunar surface profile, generation of surface images, and measurement of galactic noise in the lunar environment. A multiple-function, three-wavelength radar was designed, built, and flown on Apollo 17 in December 1972 to attain these objectives. The flight hardware consisted of two distinct coherent radar subsystems, one operating at two frequencies in the HF band and the second on a single VHF channel. Both radar subsystems shared a common data storage device, which was an optical recorder. Table 1.7 shows the characteristics

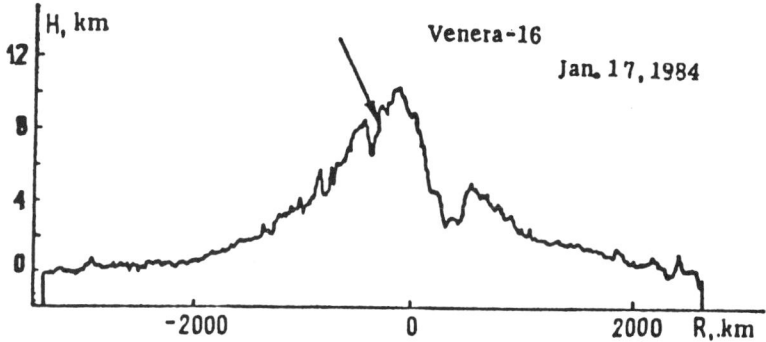

Figure 1.12 Polyus-SAR and altimeter displays [45].

Table 1.7
Apollo 17 S-209 Lunar Sounder Radar System [46]

Parameter	HF_1	Mode HF_2	VHF
Wavelength (m)	60	20	2
Estimated Depth (m)	1300	800	160
Center Frequency (MHz)	5.266	15.8	158
RF Bandwidth (MHz)	0.5333	1.6	16.0
Pulsewidth (μs)	240	80	8.0
Range Resolution, free space (m)	300	100	10
Transmitter Peak Power (W)	130	118	95
Transmitter Average Power (W)	12.4	3.7	1.5
Effective Antenna Gain (dB)	−0.8	−0.7	+7.3

Table 1.7 cont'd.

Noise Figure (dB)	11.4	11.4	10.0
PRF $(s)^{-1}$	397	397	1984
AGC Gain Range (dB)	12.1	12.1	13.9
Recorder:			
Duration (μs)	600	600	70
Type	Optical		
Film Length	195 m		
Width and Type	70-mm Kodak 3400		
Film Capacity	10 hr		
Weight	55 kg		
Prime Power	100 W dc, 17 VA, 400 Hz		
Radar System Weight	49 kg		
Radar Prime Power	103 W		

of the radar and the recorder subsystems. The HF subsystem radiated signals, HF_1 and HF_2, on alternate pulses by way of a center-fed dipole antenna. The VHF subsystem radiated through a Yagi antenna. Each of the three frequency bands used an LFM pulse, and pulse compression was performed optically in the data processor. The antenna subsystems limited the FM bandwidth to 10% of the carrier frequency. The duration of FM sweep in each system was set to maintain a time-bandwidth product of 128. Automatic gain control (AGC) was incorporated at all three frequencies to preserve large dynamic range in the receiver and signal processing systems. The gain setting was allowed to change no more than once every 30 seconds, and henceforth in discrete steps. Variations up to 12 dB at each frequency were permitted. During the Apollo 17 flight, significant variations in return power were observed over some regions of the lunar surface. The S-209 radar weighed 49 kg and required a prime input power of 103 W.

The performance of the radar was determined by aircraft tests and in the Apollo 17 mission. During August and September 1972, a prototype radar was flown in a KC-135 aircraft with the objective of validating the hardware and exercising the data analysis system. Data runs were made over deserts in the southwestern US and over ice fields in Greenland. Aircraft altitude limitations prevented data collection with the HF radar, but VHF data indicated subsurface reflections at depths of 120 m below the Greenland ice when it is assumed that the dielectric constant of the ice is 3.1. Measurements on lunar material brought to earth during previous Apollo missions indicated loss tangents on the order of 0.025 at frequencies of 5 MHz, 0.01 at 15 MHz, and 0.003 at 150 MHz. For these loss tangents, the decay of energy through the lunar material is 3.4 dB/μs at 5 MHz, 4.1 dB/μs at 15 MHz, and 12.3 dB/μs at 150 MHz. Estimated depth of penetration into the lunar subsurface was 160 m at VHF and 1300 m at HF. In the SAR mode, the swath width was 40 km and the map resolution was 10 m. The ALSE system development and performance results provided NASA with knowledge of system design, data analysis, and interpretation methods that proved to be valuable for planning future remote sensing systems.

1.5 SBR SYSTEMS OF THE FUTURE

During the 1965–1987 time period, more than 50 contractors prepared many reports describing future possible SBR systems for a variety of missions. Estimates for improved technology, particularly that of antennas in space, have been reported. Examples of Types I, II, and III SBR configurations have been selected and are described in the following paragraphs. The OMV rendezvous radar and a space station tracking system are described as examples of the future for Type I SBR. Radars for remote sensing missions are discussed as examples of Type II SBR. Examples of concepts for Type III SBR systems are described, including (1) a global air traffic surveillance and control system of satellites using SBR; (2) a US military SBR satellite system that provides basic surveillance missions for fleet and air defense of CONUS; (3) SBR to support the Strategic Defense Initiative (SDI) mission; and (4) a high-resolution mapping SBR to support the proposed United Nations International Satellite Monitoring Agency.

1.5.1 Future Rendezvous Radar Missions

OMV

All US satellite rendezvous missions have been performed by manned vehicles. In the foreseeable future, the majority of rendezvous missions will be conducted by unmanned vehicles such as the orbital maneuvering vehicle. The planned list of missions for the OMV includes: (1) large observatory servicing at the shuttle; (2) payload placement; (3) payload retrieval; (4) payload reboost; (5) payload deboost to re-entry; (6) payload viewing; (7) subsatellite mission; (8) multiple payload mission; (9) *in situ* servicing mission; (10) STS to space station transfer; and (11) base support. Details of these missions may be found in NASA's OMV request for proposal [16]. The initial design of the OMV is modular to permit upgrading its capability to operate from the space station and to accommodate the following growth missions by addition of appropriate kits or elements to the system: (a) logistical support; (b) debris collection mission; (c) extended on-orbit operation; (d) satellite buildup; (e) satellite refueling; (f) servicing mission; and (g) space station reboost.

A low-cost and lightweight rendezvous radar will be used to perform the above-mentioned future OMV missions. Such an OMV radar is in development [17] and has the major performance characteristics shown in Table 1.8. The *ren-*

Table 1.8
OMV Radar Characteristics [17]

Detection Performance	
P_d:	99%
Target:	1 m²; Swerling I
Range:	4.5 nmi
False-Alarm Rate:	1/hr
Search Scan:	± 20° cone
Scan Time:	5 min
Track Performance	
Angle Accuracy (3Σ):	20 mrad
Range Accuracy (3Σ):	Greater of 20 ft or 2% of range
Range-Rate Accuracy (3Σ):	Greater of 0.1 ft/s or 2% of range rate
System Parameters	
RF:	9.5–9.8 GHz
PRF:	6.67 kHz
Pulsewidth:	0.05, 0.2, 1.5, 15 μs
System Noise Temperature:	≈900 K
Antenna	
Type:	Planar slotted array
Size:	14 in × 15 in
Depth:	1 in overall
Gain:	30.5 dB at 9.65 GHz
Beamwidth:	5.0°
Polarization:	Linear
Transmitter	
Type:	GaAs FET
Peak Power:	2 W
Gain:	≈30 dB
Receiver	
Type:	3-channel monopulse
Noise Figure:	< 4 dB; GaAs FET LNA
System Physical	
Deployed Assembly Weight:	26 lbs
Inboard Assembly Weight:	50 lbs (redundant total)
Electronics Volume:	≈2 ft³ (redundant total)
Prime Power:	< 60 W

dezvous radar set (RRS) will be an X-band, coherent, range-gated, pulsed doppler radar with redundant electronics and gimbal motor windings. The antenna is a planar slotted array with 30 dB gain at 9.6 GHz and 5° beamwidth. The OMV system computer initiates the acquisition-search function to permit target detection at a 4.5 nmi range. Monopulse tracking is performed to within a minimum range of 35 ft. Peak power is programed over a 50 dB range during the rendezvous maneuver to minimize the RF radiation intensity on sensitive targets. Pulse frequency agility is employed, and up to 30 carrier frequency changes in 10 MHz steps over the 300 MHz operating band are used to decorrelate Swerling I target fluctuations. At each dwell, 128 pulses are coherently integrated in the fast Fourier transform (FFT) processor prior to noncoherent integration of the outputs. Up to 30 FFT outputs can be integrated. Detailed discussion of the OMV radar is given in Chapter 6 [17].

Space Station Tracking System [47]

Initial configuration of the Space Station will have limited tracking system requirements that include tracking of cooperative vehicles within a 37 km control zone, based on the assumption that all vehicles will provide accurate position and velocity data to the Space Station tracking system by way of the space-to-space link. Automatic tracking of an extravehicular activity astronaut is not required. For growth configurations of the Space Station, the tracking system must expand to meet additional requirements. More coorbiting vehicles, noncooperative or disabled vehicles, automatic tracking of EVA astronauts, and sensors for berthing-docking operations will require additional tracking capabilities within the system. Some type of short-range radar may be required to track vehicles that do not have a positioning (GPS) capability or have been disabled. Results of preliminary tradeoffs on multiple-target tracking (MTT) radars indicate [48] that either a Ka-band or an X-band phased array radar is the preferred approach.

1.5.2 SBR in Future Remote Sensing Missions [21]

SBR will participate in many remote sensing missions for observation of the earth and planets. The SIR series will continue to form the core of the US program with flights of the SIR-C and SIR-D. These radars will have the capability to image the earth's surface by using all polarization states (HH, VV, and HV) and at least two frequency bands (L- and C-band). A third frequency band may be flown by using an X-band imaging radar developed in the Federal Republic of Germany. The SIR-D mission is also planned for early in the 1990s when a four-frequency capability (L-, C-, X-, K-bands) will be used with electronic scanning beams. These radars will also develop the technology necessary for long duration orbiting SAR on the Earth Observing System (EOS).

A number of space-based SAR missions are planned by the European Space Agency (ESA), Japan, and Canada. ESA selected Thomson-CSF to design an SAR at C-band with 30 m resolution and a 100 km swath. ESA missions will focus on long-term oceanic investigations. The ERS-1 satellite will also include a six-frequency imaging microwave radiometer, a dual-frequency scatterometer for oceanic wind direction and velocity, and a radar altimeter for sea state observations. The National Space Agency of Japan is planning an earth resources satellite, which would carry an L-band SAR with 25 m resolution and 75 km swath width. The Japanese satellite mission will focus mainly on geological mapping, primarily surface feature morphology. The Canadian RADARSAT, scheduled for launch in 1990, will employ use a C-band space-based SAR primarily for monitoring polar ice dynamics for use in ship routing; the SAR will have a 200 km swath width.

In planetary exploration areas, SBR imaging systems are key elements for exploration of two continuously cloud-covered bodies, Venus and Titan. In the exploration of Venus during the late 1970s, a radar sensor on the Pioneer Venus Orbiter provided low-resolution (40–100 km) images of the planet. More recently, a USSR Venera satellite [45] produced radar images of part of the planet's northern hemisphere with a resolution of 1300 m. In the near future, a sophisticated radar sensor will be placed in orbit around Venus as part of the US Venus radar mission with the objective of providing global coverage with a resolution of 150 m. In the exploration of Saturn's satellite, Titan, the larger distance to the earth will impose a narrow limit on the data-rate transmission, which directly affects the mapping coverage and resolution. The plan is to put the spacecraft into orbit around Saturn and the spacecraft will fly by Titan on selected orbits. The flybys will be targeted in such a way that during each a different region of Titan will be mapped with a space-based SAR. Preliminary mission scenarios contain about 20 flybys. The Titan radar mapper will have a very wide swath, 600–800 km, to obtain a global map during the small number of flybys. Real aperture imaging will provide resolution of 6–40 km. A synthetic aperture mode can be used to observe limited regions with a resolution of about 200 m. The Titan radar mapper is planned for launch in the mid-1990s as part of the Cassini Mission by JPL.

Other missions using radar are planned for ocean scatterometry and altimeters. Scatterometers are used to obtain accurate measurement of global surface winds for oceanography and meteorology. Scatterometers are expected to obtain wind speed with errors of about 2 m/s. The NSCAT instrument plans to use six fan-beam antennas to obtain wind direction error of less than 16°. SBR altimeters, such as those planned for the ocean topography experiment (TOPEX), expect to measure altitude with an error of 5 cm from a 1300 km polar orbit inclined 65° over the ocean. Over the solid surface of Mars, a 37 GHz altimeter on the Mars orbiter mission expects to gather global high-resolution topographic mapping data with a height resolution of 15 m.

The Earth Observing System [49]

The overall goal of the Earth Observing System (EOS) is to advance the scientific understanding of the entire earth system on the global scale through developing a deeper understanding of its components and the interactions among them, and how the earth system is changing. International space station elements include the following satellites in polar and equatorial orbits: (1) a NASA EOS platform at 824 km sun-synchronous, 1:30 PM equator crossing time, ascending node orbit; (2) an ESA platform at 824 km, sun-synchronous, 10:00 AM equator crossing time, descending node orbit; and (3) the manned space station in a 335–460 km, 28.5° inclined orbit. Launch vehicles include the Titan IV and the STS by NASA, the Ariane V by ESA and the Japanese H-2. The NASA polar platform parameters are shown in Table 1.9. Instruments planned for the satellites include radar, radiometers, IR, optics, and UV sensors. These sensors will measure parameters such as winds, clouds, rain, liquid moisture content, geologic parameters, and ocean currents. Radars will be used to make atmospheric and geological observations. Two of the radars proposed are TRAMAR (*tropical rain mapping radar*) and LORRA (*land, ocean, and rain radar altimeter*). Some of the parameters for these radars are shown in Tables 1.10 and 1.11 [50]. The LORRA was designed for deployment on the EOS, and the TRAMAR was designed for deployment on the manned space station.

Table 1.9
NASA Polar Platform Parameters [49]

Orbit:	824 km polar sun-synchronous 1:30 PM equator crossing time, ascending node
Launch Date:	October 1995
Launch Vehicle:	Titan IV
Platform Size:	7.5 m (L) × 4 m (W)
Platform Mass:	3500 kg
Power:	3.5 kW average to payload
Data Relay via TDRSS:	300 Mb/s (max)
On-board Data Storage Capacity:	10^{12} bits
Data Recording Rate:	300 Mb/s (max)
Playback Data Rate:	300 Mb/s (max)
Direct Downlink Data Rate:	100 Mb/s (max)
Command Uplink Rate:	100 kbs
Pointing Accuracy:	270 arcsec
Pointing Knowledge:	90 arcsec
Pointing Stability:	10 arcsec

Table 1.10
TRAMAR Radar Parameters [50]

Function:	Tropical rain mapping radar with 2 simultaneous beams
Antenna Beams:	Narrower beam steps within broad X-band beam from one side of the broad beam to the other; then, both beams shift to a new beam position
Frequencies:	9.70 GHz and 24.1 GHz
Bandwidth:	6 MHz
Antenna Gains:	51.6 dBi at 9.70 GHz
	54.5 dBi at 24.1 GHz
Antenna Beamwidths:	0.255° × 0.506° at 9.7 GHz
	0.103° × 0.506° at 24.1 GHz

Table 1.10 cont'd.

Scan Angle:	± 50° cross track off nadir
Minimum Angle Step Size:	0.118° at 9.7 GHz
	0.0475° at 24.1 GHz
Minimum Dwell Time:	70 µs
Switch Time:	5 µs
Pointing Knowledge:	0.011° in scan direction
	0.047° in orthogonal direction
Polarization:	Linear
Sidelobes:	−30 dB at 9.7 GHz
	−27 dB at 24.1 GHz
Transmitter	
Peak Power:	300 W at 9.7 GHz; 200 W at 24.1 GHz
PRF:	2184–5000 pulses/s
Pulse Duration:	70 µs or 20 µs
Modulation:	LFM 4.5 MHz

Table 1.11
LORRA Radar Parameters [50]

Function:	Land and ocean modes measurement of rain, altitude, topography	
Frequency:	35.55 GHz	
Antenna Bandwidth:	300 MHz	
Antenna Polarization:	Linear	
Transmitter		
Peak Power:	200 W at antenna waveguide flange	
PRF:	2, 12, or 18 kHz	
Pulse Duration:	15 µs	
Burst Mode:	5.5 ms burst with 5.5 ms interburst	
Modulation:	LFM 50, 100, or 300 MHz on each pulse	
	Land Mode	*Ocean Mode*
Modulation:	LFM 50, 100, or 300 MHz	LFM 300 MHz
PRF:	18 kHz	2 or 12 kHz
Antenna Gain:	58.7 dBi	49.4 dBi
Beamwidth:	0.046° × 0.44°	0.041° × 0.41°
Scan:	± 6° cross track	Form beam at 0°, + 1.74°, + 3.47°, −1.74°, or −3.47°
Switch Time:	5 µs	10 µs
Dwell Time:	11.0 ms	—
Minimum Step:	0.0264°	—
Pointing Knowledge:	0.003°	0.03°
Pointing Accuracy:	—	0.03°
Antenna Peak Sidelobe:	−25 dB	−25 dB

1.5.3 SBR Concept for Global Air Traffic Surveillance

In 1982, a concept was proposed for a constellation of 14 satellites to provide tracking of aircraft for global air traffic surveillance and control [12]. *Air traffic control* (ATC) is increasingly a matter of global concern, and the explosive growth of aircraft density in, around, and between major metropolitan areas in Europe and North America is common knowledge. If a United Nations organization were responsible for ATC for 120 to 130 nations in the world, as many as 84,000 commercial aircraft could conceivably require ATC in the twenty-first century. A rosette constellation of SBR satellites at an orbital altitude of 5600 nmi (10,371 km) in a 14/14/12 Walker orbit [31] inclined at 49.4° provides continuous worldwide visibility by at least two satellites simultaneously. Figure 1.2 above shows a snapshot of the 14-satellite constellation. Each satellite provides radar coverage between grazing angles from 3 to 70°. The major subsystems in the satellite include (1) radar, (2) communications, (3) guidance and control, and (4) electrical power subsystem. Details of these subsystems are found in [12], the radar parameters are shown in Table 1.12, and the T/R module characteristics are given in Table 1.13. The 10 GHz receive-only module is similar, except that no transmitter amplifier function is contained in the module and the input power is reduced by the 60 mW input normally provided to the transmitter. Performance of the proposed radar is shown in Figure 1.13.

Table 1.12
Radar Parameters [12]

Antenna	
Type:	Corporate-fed, active, phased array
Diameter:	100 m
Frequency:	2 GHz
Wavelength:	0.15 m
Polarization:	Circular
No. of elements:	576, 078
No. of Modules:	144, 020
Element Spacing:	0.7244 λ
Beamwidth:	1.83 mrad
Directive Gain:	66.42 dB
Maximum Scan Angle:	22.4°
Receiver	
Type:	Distributed, solid-state, monolithic T/R module
Bandwidth:	500 kHz
System Noise Temperature:	490 K
Compressed Pulsewidth:	2 μs

Table 1.12 cont'd.

Transmitter	
Type:	Distributed, solid-state, monolithic T/R module
Peak Power:	22.33 kW
Pulsewidth:	2000 µs
Maximum Duty Cycle:	0.20
Frequency:	2 GHz
Signal Processor	
Type:	Digital
Input Speed:	50 Mwords/s

Table 1.13
T/R Module Characteristics

Type:	Solid-state monolithic
Frequency:	2 GHz ± 2 MHz
Peak Power:	0.155 W
Average Power:	15 mW (max. search function)
	15 mW (max. track function)
Ripple Across Band:	0.3 dB amplitude
	3.7° phase
Module-to-Module Deviation:	Same as ripple
Receiver Bandwidth:	500 kHz
Receiver Gain:	28 dB (three stages)
Receiver Noise Figure:	3 dB
Phase Shifter:	
Type	Monolithic SOS IGFET switches
No. of Bits	4 bits
Switch Power	< 1 mW
Overall Module Efficiency:	50%
Size:	2.54 cm × 2.54 cm × 1.53 mm
Weight:	5 g
Input Power:	< 71 mW

1.5.4 Military SBR Systems

Although the US has been studying SBR and developing its technology for many years and the USSR has had a space-based military radar system known as Rorsat in operation for many years, the US is still apparently decades away from fielding an operational military SBR [10]. Late in 1983, the Pentagon took an important step toward fielding an SBR that can detect and track strategic and tactical targets on the oceans, in the air, and on land when Deputy Secretary of Defense Paul

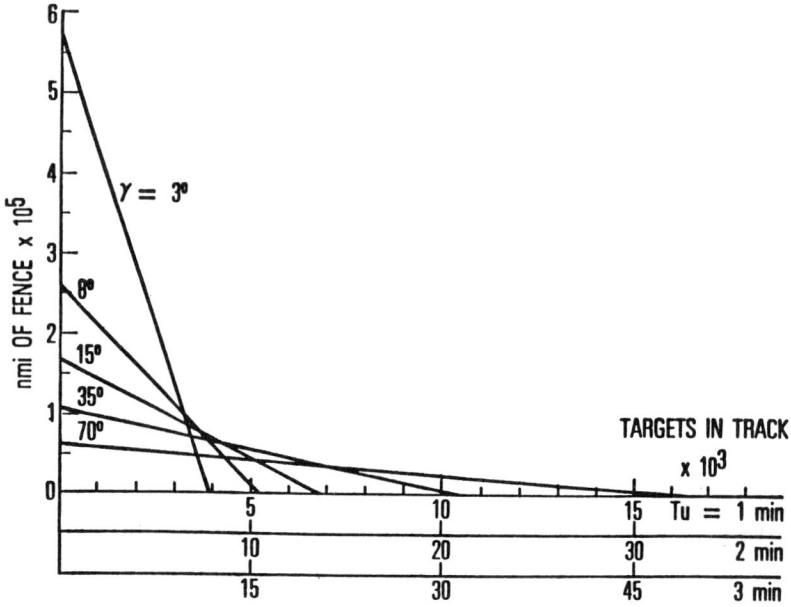

Figure 1.13 Search-track trade-off *versus* geometry and revisit time.

Thayer settled a long-festering dispute within the national security community over technological approaches to such a system. The Defense Department decided in favor of a long-term technology program recommended by the Air Force and some elements of the Navy. A short-term concept based on a Lincoln Laboratory design was considered to be confined mainly to detection and tracking of surface ships. It was judged to require excessive amounts of power and provide only limited performance in low earth orbits (below 1000 km). The long-term technology program utilizes advanced technologies, but probably will not be available before the year 2000. Three pivotal technologies that are being developed by DARPA and R&D elements of all three services included: (1) T/R modules; (2) advanced on-board signal processors; and (3) a lightweight, highly efficient, on-board power and power distribution system.

Security classification has prevented specific details of military SBR configurations from appearing in the open literature. Ulsamer [10] indicates that phased arrays would be used that are some 30 m in diameter and contain between 50,000 and 90,000 T/R modules. Brookner and Mahoney [11] derived a satellite radar architecture for performing the basic surveillance missions for fleet and air defense of CONUS. (Their work was performed under subcontract to Lockheed Missiles and Space Company, Sunnyvale, California. The prime contract was sponsored by the Naval Electronics System Command.) Both bistatic and monostatic configurations were considered. Orbit altitude, satellite constellation size, radar frequency, and power-aperture trade-offs were analyzed, and a number of antenna configurations were examined. The system selected by Brookner and Mahoney was an L-band, corporate-fed phased-array radar in orbit constellations of three to twelve satellites at altitudes from 600 to 2000 nmi. At the highest orbital altitude, a 10 × 30 m phased array that contains 15,000 radiating elements and modules was designed. The modules delivered an average power of 6 kW and the radar required a prime power of 30 kW.

1.5.5 Military SBR Applications

Both near-term and far-term SBR concepts are possible for use in military missions to perform precision tracking and discrimination functions. Each proposed radar concept would operate at one of the millimeter-wave (MMW) absorption bands, such as 60 GHz or 118 GHz, to decrease clutter and interference effects from the earth. The choice of such MMW frequencies is also based on the following: (a) MMW operation allows narrow antenna beamwidth and low sidelobes; (b) MMW operation requires lower transmitter power than microwaves; (c) MMW operation allows use of a small antenna, thereby yielding a lightweight SBR design; and (d) resolution and accuracy measurements are improved at MMW.

The performance requirements for a near-term design are modest compared to the far-term requirements, as shown in Table 1.14. The radar concept employs

Table 1.14
Military Radar Requirements

	Near-Term	*Far-Term*
Function	Precision Tracking	Track and Discrimination
In Operation	1991	1999
Technology	Existing	New
Range	370 km	2000–3000 km
Target RCS	−20 dBSM	−30 dBSM
Number of Targets	< 10	> 1000
Target Rate	6/s	600/s
Coverage	2π steradian (mechanical)	± 60° (electronic)
Orbit	380 km	> 1500 km

a reflector antenna on gimbals with an offset phased array feed [51]. This will provide 2π steradian coverage about a selected point. A four-meter diameter reflector will provide a signal-to-noise ratio (SNR) of 15 dB on a 0.01 m^2 target at a range of 370 km when the transmitter peak power is 10 kW. Six tracking measurements per second can be obtained when the average power is 300 W. A transmitter tube (such as the Varian *extended interaction amplifier*, EIA [52]) can deliver such a power level at 60 GHz with a weight of 4.5 kg. Radar system weight is estimated at 37 kg. Tracking accuracies are estimated to be range of 2 cm, range rate of 4 cm/s, acceleration of 75 m/s, and angle 0.17 of mrad when the dwell time is 5 ms. Table 1.15 shows a summary of the SBR radar system performance and parameters for the near-term concept.

Table 1.15
Near-Term Military Radar Concept Performance and Parameters

Performance	
Coverage:	2π steradians
Range:	370 km
SNR:	15 dB
Target:	-20 dBSM
Number of Measurements:	6/s
Track Accuracy (5 ms Dwell):	
Range	2.22 cm
Range Rate	4.44 cm/s
Acceleration	75.9 m/s
Angle	0.17 mrad
Parameters	
Antenna Type:	Gimballed reflector with offset feed
Beamwidth	1.6 mrad
Diameter	4 m
Transmitter Type:	Varian EIA
Frequency	60 GHz
Power (peak; ave)	10 kW (peak); 300 W (ave)
Receiver Type:	HEMT
System Noise Temperature	1000 K
Bandwidth	600 MHz (max)
Prime Power:	1500 W
Weight:	37 kg

The far-term military SBR system concept uses a large phased array antenna and a *free-electron maser* (FEM) transmitter. The antenna is an outgrowth of the Air Force's large radar array technology [53] studies. The FEM transmitter is a TRW development [54–56], which provides unique capabilities, including (a) very high average and peak power in the MMW band, (b) large instantaneous bandwidth, (c) electronic tuning ability over an octave or more, (d) high-efficiency amplifier operation, and (e) compact form, suitable for space and airborne operation. The configuration of this concept is shown in Figure 9.1 of Chapter 9 [57]. The radar antenna and transmitter parameters are shown in Table 1.16.

1.5.6 SBR for the Proposed United Nations International Satellite Monitoring Agency

The delegation from France to the United Nations has proposed the establishment

Table 1.16
Far-Term Military SBR Antenna and Transmitter Concept

Transmitter	
Type:	Free-Electron Maser (FEM)
Frequency:	60 GHz ± 5%
Peak Power:	10 MW
Average Power:	100 kW
Gain:	43 dB
Beam Voltage:	715 kV
Beam Current:	250 A
Efficiency dc-RF	37%
Wiggler Length:	99 cm
Tube Weight:	< 100 kg
Waveguide Output:	2.5 cm
Antenna	
Type:	Space-fed phased array F/D = 1
Size:	7 hex. panels—19.9 m × 20.7 m (overall)
Beamwidth:	227 μrad
Gain:	73 dB (area)
No. of Elements:	14,300,000
Illumination:	$\cos^2$ plus uniform-space taper
Surface Error:	< 7.6 mm (rms)
Element Type:	Microstrip
Phase Shifter:	MMIC
Bits	5 bits
Size	2.54 × 1.58 × 0.38 mm
Weight:	3457 kg

of an International Satellite Monitoring Agency (ISMA). Several resolutions have been passed by the UN General Assembly and a study report concerning ISMA has been issued [14]. Using the requirements given in the ISMA report (Table 1.17 shows resolution requirements), a concept for an SBR configuration has been designed and was described at the 1984 Military Microwaves Conference [13]. The satellite includes radar, electro-optical, infrared, and other sensors which operate in low earth orbit and provide observations that have various coverages and resolutions. The SBR is a spotlight SAR that operates in the millimeter-wave band. The antenna is hybrid reflector–phased array on gimbals that essentially provides 2π steradians of coverage with respect to the satellite nadir. During a typical pass over the area to be observed, a total of 80 maps can be collected using electronic scanning; each map has a size of 3740 × 430 m. Resolution is 15 cm when the integration time for each map is 2.3 s. The SBR characteristics are given in Table 1.18. The antenna has a diameter of 8.54 m and a beamwidth of 0.5 mrad at the design wavelength of 3.2 mm. The transmitter output stage is a parallel combination of Varian EIAs that provide a total average power output of 350 W. The receiver system noise temperature is 2000 K with a 1 GHz bandwidth, digital processing of the signal is performed, and memory size is 58.5 Mb. The data are stored on a

Table 1.17
Ground Resolution (in meters) Required for Treaty Verification and Crisis Monitoring [14]

Object	Detection	Recognition	Identification	Description
Bridges	6	4.5	1.5	0.90
Radar	3	0.9	0.3	0.15
Radiocommunications	3	1.5	0.3	0.15
Material Depots	1.5	0.6	0.3	0.25
Troop Units or Bivouacs	6	2.1	1.2	0.30
Air Base Equipment	6	4.5	3	0.30
Artillery and Rockets	0.9	0.6	0.15	0.05
Aircraft	4.5	1.5	0.9	0.15
Headquarters	3.0	1.5	0.9	0.15
Ground-to-Ground Missile and Anti Aircraft Sites	3	1.5	0.6	0.30
Medium-sized Surface Vessels	7.5	4.5	0.6	0.30
Vehicles	1.5	0.6	0.3	0.05
Land Mine Fields	9	6	0.9	0.025
Ports	30	15	6	3
Coasts and Landing Beaches	30	4.5	3	1.5
Marshalling Yards and Railways Shops	30	15	6	1.5
Roads	9	6	1.8	0.6
Urban Areas	60	30	3	3
Military Airfields	—	90	4.5	1.5
Submarines on the Surface	30	6	1.5	0.9

digital recorder for transmission to an ISMA ground station. The recorder has a capacity of $35 (10)^{11}$ bits, a data-rate capability of 60–300Mb/s with a BER of 10^{-6}. The estimated weight of the SBR is about 310 kg and requires prime power of 1700 W.

1.6 SBR ISSUES

Critical issues in the development of SBR include (1) system cost, (2) system survivability and vulnerability, (3) system calibration, (4) antenna deployment and distortion, (5) on-board processing, (6) nuclear prime power, (7) thermal control, (8) clutter and interference, and (9) launch vehicle capabilities. A succinct treatment of system cost, survivability and vulnerability, nuclear prime power, system calibration, launch vehicle capabilities, and clutter and interference is given here. Details of these and other issues are contained elsewhere in this book. For example, discussion of system survivability and vulnerability is found in Chapter 3 [58].

Table 1.18
ISMA Radar Parameters [13]

Antenna	
Type:	Reflector with offset-fed phased array
Size:	28 ft (8.54 m) diameter
Beamwidth:	0.5 mrad
Gain (area):	78.5 dB
F/D:	0.8
Transmitter	
Type:	EIA tubes in parallel
Frequency:	95 GHz +2%
Modulation:	Pulse +1 GHz LFM
Waveform:	Burst
Burst Length:	2000 μs
No. of Pulses/Burst:	10
Subpulse Width:	100 μs
No. of Bursts/s:	170
Average Power:	350 W
Receiver	
Type:	GaAs FET
System Noise Temp.:	2000 K
Bandwidth:	1 GHz
Signal Processor	
Type:	Digital
No. of Bits/Range Cell:	6 bits
Memory:	58.5 Mbits

Clutter and interference are covered in Chapter 11 [59]. System calibration is treated in Chapter 12 [42]. Antenna deployment and distortion are covered in Chapter 16 [60]. On-board processing is discussed in Chapter 14 [61]. Nuclear prime power is jound in Chapter 15 [62]. Thermal control considerations are treated in Chapter 9 [57].

1.6.1 SBR System Costs

The author uses a cost estimating ratio for SBR satellites of $64,000 per kilogram in 1988 dollars, which is based on informal study of many satellites that have been placed into orbit. The SEASAT cost $40,000 per kilogram in 1978 dollars. Assuming six percent average inflation since 1978, the SEASAT SBR cost in 1988 dollars would be $64,000 per kilogram. Launch costs are not included and are dependent on the launch vehicle. Informal study of many satellite launches has resulted in the data shown in Figures 1.14 and 1.15, which give launch cost for several types of vehicles when launched from two US launch sites, the Eastern Test Range (ETR) and the Western Test Range (WTR). We can see that polar orbit costs are greater than launches from the ETR due east, and it is more economical (on a dollars per pound basis) to launch large payloads on STS and Titan class vehicles.

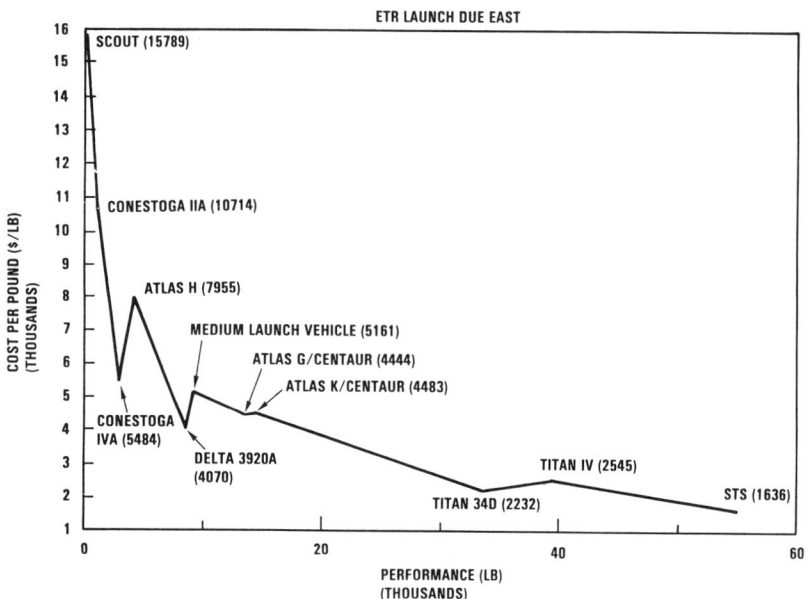

Figure 1.14 Cost per pound to low earth orbit—ETR launch due east.

System cost reduction will be possible if T/R modules are low in cost. One possible way in which this can be done is by mass production of the modules using MMIC technology and automatic testing techniques. During the 1970–1975 time period, Motorola mass produced a printed circuit radar fuze for 40 mm shells. Several million "radars" were automatically produced and tested at an average cost of $3 each. If the SBR module cost were $20, a typical large SBR could require about $2.5 million worth of modules. Considering the total system costs, this would be a reasonable price to pay. However, this possibility must be demonstrated. Typical T/R module assembly and test cost is half the module cost. Low system

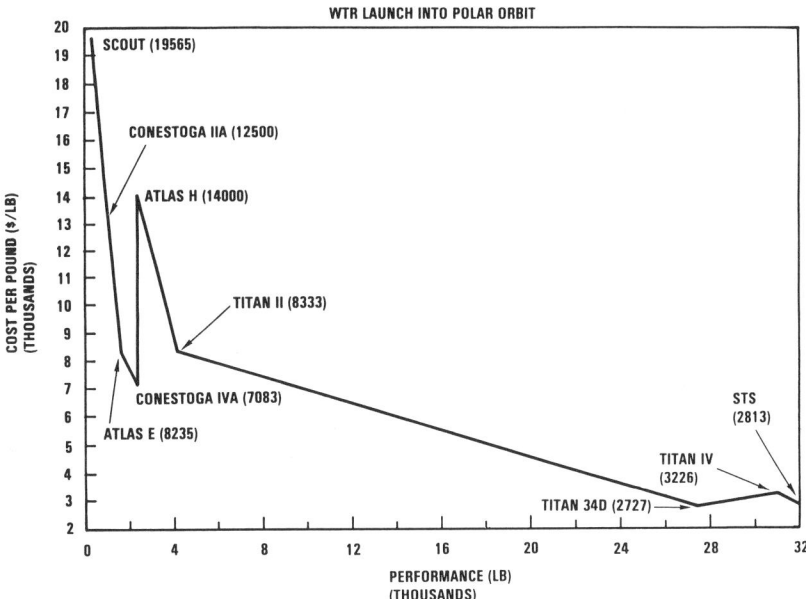

Figure 1.15 Cost per pound to low earth orbit—WTR launch into polar orbit.

costs will also result if the overall SBR system weight is kept as low as possible and automatic assembly and test techniques are used.

1.6.2 Survivability and Vulnerability

SBR system survivability and vulnerability must be demonstrated and tested. The natural space environment will cause a significant total radiation dose on a T/R module, depending on its shielding. Table 1.19 is a summary of the total dose for a five-year period for circular orbits at altitudes of 450, 900, and 5600 nmi [38]. The T/R module in the analysis has an area of one square inch, and we have assumed the total dose values will be double that expected in order to account for particle radiation which penetrates both sides of the module package. Some shielding may be provided by the chip substrate, but this has been ignored. A hardness level of $5(10)^6$ rad (Si) appears achievable in the 1985–1990 time period.

1.6.3 Nuclear Prime Power

Nuclear prime power system development depends on the successful test of a heat-pipe-cooled reactor that operates at high ($\approx$1500 K) temperatures. The current technology appears to be close to that required. In 1980, a two-meter long molybdenum heat pipe was constructed and tested [68]. The first test series in a quartz vacuum tube was performed at 1400 K. The heat pipe was tilted up to 60° to demonstrate performance against gravity.

1.6.4 SBR System Calibration

A SBR has unique calibration problems. For example, the antenna pattern measurements are frequently made at the far-field (Fraunhofer) range of $2D^2\lambda^{-1}$ from the antenna. For most antennas, these measurements give a reasonable approximation to the antenna pattern at all longer ranges. In particular, the measured gain in the main lobe is approximately 99% of the true gain. Sidelobe measurements are also quite good for all sidelobes beyond the first two. However, for large antennas operating at high frequencies, the far-field distance is very great. For example, the far-field distance for a 50 m diameter antenna operating at a 0.1 m wavelength is 50 km. Furthermore, the error in measurements of the first two sidelobes made at the Fraunhofer range is quite large for antennas specifically designed to have low near-in sidelobes. This error is due to the phase curvature of the wavefront at the antenna when used as a receiver. Therefore, new near-field antenna measurement techniques may need to be used for SBR calibration.

Table 1.19
Space Radiation Environment Summary [38]

SBR Orbit (nmi)	Five-Year Total Dose in rad (Si) Al Shielding thickness		
	15 mils	25 mils	50 mils
450	$2(10)^5$	$6(10)^4$	$2(10)^4$
900	$2(10)^6$	$4(10)^5$	$2(10)^5$
5600	$6(10)^6$	$4(10)^6$	$(10)^6$

1.6.5 Clutter and Interference

The SBR performance is significantly dependent on clutter and interference, either intentional or unintentional. To illustrate the magnitude of the clutter problem, consider the ATC radar described in Section 1.5. When the grazing angle is 70° and the reflectivity of the ground is -15 dB, the main beam clutter cross section is $+57$ dBSM. If the desired radar performance requires that a target with an RCS of $+13$ dBSM will have a signal-to-clutter ratio of 25 dB, the main beam clutter cancellation ratio must be at least 69 dB. Therefore, SBR performance requires clutter cancellation ratios that are large. A review of present capabilities in clutter suppression and cancellation indicates that the current standard is greater than 69 dB using pulsed doppler radar techniques. Barton [64] indicates clutter cancellation ratios up to 90 dB can be obtained by using pulsed doppler and displaced phase center antenna (DPCA) techniques.

Interference will primarily enter the SBR antenna through the sidelobes because the beamwidth is narrow. This interference can be either intentional noise jamming or unintentional from other radars. The effects of interference can be reduced to acceptable levels if adaptive sidelobe cancellation and sidelobe blanking techniques are used.

1.6.6 Launcher Vehicle Capabilities

The most probable launch vehicle for the SBR is the STS. Therefore, we must consider the STS capabilities to put various payloads that include one or more SBR satellites (and the propulsion systems to place them into the desired orbits). Figure 1.16 shows the STS cargo weight as a function of orbit inclination angle for various circular orbital altitudes and on-orbit velocity increments of the *orbital maneuvering system* (OMS). Note that 64,000 lbs can be delivered to a 100 nmi circular orbit, inclined at 50°, from the Kennedy Space Center launch site in Florida. If each SBR weighed 9500 lbs, three SBR satellites could be placed into orbit by one STS with 35,500 lbs in propulsion for orbital transfer.

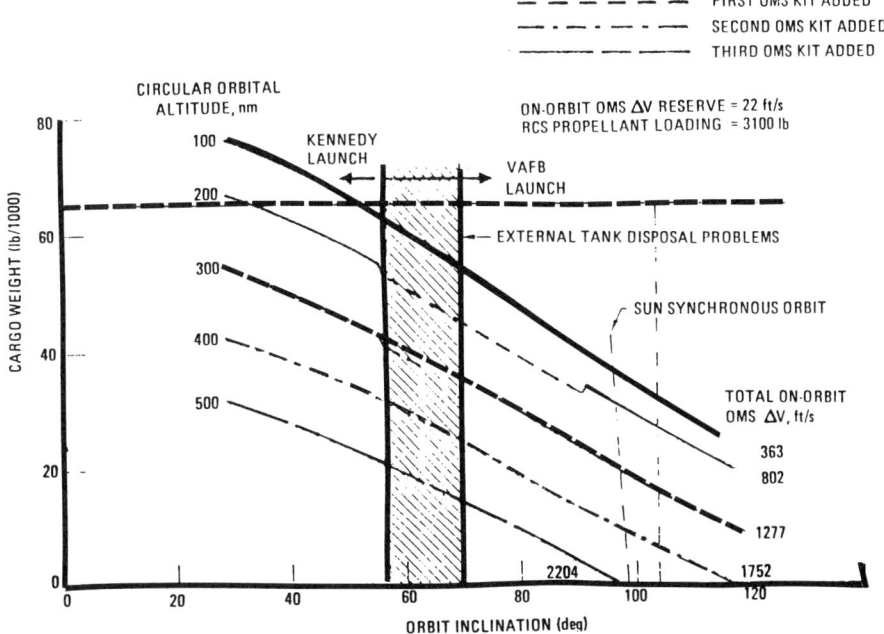

Figure 1.16 STS cargo weight *versus* inclination for various circular orbit altitudes (delivery only—no rendezvous).

REFERENCES

1. Hughes Aircraft Company, "Ku-band Integrated Radar and Communications Equipment for the Space Shuttle Orbiter Vehicle," Preliminary Design Review, Vol. I, March 14–24, 1978.
2. RCA Government and Commercial Systems, Aerospace Systems Division, Burlington, MA, "The Apollo LM Rendezvous Radar and Transponder," Report LTM 3300–15D, February 1971.
3. Quigley, W.W., "Gemini Rendezvous Radar," *Microwave Journal*, Vol. 8, No. 6, June 1965, pp. 39–45.
4. Fenner, R.G., and R.F. Broderick, "Spaceborne-Radar Applications," Chapter 34, *Radar Handbook*, M.I. Skolnik (ed.), McGraw-Hill, New York, 1970.
5. Elachi, C., et al., "Spaceborne Synthetic Aperture Imaging Radars: Applications, Techniques and Technology," *Proc. IEEE*, Vol. 70, No. 10, October 1982, pp. 1174–1209.
6. Elachi, C., and J. Granger, "Spaceborne Imaging Radars Probe "in depth," *IEEE Spectrum*, Vol. 19, No. 11, November 1982, pp. 24–29.
7. Williams, F.C., et al., "The Pioneer Venus Orbiter Radar," *IEEE WESCON '76*, Session 4, Los Angeles, CA, September 14–17, 1976.
8. Hofmeister, E.L., et al., "GOES-C Radar Altimeter," Vol. 1, *Data Users Handbook*, General Electric Company, Utica, NY, May 1976.
9. "Soviet Radar Records Venus Surface Imager," *Aviation Week and Space Technology*, Vol. 119, No. 17, October 24, 1983, p. 18.
10. Ulsamer, E., "Approach Set on Space Radars," *Air Force Magazine*, Vol. 67, No. 2, February 1984, pp. 17–18.
11. Brookner, E., and T.F. Mahoney, "Derivation of a Satellite Radar Architecture for Air Surveillance," *IEEE EASCON '83 Conf. Record*, Washington, DC, September 19–21, 1983.
12. Cantafio, L.J. and J.S. Avrin, "Satellite-Borne Radar for Global Air Traffic Surveillance," *IEEE ELECTRO '82 Professional Program Session Record*, Boston, MA, May 25–27, 1982.
13. Cantafio, L.J., "Space Based Radar Concept for the Proposed United Nations International Satellite Monitoring Agency," *Military Microwaves '84 Conf.*, London, October 24–26, 1984.
14. *The Implication of Establishing an International Satellite Monitoring Agency*, United Nations Publication No. E.83.IX.3, 1983.
15. Griffin, J.W., et al., "Ku-band—The First Year of Operation," *Proc. IEEE Int. Radar Conf.*, 1985, pp. 330–338.
16. NASA, "Orbital Maneuvering Vehicle," George C. Marshall Space Flight Center, AL, Request for Proposal 1-6-PP-01438, November 1985.
17. Locke, J., and L.J. Cantafio, "Rendezvous Radar," Chapter 6, *Space-Based Radar Handbook*, L.J. Cantafio (ed.), Artech House, Norwood, MA, 1989.
18. NASA Brochure, "Geodynamics Experimental Ocean Satellite Project of the Earth and Ocean Physics Applications Program," NASA Wallops Flight Center, Wallops Island, VA. 1975.
19. NASA News Release, "New Satellite to Measure Ocean Surface Topography and Sea State," No. 75–88, NASA, Washington, DC, March 31, 1975.
20. NASA, *GEOS-C Mission Plan*, TK-6340-001 Rev. 3, NASA Wallops Flight Center, Wallops Island, VA. December 18, 1974.
21. Carver, K.R., C. Elachi, and F.T. Ulaby, "Microwave Remote Sensing from Space," *Proc. IEEE*, Vol. 73, No. 6, June 1985, pp. 970–996.
22. McCandless, S.W., "Synthetic Aperture Radar in Space," Chapter 4, *Space-Based Radar Handbook*, L.J. Cantafio (ed.), Artech House, Norwood, MA, 1989.
23. Lund, T.J., "Radar Altimeters for Space Vehicles," Chapter 7, *Space-Based Radar Handbook*, L.J. Cantafio (ed.), Artech House, Norwood, MA, 1989.
24. Moore, R.K., "Scatterometers and Other Modest Resolution Radar Systems," Chapter 8, *Space-Based Radar Handbook*, L.J. Cantafio (ed.), Artech House, Norwood, MA, 1989.

25. Meneghini, R. and T. Kozu, *Weather Radar in Space,* Artech House, Norwood, MA, to be published 1990.
26. Cantafio, L.J., "Satellite-Borne Radar," Lecture, Advanced Radar Technology Short Course, Technology Service Corporation, San Diego, CA, April 22, 1983.
27. Wolverton, R.W., *et al.*, *Flight Performance Handbook for Orbital Operations*, John Wiley and Sons, New York, 1961.
28. Lüders, R.D., and L.J. Ginsberg, "Continuous Zonal Coverage—A Generalized Analysis," "AIAA Paper No. 74-842, *AIAA Mechanics and Control of Flight Conference*, Anaheim, CA, August 5-9, 1974.
29. Emara, E.T., and C.T. Leondes, "Minimum Number of Satellites for Three-Dimensional Continuous Worldwide Coverage," *IEEE Trans. Aerospace and Electronic Systems*, Vol. AES-13, No. 2, March 1977, pp. 108-111.
30. Ballard, A.H., "Rosette Constellation of Earth Satellites," *IEEE Trans. Aerospace and Electronic Systems*, Vol. AES-16, No. 5, September 1980, pp. 656-673.
31. Walker, J.G., "Continuous Whole Earth Coverage by Circular Orbit Satellite Patterns," Royal Aircraft Establishment Technical Report 77044, March 24, 1977.
32. Beste, D.C., "Design of Satellite Constellations for Optimal Continuous Coverage," *IEEE Trans. Aerospace and Electronic Systems*, Vol. AES-14, No. 3, May 1978, pp. 466-473.
33. Harney, E.D., *Space Planners Guide*, USAF Systems Command, U.S. Government Printing Office, Washington, DC, Pub. 0-774-405, 1965.
34. Chobotov, V.A., "Orbital Mechanics Considerations for Space-Based Radar," Chapter 2, *Space-Based Radar Handbook*, L.J. Cantafio (ed.), Artech House, Norwood, MA, 1989.
35. Kendrick, J.B. (ed.), *TRW Space Data*, 3rd Edition, 1967.
36. Schultz, J.L., and P. Nosal, "Space-Based Radar," *Horizons*, Grumman Aerospace Corporation, Vol. 15, No. 1, 1979, p. 10.
37. Fawcette, J., "Large Radar Satellite Proposed," *Microwave Systems News*, Vol. 8, No. 9, September 1978, pp. 17-20.
38. Mrstik, A.V., *et al.*, "RF Systems in Space—Space-Based Radar Analysis," General Research Corporation, RADC TR-83-91, Vol. II, Final Technical Report, April 1983.
39. Ludwig, A.C., *et al.*, "RF Systems in Space—Space Antennas Frequency (SARF) Simulation," General Research Corporation, RADC TR-38-91, Vol. I, Final Technical Report, April 1983.
40. Ludwig, A.C., *et al.*, "Functional Requirements for The SEASAT-A Synthetic Aperture Radar System," Jet Propulsion Laboratory, Pasadena, CA, FR No. FM511774 (Rev.), August 2, 1976.
41. Brejcha, A.G., L.H. Keeler, and G.G. Sanford, "The SEASAT-A Synthetic Aperture Radar Antenna," *Synthetic Aperture Radar Technology Conference*, Las Cruces, NM, March 8-10, 1978.
42. Cantafio, L.J., "Space Antenna Technology," Chapter 12, *Space-Based Radar Handbook*, L.J. Cantafio (ed.), Artech House, Norwood, MA, 1989.
43. Kalmykov, A.I., *et al.*, "Side-Looking Radar of Kosmos-1500 Satellite," *Issledovaniye Zemli Iz Kosmosa*, No. 3, May-June 1985.
44. Anon., "Soviets Plan to Launch New Spacecraft," *Aviation Week and Space Technology*, Vol. 127, No. 16, October 19, 1987, p. 27.
45. Bogomolov, *et al.*, "Venera 15 and 16 Synthesized Aperture Radar in Orbit Around Venus," *Izvestiva Vvsshikh Uchebnvkh Zavedennii, Radiofizika*, Vol. 28, No. 3, March 1985, pp. 259-274.
46. Porcello, L.J., *et al.*, "The Apollo Lunar Sounder Radar System," *Proc. IEEE*, Vol. 62, No. 6, June 1974, pp. 769-783.
47. Dietz, R.H., "Space Station Communications and Tracking Systems," *Proc. IEEE*, Vol. 75, No. 3, March 1987, pp. 769-783.
48. Tu, K., *et al.*, "Space Shuttle Communications and Tracking System," *Proc. IEEE*, pp. 356-370.
49. NASA, Announcement of Opportunity—The Earth Observing System (EOS), January 19, 1988.

50. Malibu Research and TRW, "Lorra/Tramar Design Feasibility Study," May 11, 1988.
51. Skahill, G., "A Dual Reflector Antenna Scans Many Beamwidths without Loss of Gain, Resolution or Sidelobe Level," *Microwave Journal*, Vol. 31, No. 3, March 1988, pp. 129–139.
52. Roach, B., *et al.*, "Ka-band Radar—Which Tube to Choose?" *Microwave Systems News and Communications Technology*, Vol. 17, No. 7, July 1987, pp. 90–96.
53. Schneible, R., "Large Radar Array Technology," RADC Briefing to SDI Architecture Contractors, December 1985.
54. Boehmer, H., *et al.*, "Conceptual Design of a Free Electron Maser with Electron Pre-bunching," *Bull. Am. Phys. Soc.*, Vol. 30, 1985, p. 1540.
55. Boehmer, H., *et al.*, "The TRW Free Electron Maser," *Bull. Am. Phys. Soc.*, Vol. 31, 1986, p. 1482.
56. Arnush, D., "Free Electron Maser Development at TRW," *Joint DoE/DoD High-Power MMW Fast-Wave Devices Workshop*, April 8, 1987.
57. Herold, L.M., and M.S. Bushby, "Thermal Control for Space Based Radar," Chapter 9, *Space-Based Radar Handbook*, L.J. Cantafio (ed.), Artech House, Norwood, MA, 1989.
58. Knepp, D.L., and J.T. Reinking, "Space Environment and Effects on Space-Based Radar," Chapter 3, of *Space-Based Radar Handbook*, L.J. Cantafio (ed.), Artech House, Norwood, MA, 1989.
59. Andrews, G.A., and K. Gerlach, "Space-Based Radar Clutter and Interference," Chapter 11, *Space-Based Radar Handbook*, L.J. Cantafio (ed.), Artech House, Norwood, MA, 1989.
60. Kovalcik, E., "Space Structures," Chapter 16, *Space-Based Radar Handbook*, L.J. Cantafio (ed.), Artech House, Norwood, MA, 1989.
61. Swartzlander, E.E., Jr., "On-Board Radar Signal Processors," Chapter 14, *Space-Based Radar Handbook*, L.J. Cantafio (ed.), Artech House, Norwood, MA, 1989.
62. Boretz, J.E., "Prime Power Systems in Space," Chapter 15, *Space-Based Radar Handbook*, L.J. Cantafio (ed.), Artech House, Norwood, MA, 1989.
63. Emigh, C.R., "Reactor Technology, January–March 1980," Los Alamos Scientific Laboratory (LASL), Progress Report LA-8403-PR-UC-80, June 1980.
64. Barton, D.K., "A Half Century of Radar," *IEEE Trans. Microwave Theory and Techniques*, Vol. MTT-32, No. 9, September 1984, pp. 1161–1169.
65. McGoogan, J.T., *et al.*, "The S-193 Radar Altimeter Experiment," *Proc. IEEE*, Vol. 62, No. 6, June 1974, pp. 793–803.

Chapter 2
ORBITAL CONSIDERATIONS FOR SPACE-BASED RADAR
V.A. Chobotov
Palos Verdes Estates

2.1 KEPLER'S LAWS

Orbital mechanics owes its origins to the observations of Tycho Brahe (1546–1601), the computations of Johannes Kepler (1571–1630), and the explanations of Isaac Newton (1642–1727). Kepler's three laws reworded to apply to geocentric orbits are:

1. Each satellite moves in an elliptical orbit with the center of the earth's mass as a focus;
2. The radius vector drawn from the center of the earth to the satellite sweeps out equal areas in equal amounts of times;
3. The square of the satellite's period of revolution is proportional to the cube of the semimajor axis of the satellite's orbit.

These laws assume that all bodies can be treated as point masses, and that each body is attracted solely by one large mass. Thus, the two-body Keplerian motion ignores the effects of nonsphericity of the attracting body and influence of other bodies on the motion of the two-body system.

2.2 ORBIT EQUATIONS

Keplerian motion is represented by the solution of Newton's law of gravitation equation:

$$\ddot{\mathbf{r}} + \frac{\mu \mathbf{r}}{r^3} = 0 \qquad (2.1)$$

where $\ddot{r}$ is the acceleration of a mass point m in a gravitational field of a mass point M. The gravitational parameter $\mu = GM$, where G is the universal gravitational constant.

The solution of (2.1) is a convenient expression for Kepler's first law. It is the polar equation for an orbit:

$$r = p/(1 + e\cos\theta) \tag{2.2}$$

where

r and θ = polar coordinates,
p = semilatus rectum or semiparameter of the ellipse,
e = eccentricity.

Kepler's second law may be written as

$$H = r^2\dot{\theta} \tag{2.3}$$

where

H = angular momentum,
$\dot{\theta}$ = time rate of change of θ.

Kepler's third law is

$$T = 2\pi\sqrt{\frac{a^3}{\mu}} \tag{2.4}$$

where

T = periodic time,
a = semimajor axis (see Figure 2.1).

The classification of orbits in terms of the eccentricity e is as given by Table 2.1.

Equation (2.1) also has two integrals (constants) of motion, the energy integral U:

$$\frac{v^2}{2} - \frac{\mu}{r} = U \tag{2.5}$$

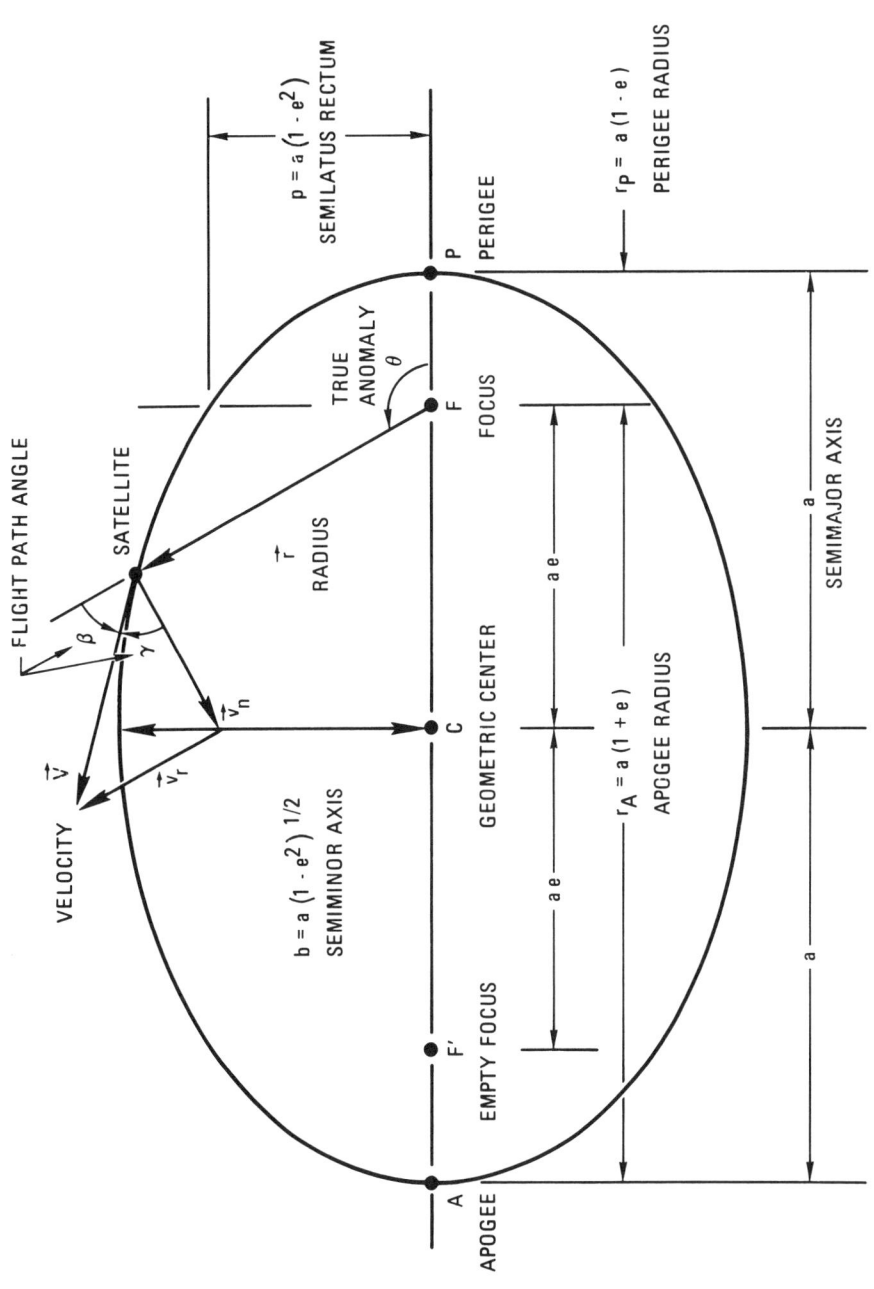

Figure 2.1 Ellipse geometry.

Table 2.1
Orbits Classified by Eccentricity

e	Orbit
0	Circle ($a = r$)
<1	Ellipse ($a > 0$)
1	Parabola ($a \approx \infty$ (undefined))
>1	Hyperbola ($a < 0$)

and the angular momentum integral H:

$$H = \sqrt{\mu p} \tag{2.6}$$

where $U = -\mu/2a$ and $p = a(1 - e^2)$.

Hence, the Keplerian velocity at any point in the orbit may be expressed as

$$v = \sqrt{\mu\left(\frac{2}{r} - \frac{1}{a}\right)} \tag{2.7}$$

The constancy of U and H reflects the fact that the Keplerian motion occurs in a plane that is fixed in inertial space.

For circular orbits, (2.7) yields the orbital velocity:

$$v_c = \sqrt{\frac{\mu}{r}} \tag{2.8}$$

Escape velocity results when $a \Rightarrow \infty$. The parabolic escape velocity then is

$$v_{esc} = \sqrt{\frac{2\mu}{r}}$$

$$= \sqrt{2}v_c \tag{2.9}$$

Several useful relationships for Keplerian orbits are

$$a = (r_a + r_p)/2 = \text{semimajor axis}$$

$$r_a = a(1 + e) = \text{apogee radius}$$

$$r_p = a(1 - e) = \text{perigee radius} \tag{2.10}$$

2.3 TIME OF FLIGHT

If the position of a satellite is desired at a specified time t, it can be found from

Kepler's equation:

$$M = n(t - \tau)$$
$$= E_t - e \sin E_t \qquad (2.11)$$

where

τ = time of perigee passage,
M = mean anomaly,
E_t = eccentric anomaly,
$n = \sqrt{\mu/a^3}$ = mean motion.

The eccentric anomaly E_t is illustrated in Figure 2.2. The true anomaly θ can be determined from

$$\tan\frac{\theta}{2} = \left[\frac{1+e}{1-e}\right]^{1/2} \tan\frac{E_t}{2} \qquad (2.12)$$

Conversely, if the time t of travel from one point on the ellipse to another point is desired, θ can be found from (2.11) by an iterative solution.

2.4 COORDINATE SYSTEMS

2.4.1 ECI Reference Frame

Several coordinate systems are used in describing the motion of an artificial satellite of the earth. The *earth-centered inertial* (ECI) coordinate system has the x and y axes in the earth's equatorial plane with the x-axis directed in the vernal equinox direction. The z-axis points in the direction of the North Pole as illustrated in Figure 2.3.

The two angles α and δ, and the radial distance r, needed to specify the position of an object in the ECI reference frame, are defined as follows:

α (right ascension)—The angle measured eastward in the plane of the equator from a fixed intertial axis in space (vernal equinox) to a plane normal to the equator (meridian) which contains the object; $0° \leq \alpha \leq 360°$.

δ (declination)—The angle between the object and equatorial plane measured (positive above the equator) in the meridional plane which contains the object; $-90° \leq \delta < 90°$.

r (radial distance)—The distance between the origin of the coordinate system and location of a point (object) within the coordinate system $r \geq 0$.

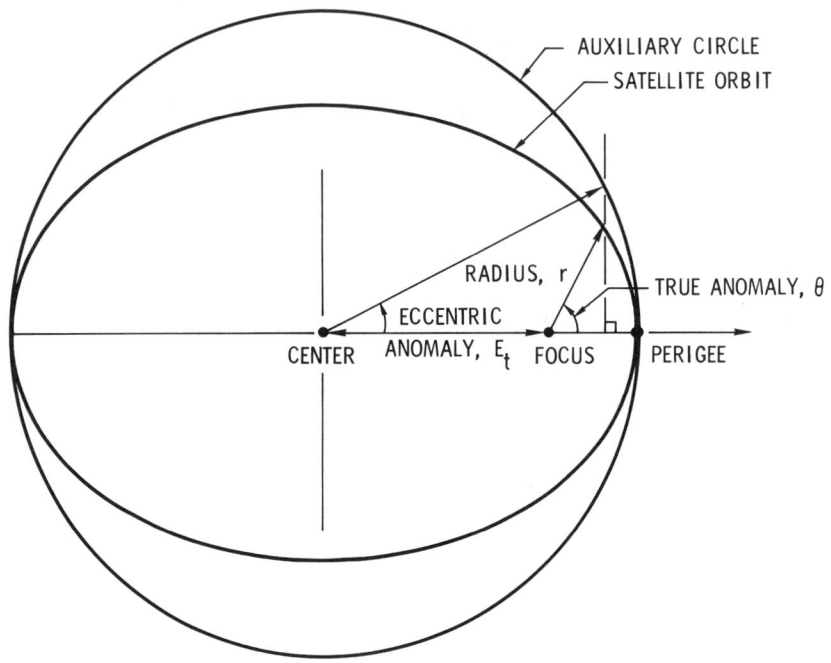

Figure 2.2 Definition of eccentric anomaly.

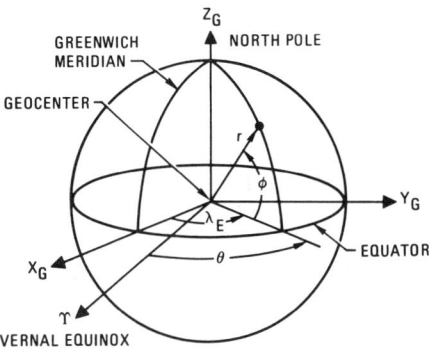

Figure 2.3 Earth-centered inertial (ECI) system.

2.4.2 Latitude-Longitude Coordinate Frame

The geographic coordinate system shown in Figure 2.4 is used to locate objects in terms of the latitude, longitude, and altitude of an object. The geocentric latitude

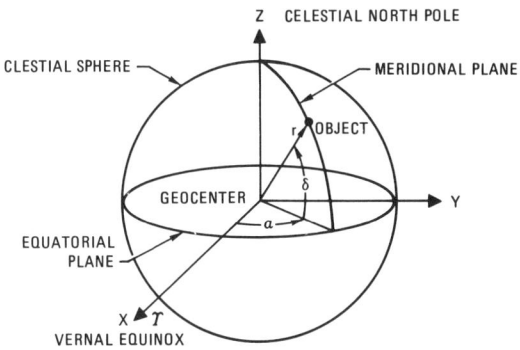

Figure 2.4 Latitude-longitude coordinate system.

and longitude are defined as follows:

ϕ (geocentric latitude)—The acute angle measured perpendicular to the equatorial plane between the equator and a ray connecting the geocenter with a point on the earth's surface; $-90 \leq \phi \leq 90°$.

λ_E (east longitude)—The angle measured eastward from the prime (Greenwich) meridian in the equatorial plane to the meridian containing the surface point; $0° \leq \lambda_E \leq 360°$.

2.4.3 Azimuth-Elevation Coordinate Frame

An observer standing at a particular point on the surface of a rotating planet sees objects in a rotating coordinate system. In this system, the observer is at its origin and the fundamental plane is the local horizon (Figure. 2.5). Such a coordinate system is referred to as a *topocentric system*. Generally, the principal axis or direction is taken as pointing due south. Relative to an observer, the object is in a meridional plane, which contains the object and passes through the zenith of the observer.

2.5 ORBITAL ELEMENTS

The motion of a satellite around the earth is described mathematically by three scalar second-order differential equations. The integration of these equations of motion yields six constants of integration. These constants of integration are known as the *orbital elements*.

The Keplerian orbital elements are often referred to as classical or conventional elements, and they are the simplest and easiest to use. This set of orbital

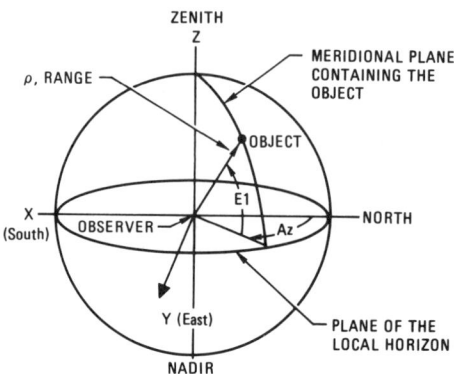

Figure 2.5 Azimuth-elevation topocentric coordinate system.

elements can be divided into two groups: the *dimensional elements* and the *orientation elements*.

The dimensional elements specify the size and shape of the orbit, and they relate position in the orbit to time (Figure 2.1). The elements are as follows:

a = semimajor axis, which specifies the size of the orbit;
e = eccentricity, which specifies the shape of the orbit;
τ = time or perigee passage, which relates position in orbit to time;
  τ is often replaced by M, the mean anomaly at some arbitrary time, t. The mean anomaly is a uniformly varying angle.

The orientation elements specify the orientation of the orbit in space (Figure 2.6). They are as follows:

i = inclination of the orbit plane with respect to the reference plane, which is taken to be the earth's equatorial plane for satellite orbits ($0° \leq i \leq 180°$). For $0° \leq i \leq 90°$, the motion is *direct*; for $90° < i < 180°$, the motion is termed *retrograde*.
Ω = longitude of the ascending node (often shortened simply to *node*); Ω is measured counterclockwise in the equatorial plane from the direction of the vernal equinox to the point where the satellite makes its south-to-north crossing of the equator ($0° \leq \Omega < 360°$).
ω = argument of perigee; ω is measured in the orbital plane, in the direction of motion, from the ascending node to perigee ($0° \leq \omega < 360°$).

The angles i and Ω specify the orientation of the orbital plane in space. The angle ω then specifies the orientation of the orbit in its plane. The argument of

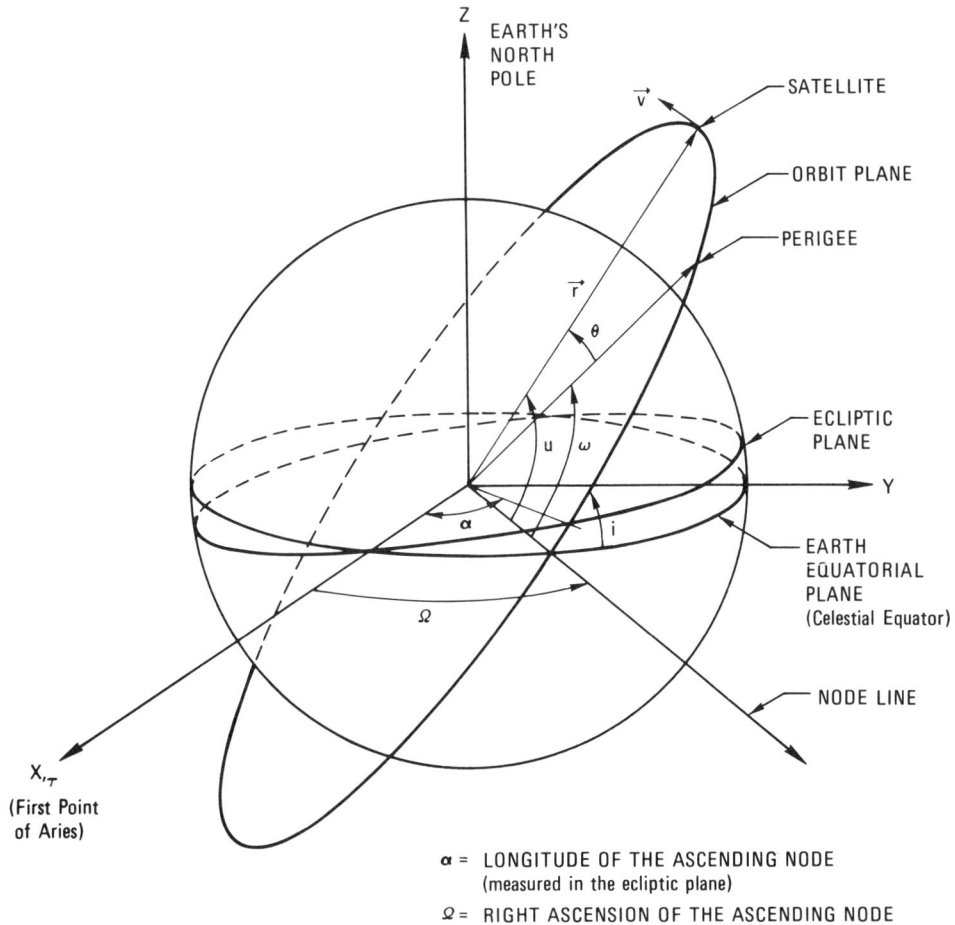

Figure 2.6 Orientation of orbit in space.

latitude u defines the position of the satellite relative to the node line.

Still another system of specifying the satellite state vector involves the scalar quantities of

v = velocity,
r = radius,
Ω = node,
γ = flight path angle, (or $\beta = \pi/2 - \gamma$),
δ = geocentric latitude (or ϕ),
A_z = azimuth of v from true north.

Various sets of elements are used in orbit determination; the inertial rectangular ($x, y, z, \dot{x}, \dot{y}, \dot{z}$) and the spherical ($\alpha, \delta, \beta, A_z, r, v$), previously introduced, are examples of orbital elements. Note that both sets of elements give the satellite's position as a point in space at any specific time. Examples of retrograde ($i > 90°$) and posigrade ($i < 90°$) orbits are illustrated in Figure 2.7. In a posigrade or direct orbit, the satellite moves eastward at the northbound crossing of the equator. The satellite moves westward in retrograde orbits.

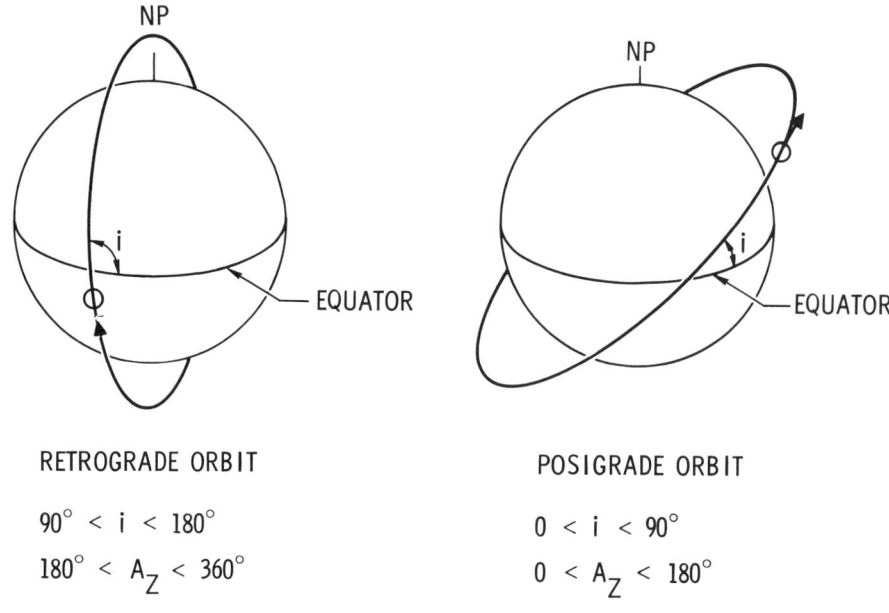

RETROGRADE ORBIT

$90° < i < 180°$
$180° < A_Z < 360°$

POSIGRADE ORBIT

$0 < i < 90°$
$0 < A_Z < 180°$

Figure 2.7 Earth orbits.

2.5.1 Summary of Orbit Equations

Eccentricity

$$e = 1 - \left(\frac{h_p + R_e}{a}\right) \text{ or } e = \frac{h_a - h_p}{2R_e + h_a + h_p} \tag{2.13}$$

Semimajor Axis   *Semiminor Axis*

$$a = R_e + \frac{h_p + h_a}{2} \quad \text{and} \quad b = (R_e + h_p)\sqrt{\frac{1+e}{1-e}} \tag{2.14}$$

Period

$$P = \frac{\pi a^{3/2}}{7500} \quad \text{(min), where } a \text{ is in nmi} \tag{2.15}$$

Average Angular Velocity

$$n = \frac{2\pi}{P} \quad \text{(rad/min)} \tag{2.16}$$

True Anomaly

$$\theta = \cos^{-1}\left(\frac{\cos E_t - e}{1 - e\cos E_t}\right) \tag{2.17}$$

Eccentric Anomaly

$$E_t = \cos^{-1}\left[\frac{e + \cos\theta}{1 + e\cos\theta}\right] = 2\tan^{-1}\left[\left(\frac{1-e}{1+e}\right)^{1/2}\tan\frac{\theta}{2}\right] \tag{2.18}$$

Mean Anomaly

$$M = nt = E_t - e\sin E_t \tag{2.19}$$

Time

$$t = \frac{1}{n}(E_t - e\sin E_t) = \frac{MP}{2\pi} \tag{2.20}$$

Altitude

$$h = a(1 - e\cos E_t) - R_e \tag{2.21}$$

Also,

$$r = R_e + h = \frac{(R_e + h_p)(1 + e)}{1 + (e\cos\theta)} \tag{2.22}$$

Velocity (at any point on ellipse with instantaneous altitude of h)

$$v = \sqrt{\mu\left(\frac{2}{h + R_e} - \frac{1}{a}\right)} \tag{2.23}$$

where μ = gravitational constant.

Flight Path Angle

$$\gamma = \tan^{-1}\left(\frac{e\sin\theta}{1 + e\cos\theta}\right) \tag{2.24}$$

The quantities h_p, h_a, R_e, a, and h may be expressed in feet in the above formulas, where R_e is the radius of the earth. Altitudes h_p and h_a are all measured from the surface of the central body; V is in feet per second; P and t are in minutes; $\mu = 1.40764545 \times 10^{-16}$ ft³/s².

2.6 GRAVITATIONAL PERTURBATIONS

2.6.1 Earth's Oblateness Effects

The principal perturbation effects in a near earth orbit are due to the earth's oblateness and its triaxiality. The earth's oblateness effects, or those caused by the J_2 zonal harmonic coefficient in the gravitational potential function, are a steady precession or regression, Ω, of the orbital plane about the earth's polar axis and the apsidal rotation ω about the normal to the orbital plane. The nodal regression, or westward motion of the orbital plane, occurs when $i < 90°$, as illustrated in Figure 2.8, where S denotes the interval between two successive crossings of the equator.

The motion of the node Ω occurs because of the added attraction of the earth's equatorial bulge, which introduces a force component toward the equator. The resultant acceleration causes the satellite to reach the equator (node) short of the crossing point for a spherical earth. This effect may also be regarded as a gyrodynamic precession due to a torque acting on the satellite angular momentum

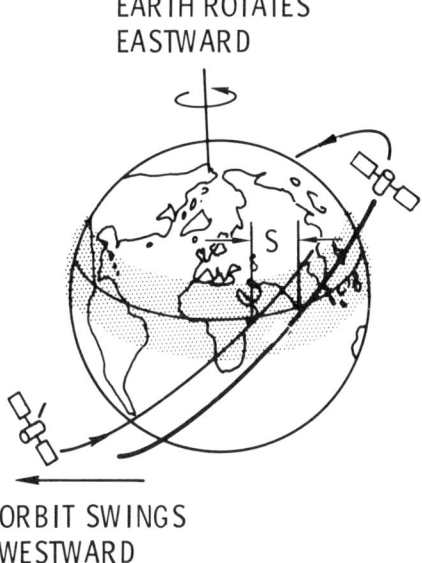

Figure 2.8 The Gravitational pull of the earth's equatorial bulge causes the orbital plane of an eastbound satellite to swing westward.

vector caused by the additional attraction of the earth's equatorial bulge. The secular nodal regression rate can be numerically evaluated to first order in the dominant oblateness parameter J_2 from

$$\dot{\Omega} = \frac{9.9639}{(1-e^2)^2}\left(\frac{R_e}{R_e + \bar{h}}\right)^{3.5} \cos i \left(\frac{\text{degrees}}{\text{mean solar day}}\right) \quad (2.25)$$

where

$$\bar{h} = \frac{h_a + h_p}{2}, \quad e = \frac{h_a - h_p}{h_a + h_p + 2R_e}$$

and R_e is the equatorial radius of the earth. As noted previously, the node regresses for direct orbits ($0° \leq i < 90°$) and advances for retrograde orbits ($90° < i \leq 180°$). Furthermore, there is no nodal regression to first order for polar orbits. Figure 2.9 illustrates the nodal regression rate *versus* inclination for various average altitude values.

The secular motion of the perigee (line of apsides) is because the force is no longer proportional to the inverse square of radial distance, and the orbit consequently is no longer a closed ellipse, as illustrated in Figure 2.10.

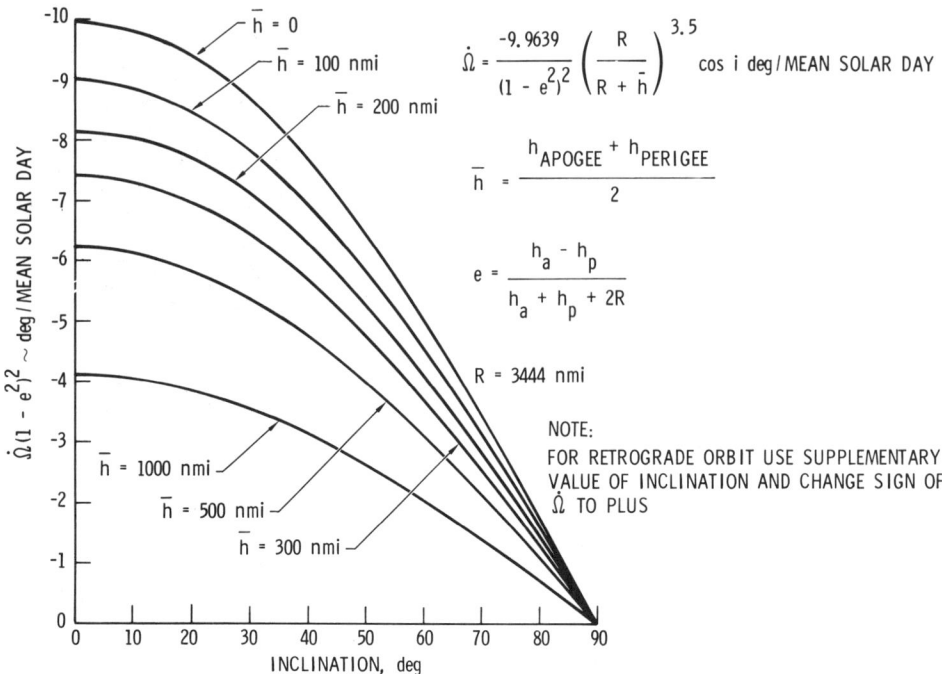

Figure 2.9 Regression rate due to oblateness *versus* inclination for various values of average altitude.

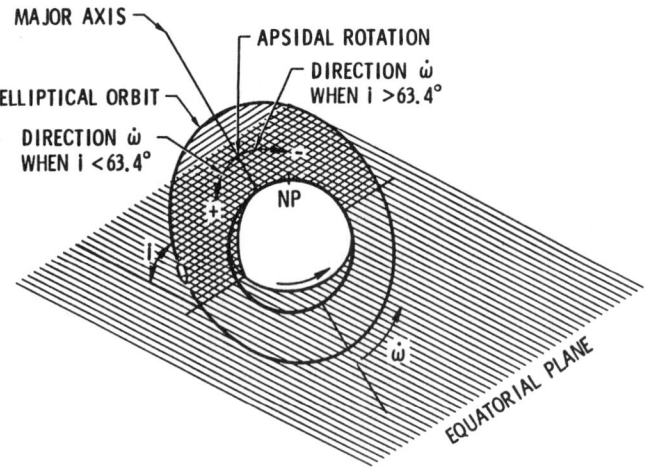

Figure 2.10 Illustration of apsidal rotation.

Table 2.2
Constants and Conversion Factors*

Earth Gravitational Constant
$$\mu = 5.53043822 \times 10^{-3} \text{ER}^3/\text{min}^2$$
$$= 3.98600800 \times 10^5 \text{ km}^3/\text{s}^2$$
$$= 1.40764545 \times 10^{16} \text{ ft}^3/\text{s}^2$$
$$= 6.27502150 \times 10^4 \text{ (nmi)}^3/\text{s}^2$$
$$= 8.13242786 \times 10^{11} \text{ (nmi)}^3/\text{hr}^2$$

Mean Equatorial Earth Radius (ER)
$$\text{ER} = 2.09256398 \times 10^7 \text{ ft}$$
$$= 6.37813500 \times 10^3 \text{ km}$$
$$= 3.44391738 \times 10^3 \text{ nmi}$$

Polar Radius (PR)
$$\text{PR} = 3432.37068 \text{ nmi}$$

Rotational Rate of the Earth
$$\omega_e = 2.62516146 \times 10^{-1} \text{ rad/hr}$$
$$= 4.37526910 \times 10^{-3} \text{ rad/min}$$
$$= 7.29211515 \times 10^{-5} \text{ rad/s}$$
$$= 1.50410672 \times 10^1 \text{ °/hr}$$
$$= 2.50684454 \times 10^{-1} \text{ °/min}$$
$$= 4.17807423 \times 10^{-3} \text{ °/s}$$

Flattening of the Earth
$$f = \frac{\text{equatorial radius} - \text{polar radius}}{\text{equatorial radius}} = 3.3528 \times 10^{-3}$$
$$= 1/298.26$$

First-Order Coefficients for Earth Oblateness and Equatorial Ellipticity
Zonal Harmonics:
$$J_2 = 1.08261579 \times 10^{-3}$$
$$J_3 = -2.53881 \times 10^{-6}$$
$$J_4 = -1.65597 \times 10^{-6}$$
Sectorial Harmonic:
$$J_2^{(2)} = 1.818300746 \times 10^{-6}$$

Characteristics of a "Grazing" Orbit at the Earth's Equatorial Radius
Velocity = 25936.2550 ft/s
Period = 84.4889851 min
= 1.40814975 hr

Conversion Factors
1 ft = 0.3048 m
1 nmi = 1852 m
= 6076.11550 ft
1 km = 0.53995680 nmi
= 3280.8399 ft
1 rad = 57.295779513°
π = 3.141 592 653 590 . . .

Time Conversions
1 mean solar day = 86,400 ephemeris seconds
1 mean sidereal day = 86,164.09054 ephemeris seconds
1 tropical year = 365.2421988 mean solar days

*The values for the constants given in this section are those included in the 1972 Geodetic System Model.

The apsidal rotation rate $\dot{\omega}$ can be expressed as

$$\dot{\omega} = \frac{9.9639}{(1-e^2)^2}\left(\frac{R_e}{R_e + \bar{h}}\right)^{7/2}\left(2 - \frac{5}{2}\sin^2 i\right)\left(\frac{\text{degree}}{\text{mean solar day}}\right) \quad (2.26)$$

The apsidal rate is shown in Figure 2.11 as a function of orbital plane inclination i with $\bar{h}$ as a parameter. Note that $\dot{\omega} = 0$ when $i = 63.4°$ or $116.5°$, corresponding to these critical inclinations.

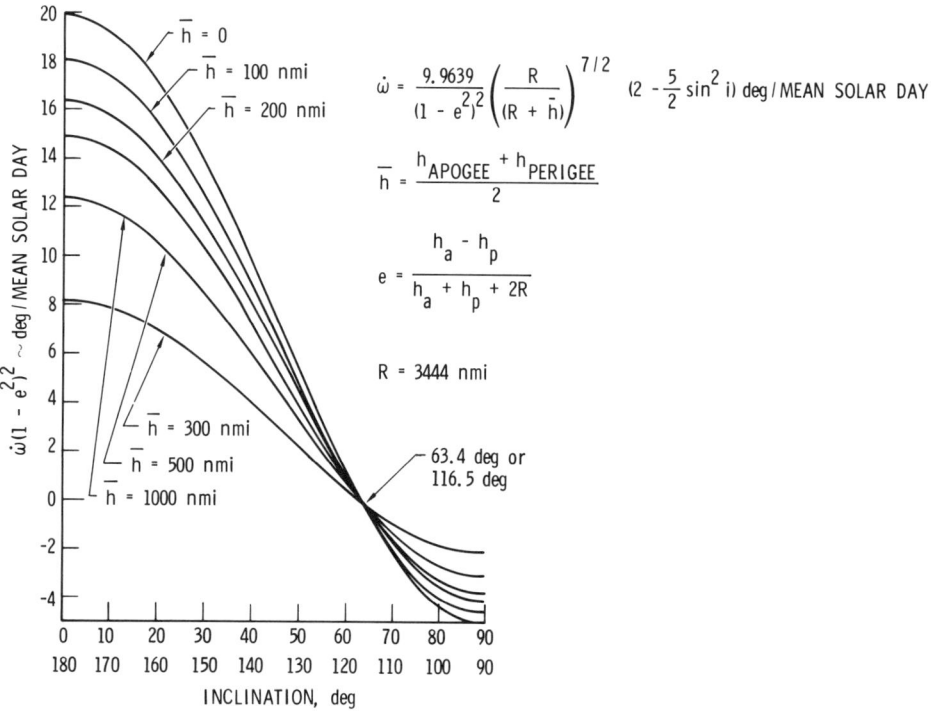

Figure 2.11 Apsidal rotation rate due to oblateness *versus* inclination for various values of average altitude.

2.7 ORBITAL SYSTEMS

Orbital systems are groups or constellations of satellites dedicated to the performance of a mission. The mission may be to communicate continuously between any two points on the globe, to provide accurate navigational fixes to various users, or to observe global weather conditions or other events in space and on the surface of the earth. The establishment of such systems requires a thorough and complete knowledge of the orbital characteristics and the technical and economic problems involved in launching satellites into mission orbits.

A constellation of satellites may be configured in many ways. An extensive parameter search usually must be made to examine factors such as altitude, total number of satellites required, launch injection parameters, and the relationships among the satellites in the constellation. The possibility of satellite failures must also be taken into account. The inoperative satellite mode may then be used to test the constellation for a degraded mode of system operation.

Satellite systems can be designed for global or regional coverage. Global coverage implies a continuous visibility of at least one satellite everywhere on the globe. Regional coverage, however, suggests that the satellite visibility (and hence mission performance) is concentrated over a specified region on the earth. An example of a regional system is the synchronous equatorial satellite, which can be placed at a specified longitude near or on the equator for transmission of messages from points east and west of the satellite, but not over the entire globe.

An example of a global system could be a number of synchronous equatorial satellites equally spaced along the equator or in inclined orbits. A continuous visibility of one or more satellites over most of the earth's geography would be ensured by such a system.

The study of orbital systems often involves examination of the satellite ground traces, initially and as a function of time to determine their evolution due to various orbital perturbations. The ground traces could be defined as the intersections of the satellite radius vector with the earth's surface. The establishment of the ground traces at specified locations on the earth's surface involves determination of the launch time and launch window. These and related problems will be discussed in this section.

2.7.1 Launch Window Considerations

2.7.1.1 Launch Azimuth

The launching of a satellite into an orbit with inclination i requires the launching in an azimuth direction Az defined by the formula

$$\sin Az = \frac{\cos i}{\cos \phi_0} \tag{2.27}$$

where ϕ_0 is the launch site latitude. For a due east launch (Az = 90°), $i = \phi_0$, or the orbital inclination is equal to the launch site latitude. Differentiating (2.27) and solving for di, we obtain

$$(\mathrm{d}i) = \frac{-\cos\phi_0 \cos\mathrm{Az}\,\mathrm{dAz}}{\sin i} \tag{2.28}$$

which shows the sensitivity of inclination (di) due to an error in azimuth (dAz). Because of range safety reasons the azimuth may be constrained between specified limits as shown in Figure 2.12, for example.

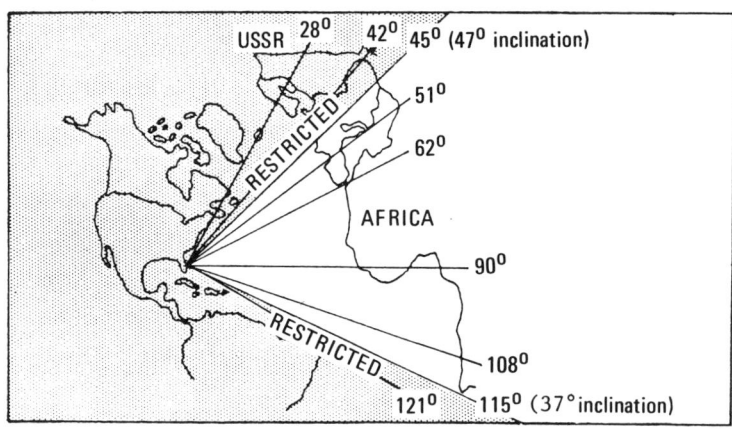

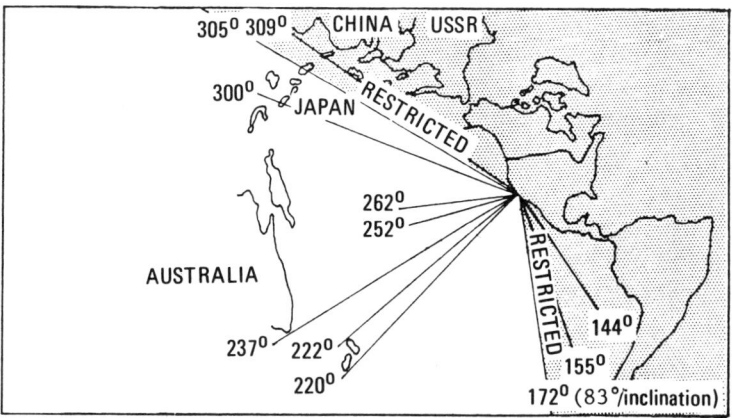

Figure 2.12 Geographic launch constraints.

2.7.1.2 In-Plane or Out-of-Plane Ascent

The most efficient launch is an in-plane ascent, characterized by a wait on the ground until the launch site lies in the mission orbital plane, which is fixed in inertial space. A plane change maneuver is not required, unless the orbit inclination is less than the launch site latitude, or the launch azimuth constraint is such that a "dog leg" maneuver is needed. A direct ascent to a parking or mission orbit may be made with or without a "phasing" orbit to place the satellite in a specified geographic location. The phasing problem, where the parking and mission orbits are not coplanar, may involve excessive waiting times, which can be somewhat reduced by "lofted" or other trajectories.

The out-of-plane launch is a form of direct ascent when the phasing problem can be solved by a wait on the ground. The launch window geometry is shown in Figure 2.13.

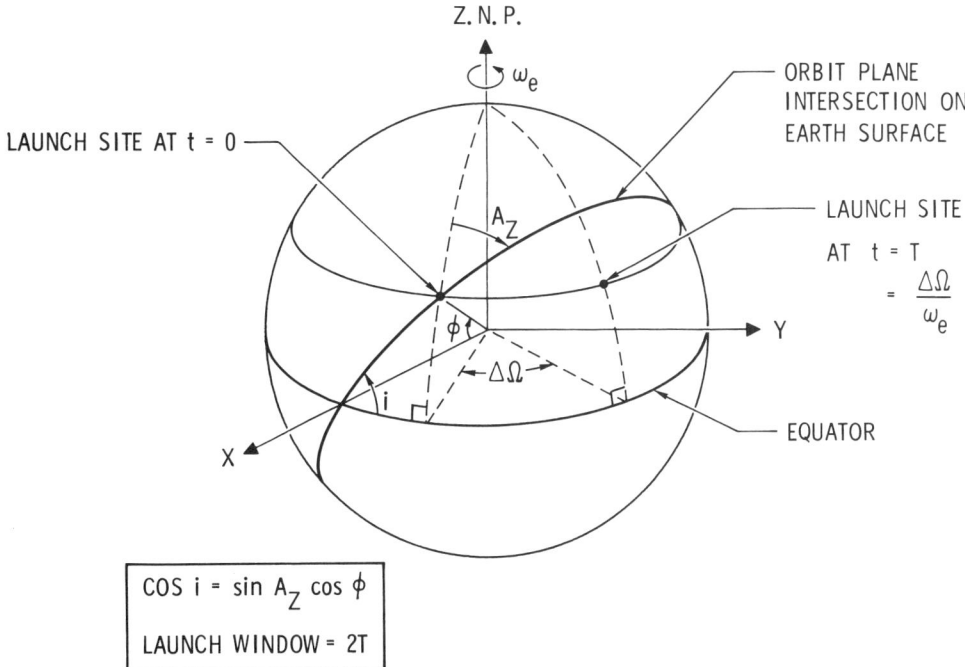

Figure 2.13 Launch window geometry.

The launch window can be defined as the time:

$$LW = 2\Delta\Omega/\omega_e \qquad (2.29)$$

where $\Delta\Omega = \omega_e t$ is the maximum permissible nodal increment consistent with the plane change angle ϵ required to transfer from the launch orbit plane to the mission orbit plane. The latter can be determined from the equation:

$$\epsilon = \cos^{-1}(\cos^2 i + \sin^2 i \cos\Delta\Omega) \qquad (2.30)$$

The longitudinal separation $\Delta\Omega$ is the same at all latitudes.

In general, a graph of ϵ *versus* time is as shown in Figure 2.14, where the plane change capability ϵ of the launch vehicle defines the launch window.

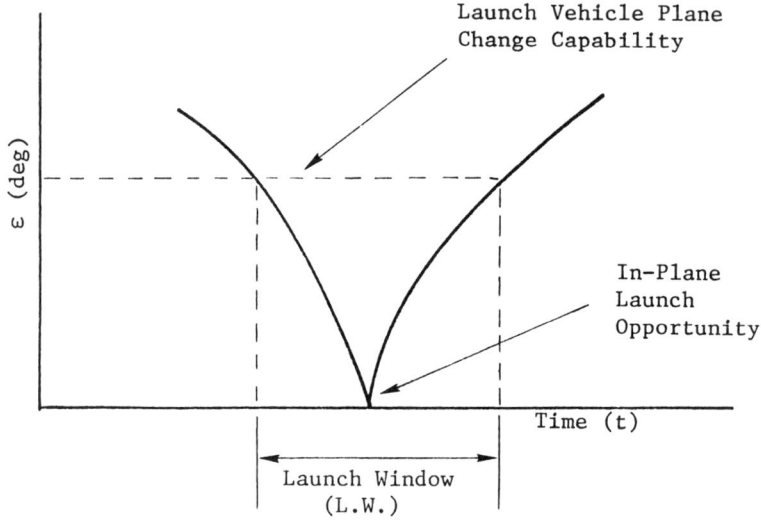

Figure 2.14 Launch window.

2.7.2 Ground Trace Considerations

2.7.2.1 *General Characteristics*

Circular figure-eight or "eggbeater" ground traces can be obtained by using satellites with 12- or 24-hour periods at different inclinations to the equator. The use of several orbital planes equally spaced in node can result in several satellites moving in the same ground trace. Two examples of this are shown in Figures 2.15 and 2.16, which show the circular and eggbeater types of ground traces with several satellites in each. Figure 2.15 is for a regional system with four satellites, and Figure 2.16 is for a global navigation system employing twenty satellites.

Other orbital systems involving up to eight 12-hour satellites in each of three orbital planes, equally spaced, have been found useful for global navigation purposes as in the Global Positioning System (GPS), for example.

One measure of performance for a navigation satellite system is the factor called *geometric dilution of precision* (GDOP), which is a measure of how satellite

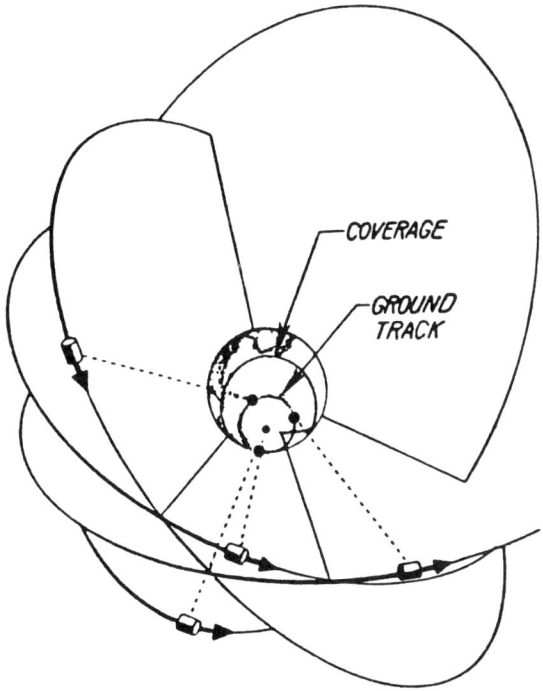

Figure 2.15 Regional navigation satellite system concept.

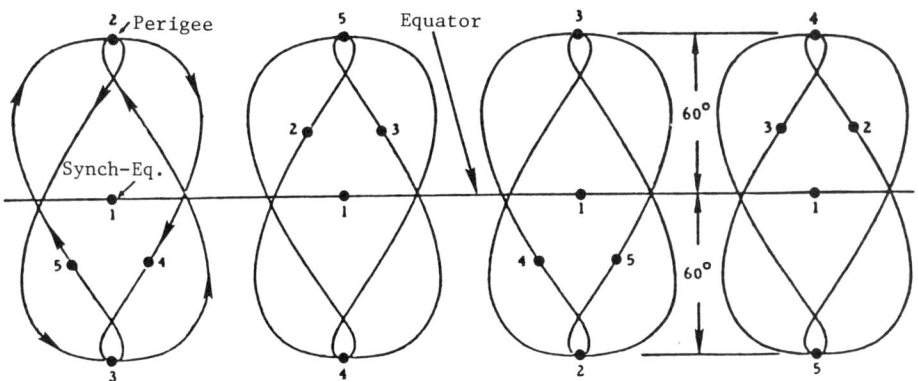

Figure 2.16 The "eggbeater" NAVSAT configuration.

geometry degrades accuracy. The magnitude of the ranging errors to a minimum of four selected satellites combined with the geometry of the satellites determines the magnitude of the user position errors in the GPS navigational fix. The four "best" visible satellites are those with the lowest GDOP [1]. Thus,

$$\text{GDOP} = \sqrt{(\text{PDOP})^2 + (\text{TDOP})^2} \qquad (2.31)$$

where

PDOP = Ratio of radial error in user position, 1σ, in three dimensions to range error, 1σ;
TDOP = Ratio of error, 1σ, in the range equivalent of the user clock offset to range error, 1σ.

2.7.2.2 Perturbation Effects

Earth's nonsphericity, solar radiation pressure, and solar or lunar gravitational attraction tend to alter the satellites' orbital elements in time. An example of how the ground trace can change due to the earth's gravitational harmonics is shown in Figure 2.17, where the initial and a four-year ground trace are illustrated.

The results are for a $Q = 1$ eccentric ($e = 0.27$) orbit, the period of which is $23^h55^m59.3^s$ with an initial inclination $i_0 = 28.5°$. The perturbations are due to the principal gravitational harmonics and the solar or lunar gravitational accelerations.

2.7.3 Highly Eccentric, Critically Inclined $Q = 2$ Orbits (Molniya)

This discussion is devoted to examining the characteristics of a very specialized orbit, the $Q = 2$ orbit, which is highly eccentric and critically inclined with apogee located over the Northern Hemisphere. This type of orbit has the ability to observe vast areas of the hemisphere for extended periods of time each day. Two properly phased spacecraft located in two ideal ground-track locations (ground tracks repeat daily) will typically continuously view 55–60% of the Northern Hemisphere, centered at the North Pole, as illustrated in Figure 2.18.

The following paragraphs examine each of the classical orbital elements and optimal values are assigned to those elements for which such assignment applies.

2.7.3.1 Semimajor Axis

The semimajor axis, a, is determined by calculation to be approximately 14,338 nmi. This value ensures that the ground track of the orbit will remain fixed relative to the earth and repeat itself daily ($Q = 2$, an integer, ensures this). Operationally, forces tend to alter a (e.g., drag, solar pressure, tesseral harmonics)

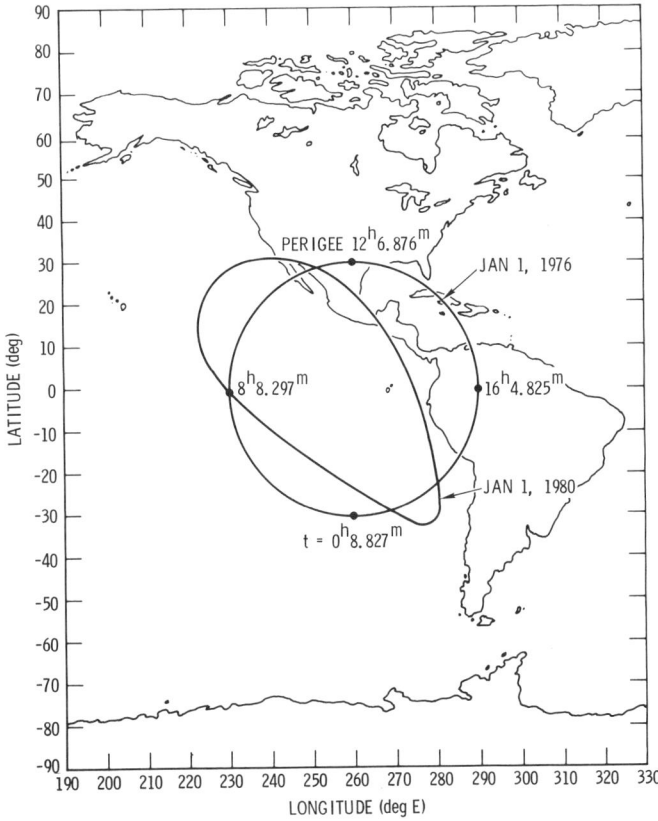

Figure 2.17 Initial circular and perturbed ground traces for a $Q = 1$ eccentric orbit ($e = 0.27$) at 28.5° inclination.

and fuel must be expended by the spacecraft to make periodic corrections to a.

When $a \approx 14338$ nmi, the Keplerian period is 11.967 mean solar hours, which is one-half of the quantity 360° (one complete earth rotation on its axis relative to the stars) divided by 15.041067 °/hr (the earth's rotation rate). So, to be a repeating groundtrack, $Q = 2$ orbit, the period is not 12 hours (one-half a mean solar day), but 11.967 hours, which is one-half of a mean sidereal day.

2.7.3.2 Eccentricity

The value of eccentricity (e) will vary over a typical mission lifetime mainly due to solar and lunar gravitational perturbations. Eccentricities ranging from 0.69 to

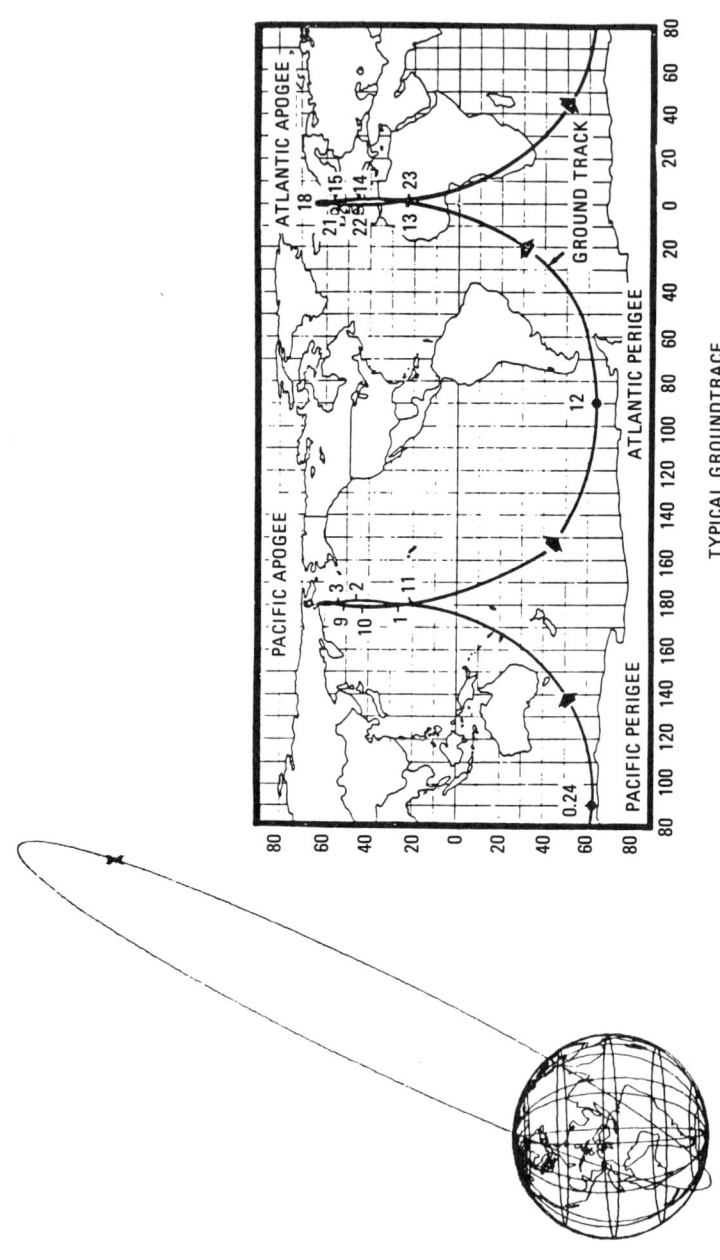

Figure 2.18 $Q = 2$ (Molniya) type of orbit.

0.74 are typical of this orbital type. The initial mission value of e is dictated by launch date, duration of mission, minimum acceptable value of h_p, ground-track location, and initial right ascension of ascending node (Ω). An adequately accurate computer program must be used to ensure that the spacecraft will not re-enter the atmosphere at a premature time.

2.7.3.3 Inclination

For observation of Northern Hemisphere regions, the inclination should be high. To maintain apogee at the northernmost point for long periods of time, the inclination must be approximately 63.44° (the critical inclination). Inclinations higher than 63.44° will force the line of apsides to rotate in a direction opposite that of satellite motion. Inclinations lower than 63.44° will rotate the line of apsides in the same direction as the satellite motion. These statements are valid only for direct orbits. The arguments are reversed for retrograde motion.

2.7.3.4 Right Ascension of Ascending Node

The node of the orbit can take any value, from zero to 360°. The value chosen essentially pinpoints the time of day of launch. At launch, the node (or time of day) must be carefully chosen so that lunar and solar perturbations do not act to reduce h_p below an acceptable level during the nominal duration of mission.

2.7.3.5 Argument of Perigee

To maintain optimal visibility of the Northern Hemisphere, ω should be maintained as close to 270° as possible. Selecting $i = 63.44°$ ensures that ω will drift away from 270° by less than $\pm 5°$ for mission lifetimes on the order of five years.

2.8 SATELLITE-BASED RADAR SYSTEMS

2.8.1 Global Coverage

Space-based radar systems consist of multisatellite constellations which provide continuous worldwide visibility by one or more satellites. Separation of observed objects from clutter by doppler processing techniques requires that at least two satellites be in view of the object at the same time. This ensures that there is a component of the object velocity vector in the direction of the radar that allows its signal processor to separate the object and clutter signals.

Space-based radar systems generally consist of satellites in circular orbits at fixed altitudes and orbital inclinations. The satellite's region of coverage (visibility) on the earth's surface is annular in shape with a visibility gap below the satellite (i.e., a nadir hole). The sensor (radar) is assumed to have elevation angle limits of β_1 and β_2, measured from the local horizontal and the earth's surface, respectively (see Figure 2.19). The former is atmosphere-constrained while the latter is signal clutter controlled. This results in earth coverage from each satellite with an earth-central angle $\Delta\theta = \theta_1 - \theta_2$, where θ_1 and θ_2 are determined by the relation:

$$\theta = \cos^{-1}\{(\cos\beta)/[1 + (h/R_e)]\} - \beta \qquad (2.32)$$

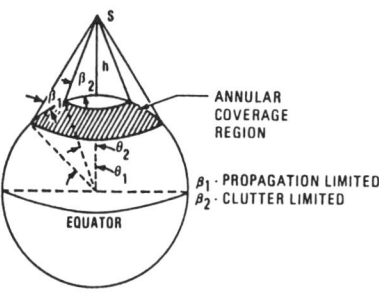

Figure 2.19 Single-satellite coverage.

where

R_e = earth radius;
h = orbit altitude above earth surface.

An infinite number of β, h pairs can thus be obtained from (2.32) for the same value of θ, as illustrated in Figure 2.20, where θ is plotted as a function of altitude for several values of β.

2.8.2 Continuous Global Coverage—Polar Constellations

2.8.2.1 Single Coverage

A preliminary estimate of the number of satellites and polar orbital planes for continuous visibility of at least one satellite over specified latitudes or the entire earth can be obtained from the results of references [2] and [3]. Thus, for example, the results of Figure 2.20 [2] for arbitrarily phased polar constellations show the allowable orbital altitudes for $\beta_1 = 3°$; $\beta_2 = 50°, 60°, 65°,$ and $80°$; and the total

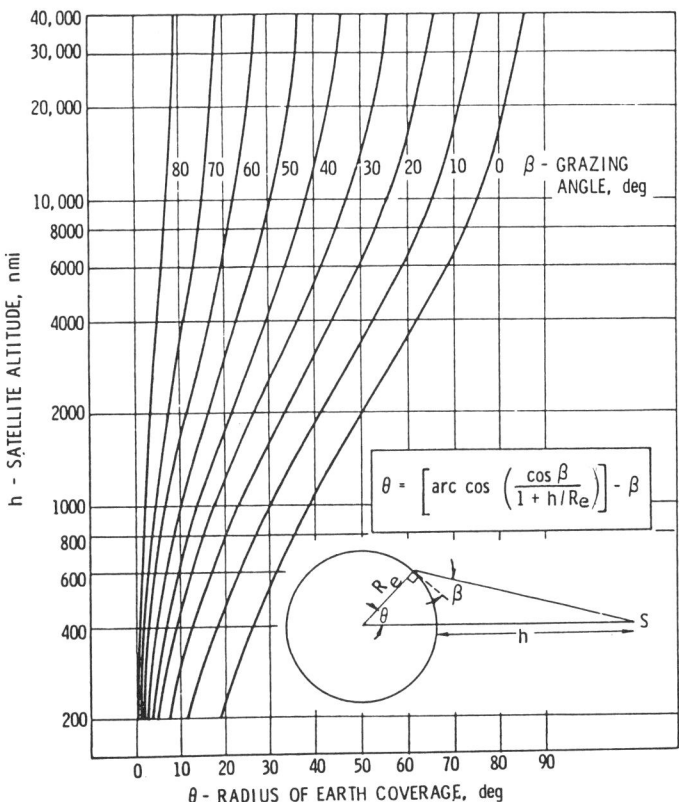

Figure 2.20 Altitude *versus* radius of earth coverage.

number of satellites per plane $N = 5, 6, 7$, and 8. The results in Figure 2.21 satisfy the conditions for nadir hole fill:

(1) $N \geq 5$
(2) $\theta_2 \leq \pi/N$
(3) $\theta_1 \geq \arccos(\cos\theta_2 \cos 2\pi/N)$

The results in Figure 2.21 show that a five-satellite circular polar orbit at an altitude of 11,500 nmi will provide continuous coverage of all earth's regions above 20°N latitude and below 20°S latitude if a grazing range of $3° \leq \beta \leq 60°$ is permissible.

Figure 2.21 also shows that an eight-satellite single-orbit coverage above 45° N latitude is obtained at an altitude of 2170 nmi. Two such orbits orthogonal to each other will provide continuous global coverage. These constellations may

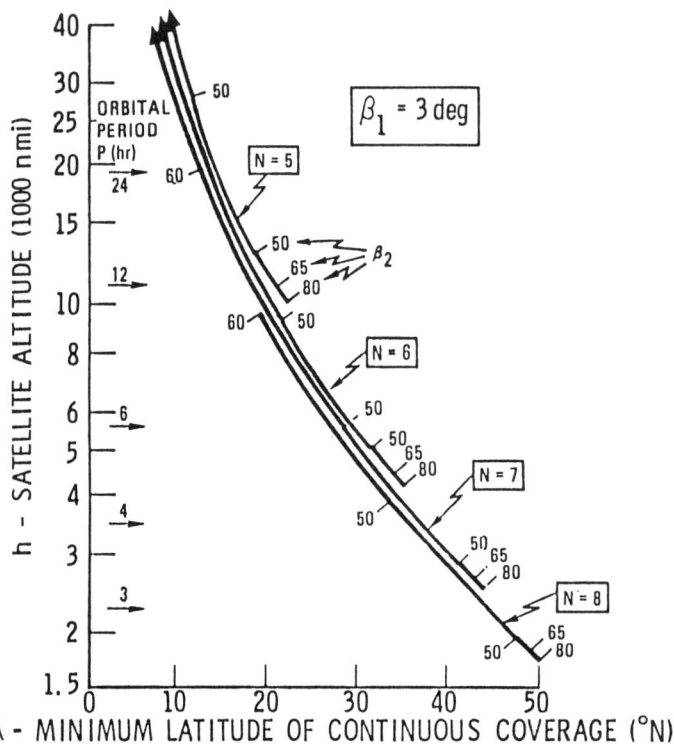

Figure 2.21 Continuous polar cap coverage.

be composed of two orthogonal polar orbits, or one polar and one equatorial orbit. Choice of orthogonal polar orbits will provide double continuous coverage above 45° N latitude and single continuous coverage below 45° N latitude. Both of these constellations require a total of 16 satellites at 2170 nmi altitude.

2.8.2.2 Double Coverage

The results of [2] and [3] show that the nadir holes of all satellites in a single orbital plane will be doubly covered by other satellites in that orbital plane if

$$\theta_1 \geq \theta_2 + 2\pi/s_2 \tag{2.33}$$

where s_2 is the number of satellites in each orbital plane. Therefore, it is necessary that

$$s_2 \leq 2\pi/\Delta\theta \qquad (2.34)$$

where

$$\Delta\theta = \theta_1 - \theta_2$$

The central angle half-width of double continuous coverage from each orbital plane of satellites is given by

$$C_2 = \cos^{-1}[\cos\theta_1/\cos(2\pi/s_1)] \qquad (2.35)$$

where s_1 is a second constraint on s, which must be satisfied from the requirement that the half swath widths of double coverage from each orbital plane of satellites must be contiguous at the earth's equator. Thus, if p is the number of orbital planes,

$$2pC_2 \geq \pi \qquad (2.36)$$

which yields

$$s_1 \geq 2\pi/\cos^{-1}[\cos\theta_1/\cos(\pi/2p)] \qquad (2.37)$$

The values of s_1 in (2.37) and s_2 in (2.34) can be computed for any altitude h and the number of orbital planes. For constellation efficiency, $s_1 \approx s_2$, where only integer values of s are used. For example, if $\beta_1 = 3°$, $\beta_2 = 70°$, a preliminary number of "efficient" constellations at different altitudes can be identified as in Table 2.3.

Table 2.3
Polar Orbit Double Coverage

$(p \times s_2)$	=	T	h(nmi)
2 × 7	=	14	5360
2 × 9	=	18	3450
3 × 9	=	27	2080
4 × 10	=	40	1450
5 × 12	=	60	900

Note that (2.36) is a conservative estimate of the required number of orbital planes by using arbitrarily phased polar constellations. This value was used to expedite the rapid generation of preliminary data. A less conservative constraint equation, obtained from [3], but one that is more difficult to solve analytically, is given by

$$p(C_1 + C_2) = \pi \tag{2.38}$$

where

$$C_j = \arccos\left[\cos\theta_1/\cos j\pi/s_1\right]$$

and will result in a more efficient constellation with fewer total numbers of satellites. Using optimally phased polar constellations (as identified in [4]) can further reduce the required total number of satellites by using the constraint relation:

$$(p - 1)(\theta + C_2) + (C_1 + C_2) = \pi \tag{2.39}$$

2.8.3 Inclined Walker Constellations

Another preliminary approach to estimating the required number of satellites for double global coverage is to use inclined orbits. For example, *Walker constellations* consist of circular orbits with particular orbital inclinations, as defined in Figure 2.22 [5].

Selected Walker constellations are shown in Tables 2.4 and 2.5 for the cases of $\beta_2 = 0$ and $\beta_2 = 90°$ (no nadir hole). A computer view of a 5/5/1 Walker constellation is shown in Figure 2.23, and its ground traces are indicated in Figure 2.24.

Some Walker constellations that have no nadir hole can be found to provide double global coverage when nadir holes are used. Thus, for example, a Walker 14/14/12 constellation at 49.4° inclination and an altitude of 5600 nmi appears to provide continuous worldwide visibility (coverage) by at least two satellites. Optimal constellations (with minimum total number of satellites) can generally be identified by computer simulation.

CONSTELLATIONS DENOTED BY T/P/F

    T = TOTAL NUMBER OF SATELLITES
    P = NUMBER OF ORBIT PLANES
    F = PHASING PARAMETER

T SATELLITES ARE EQUALLY DIVIDED AMONG P PLANES

P PLANES ARE EQUALLY SPACED IN Ω

SATELLITES WITHIN A PLANE ARE EQUALLY SPACED IN ARGUMENT OF LATITUDE

F $\times \frac{360°}{T}$ = PHASING DIFFERENCE BETWEEN SATELLITES IN ADJACENT PLANES. SEE SKETCH

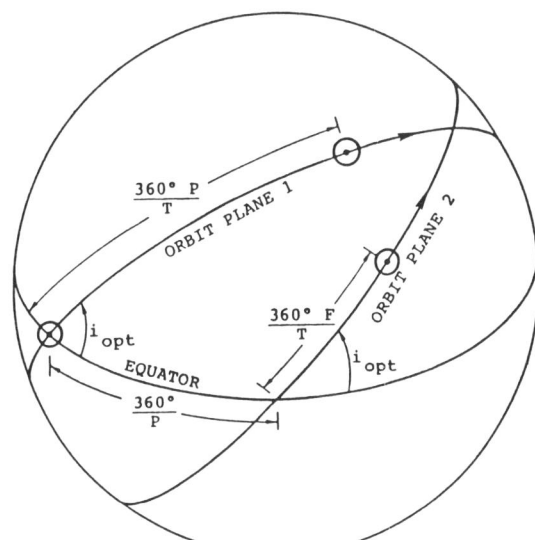

i_{OPT} = OPTIMIZED INCLINATION THAT IS COMMON TO ALL PLANES

R_{MAX} OR MIN d_{MAX} = COVERAGE CIRCLE RADIUS REQUIRED BY A GIVEN CONSTELLATION. FOR A GIVEN ELEVATION ANGLE, R_{MAX} DEFINES THE MINIMUM COMMON ALTITUDE FOR ALL THE SATELLITES IN THE CONSTELLATION.

D_{MIN} = MINIMUM EARTH CENTRAL ANGLE BETWEEN ANY TWO SATELLITES IN THE CONSTELLATION

Figure 2.22 Walker's notation and constellation definition.

Table 2.4
Walker Constellations Providing Continuous Global Coverage ($\beta_1 = 0$, $\beta_2 = 90°$), Single Coverage

Pattern T/P/F	$R_{max,1}$ (degrees)	i_{opt} (degrees)	D_{min} (degrees)	Minimum Altitude (nmi)	Reference
5/5/1	69.15	43.66	60.86	6234	Walker [6]
5/5/3	75.5	51.8	82.2	10,311	Walker [7]
6/6/4	66.42	53.13	73.7	5165	Walker [7]
6/2/0	66.72	52.24	35.7	5269	Walker [6]
7/7/5	60.26	55.69	57.0	3499	Walker [6]
7/7/1	60.5	48.0	42.5	3550	Walker [7]
8/8/6	56.52	61.87	56.3	2799	Walker [7]
8/2/1	56.9	48.2	29.5	2863	Walker [6]
9/9/7	54.81	70.54	43.1	2532	Walker [7]
9/9/1	57.28	49.88	—	2927	Mozhaev [8]
9/9/2	57.9	61.3	26.0	3037	Walker [7]
9/3/0	61.9	70.5	33.6	3868	Walker [6] Mozhaev [8] Ballard
10/10/7	51.535	47.93	0	2093	Webb
10/5/2	52.231	57.11	46.62	2179	Walker [6]
10/5/1	52.3	47.4	14.8	2188	Walker [7]
10/10/2	52.5	48.8	37.9	2213	Walker [7]
10/2/0	53.2	47.7	24.0	2305	Walker [7]
11/11/4	47.61	53.79	49.0	1664	Walker [7]

Table 2.5
Walker Constellations Providing Continuous Global Coverage ($\beta_1 = 0$, $\beta_2 = 90°$), Double Coverage

Pattern T/P/F	$R_{max,2}$ (degrees)	i_{opt} (degrees)	D_{min} (degrees)	Minimum Altitude (nmi)	Reference
7/7/2	75.97	61.81	37.1	10,762	Walker [7]
8/8/2	71.0	57.1	9.6	7134	Walker [7]
8/8/6	74.0	58.0	61.3	9051	Walker [7]
8/8/5	74.2	56.5	0	9205	Walker [7]
9/3/2	66.2	62.1	24.2	5090	Walker [7]
9/3/0	66.8	65.5	29.4	5298	Walker [7]
10/10/2	64.1	61.6	21.2	4441	Walker [7]
10/5/2	65.151	52.56	52.8	4752	Walker [7]
10/10/7	65.173	62.83	0	4758	Webb [9]
10/10/8	65.5	49.4	46.7	4861	Walker [7]
10/2/0	73.1	44.4	25.5	8403	Walker [7]
11/11/9	62.0	52.7	44.1	3892	Walker [7]
12/3/1	56.6	57.0	18.2	2812	Walker [7]

Table 2.5 cont'd.

12/6/2	56.6	54.0	0	2812	Walker [7]
12/3/2	56.7	58.5	23.9	2829	Walker [7]
12/12/10	59.3	56.8	42.2	3302	Walker [7]
12/2/0	63.7	45.7	0	4329	Walker [7]
12/2/1	64.3	45.0	21.1	4498	Walker [7]
13/13/3	54.7	52.8	38.0	2516	Walker [7]

WALKER 5/5/1 AT H=10900 NMI, I=43.7

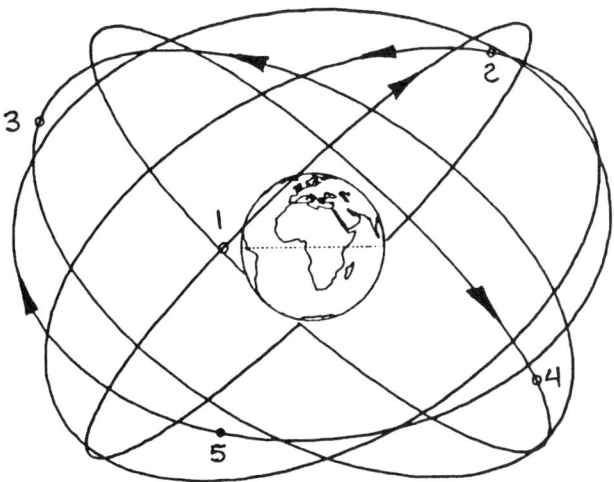

Figure 2.23 Orbital configuration.

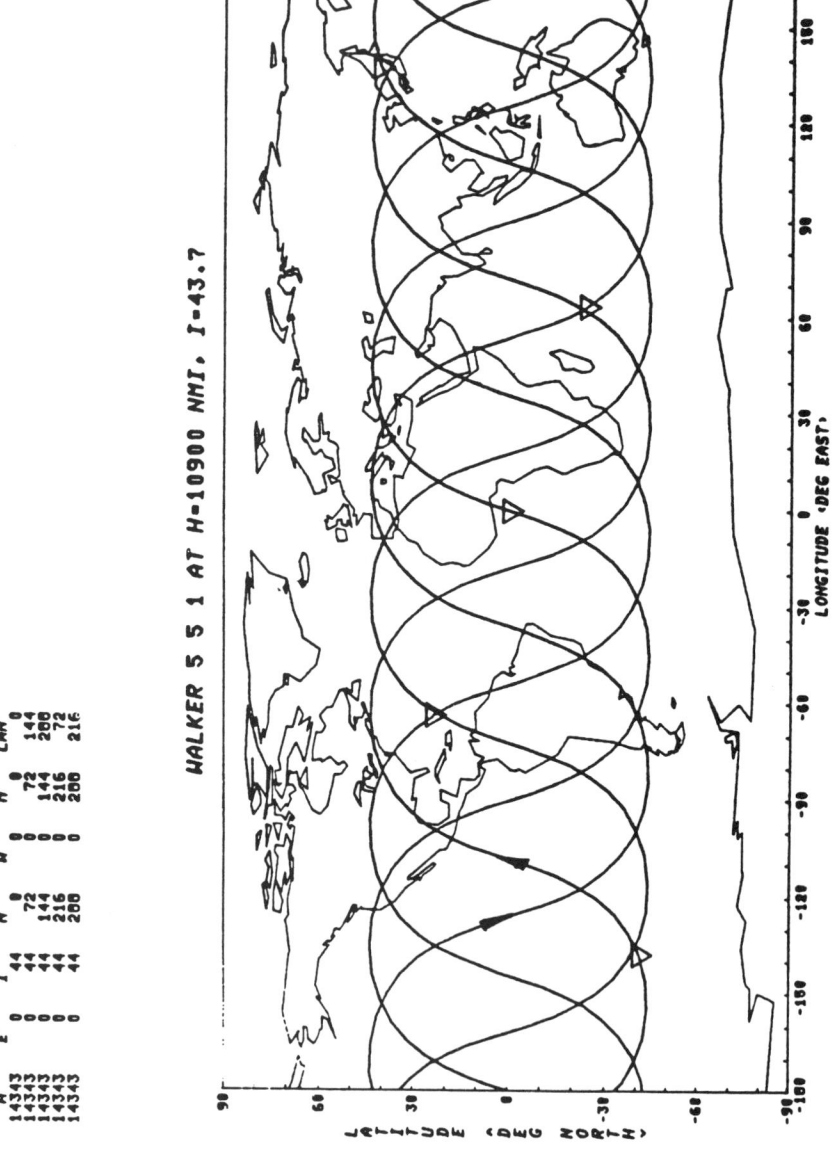

Figure 2.24 Ground trace.

REFERENCES

1. Milliken, R.J., and C.J. Zoller, "Principle of Operation of Navstar and System Characteristics," *Navigation: Journal of the Institute of Navigation*, Vol. 25, No. 2 (Summer 1978), pp. 95–106.
2. Rider, L., "Nadir Hole-Fill by Adjacent Satellites in a Single Orbit," Technical Note, *J. Astro. Sci.*, Vol. 28, No. 3, July–September 1980, pp. 299–305.
3. Adams, W.S., and L. Rider, "Circular Polar Constellations Providing Continuous Single or Multiple Coverage Above a Specified Latitude," *J. Astro. Sci.*, Vol. 35, No. 2, April–June 1987, pp. 155–192.
4. Rider, L., "Optimized Polar Orbit Constellations for Redundant Earth Coverage," *J. Astro. Sci.*, Vol. 33, No. 2, April–June 1985, pp. 147–161.
5. Chobotov, V.A., C.C. Chao, H.K. Karrenberg, T.J. Lang., and J.Y. Miyamoto, "Orbital Mechanics," class notes, Aerospace Corporation, El Segundo, CA 90274.
6. Walker, J.G., "Circular Orbit Patterns Providing Continuous Whole Earth Coverage," Royal Aircraft Establishment Technical Report 70211, November 1970.
7. Walker, J.G., "Continuous Whole-Earth Coverage by Circular-Orbit Satellite Patterns," Royal Aircraft Establishment Technical Report 77044, March 1977.
8. Mozhaev, G.V., "The Problem of Continuous Earth Coverage and Kinematically Regular Satellite Networks, II, *Kosmicheskie Issledovaniya* (in Russian), Vol. 11, No. 1, January–February 1973, pp. 59–69. Translated in *Cosmic Research*, Vol. 11, No. 1, January–February 1973, pp. 52–61.
9. Webb, E., Lockheed, Private Communication.

SELECT BIBLIOGRAPHY

A. *Books on Orbital Mechanics*

1. *American Ephemeris and Nautical Almanac*, US Government Printing Office, Washington, DC (published annually).
2. Moulton, F.R., *Celestial Mechanics*, Macmillan, New York, 1914.
3. Escobal, P.R., *Methods of Orbit Determination*, John Wiley and Sons, New York, 1965.
4. Smart, W.M., *Celestial Mechanics*, Longmans, Green, and Company, London, 1953.
5. Danby, J.M.A., *Fundamentals of Celestial Mechanics*, Macmillan, New York, 1962.
6. Roy, A.E., *The Foundations of Astrodynamics*, Macmillan, New York, 1965.
7. Winter, A., *The Analytical Foundations of Celestial Mechanics*, Princeton, 1947.
8. Ehricke, K.A., *Space Flight: Vol. I, Environment and Celestial Mechanics*, 1950, and *Vol. II, Dynamics*, Van Nostrand, New York, 1962.
9. Seifert, H., ed., *Space Technology*, John Wiley and Sons, New York, 1959.
10. Wolverton, R.W., ed., *Flight Performance Handbook for Orbital Operations*, John Wiley and Sons, New York, 1963.
11. Smart, W.M., *Textbook on Spherical Astronomy*, Fifth Edition, Cambridge University Press, Cambridge, 1962.
12. Hoelker, R.F. and R. Silber, "The Bi-Elliptic Transfer Between Circular Coplanar Orbits, *Ballistic Missiles and Space Technology*, Vol. 3, Pergamon, New York, 1961.
13. Leitmann, G. (ed.), *Optimization Techniques with Applications to Aerospace Systems*, Academic Press, New York, 1962.

14. Batin, R.H., *An Introduction to the Mathematics and Methods of Astrodynamics*, AIAA Education Series, 1987.

B. Papers on Earth Coverage

1. Walker, J.G., "Continuous Whole Earth Coverage by Circular Orbit Satellites, Royal Aircraft Establishment Technical Memorandum Space 194, April 1973.
2. Walker, J.G., "Some Circular Orbit Patterns Providing Continuous Whole Earth Coverage," *Journal of the British Interplanetary Society*, Vol. 24, 1971, pp. 369–384.
3. Mozhaev, G.V., "The Problem of Continuous Earth Coverage and Kinematically Regular Satellite Networks, I, *Kosmicheskie Issledovaniya* (in Russian), Vol. 10, No. 6, November–December 1972, pp. 833–840.
4. Walker, J.G., "Satellite Patterns for Continuous Multiple Whole-Earth Coverage," *Maritime and Aeronautical Satellite Communication and Navigation*, IEE Conf. Publ. 160, March 1978, pp. 119–122.
5. Ballard, A.H., "Rosette Constellations of Earth Satellites," *IEEE Transactions on Aerospace and Electronic Systems*, Vol. AES-16, No. 5, September 1980, pp. 656–673.
6. Walker, J.G., "Coverage Predictions and Selection Criteria for Satellite Constellations," Royal Aircraft Establishment Technical Report 82116, December 1982.
7. Draim, J.E., "Three- and Four-Satellite Continuous Coverage Constellations," *AIAA Journal of Guidance, Control, and Dynamics*, Vol. 22, No. 6, November–December 1985, pp. 725–730.
8. Draim, J.E., "A Common Period Four-Satellite Continuous Global Coverage Constellation," AIAA Preprint 86-2066-CP, presented at AIAA/AAS Astrodynamics Conference, Williamsburg, VA, August 18–20, 1986.
9. Beste, D.C., "Design of Satellite Constellations for Optimal Continuous Coverage," *IEEE Trans. Aerospace and Electronic Systems*, Vol. AES-14, No. 3, May 1978, pp. 466–473.
10. Emara, E.T., and C.T., Leondes, "Minimum Number of Satellites for Three-Dimensional Continuous Worldwide Coverage," *IEEE Trans. Aerospace and Electronic Systems*, Vol. AES-13, No. 2, March 1977, pp. 108–111.
11. Harney, E.D., Space Planners Guide, USAF Systems Command, US Government Printing Office, Washington, DC, Pub. No. 0-774-405, 1965.
12. Draim, J.E., "A Six-Satellite Continuous Global Double Coverage Constellation," AAS/AIAA Astrodynamics Specialist Conference, Paper No. AAS 87-497, 1987.
13. Lang, T.J., "Symmetric Circular Orbit Satellite Constellations for Continuous Global Coverage," AAS/AIAA Astrodynamics Specialist Conference, Paper No. AAS 87-499, 1987.
14. Lang, T.J., "Orbital Constellations Which Minimize Revisit Time," AAS/AIAA Astrodynamics Specialist Conference, Paper No. 83-402, 1983.

Chapter 3
IONOSPHERIC ENVIRONMENT AND EFFECTS ON SPACE-BASED RADAR DETECTION
Dennis L. Knepp and J. Todd Reinking
Mission Research Corporation

3.1 INTRODUCTION

The natural ionosphere can produce a variety of disturbances to radar signal propagation. Disturbances due to mean or very large scale ionization include attenuation, phase shift, time delay, dispersion, polarization rotation, refraction, and multipath. In addition, relatively small-scale ionospheric structure in the propagation medium can cause *signal scintillation*, which comprises essentially random fluctuations in the received signal phase, amplitude, angle of arrival, and other signal properties. The effects of mean propagation disturbances have been the subject of many studies [1–2] and are well known. The effects of scintillation on the detection performance of space-based radar (SBR) form the subject of this chapter.

Electron density structure in the natural ionosphere can produce random variations in the amplitude and phase of a propagating wave, even at frequencies in the gigahertz range [3, 4]. These rapid variations in signal phase, amplitude, and angle of arrival are called *scintillations,* and they are often observed over satellite links through the ambient ionosphere at VHF and UHF [5]. Measurements taken with the ALTAIR radar in the Pacific Test Range are known to give severely disturbed scintillation at 156 and 415 MHz [6]. Strong scintillation is occasionally observed at frequencies as high as C-band [7]. Because even small fluctuations in the received signal can degrade performance, the effect of scintillation must be considered in the design of a space-based radar system intended to operate through an ionospheric channel.

The severity of the fluctuations depends on the irregularity of the ionization structure and on the radar geometry and frequency. If the propagation environment is highly disturbed and the radar frequency is not sufficiently high, worst-case scintillation may occur, wherein the received signal quadrature components are uncorrelated Gaussian variates, after one-way propagation over the severely disturbed path. Worst-case (i.e., Rayleigh amplitude) scintillation is likely to occur if the ionosphere is highly disturbed, such as, for example, by high-altitude nuclear explosions [8, 9] or by chemical releases [10]. An increase in the radar frequency or a lessening of the ionization irregularity can lead to a decreased disturbance in the received signal with a corresponding change in the signal's statistical description. In general, however, a signal at one frequency (or time) may show some statistical correlation to a signal received at a different frequency (or time). The effects of signal decorrelation with time and frequency at all levels of scintillation severity are important to any VHF through X-band radar system that must operate through an ambient or disturbed ionospheric channel.

Previous authors [11–13] have studied the performance of M out of N radar detection for the case of an undisturbed propagation channel. In an earlier work [14], we reported the effect of strong (Rayleigh) fading on radar noncoherent detection performance, assuming various radar geometries and target models. The effects of frequency correlation were also studied in this work, but the method used to model frequency correlation was limited to a single coherence function.

The effects of scintillation on monostatic radar employing *double-threshold* (M out of N) detection or noncoherent integration are the subjects of this chapter. (In general, noncoherent integration provides superior performance; however, double-threshold detection is of interest because it affords a radar design simplification.) In particular, we develop methods to study the effects of varying levels of scintillation on the probability of detecting targets of constant cross section as well as targets having a cross section that follows the Swerling II model. In addition, we develop simulation methods to study the effects on detection probability of correlation of the propagation channel over time and frequency.

This chapter uses the Nakagami-m probability distribution [15] to describe the statistics of the received signal power for the monostatic radar geometry. This simple, one-parameter distribution correctly describes both the weak and strong scattering limits of the actual probability distribution of scintillation [16] and has been found to give a very good fit to observed scintillation data [17] over a wide range of scintillation conditions. The Nakagami-m distribution, as used here, permits the study of radar performance over the entire range of possible scintillation conditions.

When the two extreme assumptions on correlation of propagation effects are made (i.e., received signal statistics due to propagation fluctuations for different bursts within a look are either independent, or otherwise completely correlated), expressions may be developed for detection probability which are readily evaluated by numerical methods. For the case of partially correlated bursts, Monte Carlo methods are required to determine the effect on detection probability. The simulation methods developed here allow a completely general specification of burst-to-burst correlation and may be used to study the effects of scintillation on the performance of frequency-agile radar.

In this chapter, a coherent pulse train is referred to as a burst, and a number of bursts combined noncoherently in some fashion are called a *look*. We will assume that there is no signal decorrelation over the duration of a burst, and therefore no coherent integration loss [14].

3.1.1 Ionization Irregularity Description

Figure 3.1 depicts the geographic distribution of ionospheric scintillation as it is currently known [18]. The severity of the scintillation is indicated in the figure by the density of the cross-hatching. Most severe ionization irregularities occur in a geographic region of about 20° of latitude north and south of the geomagnetic equator. Equatorial scintillation is observed during a period of approximately 8 PM local time to about 2 AM, and may regularly take the form of saturated Rayleigh amplitude fading at VHF and UHF, or occasionally at L-band. Equatorial scintillation is known to be more severe than mid- or high-latitude scintillation.

At high latitudes, scintillation occurs within the auroral region and over the polar cap. At present the occurrence of auroral scintillation is less understood than equatorial scintillation. However, auroral scintillation is known to be more irregular than equatorial scintillation and can occur at any time during the local day or night. Increased magnetic activity brings higher levels of scintillation and can cause changes in the location of scintillation boundaries at high latitudes. The severity of scintillation can be measured by the *scintillation index, S_4*, which characterizes the depth of fading.

An example of the hourly and seasonal distribution of amplitude scintillation is shown in Figure 3.2 [19] in terms of average contours of the scintillation index (SI). This measurement of scintillation severity can easily be obtained from strip charts. The innermost contour shown corresponds to a value of SI of 70 or roughly a value of S_4 of 0.36, signifying 8 dB peak-to-peak fading. The data were obtained from six years of cumulative observations of ATS-3 from Huancayo, Peru. The figure shows that scintillation at this location is most severe in the months of February through October and during the hours immediately preceding local midnight. Similar measurements from other geographic locations indicate the presence of a longitudinal dependence on seasonal scintillation activity. Although average results of this type are extremely useful, they are insufficient for designing an SBR to operate during signal scintillation. SBR design and evaluation are best served with measurements of the distribution function of S_4 so that the designer can determine the percentage of time when peak-to-peak signal fluctuations exceed a certain level.

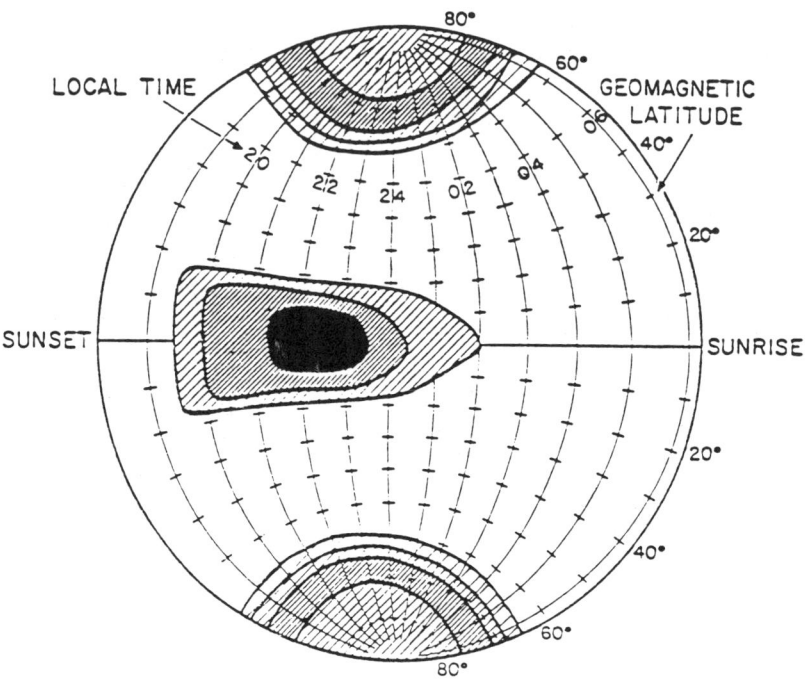

Figure 3.1 Geographic distribution of ionospheric fading.

The DNA Wideband Satellite experiment has assembled the most extensive collection to date of phase coherent scintillation data [5]. Figure 3.3 [20] shows the seasonal dependence of scintillation at 137 MHz as measured during Wideband passes observed from Ancon, Peru. The Wideband orbit is sun-synchronous so that the satellite always passes over the equator at a constant local time of about 11:30 PM. Thus, from a fixed ground location, three or four passes can be observed during each night of operation. In addition, each pass takes about 15 minutes. For the purposes of data analysis, this time period is divided into many 10 s segments. The S_4 scintillation index is obtained once for each 10 s segment giving many measurements of S_4 during the night. Figure 3.3 shows the percentage of measurements of S_4 at Ancon, where the value of 0.3 was exceeded at 137 MHz. This value corresponds to about 6 dB peak-to-peak fading.

Values of scintillation index depend on geometry, frequency, and the ionization irregularity structure. Results from the DNA Wideband satellite should be quite appropriate for the design of systems that might use high inclination orbits because the Wideband satellite was in an almost polar orbit at an altitude of 1000 km. Therefore, these measurements may be used directly for satellites in similar

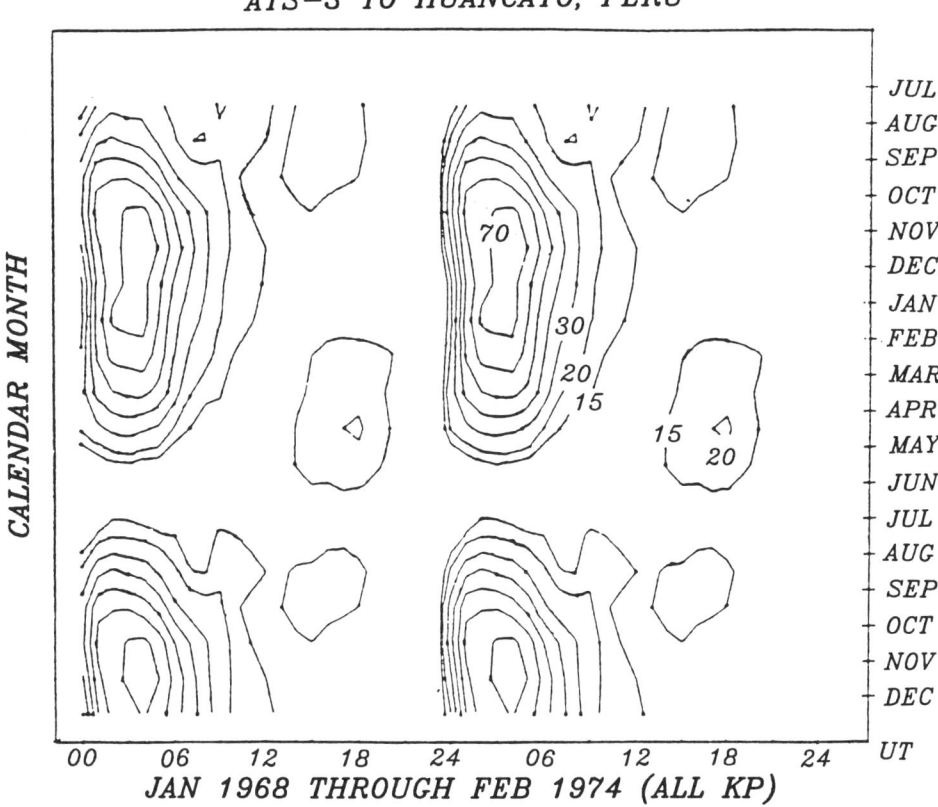

Figure 3.2 Average scintillation (SI) contours at 136 MHz obtained over a six-year period. Universal time (UT) shown is five hours less than local time (LT); 05 UT is 24 LT.

orbits, but need to be scaled to apply to different radar transmission frequencies and satellite altitudes.

Figures 3.4 and 3.5 [21] show Wideband data taken on a particularly severe pass over Ancon, Peru on December 16, 1976. Figure 3.4 shows values of the S_4 scintillation index obtained at 138, 379, and 447 MHz during a brief portion of a satellite pass. A measurement of S_4 is obtained by averaging over each 10.5 s of data, giving 16 measurements at each of the three frequencies during the 2.8-minute time period shown. During the beginning of this period propagation conditions were quite severe with worst-case Rayleigh fading (unity S_4) observed for VHF and the two UHF frequencies shown. However, a few minutes later, the scintillation severity is observed to decrease somewhat with very little scintillation

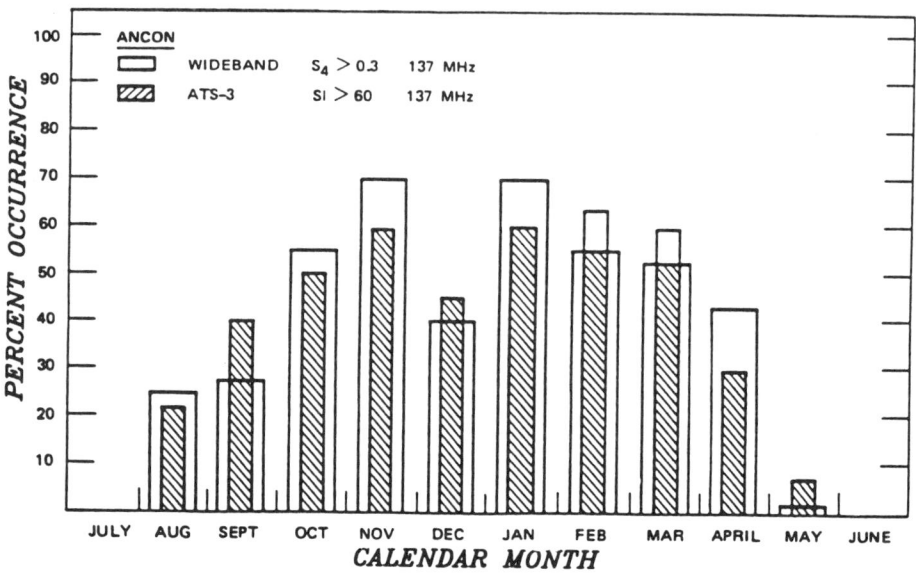

Figure 3.3 Seasonal dependence of scintillation severity at Ancon, Peru.

(small S_4) for UHF at 23:51, although at VHF scintillation is still severe.

Figure 3.5 shows values of the cross-correlation coefficient of the received intensity or power as obtained from the first six 10.5 s intervals shown in the preceding figure. The cross-correlation coefficient is shown for correlations between the lowest Wideband UHF tone at 379 MHz and the tones at 390, 413, 436, and 447 MHz. From the figure it is evident that the tones are well decorrelated for the two earliest measurements at 23:48:29 and 23:48:50 for the 379–413 MHz frequency pair. For later times during this Wideband pass, the UHF frequency tones become well correlated across the spectrum from 379–447 MHz as shown by the curve at 23:49:11. The data here show an example where the tones at 379 and 413 MHz were decorrelated. The coherence bandwidth is proportional to transmission frequency to the fourth power so that if this 34 MHz frequency separation were scaled to apply to a VHF radar at 200 MHz, the resulting coherence bandwidth would be about 4 MHz. Given the scintillation conditions observed during these measurements, the propagation channel thus would effectively decorrelate for VHF frequencies separated by more than 4 MHz. A good radar design could take advantage of this decorrelation with transmission frequency by noncoherently combining the radar returns from independent samples of the fading channel.

The mechanisms accounting for production of irregularities have been the subject

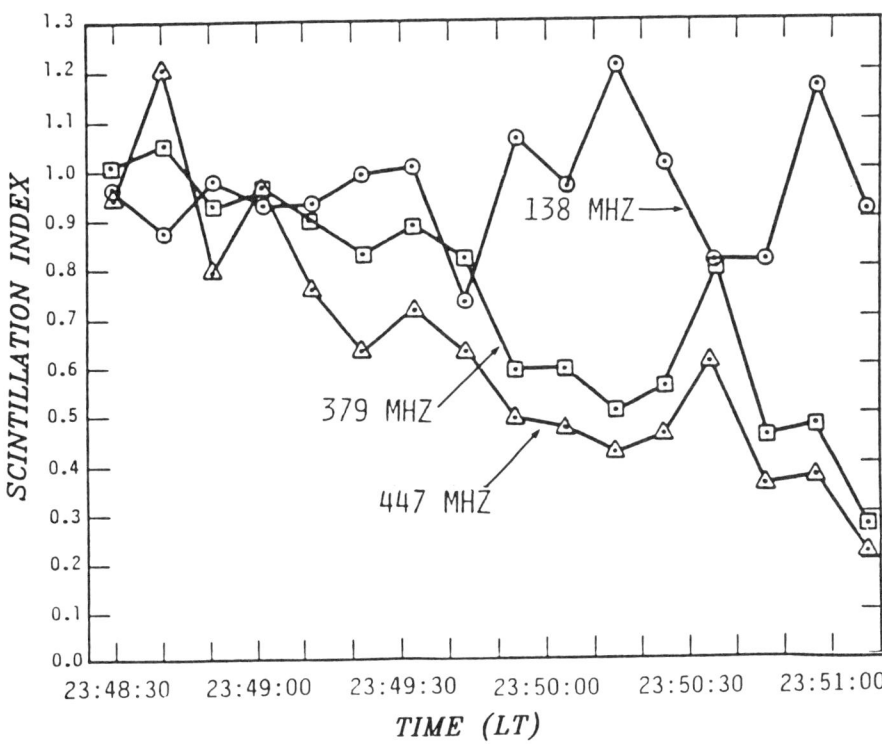

Figure 3.4 Simultaneous VHF and UHF scintillation observed during the DNA Wideband satellite experiment.

of many research papers over the last decade and are now reasonably well understood [22, 23]. In the equatorial regions, the dominant mechanism for production of large scale structure or equatorial spread-F is the collisional Rayleigh-Taylor instability. The emerging view is that high-latitude irregularities in electron density are produced primarily by the gradient-drift (or $\overline{E} \times \overline{B}$) instability.

Although other forms have been observed, the power spectral density of the electron-density structure is well represented by a two-component power-law form with the irregularities highly elongated along the direction of the earth's magnetic field. The precise values of the parameters that describe the spectrum have been investigated by a number of researchers. However, the current belief is that the spectrum behaves as K^{-4} for spatial wavenumbers (K) greater than the freezing scale, and flattens to between $K^{-2.3}$ and $K^{-2.7}$ for scale sizes ranging from the freezing scale to the outer scale. The outer scale ranges from 10–100 km and the freezing scale ranges from 150–1000 m. One view of the irregularity structure causing scintillation is a thick layer (~200 km) extending upward from the base

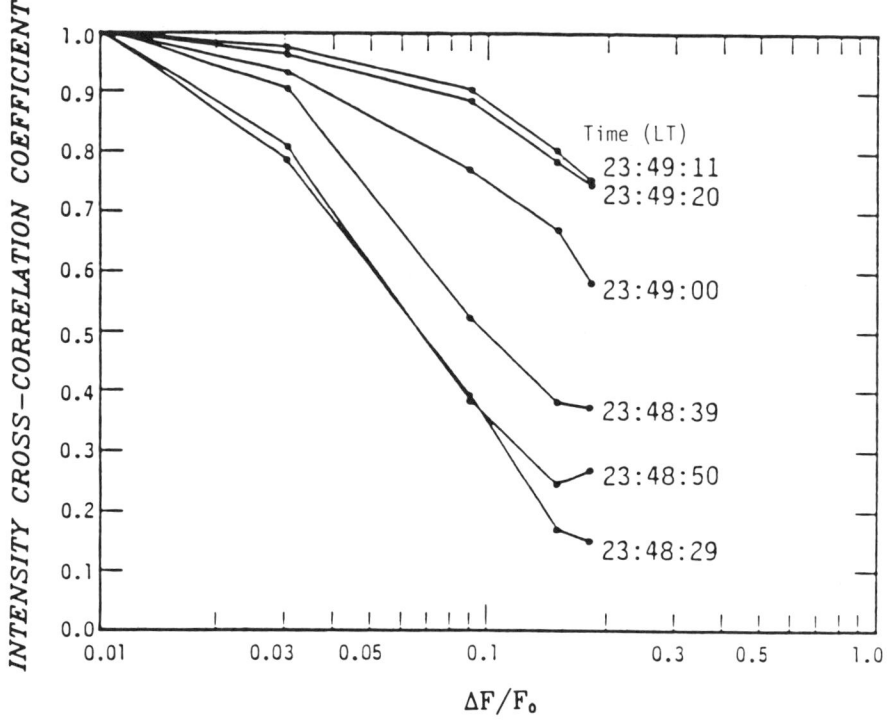

Figure 3.5 Cross-correlation of power at different UHF frequencies during the DNA Wideband satellite experiment.

of the F-region and having electron density fluctuations on the order of 50–100% of the mean electron density [24].

3.2 RECEIVED SIGNAL DESCRIPTION

If the radar signal passes through a disturbed propagation channel subject to electron-density irregularities, the received signal amplitude may experience fluctuations or scintillation. Knowledge of the probability distribution of the received signal is essential to determine radar system performance. Since the 1950s, many authors have investigated the probability distribution of the received signal after one-way propagation through turbulence (Knepp and Valley [16] and references therein). A large number of probability functions that describe various aspects of the received scintillated signal are possible. Many distributions correctly describe the weak-scattering regime, and parametrically extend to the strong-scattering limit. The joint Gaussian distribution is generally the most useful to describe higher moments of the received electromagnetic field. However, in a definitive paper, Fremouw, *et al.* [17] used measurements to show that the Nakagami-*m* distribution is clearly superior to the log-normal, generalized Gaussian, and two-component Gaussian distributions to describe the probability density function of the received power for the case of radio wave scattering. This conclusion is based on a comprehensive analysis of the large body of multispectral scintillation data collected during the DNA Wideband Satellite experiment.

3.2.1 Received Signal First-Order Statistics

In fading conditions, a convenient way to write the received signal power is [14]:

$$S_r = S_0 S \sigma / \langle \sigma \rangle \tag{3.1}$$

In this expression for the received power, S_0 is the mean signal power received from the target, S is the fractional change in the signal power due to variations caused by the propagation channel, and $\sigma/\langle\sigma\rangle$ is the fractional change in the signal power caused by target cross section fluctuations. The factor S_0 contains the mean signal level; therefore, the mean value $\langle S \rangle$ is unity. Because the mean signal level has no statistical variation, the received signal power fluctuations can be expressed as the product of the effects of fluctuations due to propagation disturbances and target variations.

3.2.2 Target Statistics

Consider the case of a radar that transmits bursts separated by a large interval relative to the decorrelation time of the target cross section variations. A similar situation arises for the case where the radar transmission frequency is sufficiently changed from burst to burst so that the target cross section is decorrelated from burst to burst. In either case, the target cross section variation can often be assumed to be described by a Swerling II model, which applies to the case where the cross section varies independently from burst to burst. This nomenclature differs from the original Swerling II convention only in the replacement of Swerling's "pulse" by the term "burst" used here. The probability density function for target cross section variation is given by the expression:

$$p(\sigma) = \frac{1}{\langle\sigma\rangle} \exp(-\sigma/\langle\sigma\rangle), \quad \sigma \geq 0 \tag{3.2}$$

where $\langle\sigma\rangle$ is the mean value. The angle brackets denote a stochastic average. A target with constant cross section is of interest to certain defense radars, and results are also included for this case.

3.2.3 Propagation Channel Statistics

To have complete characterization of the first-order statistics at the radar receiver, we also require the statistics of the signal fluctuations caused by propagation through a disturbed channel. Nakagami-m statistics have been shown [17] to be useful for describing the range of possible scintillation conditions from weak to strong scattering. Under this description, the probability density function of fading on the one-way propagation channel is controlled by the m-parameter defined by the equation:

$$\frac{1}{m} = S_4^2 = \frac{\langle(P - \langle P \rangle)^2\rangle}{\langle P \rangle^2} \tag{3.3}$$

In this equation, S_4 is the scintillation index and it is a measure of the severity of the power fluctuations. The quantity P is the received power on a one-way propagation path. The S_4 index is thus the normalized standard deviation of the received power on a one-way propagation path where the transmitted power is constant. Values of S_4 generally range from a minimum of zero, signifying constant power or no scintillation, to a maximum of unity, indicating worst-case (Rayleigh) fading of amplitude where the in-phase and quadrature components of the received one-way signal are uncorrelated Gaussian variates. Values of S_4 greater than unity have been observed, but these indicate the presence of signal enhancements due to focusing. Because S_4 has been found to be a much more useful parameter to describe scintillation, the scintillation index will be used to quantify the scintillation severity instead of m for the remainder of this chapter.

For a one-way propagation path the probability density function of received power is given by [15]:

$$p_1(S) = \frac{m^m S^{m-1}}{\Gamma(m)\langle S \rangle^m} \exp\left(\frac{-mS}{\langle S \rangle}\right), \quad S \geq 0, \text{ one-way} \tag{3.4}$$

where the subscript 1 refers to the one-way propagation path. For the case of monostatic radar operation, the transmitter and receiver are collocated so that the signal propagates twice over the same path, passing through identical irregularities. In this case, the received voltage is proportional to the square of the voltage after one-way propagation [25]. The probability density function for the received power may be obtained from (3.4) by using the transformation $Q(\text{monostatic}) = S^2$ (one-way) with the result:

$$p_2(Q) = \frac{[m(m+1)]^{m/2} Q^{m/2-1}}{2\Gamma(m)\langle Q \rangle^{m/2}} \exp\left(-\sqrt{\frac{m(m+1)Q}{\langle Q \rangle}}\right), \quad Q \geq 0, \text{ two-way} \tag{3.5}$$

Note that the Nakagami-m density is a generalization of the chi-square probability density function. The chi-square density can be obtained from (3.4) by replacing the m-parameter with $n/2$ and performing the change of variable $y = nS$ on the resulting equation. In this equation, y is the chi-square variate and n is the number of degrees of freedom which is restricted to positive integer values.

It is straightforward to compute the relationship between the value of S_4 measured on the monostatic radar, two-way path as a function of m on the one-way path as S_4^2 (two-way) $= (4m^2 + 10m + 6)/(m(m + 1)^2)$. Thus, under Rayleigh statistics, the maximum value of S_4 on the two-way path, corresponding to a monostatic radar geometry, is $\sqrt{5}$. Figure 3.6 shows the cumulative distribution of the received power on a round trip propagation path that characterizes a monostatic radar geometry consisting of two one-way propagation paths. The cumulative distribution is shown as a function of the value of S_4 on each one-way propagation path. For worst-case scintillation in the monostatic radar geometry, we see from the figure that the probability of a 10 dB or greater fade is about 40%. This high probability of fading is reflected in degradation of various aspects of radar performance, including reduced radar target detection performance.

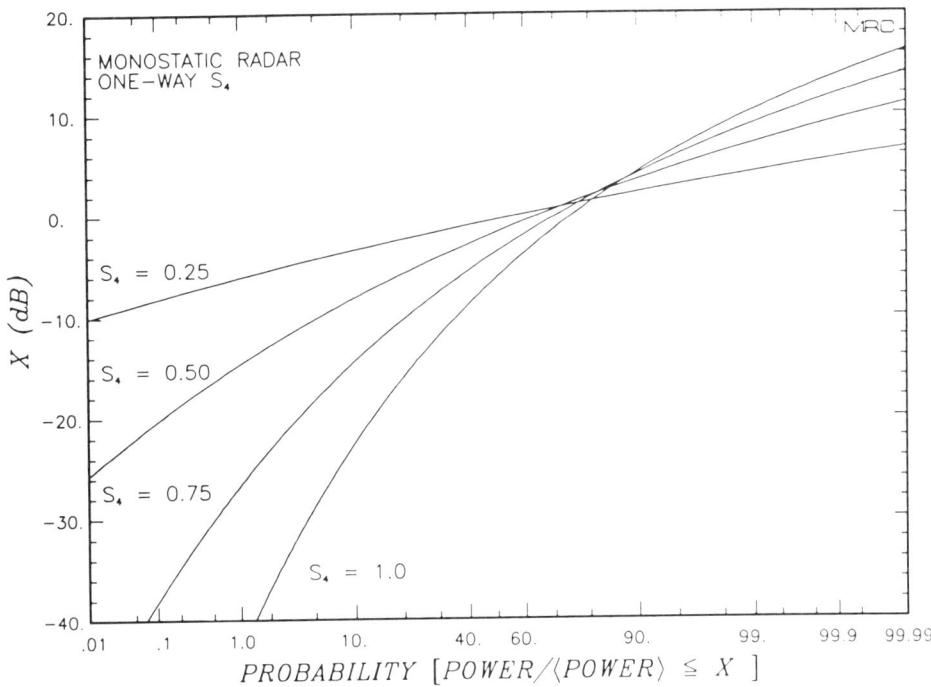

Figure 3.6 Cumulative probability distribution of received two-way power as a function of S_4 on the one-way propagation path.

3.2.4 Propagation Channel Coherence

The above discussion fully describes the first-order power statistics to be expected after propagation of a radar signal through a disturbed ionospheric channel.

The second-order fading statistics are specified by the correlation function of the received complex voltage. For the case of one-way propagation of an initially constant amplitude signal through a severely disturbed ionospheric channel, the autocorrelation function of the received voltage is given as the two-position, two-frequency mutual coherence function. The effective velocity of the radar signal's line of sight through the ionospheric irregularities can be used to convert the spatial coordinates of the mutual coherence function into temporal coordinates, thereby obtaining the correlation function of signal fluctuations due to ionospheric irregularities.

For worst-case Rayleigh fading, the correlation function of the received complex voltage $E(t, f)$ has the form:

$$\langle E(t + \tau, f + f_d) E^*(t, f) \rangle = |E_0|^2 \exp\left(\frac{-\tau^2/\tau_0^2}{1 + if_d/f_{coh}}\right) (1 + if_d/f_{coh})^{-1} \quad (3.6)$$

where τ_0 is the decorrelation time (the fading rate is $1/\tau_0$) for fluctuations over a one-way propagation path. The actual value of τ_0 is a function of radar geometry and the irregularity structure and intensity of the disturbed ionospheric channel. Large values of τ_0 correspond to slow fading conditions and small values correspond to fast fading.

As a concrete example, consider the case of a radar and target separated by a layer of ionization such as may occur in the case of an SBR observing a target near the ground. For a K^{-4} *in situ* power spectrum of three-dimensional ionization irregularities between outer scale L_0 and inner scale l_i, the decorrelation time is

$$\tau_0 = \sqrt{2} \, L_0 / \sqrt{\ln(L_0/l_i)} \sigma_\phi v_L \quad (3.7)$$

where v_L is the velocity of the line of sight through the center of the ionized layer, σ_ϕ^2 is $2(r_e \lambda)^2 L_0 L \overline{\Delta N_e^2}$ rad^2, λ is the RF wavelength, r_e is the classical electron radius (2.82×10^{-15} m), L is the thickness of the ionized layer, and $\overline{\Delta N_e^2}$ is the variance of electron-density irregularities.

We assume that τ_0 is large with respect to the duration of the transmitted pulse. The received signal is then coherent during the pulse duration, which is typically on the order of several tens of microseconds.

The channel coherence bandwidth is a measure of the maximum bandwidth available in the propagation channel over which it is possible to transmit a signal without undesired pulse distortion. Thus, in a fading environment, signal spectral components separated by less than the coherence bandwidth exhibit correlated fluctuations. If the signal spectral components are separated in frequency by an amount greater than the coherence bandwidth, different spectral components will undergo uncorrelated fading. This distortion in the received signal spectrum causes the received time-domain signal to display undesired time sidelobes. However, a propagating pulse remains undistorted as long as the maximum instantaneous signal bandwidth is less than the coherence bandwidth. This is often the case for frequency-agile radar systems. In addition, frequency-agile radar signals separated in frequency by an amount exceeding the coherence bandwidth will experience nearly independent fading. Herein we assume that the channel coherence bandwidth is sufficiently large compared with the radar pulse bandwidth so that time-domain distortion of the received pulse does not occur.

For a K^{-4} *in situ* power spectrum of three-dimensional ionization irregularities and one-way signal propagation path geometry with the radar and target on opposite sides of a scattering layer, the coherence bandwidth is given by

$$f_{coh} = \frac{\pi c(z_t + z_r)L_0}{r_e^2 \lambda^4 \ln(L_0/l_i) z_t z_r \overline{L \Delta N_e^2}} \tag{3.8}$$

In this expression, c is the velocity of light in a vacuum, z_t is the distance from the transmitter to the center of the ionized layer, z_r is the distance from the target to the center of the ionized layer, and $z_t + z_r$ is the total one-way propagation distance. The relationship between one- and two-way propagation paths is discussed in Knepp [25]. To obtain the appropriate description of a two-way monostatic radar propagation geometry, reduce the values of τ_0 and f_{coh} on the one-way path by a factor of $\sqrt{2}$.

In general, the signal power fluctuation due to variations caused by the propagation channel, S, exhibits correlation over both time and frequency as discussed above. However, in this work only the burst-to-burst correlation is of interest. Therefore, the resulting correlation function may be modeled as a frequency-coherence function given by

$$C_{SS}(f_d) = \langle S(f)S(f + f_d) \rangle - \langle S \rangle^2 \tag{3.9}$$

where S is assumed to be a stationary process over the frequency range of interest.

Consideration of channel coherence is important for the assessment of the effectiveness of burst combining using a frequency-agile radar in a fading channel. If the separation of burst transmission frequencies is great enough to ensure that $S(f)$ is uncorrelated, radar performance will be greatly enhanced through burst combining. However, if the received power shows correlation from burst to burst, then accounting for C_{SS} is important in the calculation of detection probability.

3.3 RADAR SYSTEM CHARACTERISTICS

In this chapter, we assume that an SBR is required to operate through a disturbed ionospheric channel for detecting and tracking targets near the earth's surface. Thus, as compared to a ground-based radar with similar functions, an SBR has several limitations. First, targets are detected and tracked at very long ranges. Second, available on-board transmitter power is relatively low. Third, because of the great target ranges involved, vast areas of the earth's surface are illuminated with resultant large clutter returns, even with a very narrow antenna beamwidth.

The first two points imply low received signal-to-noise ratio (SNR) per pulse and therefore require long integration times. However, the cross section of a moving target remains constant or coherent for only a few tens of milliseconds because of target motion and resulting constructive and destructive interference between many scattering centers. Hence, during a radar look, the total energy transmitted at a target is divided into a number of bursts. Each burst consists of some number, n, of pulses, which are coherently integrated. Bursts are transmitted at a different frequency in the case of a frequency-hopping radar. The detected amplitude of all the bursts that form the total radar target look are noncoherently combined in a postdetection integration process. The resulting signal power is compared to a threshold level to decide whether a target has been detected during the radar look.

One reason for choosing a waveform consisting of many coherently related pulses is that the radar transmitter may be power-limited and therefore unable to generate sufficient energy in a single pulse. Thus many pulses with low power are coherently transmitted, then coherently integrated on reception to achieve high SNR. A second advantage in using a coherent pulsed radar waveform is that doppler processing techniques may be applied to the pulses composing a single burst to achieve clutter suppression.

The use of frequency hopping can produce independent samples of the target cross section from burst to burst. Depending on the target geometry and motion relative to the radar, the target cross section from burst to burst may also be independent. In either case, a Swerling II cross section model applies. A nonfluctuating target serves as a convenient reference for computation.

When the duration of a burst is short enough that the received signal power remains essentially constant within it, the fading is referred to as *slow fading*. The slow fading assumption will apply for all cases considered in this chapter.

Two types of radar detection techniques are considered here, *double-threshold detection* and *noncoherent integration*. In both techniques, the first stage is quadrature detection of the radar return from a transmitted burst. The quadrature detector output is a measure of received power. For the M out of N detection scheme, N bursts are transmitted to form a look. If the power received from M or more bursts exceeds a predetermined threshold, a detection is declared. In the case of noncoherent integration, the power measured from each burst in a look is summed and, if the resultant sum exceeds a threshold, a detection is declared.

A phased array radar system having sufficient beam-pointing flexibility might use a modified M out of N detection scheme. For example, if a hit is declared on the first M bursts, the rest need not be transmitted. For a radar that must detect multiple targets, additional bursts may be transmitted to a specific location upon a threshold crossing of the return from the first burst.

3.4 PROBABILITY OF DETECTION

The goal of this section is to develop methods to obtain the probability of detection for a radar operating in the presence of scintillation. Mathematical expressions, which are readily evaluated using numerical methods, are found for the detection probability when constraints are placed on the burst correlation function of the random process S_i, which is the fractional change in signal power in burst i due to variations caused by the propagation channel. For the remaining cases, the detection probability is calculated by using Monte Carlo simulation methods. The simulation method presented here is general enough to include any coherence function for S.

Expressions that may be readily evaluated numerically are developed for the following cases: M out of N detection of both constant and Swerling II targets when the received power S is assumed independent from burst to burst; M out of N for both constant and Swerling II targets when S is assumed identical for each burst in a look; and noncoherent integration of bursts for a Swerling II target when S is assumed identical for each burst in a look. In all cases, S is assumed to follow a Nakagami-m amplitude distribution.

Monte Carlo methods developed here may be used to determine the probability of detection for both M out of N detection and noncoherent integration for a Swerling II target. The received signal power fluctuation due to propagation for each burst, S_i, follows an arbitrary correlation function and a Nakagami-m amplitude distribution.

3.4.1 Receiver Model

Under slow fading conditions, let the signal amplitude and phase during a received radar burst be constant so that the in-phase and quadrature voltage components are given by the expressions:

$$i = a \cos \phi + n_i \qquad (3.10)$$

$$q = a \sin \phi + n_q \qquad (3.11)$$

where i is the voltage (signal plus noise) of a single burst in the in-phase channel and q is the voltage of a single burst in the quadrature channel. The total average signal power is $\langle i^2 \rangle + \langle q^2 \rangle = S_r$. The noise is assumed to be additive white Gaussian noise with a probability density function of

$$p(n_i) = \frac{1}{\sqrt{2\pi}\,\sigma_N} \exp\left(-\frac{n_i^2}{2\sigma_N^2}\right) \tag{3.12}$$

with a similar expression for $p(n_q)$. The total noise power per burst is

$$\langle n_i^2 \rangle + \langle n_q^2 \rangle = 2\sigma_N^2 \tag{3.13}$$

so that the SNR per burst is $S_r/2\sigma_N^2$. Now, the output voltage from a quadrature detector giving $w = (i^2 + q^2)^{1/2}$ has the well known Rician probability density function conditioned on S_r:

$$p(w|S_r) = \frac{w}{\sigma_N^2} \exp\left\{\frac{-(w^2 + S_r)}{2\sigma_N^2}\right\} I_0\left(w\sqrt{S_r}/\sigma_N^2\right) \tag{3.14}$$

where I_0 is the modified Bessel function. The above density function may also be conditioned jointly on S and σ through the use of (3.1). For a Swerling II target, the probability density function of the target cross section is given by (3.2). Because S and σ are independent variables, a conditional density with respect to S alone is obtained by

$$p(w|S) = \int_0^\infty p(\sigma) p(w|S_r = [S_0 S \sigma / \langle \sigma \rangle]) \, d\sigma \tag{3.15}$$

where $S_r = S_0 S \sigma / \langle \sigma \rangle$. The integral expression in (3.15) easily reduces to the conditional density for w. Thus,

$$p(w|S) = \frac{w}{\sigma_N^2 (1 + S_0 S/2\sigma_N^2)} \exp\left[\frac{-w^2}{2\sigma_N^2(1 + S_0 S/2\sigma_N^2)}\right], \quad \text{Swerling II target} \tag{3.16}$$

3.4.2 Nonfluctuating Target

For a nonfluctuating target, the probability density function of the target cross section can be represented by the Dirac delta function $\delta(\sigma - S_0)$. In this case, the integration in (3.15) gives

$$p(w|S) = \frac{w}{\sigma_N^2} \exp\left\{\frac{-(w^2 + S_0 S)}{2\sigma_N^2}\right\} I_0\left(w\sqrt{S_0 S}/\sigma_N^2\right), \quad \text{constant target} \qquad (3.17)$$

Note that the mean received power for both the Swerling II and the constant target is $S_0 S + 2\sigma_N^2$ as it should be.

3.4.3 Single-Burst Detection

The quadrature detector output is compared to a preset threshold at the end of a burst interval, and a "hit" or "miss" is declared. The threshold is set on the basis of the noise alone so that the probability of false alarm is small. The false-alarm probability for detection of a single burst is given by

$$p_{fa} = \int_{th}^{\infty} p(w|S=0)\,dw \qquad (3.18)$$

The quantity p_{fa} is simply the single-burst probability that the noise amplitude alone exceeds the threshold. The above equation is integrated to obtain the familiar result:

$$th = \sqrt{-2\sigma_N^2 \ln(p_{fa})} \qquad (3.19)$$

Thus, the threshold amplitude is expressed as a function of the rms noise amplitude and the desired probability of false alarm for a single burst. The threshold is the same for all channel and target models because the noise is independent of target and channel statistics.

Equations (3.15) and (3.19) may be used to find the single-burst probability of detection as a function of S, the fractional change in signal amplitude due to scintillation:

$$\begin{aligned} p_d(S) &= \int_{th}^{\infty} p(w|S)\,dw \\ &= p_{fa}^{(1+S\langle \text{SNR}\rangle)^{-1}}, \quad \text{Swerling II target, or} \qquad (3.20) \\ &= Q\left(\sqrt{-2\ln(p_{fa})},\sqrt{2S\langle \text{SNR}\rangle}\right), \quad \text{constant target} \end{aligned}$$

where $\langle \text{SNR}\rangle = S_0/2\sigma_N^2$ is the average signal-to-noise ratio at the amplitude detector output and Q is the Marcum-Q function [26]. When $S = 1$, (3.20) is the familiar formula for detection probability of a Swerling II target.

3.4.4 M out of N Detection

In the case of *M* out of *N* or double-threshold detection, to obtain simple analytic expressions is straightforward for the overall probability of detection in terms of the probability of detection per burst, provided that the radar returns from each burst are independent.

Now, assume that the probability of detection per burst is given by p_d. To find the overall probability of detection, P_D, let $p_b(i, j)$ be the probability of obtaining exactly *i* detections in *j* bursts. Then, the probability of obtaining *M* or more detections out of N bursts is the sum:

$$p(M, N) = p_b(M, N) + p_b(M + 1, N) + \ldots + p_b(N, N) \quad (3.21)$$
$$= \sum_{k=M}^{N} P_b(k, N)$$

but because p_b is given by the binomial distribution [27]:

$$P_b(i, j) = \binom{j}{i} p_d^i (1 - p_d)^{j-i} \quad (3.22)$$

it is easy to obtain the probability of detection for the *M* out of *N* process:

$$P_D = \sum_{k=M}^{N} \binom{N}{k} (p_d)^k (1 - p_d)^{(N-k)} \quad (3.23)$$

In the absence of a target, the overall probability of false alarm for double-threshold detection is given by a similar expression:

$$P_{FA} = \sum_{k=M}^{N} \binom{N}{k} (p_{fa})^k (1 - p_{fa})^{(N-k)} \quad (3.24)$$

where p_{fa} is the single-burst false alarm probability.

3.4.5 Independent Bursts

Consider the case where the contribution to the received signal due to channel fluctuations is independent from burst to burst, either due to temporal or frequency offsets between bursts. Here, the probability of detection for a single burst is obtained by averaging $p_d(S)$, given by (3.20), over the distribution of channel fluctuations $p_2(S)$. Thus,

$$p_d = \int_0^\infty p_d(S)p_2(S)\,dS \tag{3.25}$$

where $p_2(S)$ is given by (3.5).

3.4.6 Constant Propagation Channel

Consider the case where the contribution to the received signal due to the disturbed channel is constant over all the bursts that constitute a look. Here, the probability of detection during fading is computed by averaging the probability of detection for constant propagation conditions over the appropriate distribution function that describes the propagation channel fluctuations. In this case, the overall detection probability may be written as

$$P_D = \int_0^\infty \left\{ \sum_{k=M}^{N} \binom{N}{k} [p_d(S)]^k [1 - p_d(S)]^{N-k} \right\} p_2(S)\,dS \tag{3.26}$$

where $p_d(S)$ is given by (3.20).

3.4.7 Noncoherent Integration—Constant Propagation Channel

For a Swerling II target and a constant propagation channel, Dana and Knepp [28] have developed a general formulation for the case of noncoherent integration that may be applied to the case of Nakagami-m statistics. The detection probability is evaluated by using the expression:

$$P_D = \int_0^\infty P_D(S)p_2(S)\,dS \tag{3.27}$$

where $p_2(S)$ is given by (3.5), and

$$P_D(S) = \frac{\Gamma\{M, th/[\sigma_N^2(1 + S\langle SNR\rangle)]\}}{\Gamma(M)} \tag{3.28}$$

$$\Gamma(a, t) = \int_t^\infty t^{a-1} e^{-t}\,dt \tag{3.29}$$

and the threshold is set by using

$$P_{FA} = \Gamma(M, th/\sigma_N^2)/\Gamma(M) \tag{3.30}$$

$\Gamma(m)$ is the usual gamma function of a single argument and $\Gamma(a, t)$ is an incomplete gamma function. Equation (3.29) is easily inverted numerically to give th/σ_N^2 as a function of the probability of false alarm and the number of bursts M.

3.4.8 Numerical Methods

Numerical evaluation of the foregoing integral expressions for probability of detection is straightforward. For the case of M out of N detection, using specified values of M, N, and the overall probability of false alarm P_{FA}, the first step is the numerical inversion of (3.24) to obtain the single-burst probability of false alarm, p_{fa}. If desired, the threshold is then obtained by using (3.19). The quantity p_{fa} is used in (3.20) to obtain $p_d(S)$ and the integrals of (3.25) and (3.26) are performed as desired. This entire process is repeated as often as desired for different values of SNR and scintillation index.

A similar procedure is used for the case of the noncoherent combining of bursts from a Swerling II target in a constant propagation channel.

3.4.9 General Correlation of Burst Returns

In the general case of a disturbed propagation channel, the radar returns from the bursts that constitute a look will exhibit some degree of correlation in time and frequency as discussed earlier. In this case, the use of numerical integration techniques becomes prohibitive because the probability of detection is expressed in the form of multivariate integrals. Therefore, Monte Carlo simulation techniques must be used to obtain the probability of detection for the case where burst-to-burst correlation is described by a general correlation function. The techniques described below are valid for general correlation functions of any form. However, in the examples to be presented, a simple Gaussian function is used. This simple one-parameter distribution is quite useful because a Gaussian function is often used to fit more complicated distributions.

The simulation is accomplished by generating random samples of the received signal power, S_r, for each burst within a look. The resulting samples may be processed directly by using either double-threshold detection or noncoherent integration. After a sufficient number of looks are processed, an estimate is obtained of the detection probability at a given SNR and probability of false alarm.

To have a valid representation of the fading channel requires that the statistics of received power follow the Nakagami-m distribution. Fortunately, as noted previously, the χ^2 number of degrees of freedom is related to the Nakagami-m parameter by the expression $m = n/2$, $n = 1, 2, 3 \ldots$ A useful range of values of S_4 may be obtained by using the available half-integer values of m.

A χ^2 variate with n degrees of freedom may easily be generated from the expression:

$$S = v_1^2 + v_2^2 + \ldots + v_n^2 \qquad (3.31)$$

where the v_i are independent, zero mean Gaussian variates with variance σ_v^2 selected so that the mean value of S is unity.

Also, all values of power S for bursts in a single look must be related according to the arbitrary correlation function C_{ss}. Write S as a function of the index of the burst, i, where there are N bursts per look and $i = 0, 1, \ldots N - 1$. Therefore,

$$C_{SS}(j) = \langle S(i)S[(i + j), \text{mod} N]\rangle - \langle S \rangle^2$$
$$j = -(N - 1), \ldots, (N - 1) \qquad (3.32)$$

is required, where the notation mod refers to the modulo remainder function defined by $(a, \text{mod } N) = r$, where $a = r + nN$ and n is any integer or zero.

Expanding (3.31) for all bursts in a look gives a set of equations:

$$S(1) = v_1^2(1) + v_2^2(1) + v_3^2(1) + \ldots + v_n^2(1)$$
$$S(2) = v_1^2(2) + v_2^2(2) + v_3^2(2) + \ldots + v_n^2(2)$$
$$\cdot \qquad \cdot \qquad \cdot \qquad \cdot \qquad \cdot$$
$$\cdot \qquad \cdot \qquad \cdot \qquad \cdot \qquad \cdot$$
$$\cdot \qquad \cdot \qquad \cdot \qquad \cdot \qquad \cdot$$
$$S(N) = v_1^2(N) + v_2^2(N) + v_3^2(N) + \ldots + v_n^2(N) \qquad (3.33)$$

Equation (3.33) may be satisfied if

$$C_{SS}(j) = 2nC_{vv}^2(j) \qquad (3.34)$$

where

$$C_{vv}(j) = \langle v_k(i)v_k[(i + j), \text{mod } N]\rangle; \quad \text{any } k, \text{ any } i \qquad (3.35)$$

The above assertion can be easily proved by using the relation [29]:

$$\langle x^2 y^2 \rangle = \langle x^2 \rangle \langle y^2 \rangle + 2\langle xy \rangle^2 \qquad (3.36)$$

where x and y are jointly normal with zero mean.

Therefore, we need to generate n Gaussian vectors, each of length N, that satisfy (3.34), which are to be used in (3.33). An efficient and flexible method for generating a vector of jointly normal variates is presented in Hurst and Knop [30].

3.5 RESULTS

In this section we give quantitative results for probability of detection for a wide range of scintillation conditions, ranging from weak to strong scattering and for various amounts of decorrelation between bursts in a look. We show that both double-threshold detection and noncoherent combining of bursts offer a degree of mitigation against the effects of scintillation when there is some diversity gain available through decorrelation of bursts in a look. Although the probability of false alarm is arbitrary in our formulation and is an input variable to the simulations, all results to be presented assume that the detection threshold is set so that the false alarm rate per look is 10^{-6}.

Figures 3.7 and 3.8 illustrate the effects of various levels of fading on a system employing a single burst for detection of a constant cross section target and a Swerling II target, respectively. The figures show probability of detection as a function of the mean SNR per burst. Comparing these figures at various values of detection probability is useful. Generally, a single-burst probability of detection of about 0.5 or greater is needed for successful tracking. For example, from Figure 3.7 at a value of detection probability of 0.7, the loss in detection sensitivity relative to undisturbed conditions is about 3.5 dB at a value of S_4 of 0.5, and it is about 11 dB at a value of S_4 of 1.0. Figure 3.8 shows similar losses in sensitivity relative to undisturbed conditions for a Swerling II target model. With the exception of very low SNR, radar performance is better for the constant target cross section model as compared with the Swerling II model due to the decreased variability in received signal power.

Figures 3.9 to 3.19 the effect of M out of N burst combining is investigated for various values of M and N. In these figures, the abscissa is the SNR per look defined as N times the SNR per burst. Thus, the curves are plotted so that various combinations of M and N may be compared on the basis of an equal amount of transmitted energy per look, assuming that all N bursts are transmitted upon each look.

For the case of a constant cross section target with no propagation channel fading, the optimum number of bursts per look is one, and there is no advantage from burst combining. When the target statistics obey the Swerling II model, however, the received signal power varies independently from burst to burst, and diversity combining is effective, as shown in Figure 3.9. At a probability of detection value of 0.8, 1 out of 2 burst combining improves detection sensitivity by about 1 dB as compared to no combining (i.e., 1 out of 1). The results presented in Figure

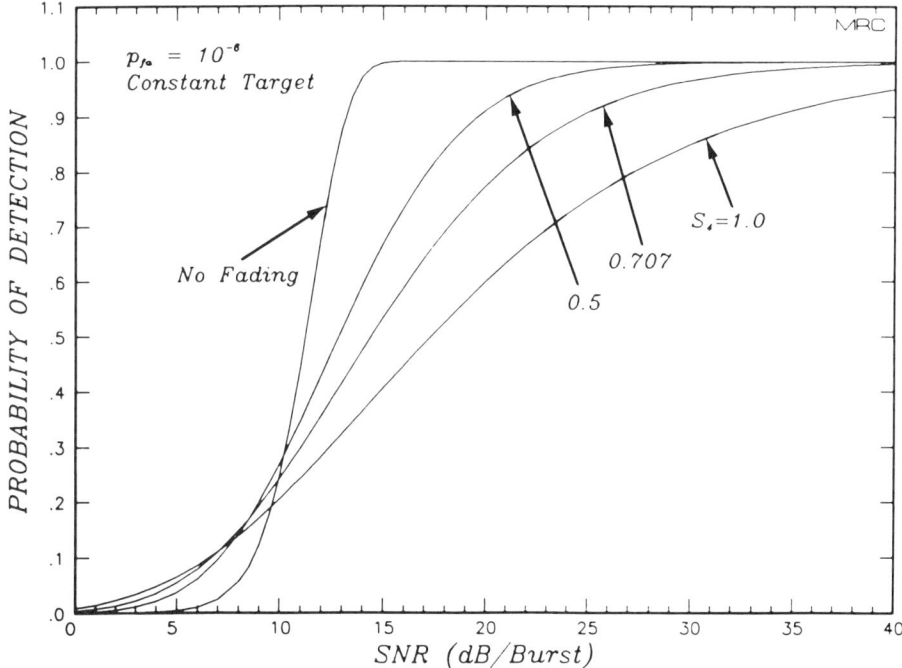

Figure 3.7 Probability of detecting a constant target with a single burst for various scintillation levels.

3.9 are well known. However, this figure is included to allow the reader to compare visually the results for detection in a benign propagation environment with the results to be presented for a fading channel.

In Figures 3.10 to 3.13, quantitative results for probability of detection are given that show the effect of various levels of scintillation severity on double-threshold detection for both target models of interest. For these results, we assume that the signal contribution due to channel fading is independent from burst to burst within a look. This case of the propagation channel disturbance is referred to as type II fading, not unlike a Swerling II target model.

Performance in worst-case scintillation, characterized by an S_4 index of 1.0 on the one-way propagation path, is considered in Figures 3.10 and 3.11. Figure 3.10 illustrates that, even for a constant target, diversity combining offers a performance improvement. An improvement of nearly 7 dB in detection sensitivity is available when the receiver operates at a detection probability of 0.75 using ? out of 8 (or 1 out of 4) detection. For the case of the Swerling II target model a shown in Figure 3.11, the use of 1 out of 8 (or 1 out of 4) combining gives a improvement of about 4 dB at a detection probability of 0.5 and, for a detectio

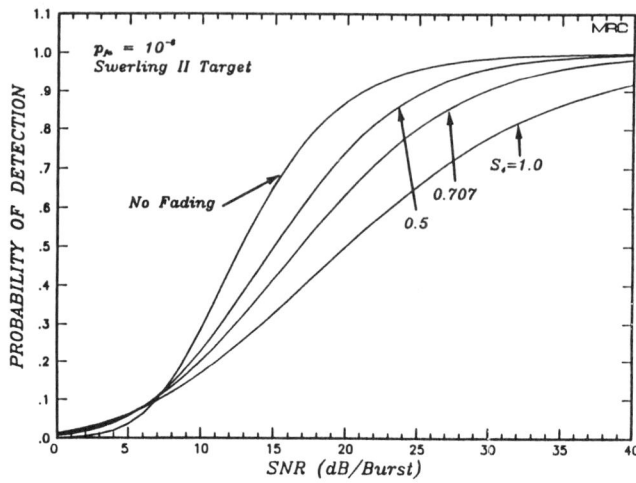

Figure 3.8 Probability of detecting a Swerling II target with a single burst for various scintillation levels.

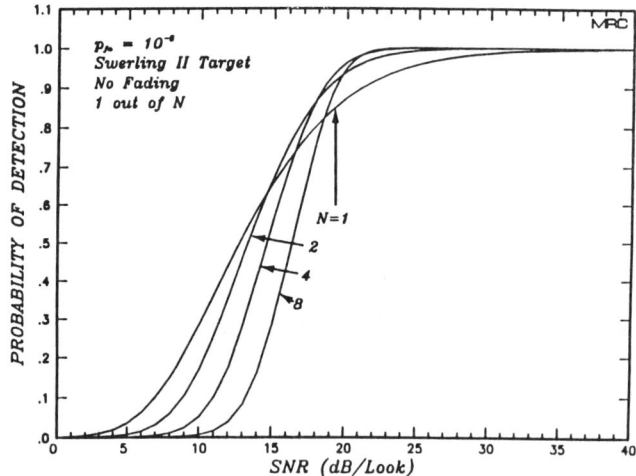

Figure 3.9 Probability of detecting a Swerling II target with no channel fading using 1 out of N combining.

probability of 0.75, the improvement is 9.5 dB.

A value of S_4 of 0.5 is generally accepted as marking the transition from weak to strong scattering. In less severe fading, with S_4 less than 0.707, there is generally no advantage to be gained from using M out of N detection against a constant

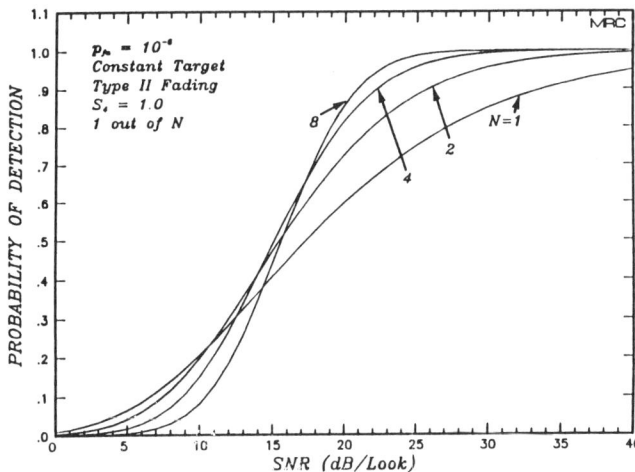

Figure 3.10 Probability of detecting a constant target with severe type II fading ($S_4 = 1.0$) using 1 out of N combining.

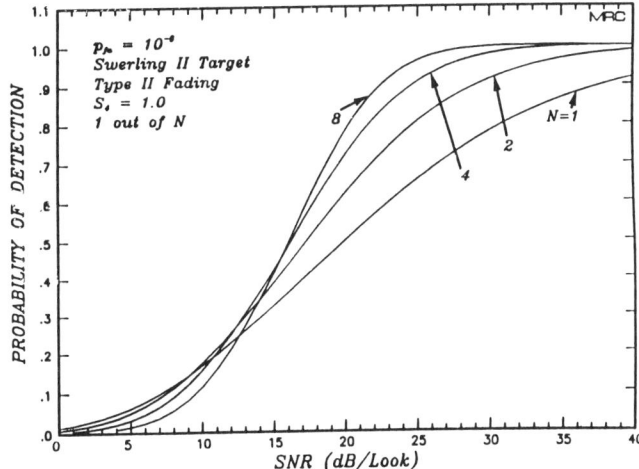

Figure 3.11 Probability of detecting a Swerling II target with severe type II fading ($S_4 = 1.0$) using 1 out of N combining.

target. For less severe scintillation, single-burst detection exhibits better performance.

Figures 3.9 to 3.11 show the effect on detection performance of variation of the number of bursts per look, N, for 1 out of N combining, for constant and

Swerling II targets. These figures illustrate that some protection against type II fading is available by using diversity combining. As an additional example illustrating that point, compare Figures 3.12 and 3.13 with Figures 3.7 and 3.8, respectively. Figure 3.12 gives probability of detection using 1 out of 8 combining for a constant target model as a parametric function of the S_4 index. Results in Figure 3.7 are for an identical case, except that only a single burst is used. Note that detection performance is not strongly affected by fading when 1 out of 8 combining is used. Also note that for a detection probability of 0.7, 1 out of 8 combining gives better performance in a strong fading environment than in a benign environment. Furthermore, performance is rather similar throughout the strong scattering regime. Of course, a large price is paid for using 1 out of 8 combining to detect a constant target in an undisturbed environment. Figures 3.8 and 3.13, for the case of a Swerling II target model, illustrate a similar situation. At a probability of detection of 0.7, a sensitivity loss of 1 dB is incurred when using 1 out of 8 combining as the propagation environment goes from no fading to worst-case fading. However, the cost for using 1 out of 8 combining in an undisturbed environment is less than 2 dB relative to single-burst detection at a detection probability of 0.7 for the Swerling II model.

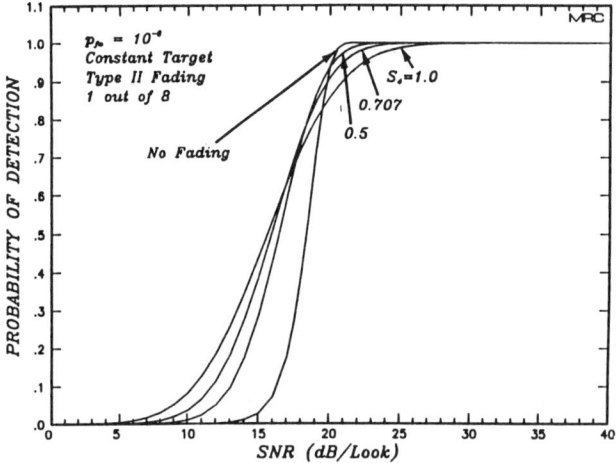

Figure 3.12 Probability of detecting a constant target for various scintillation levels during type II fading using 1 out of 8 combining.

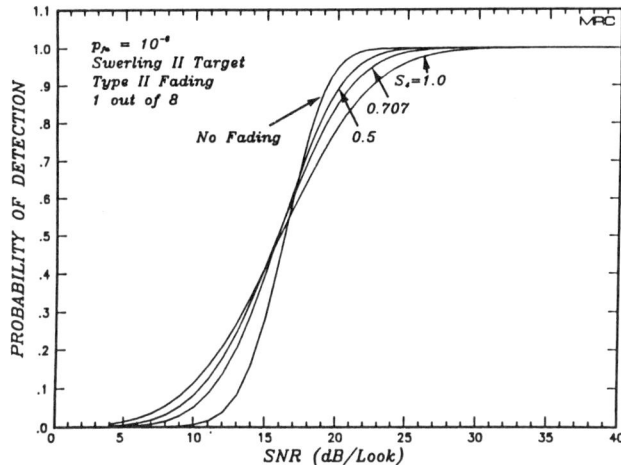

Figure 3.13 Probability of detecting a Swerling II target for various scintillation levels during type II fading using 1 out of 8 combining.

3.5.1 Type I Fading

The previous discussion has demonstrated the effectiveness of M out of N combining against type II fading, where the effects of propagation channel scintillation are independent from burst to burst. If, however, the time and frequency scheduling of bursts is such that the effect of the propagation channel is identical from burst to burst (i.e., type I fading), diversity combining is no longer effective. Figure 3.14 shows the effect of type I fading on 1 and 2 out of 8 combining for values of S_4 of 0, 0.5, 0.707 and 1.0 for a Swerling II target. A comparison of these curves to those of Figure 3.8, which gives single-burst detection performance, shows that there is generally a loss suffered from combining in a type I scintillation environment as oposed to using a single burst, for levels of scintillation severity where S_4 is greater than 0.5.

3.5.2 General Burst Coherence

To investigate the effects of partial correlation, Monte Carlo simulations are performed for a Gaussian correlation function relating power from burst to burst.

This correlation function has the form:

$$C_{SS}(n) = \exp\{-(n \ln \phi)^2\}, \quad n = 0, \ldots, N - 1 \qquad (3.37)$$

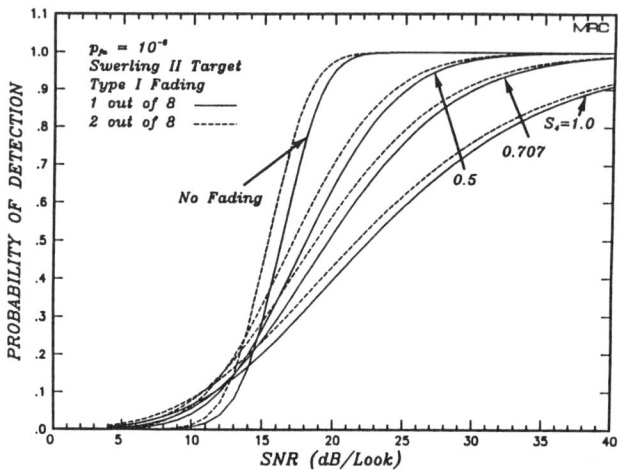

Figure 3.14 Probability of detecting a Swerling II target for various scintillation levels during type I fading using 1 and 2 out of 8 combining.

where the index n refers to burst number $n + 1$ of the N bursts in a look. The symbol ln is the natural logarithm and ϕ is a measure of the amount of correlation between bursts, with $\phi = 0$ for uncorrelated bursts and $\phi = 1$ for completely correlated bursts. Note that the Monte Carlo simulation applies to a general correlation function, and therefore can be used to study the effects of both temporal and frequency correlation between bursts. This simple Gaussian correlation function is chosen as the first application. Results obtained in this manner give a set of bursts in a look that statistically have the same correlation function for every look. Thus, the results obtained here are not strictly applicable to the case of a frequency-hopping radar, where the transmission frequency of a burst is uniformly distributed over some hopping bandwidth, giving the possibility of an arbitrary correlation function relating the bursts in a specific look. Later applications are required to address specific radar systems and propagation environments.

Figures 3.15 and 3.16 give the detection performance results for 1 and 2 out of 8 combining for a Swerling II target and values of ϕ of 0.5 (moderately correlated bursts) and 0.8 (well correlated bursts), respectively. A comparison of the solid $S_4 = 1.0$ curves of Figure 3.15 to the $N = 8$ curve of Figure 3.11 shows that there is about a 2 dB loss in detection sensitivity at a 0.7 value of probability of detection

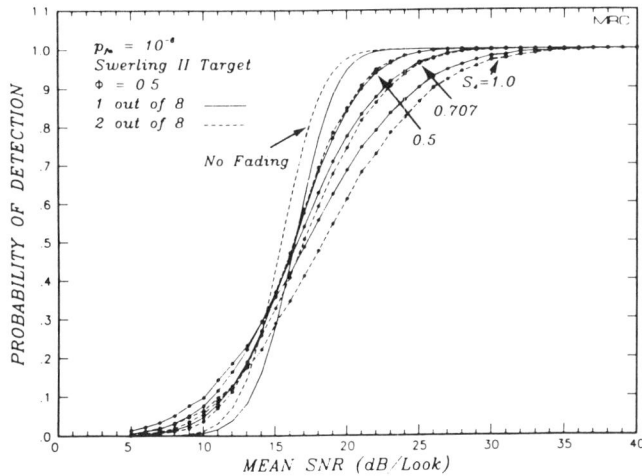

Figure 3.15 Probability of detecting a Swerling II target using 1 and 2 out of 8 for various scintillation levels, assuming Gaussian correlation of bursts ($\phi = 0.5$).

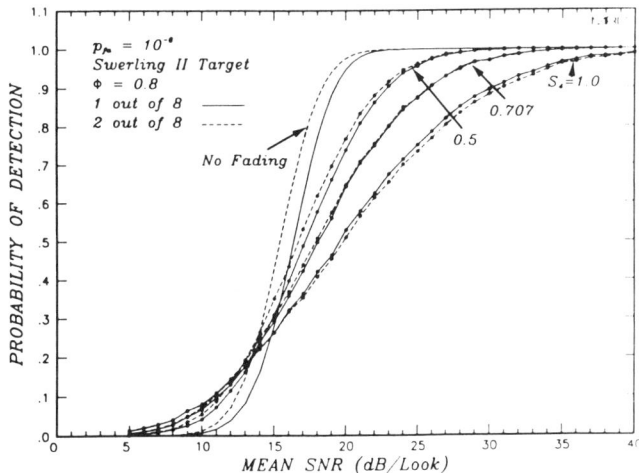

Figure 3.16 Probability of detecting a Swerling II target using 1 and 2 out of 8 for various scintillation levels, assuming Gaussian correlation of bursts ($\phi = 0.8$).

due to the moderate amount of correlation of power between bursts. For the case of greater ($\phi = 0.8$) correlation shown in Figure 3.16, the loss in detection sensitivity due to burst-to-burst correlation is about 5.5 dB at a probability of detection of 0.7 and a value of S_4 of unity. However, a comparison of Figure 3.16 with Figure 3.8 for the case of a single burst per look shows that there is some diversity gain, even in this highly correlated case, provided that sufficient power is available. In other words, only a small amount of decorrelation per burst is needed to give improved detection performance by using M out of N combining.

3.5.3 M Other Than 1

Figure 3.17 shows the effect of variation in M on detection sensitivity for the case of an undisturbed propagation channel with a Swerling II target. For a constant target (not shown), detection performance is best using 4 (or 5 out of 8), with higher values of M yielding decreased detection sensitivity. For a Swerling II target, 2 out of 8 (or 3 out of 8) provides the best detection performance.

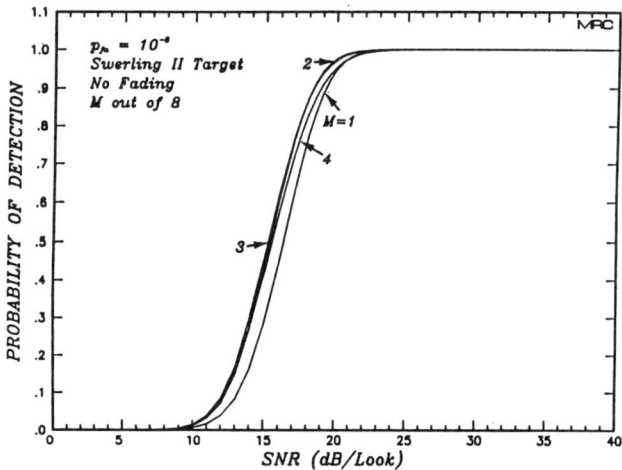

Figure 3.17 Probability of detecting a Swerling II target with no fading using M out of 8 combining.

Figures 3.18 and 3.19 show probability of detection during type II fading of different levels of severity by using 2 out of 8 combining for a constant target and a Swerling II target, respectively. These figures can be directly compared to Figures 3.12 and 3.13, which give detection performance by using 1 out of 8 combining. From the comparison, we see that 1 out of 8 yields better detection performance than 2 out of 8 during strong scintillation conditions ($S_4 > 0.5$), but the reverse is true given weaker scattering ($S_4 < 0.5$). Note, however, that this advantage is dependent on the assumption of type II fading, as we may see by comparing Figures 3.15 and 3.16 for different amounts of burst-to-burst correlation.

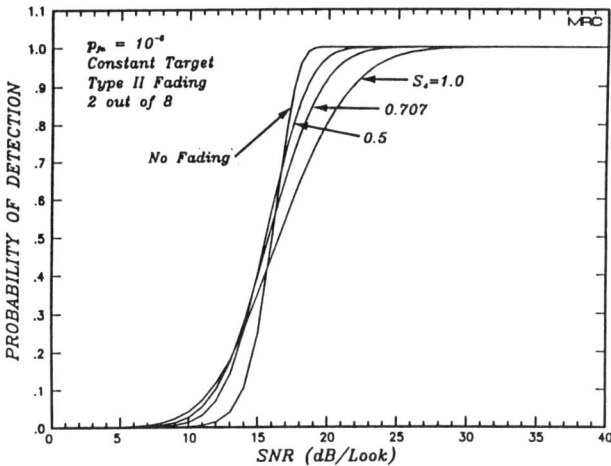

Figure 3.18 Probability of detecting a constant target during type II fading using 2 out of 8 combining.

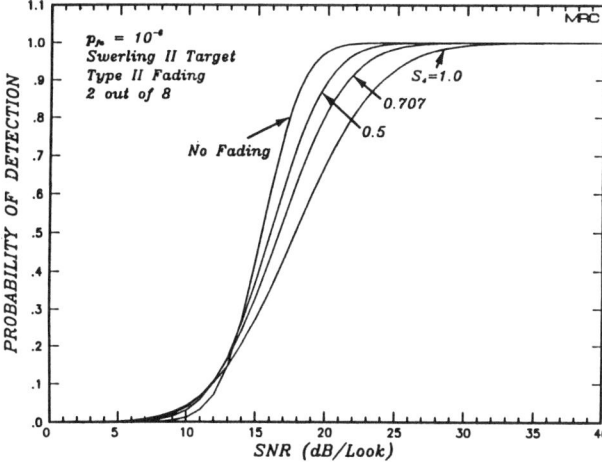

Figure 3.19 Probability of detecting a Swerling II target during type II fading using 2 out of 8 combining.

3.5.4 Noncoherent Integration

Results are also obtained in this section for detection using noncoherent integration of bursts to compare with the previous results for double-threshold detection. Double-threshold detection is sometimes referred to as *binary integration* because

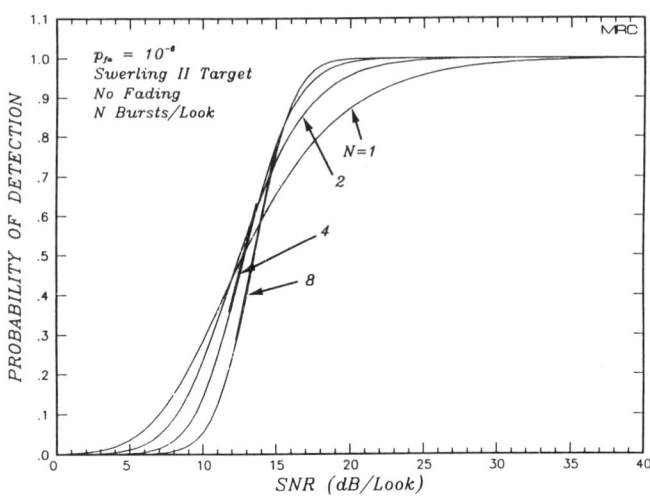

Figure 3.20 Probability of detecting a Swerling II target using noncoherent integration of bursts with no fading.

it can be considered as a modification of noncoherent integration. The modification is that the output of the quadrature detector is A/D converted by using one bit of resolution (threshold detection) prior to integration. Because some information is lost when using such a coarse measure of the quadrature detector output, we can expect a penalty to be exacted in detection sensitivity for using double-threshold detection. Results in this section illustrate the loss in detection sensitivity for Swerling II targets.

Figure 3.20 shows the probability of detection for noncoherent integration for the case of no fading and various values of N, the number of bursts per look. This figure can be compared with Figure 3.9 for double-threshold detection using 1 out of N. From Figure 3.20, we can see that for noncoherent integration and a Swerling II target, having more bursts per look is advantageous, provided that the SNR is sufficiently high to maintain the probability of detection at about 0.5 or greater, thereby avoiding the abrupt degradation in detection performance with decreasing SNR. However, from Figure 3.9, for double-threshold detection using 1 out of N in an undisturbed propagation environment, only 1 out of 2 is generally useful.

Figures 3.21 to 3.23 give results for the probability of detecting a Swerling II target by using noncoherent integration of 8 bursts per look under various conditions of scintillation severity and degrees of correlation among the bursts of a look. All three figures show curves for the case of no fading and for values of one-way S_4 of 0.5, 0.707, and 1.0. Figure 3.21, for type II fading, may be compared directly with Figures 3.13 and 3.19 for double-threshold combining by using 1 or

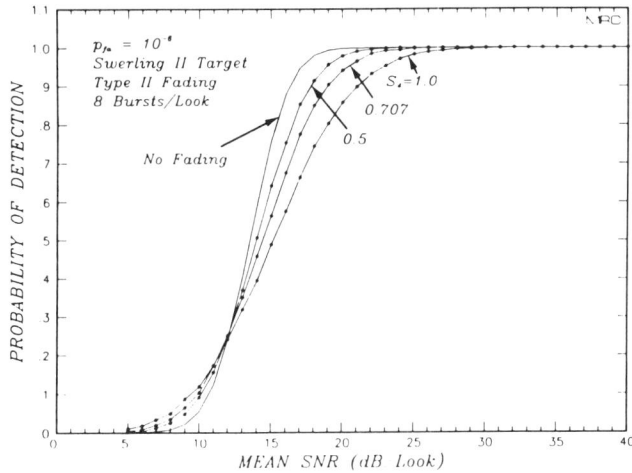

Figure 3.21 Probability of detecting a Swerling II target using noncoherent integration of 8 bursts during type II fading.

2 out of 8. In the case of S_4 equal to or greater than 0.5, 1 out of 8 gives better performance than 2 out of 8. However, comparison of Figures 3.13 and 3.21 shows that noncoherent integration of 8 bursts enjoys an advantage of about 1.5 dB relative to double-threshold detection using 8 bursts per look.

Figure 3.22 shows detection performance for the case of partially correlated bursts in a look with a value of ϕ of 0.8, and may be directly compared to Figure 3.16. This comparison indicates that noncoherent detection using 8 bursts per look enjoys an advantage of about 2 dB relative to double-threshold detection using 1 or 2 out of 8.

Figure 3.23 shows detection performance by using 8 bursts per look in a type I fading environment where the propagation effects on all bursts in a look are perfectly correlated. A comparison of this figure with Figure 3.14 again gives incoherent integration an advantage of about 2 dB relative to double-threshold detection using 8 bursts per look.

3.6 CONCLUSIONS

Severe fading has a strong effect on the target detection performance of a space-based radar. In this chapter both analytic and simulation results for the performance of double-threshold detection are compared to that of noncoherent integration of the returns from multiple bursts per look. Quantitative graphical results are given for various levels of scintillation severity ranging from no scintillation to worst-case Rayleigh fading on the one-way propagation path. In addition, the effects of

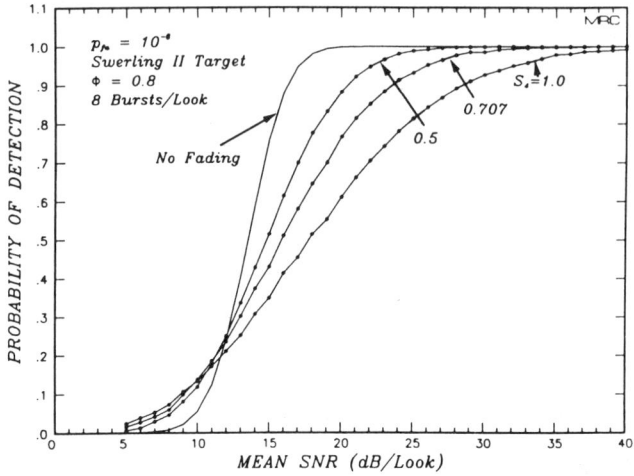

Figure 3.22 Probability of detecting a Swerling II target using noncoherent integration of 8 bursts of various scintillation levels, assuming Gaussian correlation of bursts ($\sigma = 0.8$).

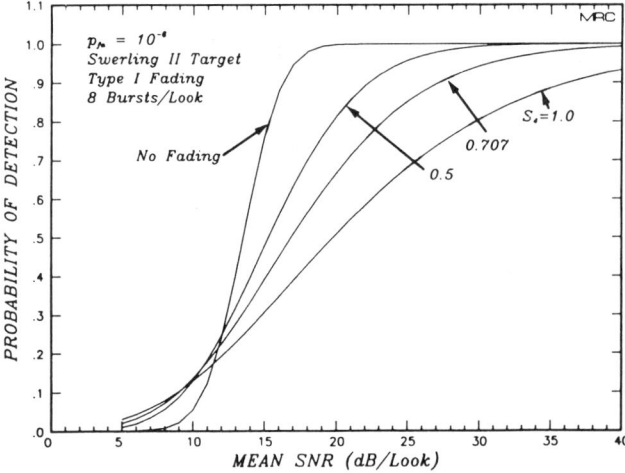

Figure 3.23 Probability of detecting a Swerling II target using noncoherent integration of 8 bursts during type I fading.

various amounts of correlation between the bursts in a look are considered through the use of Monte Carlo simulation techniques. For this work, a simple Gaussian correlation function is used to specify the burst-to-burst correlation. However, the simulation allows for an arbitrary correlation between the radar returns from the bursts in a look and includes the capability, not yet exercised, to change the burst correlation function from look to look, as required to model a frequency-hopping radar.

For a nonfluctuating target and worst-case fading, some gain is available through the M out of N combining process. However, as the level of scintillation decreases to less severe fluctuations, burst combining is to be avoided because better performance is possible with a single burst. For a Swerling II target, both double-threshold detection and noncoherent integration of multiple bursts per look offer some mitigation of the effects of fading, depending on the severity of the fluctuations and the correlation properties of the bursts of a look.

For the case of greatest interest here, the combining of 8 bursts to form a look, there is a gain of about 1.5 to 2 dB in detection sensitivity for noncoherent integration in comparison to double-threshold detection. This conclusion is based on detection performance alone, and does not consider any burst scheduling advantages of M out of N detection.

To apply the results for detection performance given in this chapter to help design a space-based radar, we need to obtain the value of scintillation index as a function of radar transmission frequency, geometery, ionospheric conditions, and time of day. The WBMOD computer code [31] contains the only worldwide model that gives this result. Unfortunately, WBMOD gives only the mean value of the scintillation index and does not yet incorporate a model of the probability density function. This function is necessary to answer questions regarding the percent of time that the scintillation index is above or below any specific value, and is therefore necessary to measure radar performance.

ACKNOWLEDGEMENT

The authors gratefully acknowledge useful discussions with G.A. Andrews of the Naval Research Laboratory, which contributed to the work discussed in this chapter. This work was supported by the Naval Research Laboratory under contract N00014-87-C-2336 and also by the Defense Nuclear Agency under contract DNA-87-C-0169.

REFERENCES

1. Budden, K.G., *The Propagation of Radio Waves,* Cambridge University Press, Cambridge, MA, 1985.
2. Lawrence, R.S., C.G. Little, and H.J.A. Chivers, "A Survey of Ionospheric Effects Upon Earth-Space Radio Propagation," *Proc. IEEE,* Vol. 52, January 1964, pp. 4–27.
3. Skinner, N.J., R.F. Kelleher, J.B. Hacking, and C.W. Benson, "Scintillation Fading of Signals in the SHF Band," *Nature (Phys. Sci.),* Vol. 232, July 1971, pp. 19–21.
4. Taur, R.R., "Simultaneous 1.5 and 4-GHz Ionospheric Scintillation Measurements," *Radio Science,* Vol. 11, December 1976, pp. 1029–1036.
5. Fremouw, E.J., R.L. Leadabrand, R.C. Livingston, M.D. Cousins, C.L. Rino, B.C. Fair, and R.A. Long, "Early Results from the DNA Wideband Satellite Experiment," *Radio Science,* Vol. 13, January-February 1978, pp. 167–187.
6. Towle, D.M., "VHF and UHF Radar Observations of Equatorial F-Region Ionospheric Irregularities and Background Densities," *Radio Science,* Vol. 15, No. 1, January-February 1980, pp. 71–86.
7. Franke, S.J., C.H. Liu, and D.J. Fang, "Multifrequency Study of Ionospheric Scintillation at Ascension Island," *Radio Science,* Vol. 19, May-June 1984, pp. 695–706.
8. Arendt, P.R., and H. Soicher, "Effects of Arctic Nuclear Explosions on Satellite Radio Communication," *Proc. IEEE,* Vol. 52, June 1964, pp. 672–676.
9. King, M.A., and P.B. Fleming, "An Overview of the Effects of Nuclear Weapons on Communications Capabilities," *Signal,* January 1980, pp. 59–66.
10. Wolcott, J.H., D.J. Simons, T.E. Eastman, and T.J. Fitzgerald, "Characteristics of Late-Time Striations Observed During Operation STRESS," *Effect of the Ionosphere on Space and Terrestrial Systems,* J.M. Goodman (ed.), US Government Printing Office, Washington, DC, 1978, pp. 602–613.
11. Schwartz, M., "A Coincidence Procedure for Signal Detection," *IRE Trans. Information Theory,* Vol. IT-2, December 1956, pp. 135–139.
12. Linder, I.W., and P. Swerling, "Performance of the Double-Threshold Radar Receiver in the Presence of Interference," ASTIA Doc. AD-11366, May 1956.
13. Walker, J.F., "Performance Data for a Double-Threshold Detection Radar," *IEEE Trans. Aerospace and Electronic Systems,* Vol. AES-7, No. 1, January 1971, pp. 142–146.
14. Dana, R.A., and D.L. Knepp, "The Impact of Strong Scintillation on Space Based Radar Design I: Coherent Detection," *IEEE Trans. Aerospace and Electronic Systems,* Vol. AES-19, No. 4, July 1983, pp. 539–549.
15. Nakagami, M., "The m-Distribution—A General Formula of Intensity Distribution of Rapid Fading," in *Statistical Methods in Radio Propagation,* W.C. Hoffman (ed.), Pergamon, New York, 1960, pp. 3–36.
16. Knepp, D.L., and G.C. Valley, "Properties of Joint Gaussian Statistics," *Radio Science,* Vol. 13, No. 1, January-February 1978, pp. 59–68.
17. Fremouw, E.J., R.C. Livingston and D.A. Miller, "On the Statistics of Scintillating Signals," *Journal of Atmospheric and Terrestrial Physics,* Vol. 42, 1980, pp. 717–731.
18. Aarons, J., "Global Morphology of Ionospheric Scintillation, II," AFCRL-TR-75-0135, USAF Cambridge Research Laboratories, March 11, 1975.
19. Hawkins, G., and J. Mullen, "Daytime Equatorial Scintillations in VHF Trans-ionospheric Radio Wave Propagation from ATS-3 at Huancayo, Peru," URSI, Boulder, CO, 1974.
20. Livingston, R.C., "Comparative Equatorial Scintillation Morphology—American and Pacific Sectors," DNA 4644T, SRI International, June 1978.

21. Knepp, D.L., *DNA Wideband Satellite Experiment—Frequency Decorrelation at UHF,* MRC-R-392, Mission Research Corporation, April 1978.
22. Keskinen, M.J., and S.L. Ossakow, "On the Spatial Power Spectrum of the $\overline{E} \times \overline{B}$ Gradient Drift Instability in Ionospheric Plasma Clouds," *Geophys. Res.,* Vol. 86, 1981, p. 6947.
23. Tsunoda, R.T., "High-Latitude F-Region Irregularities: A Review and Synthesis," *Rev. Geophys.,* 1988.
24. Basu, Sunanda, and S. Basu, "Correlated Measurements of Scintillations and In-Situ F-Region Irregularities from Ogo-6," *Geophys. Res. Lett.,* Vol. 3, 1976, pp. 681–684.
25. Knepp, D.L., "Aperture Antenna Effects After Propagation Through Strongly Disturbed Random Media," *IEEE Trans. Antennas and Propagation,* Vol. AP-33, No. 10, October 1985, pp. 1074–1084.
26. Marcum, J.L., *A Statistical Theory of Target Detection by Pulsed Radar: Mathematical Appendix,* RM-753, Rand Corporation, July 1948.
27. Feller, W., *An Introduction to Probability Theory and Its Applications,* Vol. 1, John Wiley and Sons, New York, 1957.
28. Dana, R.A., and D.L. Knepp, "The Impact of Strong Scintillation on Space Based Radar Design II: Noncoherent Detection", *IEEE Transactions on Aerospace and Electronic Systems,* Vol. AES-22, No. 1, January 1986, pp. 34–46.
29. Papoulis, A., *Probability, Random Variables, and Stochastic Processes,* McGraw-Hill, New York, 1965.
30. Hurst, R.L. and R.E. Knop, "Generation of Random Correlated Variables," *Comm. ACM,* May 1972, pp. 355–357.
31. Fremouw, E.J., and J. Secan, "Modeling and Scientific Application of Scintillation Results," *Radio Science,* Vol. 18, May-June 1984, pp. 687–694.

Chapter 4
SAR IN SPACE—THE THEORY, DESIGN, ENGINEERING AND APPLICATION OF A SPACE-BASED SAR SYSTEM
S.W. McCandless
User Systems, Inc.

This chapter identifies and describes the important functions and applications of space-based *synthetic aperture radar* (SAR). In 1978, the SEASAT satellite introduced the first SAR to space. Since then, two successful US Space Shuttle SAR exposures, SIR-A and SIR-B, have occurred.

The future is bright for US and foreign space-based SAR systems. Application candidates range from wide area search or mapping to more focused spotlight SAR or *inverse synthetic aperture radar* (ISAR) targeting modes. Few modern airborne radars operate solely in one mode. Instead, multimode systems for wide area search commonly employ real aperture systems and switch to higher resolution SAR or ISAR modes as the search narrows to specific targets. Space-based systems will soon follow this approach to multimode operation.

SARs are generally characterized by:

- Small apertures in comparison to *real aperture radar* (RAR) systems;
- Lower power requirements in comparison to RAR systems;
- A fixed antenna scan instead of the RAR raster beam scanning search pattern;
- Limited swath coverage in comparison to RAR systems;
- Moderate pulse *signal-to-noise ratio* (SNR) made possible by the large signal processing gain afforded by SARs;
- *Pulse repetition frequencies* (PRFs) bounded by range and doppler ambiguity limits.

The SAR designer must consider an ambiguity-based minimum antenna area which limits swath width, along-track resolution, or both. This increases the time that the search pattern takes. The limit on swath width per aperture requires SARs to fly higher, use multiple apertures, or both to provide wide area coverage.

Current and projected space-based SARs are data-rate limited. Consequently, range resolution, swath width, and dynamic range vie for performance within a data-rate ceiling. Considerable effort is underway to improve limiting technologies and to use more efficiently the available analog-to-digital (A/D) conversion, on-board recording and processing, and downlink capabilities.

SAR systems clearly have potential to many civil and military applications, and the following discussion considers this range of possibilities.

4.1 SPACE-BASED SAR DESIGN PRINCIPLES

For any application, radar performance must be defined in terms of the spatial and radiometric considerations necessary to resolve specific targets or scenes. This introductory section defines the fundamental relationships that govern spatial and radiometric performance, extending these relationships to simplistic radar-target interactions.

4.1.1 Basic Principles of Aperture Synthesis

SAR and a conventional real aperture radar achieve range spatial resolution (in the direction of the radar beam) in the same way by using the pulse-ranging technique. SAR is distinctive in its use of aperture synthesis to improve along-track (also called azimuth) spatial resolution. By comparison, a side-looking conventional radar obtains its resolution in this direction through the physical dimensions of its antenna. The real aperture cross-range spatial resolution is a function of radar wavelength, target range, and antenna dimension as shown by the equation:

$$\text{Resolution} = \frac{\lambda}{D_{AT}} R \qquad (4.1)$$

λ = wavelength;
D_{AT} = antenna length in the along-track or cross-range direction (for the case where the radar beam is normal, squinted 90°, to the line of flight);
R = range from the radar to the target.

This can result in unacceptable spatial performance for many applications.

A SAR "synthetically" obtains its along-track resolution by moving the real antenna in relation to the target. In the most basic sense, there must be a translation of either the target through the real beam, the real beam through the target, or a combination of both processes. Consider the string of dots in Figure 4.1 as a set of positions at each of which the SAR sends out a pulse, illuminates targets, and receives their reflections. The SAR system saves the phase histories of the responses at each position as the real beam moves through the scene, and then weighs, phase shifts, and sums them to focus on one point target (resolution element) at a time and suppress all others. The SAR image signal processing system performs the weighting, shifting, and summing to focus on each point target in turn, and then paints an image by placing the total energy response obtained in the focusing on a particular target at the position in the image corresponding to that target. Figure 4.2 illustrates the imaging geometry of the first space-based SAR, the SEASAT satellite's SAR, for illustrative purposes. Instead of a diffraction-limited resolution, the along-track resolution is independent of range and wavelength, becoming equal to half of the real antenna length D_{AT} in the direction of travel.

The length of the synthetic aperture is a function of the beamwidth of the real aperture, and becomes

$$L_{SA} \approx \frac{\lambda}{D_{AT}} R \qquad (4.2)$$

The azimuth resolution (two-way path) of the SAR becomes

$$\delta_{AT} \approx \frac{\lambda}{2L_{sa}} R \qquad (4.3)$$

Substitution of L_{SA} yields

$$\delta_{AT} \approx \frac{D_{AT}}{2} \qquad (4.4)$$

The equations shown do account for neither antenna nor processing weighting factors that will be discussed later. SARs achieve a very high signal processing gain as a result of coherent (in-phase) summation of the range-correlated responses of the radar. All of the signal returns that occur as the real beam moves through each target, as shown in Figure 4.1, can be coherently summed. In many instances, thousands of pulses are summed resulting in tens of decibels of processing gain more than the 40 dB for SEASAT SAR.

All or part of the available phase history can be coherently processed. If all of the pulses are used, the result is referred to as *single-look, one-look,* or *fully*

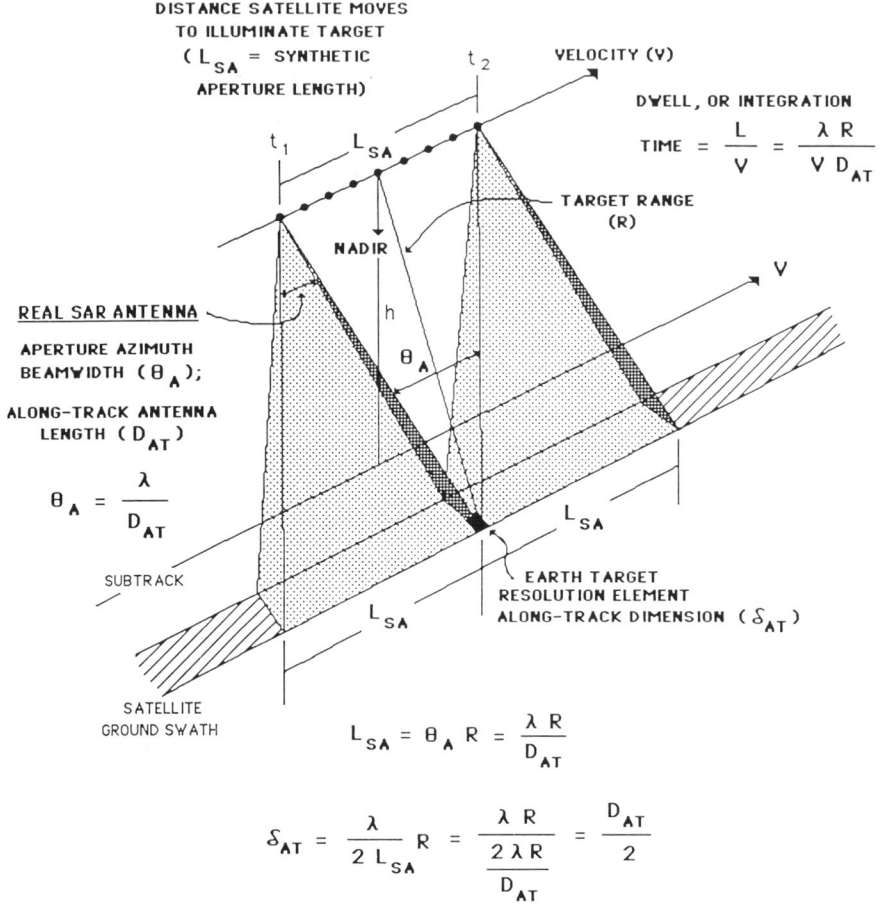

Figure 4.1 Aperture synthesis.

focused processing, achieving a spatial resolution of $D_{AT}/2$. Use of only part of the available aperture accordingly reduces along-track resolution. For example, use of half of the available aperture results in a spatial resolution of D_{AT}. This is called *two-look* processing.

As we indicated, SARs and RARs achieve spatial resolution in the direction of the radar beam by use of pulse ranging. As shown in Figure 4.3, the radar

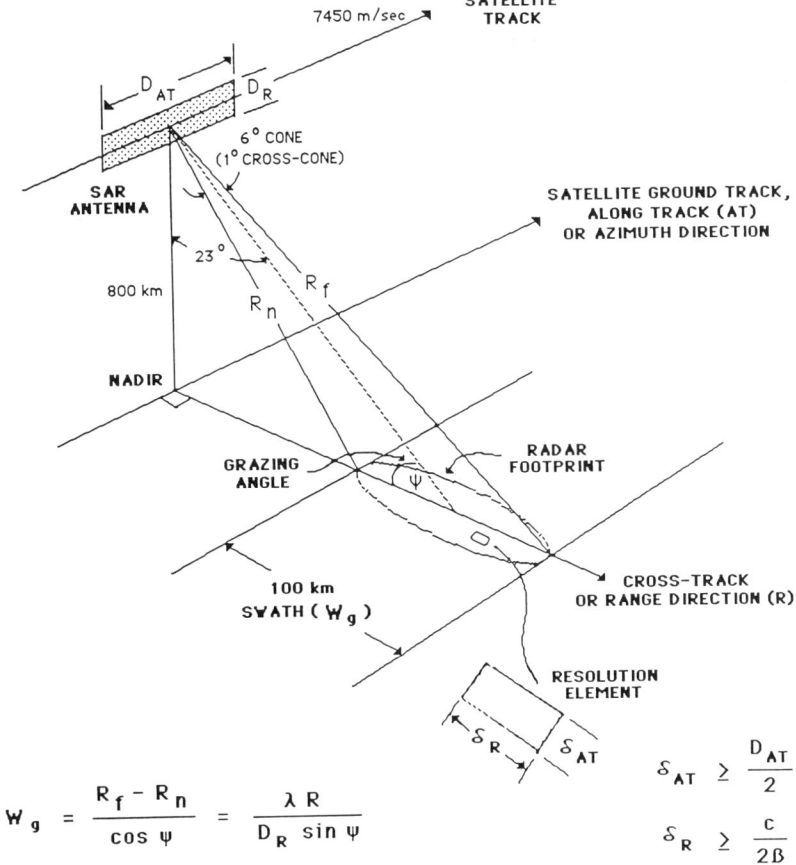

Figure 4.2 SEASAT SAR imaging geometry.

transmitting function will generate a pulse of energy. Each pulse will travel to the target area, and the discernible surface range dimension that can be recovered will be a function of the pulse length τ. The slant-range resolution is given by

$$\delta_{SR} = \tau c/2 \tag{4.5}$$

Translation to the ground plane produces

$$\delta_R = \frac{\tau c}{2 \cos\psi} \tag{4.6}$$

126

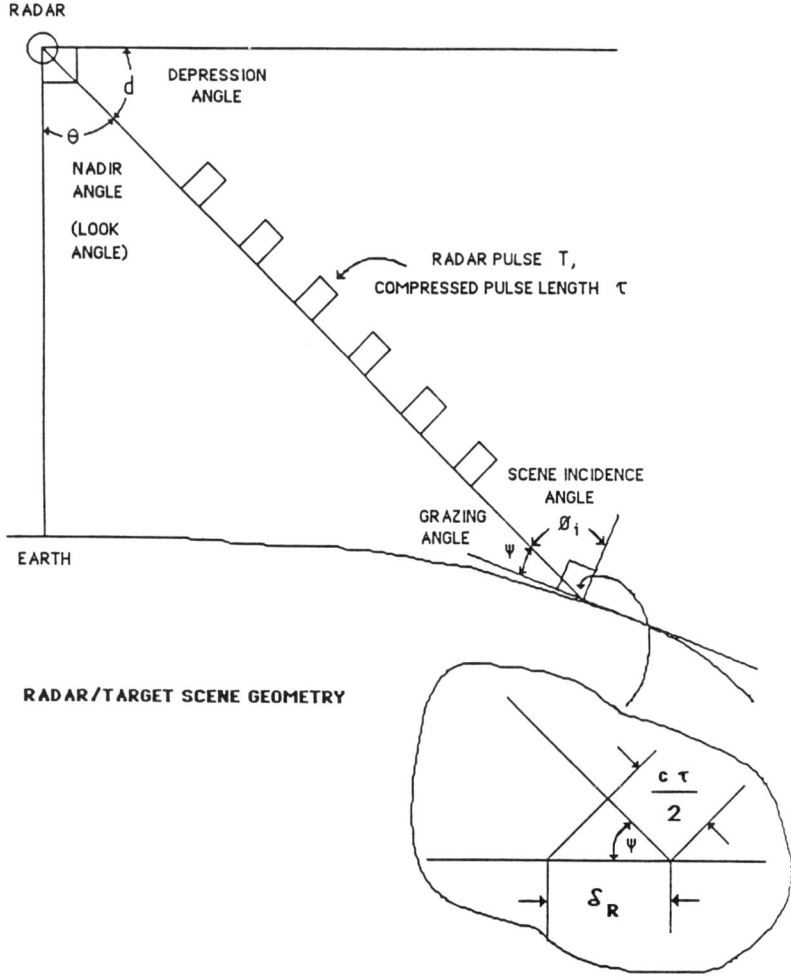

Figure 4.3 Definitions of radar angle notation and pulse ranging.

where

δ_R = ground-plane spatial resolution;
τ = effective pulse length;
c = speed of light;
ψ = grazing angle, from radar incidence vector and the earth's tangent at the point of target intercept;

or

$$\delta_R = \frac{c}{2\beta \cos\psi} \quad (4.7)$$

where

β = transmitted pulse bandwidth.

For the radar to distinguish between two features in range, the processed output resulting from these signals must appear at different times, or it will appear to be one large object.

The radar pulses apparently should be as short as possible to obtain the best resolution. However, they must also transmit enough energy to detect the reflected signals. This means that if the pulse is shortened, its amplitude needs to be increased accordingly to keep the same total energy in the pulse. However, the equipment required to transmit a very short, high-energy pulse is difficult to design and to build. Consequently, an alternate approach is used for most radar systems, including SAR. Instead of a short, constant frequency pulse, this approach uses a long, variable frequency pulse of lower amplitude.

The received signals are compressed via signal processing, and the result is that the real pulse length T yields the compressed pulse length τ, which appears in (4.5) above. The ratio T/τ is called the *pulse compression ratio*. An alternative form $T\beta$ is referred to as the *time-bandwidth product*. The compressed pulse is the result of signal return processing and appears as a $(\sin x)/x$ function.

This process is illustrated in Figure 4.4, which uses one form of encoding, called *linear frequency modulation* (LFM), sometimes referred to as *chirp*.

4.1.2 Ambiguity Relationships

4.1.2.1 Range Ambiguity Limit

Fine along-track resolution apparently can be obtained by simply making the real antenna length small. This is, in fact, the case, but ambiguity conditions place certain limits on this assumption.

The range-ambiguity limit establishes an upper bound on the PRF. An elevation view of the three-dimensional radar beam geometry is depicted in Figure 4.5. The radar beam intercepts the ground at near range R_n and far range R_f. The range limits shown are not those set by the half-power beamwidth, but rather they extend to the limit of significant two-way gain. The ground swath width is

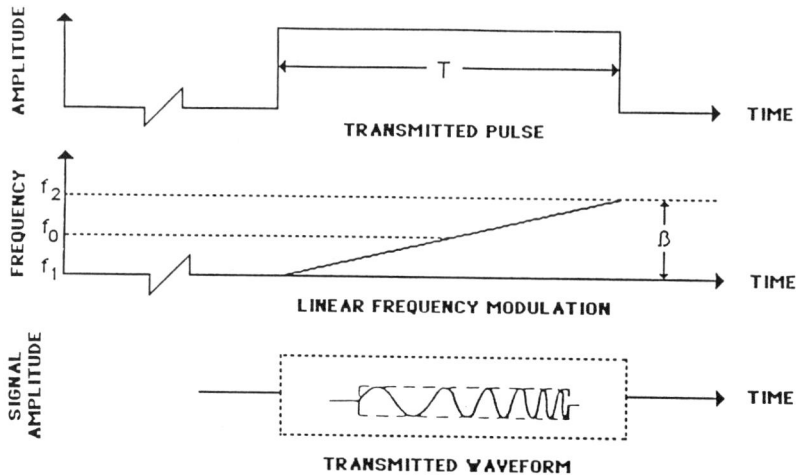
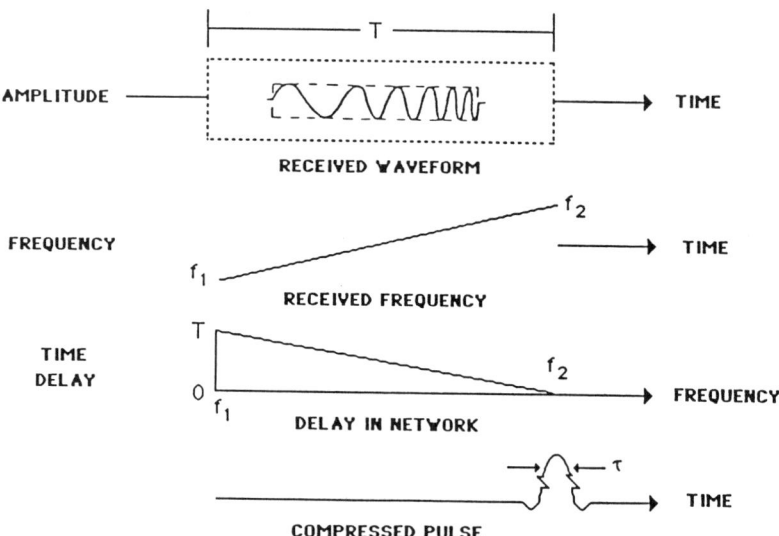

Figure 4.4 Pulse compression linear frequency modulation ("chirp" — coherent integration of radar pulses).

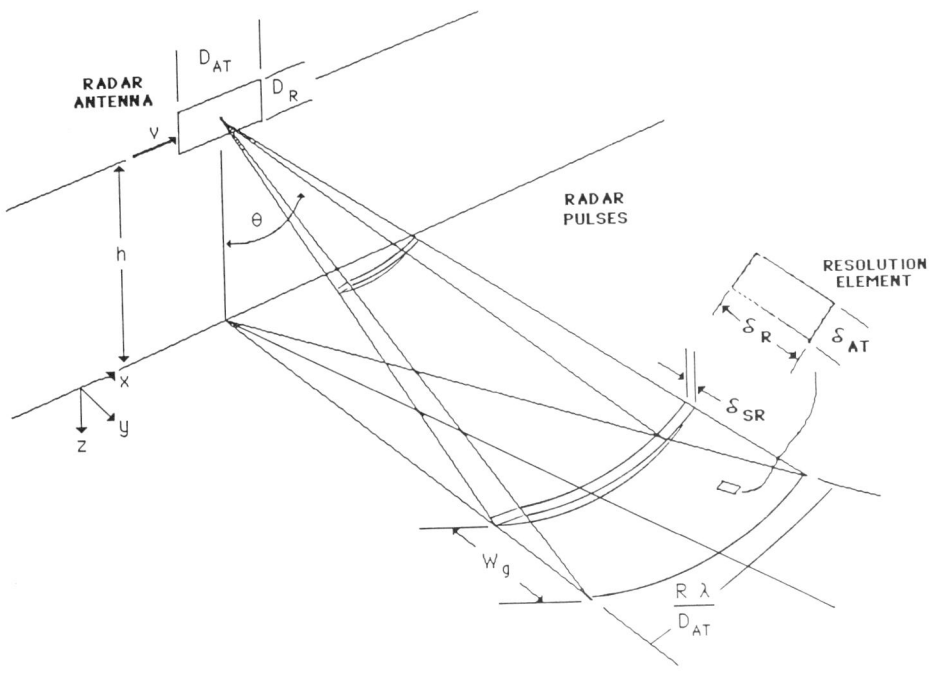

Figure 4.5 Geometry for describing SAR observations.

$$W_g = (R_f - R_n)/\cos\psi$$

or (4.8)

$$W_g = (R_f - R_n)/\sin\phi_i$$

where ϕ_i is the target incidence angle defined in Figure 4.3. The high-PRF limit to avoid receiving simultaneous return signals from two successive pulses is given by

$$\text{PRF}_{high} = 1/(2T + 2(R_f - R_n)/c) \quad (4.9)$$

where T is the real pulse length.

Figure 4.6 provides a view of the sampled return that must fit between transmissions to avoid range ambiguity. This upper PRF limit is commonly called the *range-ambiguity limit*. In practice, the highest operating PRF is set lower than this limit. For space, several pulses will be in transit simultaneously, and temporal interlacing of returns with the transmitted pulses is required to avoid eclipsing the reflected signals. As swath widths or incidence angles increase, more spacing between pulses is required, resulting in a lowering of the high PRF limit. This process will eventually conflict with a desire for improved along-track spatial resolution, which influences the doppler ambiguity relationship.

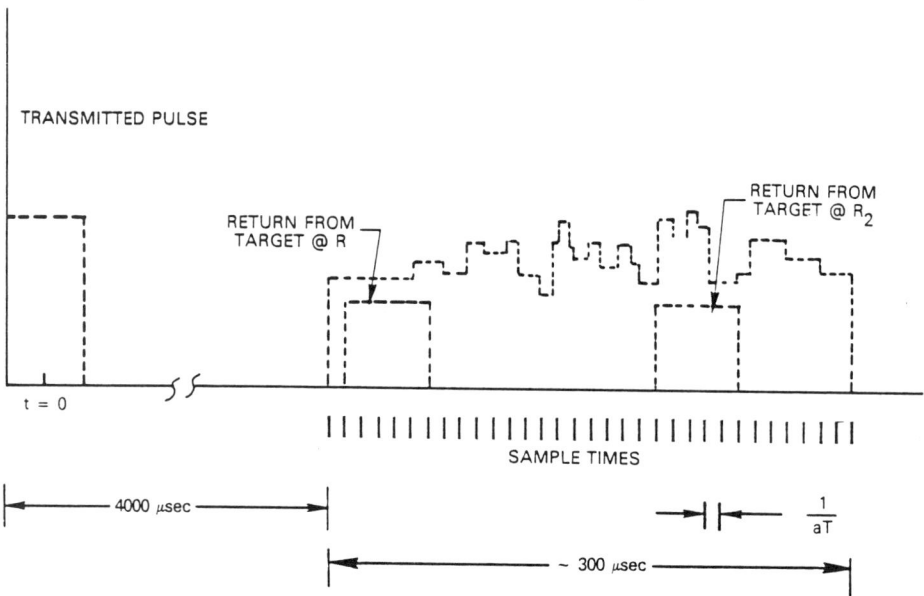

Figure 4.6 Data sampling diagram.

4.1.2.2 Along-Track Doppler Ambiguity Limit

Doppler relationships produce a low-PRF limit that is governed by a maximum phase shift of less than 2π radians from pulse to pulse for any target illuminated by the radar beam. If a 2π radian or greater phase shift occurs, there will be ambiguity in distinguishing these targets from others at the same range that cause zero phase shift (zero doppler). There are different approaches to establishing the low-PRF limit. In all cases, it is shown that

$$\text{PRF}_{\text{low}} = v/D_{AT}/2 = 2v/D_{AT} \tag{4.10}$$

where v is the platform velocity and D_{AT} is the along-track radar antenna length.

This equation implies that the transmitter must be pulsed before the space-based radar platform moves a distance equal to one-half the real SAR antenna length.

4.1.2.3 Data Rate and Ambiguities

Highlighted in the introduction was the fact that SAR designs are often data-rate limited. The data rate, or the system bandwidth, is also one of the most important determinants of a system's cost because it influences data collection, image signal processing, and the entire applications decision process.

$$\text{Data Rate} = (2\beta) \times Q \times \left(\frac{\text{Swath Width Time}}{\text{Interpulse Period}}\right) \qquad (4.11)$$

Four important performance factors are involved in determining the data rate. They are as follows:

1. $\beta = c/2\delta_R$—As range resolution improves (i.e., becomes smaller), the system bandwidth and data rate grow.
2. Q—Dynamic range requirements determine how many bits per sample (Q) are needed to preserve image data quality.
3. W_g—Swath width requirements dictate a minimum observation time to support data collection.
4. δ_{AT}—Along-track resolution determines the maximum extent of the interpulse period.

The last term in (4.11) provides some needed relief in many SAR designs because the time required to collect information from the swath $2(R_f - R_n)/c$ can be less than the interpulse period, which, in turn, must be smaller than $\delta_{AT}/2v$. Positive use of this time advantage reduces data rate and is referred to as *time expansion buffering* or *stretch processing*. This technique allowed SEASAT's data rate to be reduced from 240 Mb/s to 120 Mb/s. The conflict between swath width and along-track resolution recurs in the data-rate relationship. As δ_{AT} becomes smaller and W_g larger, the ratio of swath width time to interpulse time approaches 1, and time expansion buffering is not available to reduce data rates.

4.1.2.4 Minimum Antenna Area for SAR

The demand for improved along-track spatial resolution is in contention with achieving wide swath width because it causes the range dimension of the antenna to increase, complying with a required minimum antenna dimension for SAR.

PRF limits are established by the SAR illumination geometry, swath width, and along-track antenna length. A minimum radar antenna size can be derived by considering only the ratio of the PRF limits.

Radar beamwidth in the elevation plane is $\lambda/D_R = (W_g \cos\phi_i)/R$, where D_R is the radar antenna length in the range direction and R is the radar range. An expression for D_R can be written as

$$D_R = R\lambda/W_g \cos\phi_i \qquad (4.12)$$

Radar antenna area: $A_R = D_{AT}D_R$.

If we assume that the transmitted pulse length is small compared to the swath time $T \ll 2(R_f - R_n)/c$, an expression for D_{AT} can be derived as a function of PRF:

$$\frac{\text{PRF}_{\text{HIGH}}}{\text{PRF}_{\text{LOW}}} = \frac{D_{AT}}{4v(R_f - R_n)/c}$$

or

$$D_{AT} = \frac{\text{PRF}_{\text{HIGH}}}{\text{PRF}_{\text{LOW}}} \frac{4v(R_f - R_n)}{c} \tag{4.13}$$

The antenna area A_R required to satisfy the two ambiguity constraints can be written

$$A_R = \frac{\text{PRF}_{\text{HIGH}}}{\text{PRF}_{\text{LOW}}} \frac{4v(R_f - R_n)R\lambda}{cW_g \cos\phi_i}$$

Substituting $\sin\phi_i = R_f - R_n/W_g$:

$$A_R = \frac{\text{PRF}_{\text{HIGH}}}{\text{PRF}_{\text{LOW}}} \frac{4v\lambda R}{c} \tan\phi_i \tag{4.14}$$

The antenna area is related to the ratio $\text{PRF}_{\text{HIGH}}/\text{PRF}_{\text{LOW}}$, and this ratio is always greater than unity. If the ratio is near unity, there exists a minimum antenna area that will ensure simultaneous compliance with both ambiguity constraints:

$$A_{\text{MIN}} = \frac{4v\lambda R}{c} \tan\phi_i \tag{4.15}$$

The required antenna size increases with v, λ, R, and ϕ_i. A ratio too near unity is operationally undesirable because this provides no allowance for design and hardware margins.

4.1.3 The Radar Equations Interpreted for SAR

Single-pulse performance (expressed as a signal-to-noise ratio) is independent of the type or amount of processing gain of the radar:

$$(S/N)_P = \frac{P_t G_t}{4\pi R^2} \cdot \frac{\sigma}{4\pi R^2} \cdot \frac{A_R}{L} \cdot \frac{1}{KT_s\beta} \tag{4.16}$$

where

P_t = pulse power (peak power);
G_t = transmitter antenna gain;
R = radar-to-target range;
σ = target cross section;
A_R = effective receiving aperture area;
L = RF system losses;
K = Boltzmann's constant;
T_s = system noise temperature, K, which includes the receiver noise figure;
β = bandwidth (capable of collection of the entire transmitted spectrum).

If the radar employs pulse compression (as shown in Figure 4.4), this represents a form of processing gain, and equation is multiplied by T/τ or $T\beta$, where T is the actual pulsewidth in time and τ is the compressed pulsewidth ($\tau \approx 1/\beta$). With pulse compression, the effective signal-to-noise ratio can be greatly improved:

$$(S/N)_{CP} = \frac{P_t G_t A_R \sigma T}{(4\pi)^2 R^4 L K T_s \beta \tau} \qquad (4.17)$$

As before, the equation does not explicitly account for compression weighting for low sidelobes.

Note that $P_t T$ equals the pulse energy, J. The pulse energy is not relieved by using pulse compression. It simply allows a lower pulse power to be transmitted over a longer interval, while preserving range resolution and achieving adequate signal-to-noise performance. The next step is to introduce the signal-to-noise influence of SAR coherent integration of hundreds to thousands of individual pulses.

The number of pulses (N) available for coherent integration, as the real beam moves through the target, are

$$N = (\text{PRF}) \cdot (\text{Dwell Time}) = \frac{(\text{PRF})(L_S)}{v} \qquad (4.18)$$

where

PRF = radar pulse repetition frequency in Hz;
L_S = length of the synthetic array;
v = radar-platform velocity with respect to the target area.

$$L_S = \lambda R/D_{AT} \tag{4.19}$$

$$N = \frac{(\text{PRF})\lambda R}{v D_{AT}} \tag{4.20}$$

For a fully focused system, the along-track resolution $\delta_{AT} = D_{AT}/2$. Applying processing gain, the net signal-to-noise ratio is

$$(S/N)_F = \frac{J G_t A_R \sigma \lambda R (\text{PRF})}{(4\pi)^2 R^4 L K T_s v 2 \delta_{AT}} \tag{4.21}$$

Finally, let $G_T = 4\pi A_R/\lambda^2$ and $P_{\text{ave}} = J(\text{PRF})$. Thus,

$$(S/N)_F = \frac{P_{\text{ave}} A^2 \sigma}{8\pi \lambda R^3 L K T_s v \delta_{AT}} \tag{4.22}$$

To date, the dominant use of space-based SAR is for ground mapping (i.e., SEASAT, SIR-A, and SIR-B), in which case S is actually a clutter signal, and the performance interest becomes clutter-to-noise (C/N) ratio.

Here, $\sigma_C = \sigma_0 \delta_{SR} \delta_{AT} \sec\psi$, where σ_0 is reflectivity expressed as a normalized radar cross section (m^2/m^2); $\delta_{SR} \sec\psi$ = range resolution along the ground.

The C/N ratio for a clutter map thus becomes

$$(C/N)_F = \frac{P_{\text{ave}} A^2 \sigma_0 \delta_R \sec\psi}{8\pi \lambda R^3 L K T_s v} \tag{4.23}$$

Inspection of the S/N equation reveals that signal- or target-to-noise performance is reduced if less than the available aperture (fully focused) is used coherently during signal processing. By contrast, clutter-to-noise performance remains the same at signal processing levels less than fully focused (multilook levels). This does not mean that nothing is gained for all the trouble of focusing a SAR. When mapping, the along-track width of the clutter patch is greater for the multilook cases, thus σ_C is much greater, so the same average power allows σ_C to exceed the noise by the same amount as a focused SAR having a much smaller σ_C.

Also, working against a clutter background, the multilook system suffers in terms of recognizable point target sizes. Possibly, high clutter areas may look like targets. Designers do not have control over σ_0, but can control the size of the processed clutter patch and hence σ_C.

Point target performance (S/N) is best at the fully focused level of processing. Signal-to-clutter ratios are also greatest at the fully focused level.

4.2 END-TO-END SYSTEM DESCRIPTION

A complete SAR design must be considered as an "end-to-end" set of choices or decisions immutably linking the radar and image signal processor. Elements that must be considered include the moving satellite platform, transmitted signal, propagation effects, complex target interactions (including motion), received signals, data recovery (options include storage, playback transmission, and direct transmission), and on-board or ground-based image signal processing. A motion compensation function in the radar or signal processor must calculate the range walk and phase corrections necessary to account for satellite motion during each of the integration periods.

In most radar designs and applications, target visibility refers to the detection of targets against the competition of a total and complex environment. A space-based SAR dedicated to observing features on the earth's surface will be used for purposes of discussion.

The complex nature of target responses (statistics, fluctuation, *et cetera*) must be considered. The environment includes the unwanted responses from thermal and jamming noise; area clutter from land, ice and sea backgrounds; and volume clutter such as that produced by natural atmospheric conditions, forests, and chaff. Atmospheric perturbations and attenuation can adversely affect high frequency radars (>15 GHz). A 35 GHz system can experience clear-air atmospheric losses of from 15 to 21 dB. Rain in this situation can add 10 to 30 dB, depending on the rain rate. Lower frequencies (<1 GHz) will encounter ionospheric influences such as Faraday rotations, refraction, and scintillations. The radar must characterize each of these unwanted responses and then identify the various modes of operation in which the radar system or radar signal processor may be designed or configured to cope with varying conditions.

Final target classification and mapping modes for SARs are characterized by the following characteristics:

- Sophisticated signal processing techniques using complex, fast signal processors;
- Special radar system or signal processor designs featuring clutter cancellation or speckle reduction techniques;
- Target-specific processing using motion compensation and image-enhancement algorithms.

For space-based systems, the orbital physics of the satellite must be considered (target range, velocity, swath extent, *et cetera*). Preferential geographic coverage and coverage rate are functions of inclination and altitude, and must be assessed to determine the number of space-based platforms required to accomplish a specified target revisit interval. SARs have orbit altitude and look geometry design sensitivities and constraints (e.g., ambiguity functions, antenna dimensions, power levels), and thus cannot easily be equipped to cover a wide variety of situations.

4.2.1 System Design and Technology Considerations

The *SAR system* comprises the following elements:
- antenna;
- transmitter, including the waveform generator;
- receiver;
- signal conditioner and image processor;
- target detection and classification processing;
- system control (antenna-platform) functions.

Figure 4.7 illustrates various elements of a *total system*.

In addition to the radar, the size, weight, and cost of major satellite subsystems establish limiting or constraining conditions. Important spacecraft systems that are significantly influenced by SAR include thermal (heat dissipation), data collection (by far, the most limiting SAR support system), attitude stability, power generation, structure, and orbit maintenance (required to maintain performance altitude, especially significant for low ($\approx$ 300 km) orbits).

In addition to swath width coverage, and spatial-radiometric resolution values, typical radar design inputs include specification of the receiver noise figure, antenna noise temperature, radar transmitter and receiver RF losses and efficiencies, pulse-compression ratio, and system dynamic range. These values have been identified in the radar equations shown in Section 4.1.3, and (4.16) is repeated to support the following discussion:

$$(S/N)_P = \frac{P_t G_t}{4\pi R^2} \cdot \frac{\sigma}{4\pi R^2} \cdot \frac{A_R}{L} \cdot \frac{1}{KT_s B}$$

In general, average and peak power requirements increase with frequency for similar applications performance. System losses also increase with frequency as receiver noise figures, RF losses, and equivalent noise temperature values rise. In (4.16), receiver and antenna noise are combined in T_s, the front-end system noise temperature. Conversely, antenna dimensions usually become smaller with increasing frequency. This is a relief, considering the desire to hold antenna tolerances to something on the order of $\lambda/20$ for the two-way radar system. More detailed comments about radar parameters and performance trends are provided in the following paragraphs.

The receiver noise figure will vary as a function of λ and the type of amplifier. Smaller values of λ will generally have higher noise figures, as will tube amplifiers

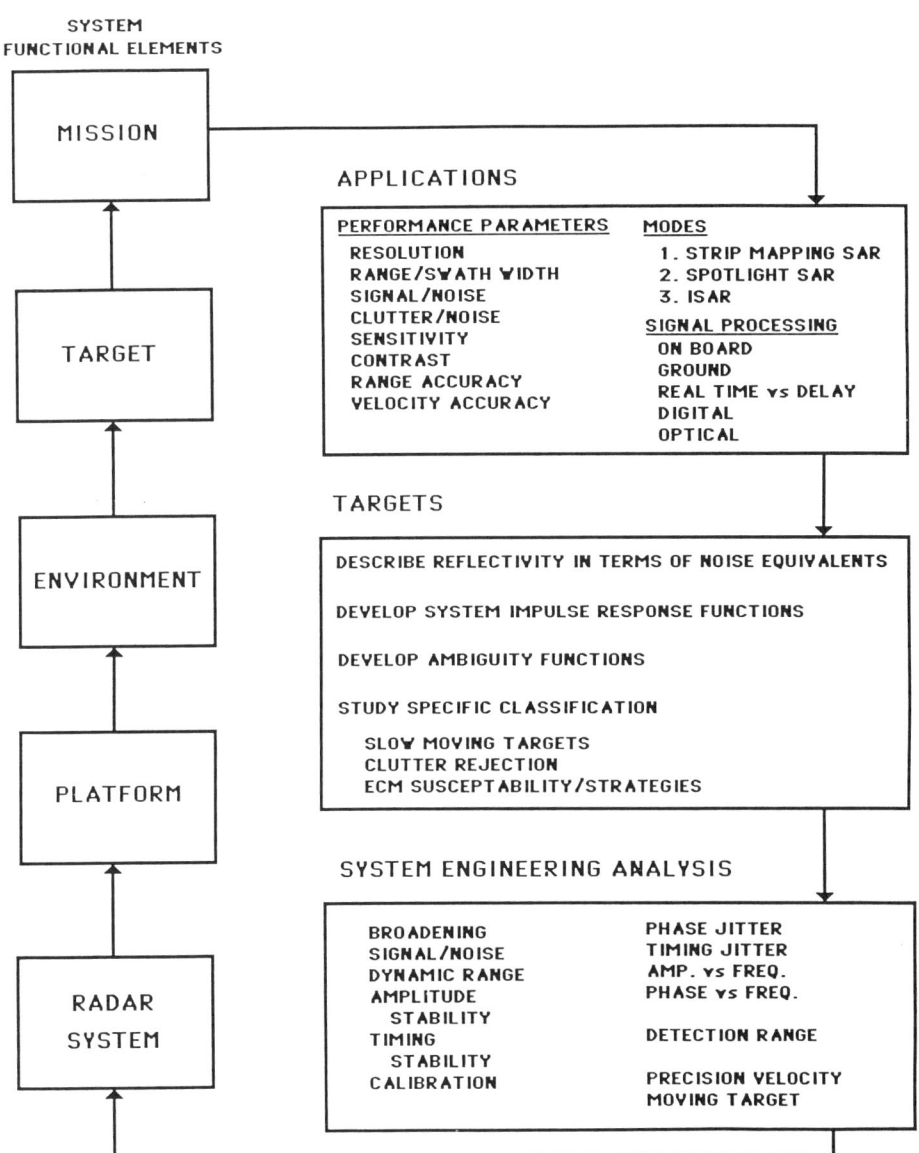

Figure 4.7 Total system.

as compared to solid-state systems. Most new space-based SAR systems are considering transmit-receive (T/R) modules, which can be physically distributed as part of the antenna (e.g., SIR-C). Tube technology is also still in use (e.g., ERS-1 and RADARSAT).

The antenna noise temperature, as used in this chapter, is a combination of the reference noise temperature (e.g., 290 K for the earth background) and antenna-to-receiver losses. A typical value for L-band systems is 500 K; for X-band systems, 600 K is more likely, and values increase with frequency.

RF losses are a function of system bandwidth, radar complexity, and size. In general, losses scale directly with frequency and bandwidth. For example, 2 dB of RF loss at L-band may become 8 dB at 35 GHz. High range resolution systems, at any wavelength, have wider bandwidths and consequently higher RF losses. In addition, wider bandwidth systems are more susceptible to ECM (i.e., barrage or brute force jamming).

The *pulse compression ratio* (PCR) is selected to keep the transmitter's peak power requirements within practical limits, while preserving the sought-after range spatial performance, δ_R. This quantity is selected by the designer. Conventional RAR and SAR systems have imposed constraints by range ambiguity that keep values in the hundreds. A typical value is the 640 PCR used for the SEASAT SAR. The radar dynamic range is designer-specified, based on the instantaneous observation dynamic range required to view the objective targets. PCR will be a function of target and background variability, and has a strong affect on the data rate created by the system, as indicated in Section 4.1.2.3. Many modern space-based SAR systems have instantaneous dynamic ranges of 25 dB or less to alleviate the data-rate problem confronted by all space-based SAR designs.

Mission design inputs cover orbital characteristics such as altitude, inclination, and mission or target requirements such as revisit intervals. Inputs are application-specific, and the designer proceeds to calculate radar values relevant to the mission on the basis of mission inputs. Calculating space and ground velocities is important to aperture dwell determination, and orbital characteristics are combined with swath extent to determine revisit intervals at specific latitudes by using both fixed and steered beam inputs.

Antenna technology paces modern SAR radar design, and is nearly as important as data collection and signal processing considerations. Radar technology inputs consider such things as the antenna design, the type of transmitter and receiver electronics that will be used, and additional losses to account for atmospheric or ionospheric variations.

Two basic antenna technologies are currently in use:

1. A corporate-fed array (planar or modified planar). This system can be pointed electronically, although some mechanical redirection also may be incorporated. Examples are SEASAT, shuttle SARs (SIR-A and SIR-B), ERS-1, and JERS-1.

2. A distributed array using T/R modules that is similar in electrical performance to the corporate-fed array. This system has the advantages of better phase control and beam steering. These systems are currently in development, and will be introduced in space by SIR-C.

Space-based SAR systems are large in comparison to other space remote sensors and have a significant effect on satellite support requirements. For example, sizing of the power system for SAR depends on the radar system power. Current systems have 5 kW peak power requirements and space designs requiring as much as 10–20 kW are being considered. The thermal subsystem weight and volume is related to the radar system peak power, and is an important part of performance in terms of the on-orbit duty cycle.

Attitude control is a function of platform stability, pointing accuracy, and radar system weight. Many airborne systems require active and passive (signal processing) motion compensation. Satellite systems are generally more stable, and can achieve good performance by using passive techniques. The key requirement for SAR is limitation, or satellite-antenna acceleration or change during an integration interval. Values of $10^{-4}°/s$ are typically required.

The data system weight and volume are a function of the radar receiver system bandwidth, duty cycle, and the type of processing being employed. Current on-board recorders can handle 60 Mbs rates with 250 Mbs recorders in the near future. Downlinks can handle rates from 80 (direct) to 300 Mbs by using relay satellites such as the NASA Tracking and Data Relay Satellite System (TDRSS).

The spacecraft structure is largely a function of the radar antenna area and weight, which can vary widely, depending on the technology used.

SAR design factors interact with one another through a series of compromises toward a feasible design. A SAR design becomes an iteration of desired performance, the practical realities of SAR physics, and the bounds of radar and satellite technology. Figure 4.8 provides an overview of how mission requirements influence system design factors.

4.2.2 System Implementations Including ISAR

There are a large number of SAR implementation and application variations. Among them are broadside strip-mapping SARs that point the beam 90° to the along-track direction. All of the civil space-based missions planned or flown to date use this configuration. A *squint mode* strip-mapping SAR would point the beam at an angle off of the broadside direction (nominally 10–80° from the along-track direction) and hold the beam fixed as it flies through the target. Figure 4.9 illustrates this geometry.

A SAR need not hold the beam in a fixed position during the target dwell or integration period. The SAR can articulate its beam as the platform flies through

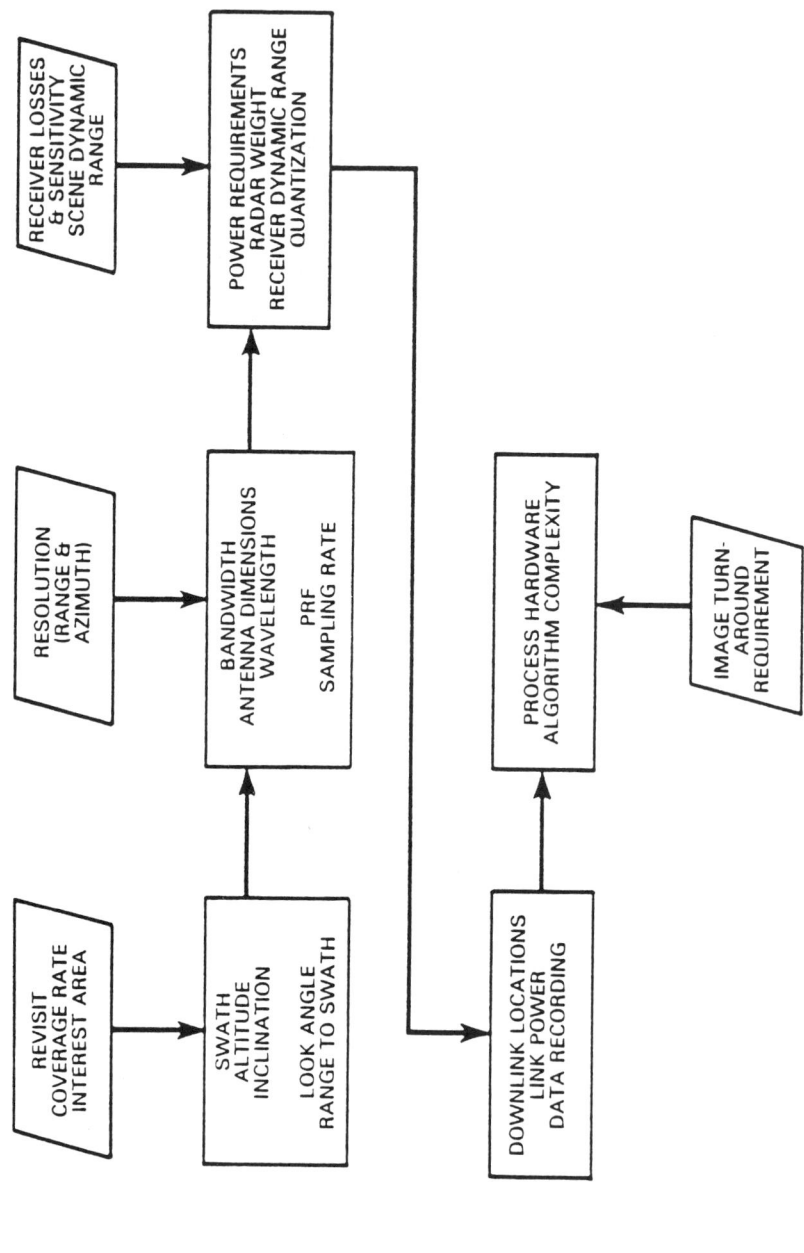

Figure 4.8 Parameter definition flow (example).

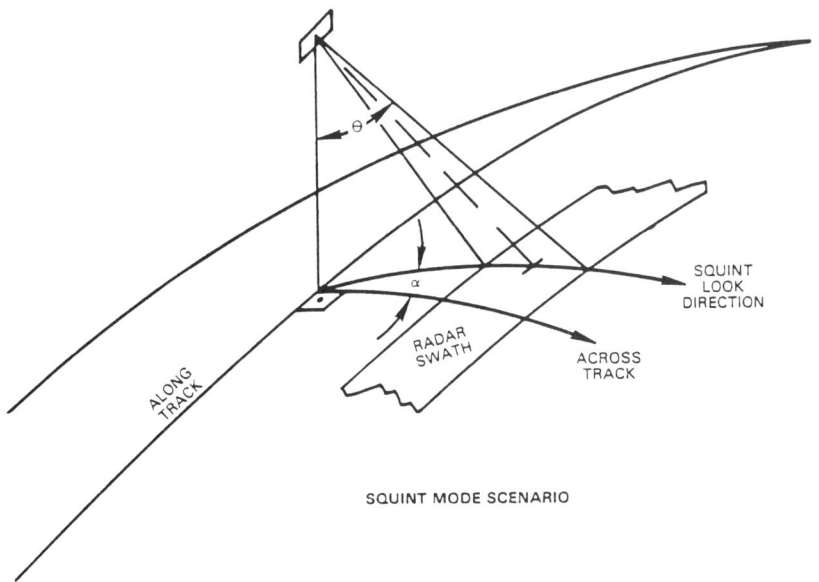

Figure 4.9 Squint mode.

the target. This increases the beam dwell period, and consequently the length of the synthetic aperture. As a result, the along-track spatial resolution is improved. This mode is referred to as *spotlight* SAR and is illustrated in Figure 4.10. Attendant with this mode of operation are narrower swath widths and added complications to the radar control interface (i.e., beam steering, variable PRFs).

Other variations on the theme include ISAR discrimination of point targets. The ISAR technique offers an approach to target classification or sorting, based on obtaining fine resolution two-dimensional images by using methods similar to those for SAR. In general, the ISAR mode offers less area coverage than SAR, but is used to produce finer resolution. Figure 4.11 provides a simplified illustration. A combination of the target's motion and the radar track defines the change in angular aspect between the *radar line of sight* (RLOS) and the target.

A simplified description of this mode would assume simple rotation of the target about one axis orthogonal to the radar incidence direction. In the simplified case, the target rotates with an angular rate of ω. The radar echo from a scattering point (P) will have a positive doppler frequency if moving toward the radar or a negative doppler rate if moving away from it. The range resolution δ_R is the same as in other modes.

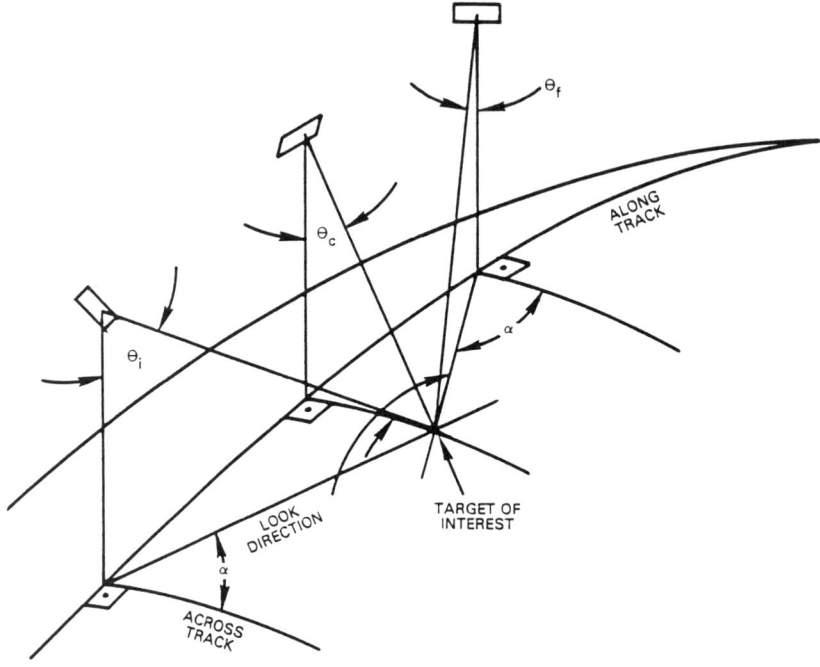

Figure 4.10 Spotlight mode.

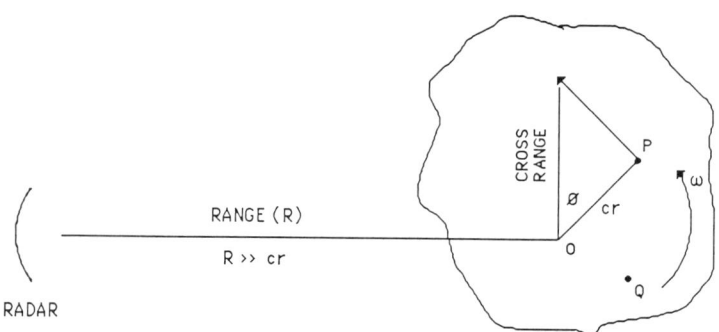

Figure 4.11 Inverse synthetic aperture radar (ISAR).

Target cross-range spatial resolution can be expressed for an aspect angle change, $\Delta\phi$, or dwell period, T_i, as

$$\delta_{CR} = \frac{\lambda}{2\omega T_i} = \frac{\lambda}{2\Delta\phi} \tag{4.24}$$

There is a problem of ambiguities in cross-range responses giving a folded effect in the resulting image. For signals sampled with an increment in angle $\Delta\theta_s$ between samples, the unambiguous interval in cross range is given by

$$A_{CR} = (\lambda/2)/\Delta\theta_s \tag{4.25}$$

Considering range cell migration, points on the target body do not move through range resolution cells during a doppler processing interval or frame time if

$$(TD/2)\Delta\phi_s < \delta_r \tag{4.26}$$

where TD is the target diameter. The number of pulses coherently integrated by the ISAR process is

$$N = \text{PRF} \times \text{Dwell Time} \tag{4.27}$$

where Dwell Time $= N \times T_i$

As a result, signal-to-noise and signal-to-clutter performance are enhanced by the ISAR mode because the clutter patch is generally smaller than the SAR mode, and the high PRF, made possible by the limited range interval being considered, increases the number of pulses available for coherent integration. Thus,

$$(S/N)_F = \frac{P_{\text{ave}} A_r^2 \sigma_T T_i}{4\pi\lambda^2 R^4 L K T_s} \tag{4.28}$$

$$(C/N)_F = \frac{P_{\text{ave}} A_r^2 \sigma_o \delta_r \delta_{CR} T_i}{4\pi\lambda^2 R^4 L K T_s} \tag{4.29}$$

4.2.3 Image Processing

In a SAR image processor there are specific operations required to convert a raw data set into an interpretable image. The raw SAR data do not constitute an image because point targets are spread out in range due to the long frequency-coded pulse and, in the along-track dimension, by the real beam moving through the point target. Figure 4.12 illustrates the raw data trace of a point target as the beam moves in the along-track or azimuth direction. As the radar moves by the target, the radar-to-target range varies, forming the curved trace shown. This translation also produces a linear FM trace in azimuth, induced by doppler, similar to the chirp transmission waveform FM trace in range.

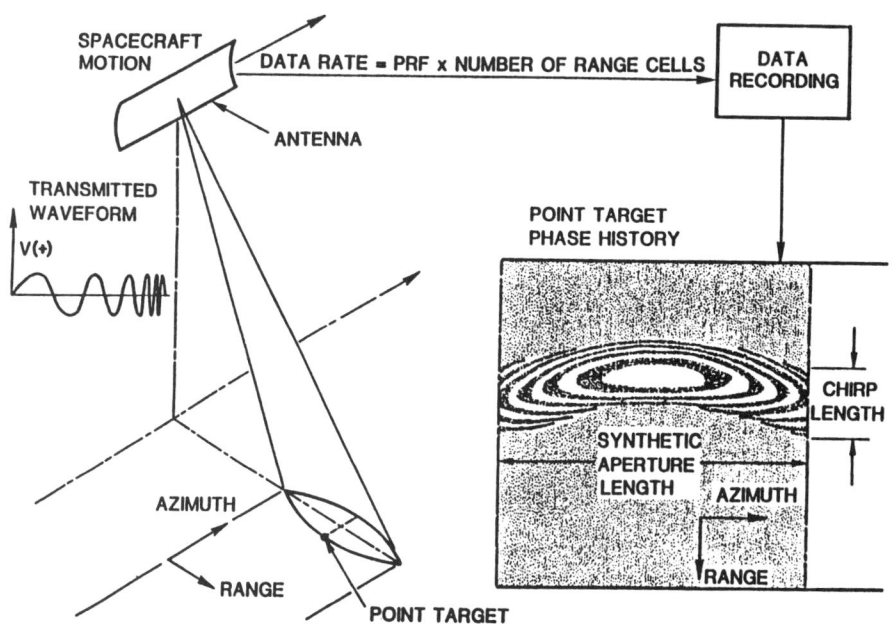

Figure 4.12 Point target echo in a synthetic aperture radar system.

The SAR signal processor compresses this distributed target information in two dimensions, range and along-track, to reproduce the image. This process is shown pictorially in Figure 4.13.

The complexity of this process would be revealed if all of the point targets (δ_R by δ_{AT} resolution elements) that exist in the beam illumination area at each pulse instance were superimposed. Actually doing so would render the figure uninterpretable and noninstructive. However, such is the situation that a SAR processor must unravel to produce an image of the sampled scene. Embedded in this process are unwanted distortions and perturbations that must be taken into account, including:

- Range effects
  - Range curvature
- Earth rotation effects

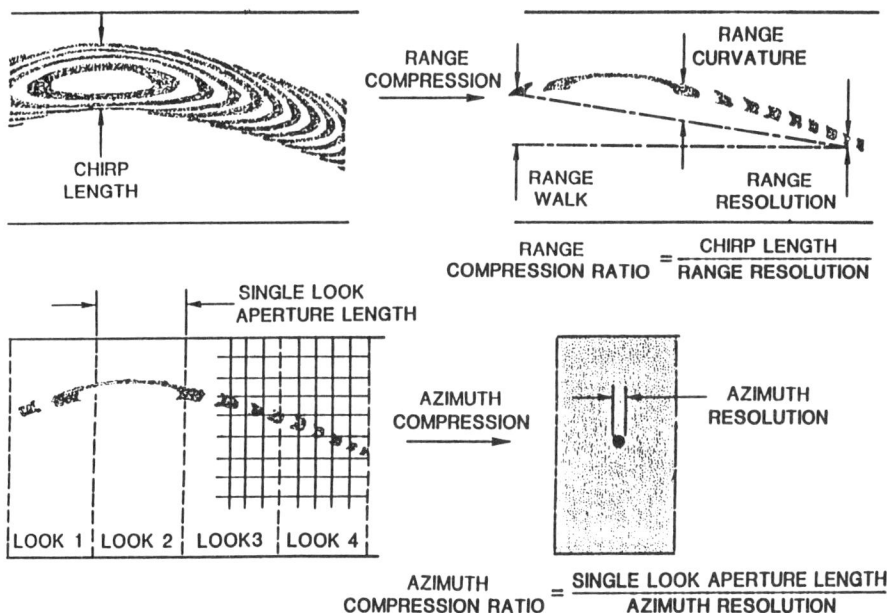

Figure 4.13 Point target compression or focusing.

- Range walk
- Squinting
• Orbit eccentricity
• Spacecraft attitude noise
• Cartographic distortion
• Topographic distortion (foreshortening)
• Shadowing and layover
• Speckle
- Radiometric resolution

After making provision in the radar or signal processor design to limit or to measure these effects, the processor must perform the following basic steps in imaging:

• Format raw radar (in-phase, quadrature (I, Q)) data
• Define the range reference function
• Perform range compression
  - Time domain: correlation
  - Frequency domain: matched filtering
• Corner turning to realign data from a range to an along-track file orientation
• Correct for earth's motion, attitude offset, *et cetera*.
• Define along-track and doppler reference functions

- Perform along-track compression
  - Time domain
  - Frequency domain
- Topographic and cartographic geometric corrections
- Calibration of radiometric values
- Form image

The culmination of these processes is shown in Figure 4.14. The range and azimuth correlation processes produce a resolution cell response shown by the two-dimensional (sin$x$)/x function in the figure. The effect of image signal processing is to subdivide the projected real beam into image pixels or resolution cells of dimension δ_R and δ_{AT}.

Range pulse compression is usually the first step in the digital processing of SAR, as shown by Figure 4.15, which illustrates the method currently used by most SAR processors, which is to process sequentially in the frequency domain in range and azimuth. Current processors for spaceborne SAR are ground-based, but space-based processors are being developed.

Major factors which may influence radar and signal processor design are listed below:

- Analog-to-digital conversion (ADC)
- Motion compensation
- Automatic gain control and sensitivity time control (AGC, STC)
- Beam-steering antenna control
- Synchronizer

The ADC is the major data interface for the radar and processor. The ultimate quality of the SAR data will be strongly driven by the ADC performance. Should it be poor, no amount of sophisticated processing would be able to restore it. Because most space-based SAR designs are data-rate limited, ADC is an essential function. For fine resolution space-based SAR, 6–8-bit ADCs with sampling rates of 100 MHz or more are needed. Block floating-point quantization or similar techniques are currently being evaluated to reduce the application limitations posed by ADC constraints. Section 4.3 discusses considerations of the *pacing technology* in more detail.

Several requirements exist for motion data to be supplied to the radar or signal processor. Among these are the following:

- Earth motion and attitude offsets
- Antenna beam-pointing
- Timing for variable PRF operation
- Spotlight SAR or scan SAR operation

As these factors suggest, a fundamental decision must be made. Should the radar calculate and compensate for its own motions, or should the data be collected and

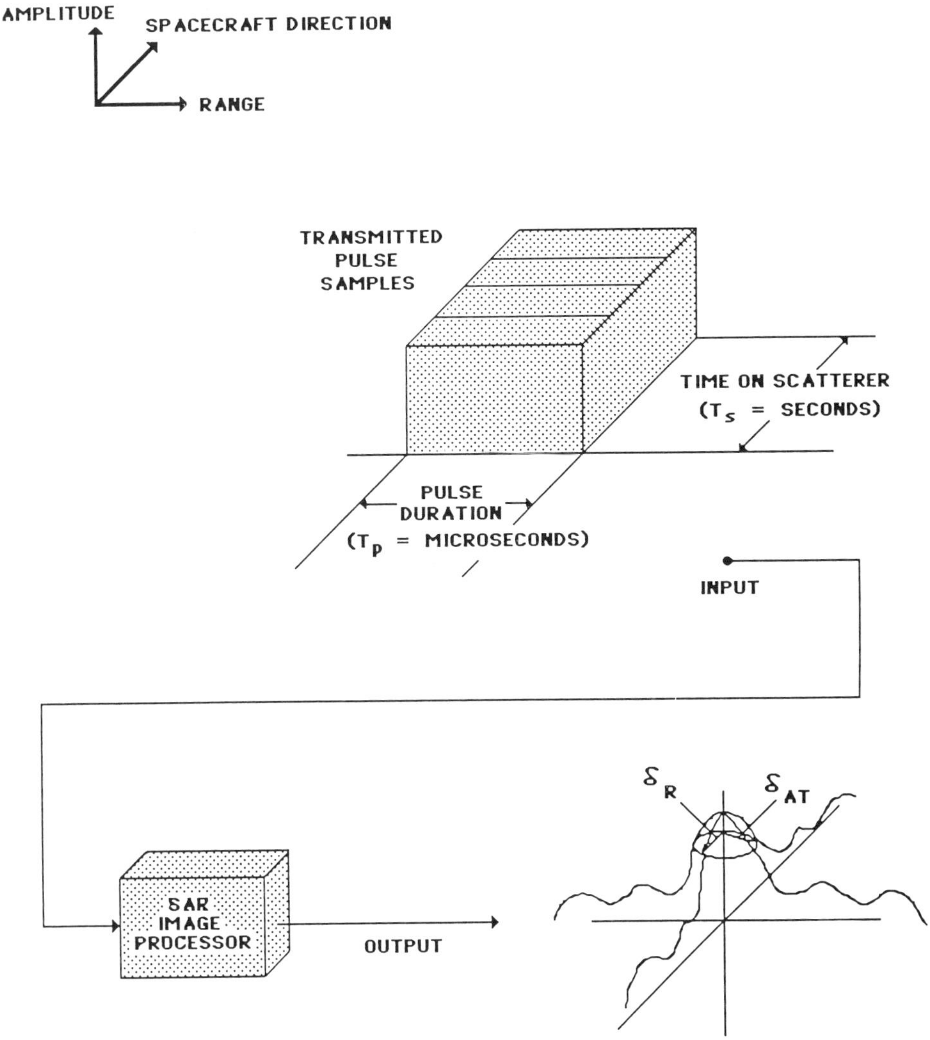

Figure 4.14 SAR processing in two dimensions.

corrections computed in the processor? Motion computations are generally handled by the processor for space-based systems.

Two *automatic gain control* (AGC) circuits will generally exist in a radar. A fast AGC is utilized to keep the receiver from saturating during jamming or when strong discrete targets are encountered. This circuit, which is controlled by the

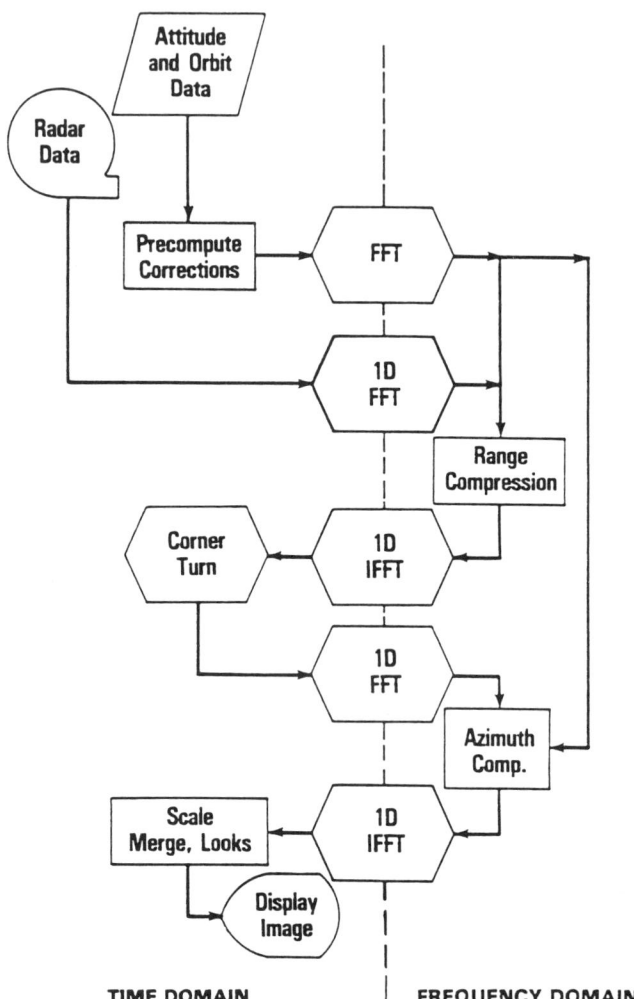

Figure 4.15 SAR digital processing.

radar, restricts the gain in all range gates. A second AGC, which may in the future be under the direction of an on-board processor, is both slower acting and more selective as to which range gates it affects. An example of the use of such an AGC would be in a space-based SAR when a strong, limited area, clutter patch exists.

Sensitivity time control (STC) is a form of AGC which anticipates the respective range dependence of the clutter and target returns, and accordingly adapts the receiver.

The signals which must pass across the AGC-STC interface from the radar to the processor include:

- The AGC setting in each active receiver channel;
- An indication of receiver saturation in any channel.

Antenna pointing commands should be supplied to the radar from an on-board processor-controller, with the controller determining the requisite phase shifts to position the beam and to provide the angular coverage.

The signals which must be passed to the radar include:

- Antenna pointing angles;
- Antenna scanning rates for scan SAR, spotlight SAR, *et cetera*.

4.3 AN ASSESSMENT OF PACING TECHNOLOGIES

Figure 4.7 provided a top-level diagram of a SAR system. A SAR consists of a real aperture radar, which has rigid phase tolerances imposed on it because it must operate as a SAR, and a signal processor capable of producing a high resolution image from the collected data. A simplified system diagram shown in Figure 4.16 can be used to focus discussions of pacing technology for SAR on critical technological functional areas and components so as to eliminate systems areas where existing technologies are adequate for space-based SAR.

Although data collection technologies are currently limiting for space-based SAR, it is difficult to categorize a projected performance requirement as an impossible technological goal based on the advances attained in the recent past and the current acceleration in direct and ancillary SAR technologies.

The following discussion begins by weeding out some obvious systems in Figure 4.16 where today's technology is more than adequate to meet applications goals. A desired 1 m^2 symmetrical resolution will be used for example purposes.

4.3.1 Nonlimiting Technologies

In general, transmitter and antenna technologies are capable of meeting space-based SAR requirements. Table 4.1 provides a tabulation of transmission bandwidth as a function of illumination geometry for a 1 m resolution in range. To achieve this resolution, the short pulse time requires pulse compression techniques to keep radar peak power and the power aperture product within technological bounds.

The transmission bandwidth to support the performance premise is well within the bounds of current technology. The need to produce a 750 MHz bandwidth, however, will limit the range of fundamental frequencies used for the radar. In

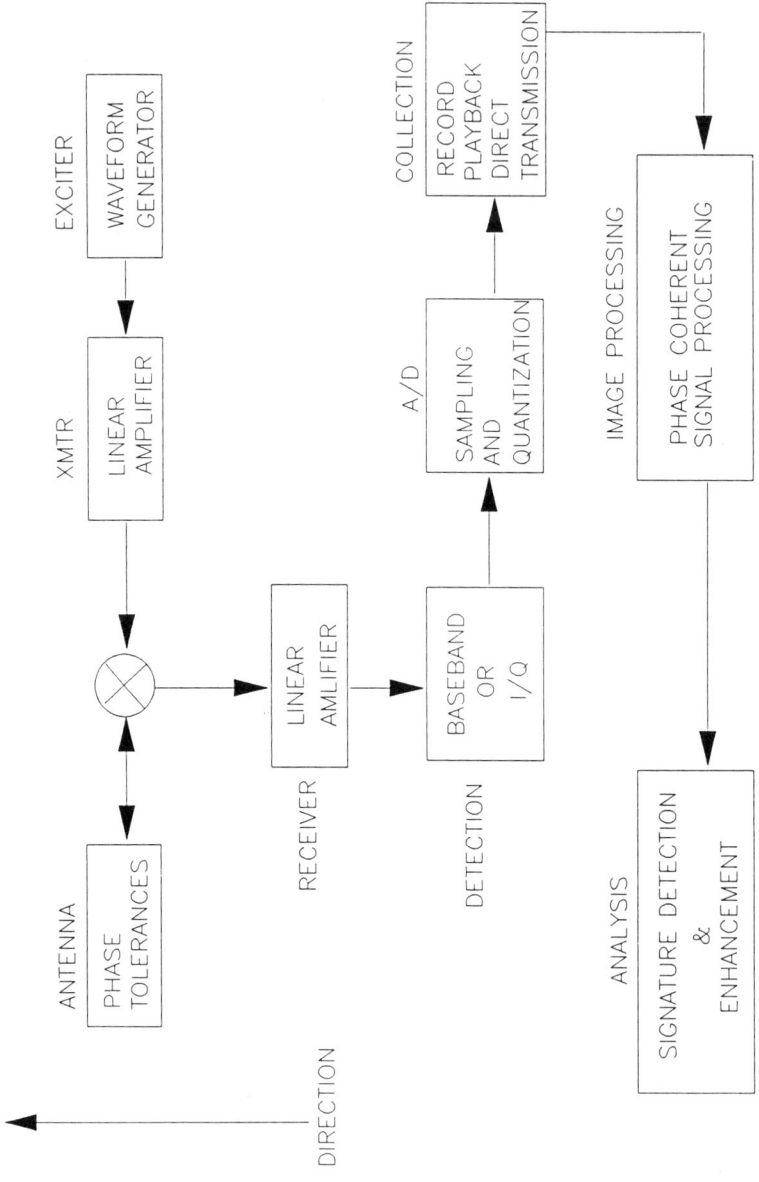

Figure 4.16 Simplified radar system diagram.

general, off-the-shelf tuned circuits and antennas can easily support bandwidths in the neighborhood of 10% of the radar fundamental frequency. In summary, the exciter does not represent a pacing technology.

Table 4.1
One-Meter Ground-Range Resolution Transmission Bandwidth as a Function of Incidence Angle ϕ
(4169-km Circular Orbit was Assumed for Purposes of Calculation)

ϕ_i (degrees)	Transmit Bandwidth (MHz)	Slant-Range Resolution (m)
10	867	.17
20	441	.34
30	300	.50
40	231	.65
50	195	.77
60	172	.87
70	160	.94
80	153	.98
90	150	1.00

SAR apertures are determined by the desire to achieve good along-track or azimuth resolution (antenna along-track dimension) and the need to avoid range and doppler ambiguities (areal considerations, in turn, drive the range dimension). Ambiguity conditions establish a lower bound on the SAR antenna area.

For a broadside strip-mapping SAR to achieve a 1 m resolution, D_{AT} must be very close to 2 m. At the same time, the designer would want D_R to be as small as possible at any satellite altitude to maximize swath coverage per aperture. Therefore, the antenna will very nearly be A_{min}. At present, antenna technologies can be characterized as shown in Table 4.2 for SAR. (The table assumes phase tolerance of $\lambda/20$ over the aperture.)

Table 4.2
Ten-Meter Antenna Technology Characterization

Technology Phase	Dimension in Wavelengths (λ)	Maximum Frequency (GHz)
Today—Reflector or corporate phased array	450	13.5
In development—Corporate-fed phased array or space-fed array	600	18.0
Projected—With a probable requirement for subaperture adjustment or phase conjugation correction	1200	36.0

The lower frequencies currently used for historic and planned space-based SAR designs are well within the range of current technology.

Current free-flying satellites use solar power systems. Projected solar-panel powered applications are in the range of 50 kW. Even higher power levels are projected for space-based nuclear reactor development. Table 4.3 tabulates power requirements and projected capabilities for space-based radar.

Table 4.3
Radar Power Limits Based on Systems Capabilities (Power in kW)

Technology	Date of Operation	System Power	Radar Power P_{avg}	$P_T(P_K)$
SEASAT (JERS-1)	1980	.4	.2	1
ERS-1 (RADARSAT)	1990	1.7	.8	5
Current Capability	1990–1995	25	12	60
50 kW Projected Solar Generators	1995–2000	50	25	120

4.3.2 Limiting Technologies

SAR sensing of ocean and polar regions began with NASA's SEASAT system in 1978. Since then, applications have continued with Space Shuttle flights. Each of these SAR systems has been limited in application due to technology-driven design constraints such as on-board recording (SEASAT, SIR-A) and data collection rate limits (SIR-A, SIR-B). Recent breakthroughs in relevant technological areas permit reevaluation of SAR system architectures and design approaches. Innovations include optical-digital on-board image processors, advanced magnetic tape recorders, and data compression algorithms. This section explores SAR pacing technology in terms of projected capabilities.

Application of space-based SAR is currently limited in the areas of data-rate conversion, storage, downlink, and image processing. These limitations can influence important performance values such as swath width, dynamic range, range spatial resolution, sensitivity, and real-time application.

As indicated in Section 4.3.1, several of the SAR subsystems can be eliminated from consideration as pacing technologies. For most historic, planned, and projected designs, the portions of the SAR from the radar transmitter through the radar receiver can be eliminated as pacing technologies.

The basic pacing technologies for space-based SAR are listed here:
- analog-to-digital conversion;
- on-board image processing;
- on-board data recording;
- data transmission to the ground.

These subsystems can be arranged in a variety of ways. Regardless of how the steps are performed, each can impose significant limits on SAR application. Image signal processing can be performed on-board (preceding the recording and downlink steps) or on the ground. Both options will be discussed.

4.3.2.1 A/D Conversion

If the data are to be processed digitally, as is the current US preference, we must sample and convert from analog to digital form, as shown in Figure 4.17.

Following the analog-to-digital conversion of the video signal produced by the SAR receiver, the digital data stream is typically time-expansion buffered to reduce the instantaneous data rate to a lower sustained rate, which is continuous over the interpulse period.

For most SAR applications that require high precision estimation of the echo power, an eight-bit ADC is needed to prevent significant saturation or quantization noise in the sampled data. Techniques have been developed for adaptive selection of the four most significant bits from the eight-bit ADC output to reduce the effective data rate. A threshold level (or exponent) is calculated to accompany a block of data based on the average power in that block. This technique, referred to as *block adaptive quantization* (BAQ), can reduce the data rate by a factor of two. The BAQ technique is being used by the NASA-JPL Magellan Venus Radar Mapper and the Shuttle Imaging Radar SIR-C, both scheduled for early 1990s operation.

A large amount of work has been devoted to another data reduction technique identified as *pre-image data compression*. However, this area is much less promising. Using noiseless (lossless) coding techniques, such as a Huffman code to eliminate redundancy in the data, will yield no more than a 20–30% reduction in data volume. This limitation is primarily due to SAR *speckle noise* effects. Lossy coding techniques for reducing the volume or rate of the raw signal data have been demonstrated with prototype systems, which implement a vector quantization fixed code-book algorithm. Prototypes are capable of handling data rates of 100 Mb/s and higher. Preliminary testing demonstrates that compression ratios of four to one can be achieved with little degradation in the resulting image quality, but this was done only for airborne X-band SAR data. More work remains to prove that this is a viable technique for spaceborne applications.

4.3.2.2 High-Rate Recorder

The high-rate recorder system for space-based SAR must be capable of recording at data rates of 50 Mb/s and above with a recording capacity of at least 9 Gbytes

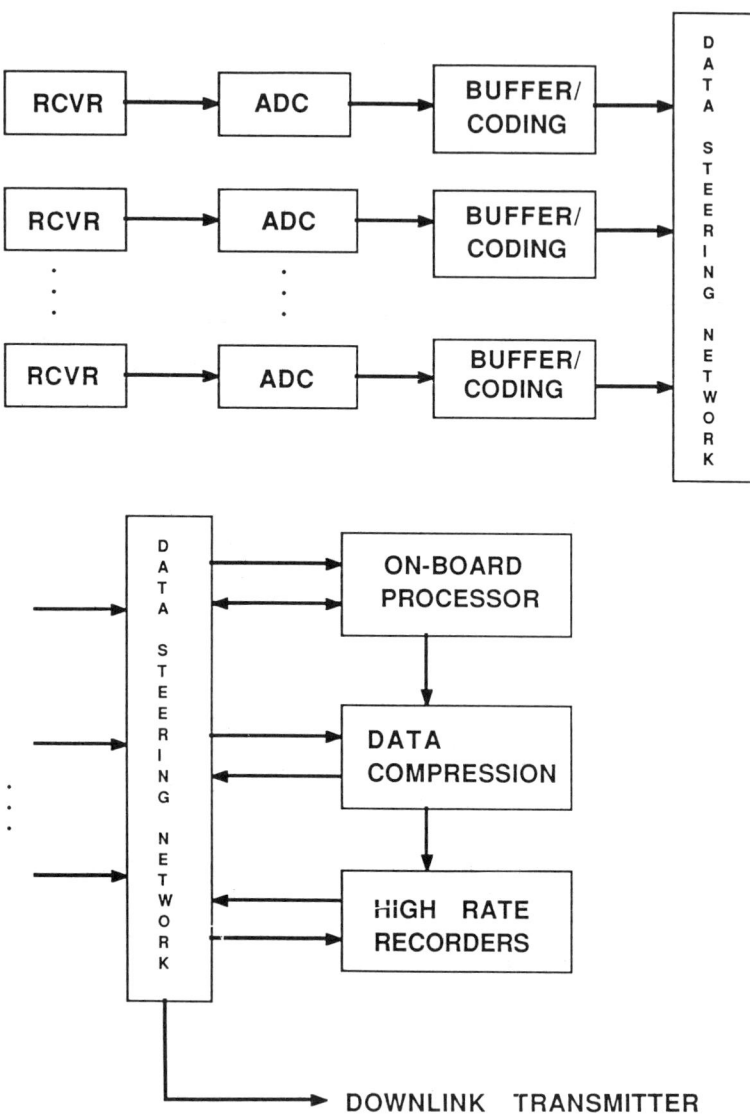

Figure 4.17 Functional block diagram of multichannel spaceborne radar system with on-board processor.

(about 20 minutes of operation) to handle even the simplest SAR system (e.g., single channel, 50 km swath). For the multiple-channel, wide swath, multipolarization SAR required to meet future observation requirements, an order of magnitude better performance is needed. To date, the most advanced recorder flown in space (manufactured by Odetics Corporation) is capable of a maximum recording rate of 60 Mb/s with a maximum capacity of nearly 10 Gbytes and a BER of 10^{-7}. This basic system has been flown on SPOT, the shuttle with SIR-B, and is planned on upcoming SAR satellites such as JERS-1. Higher rate and capacity systems are being designed and tested (by Odetics) for use on future missions.

The recording technique used in most ground recorders (e.g., Honeywell HD-96) is called *linear recording*, where as many as 42 parallel longitudinal tracks have been packed onto a one-inch wide tape. The helical recording format can realize 600 tracks per inch, producing a significant increase in packing density. A helical-format recorder (Schlumberger Industries model PV6420 (MIL STD 2179)) is capable of 240 Mb/s, a capacity of 50 Gbytes at a BER of $<10^{-10}$. This machine is being used in military combat aircraft, helicopters, and ships. It is currently undergoing testing by NASA for use with SIR-C.

Although high-rate recording systems are still a factor of two less than what is required for projected SAR systems, the compact size (.04 cu · m), low power (0.3 kW), and light weight (30 kg) of high-rate recorders permit multiple systems to be flown on a single platform. We expect that these systems will improve performance above current specifications, and the recording technology will not be a significant limitation in future spaceborne SAR systems.

4.3.2.3 Optical Image Signal Processing

An analysis of current and future technology related to the signal processing of SAR data must consider both optical and digital computing systems. In addition to the historic limitations of spaceborne processors, such as weight, power, and environmental effects (radiation shielding, outgassing), we must also consider system control, image calibration, processor flexibility, and reliability (graceful degradation) issues.

The early versions of ground-based SAR processors, some of which are still in operation, are analog (optical) systems using a laser light source with a series of lenses to perform the two-dimensional convolution processing. Film is typically used for both the input and output media. Optical processing systems feature high throughput relative to a digital processor, but are constrained in terms of the dynamic range of the film, limitations of the lenses for high resolution imaging, and swath width limitations.

Recent technological advances in the field of electro-optics have resulted in improvements of optical computing. A functional block diagram of an optical SAR processor is shown in Figure 4.18. Specific advances include available semiconductor light sources (light emitting diodes (LEDs), laser diodes) that are more reliable than previous light sources, acousto-optic devices (AOD) that have improved for input spatial light modulation use, and semiconductor detector arrays (charged coupled devices (CCDS)) that replace the film. With such technology, construction of a real-time, compact, power-efficient optical computer to perform on-board processing of SAR imagery is feasible. However, improvements must yet be made in the areas of performance, flexibility, and reliability.

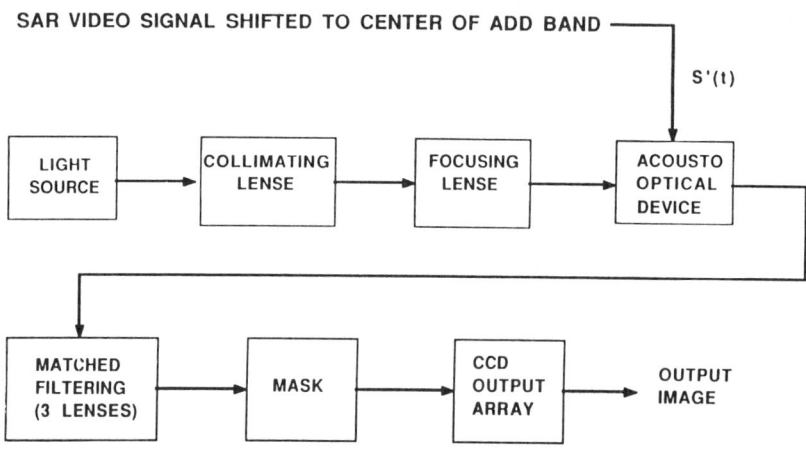

Figure 4.18 Functional block diagram of optical SAR processor.

The performance in terms of the image pixel resolution will be limited by factors such as aberrations in the optics, mechanical and electronic stability, and light source coherence. Resolutions on the order of SEASAT SAR (25 m) can be achieved with today's technology, but an order of magnitude better resolution is not currently feasible. Additional problems exist with the light source. The duration of the light source pulse must be shorter than the inverse bandwidth of the transmitted signal to avoid range smearing, and so extremely short pulses (in the 10 ns range) must be used. The use of short pulses presents a problem with coherence and gain transients that effectively degrade the resolution. This problem could be overcome if a pulsed gas laser were used, but this would only be feasible for a ground processing system. The other major area of technological improvement is in the CCD array. CCD chips are currently limited in width to 1000 elements. This limitation can be overcome by interfacing a number of chips to realize a wide range swath. However, the time-bandwidth product of the AOD may then become a constraint on the swath width. Another severe limitation of existing CCD arrays is dynamic range. Without special cooling to reduce the dark current, a reliable dynamic range of only 30 dB can be achieved, which is too small for high-precision, calibrated, SAR image processing.

Although tremendous strides have been made in electro-optical computing since the 1980s, major technological advances are still required before these systems can compete with digital computers in terms of image quality and performance. However, the potential advantages of optical systems for on-board SAR processing are tremendous. Specifically, the high-speed ADC and buffers shown in Figure 4.17 can be eliminated from the SAR system, and the signal can be routed directly to the optical processor, which, in turn, nearly instantaneously generates the image. This image is captured on a digital CCD and can be routed to a downlink processor for transmission to a ground receiver before the satellite has passed from reception range.

Considering the size of the engineering community working in optical computing today, we may reasonably assume that a high-performance spaceborne SAR processor will become technically feasible during the 1990s.

4.3.2.4 *Digital Image Signal Processing*

The trade-off in digital *versus* optical processing is typically performance for throughput. Digital signal processing is not theoretically limited in terms of dynamic range, swath width, or resolution that the processor can achieve, but rather by the quality of input signal data and the ancillary data such as platform ephemeris, attitude, and sensor calibration. The processing algorithms and techniques are sufficiently mature today such that they contribute little or no degradation to the resultant image quality. The major issue in implementing an on-board digital processor is in achieving the necessary computational power within the size, weight, and power limitations of the platform.

Recent gains in semiconductor technology, using gallium arsenide (GaAs) circuitry and advanced complementary metal oxide semiconductor (CMOS) devices produced with silicon-on-insulator (SOI) architecture, bode well for major performance gains in the near future. The potential of superconductivity will revolutionize the microelectronics industry. Superconductor research is ongoing and can greatly increase the cycle time through improved junctions.

Currently available or nearly operational technology primarily utilizes CMOS circuitry, which exhibits excellent speed and power characteristics. For spaceborne systems, a 64K radiation hardened RAM and a 16-bit radiation hardened microprocessor (80C86RH) are to be used in the Mars Observer satellite. One of the most advanced processor candidates for space use is the IBM Common Signal

Processor (CSP) developed for the Advanced Tactical Fighter (ATF) program and used in conjunction with the Westinghouse Ultra-Reliable Radar (URR). The architecture consists of a number of processing elements, each rated 125 MFLOPs contained in a module with 32 Mbytes of memory. *The system can be configured with a number of modules interfacing into a common data network, achieving a maximum performance of 1.8 GFLOPs. The CSP uses CMOS technology, does not require special cooling, and is currently constructed with 6 × 9 in boards (92 ft^3 volume). Considering that this system is within a factor of four of SEASAT's real-time processing requirement, the technology probably will soon exist for spaceborne, high-resolution, real-time SAR processing.

An alternative approach for the SAR signal processor architecture is to develop a custom chip, taking advantage of the highly repetitive nature of the SAR processing algorithm. The range processing could be performed with an analog device, such as a *surface acoustic wave* (SAW) filter, before digitization as shown in Figure 4.19. Following the ADC, an azimuth correlator chip could be designed that would not require a transition of the data from a range file to an azimuth file. The most efficient algorithm for this type of implementation is the time-domain approach, where the azimuth correlation is a convolution operation. A custom chip is required to perform resampling for the range migration correction, followed by a complex multiplier for the reference function weighting and an accumulator (complex adder with memory for one range line). This chip would be replicated for each element in the synthetic aperture followed by a multiplexer to recombine the data. This approach does not appear to represent any significant technological drivers, although this chip has not yet been fabricated.

The signal processor should also be followed by a compression subsystem, which would perform spatial compression of the SAR imagery. Studies have shown that compression ratios of 20:1 can be achieved with little degradation in the image quality. This would significantly reduce the requirement on the digital downlink and the ground data-handling systems.

4.3.2.5 *Digital Downlink and Ground Data System*

With recent technological advancements in digital telecommunication systems, wide-bandwidth digital links have become commonplace. The NASA TDRS has two 150 Mb/s communication channels for a total capacity of 300 Mb/s. To increase the link capacity or, equivalently, the system bandwidth requires an increase in power to offset the commensurate increase in noise.

Considering the potential of on-board data processing and compression, the

---

*FLOPs-Floating octal point operations; M denotes mega (million); G denotes giga (billion).

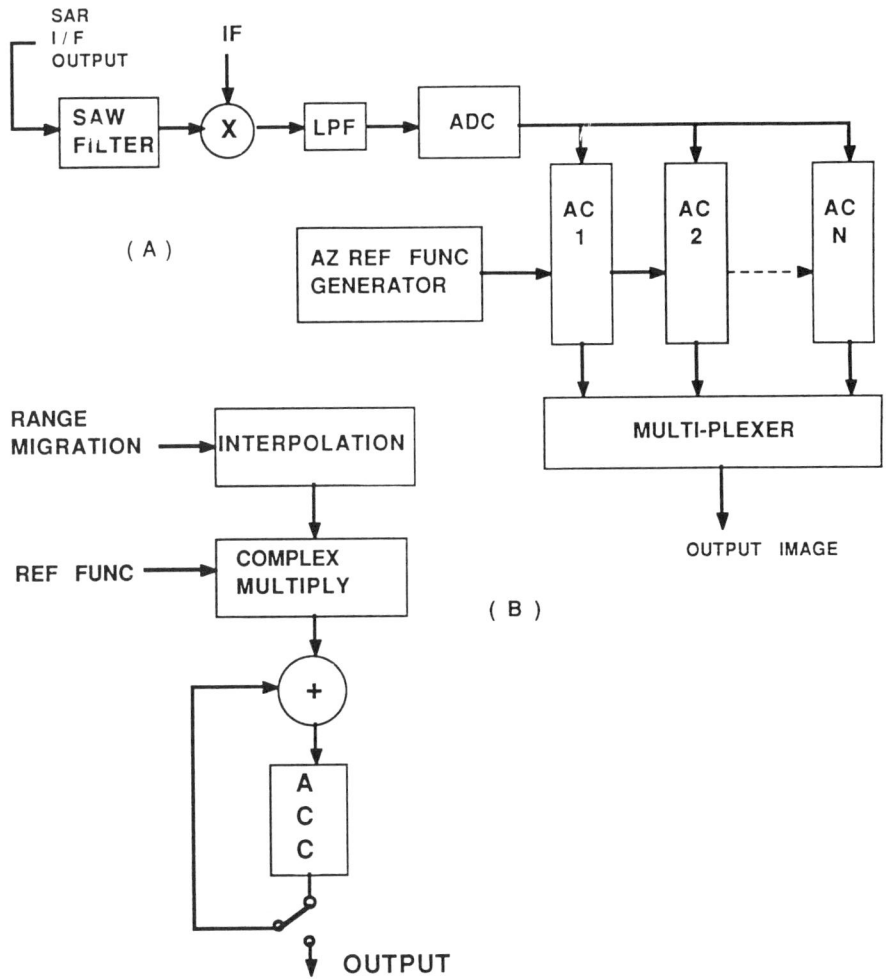

Figure 4.19 Architecture for (a) hybrid analog/digital time domain SAR correlator; (b) azimuth correlator (AC) chip.

next generation of direct downlink systems and data relay satellites should be able to handle data rates on the order of 0.5 to 1.0 Gb/s, which even after error correction coding would be equivalent to instantaneous rates of 8 Gb/s coming from the SAR's ADC. As previously reviewed, the recorder technology is currently available to capture such a data stream. A number of recorders, each with a 240 Mb/s capacity, could be interfaced in parallel to achieve the desired capacity.

We may reasonably conclude that existing integrated circuit technology and data networks (using fiber optics) guarantee the feasibility of real-time ground processing. The pertinent question then is not the feasibility of such a system. Rather, what is the optimal architecture and what systems exist or are under development? A typical ground data processing architecture is shown in Figure 4.20. A number of organizations have already built custom signal processors with GFLOP computation capabilities. Specifically, the Advanced Digital SAR Processor (ADSP) designed and built by NASA and JPL to support the Magellan Venus Radar Mapper and the Shuttle Imaging Radar (SIR-C) has demonstrated its capability of 6 GFLOPS. This system would have processed SEASAT SAR data in real time.

An architecture with dedicated pipeline processing modules, custom designed to perform a specific function, reduces system reliability because a failure typically halts all processing. This architecture is used in the ADSP, as shown in Figure 4.21, and is capable of extremely high computational rates when the pipeline is full, but relies on every module being functional for the system to be operational (i.e., no graceful degradation).

The IBM CSP represents an alternative approach using multiple identical boards for each type of processing (e.g., FFT, memory, complex interpolators), and routes the data by a high-speed switch to each board as required in the processing algorithm. The CSP architecture is shown in Figure 4.22. One drawback to this architecture is the extremely high data rates at the data transfer node. However, this system features graceful degradation at the computational board level.

A third potential architecture is the concurrent processing system such as the Massively Parallel Processor (MPP), developed by Goodyear for the NASA Goddard Space Flight Center (GSFC), or the *hypercube* developed by Cal Tech and JPL. An example of this architecture is shown in Figure 4.23. These systems consist of a large number of identical microprocessors with local memory. The microprocessors are interconnected with different configurations for each application. For example, the MPP is a planar array in which each microprocessor communicates with its nearest neighbor. The hypercube permits multiple interconnection schemes such that each processor can communicate with any other. As the microprocessor technology improves, this type of architecture becomes more desirable due to its high redundancy, system reliability, and flexibility.

4.3.2.6 Integrated System Considerations

The current technology and expected future developments in data handling and on-board signal processing were presented at a functional module level. The analysis indicates that the primary technological driver is in the spaceborne image

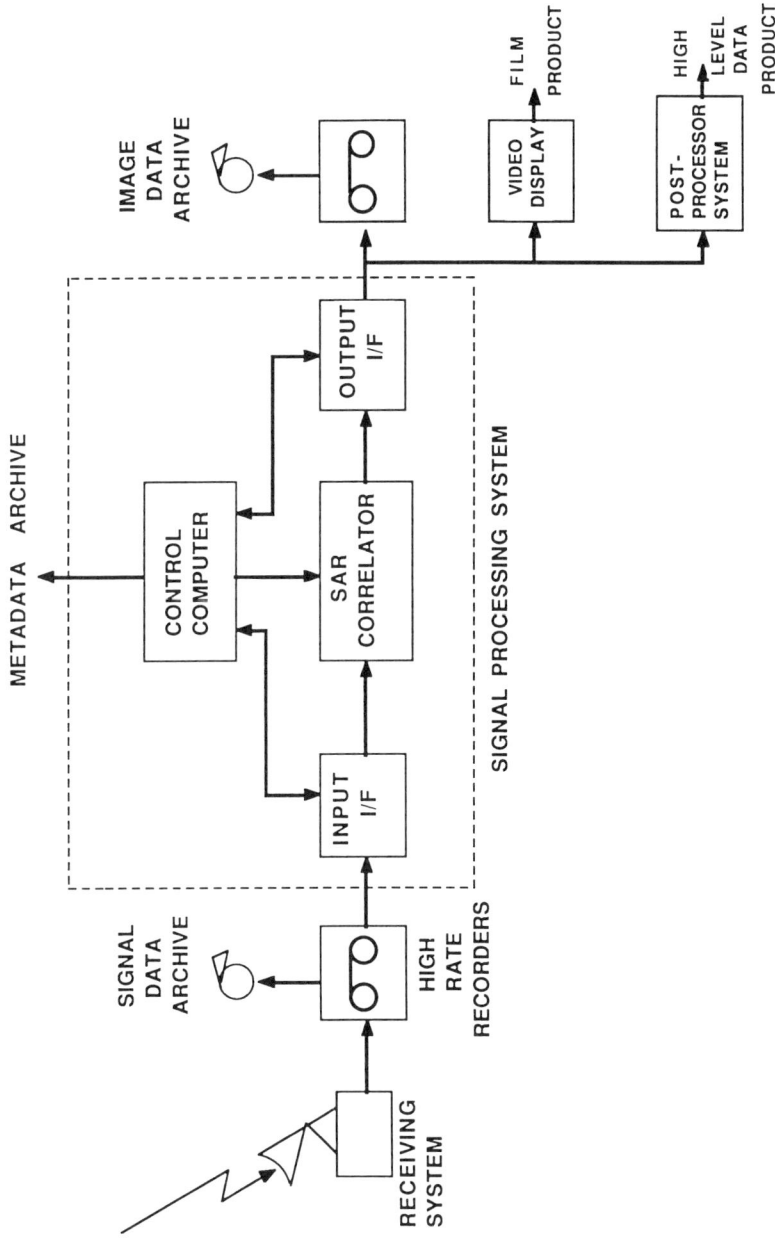

Figure 4.20 Functional block diagram of ground data system for high-rate SAR signal processing.

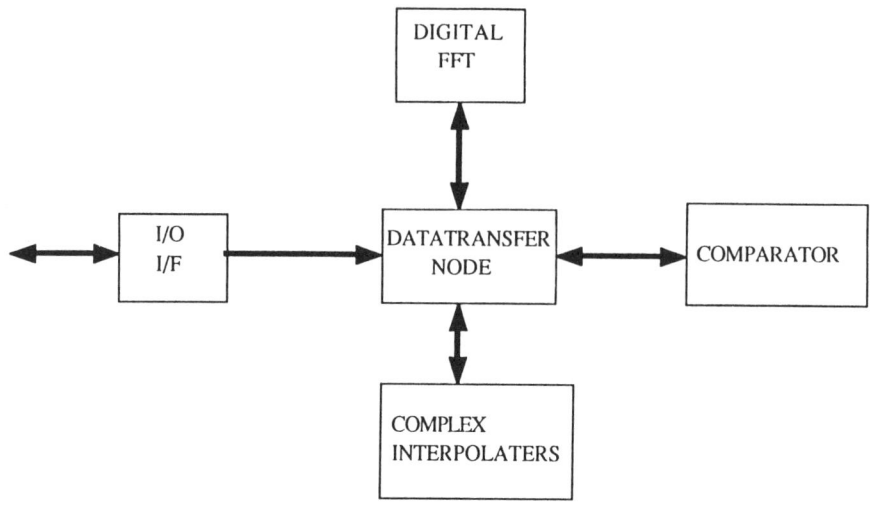

Figure 4.21 Pipeline architecture for SAR processor (control processor not shown).

signal processor. Optical, digital, or combinational hybrid designs can be considered for space-based implementations.

If an on-board processor were implemented, this system could be followed by a data compression system that would reduce both the data rate and volume by a factor of at least 20, alleviating any issues regarding the capacity of available high-rate recorders and downlink systems.

An area of lingering concern is the control of the processor and data calibration. The requirement that the SAR be capable of discerning small changes in radar cross section is an extremely difficult specification to meet in the laboratory, and more so in space. A rule of thumb is that the calibration accuracy be inversely proportional to the resolution. Thus, the higher the resolution, the poorer is the calibration performance. Typical specifications for planned spaceborne systems are on the order of 0.5 to 1.0 dB (e.g., SIR-C) for the relative calibration stability of the system. It is difficult to make a measurement to better than 0.1 dB.

A more reasonable specification to achieve for the calibration is 0.2 to 0.3 dB. This will require internal calibration signals to characterize the amplitude and phase characteristics of the system, including the antenna, over the entire signal bandwidth. This information must then be given to the processor to modify the appropriate processing parameters in near real time. High-precision global positioning satellite (GPS) and attitude determination are also required for estimation of the doppler parameters and antenna boresight. All of these elements are technically feasible, but have yet to be incorporated into an operational system.

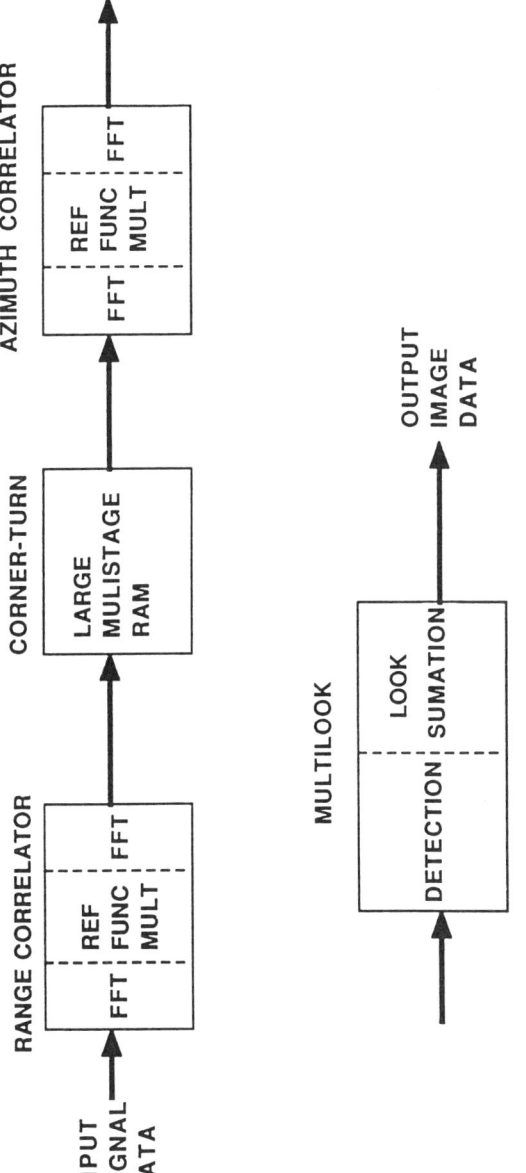

Figure 4.22 Common signal processor (CSP): each module may be configured with multiple boards and up to 32 Mbytes local RAM (control processor not shown).

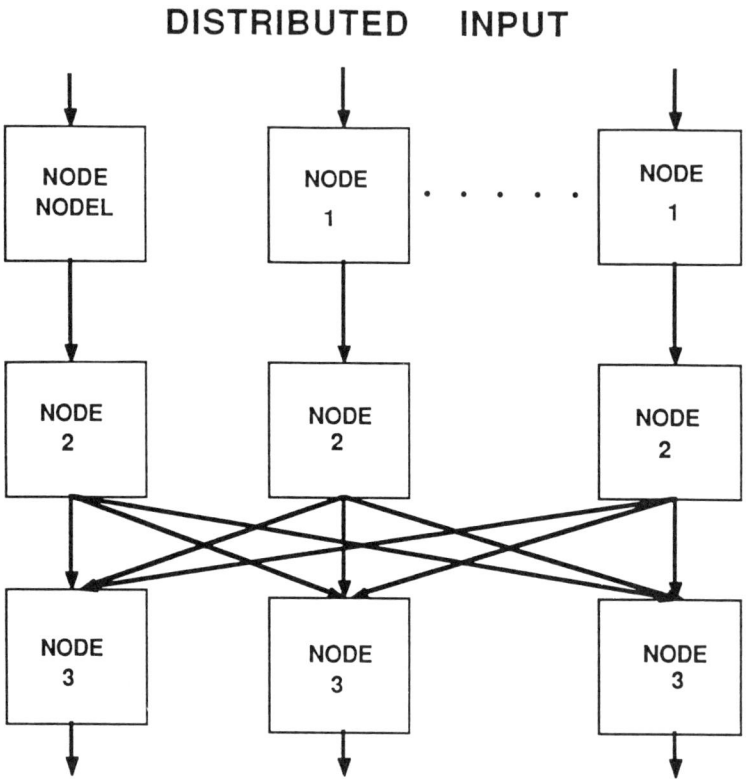

Figure 4.23 Concurrent processor architecture for SAR signal processor. Each node is microprocessor with RAM (control processor not shown).

Table 4.4 provides a summary of some of the discussion in this section. Some of the limiting technologies are shown as they are influenced by application needs such as swath width, spatial resolution, and scene dynamic range. The quest for high resolution, wide swath systems meets technological constraints at the point of data conversion, processing, and link collection. Innovative technologies and system designs will be required to overcome current limitations. Among these innovations, to have on-board, real-time processing is the best hope for a rapid improvement in system application.

Table 4.4
SAR Systems Evaluation

System	Technological Situation
Antenna	• Size is well within current limits
	• Aperture power requires modest performance trade-offs
Waveform Generation	• Current standard in bandwidth and pulse-compression ratio
Transmitter	• Current technology requiring modest power aperture performance trade-offs
Receiver-Detector	• Current generation of technology
	PACING TECHNOLOGY BEGINS HERE
A/D Converter	• Imposes significant performance limitation on range resolution, swath width, and dynamic range
Collection	• Imposes significant performance limitation similar to A/D converter
Image Processing	• Projected capabilities match projected applications. On-board processor would significantly improve system's utility
Analysis	• Perfectly matched filter awaits development

REFERENCES

1. Held, D., personal communication, California Institute of Technology, Jet Propulsion Laboratory.
2. Jain, A., "Radar Speckle Reduction in Synthetic Aperture Radar Processor by a Moving Diffuser," California Institute of Technology, Jet Propulsion Laboratory, October 1976.
3. Kuan, D.T., et al., "Map Speckle Reduction Filter for Complex Amplitude Speckle Images," Strand Image Processing Institute, Los Angeles, CA, IEEE, 1982.
4. Lee, J.S., "Speckle Analysis and Smooting of Synthetic Aperture Radar Images," Systems Research Branch, Naval Research Laboratory, Washington, DC, July 1980; rev. December 1980.
5. McCandless, S.W., and J.C. Curlander, "The Influence of Pacing Technologies on Environmental Application of Space-Based Synthetic Aperture Radar," User Systems, Inc., *Oceans-88 Proc.*, November 1988.
6. Moore, R.K., "Trade-off Between Picture Element Dimensions and Noncoherent Averaging in Side-Looking Airborne Radar," University of Kansas, *IEEE Trans. Aerospace and Electronic Systems*, Vol. AES-15, September 1979.
7. Ramapriyan, H.K., J.P. Strong, and S.W. McCandless, Jr., "Development of Synthetic Aperture Radar Signal Processing Algorithms on the Massively Parallel Processor," Space Data and Computing Division, Goddard Space Flight Center, Greenbelt, MD.
8. Tomiyasu, K., "Computer Simulation of Speckle in a Synthetic Aperture Radar Image Pixel," General Electric Co., Valley Forge Space Center, Philadelphia, PA, February 1983.
9. Tomiyasu, K., "Tutorial Review of Synthetic Aperture Radar with Applications to the Imaging of the Ocean Surface," *Proc. IEEE*, Vol. 66, May 1978.

GLOSSARY OF TERMS

Symbol or Acronym	Definition (units)
IFOV	Instantaneous Field of View (m)
λ	Radar Wavelength (m)
D_{AT} or D_R	Antenna Dimension Along-Track and in Range (m)
R	Radar Range to the Target (km)
L_{SA}	Length of the Synthetic Aperture (km)
δ_{AZ}	Azimuth IFOV (m), also called *Along-Track* (AT)
δ_R	Range IFOV (m)
τ	Effective Pulse Length (seconds)
T	Real Pulse Length (seconds)
ψ	Grazing Angle (degrees or radians)
ϕ_i	Incidence Angle, the Complement of the Grazing Angle (degrees or radians)
c	Speed of Light (m/s)
β or B	Bandwidth (Hz)
R_F	Far Range (km)
R_N	Near Range (km)
W_g	Swath Width/Scan Extent (km)
P_t	Pulse Power (watts)
G	Antenna Gain (dB)
σ or σ_0	Target or Clutter Cross Section (m² or dB, respectively)
L	RF Losses (dB)
T_s	System Noise Temperature (K)
K	Boltzmann's Constant 1.3×10^{-23} (joule/°)
PRF	Pulse Repetition Frequency (Hz)
v	Velocity (m/s)
T_i	Dwell Time (seconds)
F	Receiver Noise Figure (dB)
S/N	Signal-To-Noise Ratio (dB)
S/C	Signal-To-Clutter Ratio (dB)
w	Angular Rotation (radians/second)

Chapter 5
BISTATIC RADAR IN SPACE
P. Hartl and H.M. Braun

Universitat Stuttgart   *Dornier*

5.1 COMPARISON OF BISTATIC AND MONOSTATIC SYSTEMS

Bistatic radars are systems in which spatial separation exists between the transmitting and receiving parts. Their fundamental principles have been known since the beginning of radar technology. However, interest in bistatic radars declined early, doubtless driven by the desire of users, particularly military users, to have radars operated from a single site. This desire resulted in intensive development of monostatic radars to a very sophisticated state, whereas bistatic radars lay dormant for more than two decades until they received new interest when the development of advanced data processing technologies allowed bistatic radars to deploy their advantages.

The following brief description of imaging, bistatic, space-based radar systems provides an overview. We indicate that bistatic systems may have important capabilities for special applications.

5.2 BISTATIC SYSTEMS

5.2.1 Low Earth Orbit (LEO) Systems

With *low earth orbit* systems, both transmitter and receiver are in a low earth orbit. The following categories of bistatic LEO systems are considered:
- single-orbit systems,

- crossing-orbit systems,
- tether systems.

In single-orbit systems, transmitter and receiver satellites are placed in the same orbit, one after another (see Figure 5.1). Two separate antennas (one for the transmitting and the other for the receiving path) provide a higher transmit-receive isolation than a circulator or T/R switch commonly used in monostatic systems. Whereas in case of monostatic radars the echo reception time and, hence, the width of the swath are limited to the *pause* interval between two consecutive pulse transmissions, the bistatic configuration allows use of the *entire* pulse repetition interval (PRI) because echo reception is not blocked during pulse transmission. This leads to a larger swath width, which is independent of the transmission pulse length. Therefore, very large pulse-compression ratios, up to continuous transmission, can be applied to lower the peak power requirement or to improve the radiometric performance (gray level resolution). CW transmission can be advantageous for the overall efficiency of the high-power amplifier, especially in the case of solid-state transmitters. These advantages typically require high development costs (two satellites) and complex systems for transmitter-receiver synchronization.

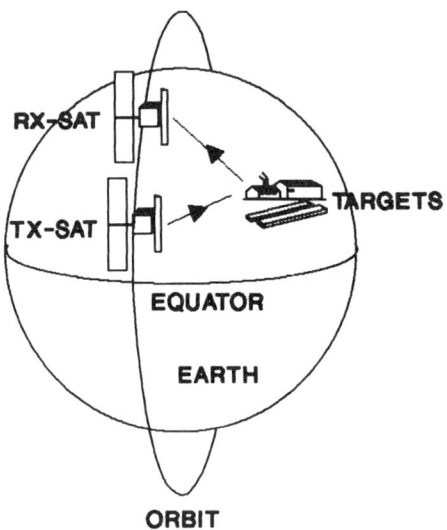

Figure 5.1 Single-orbit system.

If transmitter and receiver are carried on free-orbiting (untethered) satellites separated in the across-track or nadir direction, the system is called a *crossing-orbit system*. Figure 5.2 shows such a system for orbits with different inclinations.

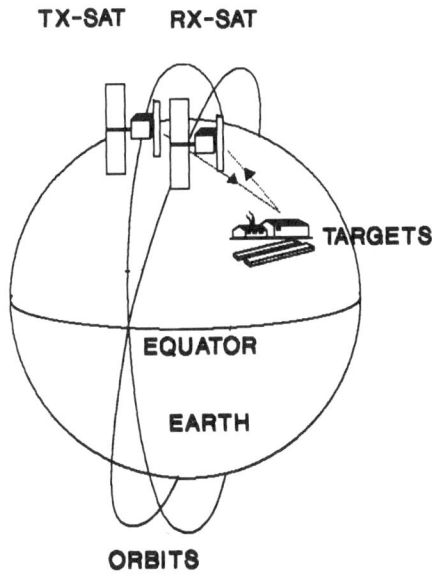

Figure 5.2 Crossing-orbit system with differently inclined orbits.

Due to orbital dynamics, the paths are crossing at the equator and the satellite distance varies from about zero at the crossing points to a maximum at the polar regions.

Figure 5.3 shows an example with orbit crossings at the poles (different ascending nodes). The system's behavior is similar to that with orbital crossings at the equator, but the maximum across-track separation appears at the equator.

Another type of crossing-orbit system is defined by spatial separation in the nadir direction (see Figure 5.4). In this case, the transmitter and receiver orbits are in the same plane, slightly eccentric, with equal eccentricity and different apogees. This results in different altitudes for transmitter and receiver, and the orbit cycle times are kept equal. A similar effect with the advantage of a constant satellite distance and a simplified radar geometry can be obtained by a tether satellite system, where transmitter and receiver satellites are attached by a cable of several kilometers in length. Tether systems are described by Yasaka and Hatsuda [12] for communication satellites. The results can be applied to radar systems as well. In this case, the synchronization signals and the radar data can be transmitted through the tether cable.

The general electrical configurations for the transmitting (TX-SAT) and receiving (RX-SAT) radars are shown in Figure 5.5. The key element of the overall system is the master oscillator of the TX-SAT. From this oscillator a beacon is sent by the direct path link to the receiver for time and phase synchronization.

170

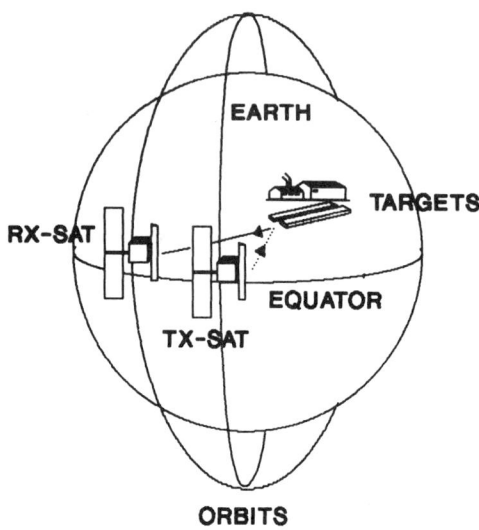

Figure 5.3 Crossing-orbit system with different ascending nodes.

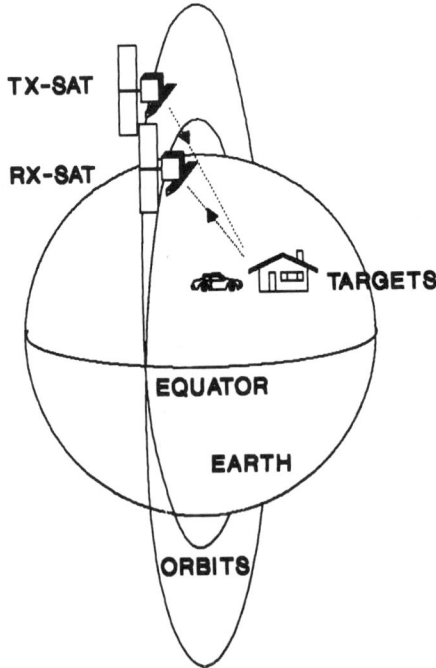

Figure 5.4 Crossing-orbit system with separation in the nadir direction.

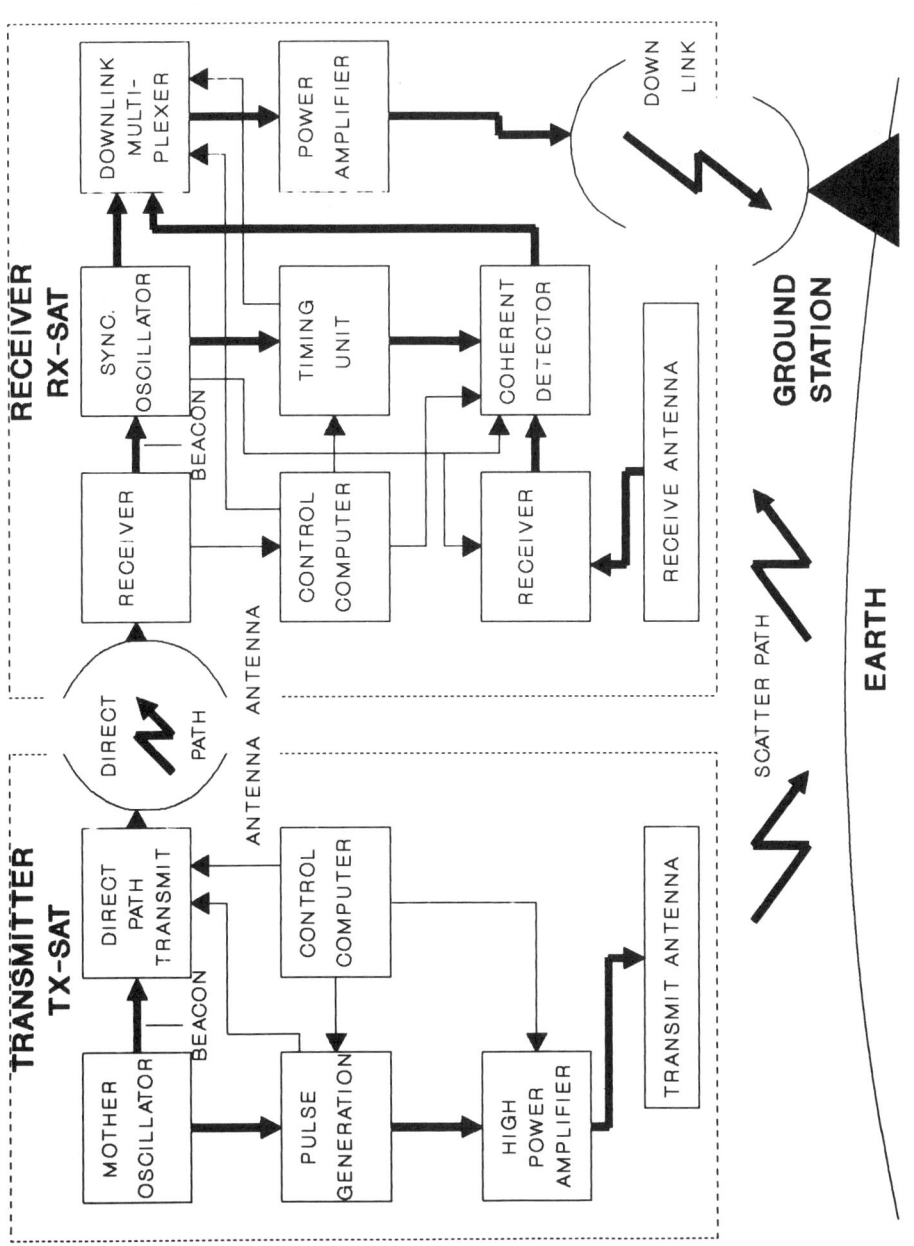

Figure 5.5 Block diagram of a bistatic radar.

Initiated by the TX-SAT control computer and based on the master oscillator, the radar pulse is generated by the pulse generation subsystem and sent to ground by the high-power amplifier and the transmitting antenna. Meanwhile, the beacon, timing references, an eventual pulse replica, and other auxiliary data are sent by the direct path to the receiver satellite. The RX-SAT control computer, having received the position data for RX-SAT and TX-SAT from the ground control station by general telecommand channels (not shown), estimates the echo time of arrival in advance as input for the timing unit. This unit controls the coherent detection of the received radar echoes. The detected signals are multiplexed with auxiliary data required for data processing on the ground and sent to receiving earth stations, either directly or by an in-orbit data relay station, which can be the TX-SAT itself or a separate communication satellite.

Crossing-orbit or tether systems are important candidates for radar stereographic imagery. In this case, the TX-SAT should additionally be equipped with a receiving channel operating in a monostatic mode, or two receivers covering the same target area simultaneously from different positions in the across-track direction. The monostatic image or the second bistatic image provides the second look required for stereo imagery. Single-orbit systems (spatial separation in along-track), however, can provide stereographic imagery only in cases of real aperture operation. Synthetic aperture methods applied to bistatic single-orbit systems destroy each kind of stereographic information within the radar data.

5.2.2 Systems with Geostationary (GEO) Transmitter

The idea of operating a radar with a powerful transmitter in a *geostationary earth orbit* (GEO) and with clusters of "small" receivers in low earth orbits was born of the desire to provide a high repetition rate of radar coverage for selected target areas on the ground. A multibeam transmitter is continuously illuminating these ground areas, or operates in a scanning mode synchronized to the receivers. Each receiver is operational as soon as it passes over these areas, and sends the received echoes back to the geostationary transmitter, which is acting as a relay satellite for these data. The receivers being carried on LEO satellites, or even aircraft, are passing by the targets closely enough to improve the link budget and to provide a radial velocity between the targets on the ground and the radar required for synthetic aperture processing.

Figure 5.6 shows a typical geostationary transmitter system, indicating that many receivers are in low earth orbits. The typical electrical layout of such a system is similar to the layout of a LEO system shown in Figure 5.5.

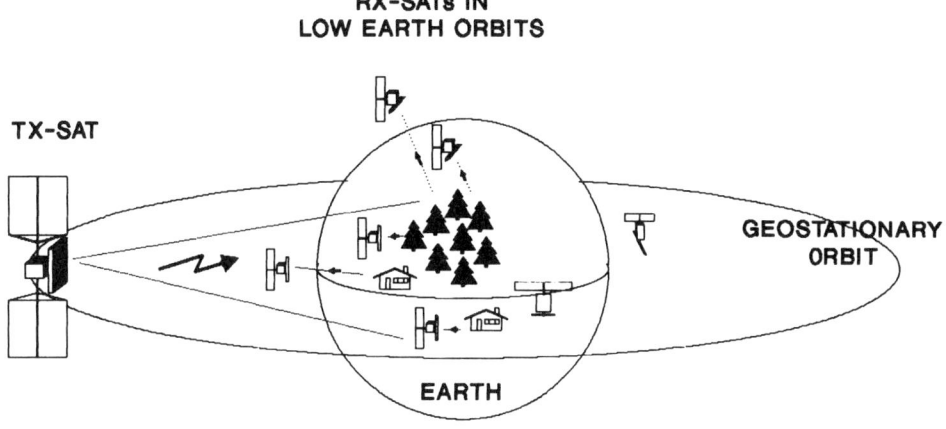

Figure 5.6 Geostationary transmitter system.

5.2.3 Parasitic Radar Systems

A BIPAR is a bistatic radar system operating with transmitters of opportunity and combining the advantages of spaceborne and airborne imaging radars. The transmitter, typically a communication satellite, is located in a geostationary orbit and continuously illuminates the area of interest on the ground. The quiet receiver is carried on an airborne platform implying a high mission flexibility, high repetition rate capability, and reduced power requirements as compared to a system with spaceborne receivers. Such a system is called *parasitic* because it uses nonradar signals from communication, navigation, or direct broadcasting satellites. The system geometry is shown in Figure 5.7.

A geostationary communication satellite is transmitting an RF signal down to earth. Within the beam of its downlink antenna, the satellite illuminates the earth's surface, and a certain portion of the RF energy is reflected back into space. The strength of the reflections depends on the reflection coefficient of the target areas on the ground, such as vegetation canopy, streets, houses, cars, and other natural or man-made targets. An aircraft that has a BIPAR receiver on-board, and is flying within the satellite's downlink antenna beam, picks up these reflections from the ground with a scanning pencil-beam radar antenna and provides real aperture radar images of that part of the earth's surface being covered by the BIPAR antenna scan range.

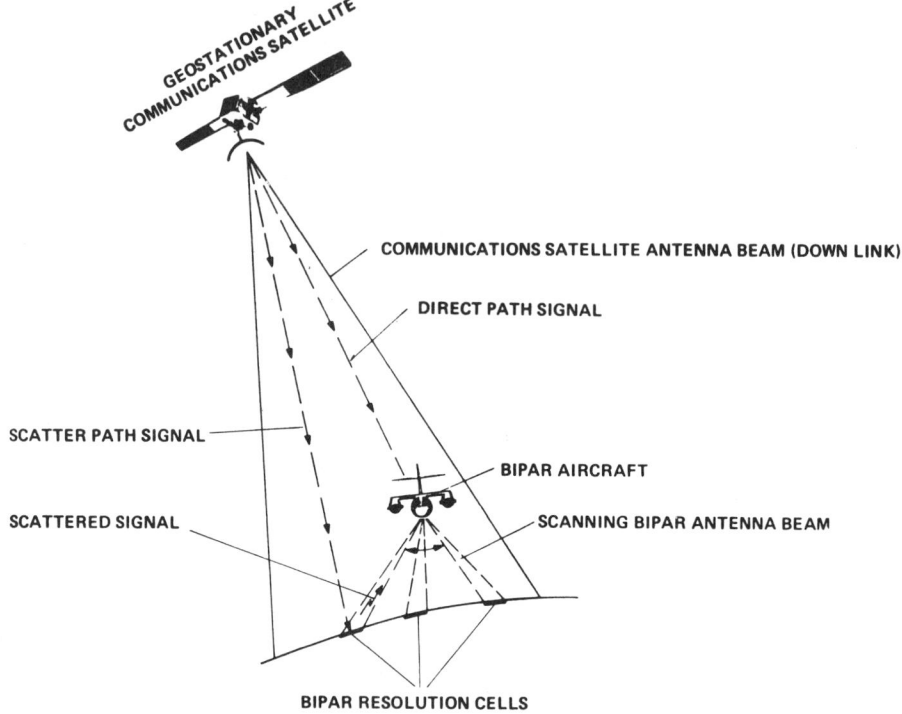

Figure 5.7 BIPAR system geometry.

In the case of low transmission power of the geostationary satellite resulting in an illumination with a low power density on the ground, the received signal reflection may be far below the BIPAR receiver noise level. Hence, correlation with a known reference is required to improve the signal-to-noise ratio (SNR). The necessary reference can be provided by receiving the direct path downlink signal with an antenna on top of the aircraft. The differences in doppler shift and round-trip delay time can easily be compensated by use of its *apriori* knowledge.

The overall block diagram of a BIPAR receiver is shown in Figure 5.8. The scattered signals are received by a scanning pencil-beam antenna and routed through a low-noise amplifier and a down-converter, which compensates for the doppler shift of the scattered signals. After appropriate gain setting, this noisy signal is fed to a coherent detector and correlated with the respective reference signal. The reference signal, received by an antenna on top of the aircraft, is routed through a low-noise amplifier, a down-converter compensating its doppler shift, and a variable delay line. The delay line allows the reference signal and the scattered signal to arrive simultaneously at the correlator's input. After integration, the signal is digitized and routed to a radar control computer. This unit analyzes the

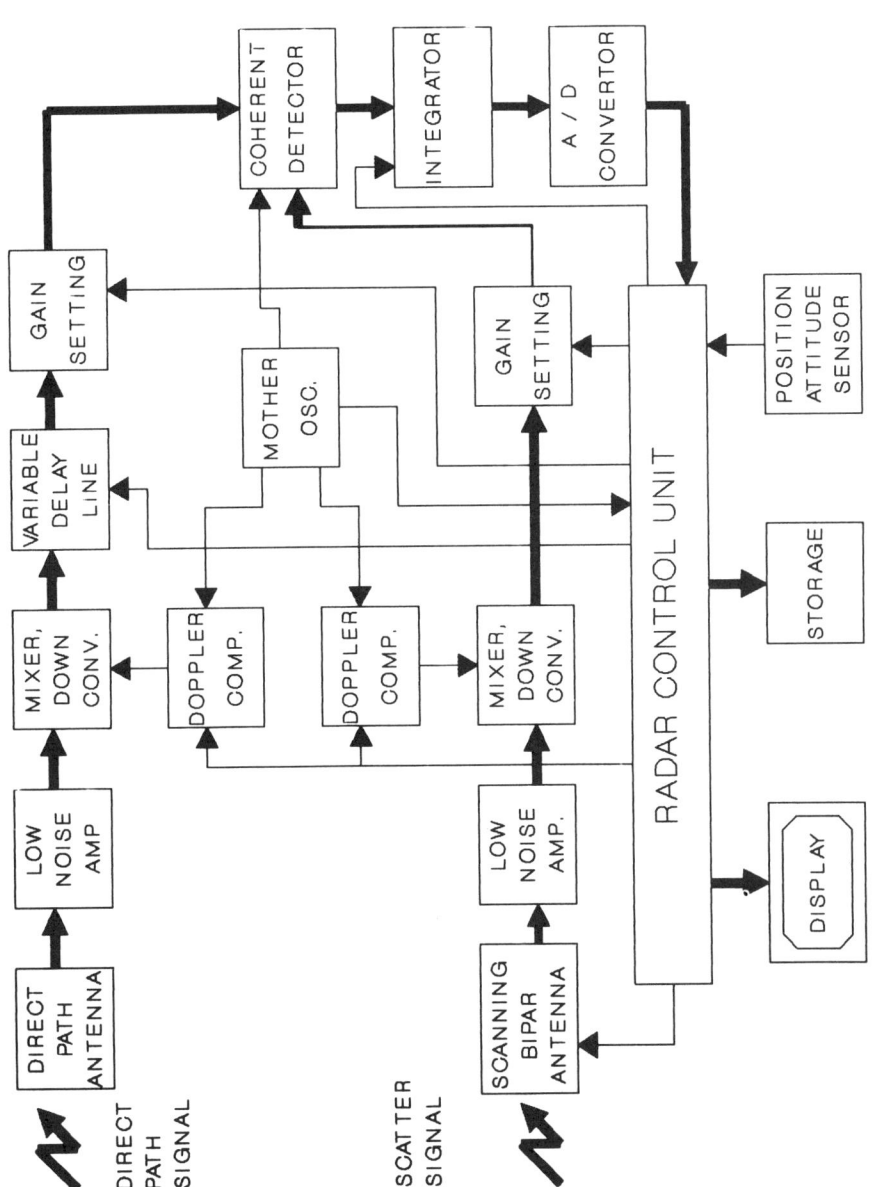

Figure 5.8 Block diagram of a BIPAR receiver.

signals and updates the gain settings. Concurrently, signals are displayed on-board as a real aperture radar image on a monitor and stored for further processing on the ground. The doppler compensation and variable time delay are driven by a radar control unit, based on data from the aircraft position and attitude sensors, on prior information about the satellite's position, downlink frequency, and the actual BIPAR antenna pencil-beam orientation.

Compared to classic monostatic and bistatic radars, we can conclude that BIPAR systems provide the following advantages:

- low power consumption,
- high operational flexibility,
- quiet and secret systems,
- low costs.

BIPAR systems are usually not optimal for radar applications, but they provide a good compromise between user requirements and available funds. BIPARs cannot replace large operational airborne and spaceborne radar systems, but in many cases they can support scientific research in areas for which operational radar systems are not available.

5.3 PERFORMANCE CONSIDERATIONS

5.3.1 Bistatic Geometry

The geometry of bistatic space radars is significantly different from any terrestrial bistatic radar geometry. Transmitters and receivers are carried on *moving* space platforms, and the targets are either located on the earth's surface or flying in the atmosphere.

A typical bistatic space radar geometry is given in Figure 5.9, showing a radar transmitter at $p_T(t)$ and its velocity vector $\mathbf{v}_T(t)$, a receiver at $p_R(t)$ with a velocity $\mathbf{v}_R(t)$, and a target on the earth's surface at $p_S(t)$ with a velocity $\mathbf{v}_S(t)$. The figure shows the direct path $\mathbf{p}_T\mathbf{p}_R(t)$ and the scatter path $\mathbf{p}_T\mathbf{p}_S(t)$ and $\mathbf{p}_S\mathbf{p}_R(t)$. The illustration identifies the nadir angles at the transmitter for the direct path (β_{TR}) and the scatter path (β_{TS}), the nadir angles at the receiver for the direct path (β_{RT}) and the scatter path (β_{RS}), the grazing angles at the target for the incoming radar signal (γ_T), the outgoing echo (γ_R), and the scatter angle (γ_S). All of these angles are functions of time.

The radar signals from the scatter path received at p_R are characterized by their difference in time of arrival between scatter path signals and direct path signals, and by their difference in doppler shift Δf_d. The propagation time for the direct path signals (T_D) and the scatter path signals (T_S) is given by

$$T_D = \frac{|\mathbf{p}_T\mathbf{p}_R|}{c} \text{ and } T_S = \frac{|\mathbf{p}_T\mathbf{p}_S| + |\mathbf{p}_S\mathbf{p}_R|}{c} \tag{5.1}$$

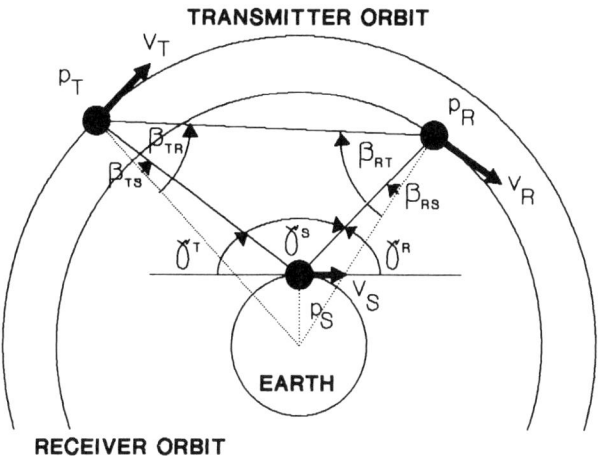

Figure 5.9 Bistatic geometry.

where c is the velocity of light ($3 \cdot 10^8$ m/s).

For a known direct path length and a measured difference in arrival time (ΔT_R):

$$\Delta T_R = T_S - T_D = \frac{|\mathbf{p}_T \mathbf{p}_S| + |\mathbf{p}_S \mathbf{p}_R| - |\mathbf{p}_T \mathbf{p}_R|}{c} \quad (5.2)$$

the scatter path length L_s can be determined by

$$L_s = |\mathbf{p}_T \mathbf{p}_S| + |\mathbf{p}_S \mathbf{p}_R| = |\mathbf{p}_T \mathbf{p}_R| + (\Delta T_R \cdot c) \quad (5.3)$$

The velocity vector of a spacecraft or a target is given by the time derivative of its path equation:

$$\mathbf{v} = \frac{d\mathbf{p}(t)}{dt} \quad (5.4)$$

The doppler shift for the scatter path is given by

$$f_{DS} = \frac{1}{\lambda}\left(\frac{d}{dt} L_s\right) \quad (5.5)$$

and for the direct path:

$$f_{DD} = \frac{1}{\lambda}\left(\frac{d}{dt}|\mathbf{p}_T \mathbf{p}_R|\right) \quad (5.6)$$

where λ is the radar wavelength.

For a known direct path and a measured doppler difference Δf_{DR} between the two signals at the receiver, the scatter path doppler is

$$f_{DS} = \frac{1}{\lambda}\left(\frac{d}{dt}|\mathbf{p}_T\mathbf{p}_R|\right) + \Delta f_{DR} \tag{5.7}$$

Imaging radars for applications in scientific earth observation generally deal with fixed or slowly moving targets on the earth's surface, implying that only two coordinates of the three-dimensional system are unknown (longitude and latitude). The third coordinate is the local earth radius. The velocity vectors of the targets are known from the earth's rotation and the local earth radius.

In an earth-centered Cartesian coordinate system, as shown in Figure 5.10, the x-axis being normalized to zero degrees longitude at time ($t = 0$), the path of a target on earth can be described by

$$\mathbf{p}_S(t) = R_S \begin{pmatrix} \cos\phi_S(t)\,\cos\delta \\ \sin\phi_S(t)\,\cos\delta \\ \sin\delta \end{pmatrix} \tag{5.8}$$

and its velocity vector by

$$\mathbf{v}_S = \frac{2\pi R_S}{T_S}\cos\delta_S \begin{pmatrix} -\sin\phi_S(t) \\ \cos\phi_S(t) \\ 0 \end{pmatrix} \tag{5.9}$$

with

$$\phi_S(t) = \frac{2\pi}{T_S}\cdot t$$

R_S is the local earth radius, T_S is the target (and earth) round-trip time ($8.64 \cdot 10^4 \mu s$ (microseconds)), δ_S is the latitude of the target, and $\phi_S(t)$ is the target angle in the equatorial plane measured from the x-axis in the east direction.

The same coordinate system as applied to a circular orbit is shown in Figure 5.11. The circular orbit can be described by the following two relations.

Transmitter Case:

$$\mathbf{p}_T = R_T \begin{pmatrix} \cos[\phi_T(t) - \omega_T]\cos\omega_T - \sin[\phi_T(t) - \omega_T]\sin\omega_T \\ \sin[\phi_T(t) - \omega_T]\cos\omega_T\cos\epsilon_T + \cos[\phi_T(t) - \omega_T]\sin\omega_T\cos\epsilon_T \\ \sin[\phi_T(t)]\cos\epsilon_T \end{pmatrix}$$

with

$$\phi_T(t) = \frac{2\pi}{T_T}(t - T_{0T}) \tag{5.10}$$

Receiver Case:

$$\mathbf{p}_R = R_R(\cos[\phi_R(t) - \omega_R]\cos\omega_R - \sin[\phi_R(t) - \omega_R]\sin\omega_R \sin[\phi_R(t)$$
$$- \omega_R]\cos\omega_R \cos\epsilon_R + \cos[\phi_R(t) - \omega_R]\sin\omega_R \cos\epsilon_R \sin[\phi_R(t)]\cos\epsilon_R)$$

with

$$\phi_R(t) = \frac{2\pi}{T_R}(t - T_{0R}) \tag{5.11}$$

T_T and T_R are the orbit revolution times. T_{0T} and T_{0R} are the times between $\phi = 0$ at $t = 0$ and spacecraft south-north equatorial plane crossings. The velocity vectors can be found by applying (5.4).

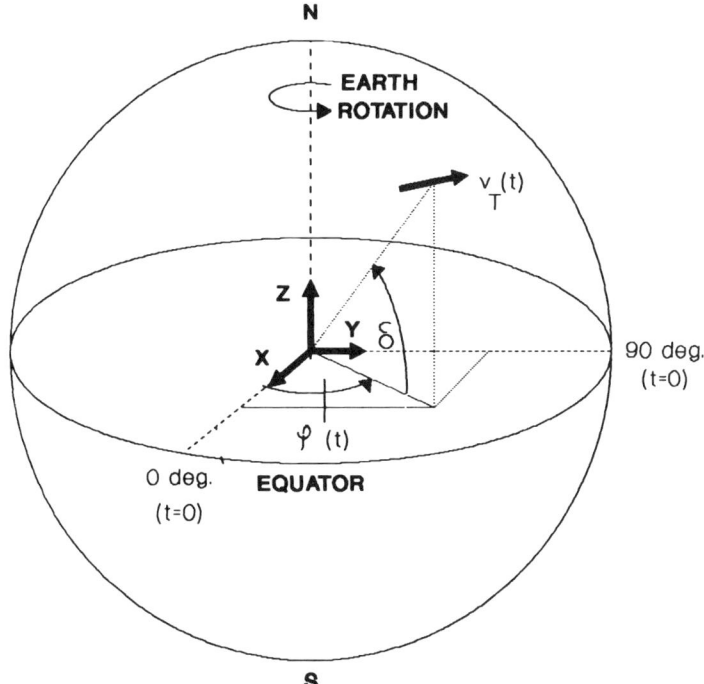

Figure 5.10 Earth-centered coordinate system.

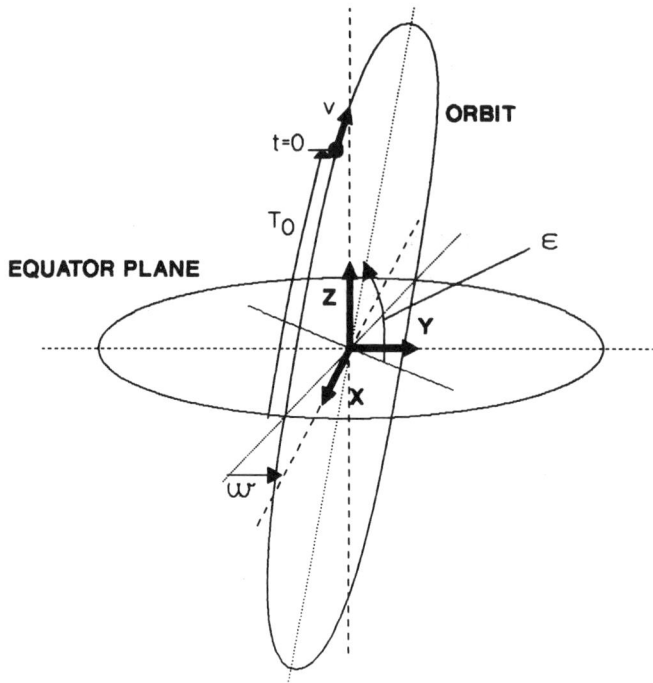

Figure 5.11 Circular orbit coordinates.

5.3.2 Spatial Performance

The spatial performance of a *real aperture down-looking radar* is described on the basis of an airborne receiver system. Spaceborne receiver systems require impractically large antennas for a reasonably fine spatial resolution. Hence, synthetic aperture methods are essential for spaceborne receivers.

The spatial resolution of a real aperture down-looking system is defined by the beamwidth of the scanning pencil-beam antenna of the receiver (see Figure 5.12). The nominal value of the spatial resolution in azimuth δ_A (flight direction) and elevation δ_E (across-track direction) at a scan angle of zero degrees (nadir) is

$$\phi_{A/E} \approx \frac{A \cdot \lambda}{L_{A/E}} \tag{5.12}$$

When the pencil beam is pointed to the outermost edge of the swath, the spatial resolution is degraded to

$$\delta_{A_{max}} \approx \frac{A \cdot \lambda}{L_A \cos(\phi_s/2)} \tag{5.13}$$

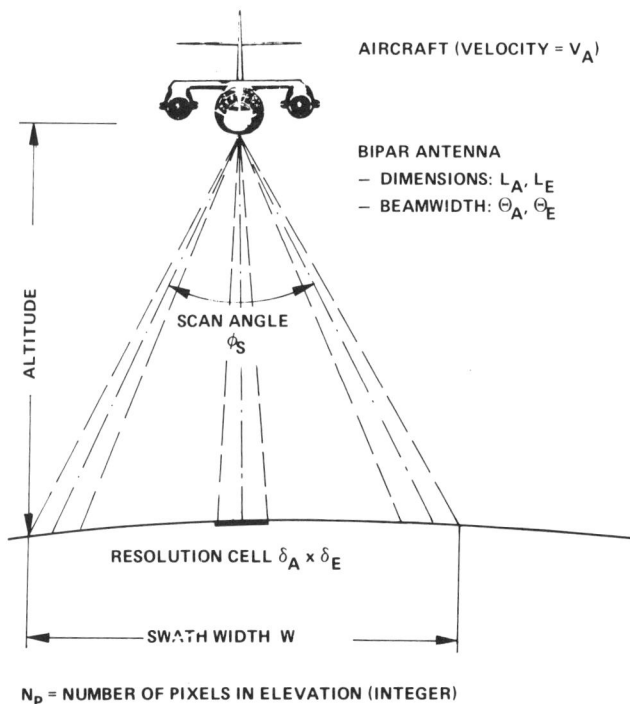

Figure 5.12 Real aperture down-looking radar geometry (receiving path).

$$\delta_{E_{\max}} \approx \frac{A \cdot \lambda}{L_E \cos^n(\phi_s/2)} \quad (5.14)$$

where A is the flight altitude, L_A and L_E are the effective antenna dimensions in azimuth and elevation, ϕ_s is the maximum scan angle between left and right edges of the swath, and n is a value of 2 for mechanical and 3 for electronical scanning of the pencil beam. Thus, the beam broadening of an electronically scanned antenna additionally widens the resolution cell in elevation being considered by n.

The width W of the swath is given by

$$W \approx A\left[2 \tan(\phi_s/2) + \frac{\lambda}{L_E \cos^n(\phi_s/2)}\right] \quad (5.15)$$

and the available pixel integration time T_i is given by

$$T_i \approx \frac{\delta_A}{v_A N_p N_L} \quad (5.16)$$

with

$$N_p \approx \left(\frac{\phi_s + \theta_E}{\theta_E}\right)$$

and

$$\theta_E \approx \frac{\lambda}{L_E}$$

where v_A is the aircraft velocity, N_L is the number of independent looks required for improvement of the radiometric resolution, and θ_E is the antenna elevation beamwidth.

In the case of a *real aperture side-looking radar*, the range resolution is no longer determined by the antenna beamwidth, but by the radar pulse length τ_s and the radar bandwidth B_{RF}:

$$\delta_R = \frac{\tau_s c}{\cos\gamma_T + \cos\gamma_R} \approx \frac{c \cdot 1.26}{B_{RF}(\cos\gamma_T + \cos\gamma_R)} \qquad (5.17)$$

γ_T and γ_R are the target grazing angles for the transmitting and the receiving paths (see Figure 5.13). The elevation beamwidth of the transmitter and the receiver antenna is matched to the radar swath width. The maximum obtainable unambiguous swath width for bistatic side-looking radars can be estimated by

$$W_{max} \approx \frac{\text{PRI } c}{\cos\gamma_{TF} + \cos\gamma_{RF}} \qquad (5.18)$$

where PRI is the pulse repetition interval, and γ_{TF} and γ_{RF} are the grazing angles at the far end of the swath. The curvature of the earth across the swath is neglected.

In this case, the receiver antenna beam should not be wider than necessary to illuminate this maximum swath to obtain a reasonable ambiguity suppression, and the transmitter antenna beam should not be smaller than required for completely illuminating the swath used.

Further improvements of the spatial resolution in azimuth (along-track direction) can be obtained by applying synthetic aperture processing. This is especially required for systems with spaceborne receivers. Two targets on the ground are distinguished by different doppler shifts in their reflected radar signals. In general, the doppler shifts of two targets are given by

$$f_{D1} = \frac{1}{\lambda}\left(\frac{d}{dt}L_{S1}\right) \qquad (5.19a)$$

$$f_{D2} = \frac{1}{\lambda}\left(\frac{d}{df}L_{S2}\right) \tag{5.19b}$$

where f_{D1} and f_{D2} are the doppler frequencies of target 1 and target 2. L_{S1} and L_{S2} are the lengths of the echo paths via target 1 and target 2.

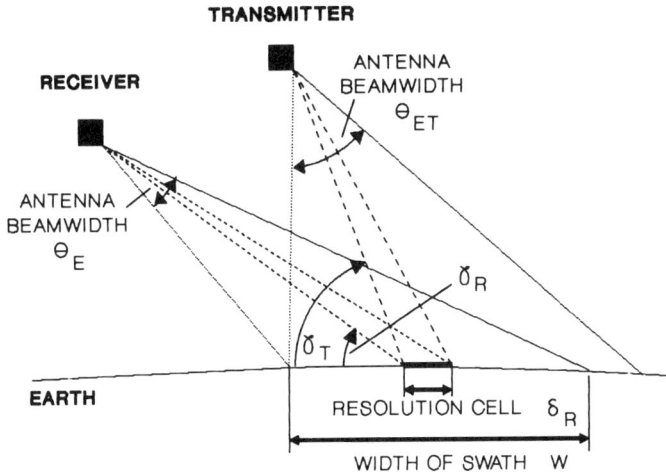

Figure 5.13 Real aperture side-looking radar geometry.

The doppler resolution δ_D is the minimum detectable doppler difference:

$$\Delta f_D = f_{D2} - f_{D1} \tag{5.19c}$$

The distance in azimuth direction between these targets is the spatial resolution in azimuth σ_A.

In the case of spaceborne imaging radars where the velocity vectors of transmitter, receiver, and targets are known and constant over the radar integration time, the doppler resolution for a given spatial resolution in azimuth can be approximated as described below.

$$f_R \approx f_T$$

$$\times \left[1 + \frac{v_T\left(\dfrac{R}{R + A_T}\right)\cos\beta_T - v_S(\cos\beta_{SR} + \cos\beta_{ST}) + v_R\left(\dfrac{R}{R + A_R}\right)\cos\beta_R}{c}\right]$$

$$\tag{5.20a}$$

where f_T is the transmitting frequency; v_T, v_S, v_R are the velocities of transmitter, target, receiver; and β_{xx} denotes the angles between the velocity vectors and the paths (not necessarily on the same plane as shown in Figure 5.14, for example). R is the local earth radius and $A_{T/R}$ is the respective satellite altitude. The correction term $R/(R + A)$ transforms satellite velocities into "footprint" velocities. This is a first-order approximation compensating for the nonlinear orbital movements of satellites and targets on earth.

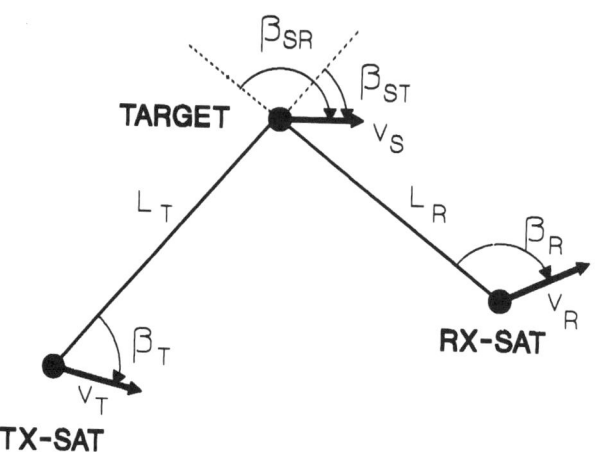

Figure 5.14 Echo path geometry.

The doppler shift of the echo path results in

$$f_D \approx f_T \left[\frac{v_T \left(\dfrac{R}{R + A_T}\right) \cos\beta_T + v_R \left(\dfrac{R}{R + A_R}\right) \cos\beta_R - v_S(\cos\beta_{SR} + \cos\beta_{ST})}{c} \right]$$

(5.20b)

To estimate the required doppler resolution δ_D, two targets ($T1$ and $T2$) can be defined, located on an isorange line with a distance of δ_A (see Figure 5.15) determining the transmitting and receiving viewing angles:

$$\Delta\beta_T \approx \beta_{T1} - \beta_{T2}$$
$$\Delta\beta_R \approx \beta_{R1} - \beta_{R2} \qquad (5.21a)$$

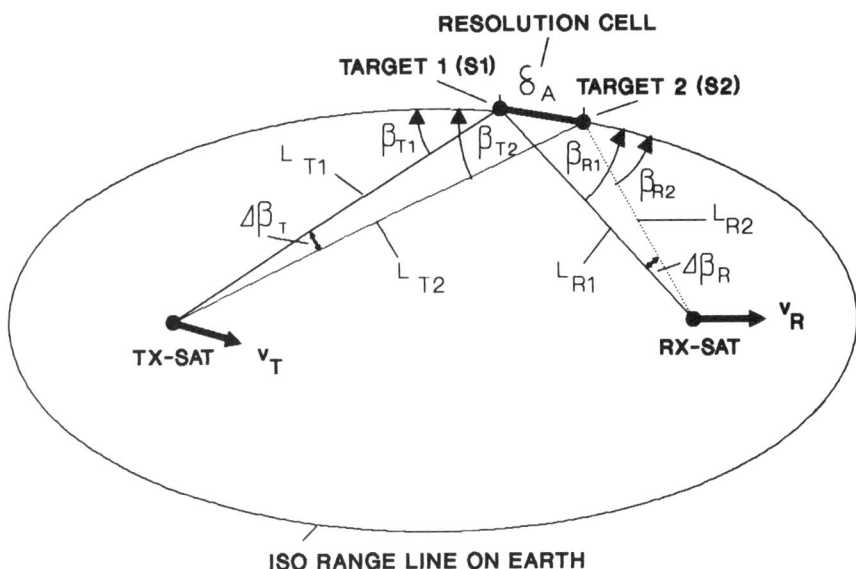

Figure 5.15 Azimuth resolution geometry.

The angles $\Delta\beta$ are very small, and therefore we let $\cos\Delta\beta$ be 1 and $\sin\Delta\beta$ be $\Delta\beta$. This leads to the following approximation:

$$\delta_D \approx \frac{1}{\lambda}\left[v_T\left(\frac{R}{R+A_T}\right)\Delta\beta_T \sin\beta_T + v_R\left(\frac{R}{R+A_R}\right)\Delta\beta_R \sin\beta_R \right.$$
$$\left. - v_s(\Delta\beta_R \sin\beta_{SR} + \Delta\beta_T \sin\beta_{ST})\right] \quad (5.21b)$$

Equation (5.21b) indicates the order of magnitude. The exact values for the doppler resolution can be found by applying (5.19), based on a three-dimensional coordinate system.

The doppler resolution is used to define the minimum required integration time per independent look T_{i1}, taking the integration efficiency η_i into consideration:

$$T_{i1} = \frac{1}{\delta_D \eta_i} \quad (5.22)$$

The total integration time of T_i for N_1 looks is

$$T_i = T_{i1} \cdot N_1 \tag{5.23}$$

and must be provided by the layout of the antennas; the radar timing takes ambiguity criteria into account.

5.3.3 Radiometric Performance

Skolnik [10] gives the bistatic radar equation:

$$P_R = \frac{P_T G_{TS} G_{RS} \lambda^2 \sigma_b}{(4\pi)^3 D_{TS}^2 D_{SR}^2 L_{PT} L_{PR} L_S} \tag{5.24}$$

where P_R is the received signal power, P_T is the transmitted power, G_{RS} is the receiver antenna gain, G_{TS} is the transmitter antenna gain, σ_b is the bistatic cross section, D_{TS} is the transmitter-to-target distance, D_{SR} is the target-to-receiver distance, L_{PT} is the propagation loss (transmitter-to-target path), L_{PR} is the propagation loss (target-to-receiver path), and L_S is the additional system loss.

If the receiver is matched to the transmitted pulse length, the receiver noise is

$$N_R = kT_R F_S B_S = kT_R F_S (\eta_p/\tau_s) \tag{5.25}$$

where k is Boltzmann's constant, T_R is the receiver temperature, F_S is the noise factor, B_S is the system bandwidth, and τ_s is the pulse length (after pulse compression, if applied).

The time-bandwidth product η_p can be approximated by

$$\eta_p = (B_S \tau_s) \approx \begin{cases} 1 \text{ for high sidelobes} \\ 1.25 \text{ for low sidelobes} \end{cases}$$

and the signal-to-noise ratio S/N after pulse compression and integration in azimuth is

$$S/N = \frac{P_R T_i f_r r_c \eta_i}{kT_R F_S B_S} \tag{5.26a}$$

where f_r is the pulse repetition frequency, r_c is the pulse compression gain, and η_i is the integration efficiency.

The integration efficiency can approximately be set to

$$\eta_i \approx \begin{cases} 1 \text{ for coherent integration} \\ \sqrt[3]{1/(T_i f_r)} \text{ for noncoherent integration} \end{cases} \tag{5.26b}$$

as shown by K. Milne [7].

If $\overline{P}_T$ is the mean transmitted power, the peak power P_T is given by

$$P_T = \frac{\overline{P}_T}{\tau_L f_r} \tag{5.27}$$

where τ_L is the transmitted pulse length (expanded pulse in pulse compression systems).

Hence, the integrated signal-to-noise ratio can be written as

$$S/N = \frac{\overline{P}_T T_i \eta_i \sigma_b \, \lambda^2 \, G_{TS} G_{SR}}{(4\pi)^3 \eta_p (kT_R) F_S \, D_{TS}^2 D_{SR}^2 \, L_{PT} L_{PR} \, L_s} \tag{5.28}$$

The radiometric resolution or gray-level resolution ρ_R of imaging radars is its ability to distinguish between two subsequent gray levels of two different targets. The resolution can be estimated by

$$\rho_R \approx 1 + \left[\frac{(S/N) + 1}{\sqrt{N_1} \cdot (S/N)}\right] \tag{5.29}$$

where N_1 is the number of independent looks. In the case of imaging radars (e.g., SAR), the radar cross section (RCS) is often expressed in terms of the normalized radar cross section σ_0:

$$\sigma_b = \sigma_{0b}(\theta_b) \cdot A_r = \sigma_{0b}(\theta_b) \cdot \delta_a \cdot \delta_r \tag{5.30}$$

where A_r is the radar resolution cell, δ_a is the azimuth resolution, δ_r is the range resolution, $\sigma_{0b}(\theta_b)$ is the normalized bistatic radar cross section, and θ_b is the bistatic or scatter angle (angle between transmitter-to-target and target-to-receiver paths).

The *bistatic* RCS is defined for two different cases. In the first one, the bistatic angle can take any value except 180°. In the other, the bistatic angle is exactly equal or very close to 180°, and is called the *forward-scatter cross section*. As described by Skolnik [10], the range of values of bistatic RCS of the first case will be of the same order of magnitude as the range of *monostatic* RCS. This does not imply that cross sections of a particular target are equal in the monostatic and bistatic cases, but on average the two will vary over comparable values.

In the forward-scatter case where the bistatic scatter angle is equal to 180°, the bistatic cross section can be many times the monostatic cross section. Skolnik [10] states that the forward-scatter cross section of a target with projected area A is

$$\sigma_{bf} = 4\pi A^2 / \lambda^2 \tag{5.31}$$

assuming that the radar wavelength is small compared to the target dimensions. This result was derived from physical optics and indicates that the forward-scatter cross section of, for example, a sphere is 36 dB greater than the backscatter cross section. However, because a forward-scatter signal has no doppler shift independent of any target velocity and the signal arrival time at the receiver is independent of the target location, space applications of foward-scatter systems are very limited.

Parasitic radar systems (called BIPARs) can be analyzed by use of a dedicated and partly simplified set of equations. Starting from a given power density p_d on the earth surface, the received power P_R can be estimated by

$$P_R = \frac{p_d G_{RS} \lambda^2 \sigma_b}{(4\pi)^2 A_R^2 L_{PR} L_s} \tag{5.32}$$

where A_R is the flight altitude.

The antenna gain G_{RS} is a function of the antenna effective area A_e and the radar wavelength. Thus,

$$G_{RS} \approx \frac{4\pi A_e}{\lambda^2} \tag{5.33}$$

The bistatic RCS ρ_b can be expressed by

$$\rho_b = \delta_A \cdot \delta_E \cdot \sigma_{0b} \tag{5.34}$$

and

$$\delta_A \cdot \delta_E \approx \frac{A_R^2 \lambda^2}{A_e} \tag{5.35}$$

The resulting equation for the received power of a BIPAR system indicates that this power is independent of the BIPAR pencil-beam antenna dimensions and the aircraft flight altitude. Hence,

$$P_R \approx \frac{1}{4\pi} p_d \lambda^2 \sigma_{0b} \left(\frac{1}{L_{PR} L_s}\right) \tag{5.36a}$$

This is due to the fact that changes in flight altitude and antenna beamwidth, implying loss and gain variations, are compensated by concomitant resolution cell size variations in the opposite direction.

The system integration gain G_i for a coherent BIPAR system is theoretically equal to the time-bandwidth product. Therefore,

$$G_i = T_i \cdot B_s \cdot \eta_i \tag{5.36b}$$

Considering the receiver noise given in (5.25), the signal-to-noise ratio can be estimated by

$$S/N = \frac{P_R G_i}{k T_R F_s B_s} = \frac{P_R T_i \eta_i}{k T_R F_s} \qquad (5.36c)$$

5.4 EXAMPLES OF BISTATIC RADARS

Despite the complexity of the design of spaceborne bistatic radar systems, they are increasingly gaining interest from the international space-based radar community. Intensive investigations and innovative research in this promising area might open up another dimension for spaceborne radar systems in future. The following examples provide an overview of the general capabilities of spaceborne bistatic radars.

5.4.1 Bistatic Parasitic Radar (BIPAR)

The transmitter of opportunity should be a geostationary communication satellite illuminating the earth's surface with a power density of -120 dBW/m^2 or a direct broadcasting (TV) satellite with about 10 times more output power. The transmitting frequency is assumed to be about 11 GHz with a bandwidth of 27 MHz.

The airborne BIPAR uses an electronically scanning antenna of 1 m × 1 m effective area with a scanning angle of ±30°. The BIPAR is flown on an aircraft at 1000 m altitude with a velocity of about 100 km/hr. The scan repetition rate is 1 Hz. It is a step scan with 40 positions within ±30°. At each beam position, three independent looks are performed, resulting in an integration time per look of about 8.3 ms. This ideally leads to a *coherent* integration gain of about 53.5 dB. The doppler spread within one resolution cell of about 30 Hz, misalignments within the time delay and doppler shift compensations, and other hardware deficiencies slightly degrade the integration gain. This must be considered within the overall system loss margin. In the case of a system with *noncoherent* integration (not considered in this example), Milne's equation (5.26b) is applicable. Values for the scatter coefficients may vary between -10 dB and $+6$ dB (1σ). These are typical values for vegetation and man-made targets.

Assuming a clear-air propagation loss of 0.5 dB, a receiver noise figure of 6 dB, and a ($k \cdot T_R$) of -204 dB/Ws and applying (5.36c), Table 5.1 shows that a signal-to-noise ratio between 0 dB and 19 dB can be obtained. An overall margin of 4.5 dB for system losses was included in this calculation.

Table 5.1
Link Calculation for a BIPAR System

+ Power density (e.g., ECS)	−120	dBW/m²
+ $(1/4\pi)$	−11	dB
+ λ^2	−31.5	dBm²
+ σ_{0b}	−10	dB – 6 dB
+ Integration gain	53.5	dB
+ Propagation loss	−0.5	dB
+ $(1/kT_R)$	204	dBWs
− Receiver noise figure	−6	dB
− System bandwidth	−74	dB/Hz
− System loss margin	−4.5	dB
Signal-to-noise ratio	0	dB–19 dB

If a direct broadcasting satellite is available as the transmitter of opportunity, the link budget will typically be improved by about 10 dB. The overall performance of this BIPAR system is summarized in Table 5.2.

Table 5.2
Performance of a BIPAR System

Spatial resolution		
nominal	27	m × 27 m
worst-case at swath edges	31	m × 41 m
Swath width	1.2	km
Signal-to-noise ratio per look	0 dB–19	dB
for a σ_{0b} of −10 dB to +6 dB		
Radiometric resolution for three looks	2 dB–3	dB
Scan repetition rate	1	Hz
Integration time per look	8	ms
Number of pixels in elevation	40	

This example shows that a relatively simple version of a bistatic radar system, comprising an airborne real aperture radar receiver picking up the ground reflections of geostationary communication satellite downlinks, is capable of achieving a satisfactory system performance.

There are several system options under discussion, including, for example, a mechanically steered or even fixed-pointed antenna. The latter antenna results in a simple system with one pixel in elevation, where the swath width is provided by several parallel flights and the signal-to-noise ratio is improved by 16 dB because of a longer integration time per look.

A further improvement could be provided by leasing one of the communication satellite transponders during BIPAR operation and the transmission of a suitably modulated signal. This obviates the need for the reference signal analysis in the BIPAR receiver because the structure of the received signal is known, and hence the signal can be detected with minor losses.

5.4.2 Bistatic Synthetic Aperture Radar (BISAR)

As another example we present a *bistatic synthetic aperture radar* (BISAR) with a TX/RX-SAT and an RX-SAT in co-orbits (see Figure 5.16). In this case, the TX/RX-SAT is a complete monostatic SAR and provides the transmission function for the bistatic system. The RX-SAT receives the radar data via the echo path and the direct path. The direct path is used for synchronization. Either satellite receives radar echoes, then detects, formats, and sends them down to the ground station on earth. The combination of monostatic and bistatic radar functions provides a higher target identification probability and better radar stereographic imaging capability.

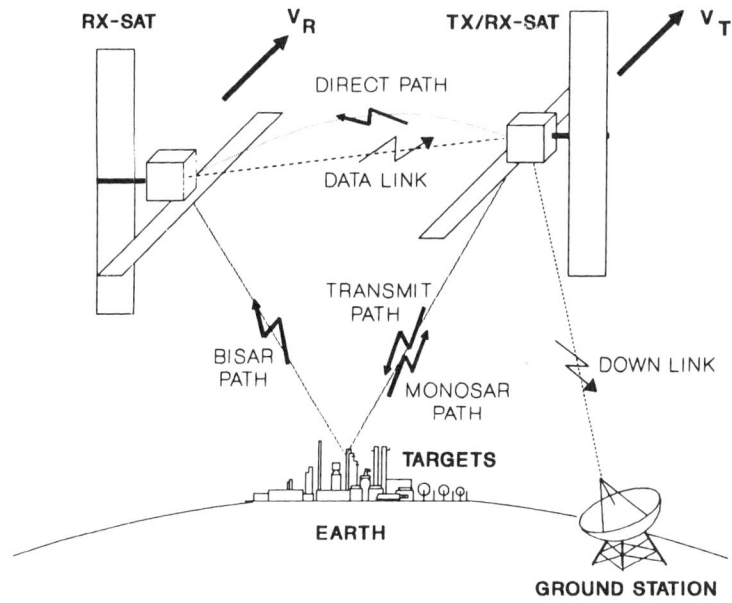

Figure 5.16 BISAR with TX-SAT and RX-SAT in co-orbits.

Let us assume an orbital configuration as previously defined in Figure 5.3 (crossing-orbits system with different ascending nodes) and shown again in detail in Figure 5.16. The system performance of a typical C-band BISAR at 600 km altitude above the equator is given in Tables 5.3 and 5.4, which are based on (5.28) and (5.29).

Table 5.3
BISAR System Parameters (Example)

Spatial resolution	30 m × 30 m
Distance between TX-SAT and RX-SAT (cross-track direction)	440 km
Slant range	640 km
Orbit altitude	600 km
Across-track side-looking operation (β_T and $-\beta_R$)	90°
Off-nadir angle of antenna beams (β_{SR} and $-\beta_{ST}$)	20°
Speed of satellites	6.5 km/s
SAR antenna dimensions	10 m × 1 m
Radar frequency	5.3 GHz
RF bandwidth	18 MHz
Doppler resolution	5.4 Hz
Integration time per look	0.2 s
Total integration time	0.8 s

Table 5.4
BISAR Link Budget (Example)

	Value	Decibel
+ Mean RF power	300 W	25 dBW
+ Integration time	0.2 s	−7 dBs
+ Minimum RCS	15 m²	12 dBm²
+ λ^2	(0.056 m)²	−25 dBm²
+ Transmitting antenna gain		40 dB
+ Receiving antenna gain		40 dB
− $(4\pi)^3$	1984	−33 dB
− (kT_R)	$(4 \cdot 10^{-21})$	+204 dB/WS
− System noise figure		−6 dB
− (Transmitting path length)²	(640 km)²	−116 dB/m²
− (Receiving path length)²	(640 km)²	−116 dB/m²
− Overall system losses		−8 dB
Signal-to-noise ratio per look		+10 dB
Radiometric resolution (four looks)		1.9 dB

The layout of the system timing and antenna parameter optimization must account for ambiguities in the time and frequency domains. For overall performance estimation, the curvature of the satellite flight paths during radar integration time can be neglected in most applications.

The term *overall system losses* covers the propagation losses on transmitting and receiving paths, the system hardware losses, and the integration efficiency.

All of these values hold for both radar channels of this system, monostatic and bistatic.

5.4.3 BISAR with a Geostationary Transmitter

As indicated in Section 5.2.2, a bistatic radar with one powerful transmitter in a geostationary orbit at an altitude of 36,000 km is an attractive system.

This transmitter illuminates the earth's surface below the receiver satellites in low earth orbits with a large, multibeam, phased array antenna and acts as a data-relay station for the radar data collected by the receiver satellites. The RX-SATs synchronize their timing by use of the direct path signal from the TX-SAT. The RX-SATs receive the echo data as long as the TX-SATs phased array antenna beam is steered to their antenna footprints. RX-SATs detect and format these data, and send them to the ground receiving station via the TX-SAT's data-relay channels (see Figure 5.17). Typical system parameters are given in Table 5.5. The resulting radiometric performance is shown in Table 5.6, using (5.28) and (5.29). The overall system losses again cover propagation losses, hardware deficiencies, and integration efficiency.

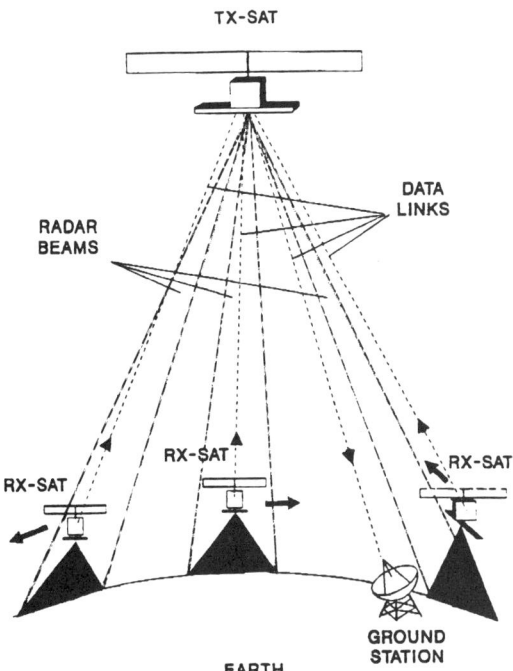

Figure 5.17 A BISAR system with one geostationary transmitter and receivers in low earth orbits.

Table 5.5
BISAR System Parameters (Geostationary Transmitter)

Spatial resolution	30 m × 30	m
TX-SAT orbit altitude (GEO)	36,000	km
RX-SAT orbit altitude (LEO)	600	km
Off-nadir angle of antenna beams	≈ 20	°
RX-SAT speed	≈ 6.5	km/s
Doppler resolution	1	Hz
Doppler bandwidth	250	Hz per look
Number of looks	3	
Integration time per look	1	s
Total integration time	3	s
TX-SAT antenna dimension	30 m × 30	m
RX-SAT antenna dimension	10 m × 2	m
Radar frequency	1	GHz
RF bandwidth	18	MHz

Table 5.6
BISAR Link Budget (Geostationary Transmitter)

	Value	Decibel	
+ Mean RF power	1 kW/beam	30	dBw
+ Integration time per look	1 s	0	dBs
+ Minimum RCS	≈ 90 m^2	20	dBm^2
+ λ^2	(0.3 m)^2	−10	dBm^2
+ Transmitting antenna gain		50	dB
+ Receiving antenna gain		34	dB
− $(4\pi)^3$		−33	dB
− (kT_0)		+204	dB/WS
− System noise figure		−4	dB
− (Transmitting path length)^2	(36,000 km)^2	−151	dB/m^2
− (Receiving path length)^2	(640 km)^2	−116	dB/m^2
− Overall system losses		−8	dB
Signal-to-noise ratio		+16	dB
Radiometric resolution (three looks)		2	dB

5.5 SUMMARY

Bistatic radar in space is a new and extremely broad area of investigation. A detailed discussion of all possible concepts would completely fill this book. Hence, this chapter concentrated on those aspects which might be important for future space-based radar developments. Monostatic and bistatic radars in co-orbits can be effectively used to measure the position of targets on earth in three dimensions (topographic measurements). A surveillance system with wide coverage and short time between subsequent observations can be realized by using one geosynchronous transmitter and many low-cost receivers in low earth orbits. Finally, a new system (parasitic radar) is discussed, where an airborne receiver uses existing downlinks of telecommunication satellites as transmitters of opportunity.

In general, the intention of this chapter is to provide the reader with a basic understanding of bistatic radar in space, its advantages, and its drawbacks. We have demonstrated that bistatic systems are viable candidates for future space-based radars and further investigation is worthwhile.

REFERENCES

1. Heilmeier, G., "Radar: Bistatic Arrays in Space?" *Microwave System News,* January 1978, pp. 39–40.
2. Pell, C., "Multi-Static Radar for Long Range Air Defense," *Microwave Journal,* Vol. 29, January 1986, pp. 171–181.
3. Montana, D.M., and R.S. Herd, US Patent, No. 4.499.468 "Range-Only Multistatic Radar System", February 12, 1985.
4. Lorti, D.C., and J.J. Bowman, "Will Tactical Aircraft Use Bistatic Radar?" *Microwave System News,* September 1978, pp. 49–54.
5. Zhou Zheng-Ou, Ding Yi-Yuan, Gong Yao-Huan, and Huang Shum-Li, "A Bistatic Radar for Geological Probing", *Microwave Journal,* Vol. 27, May 1984, pp. 257–263.
6. Ewing, E.F., "The Applicability of Bistatic Radar to Short Range Surveillance," *Radar-77,* IEE Conf. Pub. No. 55, 1977, pp. 53–58.
7. K. Milne, "Principles and Concepts of Multistatic Surveillance Radars," *Radar-77,* IEE Conf. Pub. No. 55, 1977, pp. 46–52.
8. Buchner, M.R., "A Multistatic Track Filter with Optimal Measurement Selection," *Radar-77,* IEE Conf. Pub. No. 55, 1977, pp. 72–75.
9. Pell, C., "Some System Aspects of Multistatic Radar for Long Range Air Defence," *Proc. Military Microwave Conf.,* 1984, London Microwave Exhibitions and Publishers, pp. 85–90.
10. Skolnik, M.I., *Introduction to Radar Systems,* 2nd Ed., McGraw-Hill, 1980, Chapter 14.6 "Bistatic Radar," pp. 553–560.
11. Hulbert, A.P., et al., *IEE Proc. Communications, Radar, and Signal Processing,* "Special Issue on Bistatic and Multistatic Radar," Vol. 133, No. 7, December 1986.
12. Yasaka, T., T. Hatsuda, "Geostationary Tether Satellite System and its Application to Communication Systems," *IEEE Trans. Aerospace and Electronic Systems,* Vol. 24, No. 1, January 1988, pp. 68–75.

Chapter 6
RENDEZVOUS RADAR
J.W. Locke and L.J. Cantafio
Motorola   *TRW*

The Gemini and Apollo programs each used a radar subsystem to assist in rendezvous and docking operations. The parameters and functions of the radars used in these programs have been described by Fenner and Broderick [1]. In this chapter, we discuss the rendezvous mission and describe the radar subsystem used by the Space Shuttle (Space Transportation System, STS). Then, we discuss the planned missions of the future Orbital Maneuvering Vehicle (OMV) and the radar being developed for it. We also cover the potential for interference from earth clutter and solar system noise sources.

6.1 RENDEZVOUS RADAR MISSIONS

The Gemini and Apollo programs demonstrated the first operational experience with the rendezvous maneuver. The successful performance of the rendezvous radars in these programs effectively verified that many possible missions may be performed in space. The Ku-band Integrated Radar and Communications Subsystem (IRACS), designed for the Space Shuttle, demonstrated the rendezvous, satellite retrieval, and station-keeping missions. The maiden voyage for this radar was aboard Challenger flight STS-7 on June 22, 1983. During the STS-11 flight in February 1984, the Ku-band radar assisted in checking the Man Maneuvering Unit

(MMU) operations. The radar acquired and tracked mission specialist Robert Stewart in the MMU during his 300-ft sojourn into space. The radar measured the cross section (RCS) of the MMU, which varied between 2.5 dBsm and 7.5 dBsm with acquisition at a range of 100 ft. and track out to the maximum range of 308 ft. Average velocity measured during the mission was 0.7 ft/s; peak velocity was 1.2 ft/s.

The rendezvous radar provides the tracking function for a guidance system. The rendezvous phase of the mission begins after the radar acquires the target satellite. Thereafter, the tracking function provides data on range, range rate, and the two components of the line-of-sight (LOS) inertial rate. A digital guidance computer calculates relative velocity perpendicular to the LOS by using range and angle-rate data. The closing component of velocity is obtained from the doppler frequency, or by differentiation of radar range measurements. A simplified block diagram of a typical rendezvous guidance subsystem is shown in Figure 6.1. The radar search and acquisition mode is initiated by the guidance computer. A relatively large solid angle is searched periodically until the target is acquired in range and angle. To maximize the probability of detection or acquisition, the kinematics are arranged such that a long search time is available before the target escapes from the search sector. When detection is accomplished, the search mode is stopped and the tracking mode is initiated by locking a tracking gate onto the target return. Thereafter, monopulse angle tracking of the antenna is initiated about an axis always directed toward the target. The tracking phase ends when rendezvous has been achieved within certain desired terminal accuracies on relative position and velocity. Requirements for the STS rendezvous radar are typical, and these are given in Table 6.1 [2].

During and immediately following acquisition, the relative velocity vector generally lies in the direction of the instantaneous LOS; however, a substantial error may exist, which is equivalent to a relative velocity component perpendicular to the LOS. The range at acquisition and the magnitude of the closing velocity are such that the duration of the rendezvous phase can be several minutes. A reasonably long period is essential to an accurate rendezvous because sufficient time must be allowed for smoothing of inherently noisy radar tracking data, and for correction of the measured errors. A period of as much as 10 to 20 minutes is short compared to the overall mission duration. The effect of the differential earth gravity field has been shown by Hord [3] to be negligible for tracking phase durations not exceeding 10 to 20 minutes. Furthermore, Wolverton [4] has shown that the orbital motion aspects of the rendezvous maneuver can be neglected when the rendezvous time t_R is small compared with the satellite orbital period T_o divided by 2π.

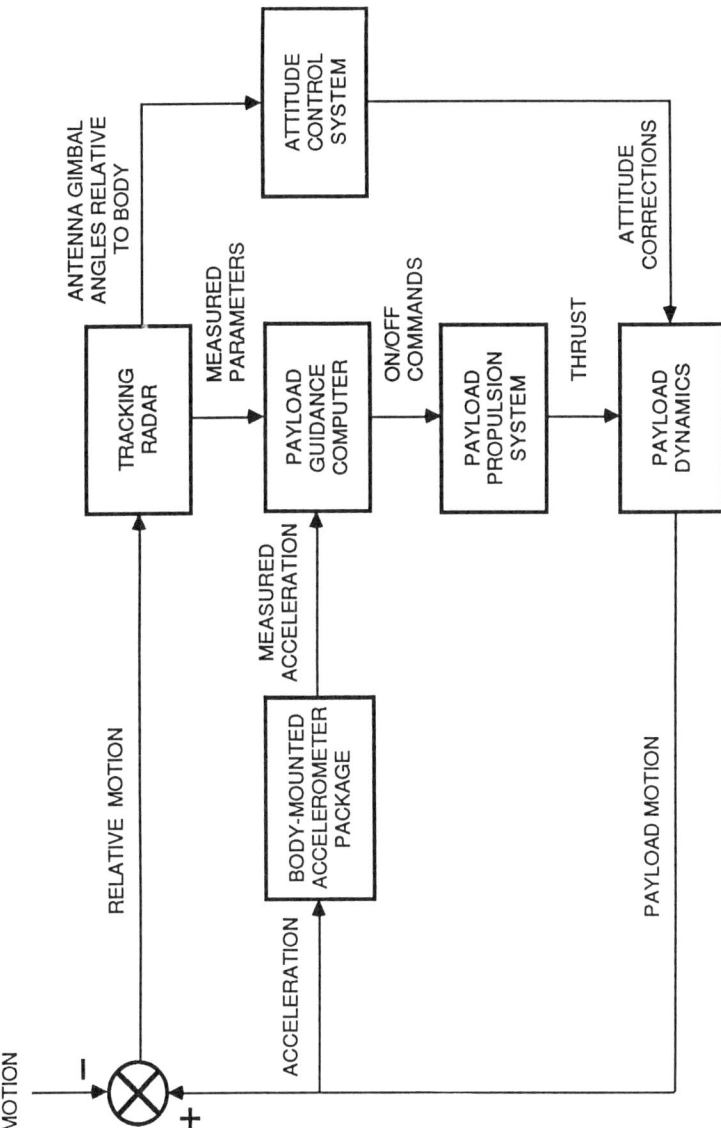

Figure 6.1 Rendezvous guidance subsystem simplified block diagram [4].

Table 6.1
STS Rendezvous Radar Requirements

Search	± 30° spiral scan
Acquisition	12 nmi on 0 dBsm, SW-I;
Track Errors	300 nmi on +14 dBm transponder
Range	±1 percent
Range Rate	1 ft/s or 1%
Angle	8 mr
Angle Rate	0.14 mr/s or 5%

6.2 SPACE SHUTTLE RENDEZVOUS RADAR

The Integrated Radar and Communications Subsystem was developed by Hughes Aircraft Company for use on the Space Transportation System orbiter vehicle. The IRACS performs both radar and communication functions for the STS. For the radar function, IRACS is a coherent, range-gated, pulsed doppler, frequency-hopping radar, which searches for, acquires, and tracks other orbiting objects and provides the spatial measurement data needed to perform rapid and efficient rendezvous with those objects. For the communication function, IRACS searches for, acquires, and tracks the Tracking and Data Relay Satellite System (TDRSS) to provide two-way communication between the shuttle and ground tracking stations.

The Ku-band IRACS operates at radio frequencies between 13.75 and 15.15 GHz, with radar operation between 13.75 and 14.0 GHz. There are two basic radar modes, *a passive-target mode* in which the target is noncooperative (in that no cross-sectional augmentation is present), and an *active-target mode* in which the target has an on-board transponder. The radar operates to a range of 12 nmi with a 1 m^2 noncooperative target in the passive-target mode, and out to 300 nmi with a +14 dBm transponder in the active mode. Submodes include an automatic search, angle and range track capability, and an external angle-control operation. Under external angle control, the antenna is positioned either by external slew commands, or by reference to inertial space or the shuttle axes. Angle, angle rate, range, and range rate measurements during automatic operation are made by the radar after track is initiated. Under external angle control, only range and range rate are measured. Table 6.2 lists the major performance and physical characteristics of the radar portion of the IRACS.

The IRACS hardware is subdivided into five major assemblies and subassemblies, which are contained in four *line replaceable units* (LRUs). The *deployed assembly* (one LRU), shown in in Figure 6.2, is mounted on the starboard longeron of the shuttle, next to the payload bay, and is extended for operation through the open payload bay doors. The deployed assembly has two major subassemblies: the *deployed mechanism assembly* (DMA) and the *deployed electronics assembly*

Table 6.2
IRACS Radar Characteristics

Detection Performance	
P_d	99%
Target	1m^2; Swerling I
Range	12 nmi
False-Alarm Rate	1/hr
Search Scan	± 30° cone
Track Performance	
Angle Accuracy (3 σ)	8 mrad
Angle Rate (3 σ)	0.14 mrad/s
Range Accuracy (3 σ)	80 ft, $R < 1.3$ nmi;
	1% of R, $1.3 < R < 4.9$ nmi;
	300 ft, $4.9 < R < 12$ nmi.
Range Rate Accuracy (3 σ)	1 ft/s, $R < 10$ nmi
System Parameters	
RF	13.75–14.02 GHz
PRF	0.3, 3, 7 kHz
Pulsewidth	0.122, 2.075, 4.15, 8.3,
	16.6, 33.2, 66.4 μs
System Noise Temperature	≈1585 K
Antenna	
Type	Parabola; prime-focus feed
Diameter	36 in
Depth	12.5 in, overall
Gain	38.4 dB at 13.8 GHz
Beamidth	1.68°
Polarization	Linear
Transmitter	
Type	TWT
Peak Power	50 W
Gain	≈44 dB
Receiver	
Type	Single-channel monopulse
Noise Figure	< 5 dB; GaAs FET LNA
System (Physical)	
Deployed Assemblies Weight	135 lbs
Radar Processor Weight	31 lbs
Electronics Volume	≈3 ft^3
Prime Power	460 W

(DEA). The other three LRUs, the *electronics assembly 1* (EA-1), the *electronics assembly 2* (EA-2), and the *signal processing assembly* (SPA), are located in the avionics bay, in the shuttle cabin.

The DMA includes the main antenna reflector, its feed, auxiliary horn antenna, and antenna gimbals, drive motors, rate gyros, digital shaft encoders, rotary joints and connecting waveguides, wires and cables. The DEA contains the frequency synthesizer, receiver, first downconverter and transmitter, all used for both radar and communication, as well as the various switches used to select the correct signal paths for the operation desired. Separate radar and communication upconverters and the radar's second down-converter are also located in the DEA.

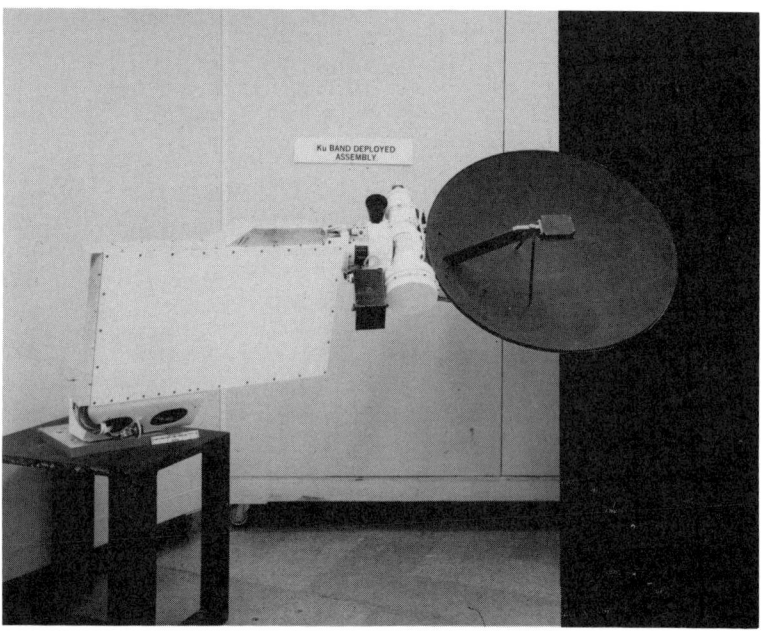

Figure 6.2 IRACS deployed assembly. (Courtesy of Hughes Aircraft Company.)

The inboard EA-2 unit contains the radar video, analog-to-digital (A/D) conversion, and the digital signal processing and control electronics. The EA-1 and SPA units are primarily devoted to communication electronics, although the angle-control electronics for both radar and communication and the system reference oscillator are contained in the EA-1 unit.

Figure 6.3 shows a functional block diagram of the IRACS. The antenna is a center-fed parabola. The five-element monopulse feed provides a sum (Σ) and two orthogonal difference (Δ) outputs. The difference outputs are time-multiplexed into a single receiver difference channel for the angle tracking operation. In the search mode, an auxilliary horn antenna is monitored by the receiver difference

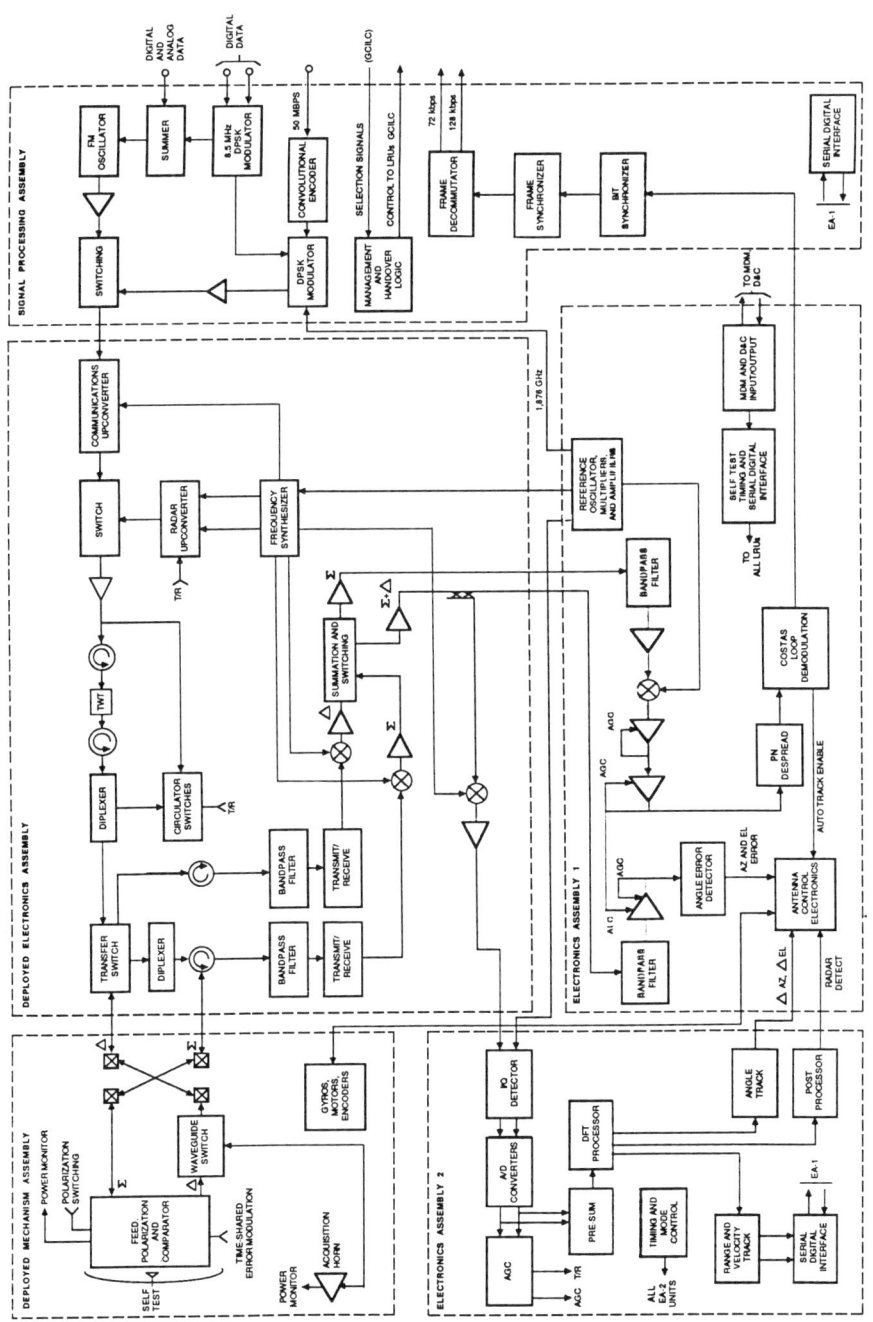

Figure 6.3 IRACS functional block diagram.

channel. The auxiliary horn signal amplitude is compared to the main antenna sum-channel signal amplitude to prevent acquisition of large targets in the sidelobes of the main antenna. The auxiliary antenna has a peak gain which is about 20 dB less than that of the main antenna. Not shown on the diagram are low-noise RF preamplifiers in the sum and difference channels. After amplification at IF, the sum and difference channels are combined into a single receiving channel for routing to the inboard electronics assemblies for further processing. The transmitter employs a *traveling wave tube* (TWT) to amplify the coherent synthesizer output to 50 W of peak power. For short-range operation (down to 100 ft), the TWT is bypassed to reduce the power on the target. Five radio frequencies are used in the radar mode to decorrelate Swerling I (slowly fluctuating) target returns, thus improving detection during search and smoothing parameter measurements for target tracking. A 16-point digital discrete Fourier transform (DFT) processor is employed for coherently integrating multiple pulse returns and to provide the highly accurate measurements of target relative velocity needed for the rendezvous mission.

During the IRACS maiden voyage into space, in June 1983, the automatic acquistion and tracking functions of the radar part of the system were successfully exercised using a mission payload satellite (SPAS 01) as a target. Operation at range from 100 to 1000 ft was demonstrated.

Long-range operation was demonstrated aboard Challenger during the STS-11 mission in February 1984. Fragments of a damaged two-meter in diameter balloon target satellite, which was released from Challenger, were tracked to a range of 114,000 ft (18.8 nmi). Also during this mission, astronaut Robert Stewart was tracked during his test of the man maneuvering unit. The IRACS radar was used on Challenger during the STS-13 mission in April 1984, where the radar aided the capture of the crippled Solar Maximum satellite. The satellite was acquired by the radar at a range of 110,000 ft (18.1 nmi) and was tracked to 88 ft. The RCS of Solar Maximum was measured as varying from 0.9 to 130 m^2 while tumbling at 1/6 rpm.[1]

Figures 6.4 through 6.8 show the predicted measurement accuracies of the IRACS as of the preliminary design review in March 1978. Figures 6.4 and 6.5 show the angle-measurement errors at the EA-2 unit outputs for the specified passive and active targets. Note that these are 1 σ errors. Figures 6.6 and 6.7 show the range-measurement errors for the specified passive and active targets. Figure 6.8 shows the doppler-derived range-rate measurement errors for the passive target. Note that this error approaches 0.1 ft/s at ranges of less than one nautical mile. Doppler-derived range rate is obtained by interpolating between adjacent *doppler filter* (DF) outputs.

6.3 FUTURE RENDEZVOUS RADAR MISSIONS

All satellite rendezvous missions have been performed by manned vehicles. In the

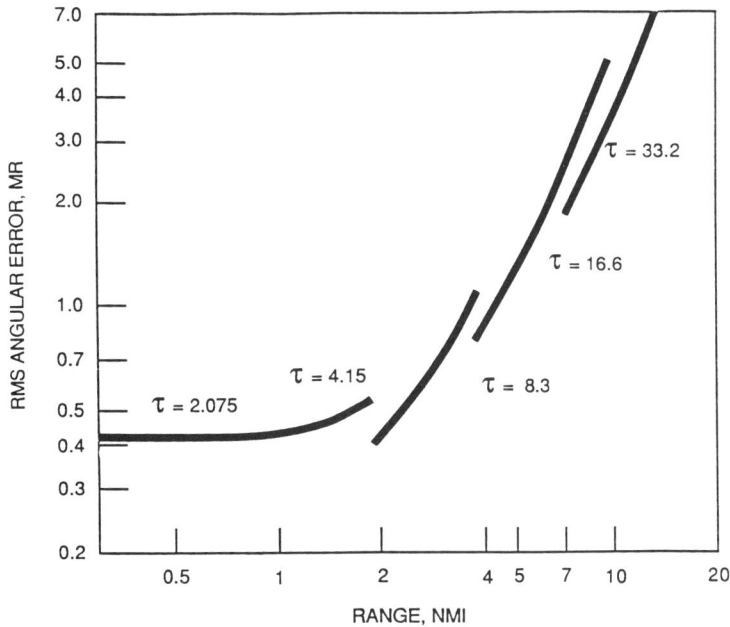

Figure 6.4 IRACS EA-2 Angle-measurement error—passive target (1σ).

foreseeable future, the majority of rendezvous missions will be conducted by unmanned vehicles such as the OMV, shown in Figure 6.9, under development for NASA by TRW. The OMV will supplement the STS capability for satellite payload delivery, retrieval, and maneuvering. The list of reference missions for the OMV design assumes an STS or space station (SS) based OMV, and includes:

1. Large observatory servicing at the shuttle;
2. Payload placement;
3. Payload retrieval;
4. Payload reboost;
5. Payload deboost to re-entry;
6. Payload viewing;
7. Subsatellite mission;
8. Multiple payload mission;
9. *In situ* servicing mission;
10. STS-to-SS and SS-to-STS module transfer;
11. Base support.

Details of these missions may be found in the OMV's RFP by NASA [5]. The design of the OMV will be modular so as to permit the upgrading of its capability

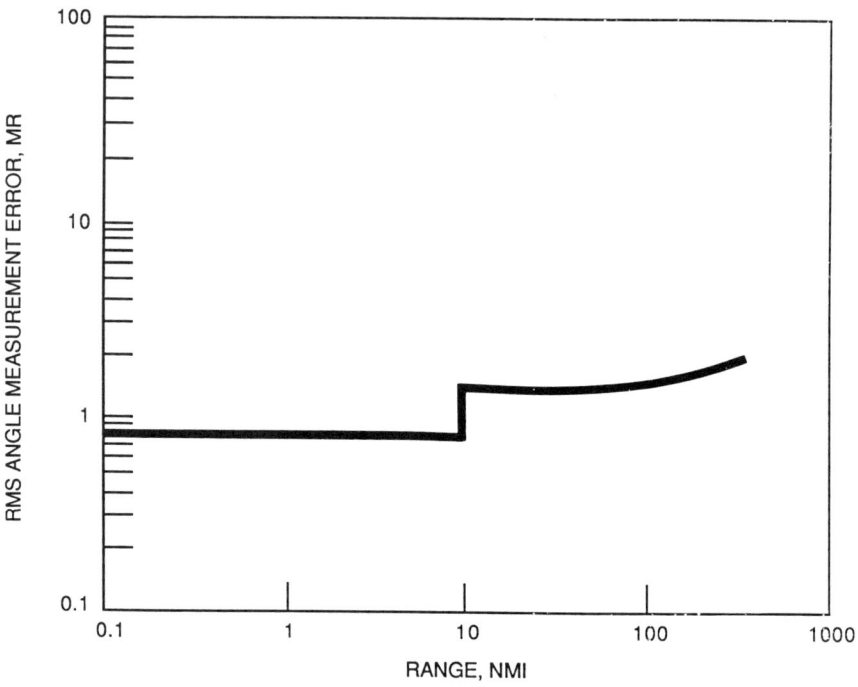

Figure 6.5 IRACS EA-2 angle-measurement error—active target (1σ).

to accommodate the following growth missions by the addition of appropriate kits or elements to the system:

1. Extended on-orbit operation;
2. Upper-stage mission;
3. Geostationary (GEO) operations;
4. Satellite refueling;
5. GEO servicing mission.

The 9000-pound, 5 ft × 15 ft diameter OMV will have limited electrical power-generating capability and will be a relatively low-cost unmanned vehicle. The rendezvous radar for the OMV will be light in weight, draw little power, and be low in cost.

6.4 OMV RENDEZVOUS RADAR

The Rendezvous Radar Set (RRS), under development by Motorola for TRW, is a coherent, range-gated, pulsed doppler, frequency-hopping radar, which supports

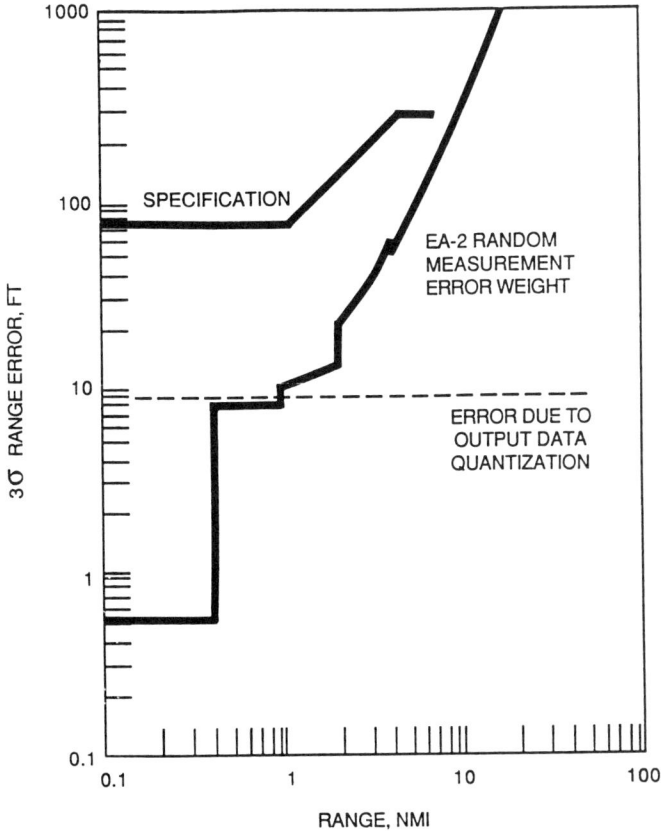

Figure 6.6 IRACS range error random component—passive target (3σ).

the rendezvous function for the OMV.

The X-band RRS operates at RFs between 9.5 and 9.8 GHz. The radar operates to a range of 4.5 nmi with a one-square-meter noncooperative target. The radar is capable of operating both autonomously and with external angle control from the OMV's on-board computer. During autonomous operation, the RRS searches an OMV specified range-angle volume and, upon target detection, acquires the target, tracks it, and makes angle, range, and range-rate measurements. During operation with OMV-designated angle position, the RRS searches, acquires, and tracks in range only, and makes range and range-rate measurements. In both modes, the antenna platform is stabilized in space by use of OMV attitude data, which are passed to the RRS from the OMV. Table 6.3 lists the major performance and physical characteristics of the RRS.

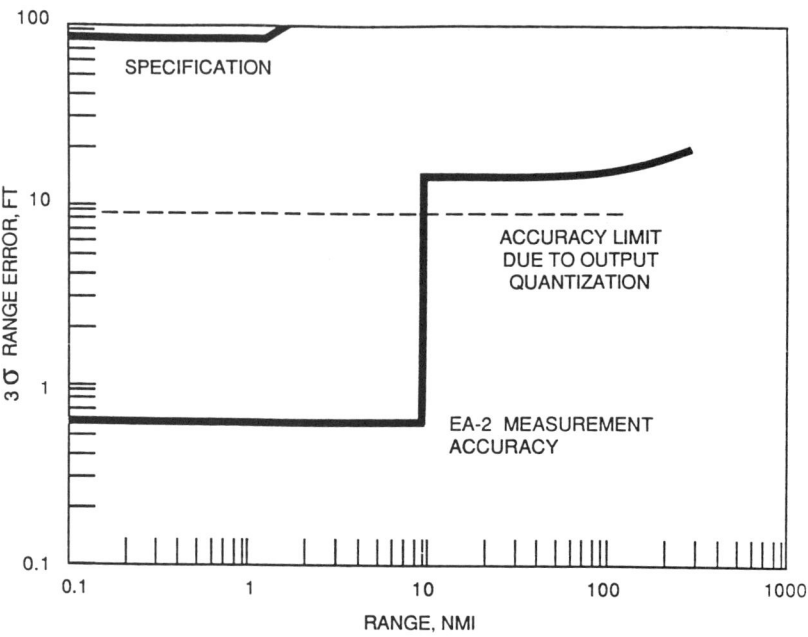

Figure 6.7 IRACS range error random component—active target (3σ).

The RRS has parallel redundant electronics hardware and gimbal motor windings. The hardware is subdivided into two major assemblies containing three LRUs. The *outboard assembly* (OA), which is one LRU, is mounted on a retractable arm which positions the OA about two feet from the OMV body (Figure 6.9). The OA includes the antenna, its antenna gimbals, drive motors, and angle position resolvers, the RF redundancy switches, and redundant *outboard electronics* (OE) modules. The OE modules each contain a GaAs FET transmitter power amplifier and three parallel GaAs FET low-noise receiver preamplifiers.

The *inboard assembly* (IA) includes two identical *inboard electronics* (IE) units, each of which is an LRU. Both IE units interface separately with the OMV computer and power systems. Only one IE/OE pair is energized by the OMV for operation.

Figure 6.10 shows a functional block diagram of the RRS. The antenna is a square planar array of waveguide-fed slot elements. A waveguide monopulse comparator network provides a sum pattern and two orthogonal difference patterns. These three output signals are separately amplified and processed in the receiver. The relative phase of the three receiver channels is monitored through use of a test signal, which is injected into the monopulse comparator. Phase correction is made in the digital postprocessor prior to the formation of angle error signals.

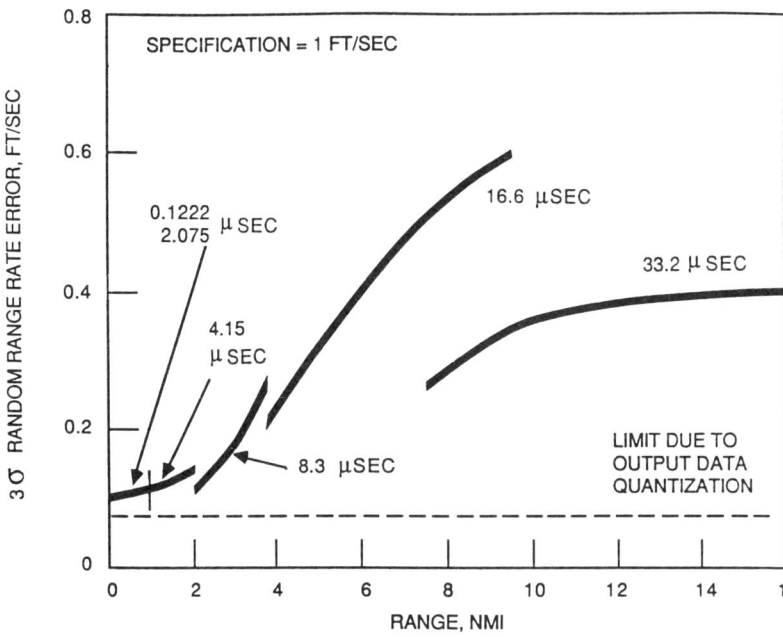

Figure 6.8 IRACS doppler-derived range-rate error random component—passive target (3σ).

The antenna is designed so that the difference pattern sidelobes are greater than those of the sum pattern in the high sidelobe regions. During the search and acquisition operations, the receiver's difference channel signal levels are compared to the sum channel's signal amplitude to prevent acquisition of large targets in the antenna sidelobes.

The location of the transmitter power amplifier and the low-noise receiver preamplifiers at the antenna minimizes the losses in the RF transmitting and receiving paths, thus minimizing the radar's peak power and antenna area needs. The RF paths between the outboard and inboard electronics are through low-loss, phase-stable coaxial cable. The transmitter peak power is 2 W. For short-range operation (down to 35 ft) the peak power is reduced by up to 50 dB to minimize the RF radiation intensity on sensitive target satellites.

Up to 30 RF, in 10 MHz steps, over the 300 MHz operating band, are used to decorrelate Swerling I target returns. At each RF, 128 pulses are integrated coherently in the FFT processor prior to noncoherent integration of the FFT records. The noncoherent integration of up to 30 FFT records and the detection and tracking functions are implemented in microprocessor firmware algorithms.

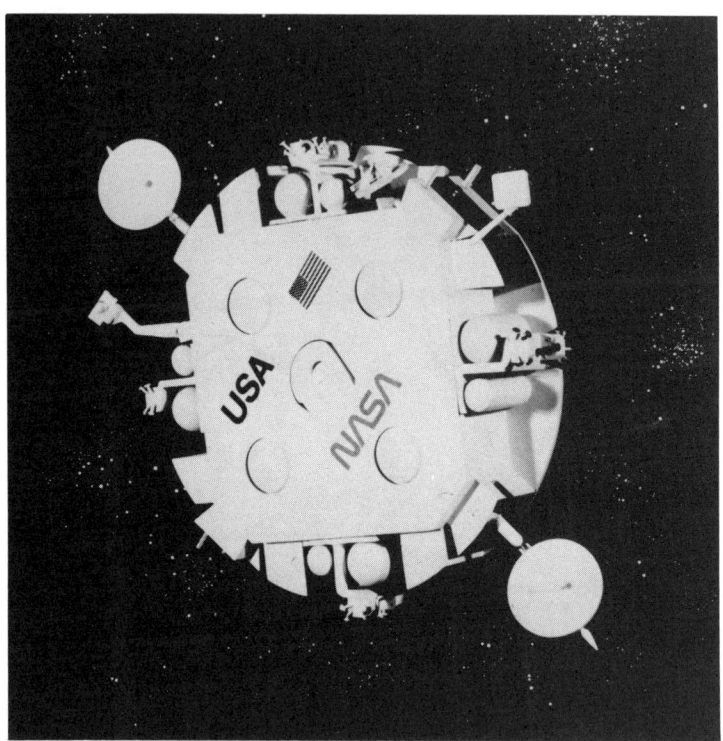

Figure 6.9 Artist's concept of orbital maneuvering vehicle with rendezvous radar outboard assembly extended (upper right). (Courtesy of Motorola and TRW.)

Table 6.3
OMV Radar Characteristics

Detection Performance	
P_d	99%
Target	1 m^2; Swerling I
Range	4.5 nmi
False-Alarm Rate	1/hr
Search Scan	±20° cone
Scan Time	5 min
Track Performance	
Angle Accuracy (3 σ)	20 mr
Range Accuracy (3 σ)	Greater of 20 ft or 2% of range
Range Rate Accuracy (3 σ)	Greater of 0.1 ft/s or 2% of range rate
System Parameters	
RF	9.5–9.8 GHz

Table 6.3 cont'd.

PRF	6.67 kHz
Pulsewidth	0.05, 0.2, 1.5, 15 μs
System Noise Temperature	≈900 K
Antenna	
Type	Planar slotted array
Size	14 in × 15 in
Depth	1 in, overall
Gain	30.5 dB at 9.65 GHz
Beamwidth	5.0°
Polarization	Linear
Transmitter	
Type	GaAs FET
Peak Power	2 W
Gain	≈30 dB
Receiver	
Type	3-channel monopulse
Noise Figure	<4 dB; GaAs FET LNA
System (Physical)	
Deployed Assembly Weight	26 lbs
Inboard Assembly Weight	50 lbs (redundant total)
Electronics Volume	≈2 ft^3 (redundant total)
Prime Power	<60 W

The use of the 128-point FFT provides the very accurate range-rate data required of the RRS. Figures 6.11 through 6.13 show the predicted 3σ measurement errors of the RRS as of the preliminary design review in December 1987.

6.5 EARTH CLUTTER EFFECTS

Certain missions may require geometries that force the rendezvous radar beam to be directed toward the earth while the radar system detects, acquires, and tracks the target of interest. There is then the potential for radar performance degradation caused by radar signal returns from the earth's surface. The operating frequencies of both the STS and OMV radars suffer little attenuation by the earth's atmosphere. The earth clutter return signal is therefore not significantly reduced by atmospheric losses. The earth clutter issue at several radar platform orbital altitudes will be considered in this section.

6.5.1 Radar Parameters

The STS and OMV rendezvous radar systems are both coherent pulsed radars operating at relatively high duty cycles. For clutter considerations, the OMV radar

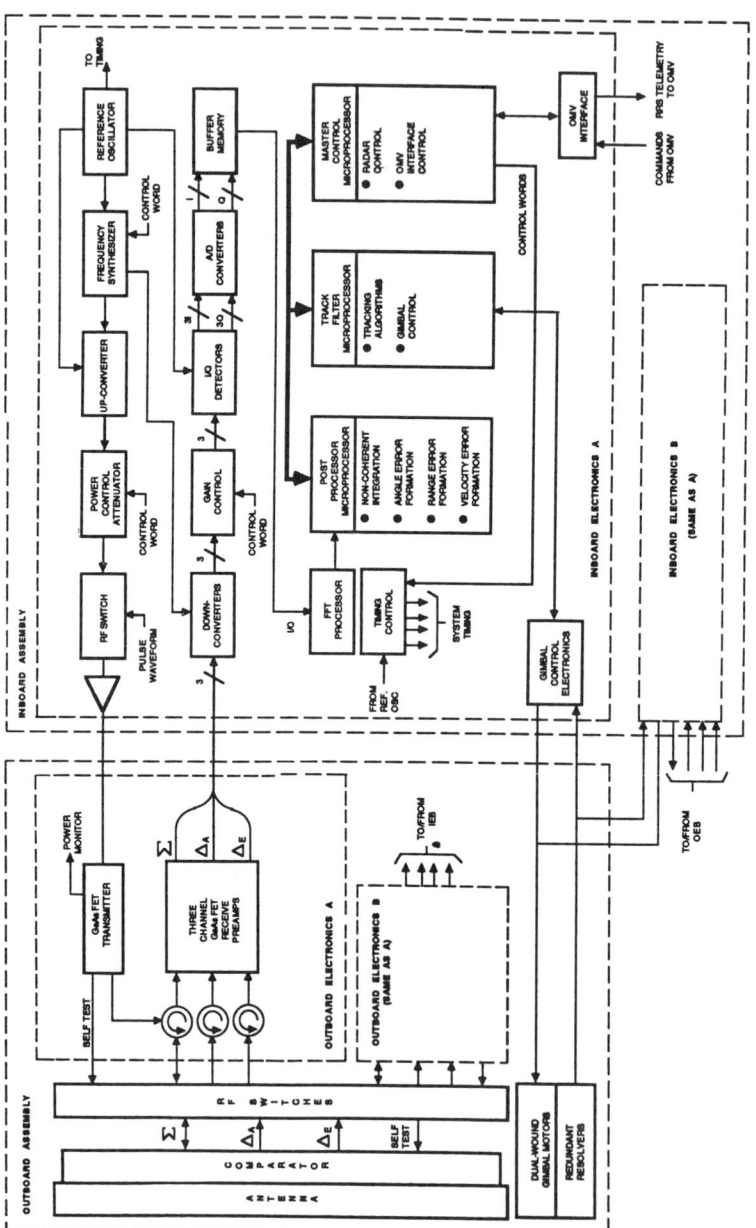

Figure 6.10 RRS functional block diagram.

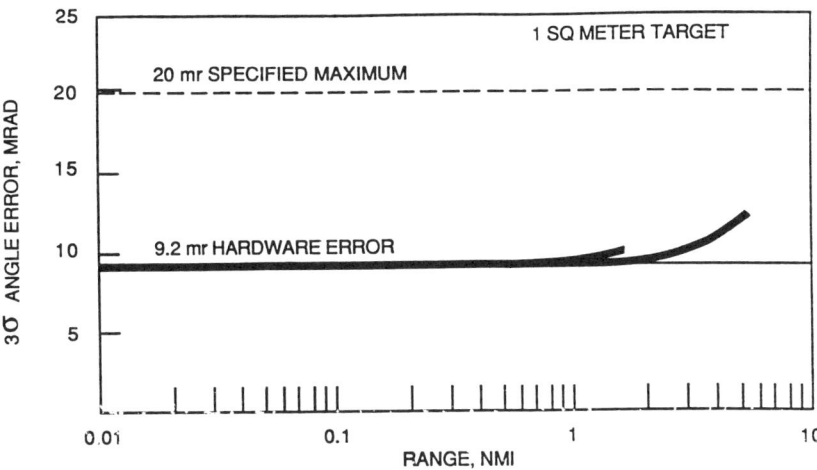

Figure 6.11 RRS angle-measurement error (3σ).

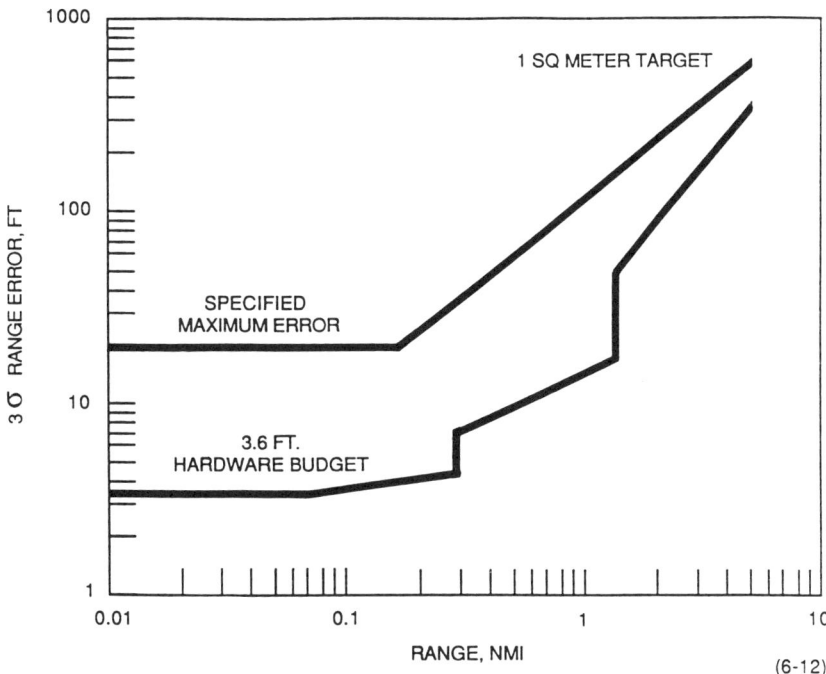

(6-12)

Figure 6.12 RRS range-measurement error (3σ).

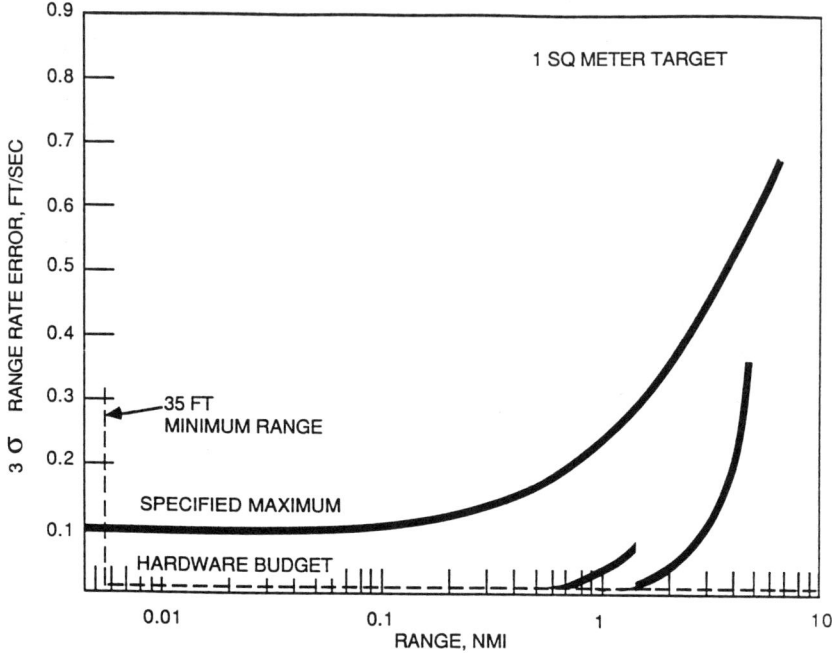

Figure 6.13 RRS doppler-derived range-rate measurement error (3σ).

parameters used are 15 μs pulsewidth, 150 μs PRI, and 5° beamwidth. STS radar parameters used are 66 μs pulsewidth, 335 μs PRI, and 1.7° beamwidth. The satellite target of interest is 4.5 nmi from the OMV radar and 12 nmi from the STS radar. The target RCS is one square meter in both cases.

It is characteristic of pulse radars operating at a constant PRI that returns from targets (or clutter), which are at a range greater than the maximum unambiguous range of the radar waveform, are not easily distinguishable from shorter range returns. For example, the return from a target, or from clutter, at a range of $R = 4.5 + (N \cdot c \cdot \text{PRI})/2$ nmi, where N is a positive integer and c is the velocity of light, would compete with the return from a target at 4.5 nmi.

The cumulative sum of these "Nth-time-around" returns from the earth's surface relative to the return from the target satellite is the subject of our discussion.

6.5.2 Radar-Earth Geometry

An understanding of the radar-earth geometry is useful for our consideration of the earth's clutter effects. Figure 6.14 shows a two-dimensional view of this ge-

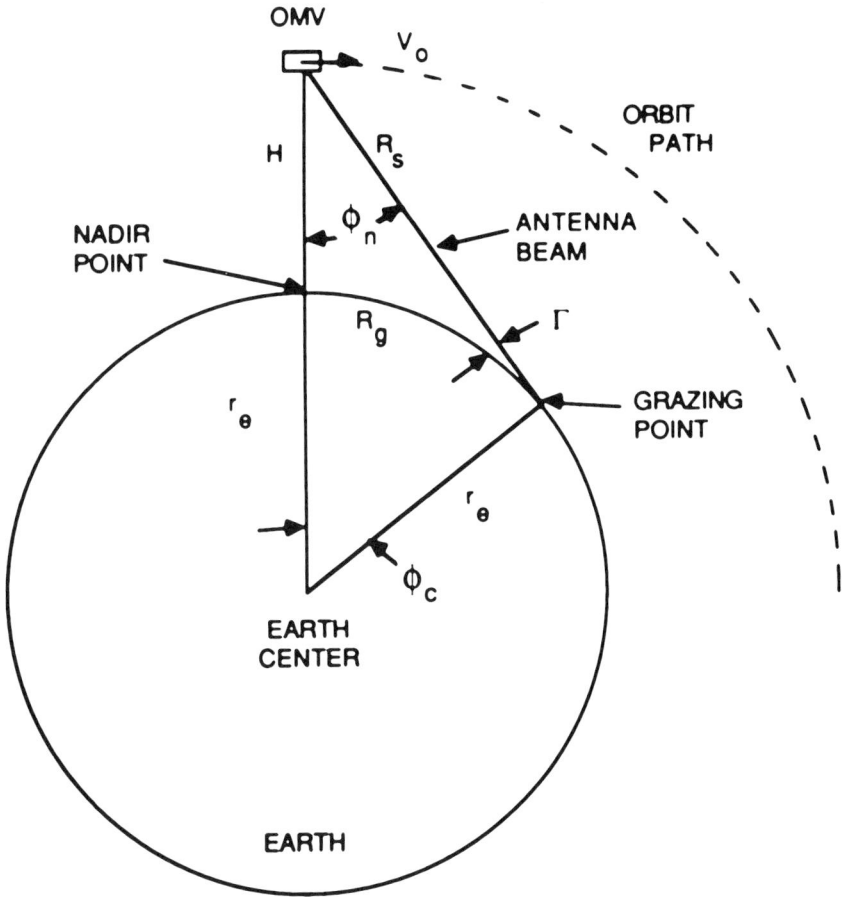

Figure 6.14 Radar-earth geometery.

ometry in the plane of the radar platform's orbit. For our discussion, the platform is in a circular orbit with an orbital altitude H and an orbital velocity V_0. The radar antenna is steered so that the beam center line is in the orbital plane (zero degrees azimuth) and intersects the earth at a distance R_s from the radar with grazing angle Γ. We will call the point on the earth's surface directly below the platform the *nadir point*. We will call the angle between the beam centerline and a line to this point (from the platform) the *nadir angle* ϕ_n. The angle between the nadir point and the grazing point, as viewed from the earth's center, will be called the *core angle* ϕ_c. The surface distance from the nadir point to the grazing point will be called the *ground range* R_g. The earth's radius r_e is 3440 nmi. The platform is R_c

(R_c = H + r_e) nmi from the earth's center. These ranges and angles have the following relationships:

$$\Gamma = \cos^{-1}(\sin\phi_n \cdot R_c/r_e) \tag{6.1}$$

$$\phi_c = 90 - \phi_n - \Gamma \tag{6.2}$$

$$R_s = r_e \cdot \sin\phi_c/\sin\phi_n \tag{6.3}$$

$$R_s = (R_c^2 - 2 \cdot r_e \cdot R_c \cdot \cos\phi_c + r_e^2)^{1/2} \tag{6.4}$$

$$R_g = r_e \cdot \phi_c \tag{6.5}$$

6.5.3 Clutter Area

The area of the earth's surface illuminated by the radar antenna beam is a function of the azimuth and elevation beamwidths, grazing angle, and range to the point of grazing. The grazing angle and range are, in turn, functions of platform altitude and radar beam nadir angle. The portion of the illuminated area of interest is that which will contribute radar returns that fall within a single range cell and degrade the detection of a true target in the cell. The projection of a range cell (in this case, one of the Nth-time-around range cells) on the earth is a function of grazing angle. A range cell of 2250 m length (15 μs pulse) projects a ground surface length of 2250/cosΓ m, as illustrated in Figure 6.15. The earth's area illuminated by the radar beam is called the *beam footprint*. The footprint outline shown in Figure 6.16 is defined by the 3-dB two-way antenna beamwidth. When the grazing angle is near 90°, the clutter area is beamwidth-limited in both dimensions, as shown in Figure 6.17(a), because the range cell projection onto the earth is longer than the elevation extent of the beam footprint. As the grazing angle decreases (i.e., as the beam is scanned out from the nadir point), the clutter area of interest becomes range-cell-limited in the elevation dimension, as shown in Figure 6.17(b). As the beam is scanned farther out, the beam footprint further elongates to include multiple Nth-time-around range cells, as shown in Figure 6.17(c).

In the beamwidth-limited case of Figure 6.17(a), the clutter area is approximately

$$A_c \approx (\pi/4)R_s^2 \cdot \theta \cdot \phi/\sin\Gamma \tag{6.6}$$

where θ is the two-way azimuth beamwidth in radians, ϕ is the elevation (or vertical-plane) two-way beamwidth in radians, and R_s and Γ are as defined earlier. The sin Γ term creates the elliptical shape of the footprint as the grazing angle is decreased.

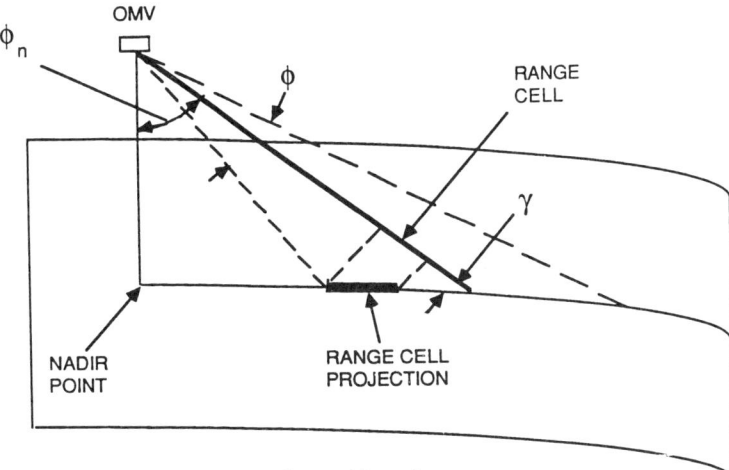

Figure 6.15 Range cell projection onto the earth's surface.

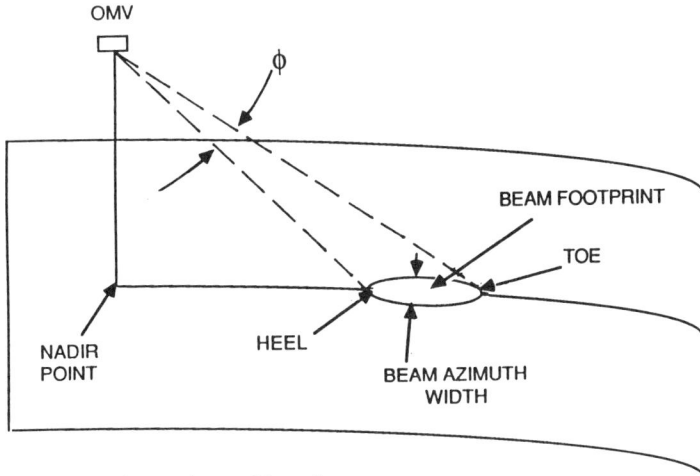

Figure 6.16 Beam footprint on the earth's surface.

In the range-cell-limited case of Figure 6.17(b), the clutter area is approximately

$$A_c = R_s \cdot \theta \cdot R_{\text{cell}}/\cos\Gamma \qquad (6.7)$$

where R_{cell} is the range cell length.

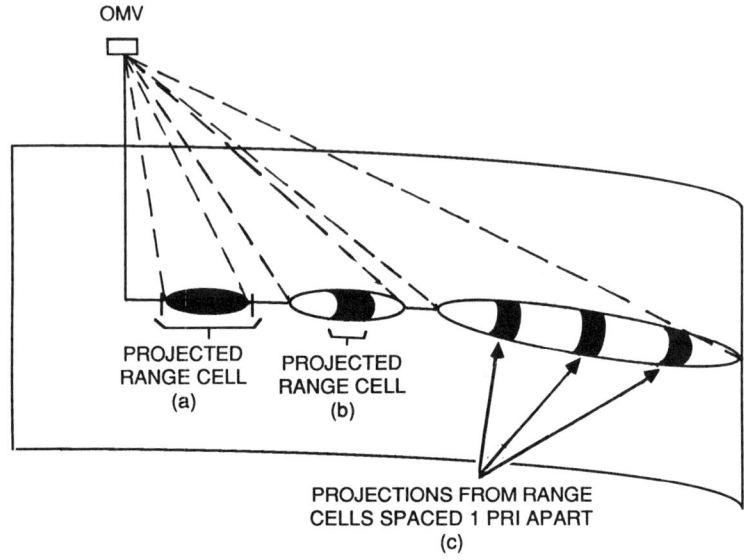

Figure 6.17 Clutter areas: (a) beamwidth-limited, (b) range-cell-limited, and (c) multiple Nth-time-around range cells.

The transition between these two cases occurs when equating (6.6) and (6.7) such that

$$\tan\Gamma = (\pi/4)\, R_s \cdot \phi/R_{\text{cell}} \tag{6.8}$$

The onset of the multiple-range-cell case of Figure 6.17(c) occurs when

$$\tan\Gamma \approx 0.5 \cdot R_s\, \phi/R_{\text{PRI}} \tag{6.9}$$

where R_{PRI} is the range projection of one PRI onto the earth's surface. At this point, if one projected Nth-time-around range cell of interest is at the center of the footprint, there will also be one at the heel and one at the toe.

A computer program was developed to simulate the geometry described above and to compute the competing earth clutter area. The program includes the effects of the earth's curvature and contributions from the antenna pattern sidelobes.[2] For the OMV radar parameters, Figure 6.18 plots competing clutter area as a function of nadir angle for platform altitudes of 100, 200, 400, and 800 nmi. The curves illustrate the effects of the three geometries described by Figure 6.17(a–c). For example, consider the 100 nmi curve of Figure 6.18. The clutter area is beamwidth-limited (and increases slightly) between nadir angles of 0° and 11°, at which point range-cell limiting starts. The clutter area then decreases until the

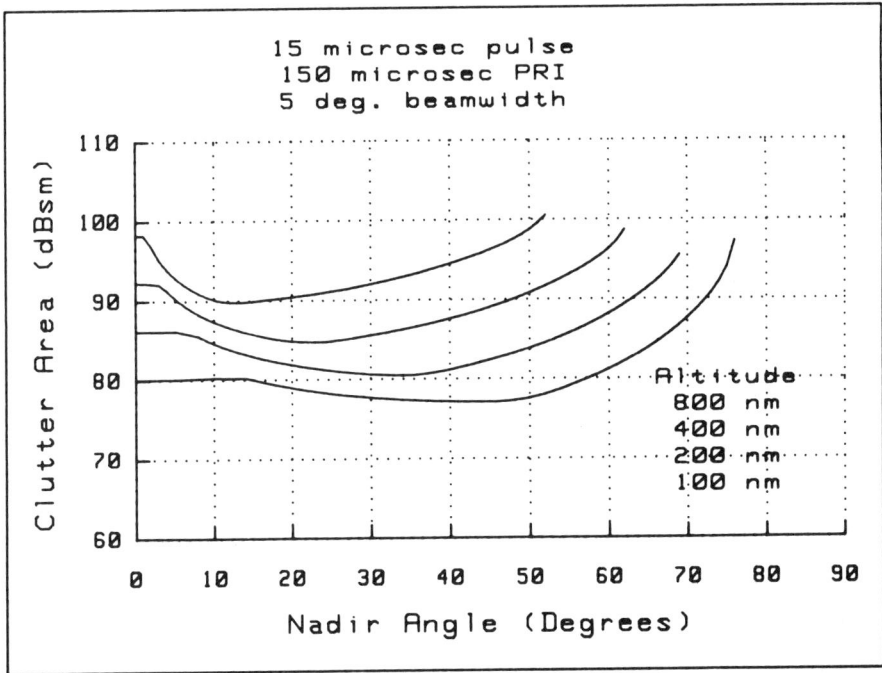

Figure 6.18 OMV radar clutter area *versus* nadir angle.

nadir angle reaches about 46°, at which point the increasing width of the clutter patch, caused by increasing R_s, counters the decreasing range-cell projection, caused by decreasing grazing angle. The clutter area hence increases slowly, with increasing nadir angle, until a nadir angle of about 59° is reached. Here, multiple range cells spaced at projected pulse repetition intervals begin entering the beam footprint and the competing clutter area increases rapidly until the beam is scanned off of the earth, at which point the curve is terminated. This sequence repeats for the higher altitude curves, but the transition points occur at ever smaller nadir angles until, at 800 nmi, the beamwidth-limited case only exists between nadir angles of 0° and 1°. These curves will obviously vary in shape and magnitude as the key radar parameters (pulsewidth, beamwidths, and PRI) are varied. Note that, for the multiple range cell portions of the curves, the computer program has corrected for slant-range differences by referring each range-cell clutter area contribution to the center of the beam footprint.

In Figure 6.19, the competing clutter area for the STS radar parameters is plotted. With the narrower antenna beam and wider pulse, the beamwidth-limited

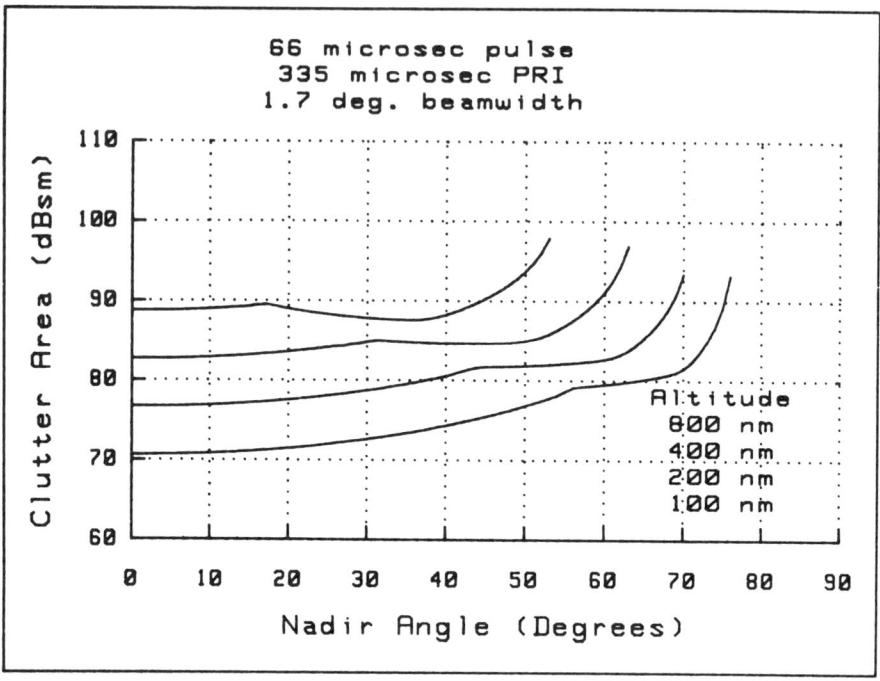

Figure 6.19 STS radar clutter area *versus* nadir angle.

condition is more dominant here, but, at the lower nadir angles, the clutter area is considerably less than that in Figure 6.18.

Before the clutter area could be compared to target area, the effects of range ratio, clutter spread in the frequency domain, and the clutter reflection coefficient should be considered.

6.5.4 Clutter-Target Range Ratio

The clutter signal strength, as measured at the radar, will be reduced relative to the target signal strength by the fourth power of the ratio of the respective ranges. For example, the return from clutter at 100 nmi must be reduced by $40 \cdot \log(100/4.5) = 54$ dB for comparison to a target at 4.5 nmi. In Figure 6.20, the curves of Figure 6.18 are plotted with the proper reduction for comparison to a target at the 4.5 nmi detection range of the OMV radar. In Figure 6.21, the curves of Figure 6.19 are plotted with the proper reduction for comparison to a target at the 12 nmi detection range of the STS radar. Note that in Figures 6.20 and 6.21 the lower

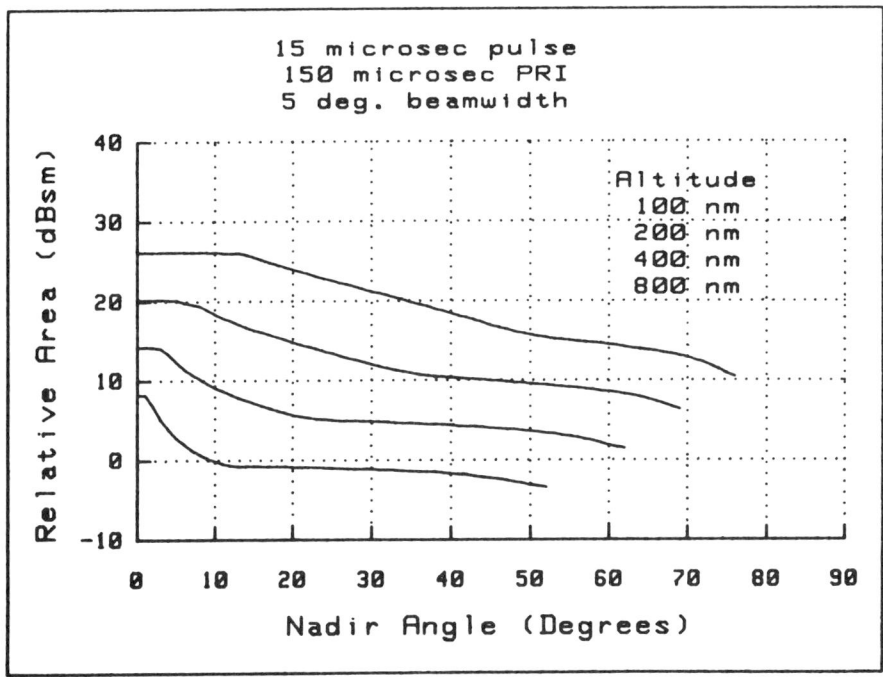

Figure 6.20 OMV radar clutter area relative to target at 4.5 nmi.

curves represent the higher altitudes. Here, we can see that the largest clutter signal returns will be at the smaller nadir angles. As we discuss in the next subsection, however, at these smaller nadir angles, the greatest benefit is derived from the spreading of the clutter signal energy in the frequency domain.

6.5.5 Clutter Frequency Spread

In a coherent pulsed radar, the frequency spectrum of the return from a point target has spectral lines separated by the PRF. Given the OMV radar's PRF of 6666 p/s, the radar processor looks for targets with velocities relative to the radar that produce doppler frequencies in the range of ±3333 Hz. At the X-band OMV radar RF this doppler frequency range corresponds to target velocities of ±170 ft/s. Returns from targets with higher relative velocities will fold ambiguously into this range of velocities and are not easily distinguishable from the target of interest. The same is true of the clutter return. However, the clutter return signal may be

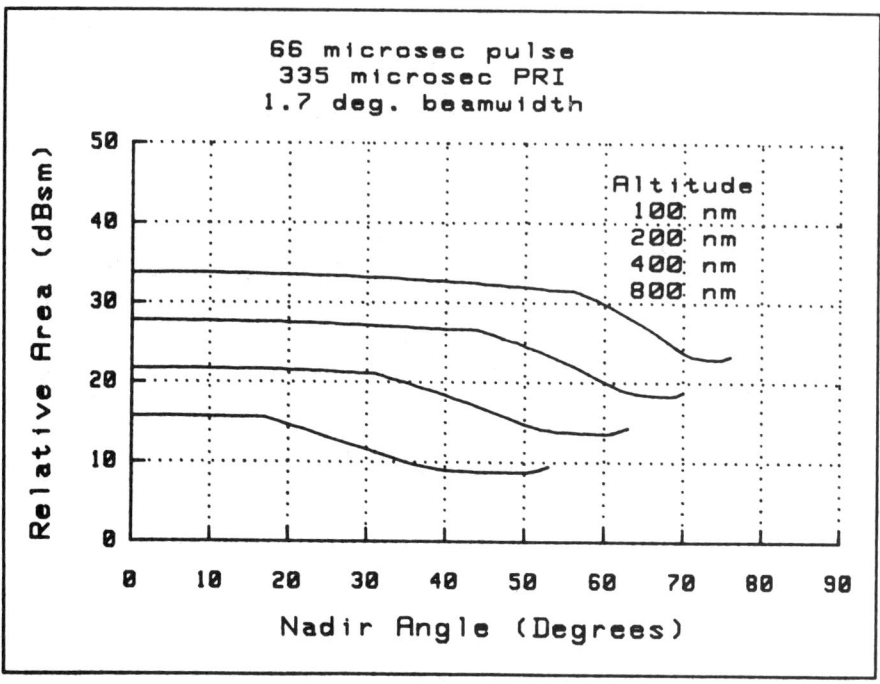

Figure 6.21 STS radar clutter area relative to target at 12 nmi.

spread over a wide range of velocities, as determined by the equations that follow, thus diluting the clutter signal energy at the target velocity of interest.

The radar is on a moving platform with an orbital velocity [4, 6] (assuming a circular orbit) calculated as

$$V_0 = [GM/(r_e + H)]^{1/2}$$
$$= 250.5/(r_e + H)^{1/2} \tag{6.10}$$

where

$GM = 62740$ nmi^3/s^2;
$r_e =$ earth radius, nmi;
$H =$ platform altitude, nmi.

The radial velocity of a point on the earth at the center of the antenna beam is

$$V_b = V_0 \cos\theta_{AZ} \sin\phi_n, \quad (6.11)$$

where

θ_{AZ} = the beam azimuth angle,[3]
ϕ_n = the beam nadir angle described earlier.

The clutter velocity at the beam center may be several times the ±170 ft/s velocity range of the radar, but the clutter return signal will fall ambiguously within this range, thus competing with the target signal. Of importance is the velocity spread of the clutter signals over the beam footprint. If the spread is significant, the target need not compete with the clutter signal energy from the entire footprint, but only with that portion of the energy which has the same apparent velocity as the target signal. The OMV rendezvous radar employs a 128-point FFT for frequency-domain signal processing and thus has 128 filters which cover the ±170 ft/s range of velocities. Therefore, if the clutter velocity spread is 340 ft/s or greater (and uniformly spread), the clutter signal in any one filter will be reduced by a factor of 128 (or 21 dB) relative to the levels shown in Figure 6.20.

The clutter velocity spread can be determined by applying (6.11) to the heel, toe, and sides of the beam footprint. Figure 6.22 shows the clutter velocity spread in the range dimension (toe minus heel)[4] as a function of nadir angle for the altitudes and radar parameters used for the previous figures. The velocity spread is higher at the lower altitudes because the orbital velocity is higher at these altitudes. The velocity spread at most nadir angles is several times the 340 ft/s unambiguous OMV radar velocity. The clutter energy will therefore be spread fairly evenly over this velocity range and be reduced by 21 dB in any one filter, as postulated earlier. The data of Figure 6.22 assume that the beam footprint is in the orbital plane. As the beam is steered out of the orbital plane, the range-dimension velocity spread will decrease (to zero for a 90° azimuth angle). However, the azimuth velocity spread (across the footprint from side to side), which is minimal for a zero beam azimuth angle (24 ft/s at 100 nmi altitude), increases as the azimuth angle increases to the maximum levels shown in Figure 6.22 to achieve similar overall results.

For the STS radar with its narrower beamwidth, the velocity spread is one third that of the OMV radar.

6.5.6 Clutter Reflectivity Coefficient

The radar reflectivity of clutter is generally defined in terms of a reflection coefficient called $\sigma°$ ("sigma zero"), which has units of m^2/m^2. The illuminated clutter area is multiplied by an appropriate $\sigma°$ value to determine the clutter RCS. The

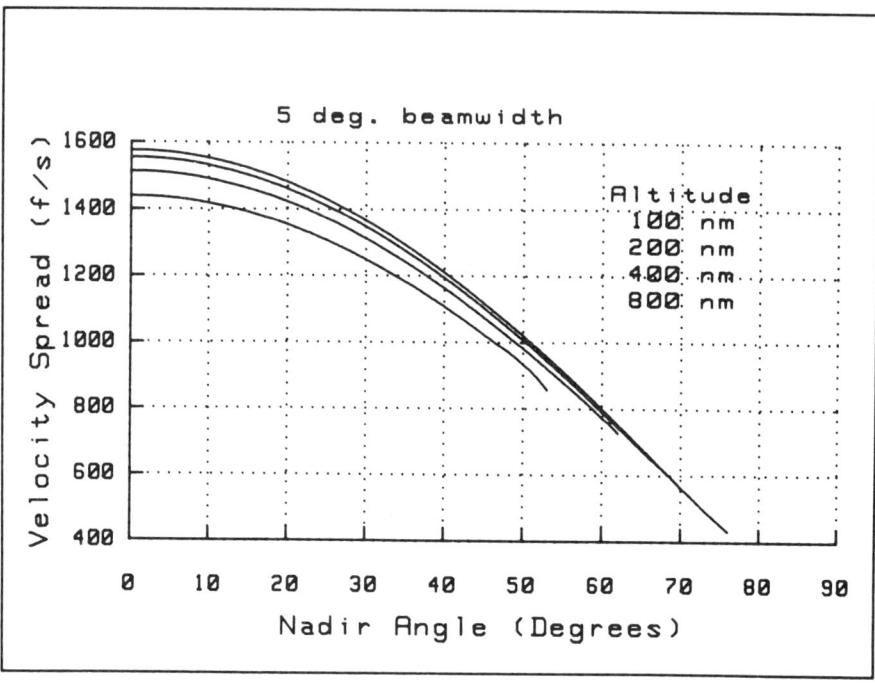

Figure 6.22 Velocity spread of OMV clutter returns over length of footprint.

curves of Figures 6.20 and 6.21 assume that the σ° of the earth clutter equals 1, or 0 dB. If σ° < 1, as is the case most of the time, the relative clutter levels will be reduced accordingly. The clutter σ° varies as a function of grazing angle, radar RF, and type of terrain. For a space-based radar in a low earth orbit, the observable clutter (earth surface) changes character rapidly. Also, the large footprint sizes of the rendezvous radars discussed in this chapter may view simultaneously several terrain types. Chapter 11 of this book discusses earth clutter as viewed from space in detail, and the reader is referred there for accurate clutter-level information.

6.5.7 Summary of Earth Clutter Effects

Consideration of the relative clutter area values from the curves of Figures 6.20 and 6.21, even when reduced by up to 21 dB to account for velocity spreading, indicates that earth clutter returns may be of concern to the rendezvous radar system designer, particularly if operation below 200 nmi in altitude is required and the target RCS is small (i.e., one square meter). Figures 6.20 and 6.21 assume target ranges of 4.5 and 12 nmi, respectively. If the target range is halved, the effects of clutter are reduced by 12 dB. The clutter RCS discussion of Chapter 11 should also be considered. If the radar orbit is such that the beam views primarily ocean areas, the clutter RCS may be low enough to mitigate any performance degradation from earth clutter. Further, the mission scenario should be considered. Accomplishing the rendezvous mission goals may be possible without steering the radar beam toward the earth, or at least toward the nadir point, until the target range is reduced.

Frequency agility can be used as a means to reject the clutter returns. For example, the coherent integration time of the radar can be reduced such that when the first earth clutter pulse return arrives at the radar the frequency has been changed. Reducing the coherent integration time while increasing the amount of noncoherent integration reduces the overall integration efficiency of the radar. Detection sensitivity can be regained by an increase in transmitter power. The primary effect of reduced coherent integration will be a concomitant reduction of doppler-derived range-rate measurement accuracy.

6.6 BACKGROUND NOISE SOURCES

The *sky temperature* (also called *antenna noise temperature*) viewed by the rendezvous radar antenna operating at X-band or Ku-band will normally be less than 10 K, contributing little to the effective noise temperature of the system. However, if the radar antenna is directed toward the sun or the earth, an increase in system noise temperature, as indicated by an increase in the receiver noise level, will be observed.

The antenna noise temperature T_a when the antenna beam is pointed at the sun, or at any solar body which does not entirely fill the beam, can be determined by [7]:

$$T_a = T_b \cdot \Omega_b/\Omega_a \qquad (6.12)$$

where

T_b = the brightness temperature of the body at the frequency of operation,
Ω_b = the solid angle subtended by the body as viewed by the antenna,
Ω_a = the solid angle of the antenna beam (one-way, 3-dB).

The contribution of the source is obviously reduced as the antenna beamwidth is increased.

The brightness temperature of the quiet sun varies approximately linearly with wavelength at microwave frequencies; about 20,000 K at 10 GHz and 14,000 K at 14 GHz. The disturbed sun brightness temperature reaches 60,000 K at 10 GHz and 30,000 K at 14 GHz[5]. The solid angle subtended by the photosphere of the sun, as viewed from a low earth orbit is 0.224 square degrees. Applying (6.12), the antenna noise temperature of the OMV radar will be increased to

$$T_{a(\text{OMV})} = 20{,}000 \cdot 0.224/20 = 224 \text{ K (quiet sun)}$$

when the antenna is directed at the quiet sun. For the 2.3-square-degree STS radar antenna beam area the antenna noise temperature is

$$T_{a(\text{STS})} = 14{,}000 \cdot 0.224/2.3 = 1360 \text{ K (quiet sun)}$$

When the antenna is scanned across the earth the beam is completely filled, and (6.12) does not apply. The full brightness temperature of the earth should be assumed. The temperature of the earth is typically 250 to 300 K [8]. The moon and planets, when (6.12) is applied, contribute less than 20 K to antenna noise temperature.[6]

The sun and earth are obviously of the most concern to a rendezvous mission controller. The antenna noise temperature adds directly to the receiver noise temperature (referred to the antenna) [7, 9]. For example, a radar system with a 5-dB receiver noise figure (referred to the antenna input) has a noise temperature (also referred to the antenna input) of 630 K. A quiet sun contribution of 1360 K would raise the total effective system noise temperature to 1990 K, an increase of 5 dB. The effect on detection performance would be a 25% reduction in range.

NOTES

1. The first year of operation of the IRACS is described by Griffin, et al. [2].
2. A Taylor-weighted antenna pattern (20 dB sidelobes, $\bar{n} = 6$) was simulated. For the geometries considered here, the sidelobe contribution to the clutter area is less than 1 dB.
3. The azimuth angle is measured in a plane which is tangent to the orbit plane at the platform position.
4. The computer program used to calculate velocity spread includes the effect of the fourth power of range when determining the footprint length. The footprint toe and heel are therefore not defined by the 3-dB width of the antenna beam, but instead they are the points at which the clutter return signal will be 3 dB less than the clutter return signal from the beam center, assuming uniform clutter reflectivity.
5. These temperatures are approximations read from Figure 8-21 of Kraus [7].
6. See Table 8-3 of Kraus [7] for detailed data on the moon and planets.

REFERENCES

1. Fenner, R.G., and R.F. Broderick, "Spaceborne-Radar Applications," Chapter 34 in *Radar Handbook,* M.I. Skolnik (ed.), McGraw-Hill, New York, 1970.
2. Griffin, J.W., H. Haddad, H.G. Magnusson, and C.L. Mohler, "Ku Band—The First Year of

Operation," *Proc. IEEE Int. Radar Conf.*, 1985, pp. 330–338.
3. Hord, R.A., *Relative Motion in the Terminal Phase of Interception of a Satellite or Ballistic Missile*, NACA TN 4399, September 1958.
4. Wolverton, R.W., *Flight Performance Handbook for Orbital Operations*, Space Technology Laboratories, Redondo Beach, CA, 1961.
5. NASA George C. Marshall Space Flight Center, AL "Orbital Maneuvering Vehicle" Request for Proposal 1-6-PP-01438, November 1985.
6. Lerch, C.S., Jr., "Satellite Surveillance Radar," Chapter 32 in *Radar Handbook*, M.I. Skolnik (ed.), McGraw-Hill, New York, 1970.
7. Kraus, J.D., *Radio Astronomy*, McGraw-Hill, New York, 1966.
8. Blake, L.V., *Radar Range-Performance Analysis*, Artech House, Norwood, MA, 1986.
9. Berkowitz, R.S., et al., *Modern Radar*, John Wiley and Sons, New York, 1965.

Chapter 7
RADAR ALTIMETERS FOR SPACE VEHICLES
T.J. Lund
Teledyne Ryan Electronics

Radar altimeters have been used in aircraft since World War II to measure clearance above the terrain. These devices supplement barometric altimeters, which measure altitude above sea level without an indication of rising terrain. Radar altimeters also provided on-board altitude measurements to the guidance system of the Surveyor and Apollo lunar landing vehicles and for the Viking spacecraft, which landed on Mars. Radar altimeters have been used to profile the terrain from an orbiting spacecraft and to provide a measure of sea state from orbit by detailed analysis of the characteristics of the received signal. Radar altimeters with resolutions of 10 cm operating in earth orbit have allowed mapping certain currents in the ocean by its surface's associated slight change in elevation.

With the advent of further exploration, and perhaps colonization of our moon and certain planets, radar altimeters will likely become an important part of future space missions. This chapter intends to give the reader some basic principles of radar altimeters, a description of various types, and a brief description of those which have already been used in space. There are two main classifications of space-based radar altimeters: those used to get altitude information for a landing spacecraft and those used for remote sensing of ocean or land terrain characteristics from orbit. In this chapter we emphasize radar altimeters that will be used on a landing spacecraft.

7.1 PRINCIPLES OF SPACE-BASED RADAR ALTIMETERS

Radar altimeters operate by directing a modulated RF signal toward the surface using a transmitting antenna, and measuring the round-trip travel time for the transmitted modulation pattern to appear at the output of the receiving antenna.

Often, but not always, the radar altimeter is required to measure the range to the nearest point of undulating terrain under the vehicle within a rather wide antenna beamwidth. Certain applications require that the average range within a narrow beamwidth be measured, while other special applications require that the radar altimeter provide a measure of the roughness of the surface below.

Radar altimeter types which have been used in space include short-pulse, linear frequency modulation, and pulse-compressed systems. In addition, biphase modulation is being used in aircraft radar altimeters. The operation and general performance of these types of radar altimeters are treated in this chapter.

7.1.1 Properties of Waveform for Radar Altimeters

The modulating waveform, time waveform, power spectrum, and the ambiguity function of common radar altimeters are shown in Figure 7.1. The ambiguity diagram is the matched filter response to the transmitted signal, and maps the response of the radar in the range and doppler planes. For radar altimeters, it is desirable for the width of the main response to be narrow in the range plane. Further, any doppler interaction with the range response should be minimized, or be such that it could be either cancelled or compensated.

Short-Pulse Radar Altimeter

The modulating and time waveforms shown in Figure 7.1 for the short-pulse radar altimeter have the same form. The pulsewidth, τ, may vary from tens of nanoseconds at low altitudes to several microseconds at high altitudes. The interpulse interval, T, which is usually selected to be longer than the round-trip time for the signal, may be in the tens of kilohertz at low altitudes and in the hundreds of hertz for high-altitude applications.

The power spectrum has discrete lines at multiples of the reciprocal of the interpulse period. The amplitude of these lines is given by the envelope function, as indicated in the figure. The frequency interval between first nulls of the spectrum is equal to $2/\tau$ in hertz.

The ambiguity diagram indicates a triangular main response with a width at the base equal to twice the two-way range, corresponding to the pulsewidth, τ. Range ambiguities exist at multiples of the range corresponding to the interpulse period. Note that there are no minor responses in the range plane, and doppler has no effect on the location of the main response peaks.

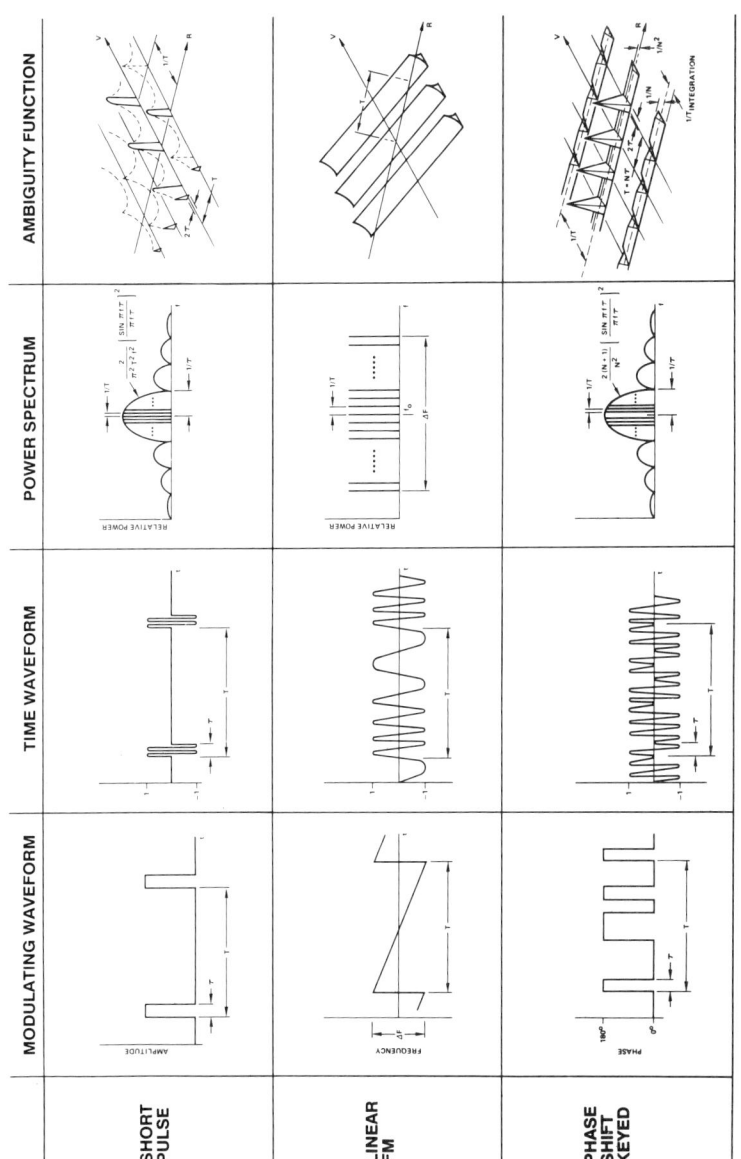

Figure 7.1 Properties of waveforms for radar altimeters.

Linear Frequency Modulation

The modulating waveform for the linear (continuous wave) frequency modulated (usually referred to as FM/CW) radar altimeter can be either saw-toothed, as shown, or triangular. The receiver is usually blanked during the short retrace time of the saw-toothed modulating waveform. The use of triangular modulation, and separate processing of the signal return on the up-sweep and down-sweep, allows doppler frequency contamination of the ranging signal to be measured and removed.

The time waveform of the transmitted signal is a linearly changing frequency from a higher to a lower and then again a rapid change to the higher. The frequency deviation of the transmitted signal is typically on the order of 100 MHz for low-altitude conditions and a few megahertz for high-altitude applications. The period of the modulation must be longer than the round-trip time at the maximum altitude. The frequency must also be chosen to yield a range of frequencies suitable for accurate processing.

The power spectrum of the transmitted waveform is a series of lines separated by the reciprocal of the modulation period with uniform amplitude over an interval of plus or minus the deviation about the center frequency.

The ambiguity diagram indicates strong coupling between range and doppler (velocity). This coupling, and ways of compensating for it, will be treated in the discussion of linear FM/CW radar altimeters.

Phase-Shift Keyed by a Psuedorandom Code

The modulating waveform of a *phase-shift keyed* (PSK) radar altimeter is a pseudorandom code. The chip width, τ, is similar to the pulsewidth of a short-pulse radar altimeter, and the code period, T, which is equal to the chip width times the number of bits in the code, is similar to the interpulse period. The modulating waveform is used to command a phase shift of either zero or 180° of the transmitted carrier.

The power spectrum of the transmitted signal is identical to that of the short-pulse radar altimeter, given the same pulse and chip widths and the same interpulse and code periods.

The ambiguity diagram indicates a triangular main response with a width at the base equal to twice the two-way range, corresponding to the chip width. Range ambiguities exist at multiples of ranges corresponding to the code period. Note that there is a low ridge response along the range axis, which extends the full range interval, due to the time sidelobes of the modulating function. Also, note that there are low-level ridge responses in the doppler plane separated by the reciprocal of the code period. The implication of this ambiguity diagram for performance is treated in the discussion of PSK radar altimeters.

7.1.2 Range Equation for Radar Altimeters

Signal Power

Equation (7.1) relates the received signal power to radar altimeter system parameters, altitude, and terrain backscatter characteristics [1]. In the case of a short-pulse radar altimeter, the equation can also be used to develop the time average shape of the echo pulse by relating the angle θ to time. The shape of the pulse, which is strongly dependent on the backscattering characteristics of the terrain, has a direct influence on the accuracy of the altitude measurement. Thus,

$$P_r = \frac{P_t G_0^2 \lambda^2}{(4\pi)^3 H^2 L} \iint [g(\theta, \phi)]^2 \sigma°(\theta) \sin\theta \cos\phi \, d\theta \, d\phi \tag{7.1}$$

where

P_t = transmitted power,
G_0 = antenna gain at peak of beam,
$g(\theta, \phi)$ = antenna pattern in θ and ϕ space,
$\sigma°(\theta)$ = Backscattering cross section per unit surface area,
λ = wavelength,
H = height above the terrain,
L = losses,
θ = angle between the vertical and a line to a point on the surface,
ϕ = angle in azimuth plane from the heading of the vehicle.

The geometry involved is illustrated in Figure 7.2 A brief explanation of the terms in the equation that may not be obvious is given in the following paragraphs.

The antenna gain, G_0, is treated as the value at the peak of the beam pattern, times an angular dependence translated from antenna coordinates to surface, based on $g(\theta, \phi)$. This translation must also take into account the pitch and roll of the vehicle.

The backscattering cross section per unit surface area, $\sigma°(\theta)$, is a measure of the radar reflectivity of the surface. It often has a strong angular dependence, which, for small angles from the nadir, can be approximated by an exponential $\sigma°(\theta) = \sigma°(0) \exp(m\theta)$ where m is the slope. Typical slopes for $\sigma°(\theta)$ are 0.1 to 0.5 dB/degree for rough land terrain and 1.0 to 1.5 dB/degree for water. Typical values of $\sigma°$ at zero degrees, $\sigma°(0)$, are 0.1 for rough land and 10 for water.

The indicated integration is carried out over an angle of 2π in azimuth angle, ϕ, and over the polar angular, θ, from zero to $\pi/2$, or as limited by the pulsewidth or beamwidth.

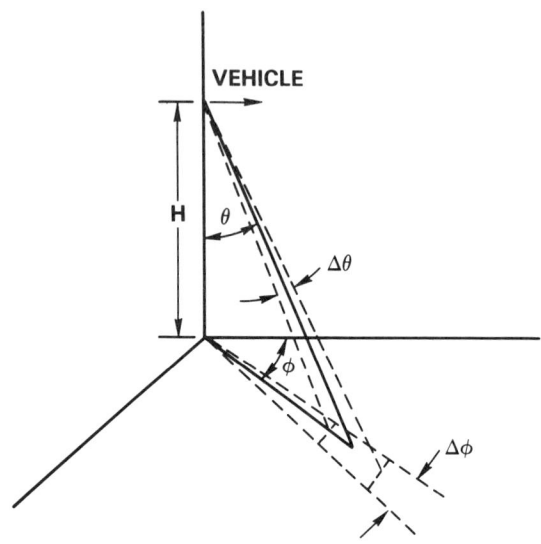

Figure 7.2 Geometry for terms used in equation (7.1).

Noise Power

We generally need to determine the signal-to-noise ratio for particular operating conditions to predict the performance of a radar altimeter. The equivalent noise power, P_n, may be expressed as

$$P_n = kT_sB$$

where

k = Boltzmann's constant,
T_s = system noise temperature,
B = bandwidth at signal processor.

The system noise temperature, T_s, may be expressed in the form:

$$T_s = T_a + T_r + LT_e$$

where

T_a = antenna noise temperature,
T_r = transmission line noise temperature = $T(L - 1)$,
T = thermal temperature,
L = loss of transmission line,
T_e = receiver noise temperature = $T_0(F - 1)$,
T_0 = reference temperature = 290 K,
F = receiver noise figure.

The radar altimeter antenna will usually be pointed toward the surface of a planet, and the antenna temperature will be essentially that of the planet. In the special case of operating over the earth, the antenna temperature will be about 290 K (17°C). If, further, the equipment temperature is maintained at about 290 K, the system noise temperature becomes equal to the commonly used expression, $T_s = T_0 LF$.

Selection of Operating Frequency

The selection of the best transmitter frequency for a particular application is strongly dependent on the antenna coverage required and practical considerations of available transmitter power as a function of frequency. If the beamwidth needs to be broad, perhaps in excess of 30°, the advantage is to use the lowest frequency practical. The reason for this is so that the capture area of a receiving antenna is related to its size, and, for a given beamwidth, a lower frequency allows the use of a larger aperture to achieve the particular beamwidth. This relationship is apparent by the λ^2 term in the equation for signal power (7.1).

However, if the requirement is to measure slant range along a narrow beam, the highest practical frequency is advantageous because, for a given size of aperture, the antenna gain increases with frequency. A radar altimeter using a broad-beam antenna is the one on the Viking Mars landing vehicle, where the beamwidth was 80° × 20°. An operating frequency of 1 GHz was selected, and a 140 W solid-state transmitter allowed operation to 450,000 ft above the planet. The radar altimeter used on the Surveyor lunar lander was required to measure slant range along the velocity vector. Hence, the altimeter used an antenna with a beamwidth of 4° at an operating frequency of 13.3 GHz.

Antenna Considerations

If a broad-beamwidth antenna is to be used to accommodate attitude variations of the spacecraft, there is an optimum antenna beamwidth for a given maximum angle between the vertical to the surface of the planet and the peak of the beam.

At the higher altitudes, the pulsewidth, or resolution cell of the radar altimeter, subtends a small angle and, to a first approximation, the antenna gain of interest is that along the vertical. The main-lobe structure of typical radar altimeter antennas can be represented closely by a Gaussian response to angles up to 1.5 times the 3-dB response points. Assuming such an antenna with an efficiency of 66%, a plot showing optimum beamwidth as a function of attitude angle is given by Figure 7.3.

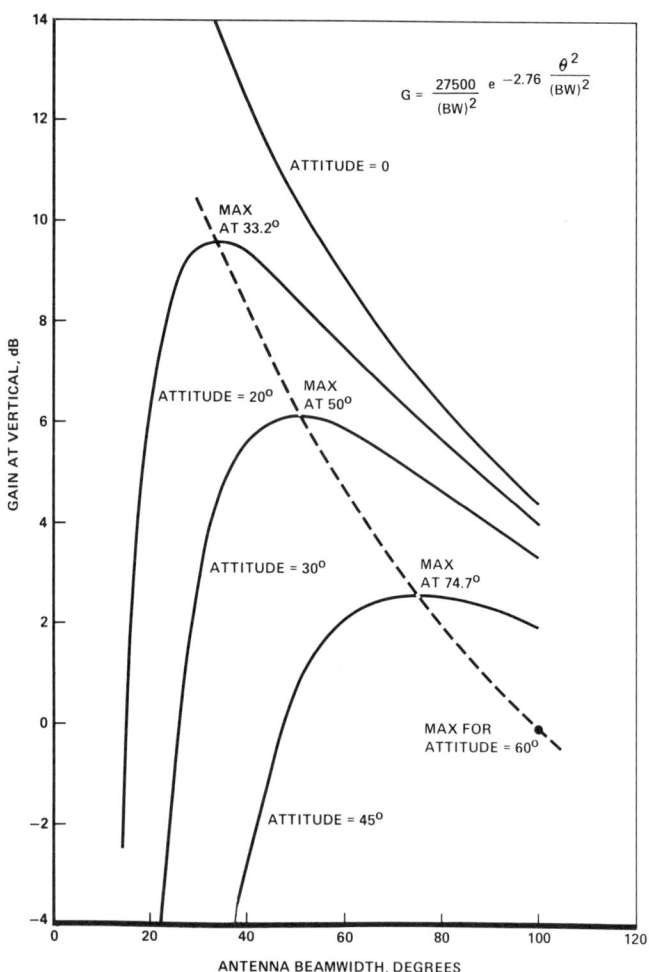

Figure 7.3 Optimum antenna beamwidth *versus* vehicle attitude.

For example, if a ±30° variation in attitude of the vehicle is to be accommodated, the optimum beamwidth is about 50°. The gain at the vertical for this antenna, at 30° attitude, is 6.1 dB.

7.2 DESCRIPTION AND PERFORMANCE OF RADAR ALTIMETER TYPES

7.2.1 Short-Pulse Radar Altimeter

7.2.1.1 Functional Description

A functional block diagram of a short-pulse power altimeter is shown in Figure 7.4. An oscillator operating at the desired frequency drives a pulsed power amplifier, which amplifies the signal to the needed power level. The higher-level power amplifier stages are usually pulsed only during transmission to minimize power consumption and to eliminate transmitter noise from leaking into the receiver during the receiving period. The short pulses of RF energy are directed toward the surface of the planet by a transmitting antenna. A detector diode coupled to the transmitter output generates a pulse in response to the RF pulse, which is used to initiate the timing sequence for measurement of the delay between transmitted and received pulses.

A portion of the signal reflected from the surface is intercepted by the receiving antenna and applied to a low-noise amplifier, which essentially sets the noise figure of the receiver. The output of the low-noise amplifier is applied to a microwave mixer, which can be excited either directly with a sample of the oscillator signal or the oscillator signal being displaced by a convenient intermediate frequency (IF). Generally, an image-rejection mixer is used to avoid foldover of amplifier noise, which otherwise raises the effective noise figure of the receiver by 3 dB. Employing an IF facilitates the use of an image rejection mixer. The signal at the output of the mixer is amplified by the IF amplifier and detected. The resulting video signal is further amplified and applied to the range tracker.

There are several implementations of range trackers, differing basically in the range-gating arrangement and method of developing the tracking error signal. Radar altimeters employing broad-beamwidth antennas require some form of leading-edge tracking of the received pulse. Types of leading-edge trackers include the broad range gate, split range gate, and a split gate with a third delayed range gate.

(1) *Broad Range Gate Followed by Adaptive Threshold Detector.* A broad range gate is positioned around the received pulse and the gated video signal is applied to a threshold detector, which is set for an acceptable false-alarm rate on noise alone. The time between the transmitted pulse and when the received pulse rises above the threshold is measured to obtain a determination of altitude. The range tracker in this case merely positions the range gate around the signal. Its dynamics do not directly affect the accuracy of the altitude measurement.

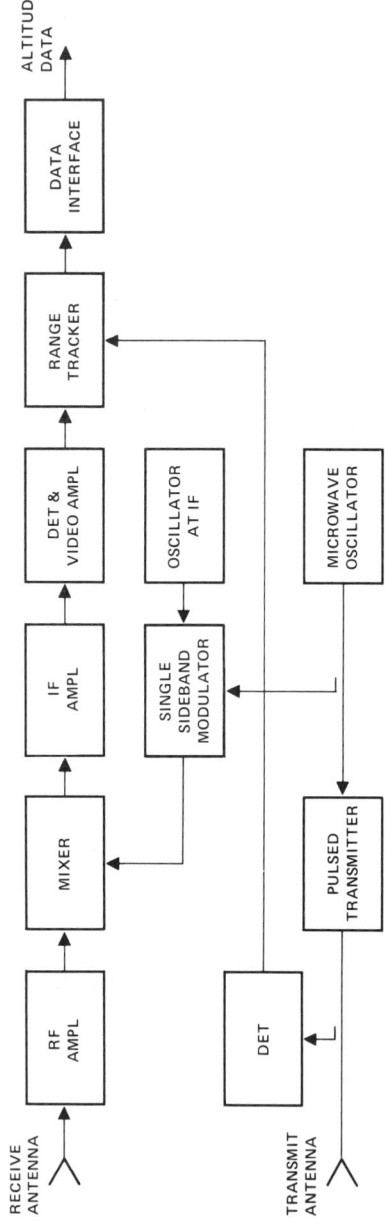

Figure 7.4 Functional block diagram of short pulse radar altimeter.

(2) *Split Range Gate Tracker.* Two contiguous range gates, each measuring approximately one pulsewidth, τ, are positioned by the tracking loop such that the amplitude of the signal in the late gate is a given factor higher than that in the early gate. The tracking point is a function of signal-to-noise ratio and the shape of the return signal.

(3) *Split Gate Tracker with a Third Delayed Range Gate.* This range tracker is similar to the broad range gate, except that a third gate is positioned one pulsewidth later in time such that it measures the amplitude near the peak of the pulse. The ratio of amplitudes allows a correction to be made in the altitude measurement for the effects of signal-to-noise ratio.

7.2.1.2 Shape of the Received Pulse

The shape of the received pulse of a short-pulse radar altimeter operating over relatively flat terrain is a function of the slope of the backscattering cross section per unit surface area with incident angle, the shape of the antenna pattern, the pulsewidth, and the square root of the pulsewidth ratio expressed in terms of equivalent distance to altitude. The basic equation for signal power (7.1) can be used to obtain the pulse shape.

There are two general cases of operation for a pulsed radar altimeter: *pulsewidth-limited* and *beamwidth-limited*. Operation at high altitudes with a short pulsewidth and wide antenna beamwidth results in the pulsewidth-limited case. Under these conditions, illustrated in Figure 7.5, the illuminated region of the surface of a planet is at first an expanding disk, and the signal strength builds up as shown in the figure. At the exact point in time when the trailing edge of the pulse intercepts the surface at the vertical to the radar altimeter, a hole forms in the center of the illuminated area, and it takes the form of a ring with expanding radius and decreasing width. The illuminated surface area remains essentially constant. Were it not for antenna pattern effects and angular dependence of the reflectivity function, the received pulse would have a relatively flat top as shown in the figure.

The angle from the vertical corresponding to the point where the leading edge of the pulse strikes the surface at the instant of time that the trailing edge also does so is

$$\theta_c = \arccos[H/(H + c\tau/2)]$$

which for small angles can be approximated by

$$\theta_c = \sqrt{(c\tau/H)}$$

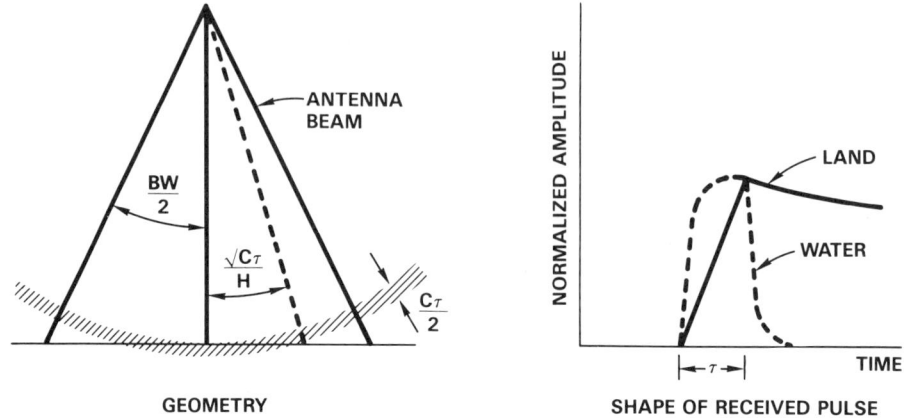

Figure 7.5 Geometry and pulse shape for pulsewidth-limited case.

If the half-beamwidth of the antenna is near this value or less and the angular dependence of surface reflectivity is small, the antenna beam pattern has an influence on the build-up of the pulse. This condition, illustrated in Figure 7.6, is referred to as being beamwidth-limited. The integral expression in (7.1), which includes antenna beamwidth effects and angular dependence of the reflectivity of the surface, automatically transits between pulsewidth-limited and beamwidth-limited conditions. Approximate equations for the signal return at the peak of the pulse for each of the limited cases are given below.

Pulsewidth-limited case:

$$P_r = \frac{P_t G_0^2 \lambda^2 c \tau \sigma°(0)}{64\pi^2 H^3 L}$$

Beamwidth-limited case:

$$P_r = \frac{P_t G_0^2 \lambda^2 \sigma°(\theta)}{32\pi^2 H^2 L} \left[\frac{1 - \cos^2\theta_0}{2} \right]$$

where $2\theta_0$ is the beamwidth of the antenna.

We may note in passing that the amplitude of the peak of the return signal pulse varies as the reciprocal of altitude cubed (H^3) in the pulsewidth-limited case and as the reciprocal of altitude squared (H^2) in the beamwidth-limited case.

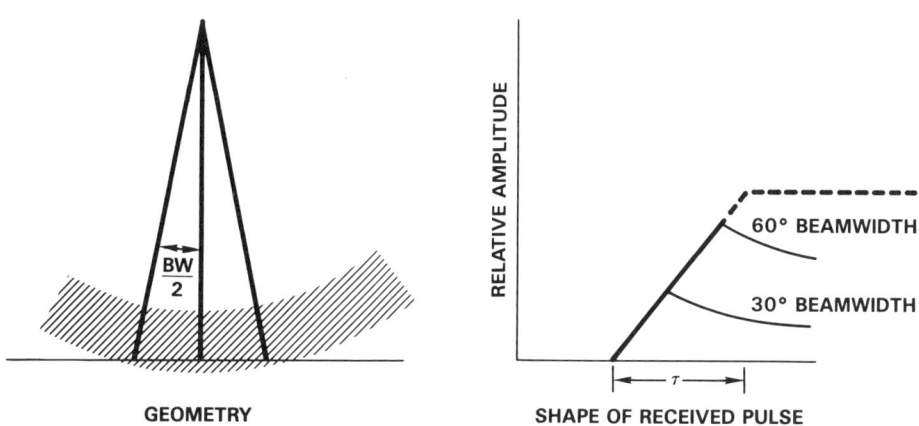

Figure 7.6 Geometry and pulse shape for beamwidth-limited case.

In the case of the antenna pointed along the vertical, the pulse shapes may be obtained by performing the integral over azimuth angle, ϕ, from 0 to 2π as the integral over polar angle, θ, is taken for increasing values of the limits of integration. These increasing angles are related to the time delay from the instant of time that the leading edge of the echo pulse is received by use of the expression:

$$t = \frac{2H}{c} \left[\frac{1}{\cos\theta} - 1 \right]$$

Typical received pulse shapes in terms of power *versus* time for the pulse-width-limited case with an antenna beamwidth of 60° and pulsewidth τ of 200 ns, for example, are given in Figure 7.7 for several different slopes of the backscattering cross section per unit surface area $\sigma°$ ranging from rough land to smooth water. The pulse amplitudes are normalized for better portrayal of the pulse shapes. These pulse shapes illustrate the fact that the largest source of altitude measurement error in most short-pulse radar altimeters is the variation in the time of thresholding on the pulse due to variation in the shape of its leading edge as the vehicle passes over different types of terrain.

The pulse shapes shown represent an average over hundreds of pulses. The individual pulses have appreciably fine structure due to the random scattering process from the surface.

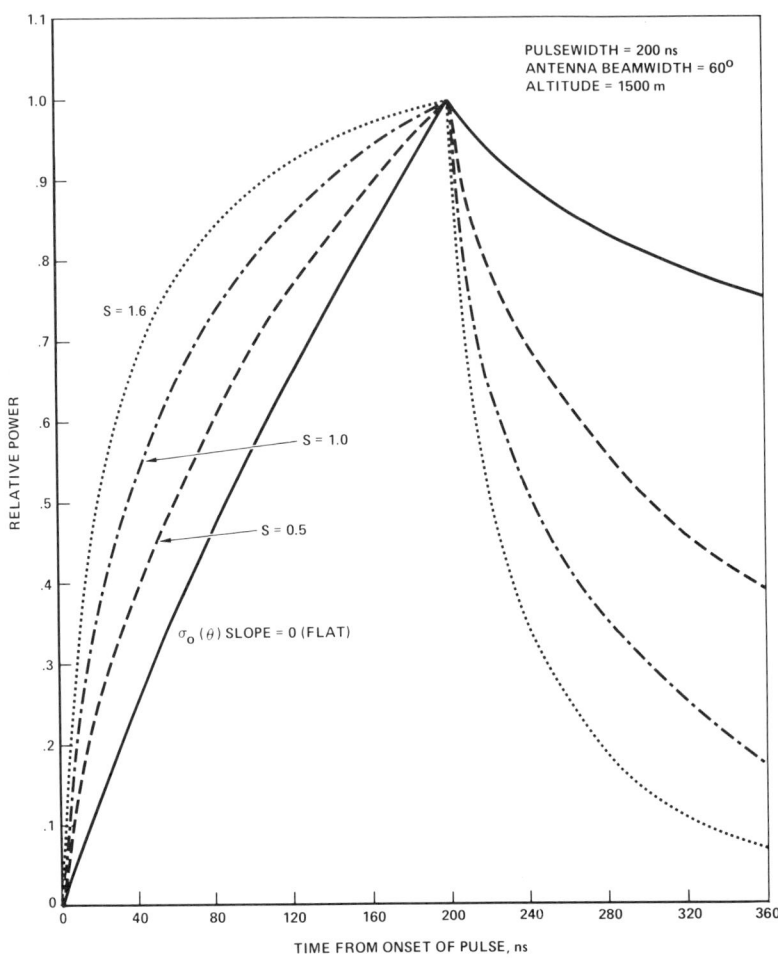

Figure 7.7 Received pulse shapes for pulsewidth-limited case for antenna beamwidth of 60°.

7.2.2 Pulse Compression and High Resolution Radar Altimeters

Pulse compression may be used for effectively developing a narrow, high-power pulse from a low-power, wide transmission interval. Pulse compression is a very useful technique to obtain short pulsewidths necessary for high resolution radar altimetry from orbital altitudes with reasonable values of peak transmitted power. Pulse compression was employed on the radar altimeters used on the Skylab and GEOS-C earth orbiting spacecraft. Those radar altimeters are described later in this chapter.

The two leading types of pulse compression are (1) *linear frequency modulation,* or "chirp," during the transmitted pulse, and (2) *phase-coded,* by PSK, within the pulse. The Skylab radar altimeter used phase-coded pulse compression and GEOS-C used chirp. A variant of the chirp pulse compression system, referred to as "stretch," uses a chirped second local oscillator (LO) in conjunction with a series of delays and bandpass filters to process the received signal. This technique allows the effective narrow pulse to be examined in very fine detail and has the potential for resolution of a few centimeters.

Detailed descriptions of the chirped and phase-coded pulse compression waveforms and associated radar equipment are given in several radar texts [2, 3]. The stretch technique is treated in [4].

A major concern for space applications of these waveforms is operation in the presence of high doppler frequencies. High doppler is a consequence of a high velocity component in the direction of the altitude measurement. The PSK type of pulse compression requires separate parallel processing channels displaced in frequency by the reciprocal of the code length. The number of channels required is a function of the total possible doppler frequency spread. The chirp type compression waveform is fairly insensitive to doppler as long as the doppler frequency is less than about a tenth of the chirp bandwidth [2].

A brief description of the stretch technique for high resolution radar altimetry is given below. A simplified block diagram of the system is given in Figure 7.8. The system uses a *dispersive delay line* (DDL) in the transmitter to develop a linear FM (or chirp) waveform at a center frequency of a few hundred megahertz. This chirped signal is up-converted to the transmitted frequency and amplified prior to transmission.

The received signal is down-converted to an IF equal to that at which the DDL operates. The signal is amplified at IF and applied to a mixer along with a signal from a DDL, which is identical to the one used in the transmitter. The receiver DDL is pulsed after a time delay approximately equal to the round-trip travel time of the pulse. The output of the mixer is a group of long pulses with a center frequency proportional to the delay between the return signal and the chirped signal generated by the receiver DDL.

The signal's spectral components are separated and stretched in time by a series of delays and bandpass filters, which can be implemented by a single surface acoustic wave (SAW) device. The outputs of the bandpass filters are summed and applied to a square-law detector. The output of the square-law detector is a series of pulses with delays corresponding to those in the SAW line, and has amplitude proportional to the strength of the signal at particular points in the leading edge of the received pulse. This pulse train can be demultiplexed, and the amplitudes at the sample points are digitized and their output comprises high resolution measurements of the shape of the compressed pulse.

A conventional altitude tracking loop with early and late gates for error sensing is used to control the delay generator, which, in turn, pulses the receiver DDL at the proper delay after the transmission pulse.

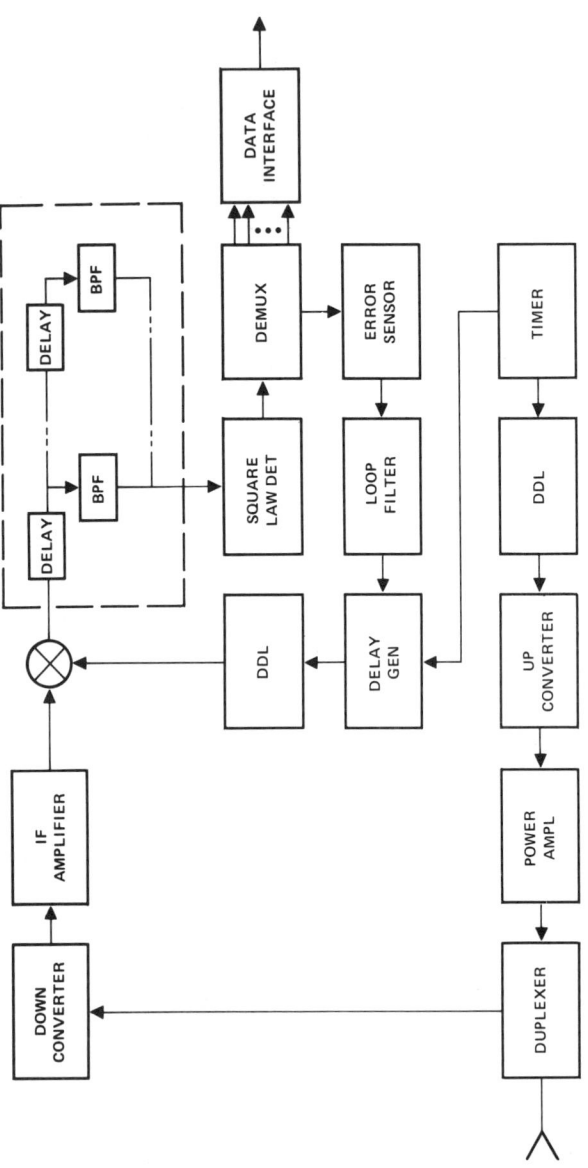

Figure 7.8 Simplified block diagram of "stretch" high resolution radar altimeter.

7.2.3 Linear Frequency Modulated Radar Altimeter

Linear frequency modulated (or FM/CW) radar altimeters are commonly used in aircraft, and they have also been used in the Surveyor and Apollo lunar landing spacecraft. The FM/CW waveform can be used with either broad-beam or narrow-beam antennas, although the type of signal processing used for these two cases is different.

Modulation waveforms typically used by FM/CW radar altimeters are illustrated in Figure 7.9. The difference between the transmitted frequency and the received frequency at any instant of time is equal to the range frequency, f_R, which is proportional to the range to the scattering element on the surface, plus the doppler frequency component, f_D, which is proportional to the velocity component in the direction of the scattering element. In applications where the velocity vector is accurately known, the doppler component of the total difference frequency can be removed in signal processing and saw-toothed frequency modulation can be used. If velocity is not known, triangular frequency modulation is usually used, and the difference frequency during the down-sweep $(-f_R + f_D)$ is subtracted from that of the up-sweep $(f_R + f_D)$ to cancel the doppler frequency and obtain twice the range frequency. When the triangular waveform is used, a time-shared frequency tracker is commonly employed to track separately the difference frequencies during the up-sweep and the down-sweep.

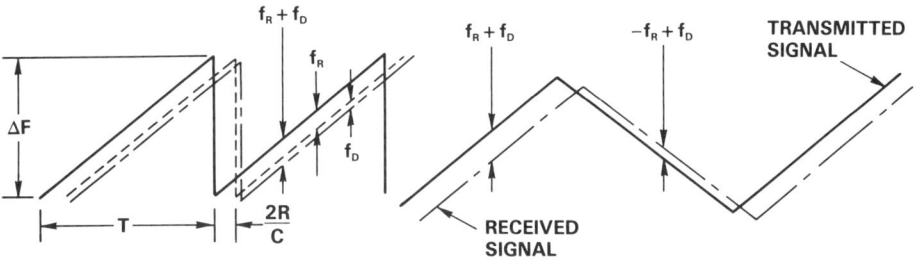

Figure 7.9 Modulation waveforms for linear FM/CW radar altimeters.

Referring to the geometry sketched in Figure 7.10, the range frequency and doppler frequency from an incremental surface area having coordinates θ and ϕ can be expressed as follows:

$$f_R = 2R\Delta F/cT = 2H\Delta F/cT \cos\theta$$

$$f_D = 2V/\lambda \, (\sin\theta \cos\phi \cos\alpha - \cos\theta \sin\alpha)$$

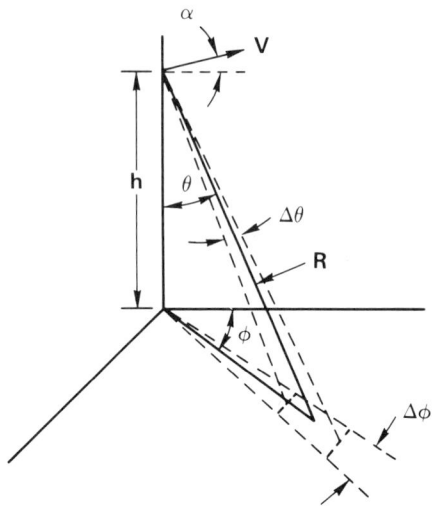

Figure 7.10 Geometry for determining characteristics of received signal in FM/CW radar altimeter.

where

R = range to surface element,
H = altitude,
ΔF = frequency deviation of transmitter,
T = frequency ramp interval,
c = velocity of propagation,
α, θ, ϕ = angles depicted in Figure 7.10,
V = vehicle velocity,
λ = wavelength.

A basic mechanization of the FM/CW radar altimeter is shown in Figure 7.11. The frequency modulator generates either a saw-toothed or triangular waveform, which is applied to a *voltage-controlled oscillator* (VCO). The VCO, which changes its output frequency linearly with the input control voltage, usually operates directly at the transmission frequency. The scale factor of the radar altimeter is directly related to the slope of the frequency modulation; thus, both the average slope and the instantaneous slope must be held to close tolerances. A linearization circuit, usually containing a SAW delay line, is used for this purpose.

The transmitter may consist of the VCO only, or a power amplifier can be used to increase the transmitted power. The transmitted energy is directed toward the surface by the transmitting antenna.

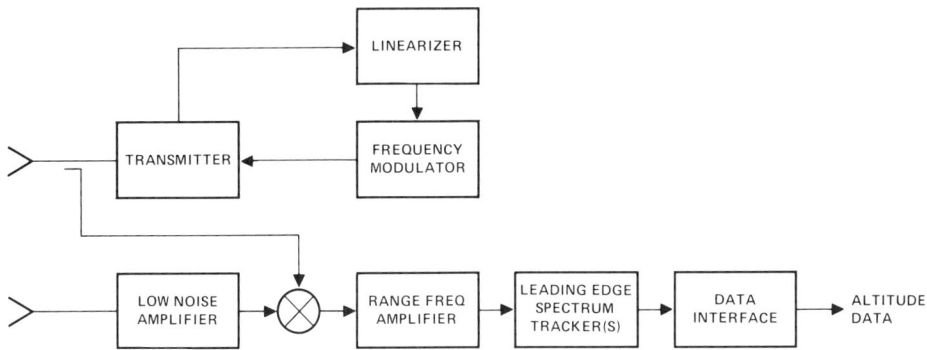

Figure 7.11 Basic mechanization of FM/CW radar altimeter.

A portion of the energy backscattered from the surface is intercepted by the receiving antenna and applied to the receiver. A sample of the transmitted signal is mixed with the received signal in the receiver, and the difference frequency is amplified and applied to the signal processor. The signal processor will be discussed separately for the narrow-beam and broad-beam configurations of the FM/CW radar altimeter.

To assess the altitude capability and altitude measurement accuracy of an FM/CW radar altimeter, we need to develop the spectra of the signal as a function of altimeter system parameters, angular dependence of $\sigma°(\theta)$, altitude, attitude, velocity vector, and antenna characteristics. The difference frequency from an incremental surface area located at the angular coordinates θ and ϕ can be expressed by

$$f(\theta, \phi) = \frac{2h\Delta F}{cT \cos\theta} + \frac{2V}{\lambda}(\sin\theta \sin\phi \cos\alpha - \cos\theta \sin\alpha)$$

The amount of signal power received from this incremental surface area is given by

$$\Delta P = \frac{P_t G_0^2 \lambda^2 \sigma°(\theta)}{(4\pi)^3 H^2} [g(\theta, \phi)]^2 \sin\theta \cos\phi \, \Delta\theta \, \Delta\phi$$

where the symbols are defined as previously.

Signal spectra may be developed by computing received power and frequency for each incremental area on the surface illuminated by the antenna beam, and then summing power within frequency resolution cells of appropriate width. This operation is best done on a computer.

Broad-Beamwidth FM/CW Radar Altimeter

Computer-generated signal spectra for an FM/CW radar altimeter with an antenna beamwidth of 60° and a frequency modulation slope resulting in a range frequency of 25 kHz for the nearest return from the surface are given in Figures 7.12 and 7.13 for various vehicle velocities. Frequency resolution cells of 100 Hz and angular resolution cells of 0.1° in θ and ϕ were used in generating the data. If a fixed frequency modulation slope is used, the range frequency varies directly with range, and at some altitude the range frequency will be 25 kHz. An alternative is to vary the frequency modulation slope such that the range frequency is held constant, and then measure the slope to determine range. The latter approach is usually used so that the range frequency can be maintained at a relatively high value.

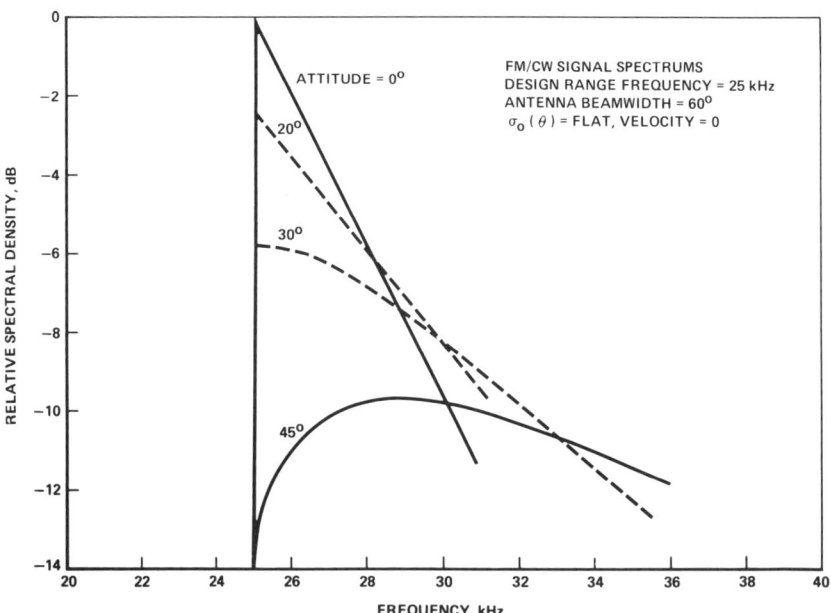

Figure 7.12 Signal spectra in an FM/CW radar altimeter at zero velocity.

Considering first the case of zero velocity, the signal spectra are shown in Figure 7.12 for vehicle attitudes of 0°, 20°, and 45° over land with flat $\sigma°(\theta)$. These spectra have the classical shape described by suppliers of FM/CW radar altimeters. In particular, the lowest frequency corresponds to the minimum range to the surface. In principle, we can configure a leading-edge frequency tracker and determine the frequency corresponding to minimum range to good accuracy, provided that the signal-to-noise ratio is adequate. The shape of the spectra suggests that we may not want to use a center-of-power type of frequency tracker for this application because the long tail on the spectrum will bias the measurement toward higher altitudes.

If the vehicle is moving with substantial velocity parallel to the surface, there is a marked influence on the location of the leading edge of the signal spectra, as shown in Figure 7.13. Unacceptable errors would result from the leading edge tracking measurement under these conditions. Note that it makes no difference if up-sweep or down-sweep modulation, or both, is used because the doppler contamination results from distributed return around the normal to the vertical. The situation worsens if the vehicle is climbing or diving.

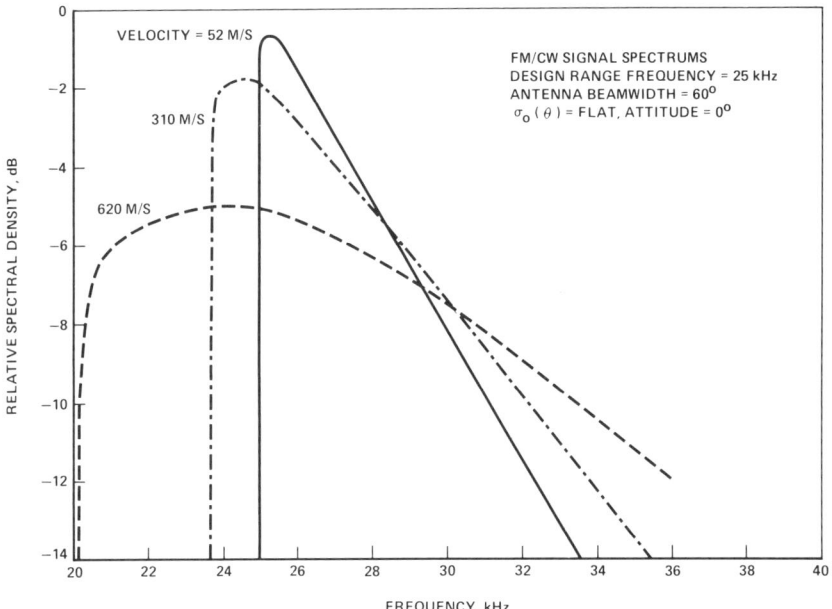

Figure 7.13 Signal spectra in an FM/CW radar altimeter at various velocities at a low-range frequency.

The effect of doppler on the leading edge of the spectrum decreases as the ratio of range frequency to doppler frequency increases. If instead of a range frequency of 25 kHz we use a steeper FM slope and achieve a range frequency of 250 kHz, the contamination of the leading edge of the spectrum becomes very small. In the case of level flight over land, the percent shift of the leading edge due to doppler at 620 m/s is only 0.2% compared to a shift of about 20% at 25 kHz. The disadvantage in using a high range frequency is that the signal density decreases because the same amount of signal power is now spread over a wider bandwidth, and the signal-to-noise ratio decreases.

The FM slope required for good accuracy for a given maximum velocity can best be determined by setting up a computer simulation using the basic equations, and plotting the resulting spectrums for conditions of interest. The shape of the spectra is influenced by the antenna pattern, attitude, and velocity of the vehicle, orientation of the velocity vector, and slope of the backscattering coefficient of the surface as well as the FM slope.

The power density of the signal is available when the signal spectra have been developed. The equivalent noise density in the system can be developed as outlined in Section 7.2.1.2. The signal-to-noise ratio can then be computed and the system parameters selected to provide the ratio required for accurate tracking of the leading edge of the spectrum.

Narrow-Beamwidth FM/CW Radar Altimeter

For those applications which do not require that altitude be measured to the closest point below the vehicle, or where it is indeed necessary to measure range to the surface along a given axis, an antenna with a narrow beamwidth is used. For example, the beamwidth used for the radar altimeter for the Surveyor spacecraft was 4° and aligned with the thrust axis.

The signal processor's requirements for the narrow-beam FM/CW radar altimeter can be established from an examination of the spectral characteristics of the signal at the output of the receiver. The signal spectra can best be developed by computer modeling, as described previously. Considerable variation in the shape of the signal spectra result from various orientations of the antenna beam relative to the normal to the surface and various orientations of the velocity vector relative to the beam axis. For example, if the narrow antenna beam were along the normal and the velocity were very low, the spectrum for an FM up-sweep and down-sweep would have the general appearance shown in Figure 7.14.

Spectral shapes are also shown for the cases of significant velocity, with the velocity vector oriented along the antenna beam and then perpendicular to the beam. The spectra shown are for the case of the scattering coefficient of the surface having a flat angular dependence. In the zero velocity case, the signal spectrum has a sharp leading edge and a somewhat more gradual trailing edge, but the bandwidth is relatively low as a result of the narrow antenna beamwidth. In the case of the velocity vector along the beam, the doppler shift is evident in the shift of the spectra from the zero velocity case. In the case of the velocity vector perpendicular to the beam, the spectra widen somewhat due to doppler, but the leading edge does not shift.

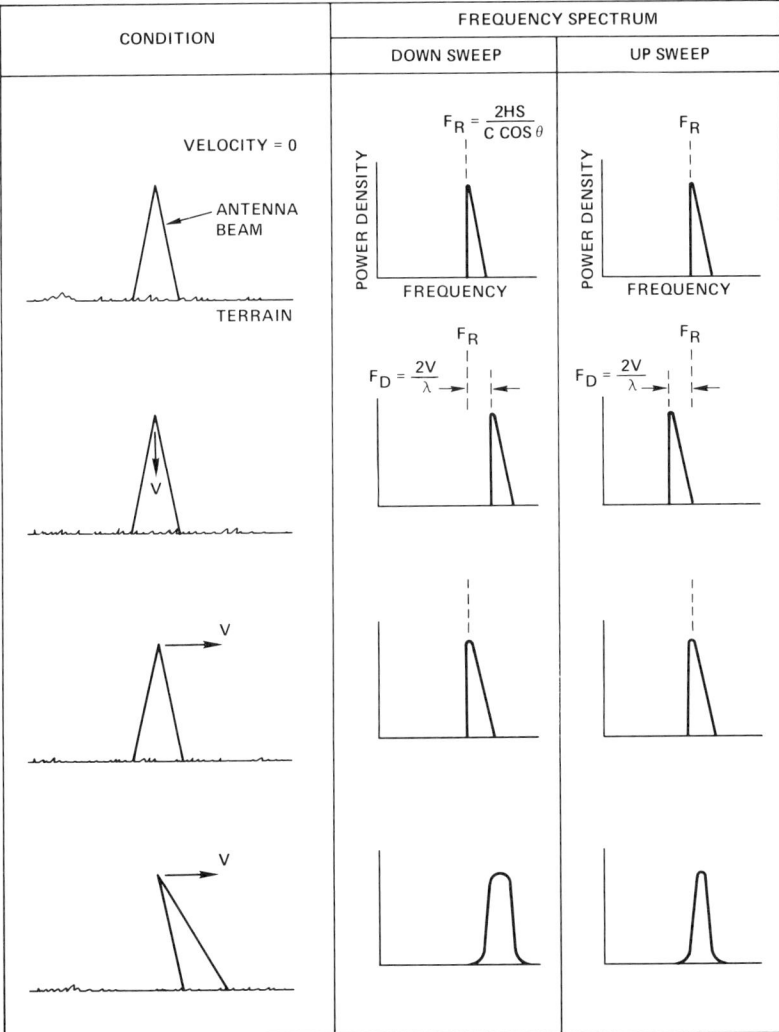

Figure 7.14 General shape of signal spectra in an FM/CW radar altimeter using a narrow-beamwidth antenna.

Signal spectra are also shown for the case of the antenna beam off normal by several times the beamwidth, and the velocity vector perpendicular to the normal to the surface. The effects of doppler adding to or subtracting from the range frequency, depending on whether an up-sweep or down-sweep is used, is apparent in Figure 7.14. Not only is average frequency of the spectrum different for the up-sweep and down-sweep cases, but the bandwidth of the spectra is also substantially different. The range frequency could be determined by using a time-shared center-of-power frequency tracker to measure the center of power of the two spectra. The average of the two frequencies would be the range frequency. Some error could be introduced in this process because the down-sweep interval would have a lower signal-to-noise density ratio and, depending on the type of frequency tracker used, the wider bandwidth interval could lead to a larger tracking error than the narrow bandwidth interval under dynamic conditions. We might note in passing that velocity of the vehicle could, in principle, be determined by subtracting the measured frequencies during the up-sweep and down-sweep. However, due to the difference in signal-to-noise density ratio, and also possible tracking errors as a function of bandwidth, the velocity determination may have substantial error.

Generally, a center-of-power frequency tracker is appropriate for FM/CW radar altimeters, which use narrow-beamwidth antennas. The bandwidths of the signal spectra are usually relatively low in comparison to the center frequencies, and the errors introduced by the nonsymmetrical spectra are usually small. A case-by-case analysis would be required to determine whether a center-of-power tracker or a more complex leading-edge tracker should be used.

7.2.4 Phase-Shift Keyed Radar Altimeter

Phase-shift keyed radar altimeters have recently been developed for use in air vehicles. PSK altimeters could also be used to provide altitude data for a spacecraft landing on a planet.

The PSK radar altimeter uses biphase modulation by a pseudorandom code and subsequent tracking of the received signal by a delayed version of the same code to measure altitude. The pseudorandom codes used are from a family of maximal length sequences. The autocorrelation function of these sequences has a single peak at time coincidence of the code and well behaved, uniform, time sidelobes when the code is displaced by one chip or greater. These sidelobes are down 20 log$N$ from the peak response, where N is the number of chips in the code. The code period and the coherent integration time is usually selected to be less than 20% of the period of the highest doppler frequency to minimize a decrease in the amplitude of the peak correlation response and an increase in time sidelobes due to doppler-induced phase reversals within the code period. If use of a coherent integration time that is short compared to the period of a doppler cycle is not possible, we must use parallel signal processor channels, offset in frequency to accommodate the total doppler spread. Another operational consideration is that isolation between the transmitting and receiving antennas is required because the waveform is continuous. Typical values of isolation are about 80 dB to allow the system to operate down to zero altitude.

There are some advantages to the PSK waveform relative to the short pulse and FM/CW waveforms previously discussed. PSK is a continuous waveform, and hence the peak power and the average power are the same, which is advantageous for the transmitter. Also, the waveform is digitally generated and processed so that both modulation and processing errors are minimized. Finally, doppler effects do not reduce the accuracy of the altitude measurement directly, although doppler effects may be apparent in signal-to-noise ratio.

7.2.4.1 Functional Description of a PSK Radar Altimeter

A functional block diagram of a PSK radar altimeter is given in Figure 7.15. The transmitter consists of an oscillator at the transmitted frequency to feed a PSK modulator. The PSK modulation is controlled by a pseudorandom code from the primary code generator. The output of the PSK modulator is amplified to the required level by a power amplifier and fed to the transmitting antenna.

A portion of the reflected signal from the terrain is intercepted by the receiving antenna and applied to the receiver. In its simplest form, the receiver consists of a balanced mixer, which is excited by a sample of the unmodulated signal from the microwave oscillator. The output of the mixer is a bipolar signal, its polarity dependent on the phase of the received signal. The bipolar video signal is amplified and limited to logic-circuit levels. A leakage nulling loop reduces the direct leakage signal from the transmitting to the receiving antenna to a value below the level of receiver noise.

A loss of about 3 dB in signal-to-noise ratio is suffered by using a single mixer channel, as shown, instead of using quadrature channels and subsequent combining. Quadrature channels may be used for high altitude applications to allow reduction in transmitted power.

The limited output of the video amplifier is applied to a tracking channel correlator and an acquisition channel correlator. Both correlators are exclusive-OR logic elements. The reference input to the tracking correlator is the tracking code, which is a delayed version of the transmitted code. The average voltage at the output of this correlator is half the amplitude of the logic-level input signals for zero correlation, and either the full amplitude of the logic input or zero for full correlation or anticorrelation, respectively. Full correlation occurs when the two codes are identical at the input to the correlator. Full anticorrelation occurs when one code is the mirror image of the other. The two states are dependent on the net RF phase of the received signal relative to the LO signal. Either state represents valid tracking data.

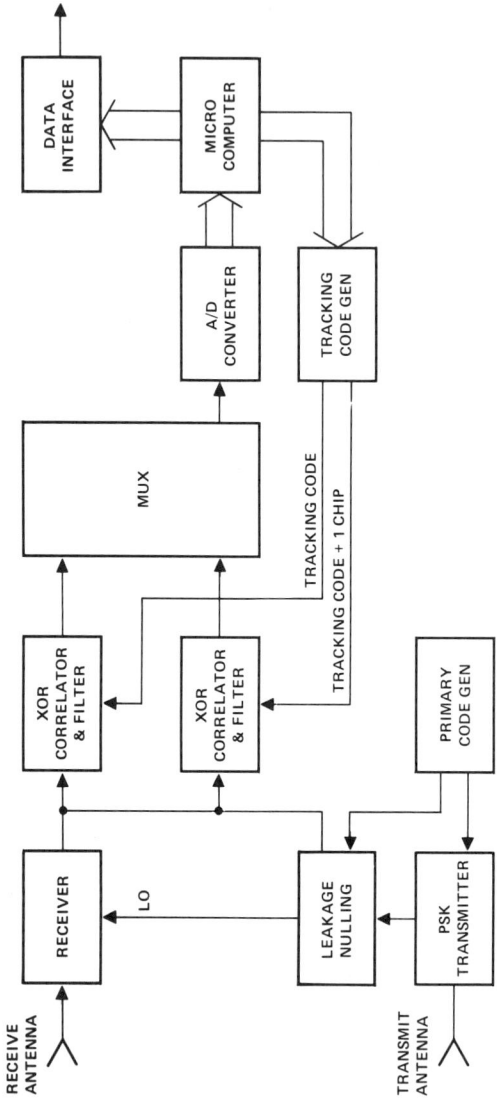

Figure 7.15 Functional block diagram of phase-shift keyed radar altimeter.

The output of the correlator is applied to an integrate-and-dump (I + D) matched filter, which integrates for the duration of the pseudorandom code, transfers the data, and then dumps to start the next interval. The output of the I + D is gathered by a sample-and-hold (S-H) circuit and routed by a multiplexer to an analog-to-digital (A/D) converter. The output of the A/D converter is applied to a microcomputer. The absolute value of the correlation, as measured by the difference between the correlator output and one-half of the logic level, is determined by the computer. It uses this information to command the tracking delay earlier or later to maintain a given level of signal correlation.

The shape of the correlation response as a function of the delay between the tracking code and the received signal is shown in Figure 7.16 for operation over land terrain using a broad-beam antenna. The tracking loop positions the delay of the tracking code such that the correlation response is just above a preset tracking threshold, as indicated in the figure. This positioning results in leading-edge tracking of the correlation response, which is the point of the nearest range to the surface. The acquisition correlator channel forms the correlation of the input signal and the tracking code delayed by one chip. The output of this correlator, which is normally near the maximum correlation of the signal, is used to ensure that the signal level is high enough for accurate tracking.

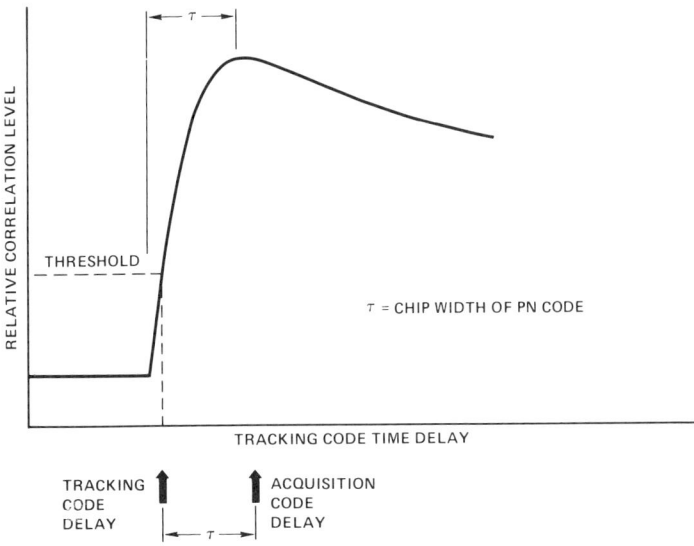

Figure 7.16 Shape of correlator output *versus* tracking code delay for land terrain.

Upon application of power or loss of signal, the computer initiates a search wherein the tracking code is stepped in one-half chip intervals until the signal is encountered. The system then switches to the tracking mode. The delay data generated within the computer while tracking is converted to equivalent altitude and formatted in the data interface circuitry into the required form for the vehicle's computer.

7.2.4.2 Performance Analysis of the PSK Radar Altimeter

The modulation used in the PSK radar altimeter yields a pulse-compressed waveform with an effective pulsewidth equal to the code chip width. The signal-to-noise ratio at a given altitude and attitude is most conveniently determined by computing the SNR for an equivalent short-pulse altimeter and subtracting processing loss unique to the PSK modulation waveform, including doppler effects, from the result. Equations for computing SNR for a short-pulse radar altimeter were given in Section 7.2.1.2

The processing loss includes the effects of hard-limiting on SNR and the effects of doppler shift on the correlation of the received signal. A loss of about 0.9 dB in SNR results from hard-limiting the signal at the output of the video amplifier.

The amplitude of the correlation peak decreases and the time sidelobes of the signal increase as the phase change due to doppler shift becomes significant over the integration time. The loss due to doppler effects for a point source target is shown in the normalized plot of Figure 7.17. For the typical case of operating over a diffuse reflecting terrain, a computer simulation can be set up to compute the composite correlation response at given delays from a field of several thousand scattering points on the surface. The antenna gain as a function of angle off the beam center and the angular dependence of the surface reflectivity function should be included in the simulation. The results of this simulation give the shape of the correlation function as well as the loss due to doppler effects. The time sidelobe levels due to doppler are also given in the resulting plots of the correlation function. The plot shown in Figure 7.16 is a result of such a computer simulation over land terrain with diffuse scattering.

The equation for SNR is evaluated by using average power for the transmitter power and the equivalent noise bandwidth of the I + D matched filter for the noise bandwidth. The effective noise bandwidth of the I + D filter is equal to $1/(2t)$, where t = sample time. The sample time is normally set as the duration of one code period, although several code periods can be integrated if the doppler frequency is low.

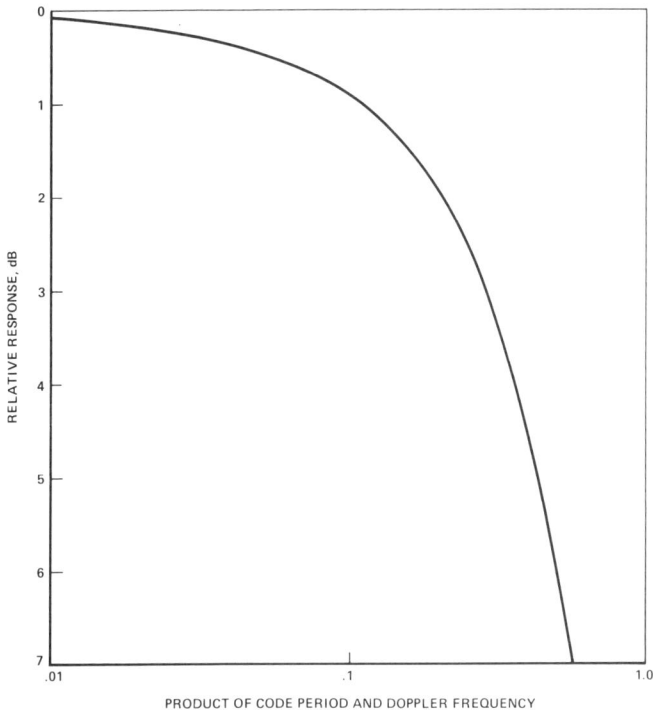

Figure 7.17 Decrease in correlation response as a function of product of code period and doppler frequency.

When the SNR and correlation curves have been developed, the accuracy of the altitude measurement can be assessed. The delay of the tracking point on the correlation function is a function of SNR and attitude of the vehicle and the magnitude and direction of the velocity vector. Unlike the short-pulse radar altimeter, there is not much shift in the shape of the correlation function with the angular dependence of the reflectivity function. As in the case of the short-pulse radar altimeter, we establish a tracking threshold at a given amount up on the correlation curve, which results in an acceptable false-alarm rate on noise alone. The delay between the transmitting and tracking codes is noted when the correlation function intersects the threshold. The error in the measured altitude above the surface can thus be predicted for each flight condition of interest.

7.3 DESCRIPTION OF RADAR ALTIMETERS USED IN SPACE

Several radar altimeters have been deployed on space missions. A chronological listing is given in Table 7.1. Brief descriptions of these instruments follow.

Table 7.1
Radar Altimeters Deployed in Space

Vehicle	Manufacturer	Altitude Range	Mission Dates
Saturn I Launch Vehicle	Teledyne Ryan	30 to 300 mi	1964–65
Surveyor Lunar Lander	Teledyne Ryan	40,000 to 14 ft	1967–68
Apollo Lunar Module	Teledyne Ryan	40,000 to 10 ft	1969–71
Skylab S193	General Electric	435 km	1973
Viking Mars Lander	Teledyne Ryan	450,000 to 50 ft	1975
GEOS-C Earth Orbiter	General Electric	843 km	1975
SEASAT-A	Applied Physics Laboratory	800 km	1978
Pioneer Venus Orbiter	Hughes Aircraft	4700 to 200 km	1978
GEOSAT	Applied Physics Laboratory	800 km	1985

7.3.1 Radar Altimeter for the Saturn I Launch Vehicle

A radar altimeter was used for range instrumentation during test firings of the Saturn I booster rockets. This altimeter reported altitude of the vehicle via telemetry during its ascent to the point of payload separation. The radar altimeter was included on five flights of the boosters. A photograph of it is shown along with a sketch of the vehicle in Figure 7.18. The rather rugged structure was designed to allow the radar altimeter to remain pressurized to avoid high-voltage breakdown as the rocket ascended out of the earth's atmosphere.

The radar altimeter weighed 11.8 kg (25.9 lbs), and it was 292 × 228 × 241 mm (11.5 × 9.0 × 9.5 in.) in size. It consumed 65 W of power at 28 Vdc [5]. The weight of the radar altimeter was not an important consideration for the test firings of the very powerful booster rockets. The Saturn radar altimeter was designed to operate at altitudes from 20 to 556 km (11 to 300 nmi). It was operated to a maximum altitude of 499 km during the rocket test program. The accuracy of the device was ±32 m.

A simplified block diagram of this radar altimeter is shown in Figure 7.19. The transmitter was a re-entrant cavity oscillator which used a planar triode tube. It was pulsed on by the modulator, which generated 3500 V pulses 1 μs long at a PRF of 144 p/s. The transmitter generated a peak power of 5 kW at a frequency of 1610 MHz. The local oscillator was a low-power cavity device that operated at 1580 MHz.

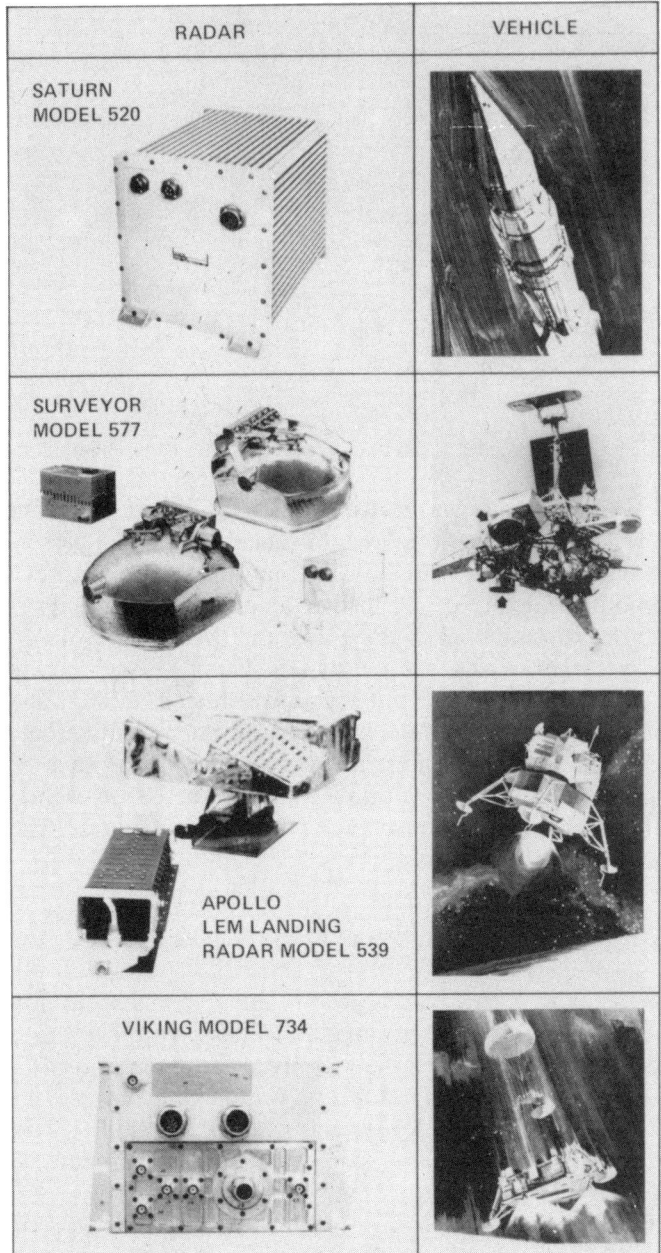

Figure 7.18 Photographs of some radar altimeters that have been used in space. (Courtesy of Teledyne Ryan Electronics.)

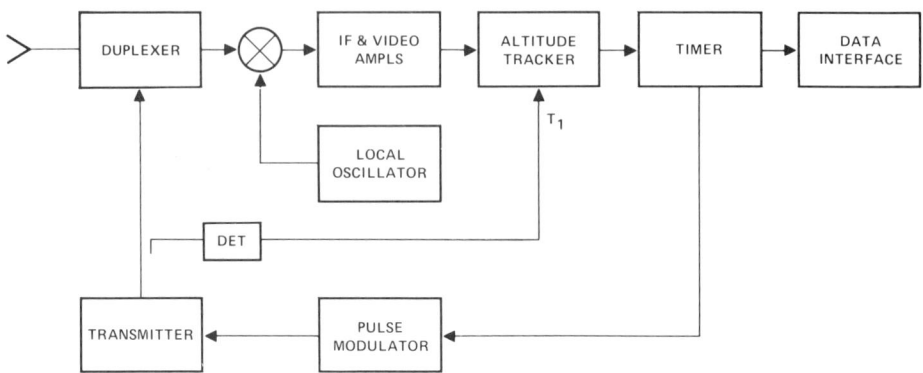

Figure 7.19 Simplified block diagram of short-pulse radar altimeter used on Saturn booster rockets.

The transmitter output was passed through a solid-state duplexer to the antenna, which was part of the rocket system. The return signal passed through the duplexer to a balanced mixer. The 30 MHz intermediate frequency at the mixer output was amplified, detected and further amplified at video. The video signal was applied to the altitude tracker along with the detected transmitted pulse, T_1.

The range tracker, which was a early-late gate type, searched for the signal, tracked it, and read it out as a counter gate which started at T_1 and ended with the detection of the return pulse. The length of the gate was therefore proportional to the rocket's altitude. The counter gate was applied to the timer module, where it was used to gate a 21 MHz clock into a counter. The counts were averaged and formatted into an 18-bit parallel binary word to represent altitude. The data update rate was 36 words per second.

7.3.2 Radar Altimeter for the Surveyor Lunar Lander

A radar altimeter with a narrow-beamwidth antenna was used on the Surveyor lunar lander to measure slant range along the velocity vector to the surface of the moon. A doppler sensor measured the velocity along the thrust axis as well as the lateral velocity components. The latter data were supplied to the control system to align the thrust axis with the velocity vector. The radar altimeter data were used along with velocity data to control the spacecraft to follow a predetermined velocity *versus* altitude profile.

The radar altimeter operated from 40,000 to 13 ft above the surface [6]. Its last important task on each flight was to provide a marking signal at an altitude of 13 ft, which was used to shut off the vernier rocket engines. This was done to minimize contamination of the lunar surface and to allow the spacecraft to settle on the surface.

Altitude measurements were made by means of a linear FM/CW ranging system operating in conjunction with a narrow-beam antenna aligned with the thrust axis. A photograph of the Surveyor landing radar, which contained both a doppler velocity sensor and radar altimeter, is shown in Figure 7.18. The accuracy of the altitude measurements was within ±4 ft plus 5% of altitude. The weight of the landing radar was 35 lbs and it consumed 590 W of primary power.

A simplified block diagram of the radar altimeter portion of the system is shown in Figure 7.20. The transmitter consisted of a 0.4 W reflex klystron, operating at a center frequency of 12.9 GHz. The reflector was modulated by a linear saw-toothed voltage waveform to generate linear frequency modulation. The resulting FM had a repetition rate of 182 Hz and a peak-to-peak deviation of 4 MHz at altitudes above 1000 ft, and 40 MHz below 1000 feet. The range scale factor, as given by $F_r = 2RS/c$, where R is the range, S is the slope of the frequency modulation, and c is the propagation velocity, is equal to about 1.5 Hz/ft at altitudes above 1000 ft, and 15 Hz/ft at altitudes below 1000 ft. In addition, a doppler shift up to 20 kHz added to the range frequency.

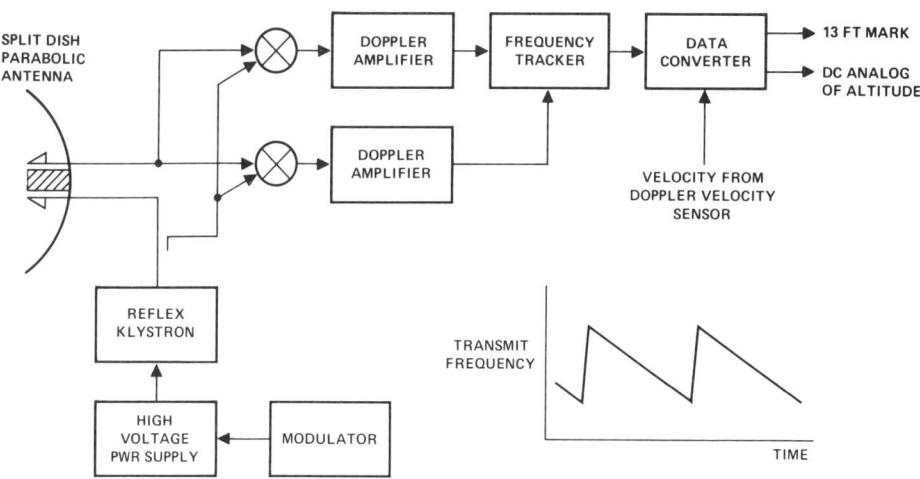

Figure 7.20 Simplified block diagram of radar altimeter used on the Surveyor spacecraft.

The antennas were of the split-dish parabolic type with two pairs of transmitting and receiving beams formed by the same antenna. One of the velocity sensor beams and the altimeter beam were generated by one antenna. Isolation between transmitting and receiving portions of the antenna was obtained by mounting the feed horns on opposite sides of a metalic septum. The beamwidth of the transmitting and receiving beams were about 4°. The received signal from the antenna was split into quadrature pairs and fed to two pairs of balanced mixers, which were excited from a sample of the transmitted signal. The difference frequency at the mixer outputs was a quadrature pair proportional to range, plus the doppler shift. These signals were amplified by means of wideband audio amplifiers, which gated out the signal return during the flyback portion of the modulation waveform.

The amplified signals were applied to a frequency tracker. It searched for the signal over the expected frequency range and, once acquired, tracked it with an accuracy of better than 0.25%. The signal acquisition program was arranged to sweep an 80 kHz band in 1.5 s and provide a probability of acquisition of 95% per sweep at SNRs of 6 dB or greater within the relatively narrow tracker bandwidth. The output of the tracker was a frequency equal to the center of power of the input signal spectrum plus a 600 kHz carrier.

The outputs of the frequency trackers in the doppler velocity sensor were combined to form a frequency term proportional to the velocity along the roll axis, which was then subtracted from the output of the altimeter frequency tracker, yielding only the range frequency. The range frequency was converted to a dc analog voltage, which was buffered and sent to the spacecraft control loops.

7.3.3 Radar Altimeter for the Apollo Lunar Module

A radar altimeter was used in conjunction with a doppler velocity sensor to provide slant-range or altitude and velocity data to the guidance computer of the Apollo 11 Lunar Module (LM). Data from an *inertial measuring unit* (IMU) was used for guidance at high altitudes prior to signal acquisition by the radar. After signal acquisition, the IMU and radar data were mixed with appropriate weighting to obtain the best altitude and velocity data. The weighting was arranged such that the IMU data dominated at high altitudes and the radar data dominated at lower altitudes.

The radar altimeter, which was a linear FM/CW type, was designed to operate from 40,000 to 10 ft [6]. Slant-range measurements were made along a narrow-beamwidth antenna, which was generally directed towards the vertical by a two-position antenna. The accuracy of the range measurement was within ±5 ft or 1.4% of altitude, whichever was smaller. (A photograph of the landing radar for the LM was shown in Figure 7.18.) It weighed 16 kg and consumed 132 W of primary power.

During the historic landing of the LM, the radar altimeter locked on to the return from the lunar surface at 44,000 ft. The altimeter provided slant-range data down to 10 ft. Armstrong and Aldrin essentially used radar data for manual control of the vehicle to a soft landing.

A simplified block diagram of the altimeter portion of the landing radar is shown in Figure 7.21. The transmitter was a frequency multiplier which generated about 0.2 watts at 10 GHz. The frequency multiplier was driven by an FM source

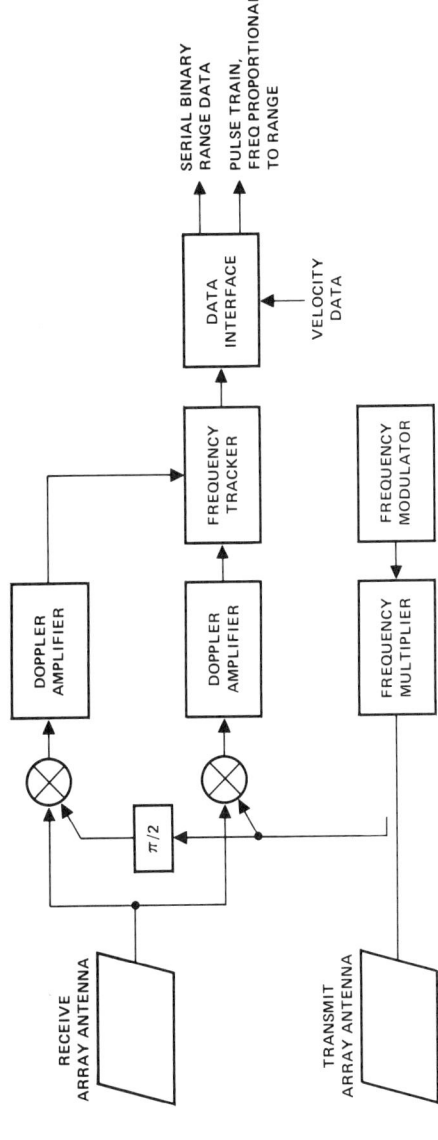

Figure 7.21 Simplified block diagram of the radar altimeter used on the Apollo Lunar Module.

operating around 100 MHz. The linear FM was saw-toothed in form with a peak-to-peak deviation of 8 MHz at altitudes above 2500 ft and 40 MHz at altitudes below 2500 ft. The repetition rate of the modulation pattern was 130 Hz.

The transmitter output was fed to a waveguide array antenna, which was interlaced with the velocity sensor array into a single aperture. A separate receiving array antenna was used. The received signal was split into a quadrature pair and applied to a pair of balanced mixers, excited from a sample of the transmitted signal. The difference frequency at the output of the mixers represented the range frequency component, $f_R = 2RS/c$, plus the doppler frequency component, $f_D = 2V/\lambda$.

The two quadrature signals were amplified by wideband, low-noise, audio amplifiers and applied to a frequency tracker. The frequency tracker searched for the relatively narrowband altimeter signal over the full frequency range, and, when acquired, tracked it with an accuracy of about 0.1%. The output of the frequency tracker was applied to the data interface unit, where the doppler frequency component of the signal was removed by using data from the velocity sensor. Range information was sent as output from the data interface unit, in serial binary form, and sent to the LM's guidance computer.

7.3.4 Radar Altimeter for Skylab S-193

A radar altimeter experiment was carried out on the Skylab S-193 mission along with radiometer and scatterometer experiments, which shared the antenna and much of the transmitter and receiver. The purpose of the radar altimeter experiment was to determine the feasibility of measuring the geoidal profile of the ocean-covered portion of the earth from orbit and also to discern wave height and wind speed. The experiment was intended to serve as a source of experimental data for the design of future satellite radar altimeter systems [7].

The antenna for the system was a parabolic dish of 44 in. diameter, gain of 42 dB, and beamwidth of 1.5° at the operating frequency of 13.9 GHz. The transmitter was a traveling wave tube amplifier, which generated a peak power of 2 kW at a PRF of 250 p/s and a pulsewidth of either 10 or 100 ns. The system used a coherent receiver containing a tunnel diode low-noise amplifier, which resulted in an overall receiver noise figure of 5.5 dB. The IF was 350 MHz, and the bandwidth was 100 MHz for the 10 ns pulse mode and 10 MHz for the 100 ns mode.

The IF amplifier output was detected and the resulting video pulses were applied in parallel to the altitude tracker and the sample-and-hold (S/H) circuits used to sample the pulse waveform. The altitude tracking loop was a three-gate type with a noise gate located before the onset of the pulse, a ramp gate, and a plateau gate. The tracker was digitally implemented with 200 MHz logic circuitry. The resolution of the altitude measurement by the altitude tracker was 38 cm (1.25 ft).

The radar altimeter could be commanded to operate in five different modes while measuring altitude. These modes were pulse shape, radar cross section, time correlation, pulse compression, and nadir alignment. All of the radar data gathered were recorded on a digital tape recorder. The types were brought back to earth for analysis.

The pulse shape mode gathered detailed pulse-by-pulse waveform information on the received signal. The received pulse was sampled at eight points with sample spacing of 10 and 25 ns for transmitted pulse lengths of 10 and 100 ns, respectively.

The radar cross section mode provided measurements of the backscattering cross section per unit surface area for all types of terrain at normal incidence and as a function of angle up to 15° degrees from the nadir.

The time correlation mode transmitted a pair of pulses with spacing variable from 1 µs to 1 ms to determine the maximum PRF at which statistically independent samples of altitude data could be obtained.

The pulse compression mode consisted of both 10 ns uncompressed and 10 ns compressed pulse operation to compare the performance and characteristics of both waveforms. The pulse compression mode used biphase modulation by a 13-bit Barker code.

The nadir alignment mode evaluated the feasibility of aligning the antenna toward the nadir by steering it until the signal strength of the gated received signal was a maximum.

The results of the altimeter experiment indicated that the mean return waveforms from the sea were in good agreement with theoretical waveforms for both 10 and 100 ns pulsewidths. The instrument's resolution of approximately 1 m clearly depicted the depression in the elevation of the surface of the sea in the vicinity of the Puerto Rican trench, which amounts to about 10 m. Vertical profiles of land masses were also made.

7.3.5 Radar Altimeter for the Viking Mars Lander

A radar altimeter system, which operated from about 137 km to 41 m (450,000 to 135 ft) above the surface of Mars, provided altitude information to the guidance computer of the Viking Lander. Two independent radar altimeters were used to provide redundancy. (A photograph of one of the two units in the redundant pair was shown in Figure 7.18.) Each unit was 254 × 203 × 152 mm (10 × 8 × 6 in) in size and weighed 4.8 kg (10.6 lbs). The input power was 26 W. The specified accuracy was 1.5% or 152 m at altitudes above 6.1 km (20,000 ft) and 4.5% plus 1.5 m at altitudes below 6.1 km.

The radar altimeter operated while the aeroshell was in place and after it was jettisoned. Consequently, two antennas were used, one mounted on the aeroshell and the other located on the bottom of the lander. An antenna switch, mounted to one of the radar altimeter units, selected one of the two antennas, and switched the selected antenna to the altimeter being used. If the in-flight self-test failed on one unit, the other would be selected. If the selected radar altimeter did not lock on to the Mars surface during entry, the computer would automatically switch to the other radar altimeter.

The performance of the radar altimeters on the two Viking spacecraft that landed on Mars were nearly identical [8]. Both locked on to the surface during the first sweep after power was turned on at an altitude of slightly less than 800,000 ft. This was a second-time-around lock, resulting from the return of the echo signal after the next pulse had been transmitted. This event was expected, and the data were used to reconstruct the trajectory. Unlock was forced at 700,000 ft and the altimeter was not allowed to relock until the altitude dropped below the unambiguous range of 450,000 ft. Acquisition again took place during the first sweep after the spacecraft had descended into the tracking range. On both flights, the radar altimeters remained locked and provided altitude data until the minimum operating altitude of 132 ft was reached. The radar altimeter data were used for continuously updating the inertial navigator during the descent. The inertial system provided altitude data for the final 132 ft.

The radar altimeter was a noncoherent pulse type, which operated at 1.0 GHz. The aeroshell antenna had a beamwidth of about 20° by 80°. The lander antenna had a nearly hemispherical pattern. With these large beamwidths the radar altimeter measured the distance to the nearest point on the surface. The minimum altitude was limited to 135 ft because a single antenna was used for transmitting and receiving.

A simplified block diagram of the radar altimeter is shown in Figure 7.22. The operating frequency was obtained by using frequency multiplication from a crystal-controlled source. This 1.0 MHz signal was applied to a power amplifier through an RF switch, which served as the pulse modulator. The power amplifier contained four power stages, the outputs of which were combined by hybrid circuits to achieve about 140 W peak power. The pulsewidth was 6.0 μs at altitudes above 20,000 ft, 700 ns at altitudes between about 2000 to 20,000 ft, and 50 ns at altitudes below 2000 ft. The transmitted pulse was applied to the antenna by use of circulator.

The echo signal received by the antenna passed through the circulator and a limiter, and was amplified by a low-noise RF amplifier. The amplified signal was applied to a balanced mixer with a local oscillator (LO) signal, which was coherent with the transmitted signal, but displaced by 75 MHz from it. The 75 MHz IF signal at the mixer output was amplified to a level set by a constant false-alarm rate (CFAR) automatic gain control (AGC) circuit. It operated from a noise-gated sample that was displaced from the signal range. The signal at the output of the IF amplifier was converted to baseband by mixing in quadrature with a second LO signal of 75 MHz. Mixing in quadrature and combining after detection was necessary to avoid a loss of about 3 dB in SNR due to the coherent relationship of

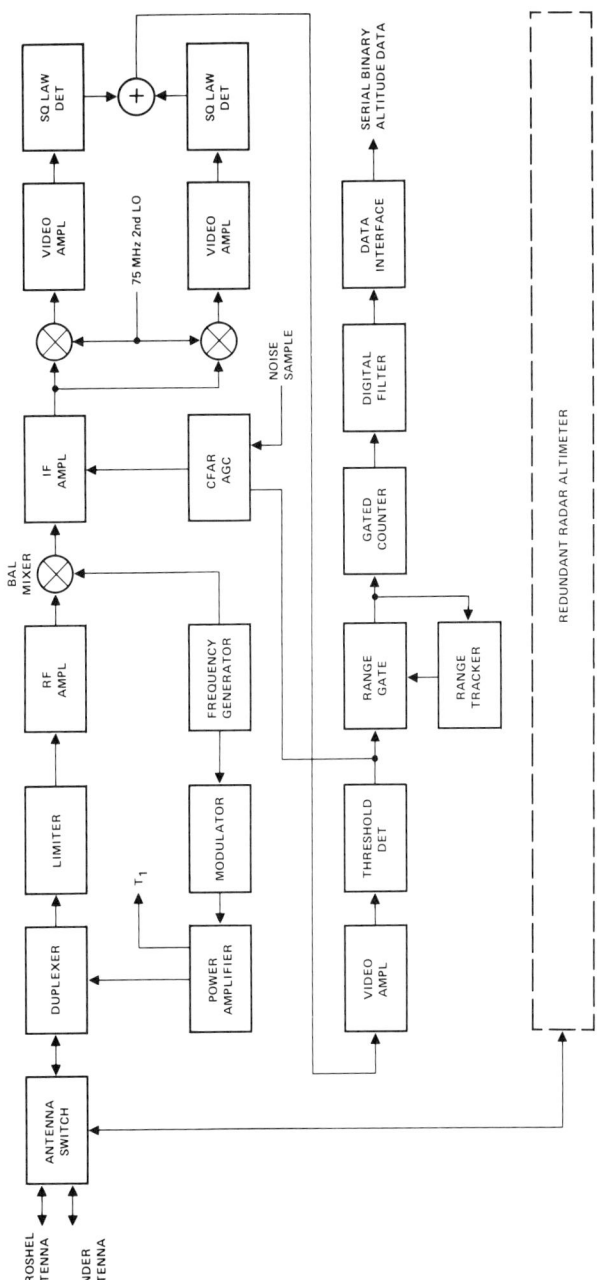

Figure 7.22 Simplified block diagram of the radar altimeter used on the Viking spacecraft.

the transmitter and receiver and the random phase of the return signal. The outputs of the two quadrature mixers were applied to variable bandwidth low-pass filters. The bandwidth was varied to match the pulsewidths used. The filter outputs were square-law detected and then combined to form the video signal for the altitude tracker.

The altitude tracker was a conventional split-gate type, which positioned an early and late gate around the received pulse. The two gates together constituted the range gate. Altitude was measured by starting a 30.6 MHz counter on the leading edge of the transmitted pulse, T1, and stopping it on the leading edge of the echo signal. The resulting count was used to load a shift register, the serial output of which was applied to a digital filter. A direct measurement of range delay with the high-speed counter was used, rather than taking data from the tracking loop, to avoid data lag due to the loop filter during the high dynamics of the mission. The 10 m quantization of the delay measurement by the 30.6 MHz clock was reduced to less than 1 m by a digital filter that operated on the altitude data before being sent to the spacecraft. The digital filter also generated altitude rate information, which was provided along with altitude on the output data bus.

7.3.6 Radar Altimeter for the GEOS-C Spacecraft

A unique radar altimeter was flown on the earth-orbiting GEOS-C spacecraft. Its mission was to demonstrate the feasibility of mapping the topography of the ocean surface to an absolute accuracy less than five meters and relative accuracy of less than one meter, and determining wave height of the ocean from measurements of the shape of the echo pulse [9].

A photograph of the GEOS-C Radar Altimeter is shown in Figure 7.23. The device weighed 68 kg and occupied a volume of 0.12 m^3. The antenna, which is at the top of the picture, was a parabolic dish of 0.6 m diameter.

The radar altimeter had two basic modes of operation: global and intensive. There were several submodes in each. The global mode provided altitude measurements with a standard deviation of less than 50 cm about a fitted mean at a data rate of one per second. Global operation also provided measurement of the backscattering cross section per unit surface area of the ocean to an accuracy of 1 dB. An internal calibration submode provided an absolute accuracy of one meter for the altitude measurements. The system operated as a noncoherent pulsed radar altimeter in the global mode.

The intensive mode provided altitude measurements with a standard deviation of 60 cm about a fitted mean at a data rate of ten per second. Intensive operation also provided backscattering cross section measurements and internal calibration to the same accuracy as in the global mode. The intensive mode contained a high-speed waveform sampling submode, which divided the received pulse into 16

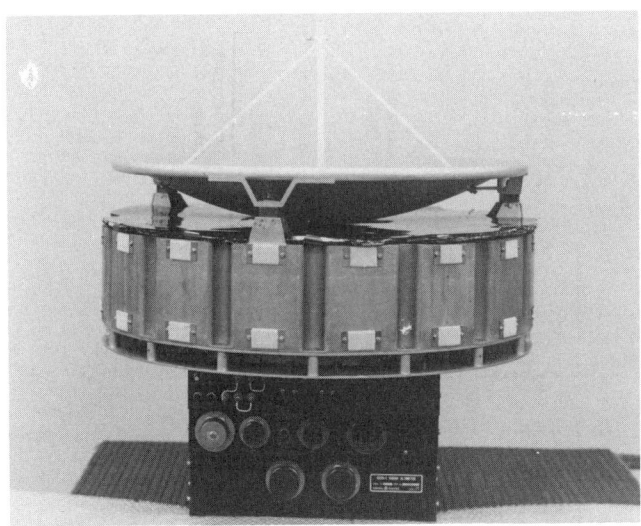

Figure 7.23 Photograph of the radar altimeter used on the GEOS-C spacecraft. (Courtesy of Thomas Godbey, General Electric Company.)

sample intervals and measured the amplitude of the pulse waveform in each. The data were subsequently used to determine wave height of the ocean. The system operated as a pulse-compressed radar altimeter in the intensive mode. A linear FM (chirp) pulse compression of a factor of about 100 was used. The resulting compressed pulse had a width of 12.5 ns.

The waveform sampler used in the intensive mode had cell widths of 12.5 ns and spacing between cells of 6.25 ns. The time position of the sample gates was fixed with respect to the altitude tracking gates. The outputs of the 16 sample gates were provided on the basis of individual pulses so that individual pulse return shapes could be reconstructed as well as the average pulse shape. Wave height can be inferred from the pulse shapes because, for a smooth sea, the rise and fall of the echo pulse are relatively steep whereas, for a rough sea, both the rise and fall are more gradual.

A simplified block diagram of the GEOS-C radar altimeter is given in Figure 7.24. The system operated at a frequency of 13.9 GHz. The 0.6 m diameter parabolic antenna provided gain of 36 dB and 3-dB beamwidth of 2.6°. The transmitter used for the global mode was a magnetron, which generated 2 kW of power at a pulsewidth of 1 μs during acquisition and 200 ns while tracking. The PRF was 100 Hz during acquisition, and a burst of 16 pulses separated by 204.8 μs was transmitted at the 100 Hz rate while tracking. The burst waveform yielded a higher PRF to improve the performance of the altitude tracker while allowing the altimeter to operate in an unambiguous manner.

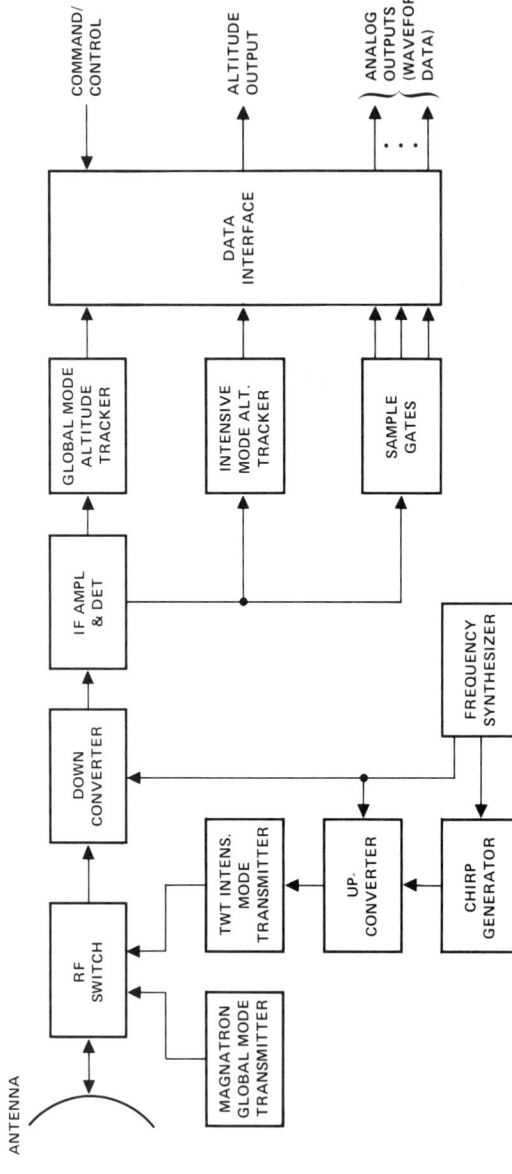

Figure 7.24 Simplified block diagram of GEOS-C radar altimeter.

The transmitter used for the intensive mode was a traveling wave tube amplifier, which generated 2.5 kW at a pulsewidth of 1.2 µs and PRF of 100 Hz. A linear chirp signal was generated at 300 MHz by means of a dispersive delay line and up-converted to the operating frequency of 13.9 GHz. This chirped pulse with a duration of 1.2 µs was amplified to the 2.5 kW level by the traveling wave tube.

The traveling wave tube was not also used for the global mode because a limited amount of primary electrical power was available on the spacecraft. The power consumption for using the magnetron in the global mode was 71 W compared with 126 W in the intensive mode.

One or the other of the two transmitters was switched to the antenna port by means of an RF switch module, which also contained the transmit-receive duplexer. The received signal was down-converted to an IF of 300 MHz. In the global mode, the IF amplifier had a bandwidth of 40 MHz. It was followed by a square-law detector and the resulting video signal was applied to the global mode altitude tracker. In the intensive mode, the 300 MHz IF signal was applied to a dispersive delay line, which performed the pulse compression. A pulse compression ratio of about 100 was used, which resulted in a compressed pulsewidth of 12.5 ns. The pulse-compressed signal was applied to a square-law detector and the resulting video signal was applied to the global mode altitude tracker.

The altitude trackers for the global and intensive modes were the same, except that the time discriminator gates were different so as to match the respective pulsewidths used. An AGC function in the receiver maintained a constant average pulse amplitude at the input to the altitude trackers to improve the accuracy of the altitude measurement. The AGC dynamic range was 50 dB. The tracking loop was implemented largely with high-speed digital circuitry. Three gates were used to develop the time discriminator for the tracking function: noise gate, ramp gate, and plateau gate. These gates were 200 ns wide for the global mode and 12.5 ns wide for the intensive mode. The separation between the gates was 100 ns for the global mode and 60 ns for the intensive mode.

The output of the altitude trackers was averaged for 1.0 s for the global mode and 0.1 s for the intensive mode. The altitude measurement output was a 32-bit serial word with a significant bit weight of approximately 0.0156, 0.156, or 1.56 ns, depending on the process period commanded.

A wealth of data was provided by the radar altimeter during the GEOS-3 mission. An excellent summary of the data is given in the July 1979 issue of the *Journal of Geophysical Research* (Vol. 84, No. B8) [10], which was devoted to that subject. The radar altimeter data provided enhanced detail of the ocean geoid over most areas of the world's oceans, and it allowed determination and location of temporary departures from the geoid [10]. For example, the boundaries of the Gulf Stream and its dynamic heights were identified, and variations in elevation of the sea surface in the vicinity of ocean trenches and sea mounts were measured in good detail. These variations in the elevation of the sea surface are generally less than 10 m. In addition, analysis of waveform data allowed determination of sea state information that was in good agreement with data collected by buoy.

7.3.7 Radar Altimeter for the SEASAT-A Ocean Dynamics Satellite

The SEASAT-A radar altimeter was the third in the series of precision radar altimeters designed to detect undulations in the mean sea level from orbital altitudes. It followed the SKYLAB and GEOS-C radar altimeters previously described. It was designed to allow resolving surface topographical features of oceanographic interest including ocean currents, tides, wave heights and coastal upwellings [11]. It was designed to operate at an orbital altitude of 800 km and provide altitude measurements with a precision better than 10 cm. The "stretch" pulse compression technique discussed in Section 7.2.2 was used to achieve an effective compressed pulsewidth of 3.125 ns.

A simplified block diagram of the SEASAT-A radar altimeter is given in Figure 7.25. The antenna, which was a parabolic reflector with a diameter of 1 m, had a 3-dB beamwidth of 1.6° at the operating frequency of 13.5 GHz. The transmitter consisted of a 2 kW traveling wave tube which was driven by a chirped source. The chirped signal was generated by a dispersive delay line (DDL) operating at 250 MHz. This was followed by an up-converter and frequency multiplier to provide a chirped signal at 13.5 GHz to drive the transmitter, and a chirped output at 13.0 GHz to serve as the LO. A trigger pulse, applied to the DDL, initiated the chirped transmit pulse.

After a time approximately equal to the round-trip time to the surface, the DDL was triggered again to develop the LO signal. Because the FM slope of the LO and the received signal are the same, time delay within the received signal maps into frequency offset. Frequency filtering can then be used to resolve differences in time delay. The time delay between the triggering of the DDL for the transmitted pulse and received pulse was controlled by the altitude tracking loop.

The output of the first mixer was amplified in a wideband IF amplifier and applied to a second mixer. The second mixer removed the 50 MHz IF and provided quadrature outputs at baseband to dual A/D converters. The converter's outputs were applied to a digital filter bank, which provided 60 samples of the waveform. Each sample was, in effect, 3.125 ns wide and spaced 3.125 ns from the previous sample.

The altitude tracking function was based on leading-edge tracking of the return signal. An adaptive tracking technique was employed, which used the outputs from various filters arranged in triplets to form a three-gate tracker. For smooth sea conditions, where the leading edge of the return pulse would be very

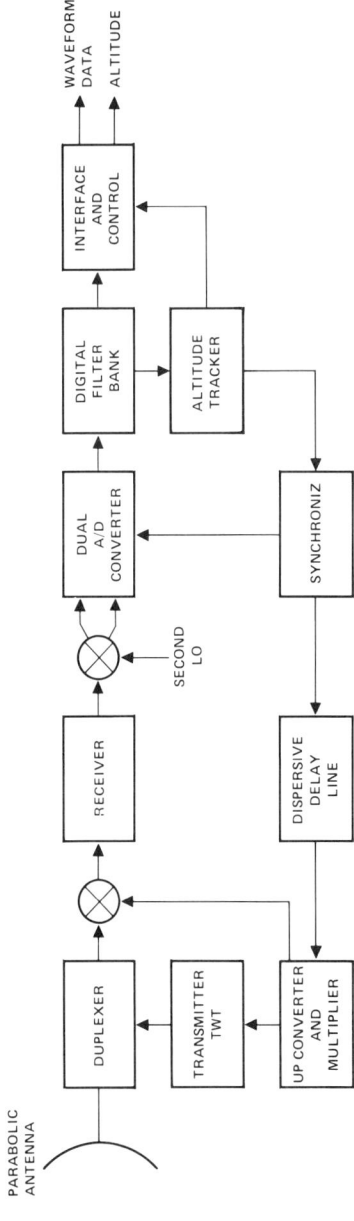

Figure 7.25 Simplified block diagram of SEASAT-A radar altimeter.

sharp, the three processing gates were the minimum width (3.125 ns). When the instrument sensed a rougher sea, evident by the leading edge being stretched out in time, more basic resolution cells were combined to form appropriately wide processing gates. As a result, the tracking point was nearly the same for a wide range of wave heights. Both altitude data and waveform sampling data were formatted in the interface as digital output.

The SEASAT-A satellite was launched on June 27, 1978 and operated until October 10, 1978 when a power failure terminated the operation of the radar altimeter. During its relatively brief operating time, the radar altimeter provided a large amount of high-quality data. The 10 cm resolution measurements allowed identification of probable sea mounts in uncharted areas of the ocean and detection of variations of the Gulf Stream and its associated eddy rings. A summary of test results is given in the April 1982 issue of the *Journal of Geophysical Research* (Vol. 87, No. C5), which was devoted to that subject.

7.3.8 Radar Altimeter for the Pioneer Venus Orbiter

The Pioneer Venus Orbiter spacecraft carried a radar that performed both imaging and altimeter functions. The spacecraft was placed in a highly eccentric orbit about Venus to meet atmospheric sampling objectives. The radar altimeter was operated at altitudes between about 4700 and 200 km [12]. The orbiter was spin stabilized with a rotational period of 12 s. Radar altimeter operation was possible for about 1 s each rotational cycle when the normal to the planet surface was within the antenna beamwidth.

The radar instrument consisted of an electronics box weighing 7.4 kg and an antenna weighing 2.3 kg. The operating frequency was 1757 MHz. The antenna, which was a reflector type with a diameter of 38 cm, had a 3-dB beamwidth of 29° by 25° and gain of 15 dB. The power consumption of the radar was 15 W.

A simplified block diagram of the radar altimeter is shown in Figure 7.26. The instrument did not contain an altitude tracking function. Instead, the altimeter measured the amplitude of the echo signal within a series of range bins, the signal then telemetered back to earth for processing.

A pulse-compressed waveform, consisting of biphase modulation of the transmitted pulse by a 55-chip pseudorandom noise code, allowed operation with a peak transmitted power of 20 W. The chipwidth was 4 μs at altitudes below 1515 km, and 6 μs at altitudes above 1515 km. The corresponding total pulsewidths were 220 and 330 μs. The peak transmitted power was 20 W.

The echo signal intercepted by the antenna was fed to a receiver. The LO frequency for the receiver was adjusted by means of a frequency synthesizer to compensate for the high doppler shift on the echo signal during portions of the trajectory. The output of the receiver was applied to a signal processor, which contained a correlator to decode the pseudorandom code within the received signal

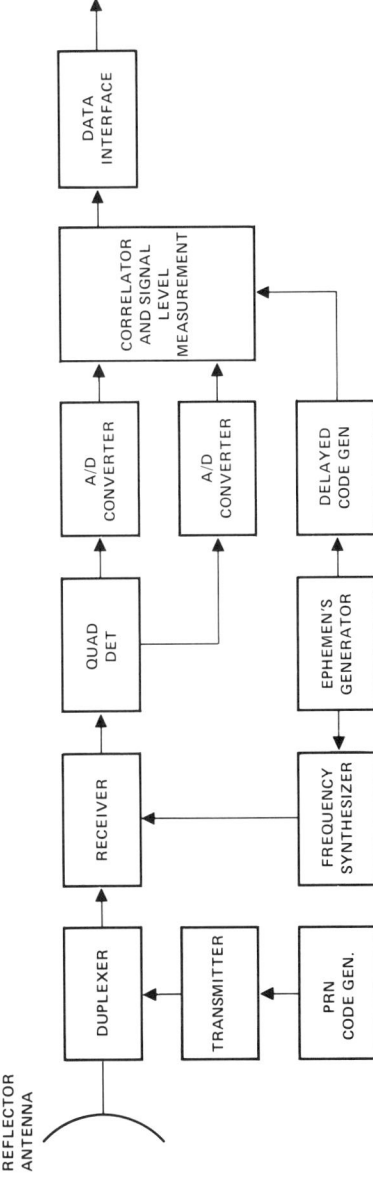

Figure 7.26 Simplified block diagram of Pioneer Venus Orbiter radar altimeter.

pulse. The correlation was performed sequentially for 64 different delays spaced at intervals of a chipwidth.

The approximate delay for the echo signal was obtained from an ephemeris generator, and the different delays were positioned above and below that value. The time delay window was wide enough to accommodate both ephemeris inaccuracy and topographic variations. The magnitude of the signal after correlation in the 64 range cells was measured. The signal magnitude for 17 delay positions, centered on the one with the strongest response, was telemetered to earth to allow construction of profile of the terrain under the spacecraft.

7.3.9 Radar Altimeter for the GEOSAT Satellite

The primary mission of GEOSAT was to provide a dense global grid of high precision radar altimeter data over the world's oceans. The data was used to refine measurements of the earth's gravitational field. To this end, height was measured to an accuracy of 3.5 cm for a 2 m significant wave height [12]. The altimeter data also yielded measurements of significant wave height to an accuracy of 10% of wave height or 0.5 m (whichever was greater) and wind speed to an accuracy of 1.8 m/s for up to 18 m/s.

The satellite was launched in March 1985. The orbital altitude was 800 km and inclination was 108°. Detailed descriptions of the GEOSAT mission, the radar altimeter, and test data are given in the Johns Hopkins APL *Technical Digest* (Vol. 8, No. 2, 1987), which was devoted to GEOSAT.

The GEOSAT radar altimeter was an improved version of the SEASAT-A radar altimeter described in Section 7.3.7. Changes included use of a 20 W, long-life, traveling wave tube for the transmitter instead of the 2 kW TWT used on SEASAT, an increase in the uncompressed pulsewidth from 3.2 μs to 102.4 μs, and a decrease in receiver noise figure from 9 to 5 dB. In addition, to accommodate the rather long uncompressed pulsewidth, digital synthesis was used to generate the chirped trasmitted pulse rather than using a SAW device.

The GEOSAT radar altimeter consisted of an RF section to which the antenna was attached, and a signal processor. The combined weight of the system was 87 kg. It consumed 146 W of primary power. The antenna was a parabolic reflector with a gain of 37.6 dB and a beamwidth of 2.0° at the operating frequency of 13.5 GHz.

As in the SEASAT radar altimeter, the stretch pulse-compression technique was used to achieve an effective compressed pulsewidth of 3.125 ns. The stretch technique transforms range delay into a frequency shift. The range-gating system then consisted of a bank of 60 contiguous digital filters, which provided an equivalent resolution of 3.125 ns (47 cm). An adaptive altitude tracker operated to center the leading edge of the return signal within the filter bank.

Altitude was measured at the time delay where the leading edge of the return compressed pulse reached half of its peak amplitude. The significant wave height was determined by measuring the slope of the leading edge of the return signal. Wind speed was determined after downlinking AGC data to the ground by computing the backscattering coefficient, and then determining wind speed from the backscattering coefficient. Digital data representing altitude, significant wave height, AGC, and the signal amplitude in each of the 60 filters were applied to the spacecraft telemetry system and downlinked to GEOSAT ground stations.

The in-orbit performance of the radar altimeter demonstrated a measurement accuracy of 3.5 cm for a 2 m signficant wave height. The specified accuracies for determining wave height and wind speed were also demonstrated.

7.4 RADAR ALTIMETERS OF THE FUTURE

Radar altimeters have benefitted from the continued increases in packaging density of electronic and microwave circuitry in recent years. Equally significant is the dramatic increase in speed and memory capability of microcomputers, making possible the setting of digital signal processing boundaries closer to the output of the microwave receiver. The highly capable signal processors have also made feasible the use of efficient pulse-compressed waveforms.

These trends toward more capable equipment of smaller size, lighter weight, and higher reliability will extend to future space altimetry. Today, for example, the entire radar altimeter, less the antenna, for the Apollo Lunar Lander could readily be implemented in a package approximately $75 \times 100 \times 190$ mm in size, approximately the same volume as the transmitter alone of that 1966 vintage equipment. By the early 1990s, using the emerging technologies of monolithic microwave integrated circuits (MMIC) and application-specific integrated circuits (ASIC), the size of an identical radar altimeter would be about $50 \times 75 \times 125$ mm. It would weigh less than 1 kg.

Anticipated advances in technology important in radar altimeters are discussed in the following paragraphs, as are the resulting benefits to space radar altimeters.

Operating Frequency

Size and weight of the antenna system can be reduced by operating at wavelengths in the millimeter and submillimeter range. For many space-based applications, there is no atmosphere or rainfall to attenuate or reflect the short wavelength transmission, which is a notable limitation of using these wavelengths for earth-based radar altimeters. There are very capable tube transmitters available in the millimeter range, which could provide considerable path loss capability for radar altimeters using small, broad-beamwidth antennas. For example, magnetrons are available at 35 GHz, which provide 2.5 kW of pulsed power and weigh only 0.16 kg. For coherent radar altimeter, klystrons are available that can provide 1 kW of pulsed power at 35 GHz. Receiver noise figures as low as 2.5 dB can be achieved at frequencies up to 60 GHz.

If the radar altimeter can function with narrow-beam antennas, relatively low transmitted power from millimeter-wave solid-state sources may be used. The antennas can be a conventional parabolic reflector or a planar or conformal array of elements. The suggestion has been made that monolithic, electronically steered antennas could be developed at millimeter-waves by using microstrip technology, where by the antenna elements are formed on semiconductor substrates of high dielectric constant, with amplifiers and switches implanted at the locations required [13]. This approach could result in a complete radar altimeter, including antennas, with an altitude capability up to 10 km, in a package size on the order of 200 × 100 × 25 mm.

Antennas

Printed or etched antennas on thin substrates are attractive for space-based applications. This class of antenna includes patch radiators on microstrip or etched slot types in stripline. These types of element can be used for either broad-beamwidth or narrow-beamwidth antennas. Parabolic reflector antennas and waveguide planar arrays are also options.

If the beam must be steered to accommodate changes in the spacecraft attitude electronically scanned antennas may be considered to avoid mechanically moving stuctures. A common form of electronically scanned antenna is comprised of a number of antenna modules, each containing a transmitter and receiver. This approach has some favorable aspects for future space-based radar altimeters because the transmitter may consist of many low-power antenna modules, and the system gracefully degrades but is still operational if a few of the modules fail.

Receivers and Transmitters

The receiver technology of today, which is practical for small radar equipment in space, can achieve noise figures of 0.6 dB at frequencies up to 12 GHz, and 2.1 dB at frequencies up to 60 GHz. Some further reduction of these values will likely be realized in the future, particularly at the higher frequencies. These relatively low noise figures will allow the space-based radar altimeter of the future to operate with lower transmitted power than those in the past.

Power GaAs field-effect transistors (FET) can produce 20 W of transmitted power at frequencies up to 8 GHz and 5 W at frequencies up to 15 GHz in a single device. Devices can be used in parallel to multiply these values. The trend toward higher available power from solid-state sources will likely continue, the result being that space-based radar altimeters of the future will use solid-state transmitters, even at orbital altitudes.

Signal Processors

Today's generation of devices for wideband IF amplifiers and high-speed A/D converters includes heterojunction bipolar transistors (HBT). These devices have been used to construct 4-bit A/D converters with sample rates up to 2Gb/s [14]. The technology of microprocessors is also advancing rapidly. Single-chip CMOS microprocessors (such as the 32-bit chip available from LSI Logic Corp.) can perform 30 MFLOPS. Microprocessors using GaAs technology (such as the 32-bit device produced by Rockwell International) will do 150 MFLOPS. One or two of these very capable devices, along with associated memory, will be able to handle much of the signal processing tasks for future space-based radar altimeters. Indeed, we are fast approaching that exciting (or frightening, depending on your point of view) time, when the output of the IF amplifier is given to the software folks and the signal processor is configured as an optimal entity, blithely applying artifical intelligence to the merest wisp of signal.

REFERENCES

1. "Minimum Performance Standards Airborne Low-Range Radar Altimeters," DO-155, prepared by RTCA ICG-2, November 1974, Appendix C.
2. Skolnik, M.I. (ed.), *Radar Handbook,* McGraw-Hill, New York, 1970, pp. 20-1 to 20-37.
3. Rihaczek, A.W., *Principles of High-Resolution Radar,* McGraw,-Hill, New York, 1969, pp. 159–224.
4. Caputi, W.J., "Stretch: A Time-Transformation Technique," *IEEE Trans. Aerospace and Electronic Systems,* Vol. AES-7, No. 2, March 1971.
5. "Operation and Maintenance Instructions, Radar Altimeter Model 520," Ryan Aeronautical Company, Report No. 52064-2, August 1963.
6. Lund, T.J., "Radar Velocity Sensors and Altimeters for Lunar and Planetary Landing Vehicles," *Proc. First Western Space Congress,* 1972.
7. McGoogan, J.T., L.S. Miller, G.S. Brown, and G.S. Hayne, "The S-193 Radar Altimeter Experiment," *Proc. IEEE,* Vol. 62, No. 6, June 1974.
8. "Entry Data Analysis for Viking Landers 1 and 2," Martin Marietta, Final Report TN-3770218, November 1976, pp. vii–4.
9. "Data User's Handbook and Design Error Analysis, GEOS-C Radar Altimeter," General Electric Company, report prepared under NASA Contract NAS 6-2619, May 1976.
10. Stanley, H.R., "The GEOS-3 Project", *J. Geophys. Res.,* Vol. 84, No. B8, July 1979, pp. 3779–3783.
11. MacArthur, J.L., "Design of the Seasat-A Radar Altimeter," *OCEANS '76 Conf. Rec.,* pp. 10B-1 to 10B-8.

12. Pettengill, D.E., D.F. Horwood, and C.H. Keller, "Pioneer Venus Orbiter Radar Mapper: Design and Operation," *IEEE Trans. Geoscience and Remote Sensing,* Vol. GE-18, No. 1, January 1980, pp. 28–32.
13. MacArthur, J.L., P.C. Marth, and J.G. Wall, "The GEOSAT Radar Altimeter," *John Hopkins APL Technical Digest,* Vol. 8, No. 2, April-June 1987.
14. Berenz, J., and B. Dunbridge, "MMIC Device Technology for Microwave Signal Processing Systems," *Microwave Journal,* Vol. 31, No. 4, April 1988, pp. 115–131.
15. McIlivenna, J.F., "Monolithic Phased Arrays for EHF Communications Terminals," *Microwave Journal,* Vol. 31, No. 3, March 1988, pp. 113–125.

Chapter 8
SCATTEROMETERS AND OTHER MODEST-RESOLUTION SYSTEMS
R.K. Moore
University of Kansas

8.1 INTRODUCTION

Spaceborne radars labeled as scatterometers and other modest-resolution space-based radars have more application than is widely recognized. These radars may be used to measure winds at sea and aloft, to map large-scale ocean features, sea ice, and soil moisture, and to measure rainfall. The radars should be viewed as complements to fine-resolution SAR systems used for many other kinds of mapping and detection.

Because of their modest resolution and relatively low power requirements, most such radars are simpler than other space-based radars. They can provide real-time readouts when needed because of their low data rates.

These radars have modest resolution requirements, and so may use antenna beamwidth to determine the size of the discriminated region. This means that they can use any combination of angle discrimination (beamwidth), range discrimination, and speed (doppler frequency) discrimination. The variety of system choices is therefore greater than for radars constrained by tight resolution requirements. Therefore, we devote considerable space here to the different methods of resolution with particular emphasis on doppler frequency contours.

Most of the modest-resolution radars are used for precise measurements of scattering coefficients. As a result, an important design consideration is providing enough integration time to allow measurements of mean amplitudes with low variance. A significant part of this is determining the rate at which independent samples become available and the number of samples needed. Hence, in this chapter we discuss in some detail the methods of measurement of amplitude and calibration of amplitude-measuring systems.

We then discuss the principles of ocean wind-vector measurement and the special scatterometers used for this purpose with emphasis on the SEASAT mission and the ESA's ERS-1. Real-aperture imaging radars (RARs), like the Soviet Kosmos 1500 series, have applications to measurements requiring only modest-resolution images, particularly over the oceans, and so RARs are also treated. Spaceborne meteorological radars have potential for rainfall measurements in areas like the tropical oceans, where almost no information is now available, so these radars are not discussed here. Winds aloft are important to meteorological modeling, but current measurements are only available where rawinsondes are sent up from weather stations. Accordingly, we discuss the potential of spaceborne radars for this purpose.

8.2 FUNDAMENTALS OF MEASUREMENT

Modest-resolution space-based radars may use nonstandard resolution techniques, whereas fine-resolution systems must use standard SAR methods. Moreover, modest-resolution systems are scatterometers; these must measure amplitudes precisely, and perhaps accurately. The primary purpose of a fine-resolution SAR is to determine spatial relationships, not amplitudes, and so this emphasis on measurement of signal strength is usually confined to scatterometers.

8.2.1 Resolution Techniques

Doppler Contours for Horizontal Motion

Because many space-based radars use doppler methods to achieve resolution, we must understand the nature of the constant doppler frequency contours (called *isodops*) on the surface before discussing resolution. Control of resolution by range and beamwidth uses methods similar to those employed in other systems, and so these are not treated here in detail.

The doppler frequency f_D at a radar for a fixed point on the ground is given by

$$f_D = 2\mathbf{u} \cdot \mathbf{R}/R \tag{8.1}$$

where $\mathbf{u}$ is the radar velocity and $\mathbf{R}$ is the vector from the radar to the stationary point. For a flat earth, this equation may be solved to show the isodops as a family of hyperbolas centered about the travel axis and reflected about its normal (negative doppler frequencies are behind the radar) (Ulaby, Moore, and Fung [1] Chapter 7). Similar isodops occur when we account for the earth's sphericity, although the equations governing these are not as simple.

Figure 8.1 shows an example of the isodops for a spaceborne radar in a 600 km circular orbit. The radar is traveling in the x direction. Away from the x-axis, the isodops for low doppler frequencies (steep depression angles) are almost straight. For high doppler frequencies the curvature is greater. For doppler frequencies of 100 kHz and less, the curves for flat earth and spherical earth are nearly the same. An example for 300 kHz, however, shows that by this frequency, the earth-curvature effect is significant.

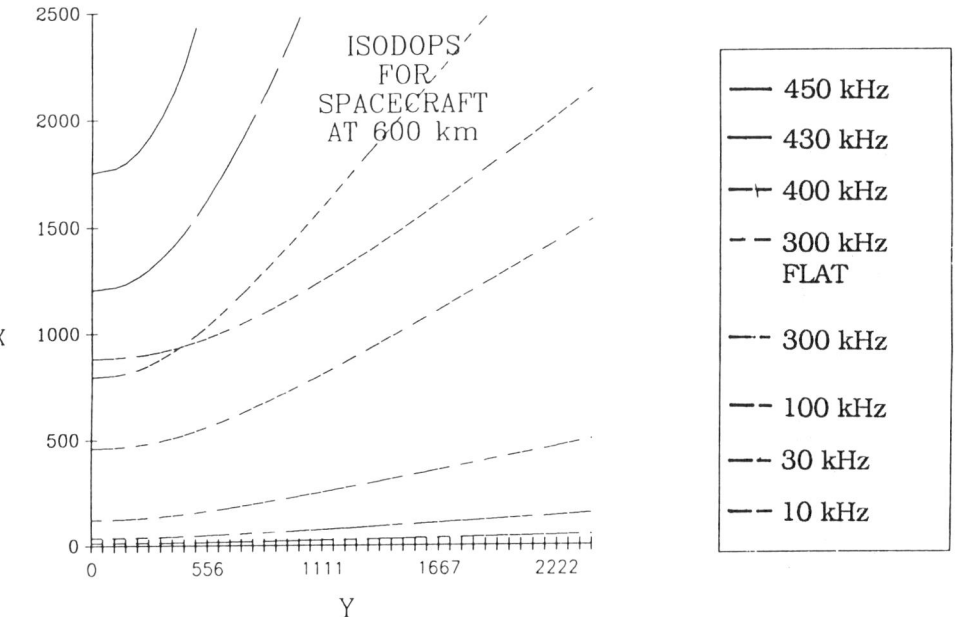

Figure 8.1 Isodops for a spaceborne radar in a 600-km circular orbit. X is distance along the orbital track and Y is distance across track. The parameter is doppler frequency in Hz. Curves are for spherical earth, but a line is shown for flat-earth calculation for 300 kHz to illustrate the difference. For small doppler frequencies, the difference between flat-earth and curved-earth calculations is negligible.

Figure 8.2 is an expanded view for Doppler frequencies up to 105 kHz. Both flat- and spherical-earth curves are plotted, but they are nearly indistinguishable. The two curves for 100 and 105 kHz are shown to indicate how a 5-kHz-wide filter could be used to obtain a resolution strip.

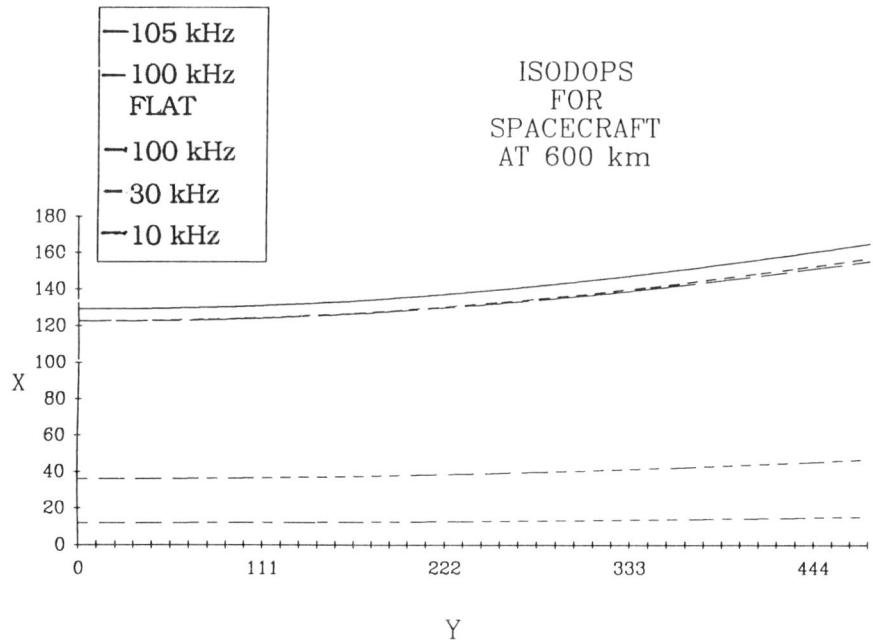

Figure 8.2 Expanded view of isodops for a spaceborne radar in a 600 km circular orbit. Coordinates are the same as in Figure 8.1. Two adjacent isodops are shown for 100 and 105 kHz to illustrate the nature of doppler discrimination on the surface (in this case, for a 5 kHz wide filter).

Comparison of Resolution Methods

Spaceborne and airborne radars have three discrimination tools available for spatial resolution: angle, range, and speed. Angle resolution is achieved by the beamwidth of the antenna. Range resolution may be achieved by pulse length, frequency modulation, or other modulation techniques. Speed resolution is attained by filtering the doppler frequencies received. Figure 8.3 shows the various combinations possible.

Figure 8.3(a) shows use of angle resolution (beamwidth) alone. Straight-CW and interrupted-CW (ICW) systems use this approach. It is the only method available for microwave radiometers. The Skylab S-193 ICW radar employed this technique.

Figure 8.3(b) shows the combination of range and angle resolution. The isorange lines are circles centered at the subsatellite point (nadir). When combined with angle discrimination, the result is the cross-hatched strip shown. This is the

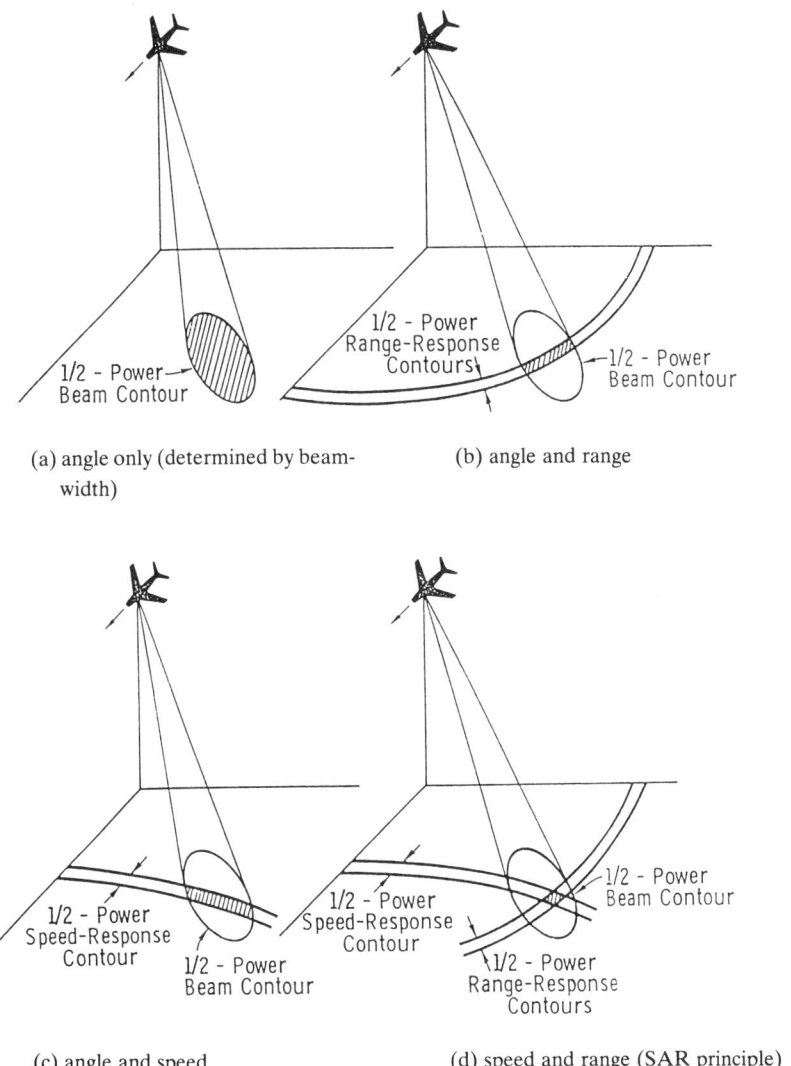

Figure 8.3 Methods of resolution possible with spaceborne radars [1].

standard method used in most ground-mapping radars other than those using synthetic aperture (SAR) methods. The Kosmos 1500 spaceborne real aperture imaging radar (RAR) and ERS-1 spaceborne scatterometer use this approach.

Figure 8.3(c) shows a combination of speed discrimination with angle resolution. This approach may be used with narrow vertical beamwidths in a side-looking mode, but with wide vertical beamwidths the antenna must be pointed away from the side to combine horizontal beamwidth with doppler filtering. The SEASAT spaceborne scatterometer used this approach.

Figure 8.3(d) shows a combination of range and speed resolution, with no use of angle resolution. This is basically the SAR approach. With a fixed filter, it is an unfocused SAR (doppler-beam-sharpening radar); if the filter is swept, we have a focused SAR. The SEASAT SAR and two Shuttle Imaging Radars (SIRs) have flown in space, but modest-resolution doppler-beam-sharpening radars have not been used.

The discussion that follows is brief. For more details the reader should consult Ulaby, Moore, and Fung [1, 2] and Elachi [3].

Range-Angle Discrimination

The geometry for range-angle discrimination of ground targets is shown in Figure 8.4(b) for plane earth. The range resolution on the ground is $c\tau_P/2 \sin\theta$, where c is the speed of light and τ_P is the pulse length. The angle of incidence θ is the same as the pointing angle for plane earth. For space-based radar, we must use the *local* angle of incidence (the angle between the incoming beam and the normal to the surface), which is different from the pointing angle with spherical earth.

The azimuth resolution is simply $\rho_h R$, where ρ_h is the horizontal beamwidth and R is the slant range. Thus, the area selected is

$$A = \rho_h R c\tau_P/2 \sin\theta \qquad (8.2)$$

We should use the two-way 3-dB beamwidth for ρ_h (one-way beamwidth at -1.5 dB).

Angle-Speed Discrimination

Figure 8.3(c) shows angle-speed discrimination. When angle and speed are combined for discrimination, the cells are seldom nearly square as they are for range-angle discrimination. They are usually diamond shaped, as shown in Figure 8.5. Because the expression for the isodops is complex, we cannot give a simple relation for the area of such cells. One pair of sides of the diamond is given by $\rho_h R$, but the length of the other pair is set by the bandwidth of the doppler filter and the geometery of the isodops. Cell areas must be calculated numerically in most cases.

This method does not apply to the side because both angle and speed discriminate in the along-track direction. An along-track fan beam, however, can be used with such a system. This approach is common with airborne scatterometers.

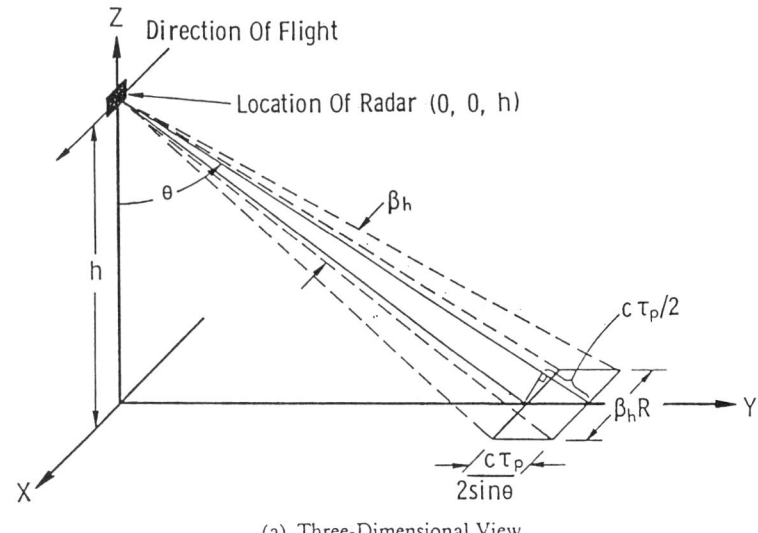

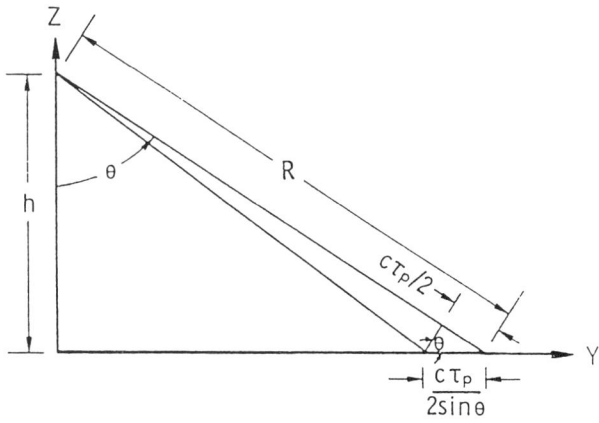

Figure 8.4 Geometry of range-angle discrimination by spaceborne radar. This is the same as for ordinary real-aperture radars [1].

For this special case, the doppler bandwidth B_D and the along-track resolution r_a are related by

$$r_a = \lambda R B_D / 2u \qquad (8.3)$$

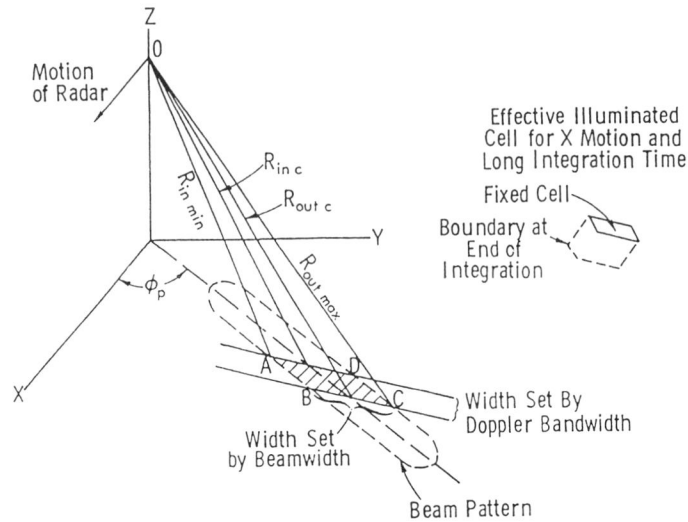

Figure 8.5 Geometry of speed-angle discrimination by spaceborne radar. Note the diamond-shaped cell. This approach is used on wind-vector scatterometers such as SEASAT and NSCATT [4].

which applies both for plane earth and low-altitude spacecraft. This is equivalent to a *vertical* beamwidth $\beta_{v(equiv)}$ of

$$\rho_{v(equiv)} = r_a \cos\theta/R$$

where θ is the local angle of incidence. The doppler bandwidth required to get a given r_a decreases as R (and θ) increases. This means that filter bandwidths must be different to get constant r_a at different angles. The reduced bandwidth at larger angles results in fewer independent samples for averaging to reduce fading effects.

The area discriminated, A, is

$$A = \lambda R^2 B_D \rho_h / 2u \qquad (8.4)$$

Range-Speed Discrimination

When range and speed are combined for discrimination (Figure 8.3(d)), the cells are also seldom nearly square. They are usually diamond shaped, as shown in Figure 8.5, but the diamonds are oriented differently than for angle-speed discrimination. Again, a simple expression is not available for the area of such cells, and so numerical calculation of the area is usually necessary. One pair of sides of the diamond is given by $c\tau_P/2 \sin\theta$, but the length of the other pair is set by the bandwidth of the doppler filter and the geometry of the isodops.

An exception is a side-looking system (unfocused SAR). For this case, the along-track resolution r_a is also related to the doppler bandwidth B_D by (8.3), but the equivalent beamwidth $\beta_{h(equiv)}$ is *horizontal*:

$$\rho_{h(equiv)} = r_a/R$$

The $\cos\theta$ factor present in the along-track case is missing, and the equivalent beam is not as narrow as the vertical beam in the angle-speed configuration.

Thus, the area discriminated is

$$A = \lambda R B_{DC} \tau_P/4u \sin\theta \tag{8.5}$$

which should be compared with (8.4).

8.2.2 Amplitude Measurement (Scatterometry)

A scatterometer is a radar calibrated to allow measurement of signal amplitudes. Thus, an imaging or tracking radar may also be a scatterometer if such calibration is present. Many radars, particularly those with modest resolution, are specifically designed to measure scattering coefficients and the term *scatterometer* is often used loosely to describe only such systems.

Although some scatterometers can measure instantaneous amplitudes, the usual applications involve measuring average signal power and converting the measurements into scattering coefficients, $\sigma°$. The scattering coefficient is a measure of the expected value of the power to be returned from a class of targets. Thus, we may have scattering coefficients for grass, corn, snow, desert, sea ice, ocean, and cities. Usually, however, a single value of $\sigma°$ will not suffice for such a class. We must specify how high the grass is, how moist it is, and, often, how rough and moist the underlying soil is, or we must express $\sigma°$ as a range for grass. Similarly, for snow, $\sigma°$ varies with depth, density, crystal size, and, especially, the amount of liquid water present (incipient melting). Similar refinements are necessary to describe each of the categories listed if $\sigma°$ is to be a single value rather than a range. Typical values of $\sigma°$ and their variations are discussed in Ulaby, Moore, and Fung ([1] Chapter 11; [4], Chapters 20 and 21), and in Long [5].

The scattering coefficient $\sigma°$ is the average radar cross section (RCS) per unit area, over an area large enough to contain many different scattering centers with random phase angles. Hence, the signal voltage fades with a Rayleigh distribution, and the *powers* of the scatterers add, producing a radar signal power which represents the average of $\sigma°$ over the resolution cell of the radar. Many independent samples of the signal must be averaged to obtain a meaningful measurement of $\sigma°$. The distribution of N averaged power measurements is the χ^2 distribution with $2N$ degrees of freedom. Figure 8.6 illustrates how this distribution narrows as more samples are integrated.

As a rule of thumb, the time between independent samples is about equal to the reciprocal of the doppler bandwidth B of the cell discriminated. For long averaging times T, and a rectangular spectrum,

$$N \approx BT \tag{8.6}$$

Spaceborne scatterometers usually measure with low signal-to-noise ratios, often considerably less than unity. For example, the Skylab S-193 scatterometer could attain 5% measurement precision with $S/N = -13$ dB. The way to get precision with low S/N is by independently measuring the noisy signal and the noise alone. Each measurement has high precision because many samples are averaged. The measured noise is then subtracted from the measured signal plus noise to obtain the signal. The results of Ulaby, Moore, and Fung [1] show that

$$\frac{\sigma}{P_s} = \sqrt{\frac{1}{N_s}\left[1 + \frac{1}{S_n}\right]^2 + \frac{1}{N_n}\left[\frac{1}{S_n}\right]^2} \tag{8.7}$$

where σ is the standard deviation, P_s is the mean signal power, S_n is the signal-to-noise ratio, N_s is the number of independent samples of signal plus noise, and N_n is the number of independent samples of noise alone. For example, with $S_n = 1 = 0$ dB and $BT = 100$, the relative precision is 22% if N_s and N_n are the same.

To obtain large numbers of independent samples, we must have wide doppler bandwidths and average for long times. The very large footprints of some spaceborne scatterometers (such as those on Skylab and SEASAT) result in both wide doppler bandwidths and relatively long times available for averaging.

Amplitude Calibration—Internal and External

Relative calibration is required for *precise* measurements, and absolute calibration permits *accurate* measurements. *Precision* depends on both the relative calibration and the number of independent samples averaged. *Accuracy* depends, in addition, on the absolute calibration. A spaceborne radar with good precision produces measurements that can be compared well with others made by the same instrument. If the measurements are not only repeatable but also absolute, systems on different

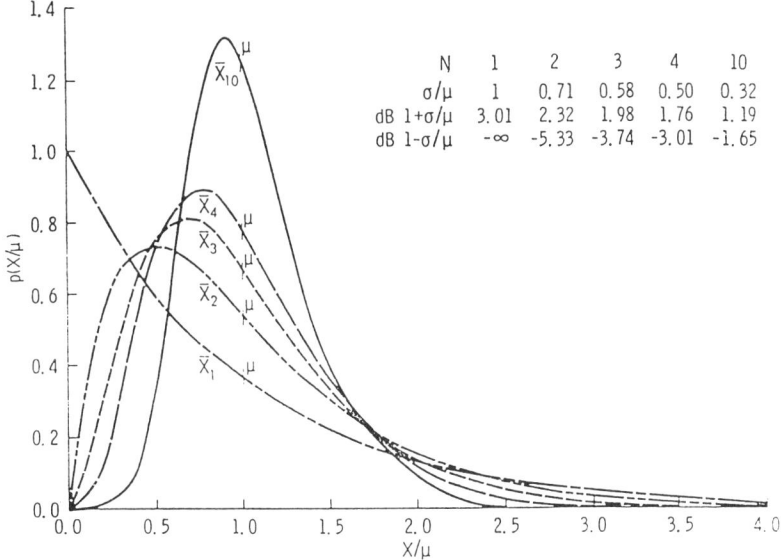

Figure 8.6 Probability density functions for N exponentially distributed variables. This is the pdf for typical fading signals with N samples averaged after detection. Note that with $N = 10$ the shape closely approaches the Gaussian shape [1].

spacecraft can be compared, and measurements and theory are directly comparable as to level. Cross-comparison of precisely calibrated instruments is possible, even without good accuracy, if they both measure the same target under the same conditions. One instrument may then be used as the *reference* standard, although neither is accurate.

The key quantity for precision is the ratio of received to transmitted power. For accuracy, we must also know the antenna gain and pattern shape as well as the pulse shape and filter frequency responses. Relative calibrations usually are *internal*. Absolute calibrations may involve *external* means.

To obtain the relative calibration needed for precision, we may either measure the transmitter power and receiver transfer characteristic separately or pass a sample of the transmitted signal through the receiver. For relative calibration, the output may be any quantity related to the desired voltage or power (e.g., digital number, centimeter on a scope, deflection of a meter, *et cetera*). The relation of this output to the actual input voltage or power is needed only if absolute measurements are the goal.

External calibration for absolute accuracy involves use of some kind of standard target of known scattering cross section. For modest-resolution space-based radars this is a problem because the cross section of the illuminated area on the surface is so large that a standard target must also have a large cross section to overcome it. Two solutions are possible: a repeater (*active radar calibrator*, ARC) or a standard area of surface.

The ARC (Brunfeldt and Ulaby [6]) is basically a receiving antenna, an amplifier, and a transmitting antenna. ARCs with radar cross sections up to 50 dB have been built. This is barely adequate to calibrate a radar with resolution of 250 × 250 m against a background with σ° of -10 dB. If it can be located in a desert with σ° of -20 dB, it may be used for a 500 × 500 m system. Another alternative is to locate it in a large lake; if calibration is accomplished under low-wind conditions, σ° may be as low as -30 dB, allowing calibration of a system with resolution of more than 1 km. For systems with more modest resolution, the ARC does not appear to offer a solution because raising the amplifier gain much more is likely to result in excessive feedback, and increasing the ARC antenna gain means that pointing must be very accurate.

Calibration against targets of known cross-sectional area can be used. This requires measurement on the ground or from aircraft of the σ° of the standard target. Moreover, the target must be homogeneous over the resolved area for the spaceborne system. For modest-resolution systems, such as real-aperture imaging radars, agricultural fields can be used in this way as they have for SARs (Dobson *et al.* [7]).

For coarser resolution systems like scatterometers used for wind measurement, a large flat forest area seems to be the only good possible standard target. Measurements over the Amazon basin aimed at antenna-pattern verification (Birrer *et al.* [8]) shows that this area is suitable, but accurate airborne scatterometer data for comparison are not now available.

8.3 OCEAN-SURFACE WIND-VECTOR MEASUREMENT

The first controlled radar measurements of the ocean were made during World War II (Goldstein [9]). These and many later measurements were aimed at learning the nature of the "sea clutter" that obscured radar echoes from ships and submarine periscopes. Interest in the use of radar to observe the sea itself started in 1964 when NASA began its earth-resources program (Moore and Pierson [10]). The desire to measure winds by using a radar scatterometer led to many airborne experiments and culminated in space-based experiments with the Skylab S-193 (Young and Moore [11]) and SEASAT scatterometer (Jones *et al.* [12]), and with operational systems for ESA's ERS-1 and the Japanese ocean-observation satellites.

Waves on the ocean are very complex. The short waves formed due to frictional drag build up first when the wind starts to blow over a calm sea; then nonlinear interactions transfer energy to waves with longer wavelengths and larger amplitudes. At some point, dissipation mechanisms balance the tendency for wave growth, and equilibrium is reached. The strength of the wind determines the equilibrium point; strong winds generate longer and higher waves. Most of the energy transfer from the atmosphere to the sea is at the very shortest wavelengths. The short waves decay rapidly and do not travel far, while the longer waves last for up to several days and travel long distances.

To understand wind measurement on the surface we must understand Bragg-resonant scattering. Figure 8.7 illustrates the Bragg resonance phenomenon. A single sinusoidal component of the surface spectrum is shown, along with an incoming plane electromagnetic wave at angle of incidence θ. The wavelength of the surface component is Λ and the radar wavelength is λ. If the excess distance from the source to each succeeding wave crest is an integer multiple of $\lambda/2$, the round-trip phase difference between signals returned from successive crests is 360°, so the signals add in phase. If ΔR is any other distance, signals from successive crests fail to add in phase and the resultant is much smaller.

If we postulate that the signal from each of the returns comes from the same part of the surface wave, the received voltage is

$$V_r = \sum_{n=0}^{N} V_0 e^{-j2kR_0} e^{-j2kn\Delta R} \tag{8.8}$$

where N is the number of wavelengths of the resonant component of the surface within the beam or within a sufficiently flat area of the long waves on which this short wave rides. This finite summation results in a sinc function for voltage, with maximum power proportional to $(N + 1)^2$. Because the average power due to adding returns from nonresonant scatterers is proportional to their number rather than the square of the power, the effect of very small scatterers can greatly exceed that of much bigger ones with nonresonant spacing but comparable numbers.

The Bragg-resonance condition is

$$k\Delta R = 2\pi\Delta R/\lambda = n\pi, \quad n = 0, 1, 2, \ldots$$

so, the Bragg-resonance condition may be written as

$$(2\Lambda/\lambda) \sin\theta = n, \quad n = 0, 1, 2, \ldots \tag{8.9}$$

where $n = 1$ is the most important. This means that for X-band ($\lambda = 3$ cm) at $\theta = 30°$, the relevant value of Λ is 3 cm. We would not expect these tiny waves (their height being much less than their length) to be important, but the Bragg resonance would make them dominant in ocean scatter. The amplitude of the radar

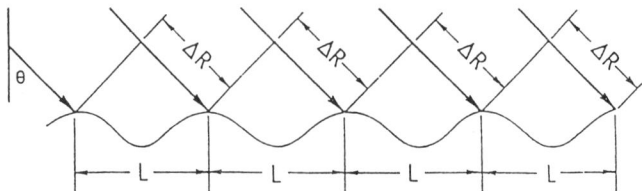

Figure 8.7 Illustration of the mechanism for Bragg scattering. Resonance occurs when $\Delta R = n\lambda/2$ so that rays from successive crests are 306° out of phase (i.e., in phase) [1].

signal is determined by the strength of the tiny ripples that are Bragg-resonant with the incoming wave and by the local angle of incidence. Hence, its average is governed by the local wind speed. Its distribution, however, depends on the slopes of the longer waves that determine the local angle of incidence.

If these long-wave slopes can be evaluated, the measurement of $\sigma°$ can be translated to a measurement of wind speed.

Many aircraft measurements have been made of the empirical variation of scattering coefficient with wind speed. A representative set at Ku-band is shown in Figure 8.8. Typically, in the range of wind speeds from about 4 to 20 ms^{-1}, the response in a given direction may be approximated

$$\sigma° = Ae^\gamma \tag{8.10}$$

where A depends on direction, polarization, and frequency. The value of γ is still in some question. At Ku-band, γ ranges from somewhat more than 1 near 20° from vertical to somewhat over 2 near 50°. Similar values are found down to C-band (Chaudry et al. [13]; Attema et al. [14]), but at L-band the values are much lower. This topic is discussed in much detail in (Ulaby, Moore, and Fung [4], pp. 1651–1675).

The azimuthal dependence of $\sigma°$ has been measured by flying aircraft in circles (Jones, Schroeder and Mitchell [15]). An example of the results is shown in Figure 8.9. The highest signals are when the radar looks upwind. A slightly lower peak occurs in the downwind direction. Minima are close to the crosswind direction, but slightly biased toward the downwind direction. A model that fits the experimental data reasonably well is of the form:

$$\sigma° = A + B\cos\phi + C\cos2\phi \tag{8.11}$$

where

$$A = a(\theta)u^{\gamma_a(\theta)} \tag{8.11a}$$

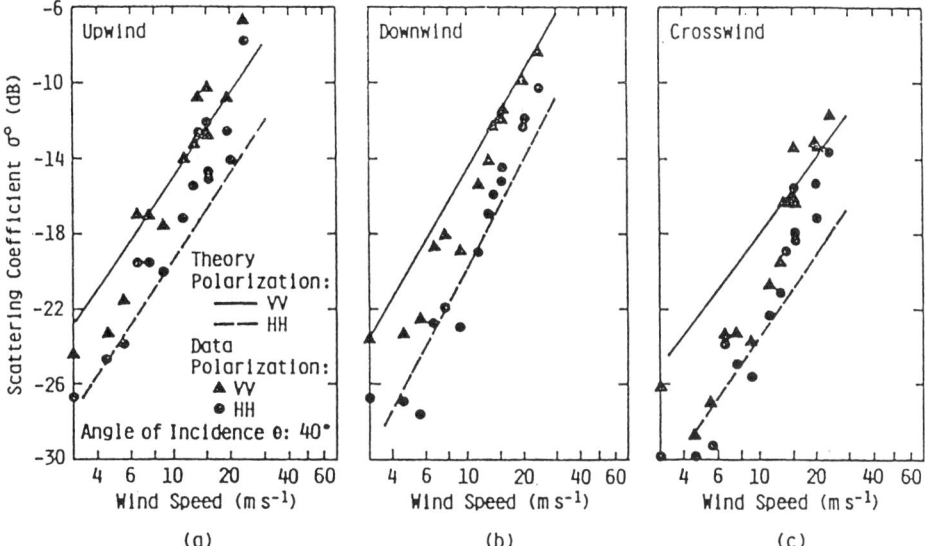

Figure 8.8 Comparison of measured values of σ° from the NASA Langley AAFE Radscat instrument with theory of Chan and Fung [21]. Note the almost-linear relation on this log-log plot, indicating a power-law relation for linear scales.

$$B = b(\theta)u^{\gamma_b(\theta)} \tag{8.11b}$$

$$C = c(\theta)u^{\gamma_c(\theta)} \tag{8.11c}$$

This variation with direction allows multibeam scatterometers to determine both wind speed and direction, but prohibits single-beam scatterometers from measuring even the speed.

Algorithms to Determine the Wind Vector

To determine the wind vector, we must have a model function of the type shown in (8.10), or, at least, a similar one. The SEASAT SASS instrument used two beams 90° apart. In this case, we can write (8.10) for beam 1 as

$$\sigma_1^\circ = A + B\cos\phi + C\cos 2\phi \tag{8.12}$$

and for beam 2 as

$$\sigma_2^\circ = A + B\cos(\phi + 90°) + C\cos 2(\phi + 90°) \tag{8.13}$$

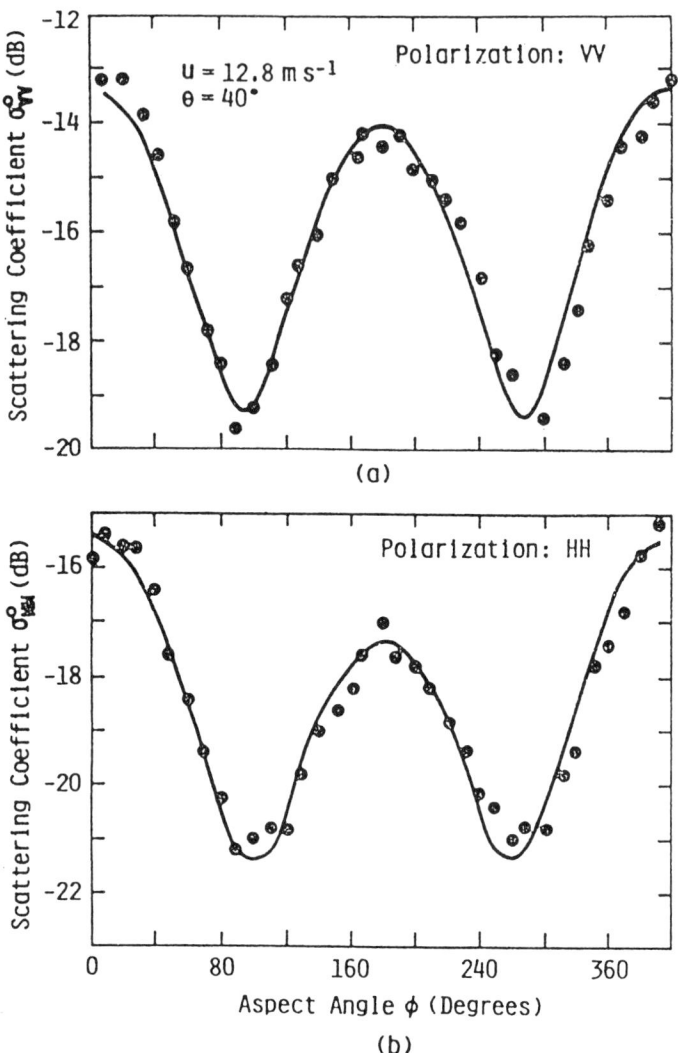

Figure 8.9 Example of circle-flight observations of the azimuthal variation of $\sigma°$ at 13.9 GHz. A regression line of the form of (8.32) is shown; $\phi = 0$ corresponds to the upwind direction.

or

$$\sigma_2° = A - B \sin\phi - C \cos2\phi \tag{8.13a}$$

These equations may then be solved for u and ϕ, but multiple solutions exist because of the periodic nature of the cosine functions.

The nature of the multiple solutions may be seen more readily if (8.12) and (8.13) are solved for u and the results are plotted. Explicit solutions are difficult to obtain, except in series form, but the form is illustrated graphically in Figure 8.10. Three cases are depicted. Lines are shown in the figure for u_1 and u_2, for assumed measured values of $\sigma°$. The intersection of the lines are the solutions of the simultaneous equations. Three cases are shown: (a) only one solution; (b) three solutions; and (c) four solutions. In Figure 8.10 (a) the maximum difference between values of $\sigma°$ measured on the two beams occurs when the reference beam is in the crosswind direction and beam 2 is in the upwind direction. This is a special case, and in general more than one solution is found. In Figure 8.10 (b) three solutions occur. This, too, is a special case, as differences in $\sigma°$ between the difference value used in (a) and that used in (b) give two solutions, whereas smaller differences give four solutions as shown in (c). For the 270° solution in (b), the speed is different than for those in the first and second quadrants.

These solution *aliases* must be resolved by other means. For SEASAT, meteorologists analyzed the patterns to resolve the aliases (Wurtele *et al.* [16]), although computer pattern recognition techniques have also been applied. Figure 8.11 shows an example of the SEASAT aliases for the JASIN experiment area in the North Atlantic.

The alias problem can be reduced by use of more beams; three-beam systems are used for the NSCAT and ESA's ERS-1 scatterometer designs (Cavanie and Offiler [17]. With a third beam, a third set of curves would be present in Figure 8.10, and the solution ambiguity would occur much less often (Schroeder *et al.* [18]). Other methods proposed would use still more beam positions, with resulting reduction in ambiguous direction results.

The situation described above is ideal in that the effects of noise are neglected. In fact, the signals are noisy and the model functions are imperfect. These combine to increase the likelihood of ambiguity and to cause errors in the solution, even when the correct alias has been picked.

Various algorithms for obtaining the wind vector solutions have been proposed. If the observed values come from widely separated spots, or if the look direction separations differ significantly from 90°, a method like maximum-likelihood or sum-of-squares (Jones, Wentz, and Schroeder [19]) must be used. The SEASAT algorithm used a procedure somewhat like a *maximum-likelihood* (ML) estimator, as proposed by Wentz, but it was really a minimum *sum-of-squares* (SOS) search procedure.

SEASAT Results

The sole purpose of the SEASAT scatterometer (Grantham *et al.* [20]) was to measure the surface wind vector over the sea. Although the measurements only

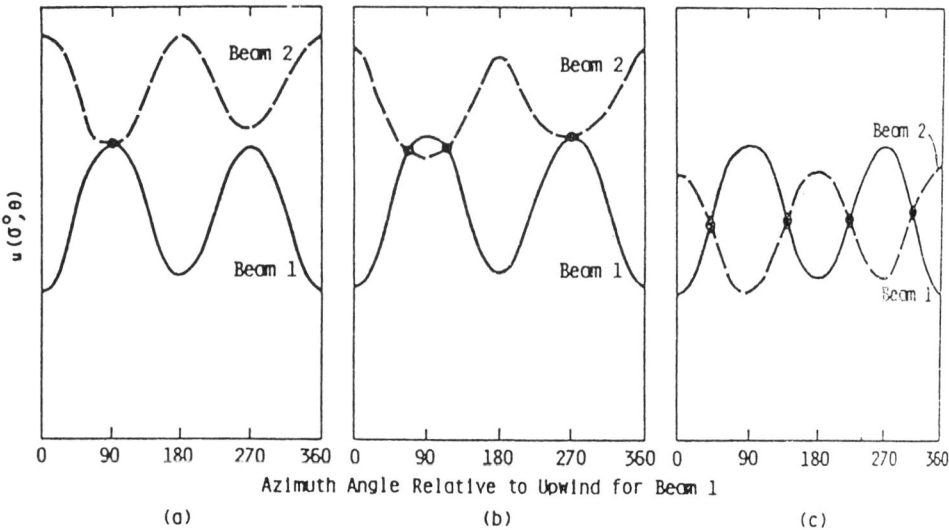

Figure 8.10 Possible solutions for the wind vector for a two-beam scatterometer with the beams orthogonal. The wind is plotted *versus* azimuth angle (relative to upwind) for pairs of measured $\sigma°$: (a) single-solution case; (b) three-solution case; (c) four-solution case [4].

lasted 99 days, an enormous quantity of data was collected. Two major experiments involving extensive surface measurements were conducted to "calibrate" the wind-vector algorithm: the GOASEX experiment in the Gulf of Alaska, and the JASIN experiment in the North Atlantic between the British Isles and Iceland. The analyses were reported in two special issues of the *Journal of Geophysical Research* ([21], 1982, [22], 1983).

On average, the SEASAT scatterometer met its goals of ± 2 m/s for speed and 20° for direction. These goals were based on use of a model function that was prepared before the mission, based on aircraft measurements. The model function was then "tuned" by using the GOASEX and JASIN results. Later, an extensive study of the statistics of the data throughout the world led to the conclusion that the model function was in error, particularly for wind speeds above 20 m/s (Woiceshyn *et al.* [23]).

One of the problems with wind vector scatterometry is answering the question, "Which wind does the scatterometer measure?" For SEASAT, the winds were supposedly normalized to the value 19.5 m above the surface in air with neutral stability. Meteorologists agree in principle, but not in detail, on how to convert surface winds to winds at this typical shipboard anemometer height. Clearly, the radar actually measures surface characteristics. Many authors believe it measures the friction velocity u_*. Recently Donelan and Pierson [24] have postulated that the actual wind measured is that $\lambda/2$ above the surface. Of course, we

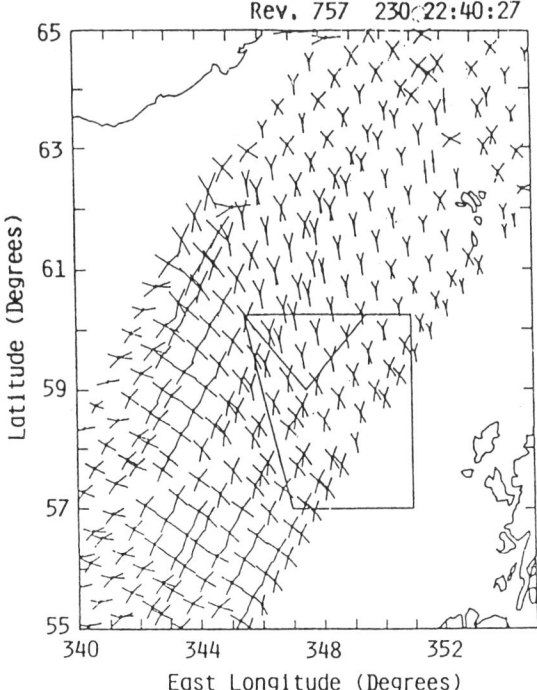

Figure 8.11 Example of wind-vector multiple solutions (aliases) for SEASAT dual-beam scatterometer. This example is in the North Atlantic; the JASIN experimental area is outlined. Lengths of the vectors are proportional to wind speed in each solution (from Wurtele *et al.* [16]).

cannot measure this wind for comparison, but the authors present methods for converting from this level to more measurable ones.

8.4 SCATTEROMETER SYSTEMS

Spaceborne scatterometer systems are primarily useful for ocean wind vector and atmospheric measurements. The latter are covered in Section 8.6. Pencil-beam (S-193) and fan-beam (SEASAT) systems have operated in space, and so both are considered here.

8.4.1 Pencil-Beam Systems

A pencil-beam system has two advantages over a fan-beam system: pencil-beam antennas have higher gain so that less power is required, and the scatterometer

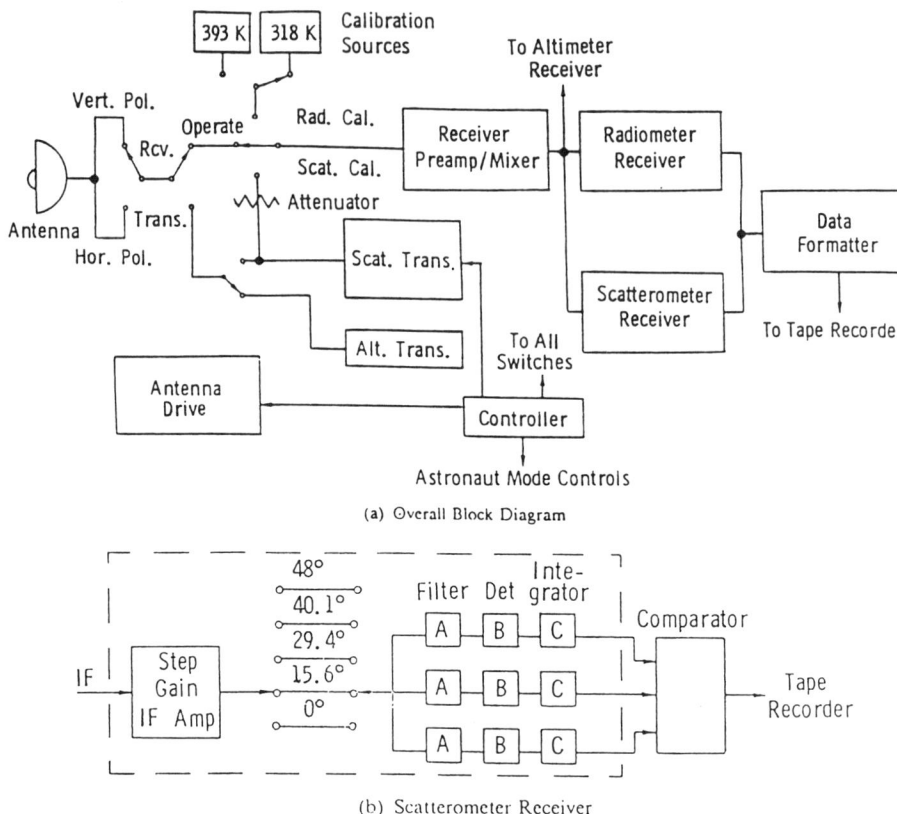

Figure 8.12 Block diagrams of the Skylab S-193 radiometer-scatterometer: (a) the overall system block diagram (the noise sources are required for radiometer calibration); (b) the scatterometer receiver. Three sets of filters, detectors, and integrators were used for each angle so that errors in pointing would not require filter bands that were too wide [1].

may be combined with a radiometer, which *requires* a pencil beam. The disadvantage of a pencil-beam system is that the antenna must be scanned, either mechanically or electrically, to achieve good coverage. Pencil-beam scatterometers for surface measurement normally operate in either CW or *interrupted-CW* (ICW) modes, attaining resolution solely by the dimensions of the beam. Rain radars, however, may need to use range resolution, even with a pencil beam.

To see the astonishingly low powers required of pencil-beam systems, we must resort to the radar equation. In its area-extensive form, this is

$$P_r = kTBFS_n = \frac{P_t G_t A_e \eta \sigma° r_x r_y}{(4\pi R^2)^2 L} = \frac{P_t A_e^2 \eta \sigma° r_x r_y}{4\pi R^4 \lambda^2 L}, \tag{8.14}$$

where P_r is received power, k is Boltzmann's constant, $T = 290$ K, B is bandwidth, F is noise figure, S_n is signal-to-noise ratio, P_t is transmitter power, G_t is transmitting antenna gain, A_e is receiving antenna aperture, $\sigma°$ is scattering coefficient, r_x and r_y are ground cell dimensions (along and across the ground track), R is slant range, η is aperture efficiency, and L is system loss. This can be converted into an equation for required power:

$$P_{t(\text{req})} = \frac{kTBFS_{n(\text{req})}(4\pi)R^4\lambda^2 L}{A_e^2 \sigma° r_x r_y \eta} \tag{8.15}$$

Here, we must know the required signal-to-noise ratio $S_{n(\text{req})}$ as well as the system parameters. With sufficient averaging, $S_{n(\text{req})}$ may be quite low. In fact, on Skylab S-193, $S_{n(\text{req})}$ was -12 dB.

To illustrate how low the required power is for such a system, we present the results for a CW X-band system (not practical, but it can be modified by dividing by duty cycle for ICW) in a 600 km circular orbit for different antenna sizes. The following assumptions are made: $\lambda = 3$ cm, $\eta = 75\%$, $F = 3$ dB, $L = 5$ dB, angle of incidence at the ground is $45°$, and $S_{n(\text{req})} = 0$ dB. The bandwidth is in every case taken to be slightly larger than the maximum value (which depends on beamwidth) at the equator. Antennas are assumed to be square arrays. Results are shown in Table 8.1.

Table 8.1
Pencil-Beam Radar Parameters

Antenna Dimension (m)	1	3	10
$P_{t(\text{req})}$ (W)	3.18	0.121	0.0032
r_x (km)	24.4	8.1	2.4
r_y (km)	34.6	11.5	3.5
Bandwidth (kHz)	16.0	5.5	1.6

The required power is very low. With a 10 m square antenna, the transmitter may require less power than the local oscillator! In practice, we cannot use a CW system because of leakage from the transmitter to the receiver, but an ICW system may be used, as on Skylab and SEASAT. The transmitter operates continuously until the first echo is received from the ground (5.4 ms for the example). It is then turned off for the duration of the echo (another 5.4 ms plus time to cross the cell, a few microseconds). Thus, the duty cycle may approach 50%. This means that the peak power must be increased by the reciprocal of the duty factor, which is somewhat more than a factor of 2 increase.

Small footprints like the one for the 10 m antenna in Table 8.1 lead to other complications. If we were to scan to achieve wide-swath continuous coverage, the rate of scanning through the cells would be so high that insufficient averaging time would be available for good $\sigma°$ measurements.

However, for many meteorological and oceanographic applications, continuous coverage is not required. A grid with, for example, 50 km or 100 km spacing is sufficient to provide inputs to the mathematical models of the atmosphere or sea surface; scanning with points spaced so widely will permit sufficient averaging time. This approach was used on both Skylab S-193 and SEASAT SASS. Multiple-beam systems could also be used to attain better coverage, but at a considerable cost in complexity.

As an example, consider the 45° circle for the table. The radius of a circle scanned at 45° from 600 km is 526.5 km. Thus, the number of cells that must be scanned for a full circle is 136 for a 1 m antenna. For continous coverage, this must be done in the time to travel 34.6 km (4.6 s). This means 33.9 ms are available for averaging of a cell. With about 16 kHz bandwidth, 542 samples may be averaged, yielding 9.6% precision for $S_n = 0$ dB. More precision could be achieved by scanning only a fraction of a circle, or by scanning slower. For instance, if a measurement is needed only every 50 km, 6.7 s are available, and the precision is 7.95%.

For the larger antennas, the beamwidths and times are correspondingly reduced. For the 3 m antenna, 409 cells are needed for full coverage and only 1.08 s are available. Clearly, special measures are needed, such as incomplete coverage or multiple beams, to allow adequate precision. Higher signal-to-noise ratios only allow a factor of 2.2 improvement in precision because of fading.

An Example—The Skylab Radscat

The Skylab S-193 Radscat (Moore *et al.* [25]) is the only pencil-beam system that has been put in space. This system combined a radar scatterometer and a radiometer. The antenna and receiver through part of the IF were common. The radiometer had a bandwidth of 210 MHz, so its IF amplifier served as a preamplifier for the radar receiver. The scatterometer had from 51 to 75 kHz bandwidth, so further filtering and amplification occurred in that part of the receiver. Separate measurements of noise level were made for using this method to achieve high precision for both parts of the instrument. Transmitter peak radiated power was 12.5 W.

Although the Skylab S-193 instrument measured radar backscatter from space, it was not capable of determining the wind vector because it could only point its antenna along one axis at a time. Thus, although it was the first wind-measuring spaceborne scatterometer, the results were subject to sizable errors because of the directional effect discussed in Section 8.3.

This instrument had many different modes, most of which involved time-sharing the radiometer and scatterometer recording. The beam could be mechanically scanned from side to side or fore and aft. In each case, both contiguous and noncontiguous modes were possible; the contiguous modes gave full coverage with less precision than the noncontiguous modes. The noncontiguous modes made observations on a 50km grid. The center of either kind of scan could be set to any one of the scan positions. Integration and dwell times for the various scan positions were adjusted to optimize precision of measurement. That is, the integration times were longer where lower signal-to-noise ratios or smaller doppler bandwidths were expected.

Figure 8.12(a) shows a block diagram of the overall system. Indicated, but not discussed, is the sharing of components with a radar altimeter. Several kinds of calibration were possible. The radiometer was calibrated with two noise generators at known effective temperatures, and this calibration could also be used with the radar if needed. The radar was normally calibrated by measuring the received noise (including that on the antenna and that from the system) after each pulse. At intervals of a few minutes, an attenuated sample of the transmitted signal was passed through the receiver. Comparison with aircraft measurements over the sea indicated accuracy no worse than 0.5 dB for the calibration.

The scatterometer receiver is shown in Figure 8.12(b). The step-gain amplifier allowed attaining a wide dynamic range. Four gain steps were available, and the system automatically selected the one with highest gain that did not cause saturation. Each of the five angles required a different doppler filter, which was followed by a square-law detector and integrator. This approach had two drawbacks: the filter bandwidths had to be 75 kHz wide, although the doppler bandwidth was about 15 kHz. Expected satellite attitude uncertainties required this. Moreover, this bandwidth requirement meant that scans could not be made effectively unless the spacecraft was in the design attitude. A better system would have been to use a tracking filter arrangement.

To illustrate the timing used in such a system, the arrangement at 48° pointing angle for a noncontiguous mode is shown in Figure 8.13. This arrangement allowed for four scatterometer polarizations and two radiometer polarizations. The scatterometer noise measurement (Scat Cal) occurred during the radiometer receiving interval. More precision could be achieved by using all the scatterometer integration time for a single polarization. The integration times were less for smaller angles and for contiguous modes. Overall specifications of the Skylab S-193 are listed in Table 8.2.

Table 8.2
Skylab S-193 Radscat Characteristics

Frequency	13.9 GHz
Antenna and Scans	
Gain	41 dB
3-dB beamwidth	1.6°
Scans:	
In-track noncontiguous (ITNC)	
Cross-track noncontiguous (CTNC)	
In-track contiguous (ITC)	
Cross-track contiguous (CTC)	
Scan angles (except CTC)	0°, 15.6°, 29.4°, 40.1°, 48.0°
CTC scan angles	±11.375° about center
Transmitter	
Peak radiated power	12.5 W
Pulsewidth	5.05 ms
PRF	125 p/s
Duty cycle	63.1%
Receiver	
First IF	500 MHz
System noise temperature	1195 K
Scatterometer Receiver	
Second IF	50 MHz
Bandwidths	51.6–72.5 kHz
Dynamic range	55 dB
Precision for $\sigma° = -30$ dB:	
Noncontiguous scans	3–7% (0.13–0.29 dB)
Contiguous scans	7–13% (0.29–0.53 dB)
Radiometer Receiver	
Bandwidth	210 MHz
Dynamic range	50–350 K
Precision	1 K (.0029 dB)

With the combined radiometer-scatterometer, the radiometer could be used to determine attenuation through the atmosphere, a standard technique for upward-looking radiometers (Wilson [26]; Haroules and Brown [27]). This is very important if the scatterometer is to be used for wind speed determination at sea, for much of the ocean is covered by clouds at any time and many of the clouds contain rain.

A windvector scatterometer proposed by Kirimoto and Moore [28] and later studied by Moore, Li, and Kennett [29] would use a pair of scanning pencil beams at angles of, for example, 40° and 50°. Each beam would scan a cone that intercepts the surface in a circle. Any point within the inner circle would be viewed from four different aspects. The authors also proposed that this system would use the same antenna for a radiometer to determine rain attenuation.

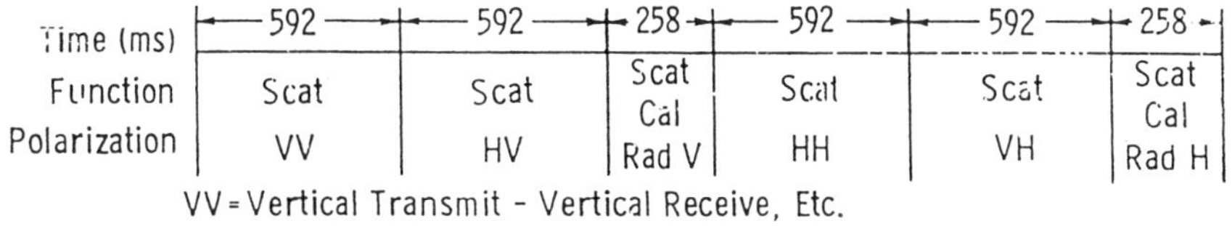

Time (ms)	⊢—— 592 ——⊣—— 592 ——⊣— 258 —⊣—— 592 ——⊣—— 592 ——⊣— 258 —⊣
Function	Scat | Scat | Scat Cal | Scat | Scat | Scat Cal
Polarization	VV | HV | Rad V | HH | VH | Rad H

VV = Vertical Transmit – Vertical Receive, Etc.

Figure 8.13 Example of timing sequence used in Skylab S-193 radiometer-scatterometer. This example is for the high-precision mode at 48°. Scatterometer integration times were smaller for smaller angles where S/N was greater [1].

8.4.2 Fan-Beam Systems

Fan-beam scatterometers allow more measurements to be made simultaneously than do pencil-beam systems. Fan-beam systems may use either range or doppler resolution in one dimension and beamwidth resolution in the other. However, fan-beam antennas have lower gains than pencil-beam antennas, so more transmitter power is needed for a fan-beam system.

We may point the fan beam ahead, to the side, or at some other angle. If it is pointed ahead, the swath covered is restricted to the width due to the narrow beam dimension, so these systems are seldom considered for space use; revisits of a target area will be too infrequent. A system with a fan-beam pointed to the side is basically the same as a real-aperture radar, which we discuss in Section 8.5. The fan-beam scatterometer for the ESA ERS-1 system, however, has one of its beams pointed to the side. As discussed in Section 8.2.1, a forward-looking fan-beam system may use either range or doppler resolution. A side-looking system may only use range resolution.

The SEASAT SASS as a Scatterometer Example

The use of multiple beams to find the vector wind was described in Section 8.3. The concept apparently arose from ideas advanced following Skylab by W.J. Pierson. The first wind-vector scatterometer was the SEASAT SASS that operated for 99 days in 1978.

The SASS was described in detail by Grantham *et al.* [20] in a special issue of the *IEEE Journal of Oceanic Engineering* devoted to the SEASAT sensor systems, and further by Boggs, Grantham, and Sweet [30]. The SASS had four fan beams, two on each side of the spacecraft, as shown in Figure 8.14. On each side, both horizontal and vertical polarizations could be used selectively or on alternating transmissions. Each cell on the surface was diamond-shaped, with a maximum dimension of about 50 km. The beams were pointed at 45° ahead of and behind the normal to the orbital plane; thus the beams were 90° apart. However,

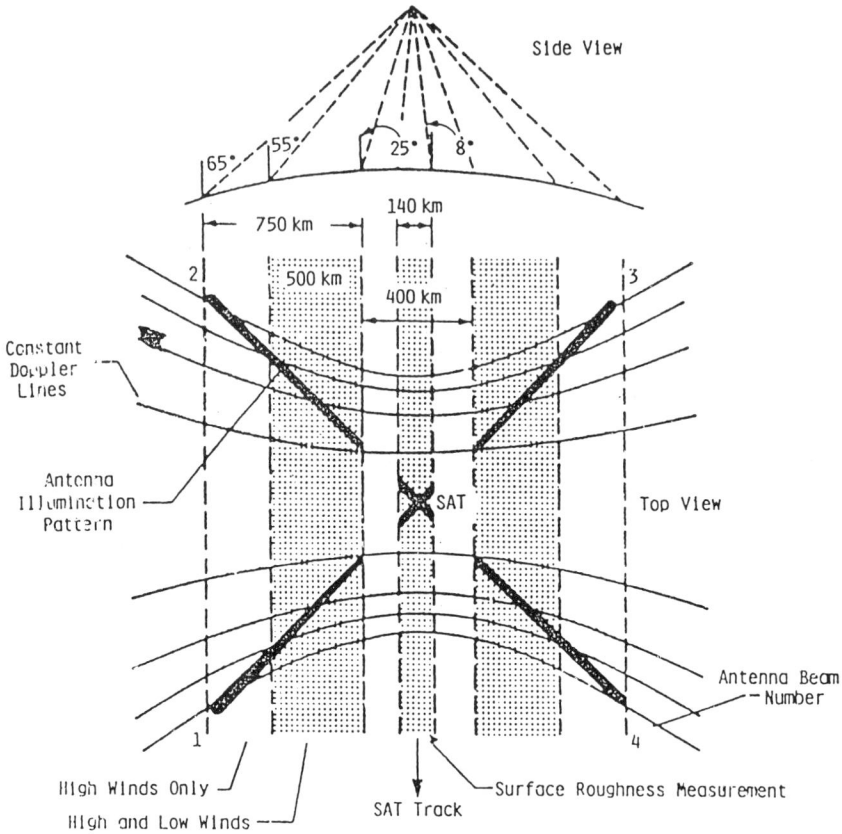

Figure 8.14 Coverage diagram for SEASAT scatterometer SASS. The region of beam overlap only extended to 55° because the spacecraft was oriented along the orbital plane rather than the ground track, so a ±3° rotation about the ground track occurred due to earth's rotation under the spacecraft. The plotted isodops can be used to show the diagonal nature of the diamond-shaped cells.

the beam orientation was related to the orbital plane (in solar inertial coordinates), rather than to the satellite ground track, thus causing a nutation about the normal to the ground track. Hence, successive observations of a point by the two beams on one side of the spacecraft were at *different* angles of incidence, resulting in complications for the wind vector processing algorithm.

This difference in incident angle is an *unnecessary complication*. If spacecraft and radar designers cooperate from the start of a project, the spacecraft may easily be oriented along the *ground track*, rather than the orbital plane. This orientation only requires enough reaction fuel to oscillate the spacecraft about ±3° per orbit, a minor matter as compared with the 360° rotation required to keep it pointed toward the earth.

The SASS operated at 14.4 GHz (2.08 cm wavelength), so the Bragg-resonant ripples ranged from 4.39 cm at 20° to 1.66 cm at 65°.

This system used an ICW transmitter as did the Skylab S-193, with the transmitter turned off to reduce feed-through while receiving. Because of the fan beam, however, the received echo came from a longer range interval, so the transmitter had to be off longer and the duty cycle had to be less. The discrimination of cells along the beam was achieved by a set of fixed doppler filters. In the ideal case, as illustrated in Figure 8.14, use of a given doppler filter for the forward and aft beams would yield the same center point for the diamond-shaped surface cells. However, because the antennas were aligned with the orbital plane and not the ground track, the rotation of the earth caused the cell centers to be displaced. Thus, the ideal situation did not prevail, and the wind vector algorithm had to account for the different locations of the cell centers. This would have been unnecessary with spacecraft orientation along the ground track.

An alternative that probably will be used in all future systems is the use of digital filters. These can be tailored by software to select identical areas for both forward and aft beams.

A block diagram of the SASS is shown in Figure 8.15. The frequency synthesizer (labeled SSS/LO) provided all of the required sinusoidal signals except the final output at 14.4 GHz, which was obtained by mixing the 14.4 GHz local oscillator signal with the 200 MHz IF signal. The carrier frequency signal was amplified in a traveling wave tube to 100 W peak power, and fed through the transmit-receive circulator to the antenna-switching matrix. The transmitted pulses were 4.8 ms wide, and the duty factor was 17% to allow adequate time for receiving signals from the complete set of ranges from that at vertical incidence to that at 60°. The transmitter power level was monitored for calibration purposes. The received signal was coupled to the mixer through a tunnel-diode amplifier (TDA) with a 5.7 dB noise figure. Provision was made for calibrating the receiver by using a noise source.

A fan-beam system cannot be used for a radiometer-scatterometer because the radiometer cell size can only be set by beamwidth. On SEASAT, the scanning multifrequency microwave radiometer (SMMR) was supposed to provide this capability. However, differences in cell size and scan pattern resulted in an ineffective rain-correction procedure (Moore, Chaudry, and Birrer [31]).

ERS-1 Example of a Spaceborne Scatterometer

The wind scatterometer (SCATT) for the ESA's Earth Resources System-1 (ERS-1) is a part of the active microwave instrument (AMI). The AMI can be operated

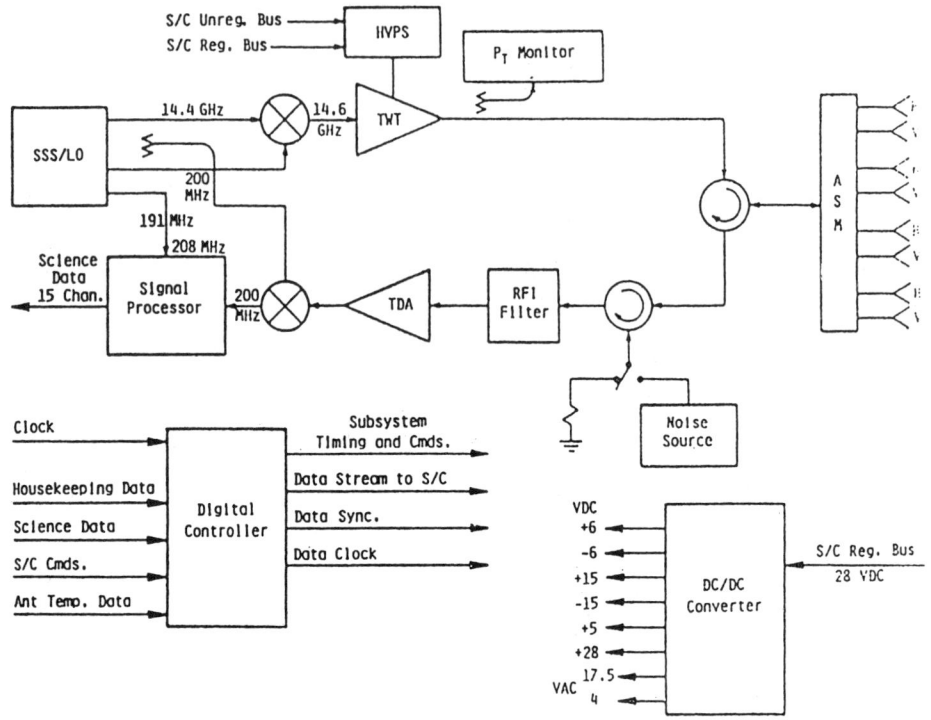

Figure 8.15 Block diagram of the SEASAT scatterometer SASS [4].

in one of two SAR modes and in the SCATT mode. For wind measurement, the AMI's performance specifications are essentially the same as those of SEASAT SASS: ±2 m/s or 10% in speed and ±20° in direction for winds of 4 to 24 m/s. The SCATT is different than the SASS in many other respects, however.

The AMI operates at C-band (5.3 GHz), whereas the SASS was at Ku-band (14.6 GHz). At this lower frequency, rain attenuation should be negligible most of the time, and no radiometer is on the ERS-1 for scatter correction or any other purpose. The AMI, including the SCATT, is a high-power pulse system. This means that the SCATT discriminates ground areas by the range-beamwidth approach, rather than the speed-beamwidth approach used on SASS. Three beams are used: two at 45° angles ahead and behind the side-looking direction and one in that direction. The three beams reduce the amount of ambiguity in direction, but the results are still ambiguous for some fraction of the time.

The SCATT has only one polarization, vertical. It has a peak RF power of 4.8 kW and a 46 W average power. Spatial resolution is about 45 km and the output is on a 50 km grid. Measurement precision is better than 8.7%.

8.5 REAL-APERTURE IMAGING RADARS IN SPACE

Imaging radars in space have great potential for describing large-scale variations in surface properties, whether on the ocean or on land. Such radars have been, and are expected to be, largely of the side-looking variety. In this configuration, the range scanning is achieved in the usual fashion by pulse length, and the "azimuth" scanning by letting spacecraft motion scan a beam pointing at a fixed angle to the spacecraft. If one wishes to have a fine resolution (pixel dimension <500 m in the along-track dimension), the synthetic aperture technique must be used, so most such systems flown or contemplated are SARs. However, the USSR has a series of spaceborne real aperture imaging radars (RARs) with 2 × 2 km pixel dimensions (Kalmykov, et al. [32]).

The SEASAT SAR was the first imaging radar flown in space (Jordan [33]). Because it was designed to study wave spectra on the sea, it had a pixel of about 25 m square. Many other oceanic features that did not require this fine resolution showed up on the images of this [21, 22, 34] and the subsequent shuttle imaging radars (SIR-A and SIR-B). Current boundaries and eddies were clearly apparent. Surface tracks of storm winds also were visible in some cases. Internal waves were clear on many images. Although the wave spectra require fine resolution, many of the other features will be visible with the much poorer resolution of a RAR, as indicated by the results from Kosmos 1500 (Shul'gin [35]).

Spaceborne imaging radars are also valuable for mapping sea ice. Even modest-resolution RARs can show the extent of the ice cover and of large open leads and polynyas. Moreover, they differentiate (at least in winter) new ice, first-year ice, and multiyear ice ([1], pp. 875–880; [4], pp. 1892–1906). Kosmos 1500 is being used operationally to provide ice information to ships' captains (Kalmykov et al. [32]), using an APT mode like that of weather satellites.

Because of the sensitivity of the radar signal to soil moisture ([1], pp. 860–863; [4], pp. 1828–1831), the modest resolution of an RAR could be useful. The mapping of moisture from a storm track across Iowa was particularly spectacular with the SEASAT SAR (Ulaby, Brisco, and Dobson [36]). This dramatic illustration of the way soil moisture affects a radar image is in Figure 8.16 [37]. This shows an area in Iowa where rain storms passed through just prior to the SEASAT overflight. The bands of high soil-moisture content are apparent. A simulation experiment (Ulaby et al. [38]) showed that a radar with a resolution between 0.1 and 1 km predicted soil moisture within 20% for 90% of the pixels in the test region. Although the spaceborne sensor of choice is a modest-resolution SAR, an RAR, particularly if combined with a microwave radiometer, can be useful in relatively flat terrains.

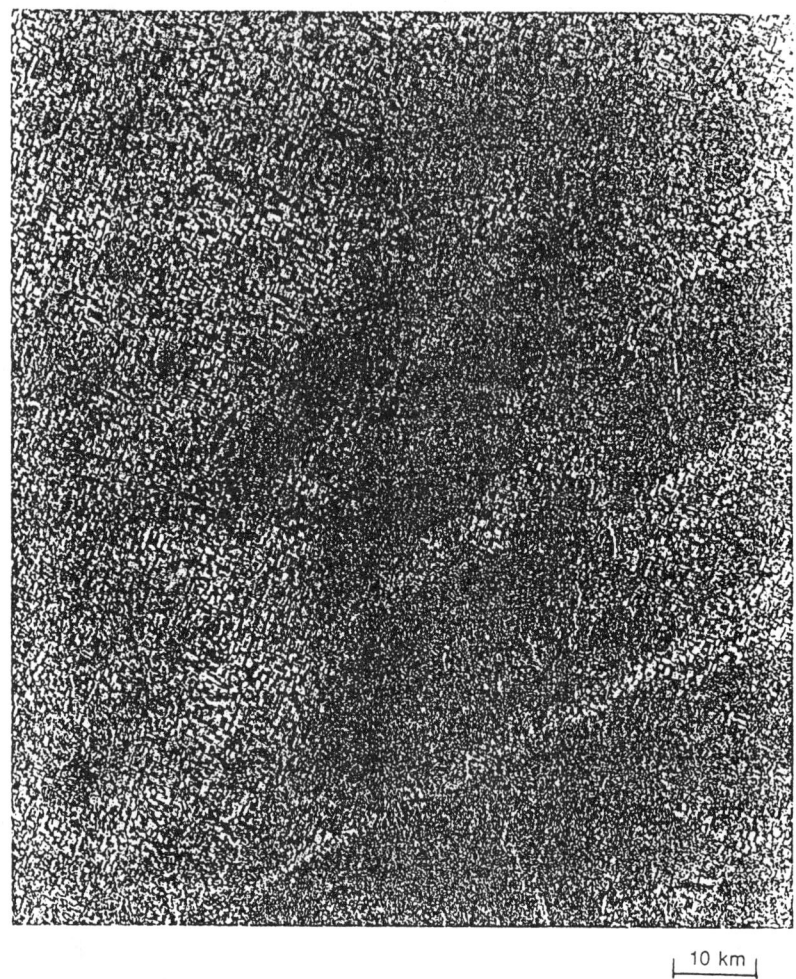

10 km

Figure 8.16 SEASAT L-band SAR image of central Iowa in the vicinity of Ames, showing patterns resulting from differences in scattering caused by soil-moisture changes due to passage of rain storms. The brighter areas are wet (from Ford, *et al.* [37]).

The radar equation for a spaceborne RAR starts in the usual form:

$$P_t = \frac{kTBFSL(4\pi)^3 R_{max}^4}{G^2 \lambda^2 \sigma° r_{x(max)} r_{y(min)}} \tag{8.16}$$

where most quantities are as defined before. Note the use of R_{max}, $r_{x(max)}$, and $r_{y(min)}$. R_{max} is at the range at the maximum angle of incidence; $r_{x(max)}$ is the along-track resolution at this maximum range; and $r_{y(\min)}$ is the ground-range resolution at this point. It is minimum because it is closest to the slant-range resolution.

This equation may be manipulated to a different form that shows dependencies directly, assuming that orbit height, swath width, maximum angle of incidence, and resolution are the selectable parameters. This form is

$$P_t = \left[\frac{2(4/3)^2 \pi k T c F S L}{\lambda^2}\right]\left[\frac{R_{\max}^2 \beta_v^2 r_{x(\max)}}{\sigma^\circ r_{y(\min)}^2}\right] \qquad (8.17)$$

where the geometric variables are in the right-hand brackets. The factor (4/3) comes from assuming 75% aperture efficiency. The speed of light is c. Because of the spherical geometry, a simple expression for β_v (the vertical beamwidth) is not possible in terms of the swath width as it is for plane geometry.

Equation (8.17) seems to imply that higher altitudes require much more power because of $R_{\max}^2$, but the product $(R_{\max}\beta_v)$ is, in fact, nearly constant. Therefore, the altitude of the spacecraft is not very important for power consumption as long as the swath width is fixed. However, the range resolution is quite important, as indicated by the $r_{y(\min)}^2$ in the denominator. Power can best be saved by degrading range resolution.

This equation has been used to calculate some examples shown in Table 8.3. Radar parameters are given at the top of the table. In each case, the resolution was selected at the outer edge of the 300 km swath and other resolutions, power, and antenna dimensions calculated from it. X-band and C-band frequencies were used. The power required is less at C-band (5 GHz), but the antennas are very long at that frequency.

Pixel areas are shown insofar as ability to interpret images depends on area (Moore [39]), rather than individual lengths. Thus, the 1000 × 500 m pixels are equivalent to pixels 690 m square. The 2000 × 1400 m pixels are about equivalent to those of the Kosmos 1500, although the standard output product for that system is about 3000 × 3000 m.

Two examples for 800 km orbital altitude show that power required is actually less than at lower altitude because of the effect of the increased antenna size on the gain. Clearly, the antenna for the 1000 × 500 m case is too long to be practical at X-band, but the 2000 × 1400 m case is realistic.

Data rates for such systems are very low, provided that on-board averaging of independent samples is achieved. For example, the 1000 × 500 m system has a data rate of only about 5200 bytes per second. Thus, RARs are quite suitable for on-board recording and dumping, and also for down-loading the images to unsophisticated ground receivers.

Table 8.3
Examples of Real-Aperture Space Radars

Outer Angle of Incidence: 60°. Swath Width: 300 km. Design $\sigma°$: -25 dB.
Noise figure: 3 dB. S/N required: 5 dB. System loss: 5 dB.

Frequency (GHz)	OUTER EDGE			350 km Orbit Height INNER EDGE			ANTENNA		
	r_x (m)	r_y (m)	Area (km²)	r_x (m)	r_y (m)	Area (km²)	P_t (kW)	Height (cm)	Length (m)
10	1000	500	0.50	653	740	0.48	26.0	8	19.6
5							6.5	16	39.1
10	1000	750	0.75	653	1110	0.72	11.6	8	19.6
5							2.9	16	39.1
10	2000	1400	2.80	1306	2073	2.70	6.6	8	9.8
5							1.7	16	19.6
10	800	1400	1.12	522	2073	1.08	2.7	8	24.5
				800 km Orbit Height					
10	1000	500	0.50	824	569	0.47	15.7	22	41.9
10	2000	1400	2.80	1648	1595	2.62	4.0	22	20.9

Power required for these 300 km swath-width systems is not unreasonable, but the antennas are a bit unwieldy. Nevertheless, these calculations show the practicality of spaceborne RARs for low orbits, provided that the applications mentioned above justify such a relatively inexpensive system.

8.6 MEASURING WINDS ALOFT FROM SPACE

Atmospheric models for global weather forecasting would be greatly enhanced if winds aloft could be measured over the oceans as with rawinsondes over land. For this reason, a Laser Atmospheric Wind Sensor is planned for NASA's EOS program to monitor winds by measuring the doppler shifts of lidar returns from aerosols in cloudfree areas (Curran [40]). Because the lidar will be scanned in a circular pattern, each point within the circle will be viewed from two different directions, so two components of the velocity can be measured.

A radar with a similar scanning pattern could measure the winds in the cloudy areas that cover much of the globe. X-band or Ku-band may be appropriate where precipitation is present, but clouds will require higher frequencies because cloud drops are very small. More power would be needed for clouds than for rain alone, but this might be available in future spacecraft with more robust power sources. Certainly, the space station will have sufficient power.

This kind of radar could also be used for measuring rain rate and surface

winds over the oceans. Its goal should be to estimate the following three quantities:
1. the echo power to establish rain rate and surface wind velocity (zero moment of the return spectrum);
2. the mean doppler frequency to establish the wind velocity (first moment of the spectrum);
3. the spread of the doppler frequency to establish the turbulent spread of wind velocity (sqare root of second moment of the spectrum).

Use of dual feeds in a reflector antenna would permit observation of each point on the inner circle from four look directions. Scanning simultaneously at, for example, 30° and 35°, would allow measurement of four velocity components. This redundant measurement could improve accuracy. This scan pattern would be similar to that for rain-rate measurements by Kirimoto and Moore [28].

Accurate determination of the doppler centroids for observed volumes will require comparison with the doppler centroids for the surface echoes. Otherwise, the small doppler shifts due to wind would be lost in changes in the doppler shifts due to spacecraft attitude changes. This occurs because the velocity component due to spacecraft motion at, for example, 35° ahead is on the order of 4300 m/s, whereas the wind-speed components to be measured are only a few meters per second. An error of 0.1° in pointing would be equivalent to 11 m/s, an intolerable value. Both the surface echo and the rain or cloud echo would be subject to the same shift due to pointing errors, and so the difference would be accurate.

At 30°, the center frequency of a stationary surface echo would vary from -250 kHz through zero to $+250$ kHz for a circular scan. Hence, such a radar would need programmed doppler filters, so the center frequency of the spectrum due to the narrow beam would be within range for rapid tracking. Because orbital parameters are reasonably well known, this kind of program can easily be incorporated. Even so, the tracking problem would not be trivial. For an antenna with 2 mrad beamwidth pointed 30° ahead, the doppler bandwidth would correspond to a velocity bandwidth of 307 m/s. Thus, if precision of ± 1 m/s is required, the center frequencies for the surface and rain spectra must be located to within about 0.3% of their bandwidth or .027% of their absolute value.

Ambiguity in doppler frequency measurement is a major problem in such a system. Presuming that we can remove the spacecraft-induced velocity by using a programmed local oscillator to beat this frequency to zero, we must still contend with the doppler bandwidth caused by the motion of the particles to be measured. For a wind speed of 100 m/s, the maximum component along a 30° line of sight is 50 m/s. At 10 GHz, doppler frequencies of ± 3333 Hz must therefore be measured. At 35 GHz, ± 11667 Hz must be measured. Even if the spacecraft pointing errors were perfectly removed, the Nyquist sampling rates would thus equal 6.67 kHz for 10 GHz and 23.33 kHz for 35 GHz. The sampling rate would need to be higher to account for errors in knowledge of the pointing angle.

The doppler spectrum could, in principle, be extracted by using a single long pulse if we merely wanted the average over all altitudes. However, this is not practical for two reasons: first, the requirements of forecasters are for wind profiles, not averages over all heights; second, the length of the pulse required for accurate estimation would exceed the time required for round-trip signal travel, so we would need to be receiving while transmitting.

For these reasons, shorter pulses must be used with a high enough PRF to sample the doppler frequency. This runs into ambiguity problems that must be resolved. The maximum *interpulse period* (IPP) is set by the travel time of the echo between surface and highest echo-producing cloud, with duration of transmitted pulse and a safety factor (guard time) added in. Thus,

$$\text{IPP} \geq 2T_{\text{trans}} + T_{\text{echo}} + T_{\text{guard}} \tag{8.18}$$

Assume an expanded transmitter pulse length $T_{\text{trans}} = 20$ μs, a detectable-cloud height of 20 km giving $T_{\text{echo}} = 154$ μs, and a guard time (including accounting for earth oblateness) $T_{\text{guard}} = 26$ μs. This results in IPP ≥ 200 μs, or PRF ≤ 5 kHz. The Nyquist criterion for measuring the doppler frequency f_D without ambiguity requires that $f_D \geq 6.67$ kHz at 10 GHz, so this criterion cannot quite be met. At 35 GHz, the situation is worse, as the Nyquist frequency is 23.67 kHz.

The ambiguity problem is also important for ground-based weather radars, and has been studied extensively (Doviak and Zrnic [41]). Some solutions proposed for this problem are (1) *phase diversity* (Zrnic and Mahapatra [42]; Sachidananda and Zrnic [43]), (2) *polarization diversity,* and (3) *PRF diversity.* With phase diversity, the phase of alternate transmitted pulses is coded differently, allowing seperation of first- and second-trip echoes. With polarization diversity, alternate pulses are transmitted with different polarizations, thereby allowing a doubling of the effective IPP. Various PRF diversity schemes have been used, as described by Doviak and Zrnic [41].

An adaptive scheme may also be used with the PRF diversity scheme. This measures the apparent doppler centroid separately for each of two PRFs. If they are the same, no aliasing has occurred. If they differ, aliasing is present. In this case, continuity with winds in areas of lower wind speeds can be used to determine the wind in the ambiguous region.

Power requirements have been calculated for the doppler wind sensor, for a few simple examples. The basic approach is the same as that used for the rain radar; indeed, this is one of the goals of the wind sensor. More power is needed for clouds, however, because the drops are much smaller. A radar such as this would be useful if it only measured winds in rain, but its value would be much greater if winds in the interior of clouds could be measured. The power levels suitable for the rain radar would seldom allow cloud measurements with adequate signal-to-noise ratio.

Use of large antennas and power in the few kilowatt range would allow wind measurement in most of the denser clouds. The peak-power level is comparable with that for the SIR-C, ERS-1, and EOS SARs. Antennas as large as 15 m diameter have been designed for use with frequencies up to Ku-band (Campbell and Belvin [44]). The peak-power level needed for cloud measurement with such an antenna is comparable with that for the SIR-C, ERS-1, and EOS SARs. This power would not allow measurements in fair-weather cumulus, but would be useful in most convective cloud systems.

Although this kind of system will be technologically challenging, it appears feasible. Hence, it will probably be built in view of the importance of wind measurement. However, for such a system (or the laser wind sensor) to be useful in forecasting, a constellation of satellites carrying the systems will be needed. In the meantime, the proof-of-concept instruments will provide valuable samples of winds aloft over tropical oceans, where they have never been measured.

8.7 SUMMARY

Modest-resolution space radars have numerous applications. Most of them are relatively simple compared with other space-based radars. Their data rates are low, so they can provide real-time readouts when needed—with or without use of data-relay satellites. Moreover, the low data rates facilitate on-board recording for subsequent dumps to telemetry stations.

A characteristic of most such radars which makes them different from others is that their applications call for precise measurement of amplitudes. They are scatterometers whether they are so-named or not! This means that they must measure fading signals with enough independent samples to reduce the variance of the measurement to reasonable levels. Some of them (the wind scatterometers, for instance) operate with signal-to-noise ratios below zero dB, which means that the integration times must be very large to allow subtracting noise estimates from signal-plus-noise estimates.

Because these radars have modest resolution requirements, they may use antenna beamwidth to determine the size of the discriminated region. This means that they can use any combination of angle discrimination (beamwidth), range discrimination, and speed (doppler frequency) discrimination. The range of system choices is therefore greater than for radars constrained by tight resolution requirements.

We have discussed applications of special scatterometers to ocean wind-vector measurement, real-aperture radars for both ocean and land applications, and meteorological radars for wind-aloft measurement. The ocean-surface wind-vector concept is proven, and operational systems such as ERS-1 are underway. The

Soviet Union has demonstrated the utility of RARs in space with its Kosmos 1500 series. Wind-vector measurement is at the conceptual study stage. By the year 2010, we can expect operational versions of all of these radars to be in use.

REFERENCES

1. Ulaby, F.T., R.K. Moore, and A.K. Fung (1982), *Microwave Remote Sensing: Active and Passive,* Artech House, Norwood, MA, Vol. 2.
2. Ulaby, F.T., R.K. Moore, and A.K. Fung (1981), *Microwave Remote Sensing: Active and Passive,* Artech House, Norwood, MA, Vol. 1.
3. Elachi, C., *et al.* (1982), "Spaceborne Synthetic Aperture Imaging Radars: Applications, Techniques, and Technology," *Proc. IEEE,* Vol. 70, No. 10, October 1982, pp. 1174–1209.
4. Ulaby, F.T., R.K. Moore, and A.K. Fung (1986), *Microwave Remote Sensing: Active and Passive,* Artech House, Norwood, MA, Vol. 3.
5. Long, M.W. (1983), *Radar Reflectivity of Land and Sea,* 2nd Ed., Artech House, Dedham MA.
6. Brunfeldt, D.R., and F.T. Ulaby (1984), "Active Reflector for Radar Calibration," *IEEE Trans. Geoscience and Remote Sensing,* Vol. GE-22, pp. 165–169.
7. Dobson, M.C., F.T. Ulaby, D.R. Brunfeldt, and D.N. Held (1986), "External Calibration of SIR-B Imagery with Area-Extended and Point Targets," *IEEE Trans. Geoscience and Remote Sensing,* Vol. GE-24, pp. 453–461.
8. Birrer, I.J., E.M. Bracalante, G.J. Dome, J. Sweet, and G. Berthold (1982a), "Signature of the Amazon Rain Forest Obtained with the Seasat Scatterometer," *IEEE Trans. Geoscience and Remote Sensing,* Vol. GE-20, pp. 11–17.
9. Goldstein, H. (1951), "Sea Echo," in *Propagation of Short Radio Waves—MIT Radiation Laboratory Series,* Vol. 13, Kerr, D.E. (ed.), McGraw-Hill, New York, Chapter 6.
10. Moore, R.K., and W.J. Pierson (1965), "Measuring Sea State and Estimating Surface Winds from a Polar Orbiting Satellite," *Proc. Int. Symp. Electronic Sensing of Earth from Satellites,* pp. R1–R26.
11. Young, J.D., and R.K. Moore (1977), "Active Microwave Measurements from Space of Sea-Surface Winds," *IEEE J. of Oceanic Engineering,* Vol. OE-2, pp. 309–317.
12. Jones, W.L., *et al.* (1982), "The SEASAT-A satellite scatterometer: The geophysical evaluation of remotely sensed wind vectors over the ocean," *J. Geophys. Res.,* Vol. 87, pp. 3297–3317.
13. Chaudry, A.H., S.P. Gogineni, and R.K. Moore (1986), "Tower-Based Broadband Backscattering Measurements from the Ocean Surface in the North Sea," *Digest IGARSS '86,* Zurich, September 1986, pp. 327–336.
14. Attema, E.P.W., A.E. Long, and A.L. Gray (1986), "Results of the ESA Airborne C-Band Scatterometer Campaigns," *Digest IGARSS '86,* Zurich, September 1986, pp. 381–387.
15. Jones, W.L., L.C. Schroeder, and J.L. Mitchell (1977), "Aircraft Measurements of the Microwave Scattering Signature of the Ocean," *IEEE Trans. Antennas and Propagation,* Vol. AP-25, pp. 52–61.
16. Wurtele, M.G., P.M. Woiceshyn, S. Peteherych, M. Borowski, and W.S. Appleby (1982), "Wind Alias Removal Studies of SEASAT Scatterometer-Derived Wind Fields," *J. Geophys. Res.,* Vol. 87, pp. 3365–3377.
17. Cavanie, A., and D. Offiler (1986), "ERS-1 Wind Scatterometer: Wind Extraction and Ambiguity Removal," *Digest IGARSS '86,* Zurich, September 1986, pp. 395–398.
18. Schroeder, L.C., W.L. Grantham, E.M. Bracalente, G.L. Britt, K.S. Shanmugan, F.J. Wentz, D.P. Wylie, and B.B. Hinton (1985), "Removal of Ambiguous Wind Directions for a Ku-Band Wind Scatterometer Using Three Different Azimuth Angles," *IEEE Trans. on Geoscience and Remote Sensing,* Vol. GE-23, pp. 91–100.

19. Jones, W.L., F.J. Wentz, and L.C. Schroeder (1978), "Algorithm for Inferring Wind Stress from Seasat A," *J. Spacecraft and Rockets*, Vol. 15, pp. 368–374.
20. Grantham, W.L., E.M. Bracalente, W.L. Jones, and J.W. Johnson (1977), "The Seasat—A satellite scatterometer," *IEEE J. Oceanic Engineering*, Vol. OE-2, pp. 200–206.
21. JGR (1982), Special issue of *J. Geophys. Res.*, Vol. 87, No. C5, April 30.
22. JGR (1983), Special issue of *J. Geophys. Res.*, Vol. 88, No. C3, February 28.
23. Woiceshyn, P.M., et al. (1986), "The Necessity for a New Parameterization of an Empirical Model for Wind-Ocean Scatterometry," *J. Geophys. Res.*, Vol. 91, pp. 2273–2288.
24. Donelan, M.A., and Pierson W.J., Jr. (1987), "Radar Scattering and Equilibrium Ranges in Wind-Generated Waves with Application Scatterometry," *J. Geophys. Res.*, Vol. 92, pp. 4971–5029.
25. Moore, R.K., J.C. Holtzman, A.C. Cook, D. Fayman, and W. Spencer (1974), "Measurement of a Microwave Antenna Pattern from an Orbiting Spacecraft," *Digest IEEE AP-S Int. Symp.*, Vol. June 1974, pp. 51–56.
26. Wilson, R.W. (1969), "Sun-Tracker Measurements of Attenuation of Rain at 16 and 30 GHz," *Bell System Tech. J.*, Vol. 48, pp. 1383–1404.
27. Haroules, G.G., and W.E. Brown, III (1969), "The Simultaneous Investigation of Attenuation and Emission by the Earth's Atmosphere from 4 Centimeters to 8 Millimeters," *J. Geophys. Res.*, Vol. 74, pp. 4453–4471.
28. Kirimoto, T., and R.K. Moore (1985), "Scanning Wind-Vector Scatterometers with Two Pencil Beams," *Proc. Conf. Frontiers of Remote Sensing of Oceans and Troposphere*, Shoresh, Israel, NASA Conf. Pub. 2303, 89, 1984 and RSL TR 0176-1.
29. Moore, R.K., F.K. Li, and R.G. Kennett (1988), "Performance of A Scanning Pencil-Beam Spaceborne Scatterometer for Ocean Wind Measurements," *Digest IGARSS '88*, Edinburgh, September 1988.
30. Boggs, D.H., W.L. Grantham, and J. Sweet (1980), "The SASS Scattering Coefficient Algorithm," *IEEE Trans. Oceanic Engineering*, Vol. OE-5, pp. 145–154.
31. Moore, R.K., A.H. Chaudry, and I.J. Birrer (1983), "Errors in Scatterometer-Radiometer Wind Measurement Due to Rain," *IEEE J. Oceanic Engineering*, Vol. OE-8, pp. 37–49.
32. Kalmykov, A.I., V.B. Efimov, A.S. Kurekin, B.A. Nelepo, A.P. Piguschin, A.B. Fetisov, B.E. Khmyrov, V.N. Tsymbal, and V.P. Shestopalov (1986), "The Radar System of the Cosmos-1500 Satellite," *Sov. J. Remote Sensing*, Vol. 4, pp. 827–840.
33. Jordan, R.L., (1980), "The Seasat Synthetic Aperture Radar System," *IEEE J. Oceanic Engineering*, Vol. OE-5, pp. 154–163.
34. Fu, L.L., and B. Holt (1982), *Seasat Views Oceans and Sea Ice with Synthetic-Aperture Radar*, Jet Propulsion Laboratory, Pasadena, CA.
35. Shul'gin, C.V. (1987), "Analysis of Cosmos 1500 Radar Images of the Ocean Surface in a Zone of Cyclones and Mesoscale Cloud Formations" (in Russian), *Issledovanie Zemli iz Kosmosa*, March–April 1987, pp. 3–11.
36. Ulaby, F.T., B. Brisco, and M.C. Dobson (1983), "Improved Spatial Mapping of Rainfall Events with Spaceborne SAR Imagery," *IEEE Trans. Geoscience and Remote Sensing*, Vol. GE-21, pp. 118–121.
37. Ford, J.P., R.G. Blom, M.L. Bryan, M.I. Daily, T.H. Nixon, C. Elachi, and E.C. Xenos, *Seasat Views North America, the Caribbean, and Western Europe with Imaging Radar*, JPL Pub. No. 80-67, November 1980, NASA-JPL, Pasadena, CA, p. 118.
38. Ulaby, F.T., M.C. Dobson, J. Stiles, R.K. Moore, and J.C. Holtzman (1982), "A Simulation Study of Soil Moisture Estimation by a Space SAR," *Photogrammetric Engineering and Remote Sensing*, Vol. 48, pp. 645–660.

39. Moore, R.K., (1979), "Trade-off Between Picture Element Dimensions and Noncoherent Averaging in Side-Looking Airborne Radar," *IEEE Trans. Aerospace and Electronic Systems,* Vol. AES-15, pp. 696–708.
40. Curran, R.J., ed. (1987), *Earth Observing System Instrument Panel Report: LAWS Laser Atmospheric Wind Sounder,* NASA, Washington, DC, Vol. IIg.
41. Doviak, R.J., and D.S. Zrnić (1984), *Doppler Radar and Weather Observations,* Academic Press, New York.
42. Zrnić, D.S., and P. Mahapatra (1985), "Two Methods of Ambiguity Resolution in Pulse-Doppler Weather Radars," *IEEE Trans. Aerospace Electronic Systems,* Vol. AES-21, pp. 470–483.
43. Sachidananda, M., and D.S. Zrnić (1986), "Recovery of Spectral Moments from Overlaid Echoes in a Doppler Weather Radar," *IEEE Trans. on Geoscience and Remote Sensing,* Vol. GE-24, pp. 751–764.
44. Campbell T.G., and W.K. Belvin (1985), "The Development of the 15-meter Hoop Column Deployable Antenna System with Final Structural and Electromagnetic Performance Results," *Digest IGARSS '85,* Amherst, MA, p. 658.

Chapter 9
THERMAL CONTROL FOR SPACE-BASED RADAR
L.M. Herold and M.S. Busby
TRW

9.1 INTRODUCTION

Thermal control of advanced space-based radar satellites will be a major issue due to the significant amounts of heat dissipated by the electrical power source, the electrical power conversion equipment, and the radar payload. The Military Space Systems Technology Plan [1] defines several military applications, where the electrical power requirements range from 50 kW to more than 500 kW. Because the overall efficiency (RF output divided by electrical power input) is typically less than 25%, the remaining electrical power is converted to heat, which must be managed by the thermal control systems. In most cases, these thermal control systems will require large, deployable radiators that will be similar in configuration to today's solar arrays and will need thermal transport systems, such as fluid loops, to acquire heat from the radar and power equipment and to transport it to the remotely located radiators. In situations where the radar is active for only portions of the orbit, thermal storage in the form of phase change materials may be included in the thermal control system as a method for saving weight and radiator area.

This chapter discusses the key requirements that drive the thermal design, the trades that lead to a selection of the preferred design approach, thermal management design data to provide performance and weight parameters for sizing the thermal management system, and a review of thermal management design integration to other major subsystems.

9.2 THERMAL DESIGN REQUIREMENTS

The mission requirements having the most influence on the thermal management system are generated from three sources:

- Radar equipment (transmitter type, phased array, *et cetera*);
- Mission parameters (orbit altitude and inclination, spacecraft orientation, operating duty cycle);
- Power train (electrical power source plus conversion equipment).

9.2.1 Radar Equipment Requirements

The payload thermal requirements will depend to a certain extent on the radar concept that is used for the mission and its operating frequency (L-band, S-band, or X-band). Typical concepts are (1) a corporate-feed phased array consisting of microwave or millimeter-wave *monolithic integrated circuit* transmitter-receivers distributed across the array, (2) an offset-feed hyperbolic reflector, and (3) a lens-feed phased array consisting of an RF-emitting *maser* transmitter illuminating phase-shifters distributed across the face of the array. Figure 9.1 illustrates the three concepts. All concepts share the need for a large, lightweight structure for focusing and directing the large number of beams required. These large structures have a common requirement of accurate dimensional stability. Because the structure will experience wide temperature excursions between full sunlight and eclipse conditions, the usual approach is to select a structural material with a very low coefficient of linear expansion (α_L), such as graphite-epoxy. Proper geometric layup of the graphite fibers can result in α_L being less than 0.18×10^{-6} cm/cm/°C, as well as a composite having other excellent mechanical properties (high modulus of elasticity and low weight). (Further discussion of the mechanical design aspects is contained in Chapter 16.)

Transmitter-Receivers (T/R)

The principal dissipative element in a space-based radar is the transmitter-receiver (T/R). The devices range from monolithic microwave integrated circuits (MMICs) to traveling wave tube-like masers. Efficiencies for MMIC devices range from a low of 10% for millimeter-wave MMICs (60 GHz and above) to a high of 30% for S-band and L-band microwave transmitters. The maser is a higher efficiency device for millimeter-wave applications ($\eta = 35\%$). The power output per MMIC module ranges from 0.1 W to approximately 2.2 W, depending on the frequency of interest. High-power radar signals are produced by combining the outputs of ten thousand to five hundred thousand modules, depending on the type of mission

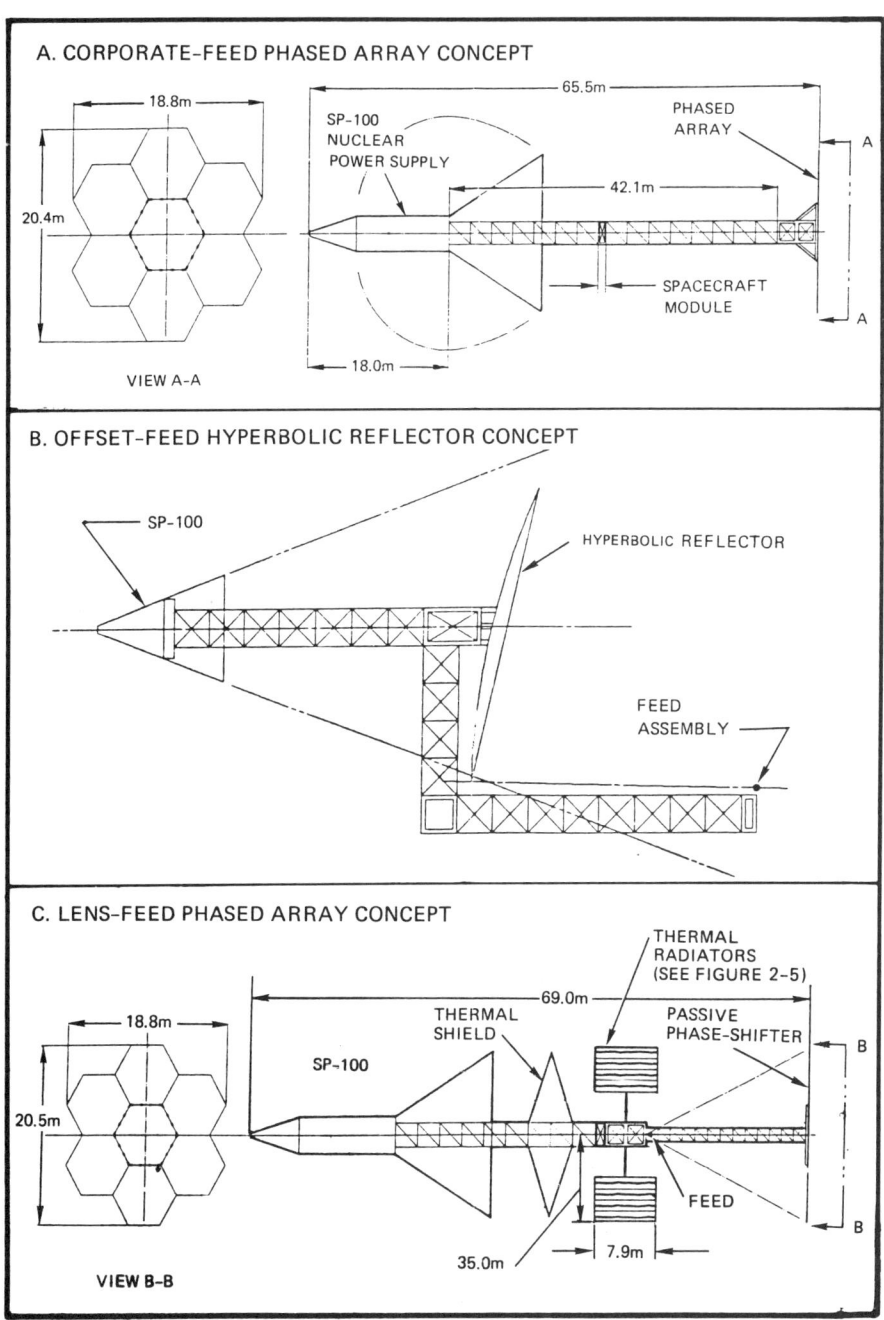

Figure 9.1 Three concepts for space-based radar.

and operating frequency. Missions requiring resolution of many targets use a larger number of modules to produce multiple beams, each of which can detect and track individual targets.

A maser transmitter emits a single, high-power beam capable of producing as much as 10 kW of RF power. The associated high thermal dissipation (≈ 18.6 kW) must be rejected along the length of its cylindrical collector, which is approximately 1.83 m (6 ft) in length by 6.35 cm (2.5 in) in diameter (Figure 9.2). This leads to a power density of 213 W/cm^2 (33 W/in^2) for the maser. Future maser transmitters are expected to operate with optimum efficiency at 250°C.

The MMIC modules, however, have a delicate transistor junction in their substrate, which must be maintained below 66°C. A large temperature gradient between the junction and module baseplate will be maintained due to the low thermal conductivity of the solid-state device. Baseplate temperatures of 40°C to 45°C will maintain the junction within acceptable limits. The MMICs are relatively stable over wide temperature ranges (zero to 50°C), and therefore do not require stringent temperature control. Because of their low thermal conductivity, the MMIC baseplate (which is only 0.39 cm^2 or approximately a half-wavelength on a side) must be embedded into the surface of a high-thermal-conductivity material. The power density then is determined by the spacing of the MMIC modules on the high-conductivity material. The module spacing, however, is dependent on the frequency of operation and the type of radar used; therefore, the frequency of operation ultimately determines the thermal power density.

The two basic forms of space-based radar using MMIC modules are the corporate-feed and space-feed phased array radars. In a corporate-feed phased array, the MMIC modules are distributed uniformly across the surface of the array (Figure 9.3). The module spacing must be maintained at approximately one-half wavelength, which indicates a tighter module spacing as the operating frequency increases. Clearly, the power density will increase as the operating frequency becomes greater. Minimum temperatures ($> -30°C$) can be maintained on the array surface by applying a small amount of power (microwatts per module) to the modules. For frequencies below 60 GHz, the power density is low enough that the phased array acts as a thermal radiator, maintaining the modules within limits (see Figure 9.3). At frequencies above 60 GHz, however, the maximum temperature of the modules is exceeded. At this point the excess heat must be transferred to a remote radiator or thermal storage must be integrated into the modules.

The space-feed array used some type of *beam-forming network* (BFN), either Rotman lens or a waveguide network, to combine the multielement RF output into coherent beams, which are directed through space to a lens array that provides the steering capability for the radar beam. The placement of the T/R modules is not constrained as is the case for the corporate-feed array. However, the closer

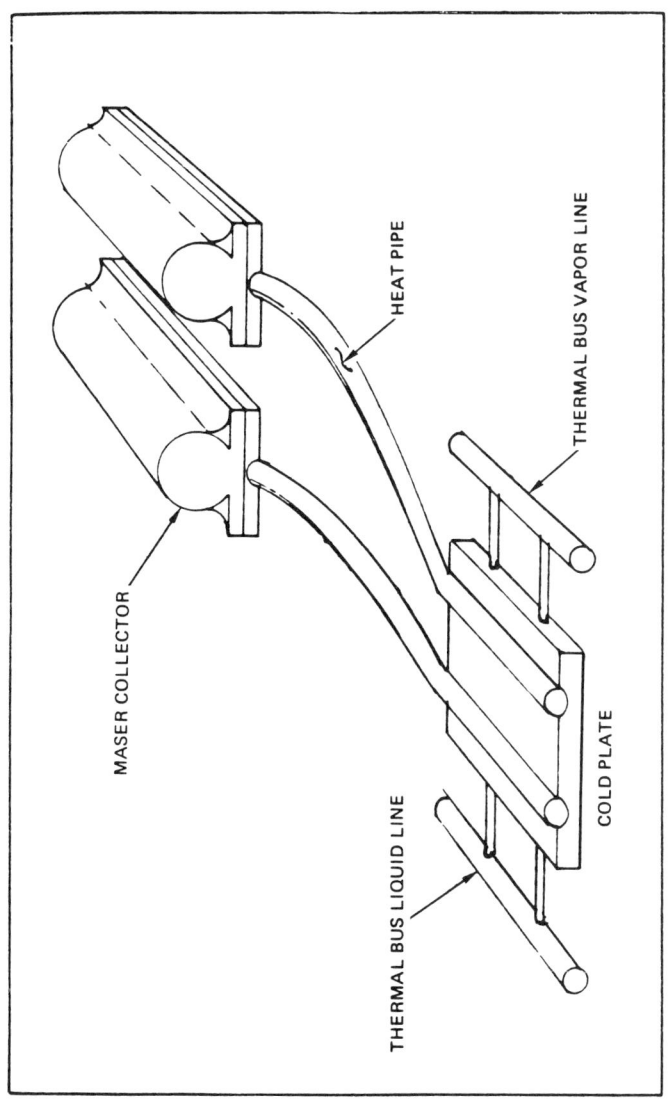

Figure 9.2 Thermal transport system for maser SBR.

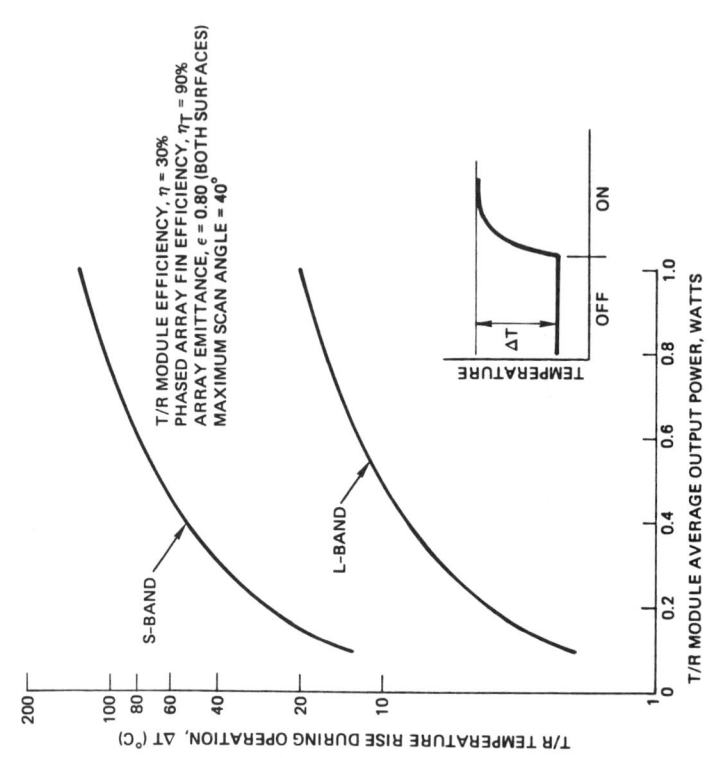

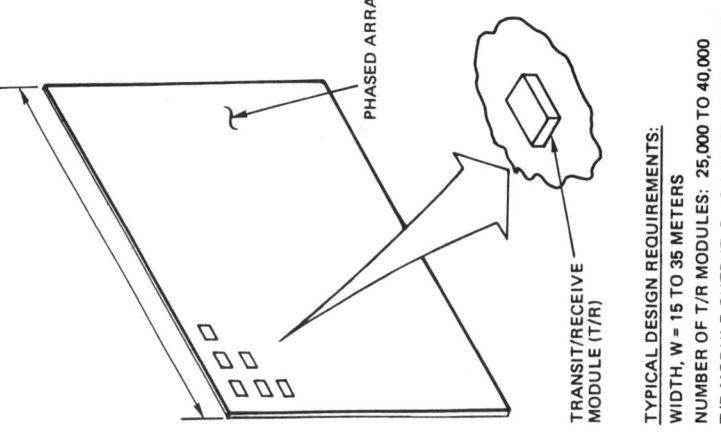

Figure 9.3 Thermal control of a phased array.

the spacing of the T/R devices, the less will be the overall loss in the system because waveguide runs are shorter. At 60 GHz, waveguide losses are 1.5 dB/m, which translate to thermal dissipation equal to 10% of the electrical input power for a short (<10m) waveguide run and 50% of the input power for a two-meter run. One possible configuration would be to mount the modules on a parabolic dish with equal-length waveguide runs to a feed located at the focus, thereby maintaining phase relations. The rear of the dish could then be used as the interface to a system of heat acquisition components.

Beam-Forming Networks

Losses in the beam-forming network (e.g., Rotman lens) will be from 1 to 3 dB (20 to 50% of the input power), and can lead to the single most difficult thermal control problem facing the thermal management system. These losses represent heat generated in the dielectric material and on the surface of the metal conductor plates. Rotman lens beam-forming networks comprise numerous disks of low-thermal-conductivity dielectric material sandwiched between metal sheets and stacked like plates in a cylindrical metal container (Figure 9.4). There are gaps between each plate so that a fluid can be passed through the container to remove waste heat because coolant tubes cannot be brought into the container. Removal of waste heat from low-power-level beamformers can be accomplished through the exterior wall of the BFN. Because of the low thermal conductivity of the dielectric plate, take care to ensure that thermal gradients do not become too severe when using this method of heat acquisition. Beam-forming networks do not contain active devices. Therefore, the maximum allowable temperature can be as high as 100°C, and possibly higher, depending on the materials and construction of the device. Integrated beamformers are limited by the amplifier's allowable temperatures.

In the microwave frequencies, there is a trade-off between BFN size and losses. Generally, the lower the frequency, the larger the beamformer will be and a lower power density results. Above 45 GHz, losses in the waveguide feeds and BFNs would be too great to allow sufficient RF output. Therefore, mirrors and lenses have been proposed for missions using 60 GHz and higher. This technique, similar to visible frequency methods, would reduce losses well below 1 dB and alleviate the thermal management problem.

9.2.2 Mission Parameters

Table 9.1 lists the key mission parameters that have a significant effect on the satellite's thermal control design. The first three parameters (orbit, spacecraft orientation, and spacecraft external configuration) define the worst-case hot and cold heating environments that each surface of the satellite will experience.

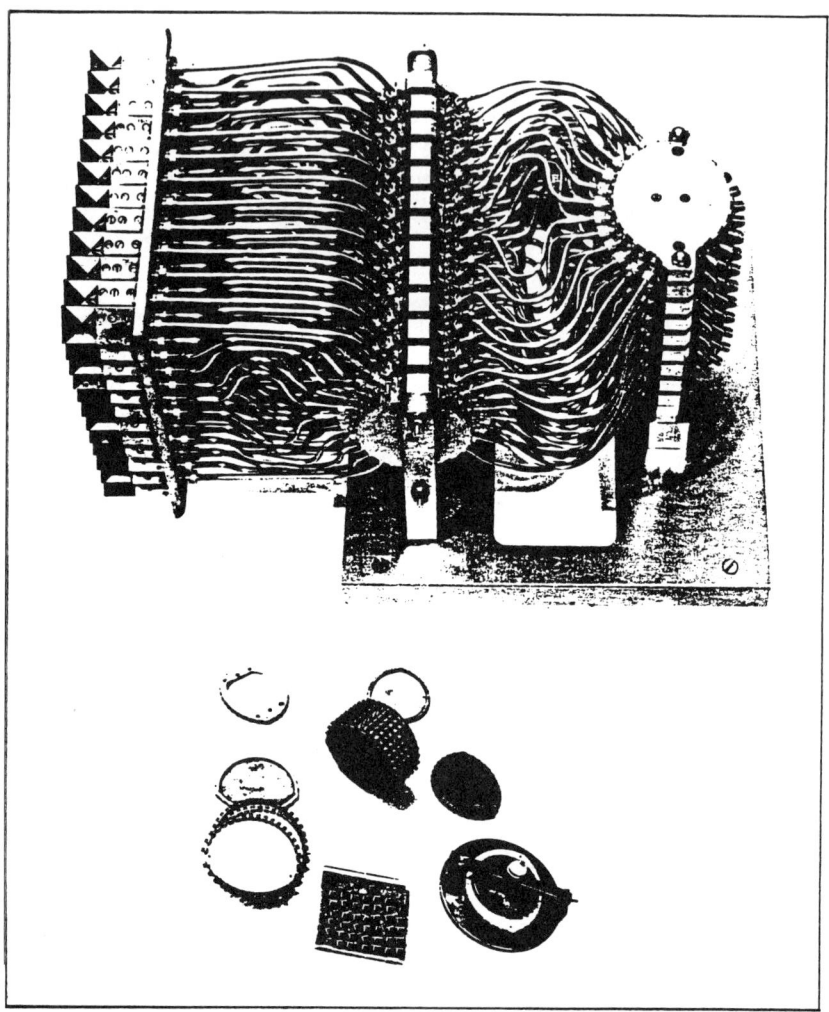

Figure 9.4 Rotman lens beamformer and internal parts.

The environmental heating is due to direct solar and earth heating, which consists of reflected solar (*earth albedo*) and earth infrared radiation. For low-to-medium altitude missions, the earth heating represents a major portion of the environmental heat load for surfaces facing the earth. Figure 9.5 shows incident earth infrared radiation heat flux to a flat panel as a function of altitude and its orientation with respect to earth. This heat flux occurs throughout the *entire* orbit, as the nighttime earth radiation is essentially the same as that of daytime. The

Table 9.1
Mission Parameters Affecting Thermal Control

1. ORBIT - altitude, inclination, eccentricity
2. SPACECRAFT ORIENTATION - orientation during all phases of the mission, and maximum body rates during spacecraft maneuvers
3. SPACECRAFT EXTERNAL CONFIGURATION - size, shape, addendages, keep-out zones. surface properties of all non-thermal control surfaces, and thruster location
4. THERMAL DISTORTION REQUIREMENTS - temperature gradients for spacecraft and appendages during all mission phases
5. LIFETIME
6. LAUNCH VEHICLE ENVIRONMENT - launch vehicle liner temperatures during ascent and pre-separation phases
7. WEIGHT, COST AND SCHEDULE CONSTRAINTS

albedo heating, which has the same spectral content as the sun, is more complicated to define because there are two additional parameters that must be defined. (See Figure 9.6.) These parameters are Θ_s and Φ_c, where Θ_s is the angle between the sun vector and a line connecting the satellite and the earth's center, and Φ_c is a longitudinal angle. A key difference between albedo and earth radiation environments is that albedo is highly transient, varying from zero on the shadowed portion of the earth to a peak at the satellite's subsolar position. Transient earth and albedo heating conditions are illustrated in Figure 9.7 for 600 and 5600 nmi altitude circular orbits. The reduction in earth heating environments with increasing altitude is apparent. The *absorbed* albedo heating can be reduced significantly by using thermal control coatings with a low solar absorptance, α_s, which also reduces the amount of direct solar heating absorbed by a surface. Thermal radiation properties of commonly used spacecraft thermal control coatings are shown in Table 9.2. There are two bounding values for solar absorptance: *beginning-of-life* (BOL) when a surface is new and clean, and *end-of-life* (EOL) when the surface has been degraded by solar ultraviolet energy, space-charge particles (electrons, protons, neutrons), and contamination. The end-of-life value depends on contamination from the satellite, the orbit-altitude, and time of exposure. Values in Table 9.2 are representative of a five-year lifetime in 600 to 6000 nmi orbits.

The solar heating (or *solar constant,* as it is usually termed) is 0.135 W/cm^2 (429 Btu/ft^2-hr) at a distance of 1.0 astronomical unit (AU) from the sun. The seasonal variations of the earth's distance from the sun cause the solar constant to vary 3.4% annually, being highest at winter solstice and lowest at summer solstice. Solar heating is the primary environmental heat load on thermal radiators and phased arrays. For a spacecraft in a high orbit, where earth heating is negligible,

Earth infrared heating, $\dot{Q}_e/A = F_{sp} \cdot E_r$

FLAT PLATE

where $\dot{Q}_e/A$ is earth radiation incident on a spacecraft surface
F_{sp} is the view factor of earth
E_r is earth radiation (nominal value is 215 to 227 W/m²)
h is the spacecraft altitude
R_p is the earth radius (3441 n. mi.)

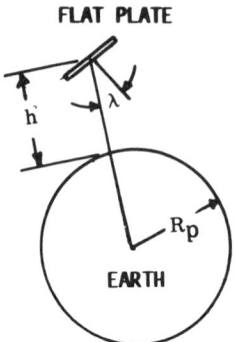

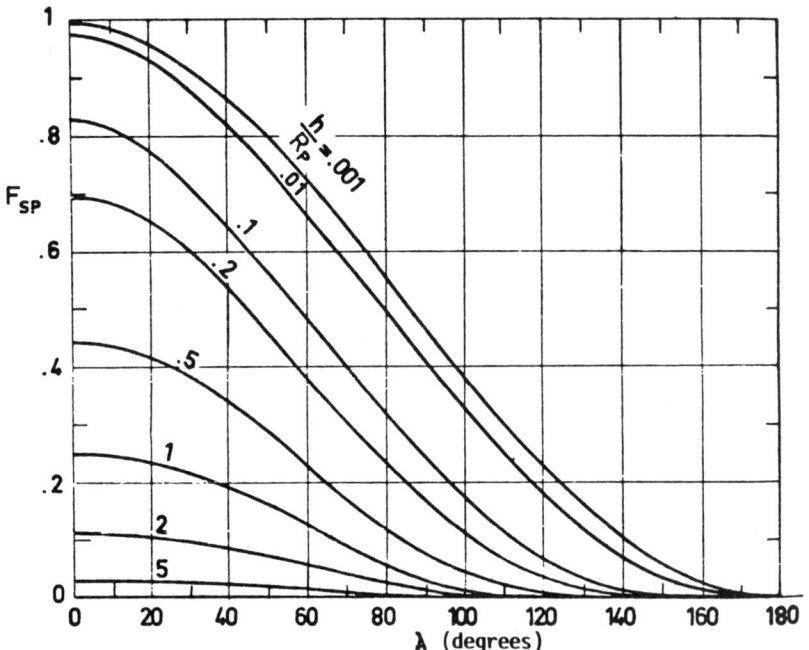

Figure 9.5 Calculation of earth infrared heating environments on spacecraft surfaces.

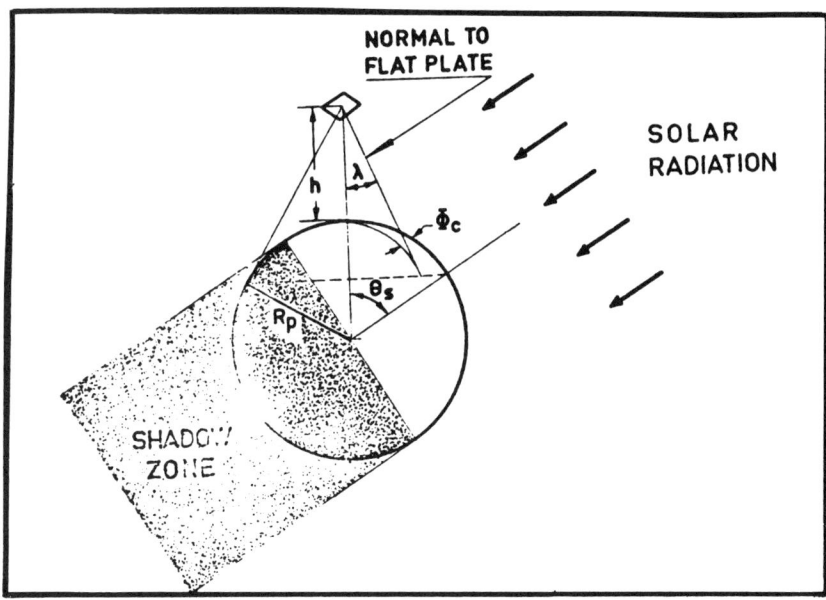

Figure 9.6 Orbital parameters and spacecraft surface orientation needed for albedo heating analysis.

the temperature of a panel (such as a radiator or phased array) can be calculated from the following expression:

$$A_1 \alpha_s G_s \cos\Theta_s + P = A_1 \epsilon_1 F_1 \sigma T^4 + A_2 \epsilon_2 F_2 \sigma T^4 \tag{9.1}$$

where

Θ_s = solar absorptance of sunlight surface;
G_s = solar constant;
P = equipment heat dissipation (w);
A_1 = area of one side of the panel;
A_2 = area of opposite side; zero if the back side does not radiate to space;
ϵ_1 = infrared emittance of surface 1;
ϵ_2 = infrared emittance of surface 2;
F_1 = view factor to space of surface 1;
F_2 = view factor to space of surface 2;
σ = Stefan-Boltzmann constant;
T = temperature (K).

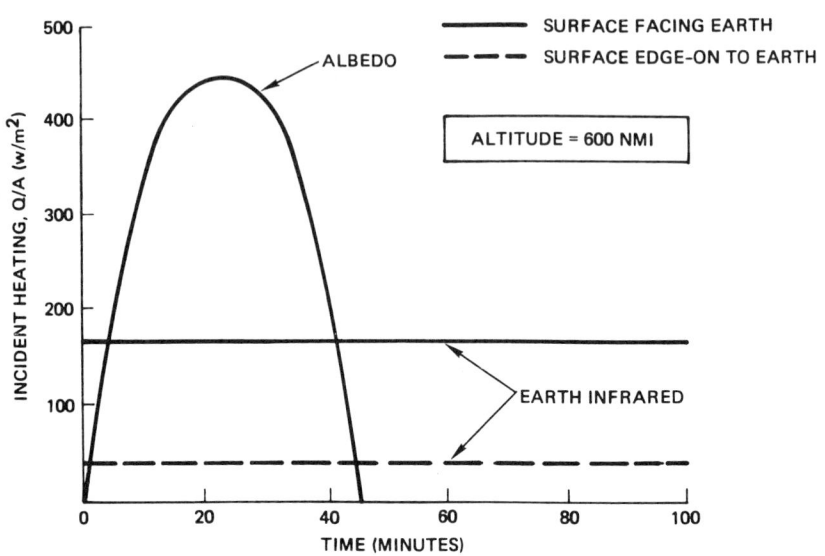
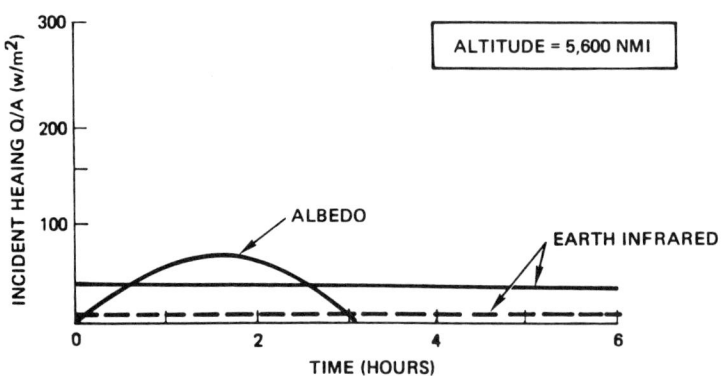

Figure 9.7 Albedo and earth heating incident on an earth-facing surface.

Table 9.2
Thermo-Optical Properties of Coatings and Paints

Material Description	Solar Absorptance (α_s)			Hemispherical Infrared Emittance (ε_h)
	Beginning-of-Life (BOL)	End-of-Life (EOL)		
		GEO*-3 Yrs	GEO-5 Yrs	
Silverized Fused Silica, Second-Surface Mirror, 8 mils thick	0.06	0.17	0.20	0.80
Silverized FEP Teflon, 5 mil thick	0.07	0.20	0.27	0.80
Aluminized FEP Teflon, 5 mil thick	0.13	0.25	0.32	0.80
Aluminized Kapton, 1 mil thick	0.36	0.54	0.66	0.61
Aluminized Kapton, 2 mil thick	0.39	0.55	0.67	0.73
Aluminized Kapton, 1st Surface	0.12	0.13	0.14	0.03
Aluminum Foil Tape 2 mil, 2 mil adhesive	0.15	0.16	0.17	0.035
Black Kapton Film, 1 mil thick	0.92	0.92	0.92	0.88
Chromized Kapton Film, 0.5 mil thick	0.70	0.70	0.70	0.70
White S-13-GLO Silicone Paint 10 mil thick	0.22	0.39	0.47	0.88
Black Z306 Polyurethane Paint, 3 mil thick	0.95	0.93	0.92	0.87
Clad 7075 Aluminum	0.25	0.26	0.27	0.04
Solar Cell w/Cover	0.69	0.71	0.72	0.86

*GEO = Geosynchronous Orbit

If a given equipment heat load, P, is to be reradiated to space at a temperature less than T, it is necessary to calculate the radiator area required $(A_1 + A_2)$. Typical results for various sun angles (Θ_s) and temperatures are illustrated in Figure 9.8.

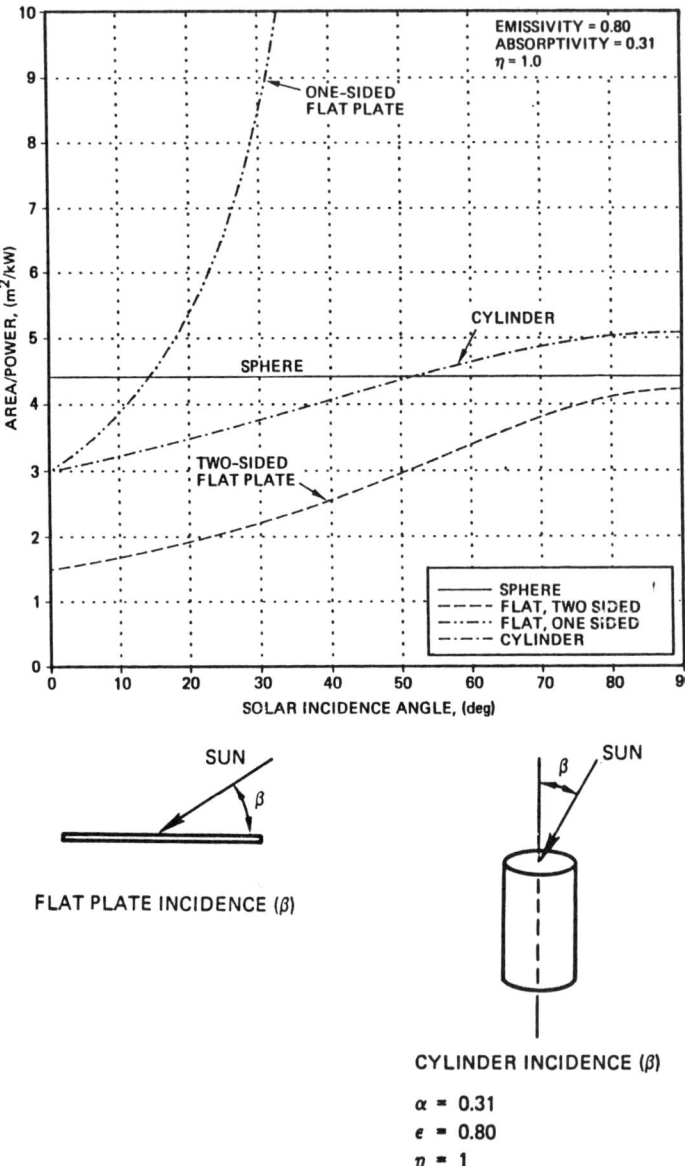

Figure 9.8 Required radiator area for $T_{\text{rad}} = 20°C$.

The spacecraft thermal design is sometimes dominated by a cold-case condition, due to either a long eclipse (in the earth's shadow), or shadowing of the sun by other appendages of the spacecraft. For example, the temperature of a lightweight phased array antenna will drop below $-50°C$ if the radar is not operating and the eclipse lasts one hour or longer. Thus, if the performance of T/R devices is adversely affected by low temperatures or high rates of temperature change, the cold-case environment must be addressed thoroughly. The maximum time in earth eclipse is shown in Figure 9.9 for various altitudes for circular orbits.

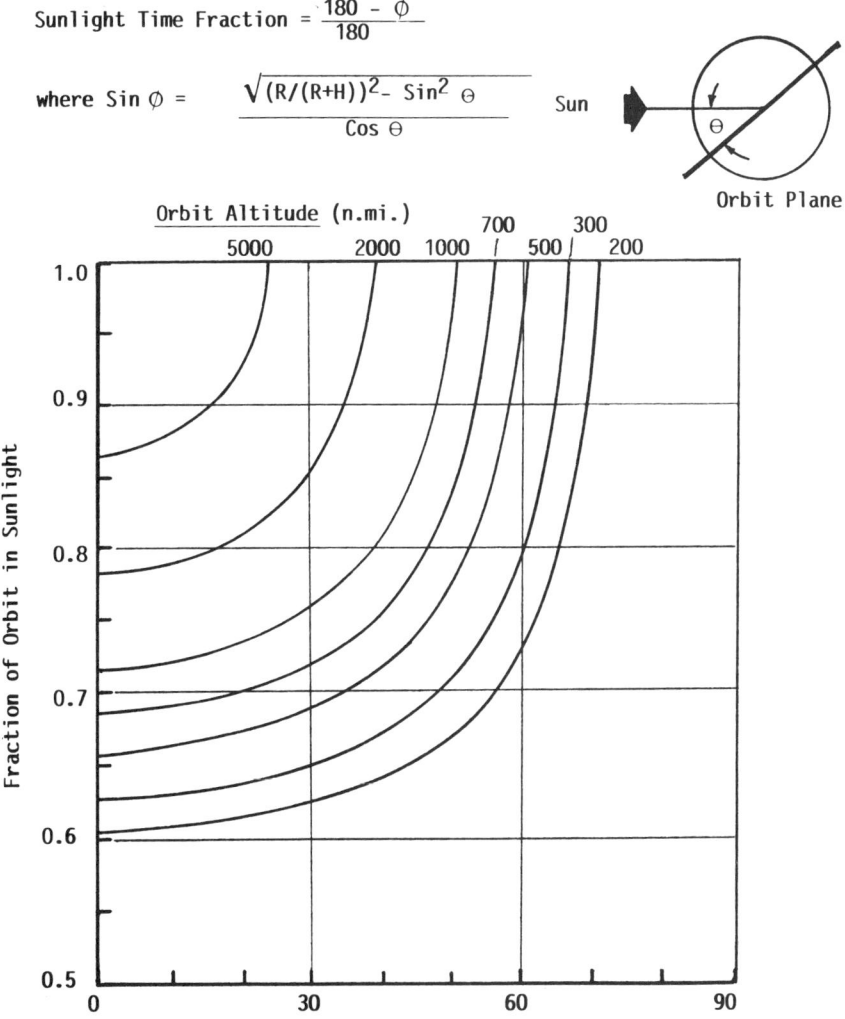

Figure 9.9 Orbit period and eclipse duration.

The operating duty cycle of the radar system has a major influence on the choice of the thermal management concept. For space-based radar satellites in midaltitude orbits (600 to 6000 nautical miles), the operating duty cycle is low (0.10 to 0.35) due to the available time over the target area. A typical operating duty cycle for a 5600 nmi orbit is characterized in Figure 9.10. The operating mode duration is about one-third of the orbit (≈ 2 hr), while the spacecraft is over the target area. The heat dissipation in the radar components, the batteries, and the power conditioning equipment is extremely high during this two-hour operating mode. During the remainder of the orbit, the payload is inactive and the electrical power system is recharged. This time can be used to advantage by the thermal system as well by reradiating heat stored in thermal storage components if they are used in the system.

9.2.3 Heat Dissipation of the Electrical Power System

There are two major heat loads that the satellite's thermal management system must dissipate: (1) the spacecraft electrical power equipment, and (2) the radar RF-emitting equipment. The electrical power system includes the power source (solar arrays, nuclear reactors, *et cetera*), power conversion, power conditioning, and power storage. A typical power chain is illustrated in Figure 9.11.

The component with the greatest heat dissipation is the power source, which may have thermal efficiency as low as 6% for the SP-100 nuclear power system to over 30% for advanced heat engine power plants. The thermal system designer must be aware of the effect that the power source's thermal control radiators will have on his design. Care must be taken to account for the radiant interchange between the high-temperature power source radiators and the spacecraft radiators as well as the phased array. The second largest heat dissipator is the power conversion and power conditioning equipment. Current electronic technology requires that these "black boxes" be maintained below 50°C for long life. The high heat loads, 10 to 20% of the spacecraft power and 65 to 90% of the radar power, and the low heat rejection temperature (50°C) will lead to large radiators. The electrical storage devices (typically batteries) have a low efficiency in their discharge mode and require radiators to reject this heat to space. Current batteries, such as nickel-cadmium and nickel-hydrogen operate at less than 20°C for long life. The batteries of the future are sodium-sulpher (NaS), which operate at 350°C; thus, the radiator size will be reduced dramatically.

Table 9.3 illustrates typical heat dissipation requirements for a space-based radar that emits 50 to 300 kW of RF power.

The resulting thermal management requirements are as follows:
- accept high heat loads generated by dissipation by electronic equipment and batteries;
- transport the heat to radiators;

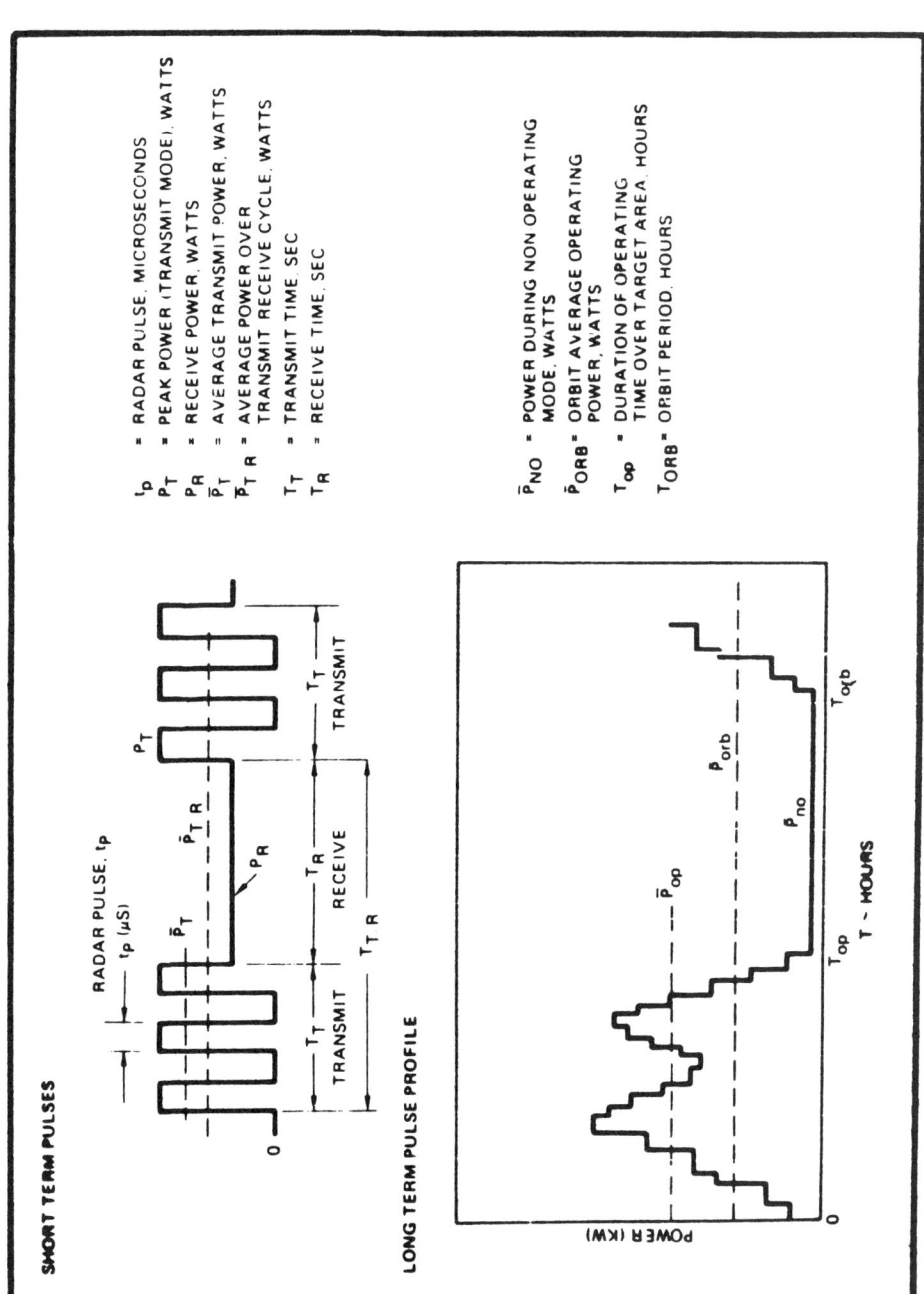

Figure 9.10 Characteristics of a space-based radar operating duty cycle.

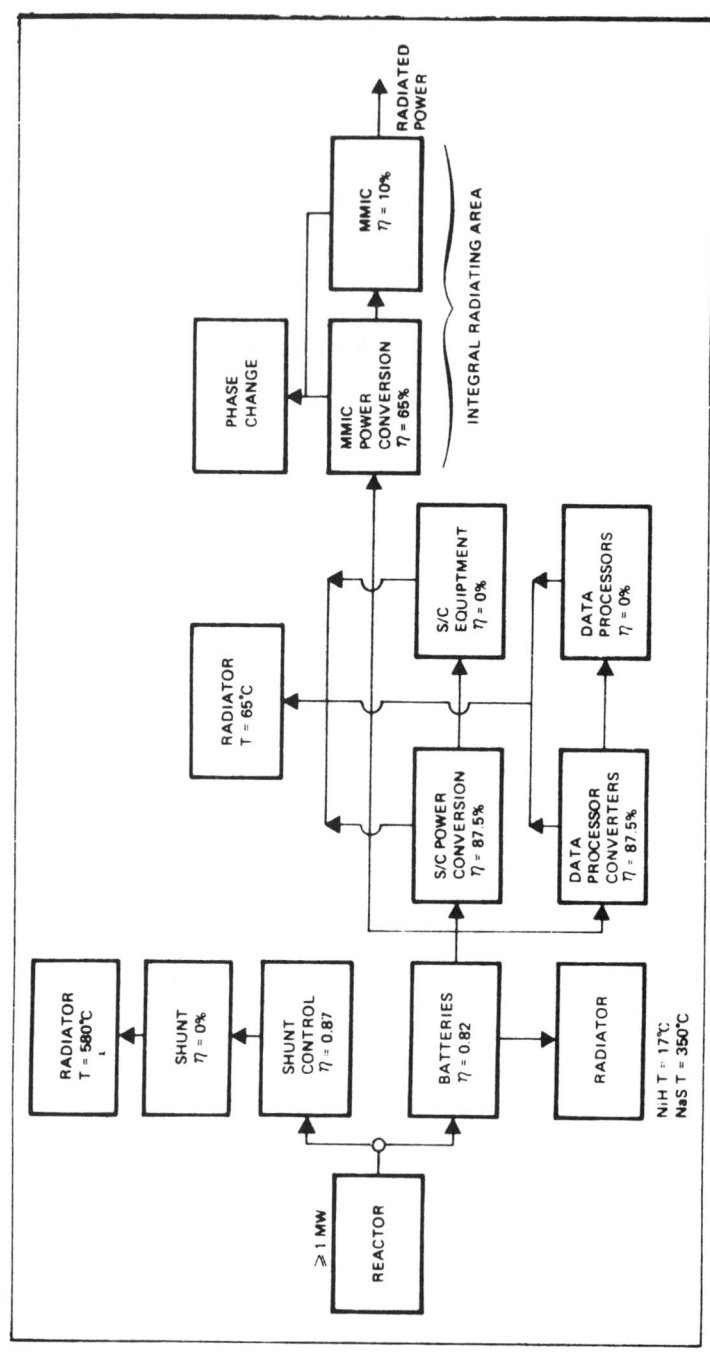

Figure 9.11 MMIC power chain.

- reject the heat to space;
- or, store most of the heat during operation and reject it to space during the nonoperating time period.

The resulting radar thermal requirements are summarized in Table 9.4. Specific orbital information and equipment powers are not shown because of their classified nature. Specific information would be obtained by the thermal systems designer from the radar systems designers on a mission by mission basis. Thermal design trades to minimize weight and radiator area are discussed in the next section.

Table 9.3
Heat Dissipation for an MMIC SBR with a Reactor-Battery Power Source

RF Radiated Power (kW)	Heat Dissipation (kW)						
	Low Duty Cycle Equipment Dissipation				Continuous Operation		
	Transmitter	Converter	Battery	Data Proc	Spacecraft	Shunt	Shunt Control
50	450	269.0	148.4	2.2	5.5	168.7	24.9
100	900	538.5	290.4	2.2	5.5	330.1	45.9
200	1800	1080.0	574.4	2.2	5.5	653.1	87.9
300	2700	1620.0	858.4	2.2	5.5	975.9	129.9

Table 9.4
SBR Thermal Requirements

Component	Maximum Operating Temperature	Thermal Dissipation	Power Density
Data processing	65°C	100 to 500 watts per processor	
Attitude control			
CMG	80°C → 100°C	<1 kW Total	
Electronics	65°C		
Communications			
Chemical laser crosslink	100°C	<1 kW	NA
TWT	80°C	<1 kW	2.3 W/cm^2
Solid-state amplifiers	65°C	<1 kW 2 kW Total	7.8 W/cm^2
Other electronics	65°C	<1 kW	
Phased Array	Figure Control	Other Requirements	
Space-feed array	λ/4	Must be RF transparent, no active devices to limit temperature swings.	
Corporate-feed array	λ/10 → λ/25	Doesn't need to be RF transparent, transmitters on array, severe surface accuracy required, transmitters generally limit temperature swings.	

Table 9.4 (cont'd)

Component	Maximum Operating Temperature	Efficiency	Thermal Dissipation
Transmitter			
MMIC	66°C junction 40 - 45°C baseplate	30% L-, S-band 10% X-band and above	0.1 to 20 W per module
MASER	250°C	30% - 35%	Up to 20 kW/tube, 5.1 W/cm^2
Beam Forming Network (BFN)	~100°C	1dB - 3dB (20% to 50% of input power)	Depends on input power
Power source			
Solar array/battery	80°C solar array 10°C NiH battery 350°C - 400°C NaS battery	18%	Mission dependent
DIPS	150°C - 260°C depending on cycle	26% - 36%	
Nuclear reactor	540°C	24% - 36%	
Power processing	65°C	88% - 95%	
Power distribution and control	65°C	90% - 98%	

9.3 TRADES TO MINIMIZE THERMAL MANAGEMENT SYSTEM WEIGHT

Weight is a precious commodity on nearly all satellites. The launch cost of a spacecraft using the Space Shuttle (Space Transport System, STS) is over $40,000 per pound. When a satellite exceeds a launch vehicle's capability, the mission cannot proceed unless a more powerful launch vehicle is found. On this basis, trades to optimize the thermal management system's weight must be given serious consideration.

Deployed thermal radiators, which appear very similar to solar arrays, will be needed to reject the high heat loads to space. A study by S. J. Mertesdorf *et al.* [1], found that thermal radiators account for about 75% of the thermal management mass. (See Figure 9.12(a).) The remainder is in the heat acquisition (cold plates) and thermal transport system (fluid loops). Radiators thus, are the major source of weight penalties, and as such the trades are generally aimed at reducing the radiator weight.

The use of thermal storage can reduce the thermal management system's weight for certain types of missions. Specifically, we know that missions with a low operating duty cycle will benefit to varying degrees by using thermal storage.

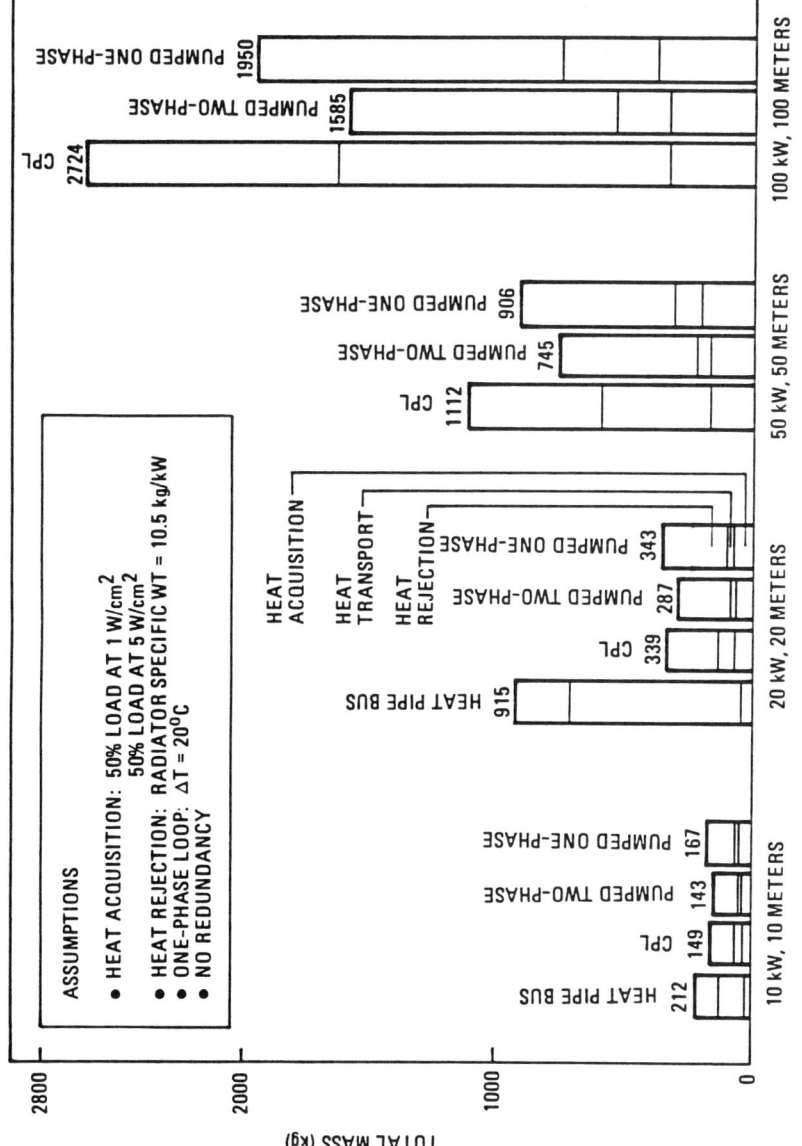

Figure 9.12(a) Weight comparisons and various types of heat acquisition and transport systems.

This duty cycle depends on the ratio between the specific radiator mass per heat rejection to the thermal storage mass per energy stored. Both of these parameters are physical constants that depend only on the performance capabilities of the radiator and thermal storage components. Therefore, the missions that will benefit from thermal storage depend on the available technology for these components and no other trades can be made. Considering the range of near-term performance parameters, operating periods up to 60 minutes result in weight savings. This operating time represents the product of the duty cycle times the orbit period. Clearly, higher altitude radar missions, due to their long orbit periods, will not benefit from thermal storage and must be sized for peak heat loads. A lower orbit (600 nmi for instance), where the operating duty cycle would be approximately 20%, would lead to an operating period of 20 to 25 minutes. A brief look at a thermal system sizing will provide some insights into the benefits of thermal storage for a mission of this type.

The orbit period for this test case is 100 minutes, of which the operating period is 20 minutes. Total system peak heat dissipation will parametrically vary between 10 kW and 100 kW. This leads to stored energy requirements of 3.3 to 33 kW-hr. Using near-term goals for two-sided, flat-plate, heat pipe radiators of 7.4 kg/kW and 14.6 kg/kW-hr for thermal storage components, Figure 9.12(b) results. The figure indicates a weight savings of 17% at all dissipation levels. Greater weight savings may be had in the future with increased performance from the thermal storage components. Further trades that can be made by the thermal designer are between components of a particular type (i.e., between heat transport buses) or between types of radiators.

9.4 THERMAL CONTROL COMPONENTS

Thermal control components that will be required for high-power SBR satellites include thermal radiators for rejecting the waste heat to space; heat acquisition and transport systems to carry the waste heat from equipment to radiators by using components, such as cold plates, heat pipes, pumped one-phase fluid loops, and two-phase fluid loops; and thermal storage devices, such as phase change materials to absorb periodic high density heat fluxes.

An overall mass summary for these components, based on current technology, was presented in Figure 9.12 (a) for waste heat loads of 10 to 100 kW, and for transport distances of 10 to 100 m. The radiator specific mass (kg/kW) was based on the use of conventional materials, such as aluminum. As more exotic materials, such as high-conductance graphite composites, are developed, the radiator specific mass will diminish somewhat, perhaps up to 50%. However, the weight estimates for the other components are not expected to drop appreciably.

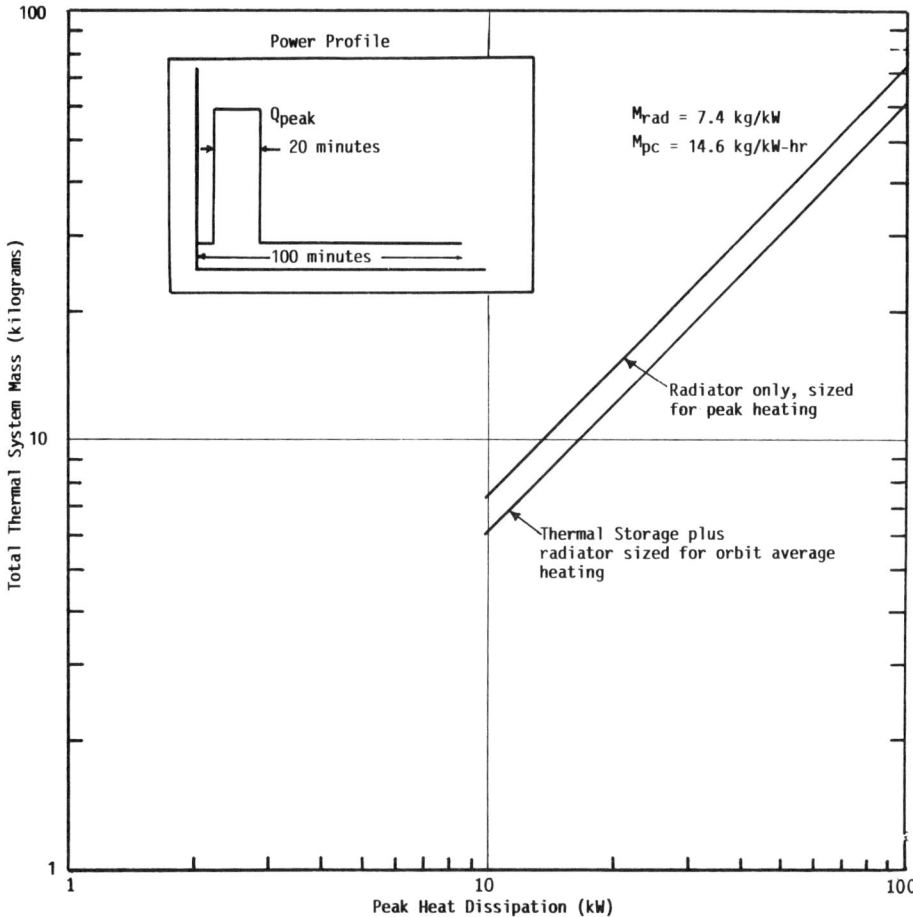

Figure 9.12(b) Weight optimization between radiator only and radiator plus thermal storage.

9.4.1 Thermal Management Systems

Recent trends in satellite technology have introduced challenging requirements for spacecraft electrical power and, correspondingly, for thermal management. Spacecraft at current power levels (1–2 kWs) employ structural panels as passive thermal radiators. As spacecraft power levels approach and exceed 10 kWs, passive thermal control techniques will be unable to accommodate the load, and various active systems such as heat pipes and fluid loops become necessary. Of critical importance is that waste heat loads from many different on-board power sources, each with its own duty cycle and temperature control limits, be integrated into the complete satellite thermal management system. Minimization of spacecraft weight and volume, while retaining simplicity in subsystem design is desirable. Thermal energy must be acquired from the heat sources and transported efficiently and reliably to the rejection site. Thermal energy acquisition devices, often known as *cold plates*, must be able to handle system transients, thermal load variations, spacecraft maneuvers, and a variety of equipment constraints.

Most thermal management system concepts for high-power satellites require an integrated thermal bus for handling thermal base or average loads. A thermal bus is a closed fluid system that consists of the following elements:

- one or more heat acquisition devices such as cold plates;
- a heat exchanger or some method of thermal communication between bus and radiator;
- transport lines connecting cold plates to the heat exchanger;
- any required control or fluid inventory management devices such as sensors, accumulators or reservoirs, and valves.

There are four types of thermal bus or closed fluid loop systems:

- heat pipe systems;
- capillary-pumped loops (CPLs);
- electromechanically pumped two-phase fluid loops;
- single-phase fluid loops.

Figure 9.13 shows a generic loop schematic.

Heat Pipe Systems

The main advantage of the heat pipe thermal bus is its inherent simplicity. Coupled with a variable conductance heat pipe radiator, the system will provide good temperature control. The basic heat pipe elements have already had considerable development effort, and their performance is well characterized. *Fixed-conductance heat pipes* (FCHP) and *variable-conductance heat pipes* (VCHP) have been used as primary thermal control components on operational spacecraft programs for both NASA and the military. Thus, various heat pipe designs have achieved a fully qualified flight status.

The main disadvantage of heat pipes is the large number required for high-power (>10 kW) systems and the resulting weight penalty. The low-pressure difference available for liquid pumping as compared with a mechanically pumped two-phase system may restrict application of heat pipes to lower powers and spacecraft for which pumping lengths are less than 15 m. The limitation on pumping distance may be overcome by cascading heat pipes and accepting a larger radiator because of interface temperature drops.

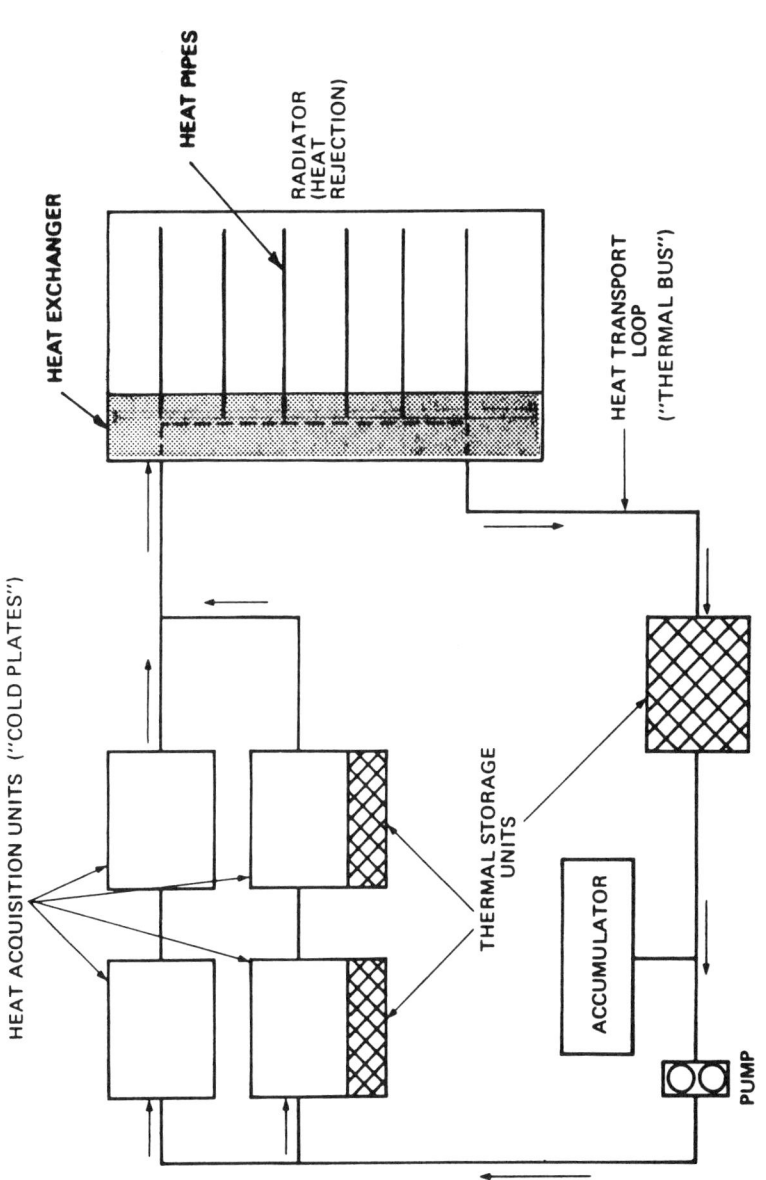

Figure 9.13 General schematic of a thermal energy acquisition and transport loop.

Capillary-Pumped Loops (CPLs)

During the 1980s, a large amount of conceptual design, hardware development, and testing of CPL components and systems was accomplished. The following summarizes the features of these efforts:

1. A CPL has few, if any, moving parts (Figure 9.14a), resulting in fewer possible mechanical failure modes. However, use of distributed electromechanical pumps in the liquid return line or use of a bellows-like accumulator may be necessary in some applications.
2. For a pure capillary system, where there are no electromechanical pumps, there is no danger to pump cavitation.
3. The heat transport capability per unit bus mass (kW/kg) is higher for a CPL than for a pumped single-phase loop.
4. Heat load-sharing is possible; for example, an evaporator that is not receiving heat from a payload can act as a condenser and maintain the payload temperature near the saturation temperature. This can result in substantial savings in auxiliary heater power.
5. Design and analysis software has been developed to evaluate steady and transient behavior of CPL systems.
6. CPL systems (Figure 9.14b) have been tested extensively in a 1-g environment.
7. Microgravity testing has been performed on CPLs in the Space Shuttle. As a result of these tests, many of the CPL's operational difficulties have been identified, and steps taken toward their solution.

Two-Phase Fluid Loops

A mechanically pumped two-phase thermal bus will ultimately have a lower mass per unit heat transported than a pumped liquid or capillary pumped loop. The two concepts for a two-phase Space Station thermal bus are fundamentally different in their management of the liquid and vapor phases. These prototype buses require much more ground and flight testing before we can consider them to be current technology.

We can identify the following advantages of two-phase loops:

1. A pumped two-phase loop is capable of transferring large amounts of heat from the payloads to the heat rejection system with low fluid mass flow rates, pressure drops, and pumping powers.
2. Unlike a CPL system, a pumped two-phase system has no rigid restriction on the total loop pressure drop. This permits the use of smaller diameter transport lines, resulting in lower transport line and accumulator masses.

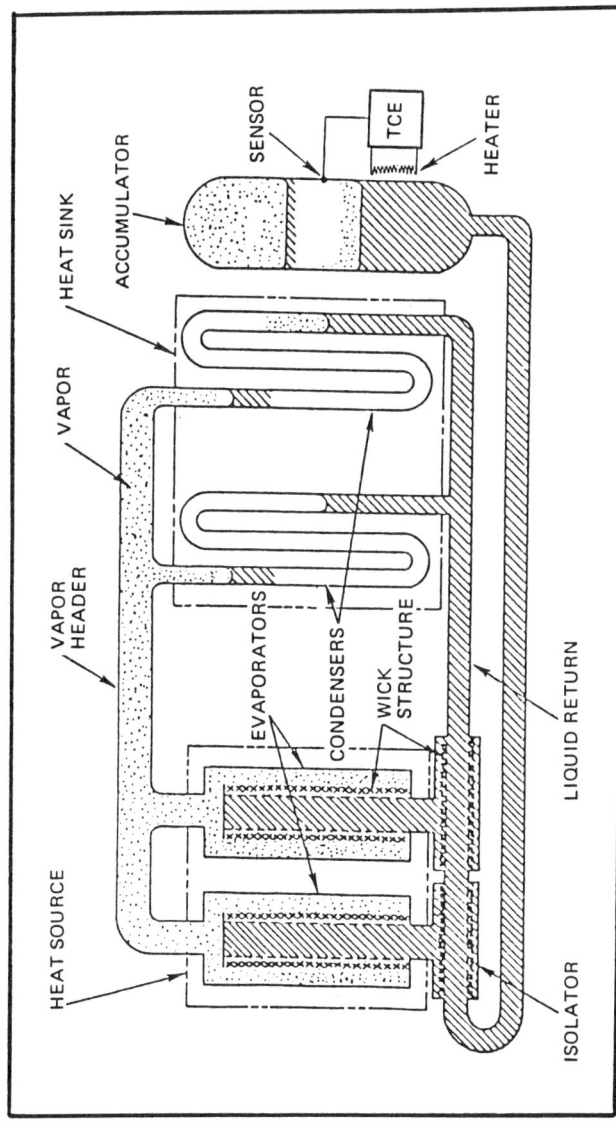

Figure 9.14(a) A typical capillary-pumped loop.

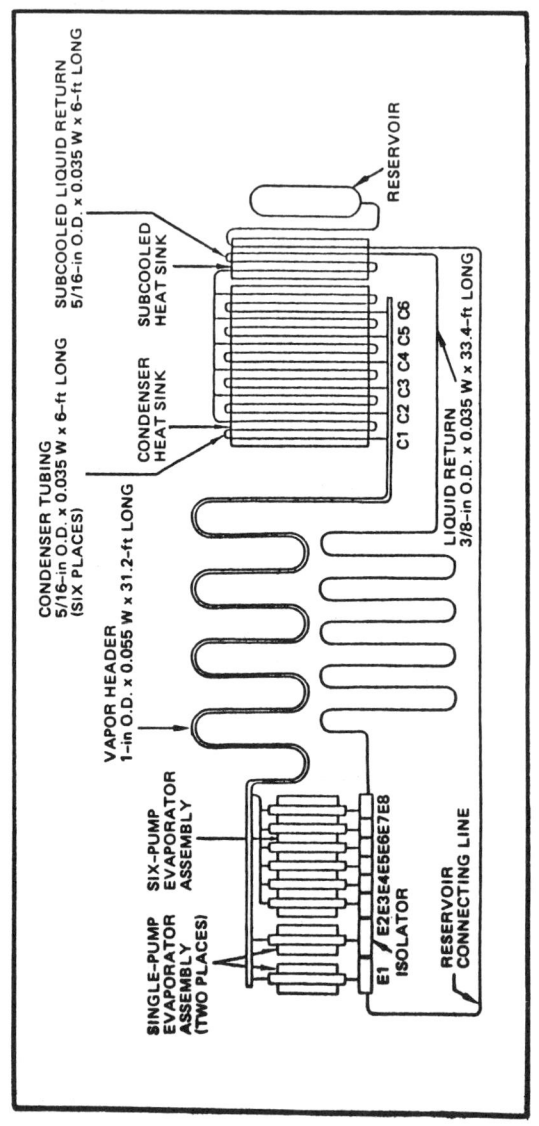

Figure 9.14(b) Schematic of CPL I and CPL II testing models.

3. A pumped two-phase thermal bus has a much lower mass per unit heat transported than a pumped liquid bus. This is because the mass flow rates, pressure drops, and pumping power required are much larger for a pumped liquid loop than for a pumped two-phase loop.
4. Heat load-sharing is possible, and so heat can be provided from the fluid to an inactive payload, resulting in large potential savings in auxiliary heater power.
5. A pumped two-phase system is nearly isothermal. For example, there is a very small fluid temperature drop from the heat dissipating payload to the radiator, resulting in considerable savings in radiator mass as compared with a pumped liquid loop.

Single-Phase Fluid Loops

Pumped single-phase loops with heat pipe radiators are very likely to become the baseline design concept for military missions with high heat dissipation requirements (10–50 kW$_e$) for the near term. Heat pipes probably will show weight and reliability advantages up to about 10 kW$_e$. Pumped single-phase loops have the advantage of being a well developed, well characterized technology, and their response to transient load and environmental conditions is easily calculated. Start-up and shutdown sequences are routine with no time delay involved.

Pumped liquid loops will remain competitive with two-phase loops if two criteria are satisfied. First, the allowable temperature drop ΔT in the loop must be fairly high, certainly larger than the 5°C specified for Space Station thermal bus requirements [3, 4, 5]. Second, the spacecraft design should provide for minimum lengths of bus plumbing; that is, the spacecraft design should arrange the dissipating equipment and radiator heat exchangers to minimize the length of the fluid lines.

Although the overall ΔT may be fairly high (30°C, for example), critical equipment can be maintained at relatively small ΔT values by their location in the loop and by appropriate flow control through their cold plates.

The utilization of a one-phase loop must be evaluated thoroughly for each military spacecraft application, but, until pumped two-phase loops are flight-qualified, one-phase loops will continue to be the preferred concept.

9.4.2 Thermal Storage

The heat rejection and transport system generally must be sized to transport and reject the maximum imposed heat load. If the peak heat dissipation rate is much larger than the orbital average rate, the radiator and bus size and weight can be reduced by use of thermal storage. It will reduce the overall heat rejection and transport system weight if that of the thermal storage device is less than the reduction of the radiator or bus weight. The magnitude of the thermal storage benefit depends not only on the spacecraft equipment heat load profile, but also on the heat rejection and transport system hardware, rejection temperature, and environmental heating profile. A report by Lockheed [6], in particular, contains the most comprehensive compilation of *phase change material* (PCM) properties and applications to date.

A systems analysis is useful for determining the maximum weight benefit of thermal storage. If a thermal management system with thermal storage does not appear to be demonstrably lighter in weight than one without thermal storage, it must be eliminated from further consideration.

For repetitive power-pulse durations of less than 30 minutes, thermal storage is beneficial in reducing the heat rejection system weight for the assumed specific masses of radiators, thermal buses, and thermal storage system. The radiator should be sized for an orbital-average heat rejection rate, rather than the pulse-mode heating rate.

The design of the thermal storage module requires considerable development because of the poor thermal conductivity of prospective PCMs. Two methods that improve the overall conductance include enhancing the PCM thermal conductivity and designing the thermal storage module with a large interface area. Enhancement of the PCM thermal conductivity can be accomplished by attaching narrow, closely spaced metal fins to the module baseplate. The conductivity of the material can be dramatically improved with as little as 10% metal (by volume), but, for maximum heat transport performance, the metal concentration should be 50%. Increasing the proportion of metal increases the weight of the module with little corresponding improvement in the thermal storage capacity. Therefore, a minimum volume percentage of metal should be chosen to achieve the required performance.

REFERENCES

1. "Military Space Systems Technology Plan, Vol. II: System Concepts and Architectures," AFSTC-TR-84-4, January 1985.
2. Herold, L.M., M.S. Busby, and S.J. Mertesdorf, "High Power Spacecraft Thermal Management Study, Final Report, Task 1. Requirements Formulation," AFWAL-TR-2121, Vol. I, August 1986.
3. Sadunas, J.A., and A. Lehtinen, "Thermal Management System Options for High Power Space Platforms," AIAA-85-1047, *AIAA Thermophysics Conf.*, Williamsburg, VA, June 1985.
4. Nason, J., "Final Report for Thermal Bus Study, Task 1," Hamilton-Standard Report No. SVHSER 8913, June 30, 1983.
5. Sadunas, J.A., "Concept Description and Trade Study for the High Efficiency Automated Thermal Control Systems Study," Rockwell International Report SSD84-0059, May 1984.
6. Hale, D.V. et al., "Phase Change Materials Handbook," Lockheed Missiles and Space Co., *NASA Contractor Report CR-61363, September 1971.*

Chapter 10
RADAR CROSS SECTION (RCS) OF SATELLITES AND OTHER SPACE-BASED TARGETS
J.W. Curtis
The Aerospace Corporation

10.1 INTRODUCTION

A radar when placed in space has many more target viewing opportunities than does a surface-based or an airborne platform. With a radar satellite platform orbit period of 90 minutes and a velocity of about 7 km/s, many targets can be viewed in a short time. Depending on the revisit time needed, use of a number of satellites (constellation) may be necessary to perform certain missions, such as forming a radar fence around a given area like a fleet of ships or the continental United States (CONUS). The space location of the radar platform has further advantages. From a military point of view, many aircraft are designed so that they cannot be seen from the ground, or at least their *radar cross section* (RCS) is substantially reduced. A radar in space can take advantage of such design by viewing these aircraft in the upper hemisphere, where a larger RCS may exist. Reducing the RCS everywhere is difficult. The sensor presence in space will force the aircraft designer into a more difficult design scenario. Frequency selection could also help the detection of space objects by taking advantage of certain absorption bands in the earth's atmosphere within which background clutter would be greatly reduced. Prominent atmospheric absorption bands are indicated in Figure 10.1, showing peak frequencies of 22.24, 60, and 118 GHz [1]. There is an additional water absorption band at 184 GHz that is not shown in the figure. This chapter deals with the RCS of various types of targets (air, surface, and space) in which such

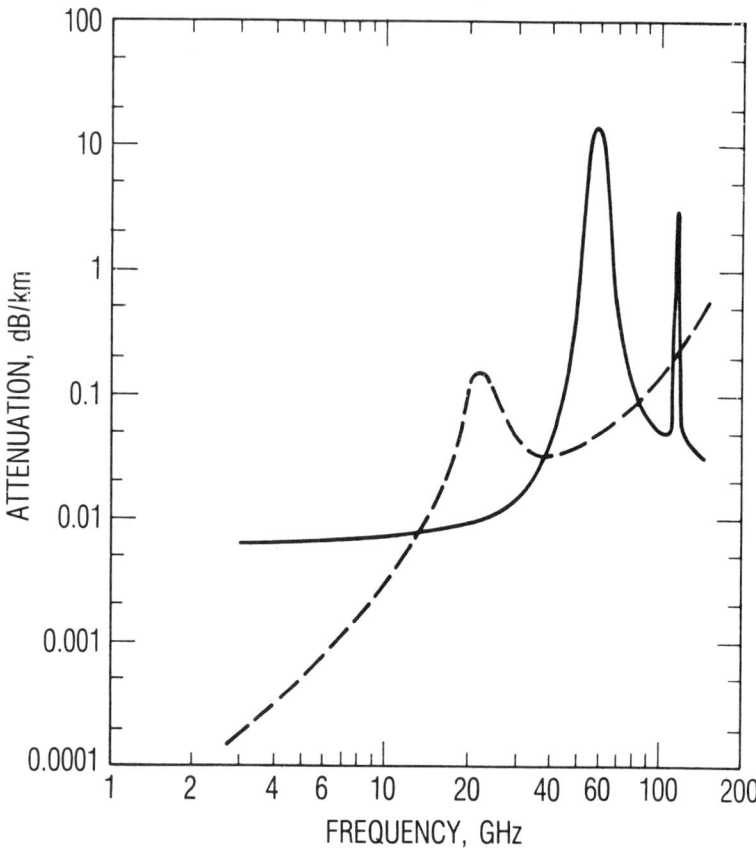

Figure 10.1 Attenuation of electromagnetic energy by atmospheric gases: dashed curve is absorption due to water vapor; the solid curve is the absorption due to oxygen (from [1]).

systems are interested, and we will discuss them as discrete targets, although they are often viewed against a background (clutter). We will not deal with large area surveillance, such as prospecting, assaying, and high resolution ground mappings, because these topics have been adequately covered elsewhere [2, 3]. Radar clutter *per se* will be discussed in the next chapter.

Before examining individual objects, we shall review some of the prediction techniques commonly used to determine the RCS of an object of interest. A number of different techniques might be applied to a given object, depending on the radar frequency. Three distinct regimes are usually of interest to RCS analysts: the *Rayleigh region* in which the object is much smaller than the wavelength, the *resonant region* in which the objects are about the same order of magnitude as the wavelength, and the *optical region* in which the object is much larger than the wavelength. These three regions are illustrated in Figure 10.2, where the RCS of a sphere is presented. In the Rayleigh region, the RCS is proportional to k^4V^2, where V is the volume and $k = 2\pi/\lambda$, with λ being the wavelength. The RCS drops off rapidly with the inverse wavelength to the fourth power. At the time of this writing, the longest wavelengths being considered for SBR are about 0.25 m (preferred frequencies are L-band and higher), whereas most objects of interest are larger. Therefore, for our purposes, we can rule out further discussion of the Rayleigh region.

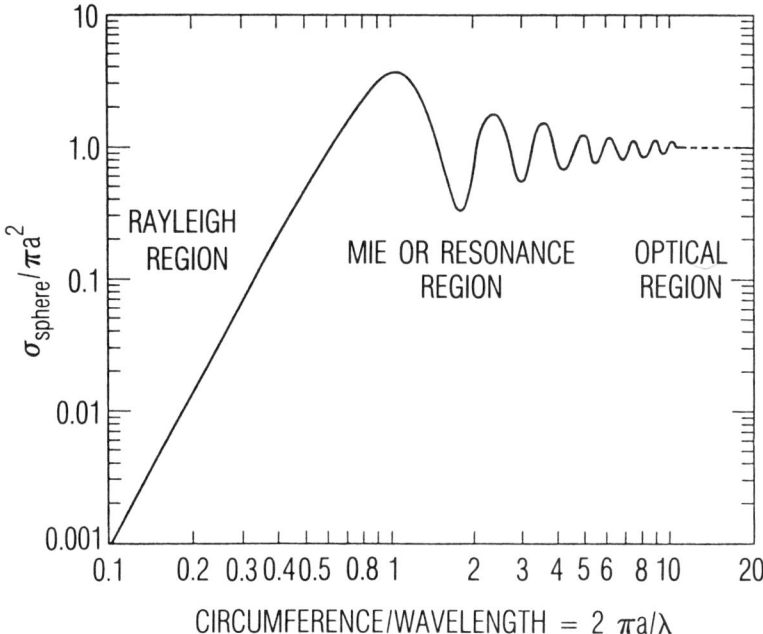

Figure 10.2 Radar cross section of a sphere.

The resonant region is not only of much more interest, but is also the most difficult to handle analytically. A powerful technique often used in this regime is the *method of moments* (MOM). When Maxwell's equations are applied in conjunction with Green's theorem to the surface of a scattering body, a set of integral field equations (Stratton-Chu [4]) results. The electromagnetic boundary conditions are then applied to the surface of the scattering body to determine a system of linear equations, which can be solved by matrix methods to generate the body currents. The scattered field can then be obtained from the body currents. These procedures are known as the method of moments. Although this technique is exact, it is limited to bodies not larger than about 10λ because, as the body size increases, the associated matrices become extremely difficult to process (invert), even for the largest computers. Most objects of interest to SBR will be in the optical or high frequency region. Modern prediction techniques in this regime include *geometrical optics* (GO), *physical optics* (PO), *geometric theory of diffraction* (GTD), and *physical theory of diffraction* (PTD). Geometrical optics assumes very short wavelengths, and is usually associated with ray tracing (direct and reflected rays). Geometric theory of diffraction is an extention of GO into shadowed regions to include diffracted rays. Physical optics retains the surface current approach of the Stratton-Chu equations, but is greatly simplified because of the higher frequency assumption although it does not account for edge or surface-wave scattering. Physical theory of diffraction extends PO to include edge effects. These high frequency diffraction theories and their extensions are discussed in more detail in standard texts on RCS prediction [4, 5]. All of the high frequency prediction methods are usually applied to reasonably simple shapes. To determine the RCS of a complex object such as a satellite or aircraft, a common method is first to break the object down into constituent parts. Each such part is then modeled as a simple scatterer with an RCS that is either known or can be calculated by one of the optical techniques. Total RCS at a given aspect angle can be estimated by

$$\sigma = \left| \sum_{j=1}^{N} (\sigma_i)^{1/2} \exp(i\phi_j) \right|^2 \tag{10.1}$$

where σ_j is the RCS of the jth component and ϕ_j is the relative phase angle associated with the jth component, calculated on the basis of its location in a body-centered coordinate system. For large objects many wavelengths in size, assume a random phase between the scatterers, in which case the expected value of RCS can be expressed as

$$\sigma = \sum_{j=1}^{N} \sigma_j \tag{10.2}$$

Probable values of RCS will then lie between $\sigma + S$ and $\sigma - S$, where S is defined by

$$S^2 = \left(\sum_{j=1}^{N} \sigma_j \right)^2 - \sum_{j=1}^{N} \sigma_j^2 \tag{10.3}$$

The utility of this method is illustrated in Figures 10.3 and 10.4, where the average RCS as determined above is compared with experimental data for a large manned aircraft [6]. We can see that such an approach, using the random phase method, is adequate, although it depends on how accurately the analysts need to model the object.

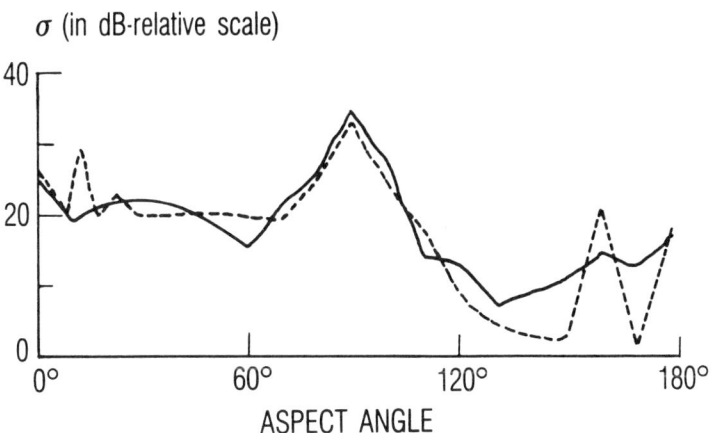

Figure 10.3 The RCS of a large manned aircraft as a function of aspect in the plane of the wings (E-vector in the plane in which aspect is measured): dashed line is theoretical (average); solid line is experimental (10° medians). © IEEE, 1965.

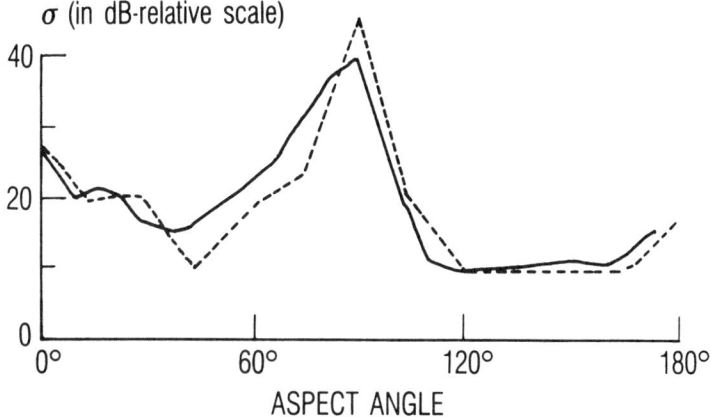

Figure 10.4 The RCS of a large manned aircraft as a function of aspect in a plane through the axis of the aircraft and normal to the plane of the wings (E-vector in the plane in which aspect is measured): dashed line is theoretical (average); solid line is experimental (10° medians). © IEEE, 1965.

10.2 SPACE TARGETS

The RCS of a satellite can be determined by model measurements, direct computation, combination of component measurements and computation, or direct measurement. Some useful measurements on satellite parts were made by Dybdal and King [7] in an anechoic chamber at millimeter wavelengths. Such knowledge can be of use in the synthesis procedure discussed in the preceding section. The millimeter wavelength region was chosen because radars in this band offer the potential of enhanced resolution and high doppler sensitivity. Typical satellites incorporate two types of targets, structural and sensor, which provide most of the return to a typical radar. In terms of structure, solar array material is very important because solar cells are present over a large part of the visible surface. Results of the chamber measurements indicate that solar panel material scatters like a perfect conductor at 93 GHz (Figure 10.5). Thermal blanket material is another important structural element. This material is composed of a layer of sheets with crinkles on the inner layers and a smooth gold-foil outer layer. The measured results seem to indicate that even the "smooth" gold layer is not flat so that the peak RCS is reduced by 20 dB below the equivalent flat plate (Figure 10.6). Of course, at this high frequency, even relatively smooth surfaces look rough. Measurements were also made on X-band and E-band horns and an optical telescope. Examples of these measurements are shown in Figures 10.7, 10.8, and 10.9. In the case of the X-band horn, the measurement frequency is 10 times the normal operating frequency and multimode propagation effects are seen. The return from a small telescope has a noticeable lobe structure with peak levels on the order of -20 dBsm when looking into the aperture. Other objects, such as a series of rough surfaces, were measured. The conclusion was that surface roughness plays a more dominant role in influencing scattering at high frequencies, and becomes more important as the target size increases and that the surface roughness lowers the peak specular amplitude. There is also some evidence that the RCS can increase at aspect angles other than specular, especially as the frequency increases. This occurs because, as mentioned above, as the wavelength decreases, a given surface becomes relatively more rough.

Good RCS models of satellites and other space objects can often be established by combining component measurement data such as obtained above with additional calculated data. Flat plates and cylinders, for instance, are commonly used as solar panels and body structures. Flat plate RCS can be determined with high accuracy, at least for smooth conductors, by using GTD methods. Results of such a calculation are shown in Figure 10.10 for a square plate [8]. Notice that, at off-normal aspect angles, large deviations between GTD and PO calculations begin to occur, showing the marked superiority of GTD. Cylindrical bodies can be analyzed in a similar manner. Figure 10.11 shows the calculated RCS of a large cylinder [9] (a commonly occurring satellite body structural member). The method of calculation used here was the PTD.

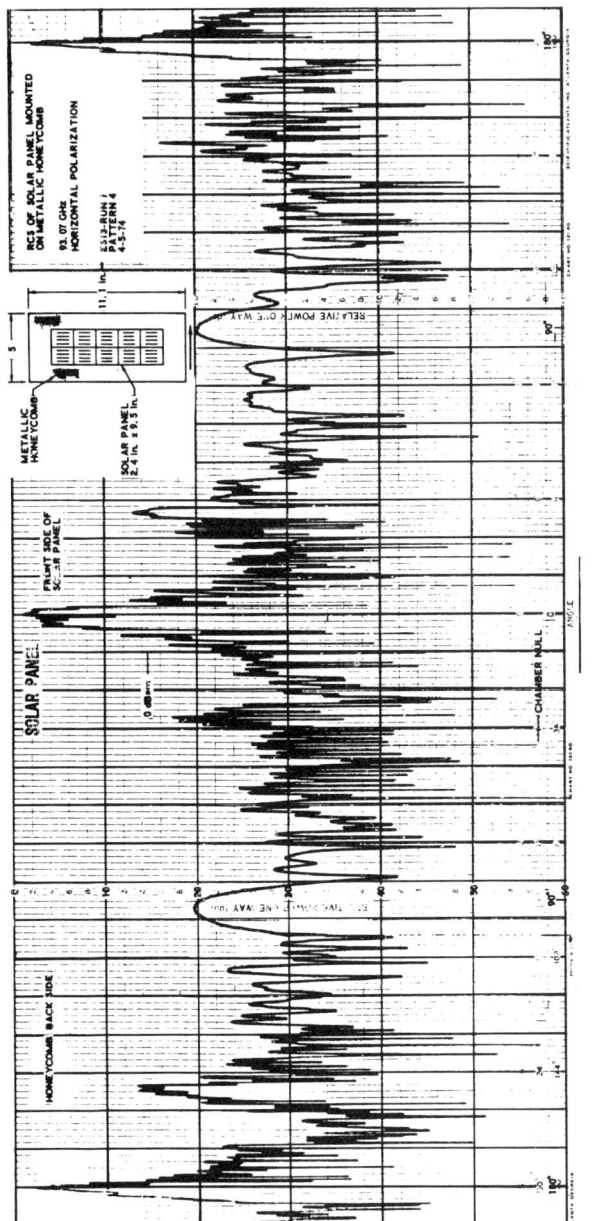

Figure 10.5 Measurement RCS of solar panel with and without back metallic reference plate. © IEEE, 1977.

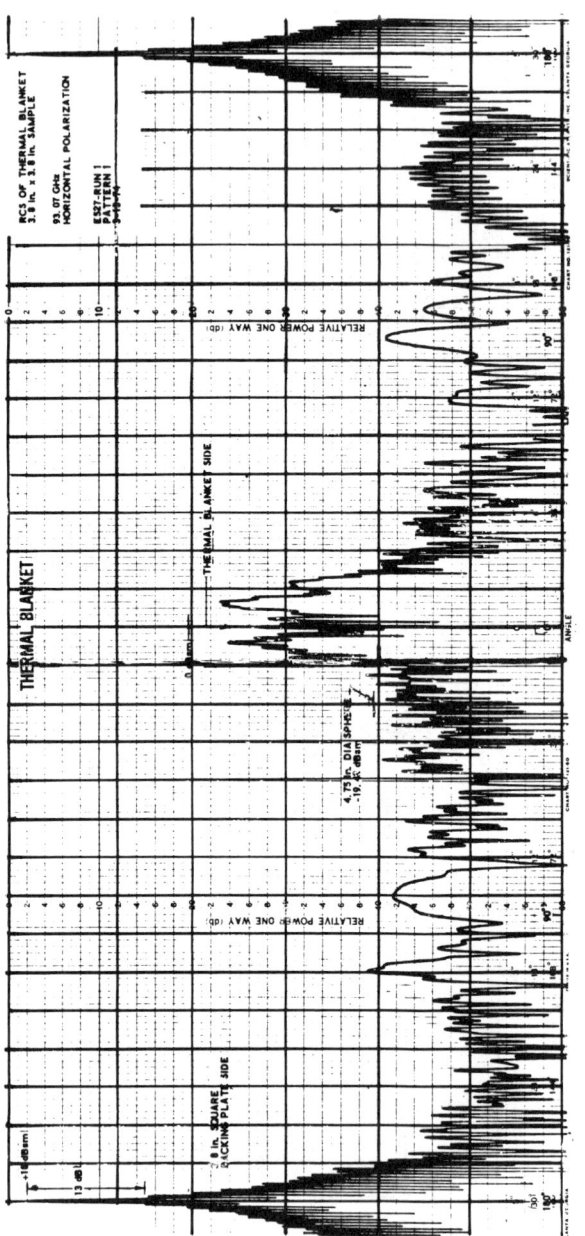

Figure 10.6 RCS of thermal blanket material with metal reference plate. © IEEE, 1977.

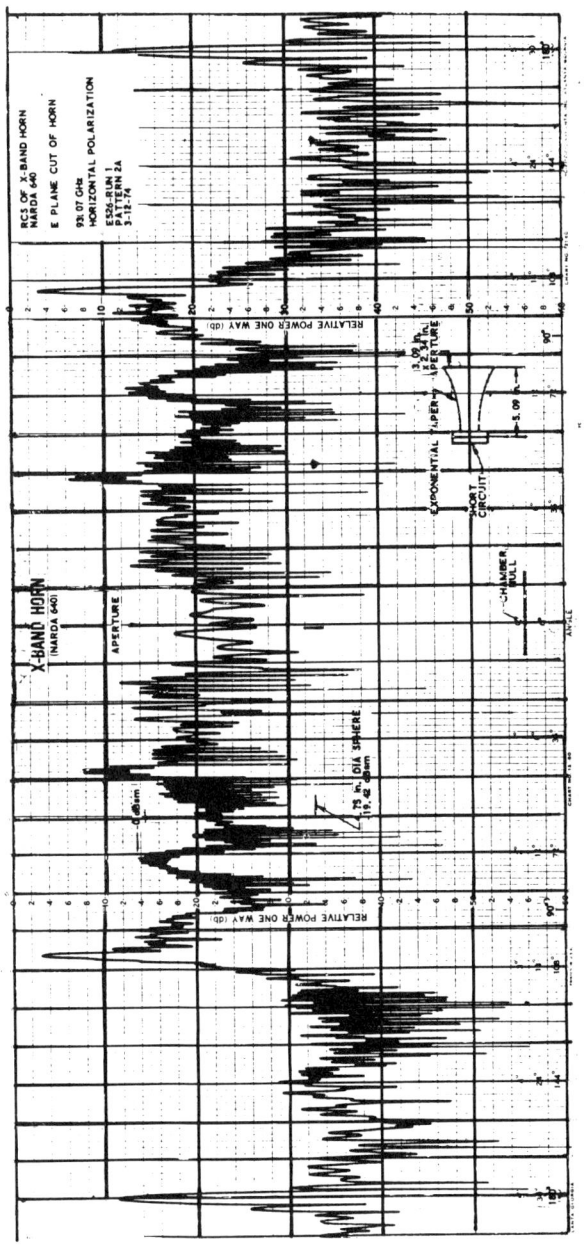

Figure 10.7 Measured 93 GHz RCS characteristics of X-band horn (E-plane cut). © IEEE, 1977.

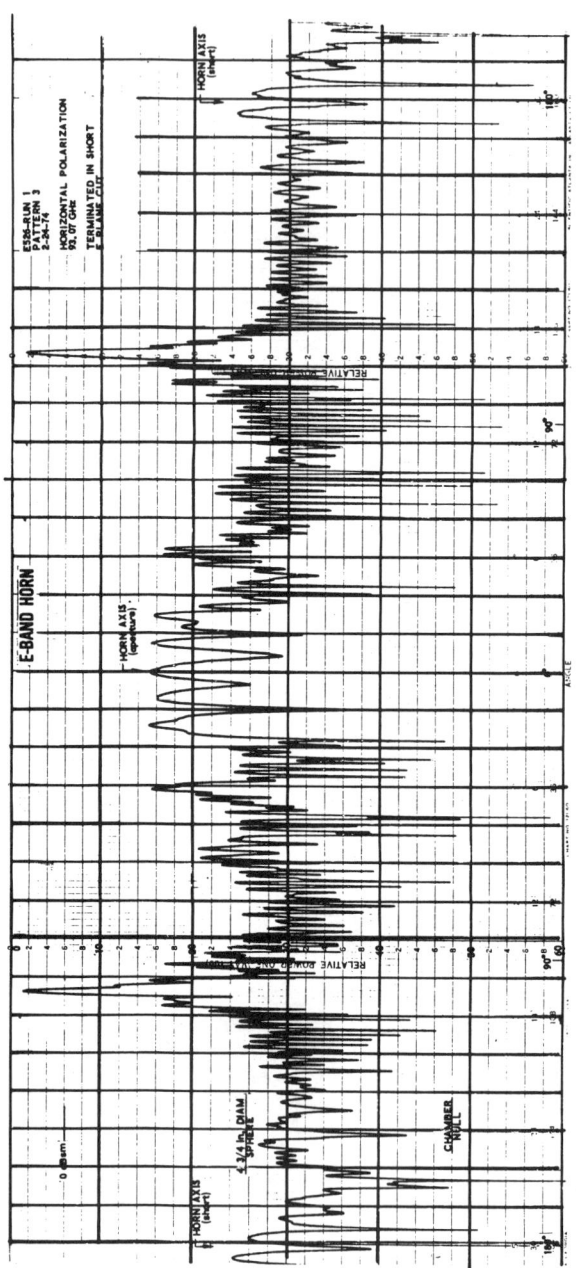

Figure 10.8 Measured 93 GHz RCS characteristics of E-band (60 to 90 GHz) horn (E-plane cut). © IEEE, 1977.

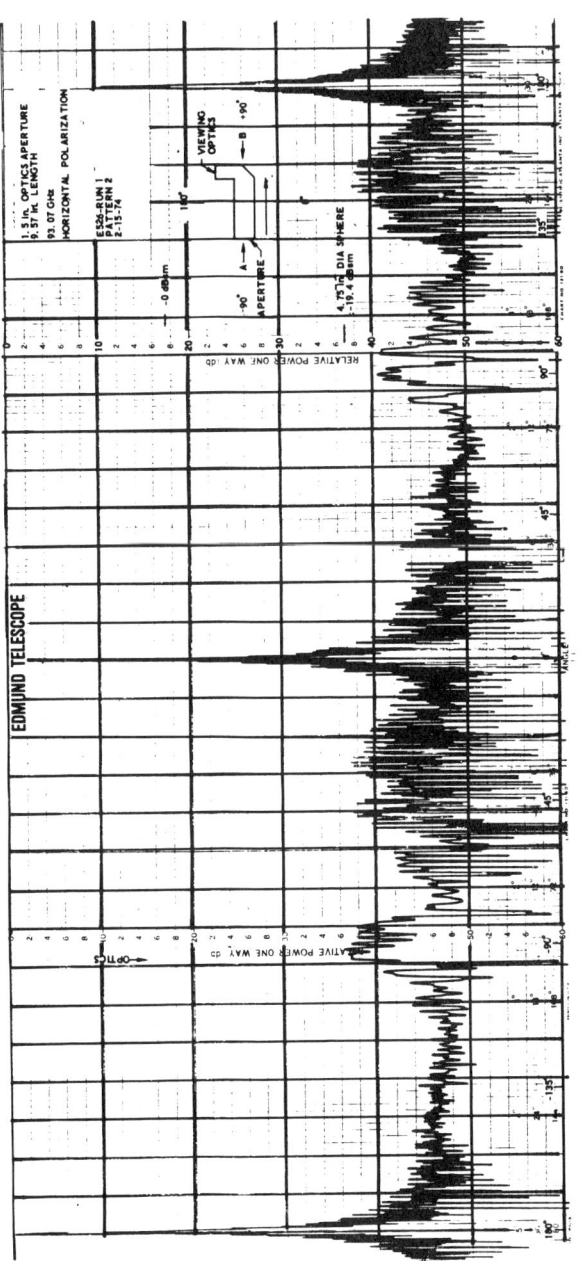

Figure 10.9 Measured RCS characteristics of telescope. © IEEE, 1977.

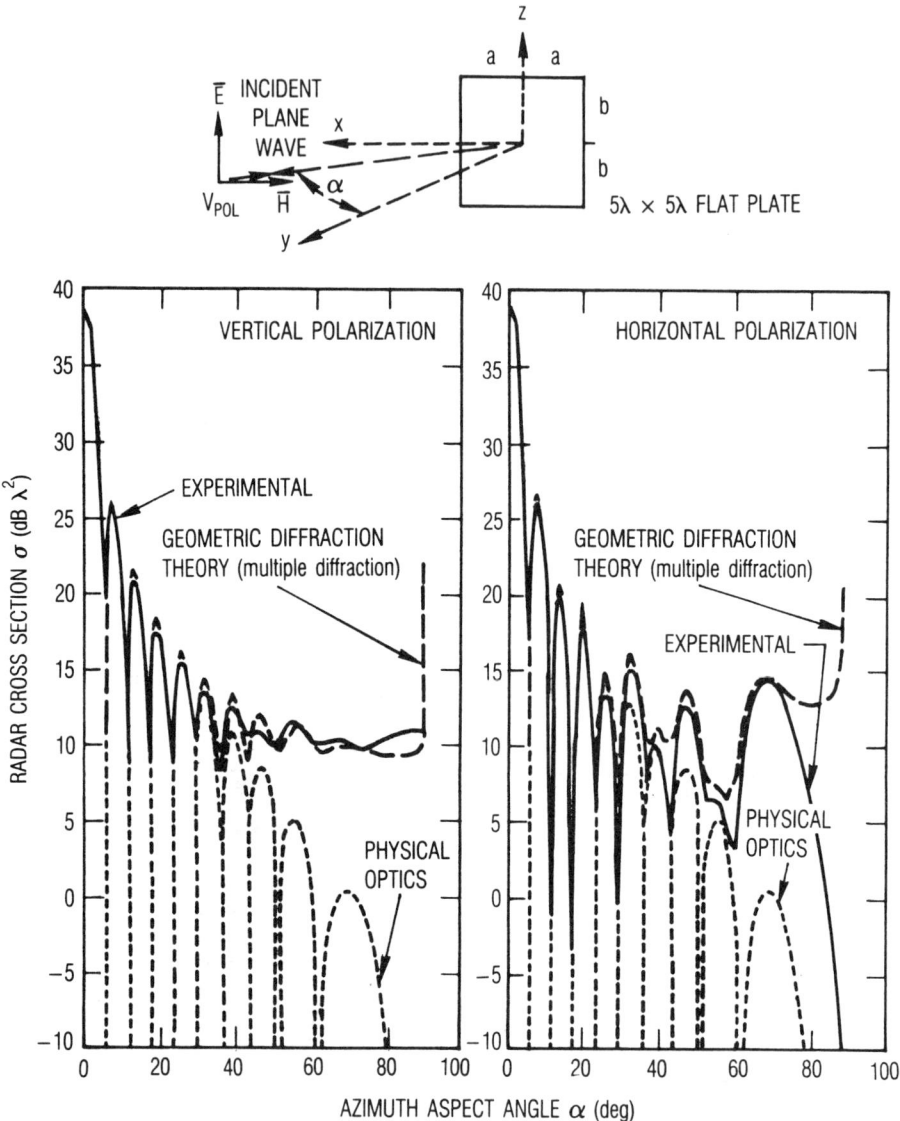

Figure 10.10 RCS patterns of a 5λ and 5λ square plate. © IEEE, 1967.

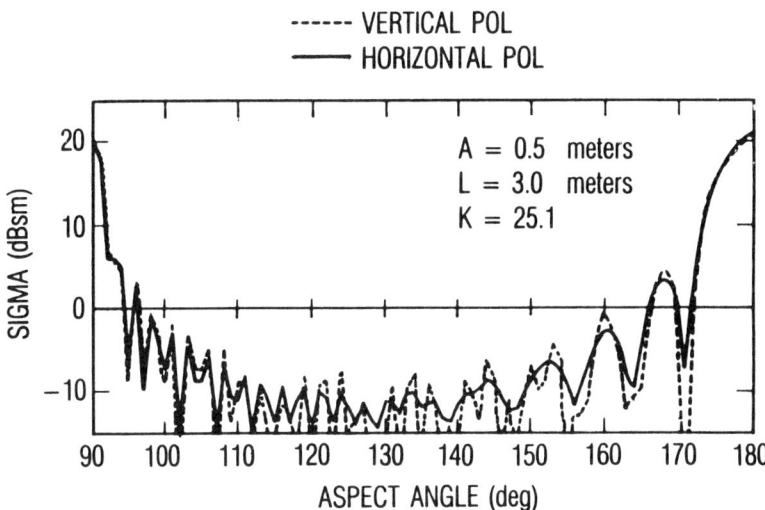

Figure 10.11 Radar cross section of a cylinder.

The backscattering property of parabolic dish antennas has been investigated by Kau [10]. His experimental data show good agreement with PO and GTD calculations at high frequencies for parabolic reflectors. At frequencies below the operating frequency, the RCS of a parabolic antenna is essentially the same as that of a paraboloidal reflector. Within or above the operating frequency, the major contributor to the RCS is the feed structure, which is to be carefully analyzed, but the upper limit for shallow dishes is still essentially that of a disk. Application of the synthesis approach, as presented above, especially the combination of measured and calculated data of components and sensor scattering, should lead to a reasonable determination of satellite RCS at aspect angles of interest.

Ground measurements, of course, have been made on numerous satellites. In many cases, the RCS has been observed to be completely dominated by the solar panel return. This return has been observed to be as high as 37 dBsm (5000 m^2) for some communication satellites. Average RCS measurement [11] made on selected satellites during the RCA TRADEX project have been included in Table 10.1. Additional RCS data are presented in Table 10.2 from full-scale model measurements [12] and other group measurements [5].

Table 10.1
Average RCS Measurements [11]

Object	Designation		Average Radar Cross Section
0029	60 Beta 2	Tiros 1	0 dBsm
0031	60 Gamma 2	Transit 1b	0 dBsm
0043	60 Zeta 1	Midas 1	0 dBsm
0049	60 Iota 1	Echo 1	+25 dBsm
0226	62 Beta 1	Tiros 4	+ 2 dBsm
0255	62 Zeta 1	OSO 1	− 3 dBsm
0309	62 Alpha-Alpha	Tiros 5	− 6 dBsm
0446	62 Beta-Mu	Anna 1b	− 5 dBsm
0740	1964-04A	Echo 2	+25 dBsm
0744	1964-05A	Saturn SA-5	+25 dBsm
1002	1965-08C	LES 1	0 dBsm

Table 10.2
Median RCS Measurements [5, 12]

Name	RCS (max) (dBsm)	RCS (min) (dBsm)	RCS (median) (dBsm)	Frequency
TIROS	39	−21	3	X-Band
NIMBUS	39	−15	6	X-Band
P706 FSV	36	−14	12	X-Band
LUNIK II	36	−21	0	X-Band
SPUTNIK III	26	−24	−1	X-Band
VOSTOK II	37	−14	6	X-Band
TELSTAR	20	−20	−6	X-Band

10.3 AIRBORNE TARGETS

As mentioned in our introduction, the SBR viewing geometry against airborne vehicles may be advantageous for a number of reasons. One suggestion is that it could be used for air traffic control [13]. From a military point of view, improved detection and tracking could be achieved because of possibly higher RCS in a target's upper hemisphere. The higher RCS could result from a viewing of the cockpit, vertical stabilizer, or a better view of the upper wing surfaces and other aircraft components that would normally be hidden from view. Typical RCS variation of a fighter aircraft in the azimuth plane is shown in Figure 10.12. The RCS increases with frequency at an azimuth angle broadside to the fuselage axis. If the

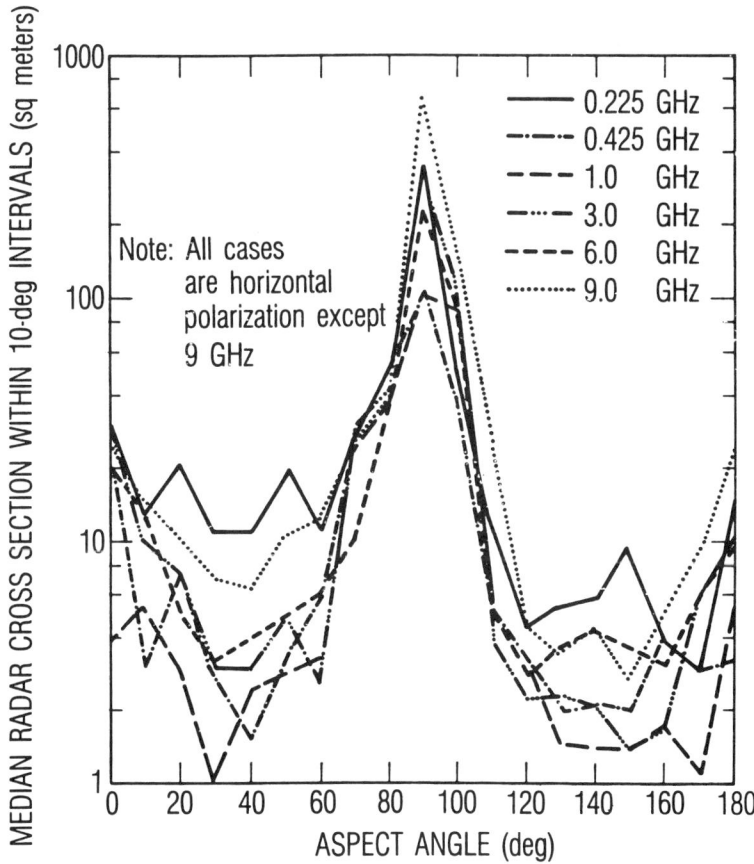

Figure 10.12 Median radar cross section of a fighter aircraft as a function of frequency (yaw plane).

RCS were measured over the top of the aircraft, the RCS could generally be expected to be even larger due to the inclusion of the wing surface to the signature in the upper hemisphere. This point is illustrated in Figure 10.13, which shows the RCS of a bomber type of aircraft in both the pitch plane and the azimuth plane. The increase in RCS in the pitch plane is probably due to the cockpit, vertical stabilizer, and wing surfaces. The peak RCS in the lower hemisphere would probably be lower. Table 10.3 lists values of RCS of several other types of aircraft at microwave frequencies [1].

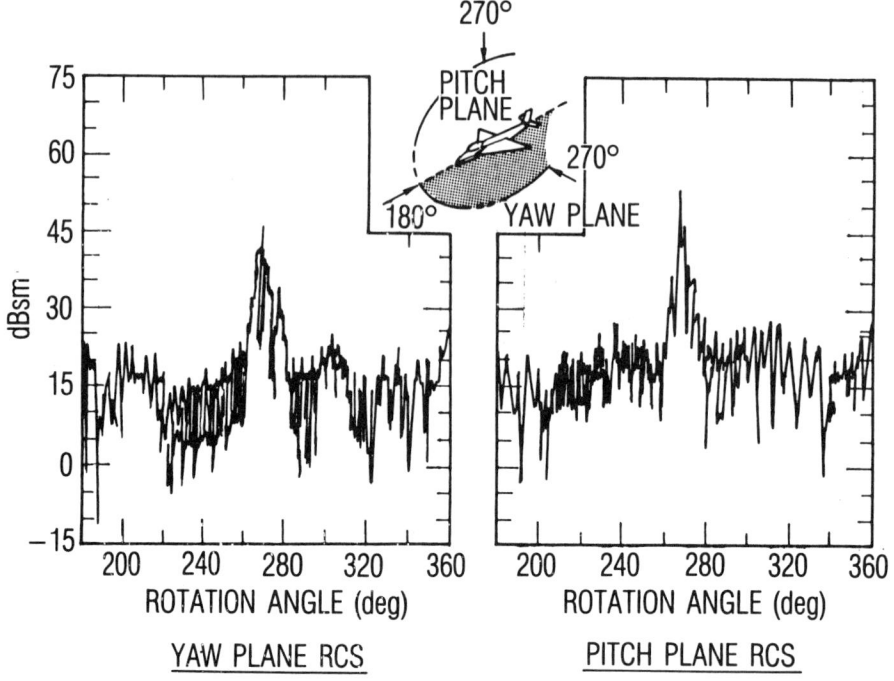

Figure 10.13 Monostatic RCS of a bomber type of aircraft.

Table 10.3
Measured Typical RCS of Several Airborne Targets [1]

Object	RCS(dBsm)
Conventional, unmanned, winged missile	−3
Small, single-engine aircraft	0
Small fighter, or four-passenger jet	3
Large fighter	7.8
Medium bomber or medium jet liner	13
Large bomber or large jet liner	16
Jumbo jet	20

The RCS discussion so far has pertained to monostatic radar geometry, whereby the transmitter and receiver are collocated. In the more general case wherein they are not collocated, we have what is called a *bistatic geometry*. The angle between the transmitter and receiver directions is known as the *bistatic angle*. Figure 10.14 shows the bistatic RCS for the case of a sphere of radius a at two polarizations [14]. We can see from the figure that as the bistatic angle increases,

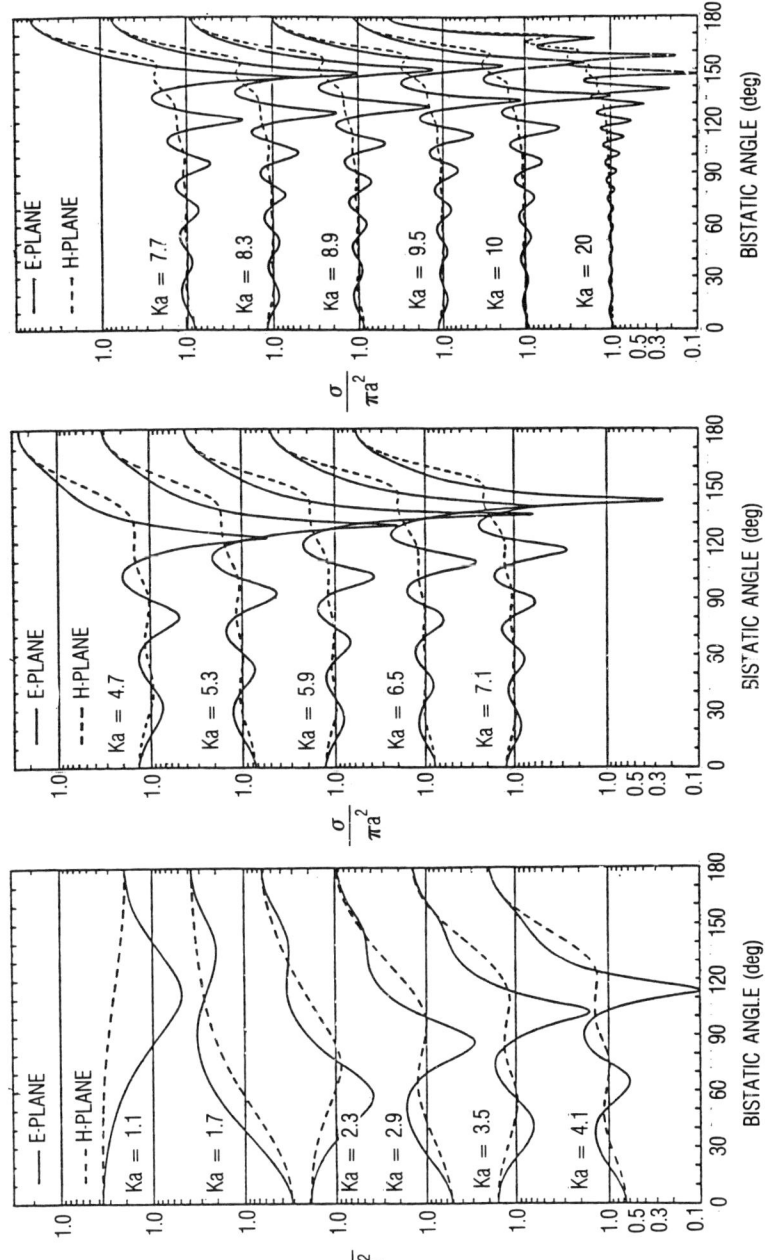

Figure 10.14 Bistatic RCS of a sphere (from [14]).

the RCS increases, especially for large values of *ka* (small wavelength). The theoretical maximum value for the bistatic RCS of a sphere is the same as that of a flat plate of the same radius, which can be several orders of magnitude greater than the monostatic RCS. Most objects have bistatic behavior similar to that of a sphere, and this enhanced RCS can possibly be exploited by separating the transmitter at a large angular distance from the receiver. This could be accomplished in the SBR case by adding a ground, airborne, or another space terminal that would create a large bistatic scattering angle. An interesting experiment using the bistatic geometry was carried out in 1977 [15], employing each of several geosynchronous military communication satellites as a transmitter. Targets were mainly commercial aircraft flying into and out of Los Angeles International Airport. The receiver was ground-based and used the X-band satellite downlink frequency to generate aircraft bistatic RCS and doppler history. Bistatic RCS on the order of 45 dBsm was observed for a DC-10. Bistatic angles for this case were about 90° or less.

10.4 OTHER TARGETS

The emphasis so far has been on the RCS of satellites and airborne targets. Reasonable estimates of the RCS of other types of targets can be made by using the synthesis methods already discussed. Another important scatterer to be considered in the synthesis method is the corner reflector. The side of a ship and the ocean surface, an aircraft wing and the fuselage, sensor boxes on a flat satellite surface, all form corner reflectors of two or more surfaces. The contribution to the total RCS for the case of a dihedral corner reflector, as shown in Figure 10.15, has been estimated [5] to be

$$\sigma \approx \frac{16\pi a^2 b^2 \sin^2(\pi/4 + \phi)}{\lambda^2} \tag{10.4}$$

and can provide a large return when viewed in the proper plane. This return can be reduced by locating the faces at some angle other than 90°. The trihedral corner reflectors are often used to enhance the RCS for detection purposes, but can occur naturally on some targets. Several types of trihedral corner reflectors are illustrated in Figure 10.16. The average RCS within the 3 dB cone of coverage for each type of trihedral corner reflector is plotted *versus* wavelength in Figure 10.17 for a series of edge lengths. The maximum, average, and 3 dB lobewidth of the RCS of each type are shown in Table 10.4.

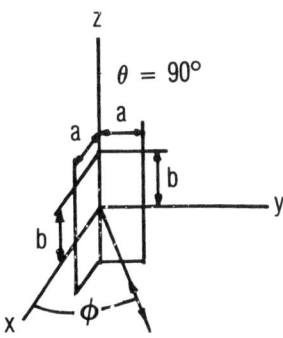

Figure 10.15 Coordinates for dihedral corner reflector.

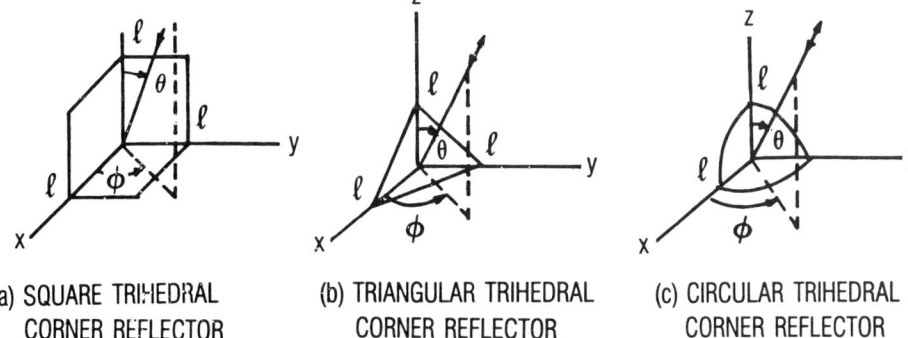

(a) SQUARE TRIHEDRAL CORNER REFLECTOR

(b) TRIANGULAR TRIHEDRAL CORNER REFLECTOR

(c) CIRCULAR TRIHEDRAL CORNER REFLECTOR

Figure 10.16 Coordinates for trihedral corner reflectors: (a) square; (b) triangular; (c) circular.

Table 10.4
Comparison of Trihedral Corner Reflectors

Trihedral Corner Reflector Type	Maximum RCS	Average RCS	Angular Coverage
Square	$\sigma = \dfrac{12\pi l^4}{\lambda_0^2}$	$\langle \sigma \rangle = \dfrac{0.7 l^4}{\lambda_0^2}$	23° cone about symmetry axis
Triangular	$\sigma = \dfrac{4\pi l^4}{3\lambda_0^2}$	$\langle \sigma \rangle = \dfrac{0.17 l^4}{\lambda_0^2}$	40° cone about symmetry axis
Circular	$\sigma = \dfrac{15.6 l^4}{\lambda_0^2}$	$\langle \sigma \rangle = \dfrac{0.47 l^4}{\lambda_0^2}$	32° cone about symmetry axis

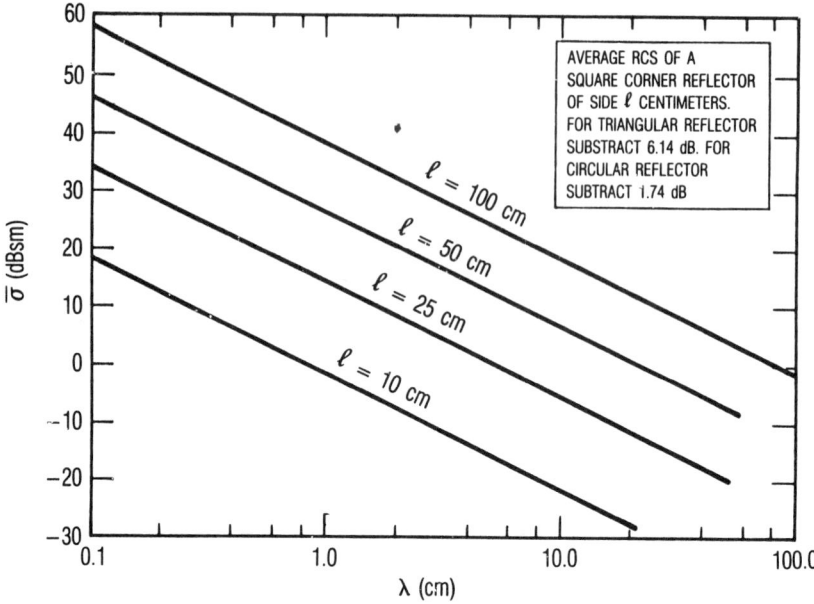

Figure 10.17 Radar cross section of a corner reflector.

There also exist experimental data on specific target types that can be scaled for application to a particular problem. A good example would be the data of Keys and Primitch [16]. This reference contains data taken on flat-backed cones, the included nose angles of which varied from 8° to 120° and their base diameters ranged from about one-third to three wavelengths. The RCS of a large number of cones can be obtained from the data by scaling. Ship targets present an especially difficult problem. Due to the multipath environment provided by the sea surface, the RCS of some ships [4] can be as high as one square mile (64 dBsm). The median RCS at grazing incidence, omitting the broadside return, can be expressed as

$$\sigma = 52\sqrt{fD^3}$$

where σ is the cross section in square meters, f is the frequency in MHz, and D is the ship full-load displacement in kilotons [17]. Some experimental results from numerous ship measurements [2] are shown in Figure 10.18 in which average RCS is plotted *versus* ship length. These results are for grazing incidence only. At higher elevation angles, such as those encountered by an SBR, a rough estimate can be made by assuming the RCS in square meters to be equal to the ship's displacement

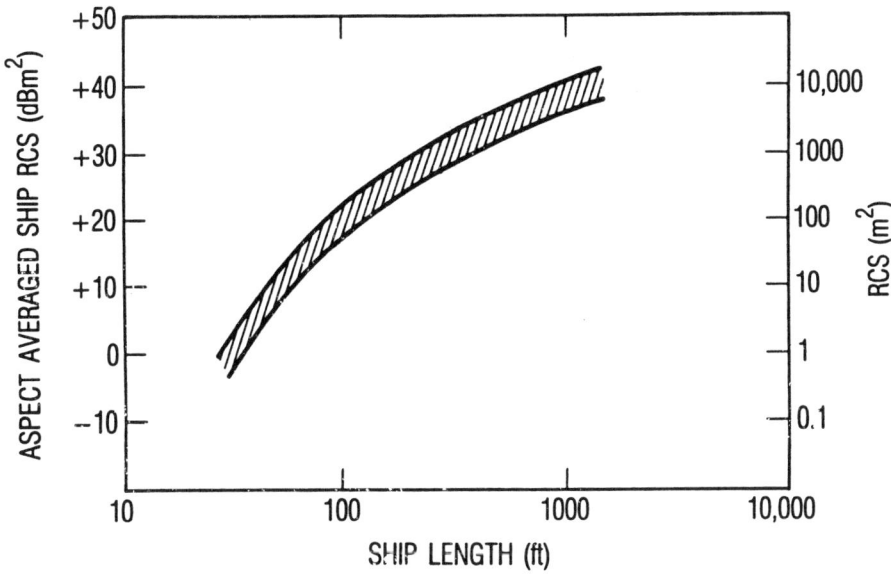

Figure 10.18 RCS *versus* ship length (from [2]).

in tons [1]. With the proper wideband waveform, the RCS profile of a ship can sometimes be generated in the range direction and used to assist in identification. The RCS of other ground objects such as tanks, trucks, and cars can be determined by either direct measurement or the synthesis method. Table 10.5 contains RCS data on some of these objects at microwave frequencies [1].

Table 10.5
Typical RCS of Ground Targets [1]

Object	RCS(dBsm)
Small open boat	−17
Small pleasure boat	3
Cabin cruiser	10
Pickup truck	23
Automobile	20
Bicycle	3
Man	0
Bird	−20
Insect	−50

10.5 RCS MEASUREMENT

In principle, measurement is the best way to determine RCS, but there are complications. Due to size limitations, for instance, scale-model measurements may be necessary. Duplication of a target's full-scale electromagnetic properties at the measurement frequency is one of the difficulties most often encountered. However, if the full-scale object is a fairly good conductor with no dielectrics or other partial conductor and the angles and surface qualities are duplicated, modeling is an excellent method of determining RCS.

Outdoor ranges have always been popular because the long measurement distances allow far-field conditions to be achieved. One of the most successful RCS measurement facilities in the US is the Radar Target Scatter Site (RATSCAT) operated by the 6525th Test Group, Air Force Special Weapons Center at Holloman Air Force Base, New Mexico. Among significant features offered by RATSCAT [18] are very broad frequency coverage and the capability to mount large targets such as full-scale aircraft. Targets as heavy as 100,000 pounds can be simultaneously measured at four frequencies at the main facility. The range of frequencies used at this complex may be between 140 MHz and 18 GHz or also at 35 GHz and 94 GHz.

Demands for making RCS measurement indoors have encouraged the development of compact ranges [19]. The far-field condition problem has been solved by using a reflector-based folding of the wave path to produce a plane wave at the target location. Such indoor ranges are especially useful for RCS reduction (stealth) studies in which aircraft or satellite components can be separately measured and their contribution minimized in the security of an indoor facility. The matter of object identification has stimulated the need for detailed characterization of RCS in terms of, for example, the exact location and description of scattering centers. This requirement has hastened the development of short-pulse or wideband RCS measurement at both indoor and outdoor facilities. Wideband synthetic imaging techniques have also been found to be useful.

After the measured data have been gathered, the problem remains as to how to apply it. At most frequencies of interest, the RCS can change radically with aspect angle, which is usually varied at a constant rate at the measurement range. When combined with the movement of the actual target in flight, the result is a dynamical variation of the RCS that an actual radar is likely to encounter. Because geometry and kinematics are not usually known beforehand, a statistical description of target RCS fluctuation is generally invoked in radar system design. Swerling [20] has suggested the well known four cases for fluctuating targets. The distribution to be used depends on whether the target is composed of many scattering centers of similar magnitudes (Swerling cases 1 and 2) or one large scattering center together with small reflectors (Swerling cases 3 and 4). Both Swerling distributions are special cases of the chi-square distribution. Ricean [21] and log-normal distributions [22] have also been suggested as being appropriate. Applying any one of these distributions to a particular target is difficult because during any observational period, the target will be viewed over a finite angular interval. If only one parameter is used to describe a target, it has been suggested [23, 24] that the median value be used with Rayleigh statistics. The median value is also useful for other reasons. It is customary to use dBsm to express RCS of measured data. The logarithm of the median in this case is equal to the median of the logarithms, which makes it a simple matter to summarize measured data over an angular interval of interest.

10.6 SUMMARY

The target upper hemisphere viewing geometry that can be achieved by a space-based radar may be exploited in some cases to improve target detection and tracking capabilities. Knowledge of the expected target RCS is necessary in the design and sizing of an SBR system. A good estimate of the RCS can be determined in a number of ways, and in this chapter we have reviewed the more important approaches. The procedure is basically to subdivide the complex target into its constituent scattering centers, then assume them to be independent scatterers and combine them, preferably in a vectorial manner. The RCS of the scattering centers, especially the simple ones (corner reflectors, flat plates, cylinders, *et cetera*), can be modeled mathematically. Other, more complex elements (antennas, blanket material, telescopes, *et cetera*) can be measured and then included in the combining process. Other approaches are use of either scale models or full-scale measurement of entire bodies of interest. Representative RCS values of typical targets have been presented to provide the reader with useful estimates.

REFERENCES

1. Skolnik, M.I., *Introduction to Radar Systems*, McGraw-Hill, New York, 1980.
2. Wehner, D.R., *High Resolution Radar*, Artech House, Norwood, MA, 1987.
3. Cimino, J.B., B. Holt and A.H. Richardson, *The Shuttle Imaging Radar B (SIR-B) Experiment Report*, NASA-JPL, March 15, 1988.
4. Knott, E.F., J.F. Schaeffer, and M.T. Tuley, *Radar Cross Section*, Artech House, Norwood, MA, 1985.
5. Ruck, G.T., *et al.*, *Radar Cross Section Handbook*, Vols. I and II, Plenum Press, New York, 1970.
6. Crispin, J.W., *et al.*, "Radar Cross Section Estimation for Complex Shapes," Proc. IEEE, Vol. 53, August 1965, pp. 972–982.
7. Dybdal, R.B. and H.E. King, *"93 GHz Radar Cross Section Measurements of Satellite Elemental Scatterers,"* IEEE Trans. Antennas and Propagation, Vol. AP-25, May 1977, pp. 396–402.

8. Ross, R.A., *"Radar Cross Section of Rectangular Flat Plates as a Function of Aspect Angle,"* IEEE Trans. Antennas and Propagation, Vol. AP-15, May 1967, pp. 329–335.
9. Ufimtsev, P. Ya., *Method of Edge Waves in the Physical Theory of Diffraction*, Radio i Svyaz, Moscow, 1962.
10. Kau, P.S.S., *Backscatter Cross Section of a Paraboloidal Antenna*, MIT Lincoln Laboratory, Technical Note 1973–39, August 1973.
11. Kendrick, J.B., *TRW Space Data*, TRW Corp., 1967.
12. Smolski, A., *Final Report on Satellite Radar Cross Section Measurements*, Report No. ESSC 63-33, Electronic Space Structures Corp., 1963.
13. Cantafio, L.J., and J.S. Avrin, *Satellite-Borne Radar for Global Air Traffic Surveillance*, IEEE Electro-82, Professional Program Session Record, Boston, MA, May 25–27, 1982.
14. King, R.W.P., and T.T. Wu, *The Scattering and Diffraction of Waves*, Harvard University Press, Cambridge, MA, 1959.
15. Avrin, J.S., *"Bistatic Space Based Radar Experimentation at The Aerospace Corporation in 1977,"* Tri-Service Multistatic Radar Workshop, RADC, Griffis AFB, NY, May 21, 1980.
16. Keys, J.E., and R.I. Primitch, *The Radar Cross Section of Right Circular Metal Cones*, Vols. 1 and 2, DRTE, Ottawa, Canada, 1959.
17. Skolnik, M.I., *"An Empirical Formula for the Radar Cross Section of Ships at Grazing Incidence,"* IEEE Trans. Aerospace and Electronic Systems, Vol. AES-10, March 1974, p. 292.
18. Garretson, H.C., III, *"Radar Cross Section Testing,"* AIAA/AHS/CASI/DGLR/IES/ISA/ITEA/SETP/SFTE 3rd Flight Testing Conf., Las Vegas, NV., April 2–4, 1986.
19. Pustai, J., *"Compact Range Technology,"* Microwaves & RF, May 1987, pp. 117–183.
20. Swerling, P., *"Probability of Detection for Fluctuating Targets,"* IRE Trans. Information Theory, Vol. IT-6, April 1960, p. 269.
21. Jao, J.K., and M. Elbaum, *"First-Order Statistics of a Non-Rayleigh Fading Signal and Its Detection,"* Proc. IEEE, Vol. 66, July 1978, pp. 781–789.
22. Pollon, G.E., *"Statistical Parameters for Scattering from Randomly Oriented Arrays, Cylinders and Plates,"* IEEE Trans. Antennas and Propagation, Vol. AP-18, January 1970, pp. 68–75.
23. Nathanson, R.E., *Radar Design Principles*, New York, McGraw-Hill, 1969, Chapter 5.
24. Wilson, J.D., *"Probability of Detecting Aircraft Targets,"* IEEE Trans. Aerospace and Electronic Systems, Vol. AES-8, November 1972, pp. 757–761.

Chapter 11
SBR CLUTTER AND INTERFERENCE
G.A. Andrews and K. Gerlach
Naval Research Laboratory

11.1 INTRODUCTION

The advantage of putting a radar in space for earth surveillance is an extension of the radar horizon so that a very large area of the earth can be searched. The major disadvantage is that this large area of the earth produces severe clutter conditions and the potential for many interfering transmitters within the space-based radar's (SBR) field of view. The fundamental design considerations for SBRs to do air or surface search are (a) the radar must have enough power-aperture product to detect the radar cross section of the targets of interest at the search rate required for the application, (b) the radar must have enough angular and range resolution to locate the target with the required accuracy or also possibly to classify the target, and (c) the radar must be capable of rejecting clutter returns from earth and interference from other electromagnetic transmissions to detect targets in the presence of these usually much larger unwanted signals.

Figure 11.1 illustrates the SBR detection problem and the sources of interference that limit the detectability of the desired target. The sensitivity or detection capability of the SBR, shown in orbit, is fundamentally limited by its internal receiver noise. This limitation can be reduced by designing more sensitive, low-noise receivers. The antenna beam is shown pointing down and generating a main-beam footprint on the earth. An aircraft is shown in the footprint. By increasing the power-aperture product, the signal level and therefore the detectability of this target can be increased without decreasing the search rate. These are the usual search radar design considerations.

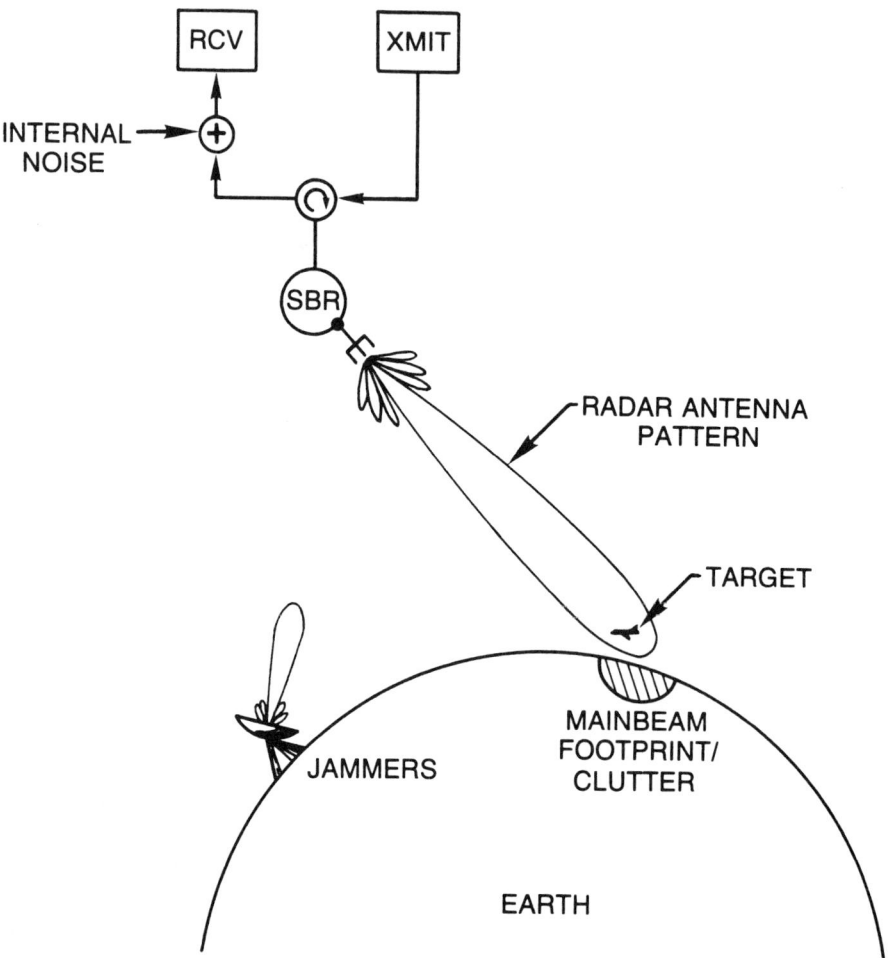

Figure 11.1 SBR search-detection scenario.

Additional factors that affect the design of a space-based search radar include the main-lobe clutter, the sidelobe clutter, and interference from earth-based transmitters within the field of view. The main-lobe clutter is the energy backscattered from the earth's surface within the footprint of the main beam and coinciding with the target returns as shown in Figure 11.1. This problem is treated by using clutter rejection techniques, similar to those used in airborne look-down radars.

The sidelobe clutter is the energy backscattered from the earth's surface outside the main-beam footprint. Because the level of this energy is determined by the level of the energy radiated through the sidelobes of the radar antenna, and the level entering the radar receiver is also determined by the sidelobe gain of the radar antenna, a very low sidelobe antenna design is a major consideration in the design of such a radar.

The interference from earth-based transmitters within the radar's field of view could be caused by other users of the radar band or adjacent bands spilling over into it. For military applications, these transmissions may be intentional jammers aimed at neutralizing the radar. Another factor in the design of space-based search radars, especially at lower frequencies, is the fading, dispersion, and backscatter of the ionosphere. (This problem is dealt with by Knepp and Reinking in Chapter 3.)

A generic SBR system configuration is illustrated in Figure 11.2, which shows the components of the system relating to clutter and interference rejection. The SBR system performance will ultimately be determined by the antenna size and accuracies that can be practically and economically achieved. Large antenna concepts that have been or are being considered for SBR applications include reflectors, dual reflectors, planar arrays, and space-fed lenses. The function of the large antenna structure is to provide the necessary aperture so that the received radar returns can be discriminated from internal noise and external interference. In addition, this antenna provides the highly accurate beam steering and the narrow pencil beam necessary for illuminating a small footprint on the earth.

The main and auxiliary antennas are respectively used to form the main channel of the radar receiver and the auxiliary channels needed for interference rejection. When sidelobe jamming is present, the outputs from the auxiliaries are used to cancel the jamming in the main channel in a procedure called *adaptive nulling*. The factors that can degrade or defeat this procedure are considered in this chapter.

The beamformers are followed by individual receivers on each channel. These receivers should be designed as low-noise front-ends with high degrees of linearity and stability in the RF and IF sections. In addition each receiver should be matched closely in its frequency response with that of the other channel receivers. Note that it is possible to perform analog adaptive nulling at radio frequencies as well as digitally at baseband. So, we have shown a feedback loop from the *interference rejection* processing to the channel receivers. This cascaded analog-digital concept has the advantage of reducing the effects of channel mismatch errors on interference rejection.

The interference rejection processing is followed by a *clutter rejection* processor, its function being to compensate for platform motion effects on the received radar returns, to reject clutter returns from the earth's surface, and to detect moving

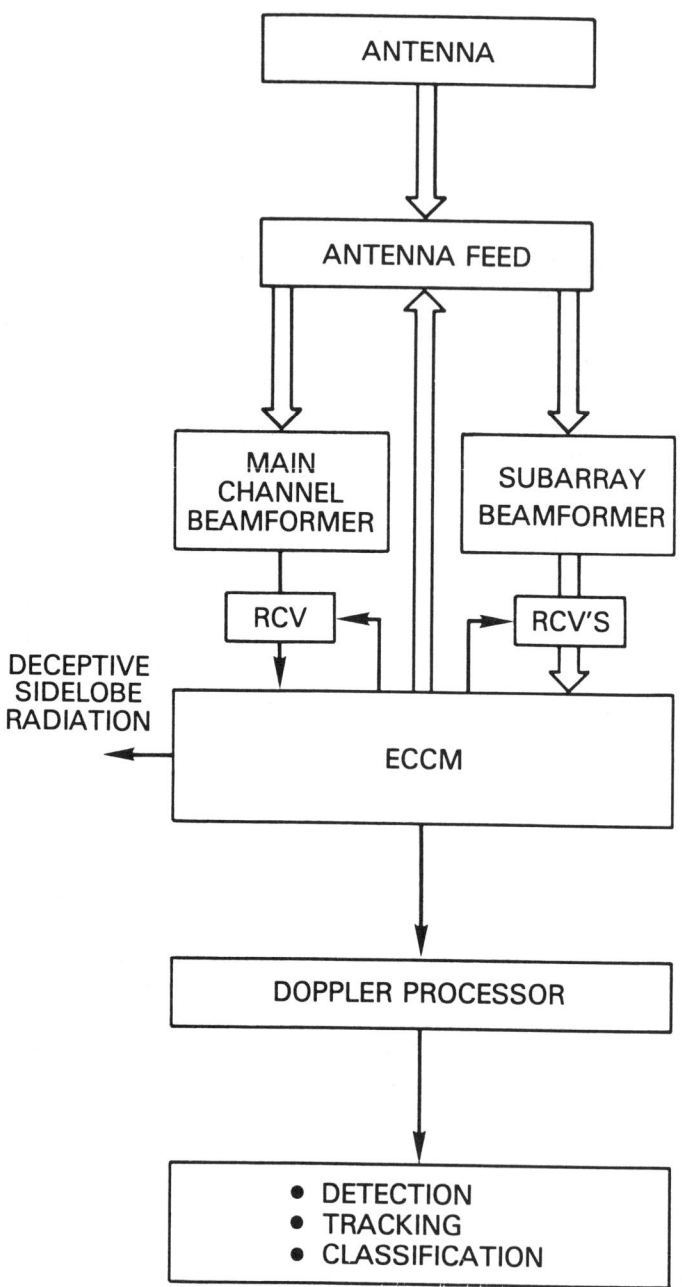

Figure 11.2 Generic SBR signal processing configuration.

targets. The techniques employed to achieve these goals have evolved for the most part from *airborne moving target indicator* (AMTI) and airborne pulsed doppler concepts. The doppler processor is followed by a postdetection processor, which develops target reports and tracks.

A major factor that must be considered in the design of an SBR of this type is the interaction of the interference and clutter processing. The presence of clutter in interference rejection processing can form nulls in the main beam, which not only cancel the clutter, but also cancel the targets. Additionally, the convergence rate of the adaptive nulling algorithms would be increased by orders of magnitude due to this clutter. The potential effects of adaptive nulling on clutter rejection are (1) increasing the composite antenna sidelobes, which would raise the sidelobe clutter level, and (2) introducing instability in pulse-to-pulse processing over the coherent integration time for clutter rejection as a result of time-varying adaptive weights. Concepts for minimizing these effects are described in this chapter.

This chapter assesses a wide range of radar techniques previously available in the open literature by translating the governing engineering equations to space geometry, adding orbital mechanical effects, and extrapolating results from other radar experience: specifically, airborne MTI and pulsed doppler radars, airborne and space-based synthetic aperture radars, and other space-based sensors such as scatterometers and radar height-finders. Fundamentals and concepts are emphasized rather then system designs. Therefore, the theory of clutter and interference characteristics and rejection techniques can be clarified and compared within the constraints of national security. We hope that the problem descriptions and engineering equations presented herein will provide the basis for further research and development in this area.

11.2 CHARACTERIZATION OF SBR CLUTTER

Backscatter from the earth's surface can be a severe limitation to a space-based surveillance radar. This section will provide models for assessing the magnitude of the problem for land, sea, weather, or aurora clutter.

11.2.1 Amplitude of Clutter

The analysis of the expected clutter environment and the radar design are commonly approached in two ways. The first, which considers the input signal-to-clutter ratio, is utilized by signal processing engineers concerned with designing clutter rejection techniques that maximize the output signal-to-clutter ratio. The second way, which considers the input clutter-to-noise ratio, is utilized by radar system engineers concerned with designing the radar so that the detection range is not degraded by clutter; hence, the radar is noise-limited, not clutter-limited.

We should consider an SBR system design from both perspectives because a feasible and affordable SBR design would involve many trade-offs. The first approach focuses on signal processing techniques and emphasizes qualitative comparisons of clutter processing techniques. The second approach is a part of system design in that it is concerned with the overall radar performance and the specific processing needs with which to maximize it. In both cases, the effect of clutter processing (which results in detection losses) on the target and the interactions between clutter processing and processing for interference rejection must be prominent considerations in system trade-offs.

The equation for signal-to-clutter ratio is [1, 2]:

$$\frac{S}{C} = \frac{\sigma_T F^4}{\bar{\sigma} F_c^4}$$

where σ_T and $\bar{\sigma}$ are the effective radar cross sections of the target and the clutter, respectively. F is the antenna pattern propagation factor for the target and F_c is the antenna pattern factor for the clutter. A monostatic radar has been assumed so that the transmit and receive pattern factors are equal.

Main-Lobe Clutter

From Figure 11.1, we can see that the pattern factor for the target and the pattern factor for main-lobe clutter are equal. Therefore, for main-lobe clutter, the received signal-to-noise ratio is simply

$$\left(\frac{S}{C}\right)_{ML} = \frac{\sigma_T}{\bar{\sigma}} \tag{11.1}$$

The effective radar cross section for main-lobe clutter backscattered from the earth's surface is given by

$$\bar{\sigma} = A_c \sigma° \tag{11.2}$$

where A_c is the area of the earth's surface illuminated by the radar and $\sigma°$ is the clutter cross section per unit area. From Figure 11.3, the illuminated area is

$$A_c = R_s \Theta_{AZ} N_a \frac{c\tau}{2} \sec\psi \tag{11.3}$$

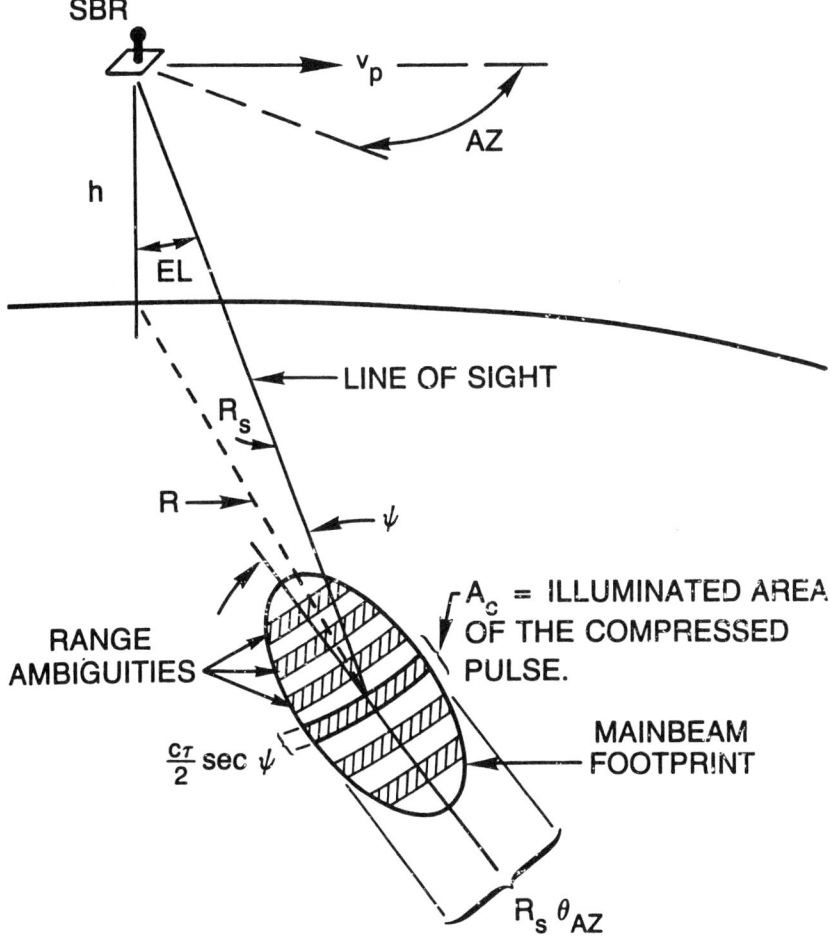

Figure 11.3 Clutter calculation geometry.

where R_S is the radar range to the surface, Θ_{AZ} is the radar horizontal beamwidth, N_a is the number of range ambiguities in the antenna footprint, c is the radar wave propagation velocity, τ is the radar pulse length (compressed pulse length for pulse-compression waveforms), and ψ is the grazing angle at which the surface is illuminated.

The radar range (or slant range, R_s) from the radar satellite to the antenna footprint on the earth's surface is given by

$$R_s = \left[R_e^2 + (R_e + h)^2 - 2R_e(R_e + h) \cos\left(\frac{R}{R_e}\right)\right]^{1/2} \tag{11.4}$$

R_e is the radius of the earth, nominally 3440 nmi (6371 km), R is the range from the nadir point (the point on the surface of the earth directly below the radar satellite) to the antenna footprint, and h is the satellite orbit height. Equation (11.4) is plotted in Figure 11.4 for several selected orbital altitudes.

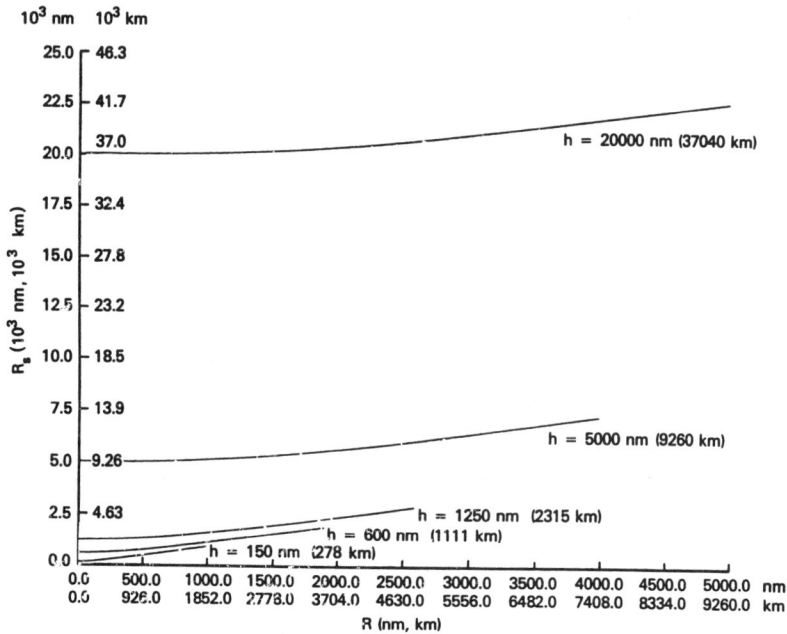

Figure 11.4 The relationship between slant range (R_S) from the SBR to the target and the range on the earth (R) from the satellite nadir point to the target; h is the orbital altitude.

The corresponding grazing angle is

$$\psi = \cos^{-1}\left[\frac{R_e + h}{R_s} \sin\left(\frac{R}{R_e}\right)\right] \tag{11.5}$$

Equation (11.5) is plotted in Figure 11.5 for several selected altitudes.

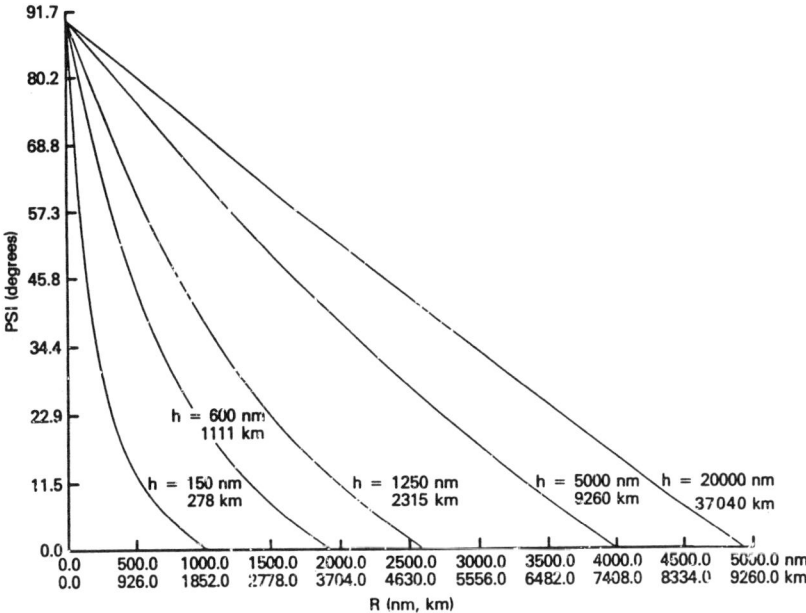

Figure 11.5 The relationship between the grazing angle (ψ) and the range on the earth from the satellite nadir point to the target; h is the orbital altitude.

The maximum range on the earth that can be viewed by an SBR of orbital altitude, h, is

$$R_{MAX} = R_e \cos^{-1}\left(\frac{R_e}{R_e + h}\right) \tag{11.6}$$

and it is plotted in Figure 11.6.

The received clutter power can be computed with a variation of the radar range equation [4]:

$$P_c = \frac{P_t G^2 \bar{\sigma} \lambda^2 F^4}{(4\pi)^3 R_s^4} \tag{11.7}$$

where P_t is the transmitted power, G is the radar antenna gain, $\bar{\sigma}$ is the effective clutter cross section given by (11.2), λ is the wavelength of the transmitted frequency, and R_s is the range from the radar to the clutter (antenna footprint).

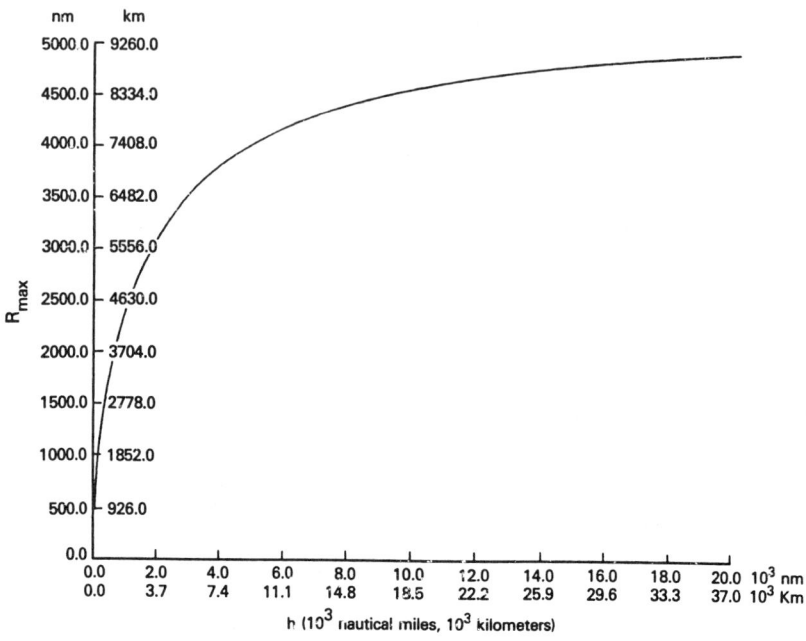

Figure 11.6 The relationship between the maximum range on the earth that can be searched *versus* the orbital altitude, h.

The clutter-to-noise ratio is

$$\frac{C}{N} = \frac{P_c}{kT_sB} \tag{11.8}$$

where P_c is given by (11.7), k is Boltzmann's constant, T_s is the system noise temperature, in K, [4], and B is the receiver bandwidth.

For volumetric clutter (rain, other meteorological particles, or chaff), the received signal-to-clutter ratio is [4]:

$$\frac{S}{C} = \frac{\sigma_t}{\eta \Omega R_s^2 \left(\frac{c\tau}{2}\right) N_a} \tag{11.9}$$

where η is the clutter-echo cross section per unit volume illuminated by the radar beam of Ω steradians.

Sidelobe Clutter

The sidelobe clutter level relative to the main-lobe clutter is determined by the integrated antenna main-lobe-to-sidelobe level. Therefore, for sidelobe clutter, the received signal-to-clutter ratio is

$$\left(\frac{S}{C}\right)_{SL} = \left(\frac{S}{C}\right)_{ML} \times \left(\frac{ML}{SL}\right)_I^{-1} \tag{11.10}$$

where $(ML/SL)_I$ is the integrated main-lobe-to-sidelobe ratio. An effective SBR in a severe clutter and interference environment must have very low sidelobes because of the effect of the antenna sidelobes on the level of sidelobe clutter and on the performance against sidelobe interference.

Clutter Cross Section Parameters

The surface clutter cross section per unit area, $\sigma°$, is a function of many parameters, such as type of terrain or sea conditions, frequency, polarization, and grazing angle, as well as radar parameters, including angular resolution, bandwidth, waveform, and clutter processing techniques. Because of the large number of variables involved and the excessive amount of data that must be taken to resolve these variables under all conditions, researchers have approached this problem by jointly developing the backscatter theory and taking data under as many conditions as possible to verify or modify the theory.

As the concepts and theory of radar evolve, the understanding of clutter must also progress. This is particularly true for space-based surveillance radars for which there is very little clutter data available. For the most part, SBR designs are based on extensions of airborne clutter data.

The space-based clutter data that are available consist of data from the SEASAT/SAR satellite, Shuttle Imaging Radars (SIR-A and SIR-B), and Skylab II scatterometer data. SEASAT, SIR-A, and SIR-B were not designed to take clutter data and were therefore not calibrated, although there was some success in using the supporting data to estimate the calibration, which could then be used to estimate the average $\sigma°$. The major limitation of the data is the low clutter-to-noise ratio, which allows only the average $\sigma°$ to be estimated and provides very little information on the statistical distributions. For an accurate prediction of the performance of an SBR, the *probability density function* (pdf) of clutter spikes must be known up to perhaps 50–70 dB above and below the mean. The Skylab scatterometer data have been analyzed to provide estimations of both mean $\sigma°$ and distributions for sea clutter [5], land, snow, and ice [6].

Long [7] has shown a general relationship between $\sigma°$ and the grazing angle, as illustrated in Figure 11.7. Three significant regions are identified in this figure: low grazing angles (0° to ψ_c), a plateau region (ψ_c to ψ_0), and near vertical incidence (greater than ψ_0). Nathanson [8] provides typical values for $\sigma°$ for sea clutter and several types of land clutter in which the grazing angle regions are divided into low (0° to 20°) and high (30° to 90° for the sea clutter and 10° to 70° for land clutter). Grazing angles greater than about 70° would usually be avoided because of the extremely high clutter returns. Lauer [9] and Tomlinson [10] have modeled SBR clutter for accurately predicting clutter amplitudes *versus* grazing as verified by existing airborne and limited space-based clutter. A plot of $\sigma°$ for sea clutter at L-band *versus* grazing angle is shown in Figure 11.8.

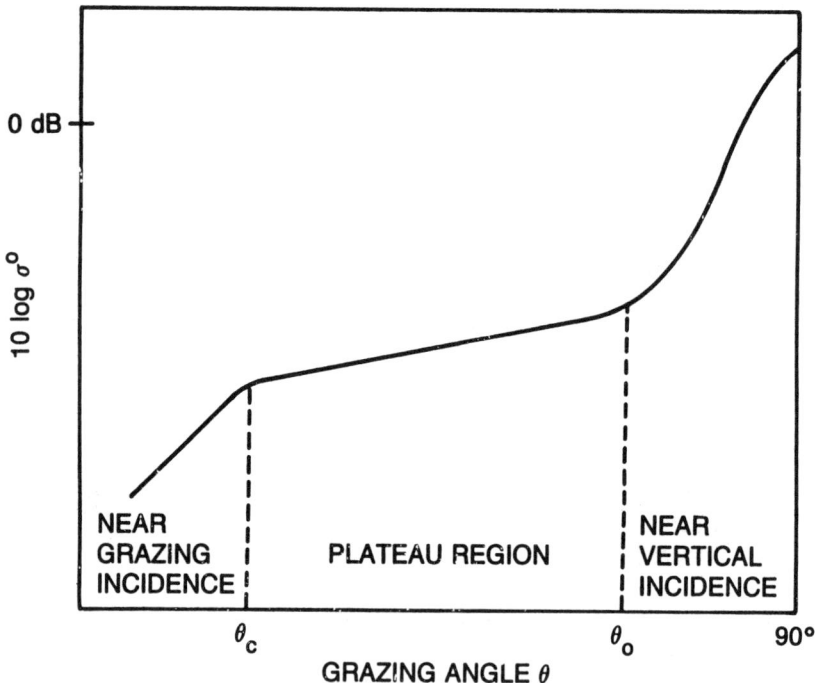

Figure 11.7 General dependence of clutter backscatter coefficient on depression angle.

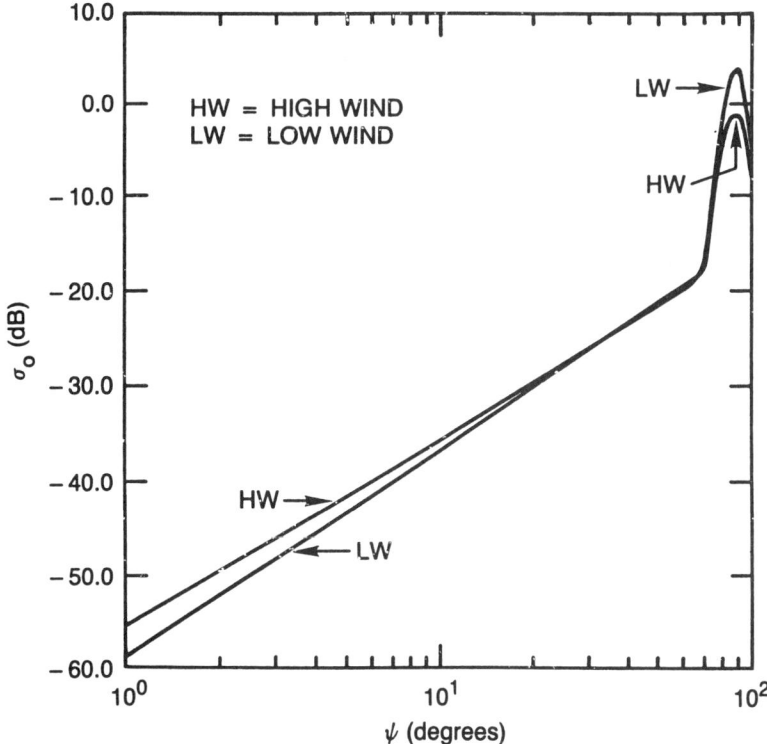

Figure 11.8 Modeled sea reflectivity *versus* grazing angle and wind conditions at L-band (from Lauer [9]).

Table 11.1 through 11.5 are compilations of typical values of clutter reflectivity for a radar in the UHF band, L-band, S-band, C-band, and X-band, respectively, and horizontal polarization. For vertical polarization, about 10 to 15 dB should be added to these values of $\sigma°$ for sea clutter at the lower grazing angles and for UHF. The 10 to 15 dB increase for vertical polarization becomes a 0 to 3 dB increase at low grazing angles for X-band [8]. Vertical and horizontal polarizations have about the same value at grazing angles of about 60° and greater for sea clutter. For land clutter consisting primarily of forests or cities, vertical polarization appears to produce about 3 to 5 dB less clutter than horizontal polarization.

Note that these values are for the reflection coefficient in dB below $1 \ m^2/m^2$. The larger this negative number the less the backscattered energy. Therefore vertical polarization at low grazing angles produces about 10 dB to 15 dB more sea clutter than does horizontal polarization.

Table 11.1
Typical Values of Clutter Reflectivity, σ°, in dB (1 m²/m²) at UHF [8]

Clutter	Grazing Angle (degrees)				
	1	3	10	30	60
Sea (SS3)	−76	−61	−50	−40	−21
Desert			−40	−40	−37
Farmland	−30		−36	−35	−32
Wooded hills	−24		−22	−19	−16
Cities	−22		−6	−9	−6

Table 11.2
Typical Values of Clutter Reflectivity, σ°, in dB (1 m²/m²) at L-Band [8]

Clutter	Grazing Angle (degrees)				
	1	3	10	30	60
Sea (SS3)	−60	−55	−48	−39	−20
Desert	−45		−40	−35	−32
Farmland	−32		−33		
Wooded hills	−35				
Cities	−30		−12	−14	−11

Table 11.3
Typical Value of Clutter Reflectivity, σ°, in dB (1 m²/m²) at S-Band [8]

Clutter	Grazing Angle (degrees)				
	1	3	10	30	60
Sea (SS3)	−48	−46	−46	−38	
Desert			−25	−31	−28
Farmland			−21		
Wooded hills	−32				
Cities			−10	−18	−15

Table 11.4
Typical Values of Clutter Reflectivity, σ°, in dB (1 m²/m²) at C-Band [8]

Clutter	Grazing Angle (degrees)				
	1	3	10	30	60
Sea (SS3)	−43	−42	−40	−37	−20
Farmland	−38		−29		

Table 11.5
Typical Values of Clutter Reflectivity, $\sigma°$, in dB (1 m²/m²) at X-Band [8]

Clutter	Grazing Angle (degrees)				
	1	3	10	30	60
Sea (SS3)	−40	−39	−37	−34	−21
Desert	−38		−26	−26	−23
Farmland	−36		−25	−21	−18
Wooded hills	−30		−25	−16	−13
Cities	−24		−14	−15	−12

Nathanson [8] and Ulaby, Moore, and Fung [12] have reported seasonal variations in terrain reflectivity as well as variations due to moisture content, particularly at higher grazing angles. In general, reflectivity decreases with snow cover and increases with leaves or grass cover [8] and moisture content of the soil or vegetation [12]. These variations can be as much as 2–10 dB at 50° grazing and decreasing to 0 dB near 0° grazing.

Volume clutter cross section per unit volume, η, can be computed for rain from [11]:

$$\eta = 6 \times 10^{-14} r^{1.6} \lambda^{-5} \tag{11.11}$$

where r is the rain rate in mm/hr and λ is the wavelength; η can be computed for chaff from [11]:

$$\eta = 3 \times 10^{-8} \lambda \tag{11.12}$$

Equations (11.11) and (11.12) were originally reported by David K. Barton (see "Radar Equations for Jamming and Clutter," *EASCON 67 Technical Convention Record,* Supplement to *IEEE Transactions on Aerospace and Electronic Systems,* Vol. AES-3, pp. 340–355).

Aurora clutter decreases at least as fast as the wavelength to the fifth power, and therefore is not usually a major consideration above UHF. Brookner [14] computed values of η for worst-case high sun-spot cycles. Typical values of η for rain, chaff, and aurora clutter are given in Table 11.6.

Table 11.6
Typical Values for Volume Clutter Reflectivity, η (m²/m³)

Clutter	Frequency Band					
	VHF	UHF	L	S	C	X
Rain (4mm/hr)			2×10^{-10}	5×10^{-9}	5×10^{-8}	5×10^{-7}
Chaff			7×10^{-9}	3×10^{-9}	1.7×10^{-9}	10^{-9}
Aurora	4×10^{-7}	10^{-8}	10^{-10}	10^{-13}		

Clutter Amplitude Distributions

The objective of clutter rejection techniques is to maximize the probability of detection of targets in the clutter background and to minimize false alarms caused by clutter residue. To achieve this, both the mean amplitude and the pdf of the clutter amplitude must be known. In recent years considerable research has been directed at identifying the "best" pdf to represent the amplitude of the major clutter types under various environmental conditions and as a function of significant radar parameters. The selected model would ideally represent the clutter in a realistic way and lend itself to analysis for predicting radar performance. In practice, the clutter data rarely match the model accurately and the database is seldom adequate to assess radar performance in all of the environments of interest. This is especially true for space-based radars. A compromise between accuracy of the model and analytical convenience is usually necessary.

The simplest and the most analytically convenient model of the clutter amplitude is the Rayleigh. When the clutter can be modeled as returns from a number of independent, random scatterers, the in-phase (I) and quadrature (Q) amplitude fluctuations of the clutter are described by the Gaussian pdf. The envelope of these fluctuations is Rayleigh distributed and the power has an exponential distribution.

Skolnik [15] and Long [7] have applied the Rayleigh pdf to sea clutter when the radar resolution cell (determined by (11.3)) is large compared to the sea wavelength and terrain clutter, such as deserts and some types of farmland. Valenzuela and Laing [16, 17] found the Rayleigh pdf to be appropriate for sea clutter from low sea states ("glassy" seas) as well as deserts and farmland. Schleher [3] suggests the Rayleigh pdf for (1) sea clutter when the radar resolution is low (pulsewidth greater than 0.5 μs) and grazing angles greater than 5° and (2) weather clutter.

When the Rayleigh conditions are met, the clutter random process is represented by $c_t = x_t \cos 2\pi f_c t - y_t \sin 2\pi f_c t$, where f_c is the radar carrier frequency; x_t and y_t are zero mean, independent, and identically Gaussian distributed processes with variance σ. These components would represent the I and Q clutter components at the radar receiver. The voltage envelope of c_t is

$$v = \sqrt{x_t^2 + y_t^2}$$

The pdf of the voltage amplitude of Rayleigh clutter is [3]:

$$P_v(v) = \frac{v}{\sigma^2} \exp[-v^2/2\sigma^2], \quad v \geq 0 \tag{11.13}$$

The clutter power is $P_c = 2\sigma^2$, which has an exponential pdf:

$$P_p(P) = \frac{1}{P_c} \exp[-P/P_c] \tag{11.14}$$

For Rayleigh clutter, the effective radar cross section (11.2) is represented by the mean of the radar cross section distribution with the exponential pdf:

$$P_p(\sigma_e) = \frac{1}{\overline{\sigma}} \exp[-\sigma_e/\overline{\sigma}] \tag{11.15}$$

The Rayleigh clutter statistics have been employed almost exclusively in clutter-rejection analyses and processing designs and assessments. An attractive feature is that it is completely defined by its variance σ^2, and target detection in clutter can be treated in the same way as Gaussian noise using the well known Neyman-Pearson hypothesis test for detection. The required signal-to-clutter ratio (S/C) for specified probabilities of detection (P_d) and false alarm (P_{FA}) can be determined by simply using the standard Neyman-Pearson P_d/P_{FA} curves for detection of signals in Guassian noise (Figure 11.9). By calculating the input S/C from (11.1) and determining the required output S/C from Figure 11.9, the required *MTI improvement factor* is simply the ratio of these.

In most clutter conditions particularly when one or more of the conditions above are not met, an assumption of Rayleigh clutter will result in an optimistic requirement for clutter rejection. As has been observed, the actual clutter has a greater probability of large values than does Rayleigh clutter [15] (i.e., the actual pdf has higher "tails" than the Rayleigh). In such cases, the log-normal pdf has been proposed to model the clutter.

The pdf describing the log-normal clutter statistics is

$$P(\overline{\sigma}) = \frac{1}{\sqrt{2\pi}\sigma\overline{\sigma}} \exp\left[-\frac{1}{2\sigma^2}\left(\ln\frac{\overline{\sigma}}{\sigma_m}\right)^2\right], \quad \overline{\sigma} > 0 \tag{11.16}$$

where $\overline{\sigma}$ is defined by (11.2), σ_m is the median value of $\overline{\sigma}$, and σ is the standard deviation of the natural logarithm of $\overline{\sigma}$.

Trunk [18] and George [19] investigated the detection of a nonfluctuating target in log-normal clutter and quantified the additional S/C needed as compared to Rayleigh clutter to achieve the same P_d and P_{FA}. Trunk proposed a median detector instead the usual mean detector as being more efficient for log-normal

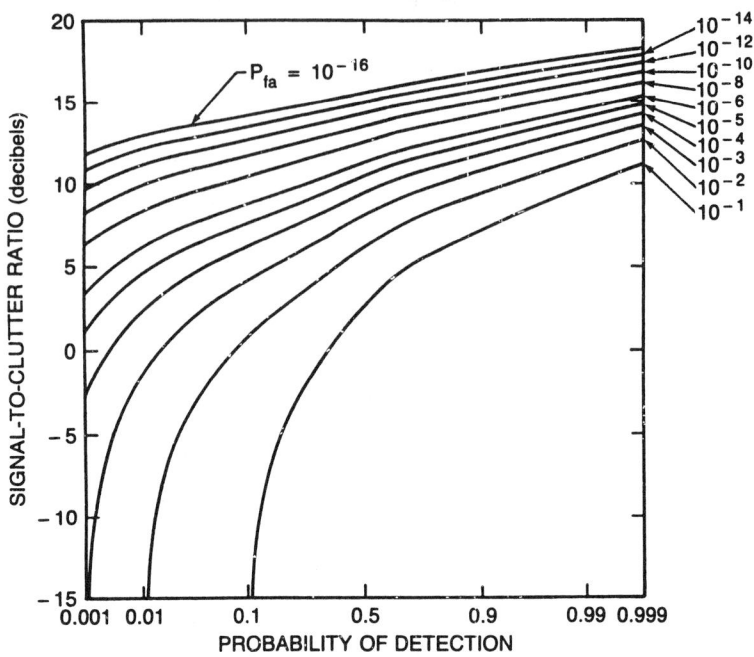

Figure 11.9 Required signal-to-clutter ratio (S/C) for Rayleigh clutter at the input terminals of a linear rectifier detector as a function of probability of detection for a single pulse, with the false-alarm probability (P_{fa}) as a parameter; a nonfluctuating signal is assumed.

clutter. An example of these results is shown in Figure 11.10. Using a mean detector for both types of clutter and a P_d of 0.9 for comparison, we see that if the number of independent pulses (N) to be integrated for detection equals 3, an additional 12 dB of S/C is needed for log-normal clutter. For $N = 29$, an additional 7 dB is needed. This would imply that an additional 7–12 dB of clutter rejection was needed for log-normal clutter than had been concluded from Figure 11.9 for Rayleigh clutter. Note that if N_c pulses were coherently processed for clutter rejection and N_D pulses transmitted during the dwell at one footprint, $N = N_d/N_c$.

Trunk showed that this 7–12 dB "loss" associated with log-normal clutter could be reduced by 1.6–2.8 dB if log-normal clutter were predicted and a median detector used instead of a mean detector (Figure 11.10). However, if the clutter is more nearly Rayleigh then the median detector is less efficient than the mean detector by a similar amount.

One commonly employed approach used to circumvent the uncertainty associated with the actual clutter distributions encountered is to provide enough clutter rejection so that the clutter residue after processing is below the system

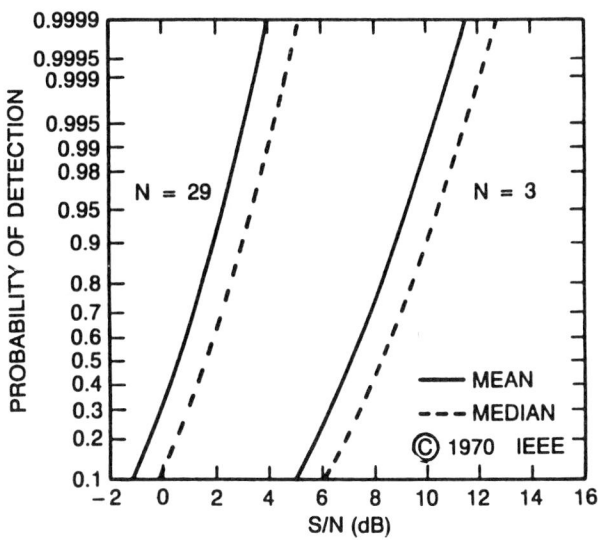

Figure 11.10 (a) Probability of detection for the Rayleigh distribution, for both a median and a mean detector and for both a small number of samples (3) and a larger number (29); signal-to-noise ratio $(S/N) = 10 \log (A^2/2\sigma^2)$, where A is the signal voltage amplitude. Probability of false alarm $= 10^{-6}$ (from Trunk [18]).

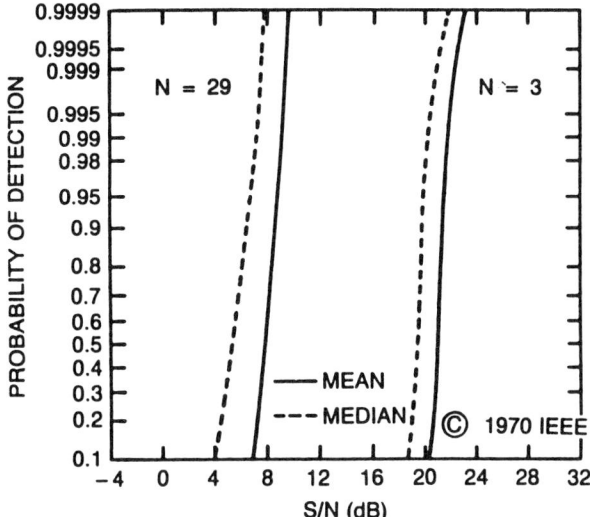

Figure 11.10 (b) Probability of detection for the log-normal distribution; signal-to-noise ratio $(S/N) = 20 \log [A/(\text{median noise value})]$, probability of false alarm $= 10^{-6}$, and the standard deviation of the log-normal distribution $= 6$ dB (from Trunk [18]).

noise level, and therefore the false alarms will be dominated by statistically more predictable noise. The procedure for doing this would be to compute the received clutter-to-noise ratio (C/N) from (11.8) and add 7–12 dB to provide for the log-normal loss. This would be a safe specification of the clutter rejection requirement, but it could be a costly overspecification if the actual clutter were more nearly Rayleigh.

For a space-based radar, another more complex but more efficient approach would be to create a worldwide clutter map by type and to predict the expected clutter statistics in a selected footprint based on the clutter type from this map [20]. The expected clutter statistics would be based on other parameters such as radar resolution, sea state, and grazing angle from the SBR to the footprint. The required clutter rejection would change as the radar was scanned to other footprints. This would mean that the coherent processing time needed for clutter rejection could be changed at each footprint position, which would change the dwell needed at each footprint position. The area coverage rate could be maximized in this way, but an adaptable processing architecture and a stepped-scan antenna capability would be needed.

The log-normal distribution is considered to be a "severe" assumption for the clutter statistics in that it predicts a need for more clutter rejection than the actual clutter for most of the data available. Other clutter distributions have been proposed to match actual clutter that falls between the optimistic Rayleigh distribution and the pessimistic log-normal. These distributions include the "contaminated" Gaussian (Trunk), Ricean, chi-square, Wiebull, an ITT model (Schleher), and a model by Moore *et al.* [3] based on space-based clutter from Skylab S-193.

11.2.2 Clutter Spectral Distributions

The magnitude of the clutter returns and amplitude distribution determine the amount of clutter rejection needed to achieve the desired target detectability, as described in the previous section. The spectral distribution of the clutter will determine the ability of doppler processing techniques to reject the clutter. In that sense, this clutter spectrum will set a ceiling on the detectability of a moving target by an SBR.

Barlow [21] proposed that the spectra of almost all types of clutter could be described by a Gaussian distribution. This distribution is usually used in the literature to assess the performance of clutter rejection techniques. The driving force for the clutter's internal motion is wind. Therefore, in virtually all types of clutter environments, the SBR would be expected to perform worse in high wind conditions (obvious exceptions include land clutter without vegetation such as deserts, some mountains, and snow or ice covered terrain). Note that the doppler spectra need not be zero-mean. Sea, rain, and snow clutters, and chaff can have a nonzero-mean spectrum, depending on the viewing angle relative to the wind direction.

This mean must be considered in the clutter processing designs. The Gaussian clutter power spectral density is given by

$$S(f) = \frac{P_c}{\sqrt{2\pi}\sigma_f} \exp[-(f - m_f)^2/2\sigma_f^2] \qquad (11.17)$$

where P_c is the total clutter power, which is proportional to σ_c (see (11.2)), σ_f is the standard deviation of the clutter frequency spread, and m_f is mean of the clutter doppler spectrum. The frequency spread is related to σ_v, the standard deviation of the clutter velocity spread, $\sigma_f = 2\sigma_v/\lambda$, where λ is the wavelength.

The effect of wind speed or sea state on sea clutter doppler spread and mean doppler shift has been a subject of intense study [2, 7, 8, 11, 15, 21–24]. First-order scattering theory appears to be successful in relating sea clutter $\sigma°$ with the directional mean-squared wave height spectrum, which can then be related to the wind level and direction [22]. However, the backscatter doppler spectrum appears to have a directional variation only in its mean and not in its doppler spread. Pidgeon [23] first noted that the sea clutter doppler spread was apparently independent of viewing angle relative to the wind velocity, and the mean doppler shift was different for horizontal and vertical polarization for low (less than 10°) grazing angle. He also observed that the mean doppler shift varied as the cosine of the viewing angle with respect to the wind velocity. Valenzuela and Laing [24] seemed to find, at least, a weak relationship between the spectral width and the viewing angle.

Barton [2] stated that there is a rough proportionality between the wind speed and the clutter velocity spread, σ_v, for both sea and rain clutter. The ratio is about eight for sea clutter and five for rain. Taking these factors into account, Figure 11.11 plots typical values for σ_v and m_v for sea, land, and rain clutter as a function of wind speed and alternatively significant wave height and sea state.

Fishbein, Graveline, and Rittenback [25] analyzed clutter data from an X-band, noncoherent, pulsed doppler radar when field tests did not demonstrate the predicted performance. They found that the clutter spectrum decreased at a much slower rate (i.e., had much higher "tails") than predicted by the Gaussian spectral assumption. They proposed a distribution that gives good agreement with the data:

$$S(f) = \frac{1}{1 + (f/f_c)^3} \qquad (11.18)$$

where $f_c = 1.33 \exp(0.1356v)$ if v is the wind speed in knots, or $f_c = 1.33 \exp(0.0692v)$ if v is the wind speed in m/s. The data were taken with horizontal polarization. Equation (11.18) is plotted in Figure 11.12 along with a best-fit Gaussian distribution for comparison.

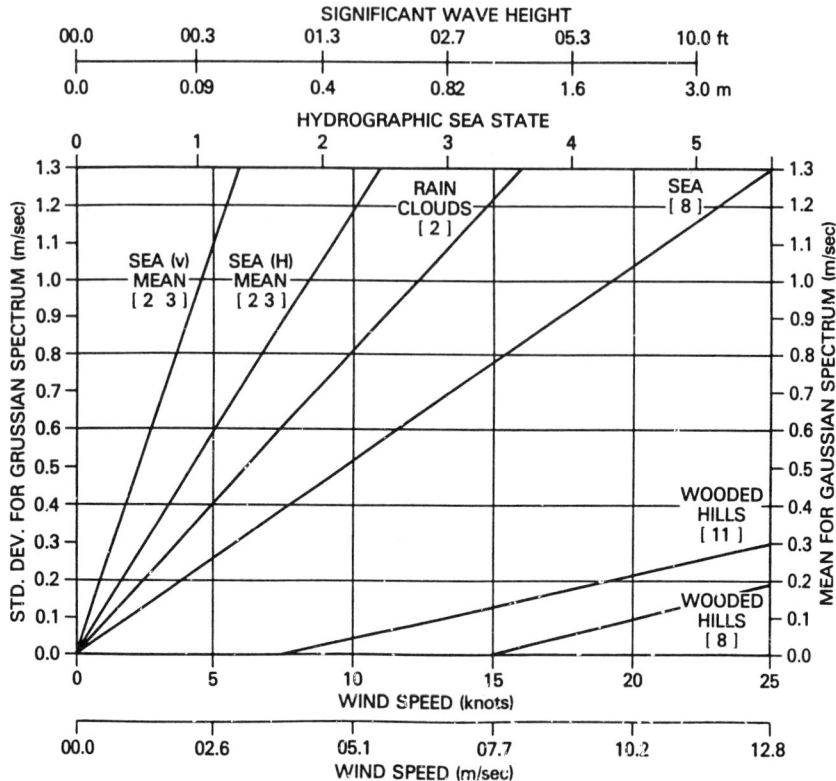

Figure 11.11 The variation in the mean and standard deviation of the velocity spectrum of several types of clutter *versus* windspeed, sea state, or significant waveheight.

A specifically designed experiment by MIT Lincoln Laboratory to measure terrain clutter spectra has been reported by J.B. Billingsley and J.F. Larrabee (see "Measured Spectral Extent of L- and X-Band Radar Reflections from Wind-Blown Trees," Lincoln Laboratory Project Report CMT-57, February 1987). Here, an exponential distribution was measured over a 60 dB dynamic range.

The composite clutter spectrum observed by an SBR is dominated by platform motion effects as described in the next section. However with the utilization of platform motion compensation techniques such as *displaced phase center antenna* and *quadratic phased modulated waveform* (also described later), the rejection of clutter may be limited by internal motion. Therefore, a better understanding of clutter spectra is needed and must be supported with enough experiments to gain

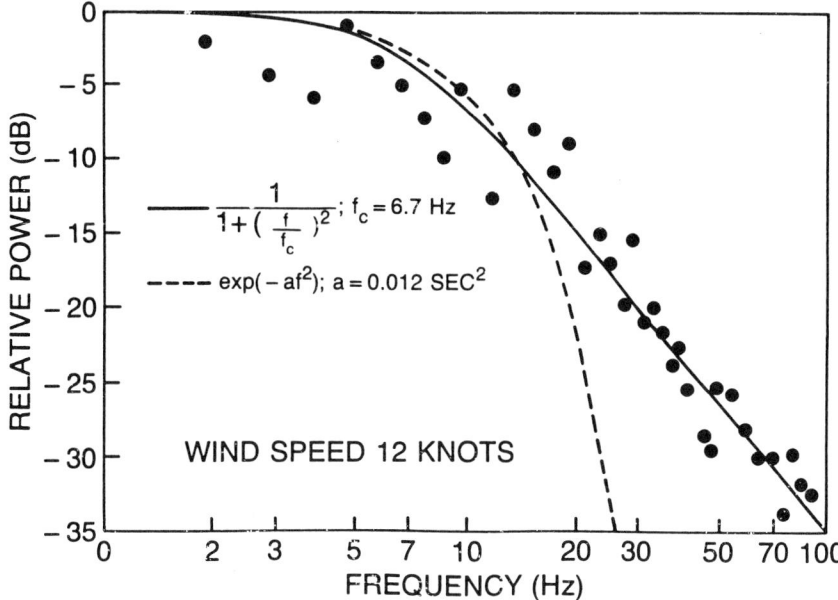

Figure 11.12 Power spectrum obtained with an X-band radar; each solid circle represents a measurement (from Fishbein, Graveline, and Kittenbach [25]).

confidence in the resolution of these issues. At this time, an assumption of a Gaussian spectrum with standard deviation and mean determined by Figure 11.11 has both a theoretical basis for many types of clutter and support by data from several experiments. The spectra defined in Figure 11.12 could be viewed as an extreme "worst-case," rather than as a design goal.

11.3 UNIQUE FEATURES OF SBR CLUTTER

An SBR must not only contend with clutter as described in the previous section, but also with certain unique space-related characteristics, such as the effects of the orbit velocity of the SBR satellite, the earth's rotational velocity, propagation anomalies, antenna pattern, and observation angles.

11.3.1 Antenna Pattern Considerations

The antenna aperture size and the transmitter power are major contributors in the radar equation to the sensitivity of the radar and, therefore, the detectability of targets of interest. The antenna size and pattern characteristics also have a profound effect on the level and characteristics of the clutter to be rejected. Therefore, although the power-aperture product is a useful figure of merit for a search radar only concerned with noise, clutter considerations make the aperture size a more important factor for an SBR designed to look down on and to search the earth's surface.

The size of the antenna aperture determines the size of the antenna footprint, which, in turn, determines the level of the main-lobe clutter, the number of range ambiguities to be encountered, and the doppler spread of the main-lobe clutter. The size of the antenna is important, but equally so are the antenna errors and tolerances, which limit the achievable antenna sidelobes. These sidelobes, in turn, limit the magnitude of the sidelobe clutter and sidelobe interference that can be rejected by the antenna pattern itself.

Antenna sidelobes are reduced by an aperture weighting, or taper function, applied to the aperture illumination function of the antenna. A random sidelobe "floor" exists, which limits the sidelobe reduction that can be achieved through weighting. The major sources of system errors that determine the random rms sidelobe floor is illustrated by the Theo C. Cheston nomograph [27] shown in Figure 11.13. The rms sidelobe levels are measured along the abscissa both relative to isotropic (dBi) and relative to the element gain (dBe). The assumed element gain is $G_e = \pi \cos\theta$. Amplitude and phase errors in the antenna, including the transmit-receive (T/R) modules for a phased array, are the fundamental limitation to the achievable sidelobe reduction.

The rms amplitude errors are measured along the first ordinate of Figure 11.13. The errors are shown in fractional rms errors (v/v) and in dB where the amplitude errors in dB are equal to 20/ln (10) times fractional rms amplitude errors (v/v). For example, to achieve rms sidelobes of −15 dBi (or −20dBe), the fractional rms amplitude errors must be less than 0.1 (or 0.87 dB), assuming contributions from other errors are negligible.

The phase errors are measured along the third ordinate and are illustrated for peak and rms values where the peak errors are $\sqrt{3}$ times the rms errors [27]. To achieve −15 dBi (−20 dBe) rms sidelobes, peak errors of less than ±10° (or 5.77° rms) would be required, assuming contributions from other errors are negligible. One source of phase errors results from the number of phase shifter bits in a phased scanned array shown on the last ordinate. The rms sidelobes (dBe) equals $\pi^2/(3 \times 2^{2P})$, for P phase shifters bits [27]. To achieve −15 dBi (−20 dBe)rms sidelobes, 5 bits would be needed; 4 bits only achieves −18.9 dBe.

Another source of errors shown in Figure 11.13, which results in higher sidelobes in a phased array, is failed radiating elements. The effect of failed elements depends on the failure mode. It the T/R module fails in such a way that it is blocked from either transmitting or receiving, the failure is less severe to the antenna pattern and its sidelobes than if the phase shifter fails while the T/R module and element continue to transmit and receive but a fixed, random phase is applied by the failed phase shifter. For a given rms sidelobe level, twice as many failures could be tolerated for blocked radiators as would be the case with failed phase shifters. To achieve -20 dBe (-15 dBi) rms sidelobes, 1% of the elements radiation can fail, but only 0.5% of the phase shifters can do so. Again, this assumes that contributions from other errors are negligible. To combine the effects of all error sources, they are usually assumed to be independent and combined insofar as the square of the total is equal to the sum of the squares of each of the individual errors.

A convenience for SBR analyses is to transform the SBR antenna pattern that is usually described in terms of an angle (ϕ) in a "horizontal" plane and an angle (ϕ) in a "vertical" plane into a range and azimuth as projected onto the earth. The transformations needed to project an SBR antenna pattern onto the surface of the earth could be computed if we selected an azimuth pointing angle, AZ, relative to the satellite velocity vector and a range, R, on the earth relative to the nadir point beneath the satellite (see Figure 11.3). The projection in the range direction is referred to as the *range scale factor*, Δ_R, and the angular projection in the azimuth direction is called the *azimuth scale factor*, Δ_{AZ}.

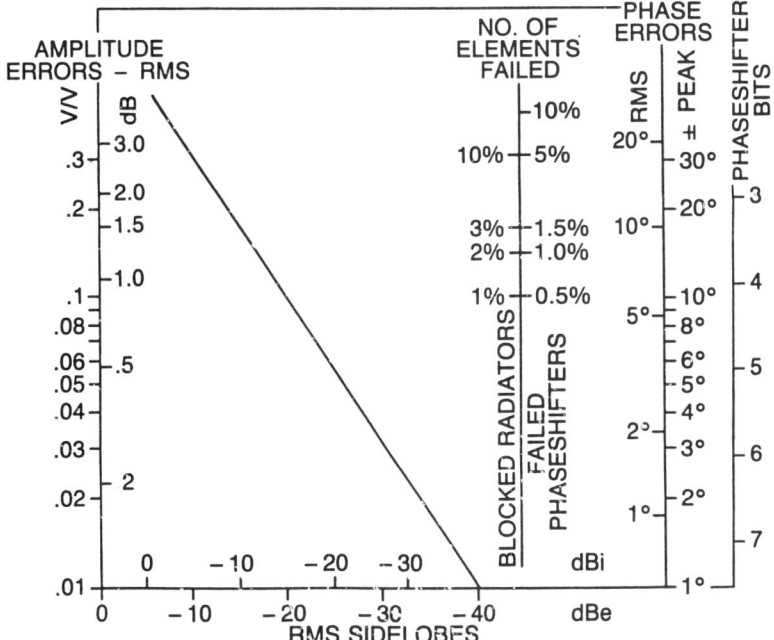

Figure 11.13 The effect of random amplitude and phase errors, quantization errors, and randomly failed antenna elements on rms sidelobes.

These scale factors are given by

$$\Delta_R = R_s \Delta\phi \csc\psi$$
$$\Delta_{AZ} = \tan^{-1}\left(\frac{R_s \Delta_\theta}{R}\right)$$
(11.19)

where R_s is given by Equation (11.4) (see Figure 11.4), Δ_θ and Δ_ϕ are the respective differential horizontal and vertical angles on the antenna pattern, and ψ is the grazing angle given by Equation (11.5) (see Figure 11.5). Here, the nose of the antenna beam is pointing in the direction of the line-of-sight (LOS) vector of Figure 11.3. The vertical angle on the antenna pattern is in the elevation plane relative to the satellite and containing the LOS vector. The horizontal angle on the antenna pattern is taken in the plane perpendicular to elevation plane and containing the LOS vector.

The antenna pattern in (R, AZ) coordinates is given by

$$G(R + \Delta_R, AZ + \Delta_{AZ}) = G(\Delta_\theta, \Delta_\phi + \Delta_{\phi_c})$$
(11.20)

and

$$\Delta\phi_c = \frac{R}{R_s}\left[1 - \sqrt{1 + \left[\frac{R_s}{R}\Delta_\theta\right]^2}\right]$$

where $\Delta\phi_c$ is the correction in the elevation plane needed to keep R constant as θ is varied. These simplified equations are applicable when the significant range of angles on the antenna pattern to be considered is small compared to the elevation scan angle, EL (see Figure 11.3). The angle, EL, is given by

$$EL = \sin^{-1}\left[\frac{R_e}{R_e + h}\cos\psi\right]$$
(11.21)

where ψ is given by (11.5). An in-depth description of SBR coordinate transformations is given in [20].

Figure 11.14 shows the pattern of a circular aperture, which is uniformly weighted on transmission and -34 dB truncated Gaussian weights on reception. These patterns are combined into a two-way pattern and projected onto the earth's surface from a satellite altitude of 5600 nmi (10,371 km) and a 3° grazing angle in Figure 11.15(a), a 25° grazing angle in Figure 11.15(b). Each pattern is truncated when it reaches the horizon. The floor is at -100 dB relative to the peak of the main beam. From this plot we can see that the sidelobes of this particular pattern

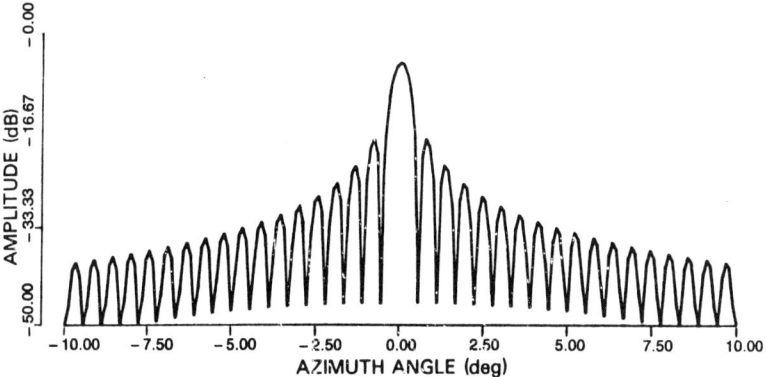

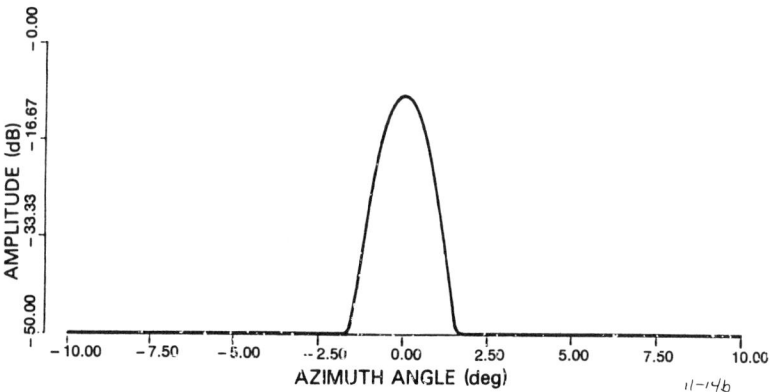

Figure 11.14 Antenna patterns for a circular aperture uniformly weighted on transmitting and −34 dB truncated Gaussian weights on receiving: (a) transmitted pattern; (b) received pattern.

are probably determined by antenna errors at angles greater than about 3° from the peak of the main lobe.

The clutter returns from each point on the earth's surface will be multiplied by this two-way antenna gain. Note that the transformations outlined above do not lead to equal incremental steps on the earth's surface when equal angular steps are taken on the antenna pattern. The clutter power received from each step on the earth will also be proportional to the size of that particular step.

The main-beam footprint on the earth's surface is usually defined at the −3 dB point on the antenna pattern (see Figure 11.16). The footprint is used to determine the level of main-beam clutter and the region to be searched for targets.

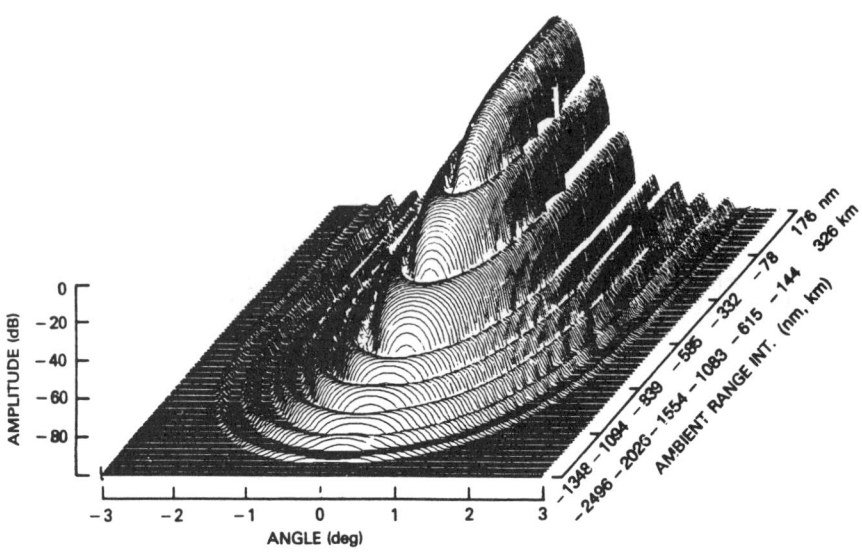

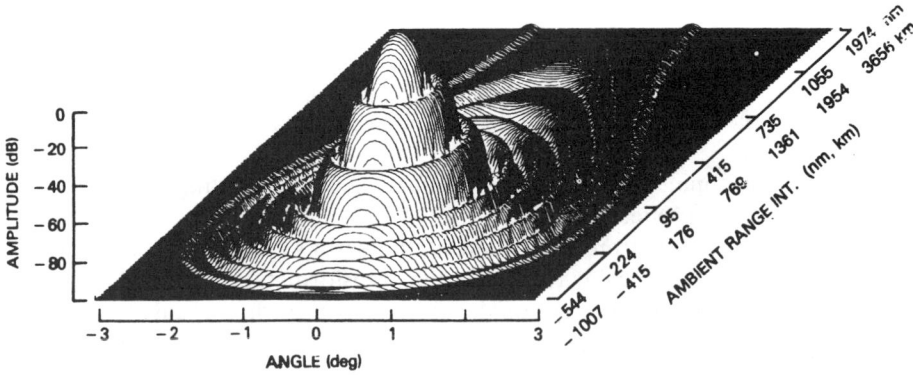

Figure 11.15 Two-way antenna pattern from Figure 11.13 projected onto the earth's surface from an altitude of 5600 nmi(10,371 km) and grazing angle of (a) 3° and (b) 26°.

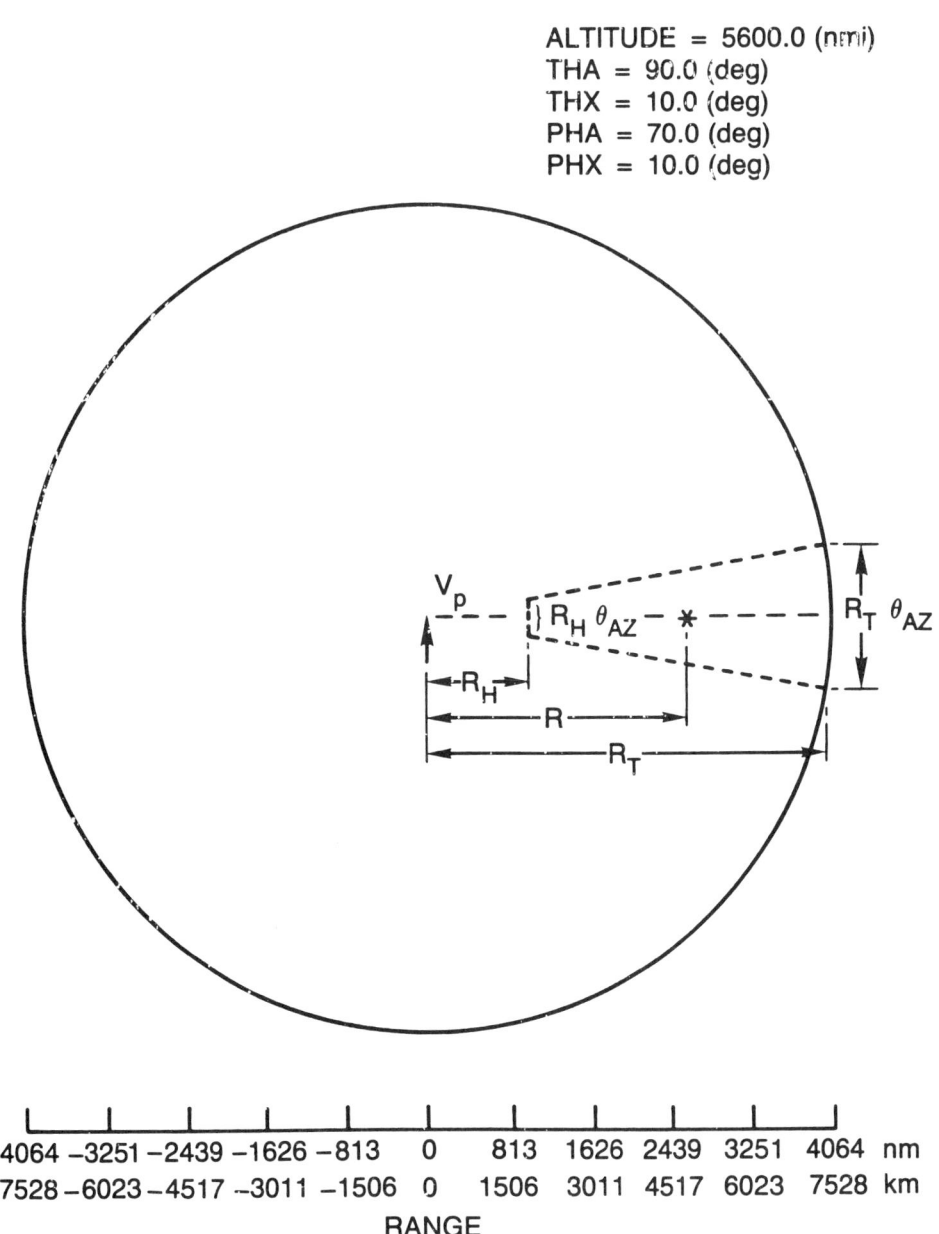

Figure 11.16 The projection of a square (10° in azimuth and 10° in elevation) window onto the earth from a 5600 nmi, (10,371 km) altitude to a grazing angle of 26° at the center of the window (∗).

The equations defining the range boundaries of this footprint for the "toe," R_T, and the "heel," R_H, are

$$R_T = R_e\left(\frac{\pi}{2} - \text{EL} - \frac{\phi_{\text{EL}}}{2} - \psi_T\right)$$
$$R_H = R_e\left(\frac{\pi}{2} - \text{EL} + \frac{\phi_{\text{EL}}}{2} - \psi_H\right) \qquad (11.22)$$

where ϕ_{EL} is the 3 dB beamwidth of the antenna pattern in the elevation plane, ψ_T and ψ_H are the grazing angles at the toe and heel of the footprint, respectively,

$$\psi_T = \cos^{-1}\left[\frac{R_e + h}{R_e}\sin\left(\text{EL} + \frac{\phi_{\text{EL}}}{2}\right)\right]$$
$$\psi_H = \cos^{-1}\left[\frac{R_e + h}{R_e}\sin\left(\text{EL} - \frac{\phi_{\text{EL}}}{2}\right)\right] \qquad (11.23)$$

The angular extent of this footprint in the azimuth plane is simply the 3 dB azimuth beamwidth, θ_{AZ}. The cross-range extent at the toe is $R_T \theta_{\text{AZ}}$ and at the heel is $R_H \theta_{\text{AZ}}$.

11.3.2 Range Resolution and Ambiguities

The SBR range resolution projected onto the earth's surface is

$$R_{\text{res}} = \frac{c\tau}{2}\sec\psi \qquad (11.24)$$

The range between range ambiguities can be approximated by

$$R_{\text{amb}} = \frac{c}{2f_r}\sec\psi \qquad (11.25)$$

for relatively high pulse repetition frequencies (PRFs), f_r, such that the grazing angles at the range ambiguities are approximately equal. The general expression is

$$R_{\text{amb}} = R_e \cos^{-1}\left[\frac{R_e^2 + (R_e + h)^2 - \left(R_s + \frac{c}{2f_r}\right)^2}{2R_e(R_e + h)}\right] - R \qquad (11.26)$$

for a selected orbit altitude, h, and a range from nadir, R.

Because a constant range from the nadir point corresponds to a circle centered at nadir, all of the range ambiguities at a given orbital altitude form concentric circles centered at the nadir point. The range ambiguities resulting from selecting $h = 5600$ nmi (10,371 km) and $f_r = 10$ kHz are illustrated in Figure 11.17. A relatively high PRF of 10 kHz might be selected to minimize the number of velocity ambiguities for the targets of interest and to maximize the range of velocities that need not compete with main-lobe clutter. However, comparing Figures 11.16 and 11.17 shows that this results in a large number of range ambiguities, N_a, within the footprint. This not only raises the average clutter level (see Equation (11.3)), but also requires that these range ambiguities be resolved.

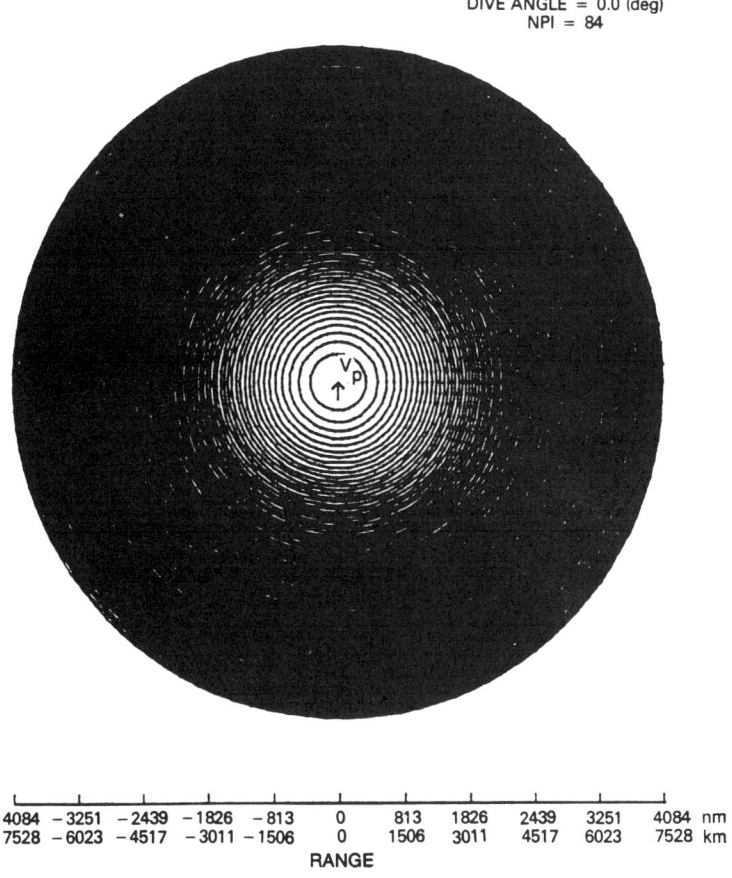

Figure 11.17 The projection of ambiguous range gates onto the earth from 5600 nmi, (10,371 km) altitude and a PRF of 10 kHz.

To have no range ambiguities within the footprint, a PRF must be selected such that $R_{amb} \geq R_T - R_H$. This leads to a limit on the PRF:

$$f_{r_{max}} = \frac{c}{2\left\{\left[R_e^2 + (R_e + h)^2 - 2R_e(R_e + h)\cos\frac{R_T}{R_e}\right]^{1/2} - R_s|_H\right\}} \quad (11.27)$$

This maximum PRF is plotted in Figure 11.18 as a function of the elevation scan angle, EL (see Figure 11.3), for several orbit altitudes, h, and 3 dB elevation beamwidths of 1° and 5°. The elevation scan angle is constrained between a minimum such that the maximum grazing angle of the heel of the footprint, ψ_H, is 60° and a maximum such that the toe of the footprint is at the horizon, $\psi_T = 0$.

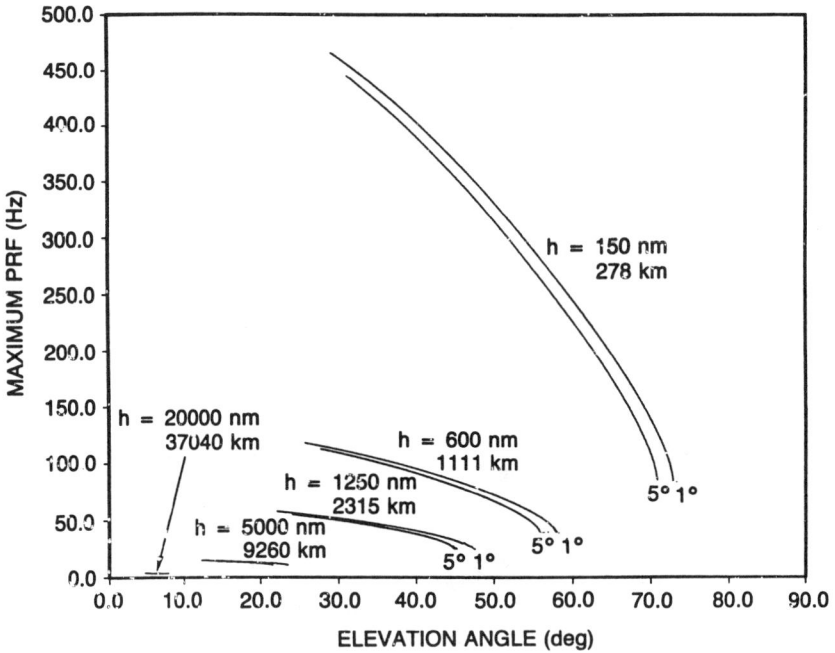

Figure 11.18 The maximum PRF for no range ambiguities within the 3 dB beamwidth of the two-way antenna pattern represented by the transmitting and receiving patterns of Figure 11.14; h is the satellite altitude.

Because of doppler ambiguities or clutter rejection considerations, a higher PRF will often be selected. The number of range ambiguities in the footprint can be calculated from

$$N_a = \frac{f_r}{f_{r_{max}}} \tag{11.28}$$

If N_a is not an integer, some range cells in the footprint (those at the heel and at the toe) will have an additional ambiguity relative to the range cells near the center of the footprint. This value of N_a should be used in (11.3) to commute the total clutter power.

11.3.3 Platform Motion Effects

For an SBR, the received clutter spectrum is caused not only by the internal motion described in Section 11.2.3, but also by the effects of the satellite velocity and the earth's rotational velocity. The relative velocity between the satellite and a particular point on the earth is a function of the locations of the satellite and that point. Therefore, the platform motion effects will be described by assuming that these two locations are known as well as the height, h, and inclination angle, η_i, of the satellite orbit.

Referring to Figure 11.19, the nadir point of the satellite, P_1, and the point of interest, P_2, are located on a spherical earth of radius, R_e, by their latitudes (α_1 and α_2), longitudes (β_1 and β_2). The range on the earth of P_2 from the nadir point, P_1, is

$$R = R_e\alpha = R_e \cos^{-1}[\sin\alpha_1 \sin\alpha_2 + \cos\alpha_1 \cos\alpha_2 \cos(\beta_2 - \beta_1)] \tag{11.29}$$

This value of R can be used in (11.5) to compute the grazing angle, ψ, at P_2. For an SBR at P_1, the azimuth scan angle of P_2 relative to north is given by

$$\beta' = \sin^{-1}\left[\frac{\cos\alpha_2 \sin(\beta_2 - \beta_1)}{\sin\alpha}\right] \tag{11.30}$$

The corresponding azimuth scan angle of the SBR relative to the orbit velocity vector is

$$AZ = \beta' + \eta_i \cos\left(\frac{\pi}{2}\frac{\alpha_1}{\eta_i}\right) - \frac{\pi}{2} \tag{11.31}$$

where η_i is the orbit inclination angle.

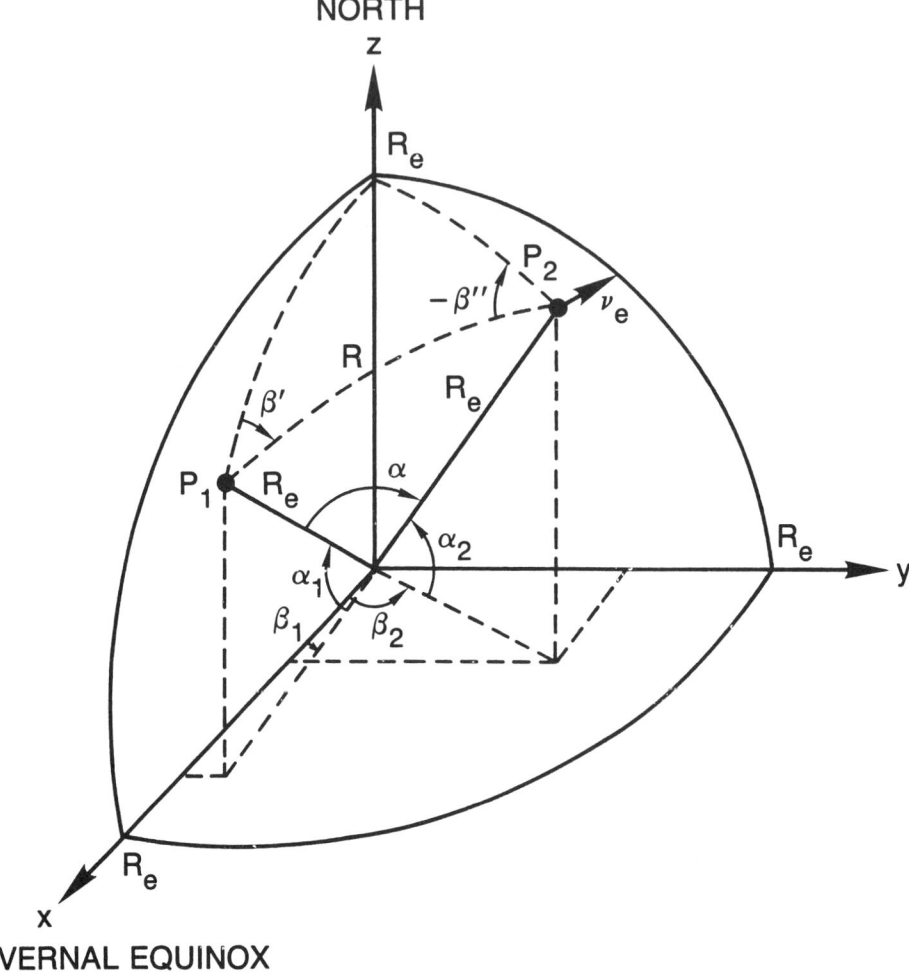

Figure 11.19 Coordinate system defining the geometry for calculating the doppler spectrum caused by platform motion and earth rotation.

From the coordinate system defined in Figure 11.3, the doppler shift relative to the SBR associated with any point on the earth is

$$f_d = \frac{2}{\lambda} [v_p(\cos D \sin \text{EL} \cos \text{AZ} + \sin D \cos \text{EL}) + v_e \sin\beta'' \cos\psi] \quad (11.32)$$

where λ is the wavelength in meters and v_p is the orbit velocity vector in meters per second given by $v_p = 629{,}575/\sqrt{h + R_e}$ m/s with h and R_e in kilometers. D is the angle of the velocity vector in the vertical plane ($D = 0$ for a circular orbit). The elevation scan angle, EL, is given by

$$\text{EL} = \sin^{-1}\left(\frac{R_e}{R_e + h} \cos\psi\right) \tag{11.33}$$

The rotational velocity of the earth at P_2 is, in meters per second, $v_e = 459 \cos\alpha_2$ m/s. The azimuth angles at P_2 are also defined relative to north; β'' is the azimuth angle of the ground range vector, R, relative to north, measured at P_2 such that

$$\beta'' = \sin^{-1}\left[\frac{\cos\alpha_1 \sin(\beta_1 - \beta_2)}{\sin\alpha}\right] \tag{11.34}$$

Isodops (curves of constant doppler shift) are computed from (11.32) and illustrated in Figure 11.20 for $h = 5600$ nmi (10,371 km), $\eta_i = 65°$, $\alpha_1 = \beta_1 = 0$ and a wavelength of 0.24 m. We also assume that doppler filters are formed by integrating 64 pulses at a PRF of 10 kHz. The resulting filter numbers for each isodop are also shown for reference. Note that a filter number occurs more than once because of the doppler ambiguities under these conditions. The platform velocity vector, v_p, points vertical in this figure. The axis of symmetry for the isodops is rotated counterclockwise relative to v_p because of the earth's rotation. The velocity vector along the ground track is also rotated counterclockwise to correspond to the axis of symmetry.

Farrell and Taylor [28] showed that the clutter spectrum for a doppler radar could be accurately described by considering three regions of the antenna pattern: the main-beam, significant sidelobe (only the first sidelobe for the airborne radar considered), and remaining sidelobe regions. The authors stated that the spreading of those remaining sidelobes over several doppler regions apparently has an averaging effect, and therefore the integrated sidelobe level can be used over all of that sidelobe region. Applying this to the antenna pattern shown in Figure 11.15, the sidelobes greater than about 3° from the main-beam peak could be represented by the integrated sidelobe level over that region.

The characteristics of the isodops in the region of the main lobe can be examined by applying a window to the isodops of Figure 11.20. A 10° window centered at an azimuth angle of 90° relative to the ground track velocity vector and a grazing angle of 26° is shown in Figure 11.21. The clutter spectrum corresponding to a range cell is found by taking a horizontal cut and applying the antenna

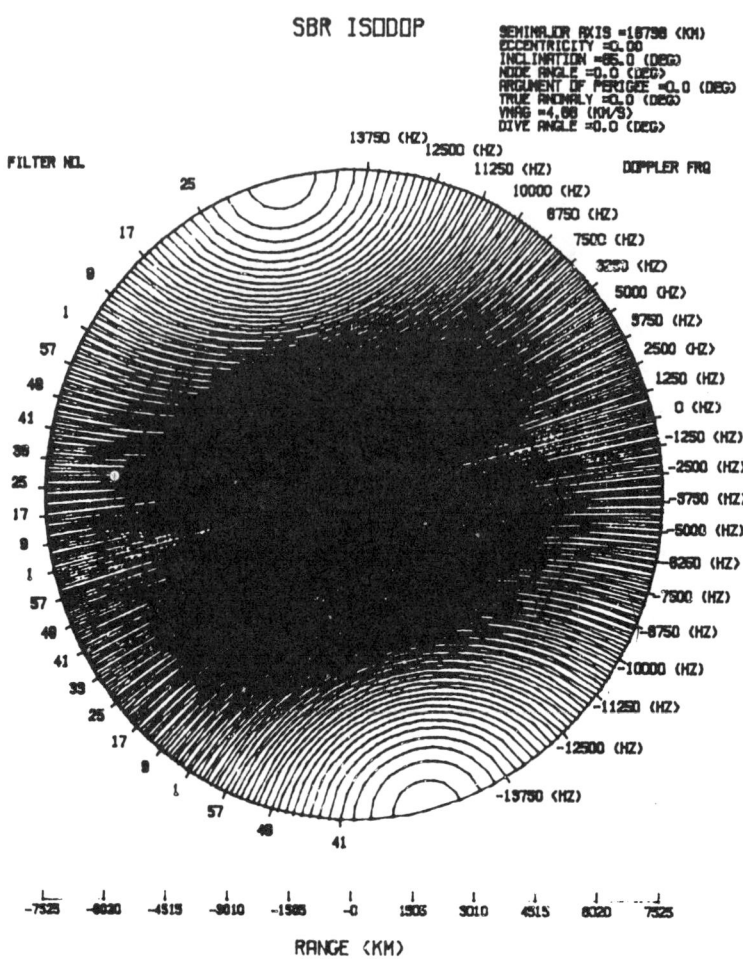

Figure 11.20 Isodops, including earth rotation, for a 5600 nmi (10,371 km) altitude SBR with an orbit inclination angle of 65° and transmitting frequency at L-band.

gain factor to each doppler (or azimuth angle) value. This results in the clutter spectrum shown in Figure 11.22, using the antenna pattern of Figure 11.15 with a random two-way sidelobe level of -70 dB.

The doppler shifts of Figure 11.21 change very slightly with range so that ambiguous range cells will have essentially the same spectrum and simply add to this spectrum without appreciably changing its shape. The spectral spread decreases somewhat as the range decreases so that the worst-case spectral spread due to platform motion at an azimuth angle of 90° is at a grazing angle of 0° (the horizon).

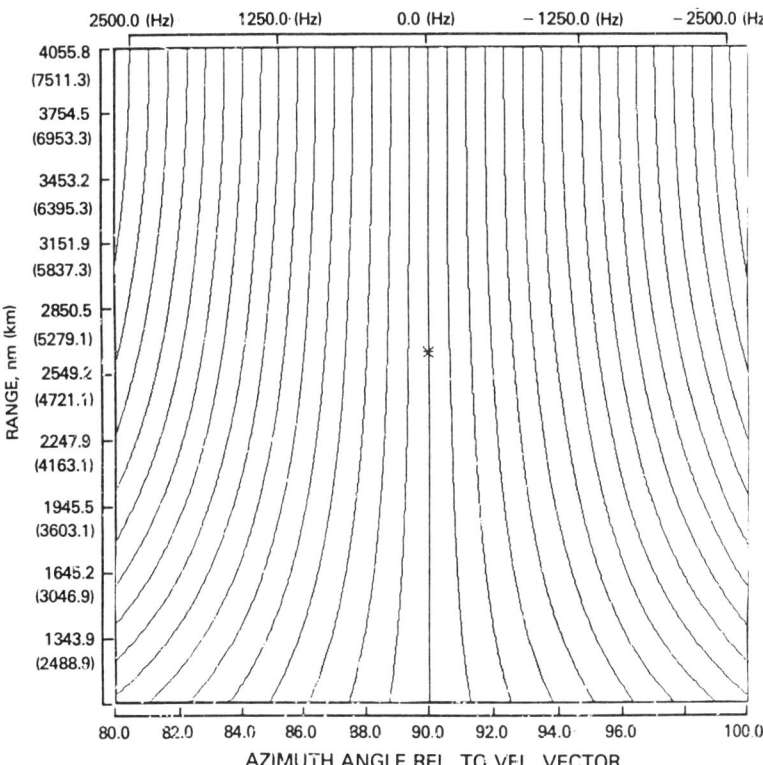

Figure 11.21 Isodops from Figure 11.20 in the 10° azimuth, 10° elevation window of Figure 11.16 pointing at 90° relative to the ground track velocity vector, and a grazing angle at the center of 26°.

When the azimuth angle relative to the ground track is changed to 0°, the isodop window shown in Figure 11.23 results. The doppler shift at the center of this window, 12.8 kHz, has been subtracted to examine the change in doppler around the peak of an antenna pattern at that point. This is equivalent to clutter-locking the radar receiver to the mean clutter doppler shift.

Notice that for this pointing angle there is almost no change in doppler for a selected range, resulting in very little doppler spread within a given range cell. However, there is a large change in doppler between range cells. Therefore, the

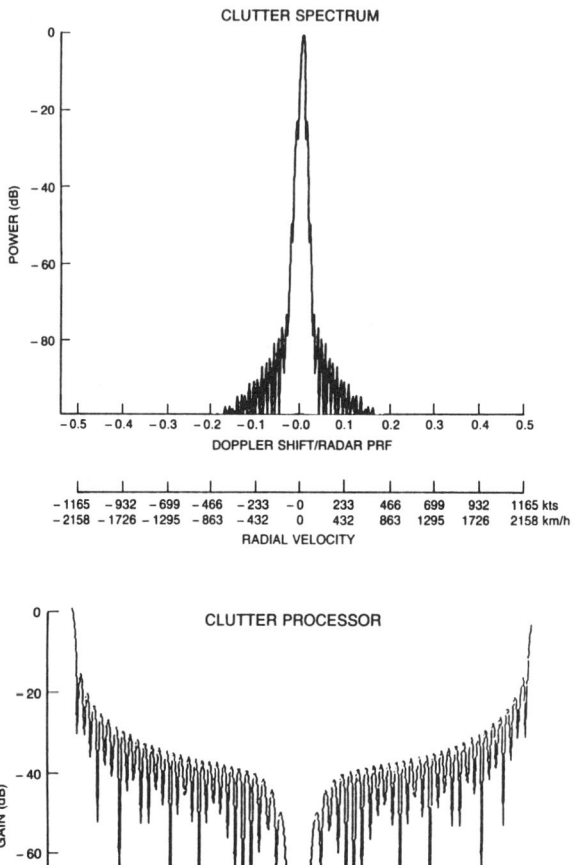

Figure 11.22 (a) Clutter spectrum resulting from the dopplers of Figure 11.21 and the antenna pattern of Figure 11.15; (b) the center filter of a 64-pulse doppler processor matched to reject the spectrum in (a). This matched clutter filter is discussed in Section 11.4.

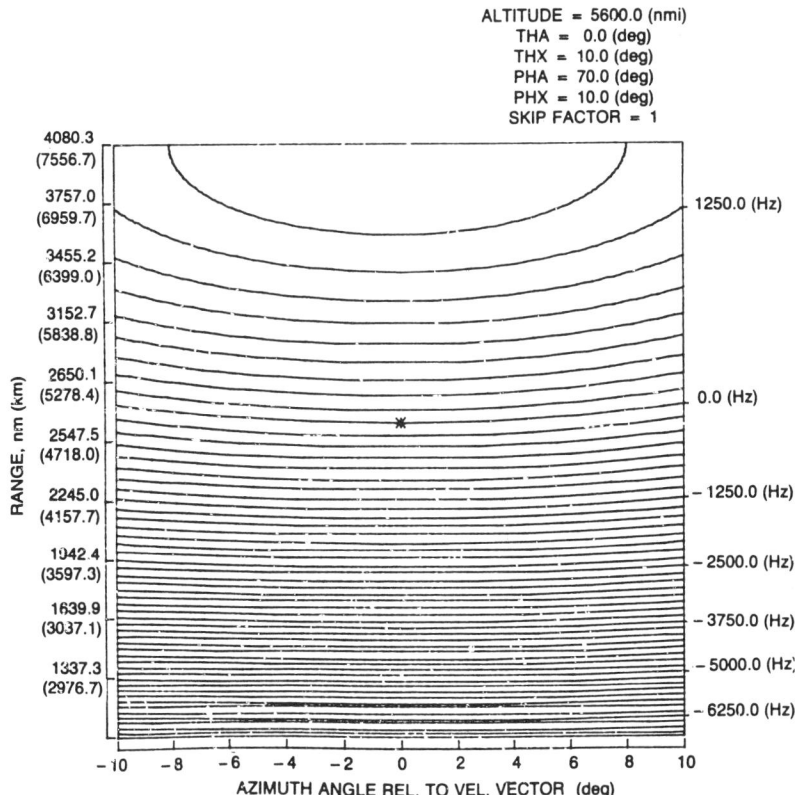

Figure 11.23 Isodops from Figure 11.20 in the 10° azimuth, 10° elevation window of Figure 11.16 pointing at 0° relative to the ground track velocity vector, and a grazing angle at the center of 26°. The mean doppler has been removed.

combining of ambiguous range cells leads to a clutter doppler spread as shown in Figure 11.24 for a PRF of 10 kHz and the antenna pattern of Figure 11.15.

The 3 dB bandwidth of the clutter spectrum can be estimated by using (11.32), which leads to

$$\Delta f_{3dB} = [(\Delta f_{AZ})^2 + (\Delta f_{EL})^2]^{1/2} \tag{11.35}$$

where the bandwidth spread in azimuth is

$$\Delta f_{AZ} = (f_d|_{AZ=(\pi/2)-\theta_{AZ}} - f_d|_{AZ=(\pi/2)+\theta_{AZ}}) \sin AZ$$

and the bandwidth in range if range ambiguities are present is

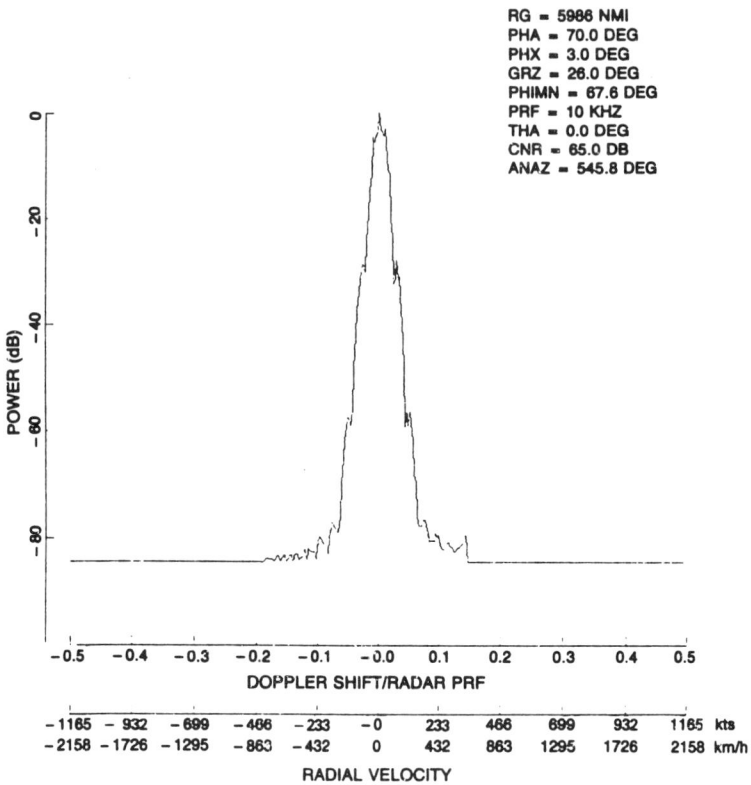

Figure 11.24 Clutter spectrum resulting from the dopplers of Figure 11.23 and the antenna pattern of Figure 11.15.

$$\Delta f_{EL} = (f_d|_{\substack{AZ=0 \\ EL \to El - \theta_{EL}}} - f_d|_{\substack{AZ=0 \\ EL \to EL + \theta_{EL}}}) \cos AZ$$

where $\to$ indicates that EL is replaced by EL + θ_{EL}.

For a step-scanned SBR, the antenna pointing angle is held constant during the coherent processing time for clutter rejection. In this case, the total clutter spectrum is found by convolving the platform motion spectrum described above with the internal motion spectrum described in Section 11.2.3. For a continuously scanned antenna, the combined clutter spectral broading due to scanning modulation and platform motion is *not* the convolution of these two effects, unless the antenna pattern is Gaussian shaped [29]. Very precise computer modeling of the effects of platform motion, scanning, and a cosine processing-window function with zero-pedestal height by O'Sullivan and Pawula [29] show that the spectral skirts do not fall off as fast as the antenna beam. In fact, the authors stated that a scanning modulation spectrum and a platform motion spectrum derived from a *Gaussian* approximation to the antenna beam convolved with each other agreed "extremely well" with the modeled results in that the spectral skirts fall off much slower than when the actual antenna pattern is used. In general, precise modeling appears to be necessary to predict clutter spectra confidently under a wide range of conditions when performance predictions of very high level clutter-rejection SBRs are needed.

11.4 CLUTTER-REJECTION TECHNIQUES

Clutter-rejection techniques applicable to an air or surface search SBR include *reducing the clutter cell size* by using high range resolution and SAR techniques (see Chapter 4), and *doppler processing* to detect moving targets and to reject "stationary" clutter. Doppler processing (or doppler-velocity filtering) generally consists of nonrecursive processing (finite impulse response filters, without feedback) and recursive processing. The nonrecursive techniques generally require more hardware, but they have become popular for SBR applications due to their superior transient response and the relatively low-cost digital implementations available. A finite transient response is particularly desirable for step-scanned antenna concepts. Optimization approaches have been developed [30] for recursive *moving target indicator* (MTI) filters, which are in widespread use in airborne pulsed doppler radar designs.

11.4.1 Radar Doppler Processors

The optimization criterion most often employed is to maximize the output signal-to-clutter ratio (S/C), but this does not always lead to the desired velocity response, especially for low velocity targets [31]. Of course, if the filters are designed to give a desired velocity response, the clutter residue generally increases so that a trade-off is needed. Both factors are usually important in SBR designs.

A *finite impulse response* (FIR) filter can be modeled as a transversal filter (Figure 11.25). If the returns from N pulses are to be processed during the antenna dwell time at a footprint, a received data vector with elements $x(nT)$, where $n = 1, 2, \ldots, N$ and T is 1/PRF, can be defined along with a weight vector with elements $a(nT)$. The elements of the input data vector are made up of a desired signal-target $s(nT)$, plus clutter $c(nT)$, plus noise $n(nT)$. Maximizing the output S/C corresponds to maximizing the improvement factor [32]:

$$I_{s/n} = \frac{a_T M_S a^*}{a_T M_N a^*} \tag{11.36}$$

where M_S and M_N represent the covariance matrices of the signal and the clutter-plus-noise, respectively. The subscript T represents the transpose of the vector or matrix and the asterisk indicates the complex conjugate. By selecting the proper representation of M_S, many FIR filters can be analyzed. M_N can be derived from the inverse Fourier transform of clutter spectrum discussed in Section 11.2.2, the platform motion spectrum described in Section 11.3.3, the spectrum of the system instabilities, and the internal noise spectrum (usually assumed "white").

To derive the optimum weight vector, all of these sources of interference must be considered together (i.e., the spectra of all sources must be jointly convolved to obtain the total spectrum). However, when the weights are known, the improvement factor against each type of interference can be computed separately by using (11.36) with the appropriate interference spectrum. The individual improvement factors are combined by [11]:

$$\frac{1}{I_{s/n}} = \frac{1}{I_{im}} + \frac{1}{I_{pm}} + \frac{1}{I_{stab}} + \frac{1}{I_N} \tag{11.37}$$

where I_{im}, I_{pm}, I_{stab}, and I_N respectively represent the improvement factors of clutter internal motion, platform motion, system instabilities, and noise.

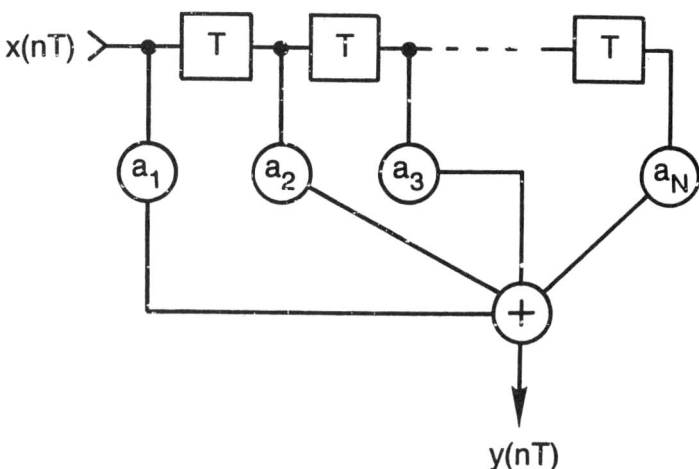

Figure 11.25 Generic transversal filter for interpulse period, $T = 1/\text{PRF}$; $a_i =$ the weights.

Emerson Weights

The solution of the optimum weight vector for (11.36) involves finding the matrix W that "prewhitens" the interference (i.e., $W_T M_N W^* = I$, the identity matrix), and finding the eigenvector that produces the largest eigenvalue of $W_T M_S W^*$ [32]. This process was used by Emerson [33] to develop the optimum weights of an MTI that maximize the output S/C for a target with uniformly distributed velocity probability over the detection range between doppler ambiguities (Figure 11.26). In this case, $M_S = I$ and the optimum weight vector is given by the eigenvector of the clutter covariance matrix, M_c, that results in the smallest eigenvalue. The Emerson weights for a three-pulse canceller give about a 2 dB larger improvement factor for a target with a uniformly distributed velocity pdf, but give much poorer performance for low velocity targets [32] than the usual binary weights that result from simply cascading cancellers [37]. In general, when $M_S \neq I$, the optimum weights are given by the eigenvector associated with the largest eigenvalue of $M_N^{-1} M_S$ [32].

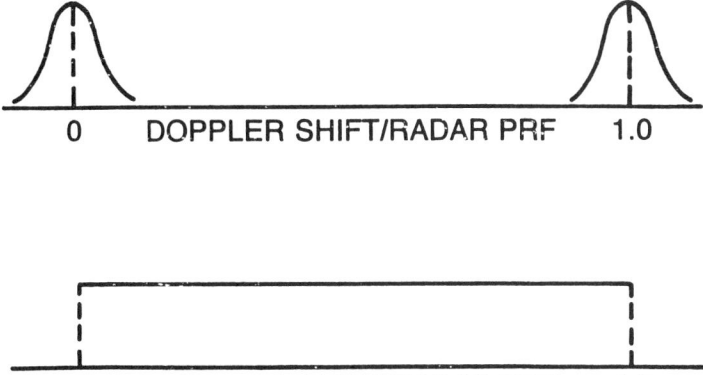

Figure 11.26 Basis for the design of a single filter to reject the clutter and detect a target with doppler shift that may be anywhere within the MTI detection region.

The Least-Mean-Squared (LMS) Algorithm

For a known signal vector, the computation of the optimum weight vector simplifies to [34, 35]:

$$a_{\text{opt}} = M_N^{-1} S^* \qquad (11.38)$$

Brennan and Reed [34] applied this algorithm to the design of a bank of doppler filters to cover the detection region. In Figure 11.27(a), eight equally spaced points in the doppler region are shown, corresponding to the eight filters of an eight-pulse processor. Each filter is optimized at one particular doppler frequency by using the above algorithm. The signal vector for a target at each doppler is known. The ith component of the signal vector at the nth point is [35]:

$$s_i = \exp\left[j\frac{2\pi(n-1)}{N}\left(i - \frac{N+1}{2}\right)\right] \tag{11.39}$$

This is a suboptimal design over the entire doppler region in that the filters are optimized only at specific points. An optimal design of an eight-pulse processor is illustrated by Figure 11.27(b), where each filter is optimized for a target with a doppler shift that is assumed to have equal probability of occurring within the interval:

$$\left(\frac{2n-1}{2NT}, \frac{2n+1}{2NT}\right), \quad n = 0, 1, 2, \ldots, N-1$$

This optimal design involves solving for eigenvalues and eigenvectors as previously outlined.

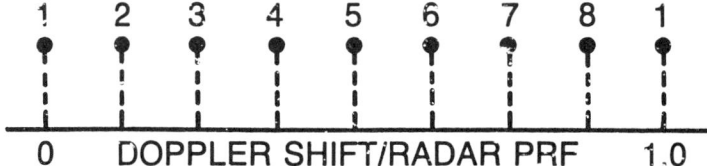

Figure 11.27 Basis for the design of a bank of filters to reject the clutter and detect a target with (a) doppler shift at one of the equally spaced points, or (b) doppler shift that may be anywhere within one of the equal intervals of the MTI detection region.

Matched Clutter Filter (MCF)

Because the algorithms for computing the eigenvalues of the above procedure converge very slowly when the clutter is highly coherent (narrow spectral width), an approximate result can be obtained by defining a signal vector for each interval [35]. The signal vector for the *n*th filter is

$$s_i = \exp\left[j\frac{2\pi(n-1)}{N}\left(i - \frac{N+1}{2}\right)\right]\frac{\sin\left[\frac{\pi}{N}\left(i - \frac{N+1}{2}\right)\right]}{\frac{\pi}{N}\left(i - \frac{N+1}{2}\right)} \quad (11.40)$$

where $i = 1, 2, \ldots, N$ and $n = 1, 2, \ldots, N$.

For example, using this signal vector for $N = 64$ and the clutter covariance matrix, M_c, derived by taking the Fourier transform of the platform motion spectrum of Figure 11.24, a bank of 64 matched clutter filters is formed. Filter number 33, with its detection region at the interval centered on the PRF/2, is shown in Figure 11.28. This filter has a maximum response in its detection region and a minimum response in the region of the clutter. This filter is actually the *N*-point FIR filter that gives the least mean squares (LMS) error approximation of the rectangular bandpass filter about this interval.

A comparison of Figures 11.24 and 11.28 shows that the shape of the filter response in the region of the clutter is the reciprocal of the clutter spectrum. Figure 11.29 shows the output of the filter with the input clutter of Figure 11.24. The clutter spectrum has been essentially "whitened," except near the signal passband. This demonstrates that the filter is "matched" to this clutter spectrum and shows a clutter *cancellation ratio* (CR) of about 49 dB. This CR combined with the ideal signal gain of SG $= 20 \log N = 30$ dB leads to an improvement factor of about 79 dB averaged over all dopplers within the filter's detection region. Another figure of merit for coherent filters of this type is its performance against noise. This filter exhibits S/N loss of only about 0.1 dB relative to the ideal gain ($10 \log N$).

With the weight vectors for the 64 filters known, the improvement factor for targets with all dopplers can be computed by using

$$I_{s/n} = \frac{|\mathbf{a}_T \mathbf{x}|^2}{\mathbf{a}_T M_c \mathbf{a}^*} \quad (11.41)$$

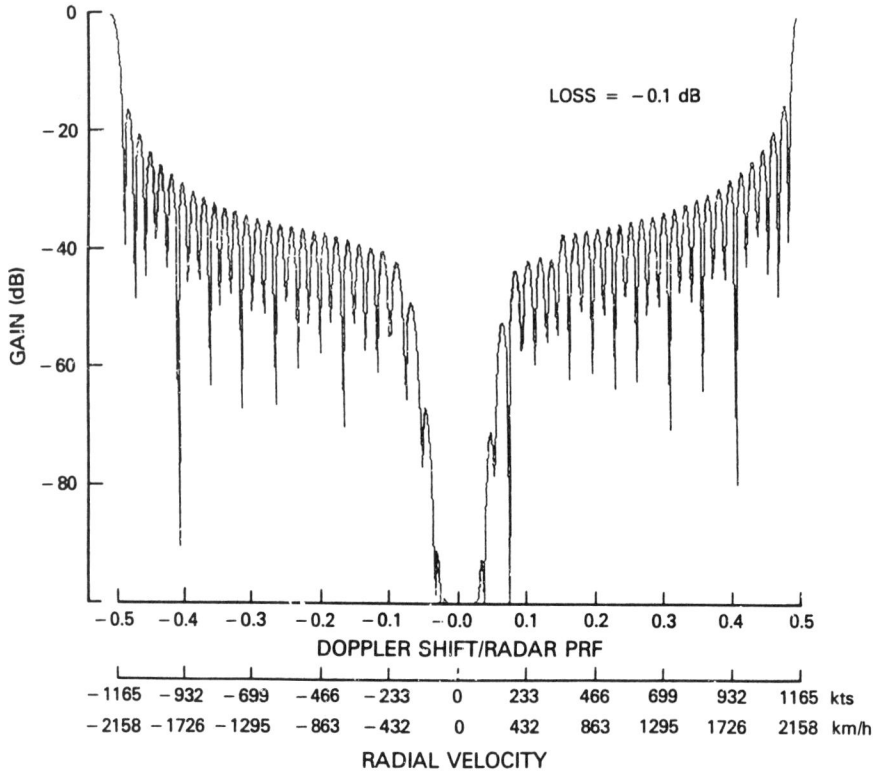

Figure 11.28 The center filter of a 64-pulse doppler processor matched to reject the clutter spectrum of Figure 11.24.

where the ith component of the signal vector $x_i = \exp[j2\pi f_d(i - 1)T]$. Figure 11.30 shows the improvement factor of the filter with the maximum response as the target doppler is varied from 0 to PRF/2. If the input S/C and the required output S/C for a desired P_d and P_{fa} are known, the *minimum detectable velocity* (MDV) is that at which the required $I_{s/n}$ is achieved, as illustrated in the figure. A comparison of MCF, FFT, and MTI cascaded with both an MCF and fast Fourier transform (FFT) is made in [36], illustrating the low losses and better detection of low velocity targets of the MCF.

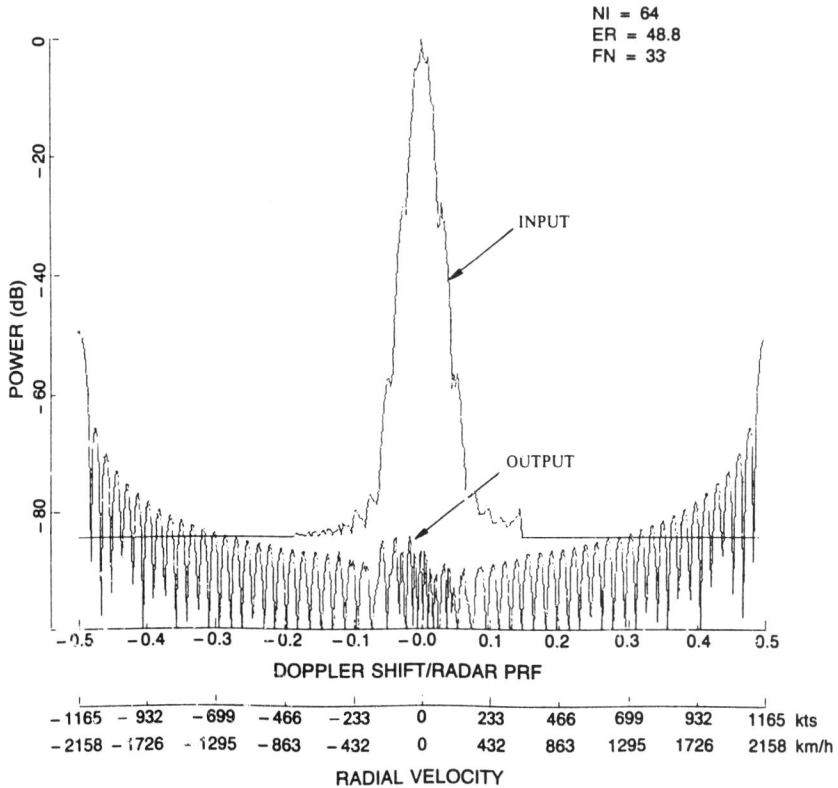

Figure 11.29 Clutter residue (output) of the filter of Figure 11.28 with the input spectrum from Figure 11.24.

Weighted FFT

The fast Fourier transform is a widely used radar doppler processor, either cascaded with MTI cancellers or weighted and used alone [37]. A heavily weighted FFT to control the filter sidelobes is a simple and appropriate concept for clutter rejection for SBR applications. The major disadvantages are the losses and broadening of the passband caused by the window function (weighting) used to reduce the filter sidelobes. The previous analysis can be applied to the FFT by defining an equivalent

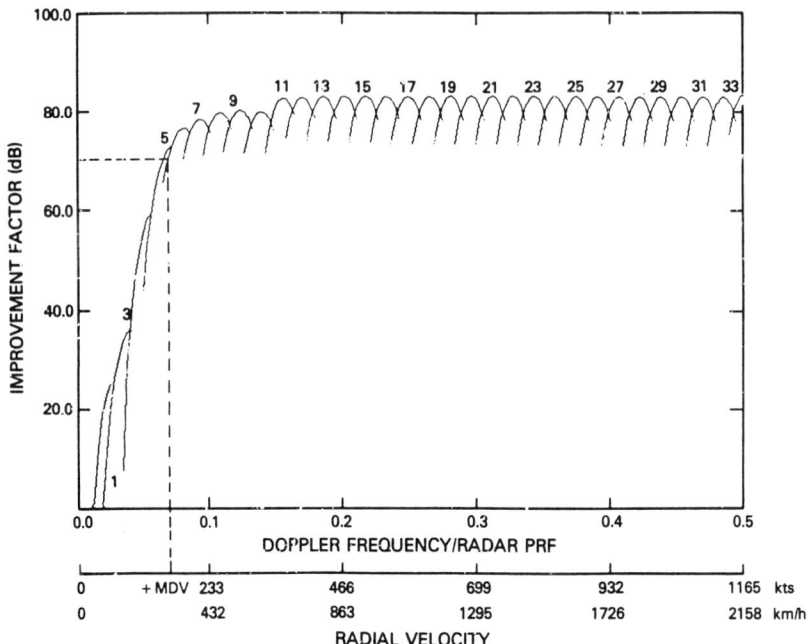

Figure 11.30 Improvement factor of the first 33 filters of a 64-pulse doppler processor matched to reject the clutter spectrum of Figure 11.24; the peaks of the filters with the maximum target response are plotted *versus* target doppler. An L-Band SBR is assumed.

weight vector. The total effective weights of the kth filter are

$$a_n = b_n \exp\left[-j2\pi\frac{(n-1)(k-1)}{N}\right], \quad n = 1, 2, \ldots, N \quad (11.42)$$

where b_n represents the weighting for sidelobe reduction. Figure 11.31(a) shows the filter centered at the PRF/2 for a 64-pulse FFT with -70 dB Chebyschev weights. Figure 11.31(b) shows the clutter residue output from this filter when the input is the platform motion spectrum shown. A clutter cancellation of 45.6 dB is shown and a "windowing" loss of 2.2 dB.

Cascaded MCF and FFT

Design analyses trading off the number of pulses processed in the MCF and the spectral characteristics of the residue indicated that, over a broad range of clutter spectral widths, the residue output from the MCF was essentially "whitened" with at most 32 or 64 pulses processed. This means that the optimal processor for further reducing this residue is simply an unweighted FFT.

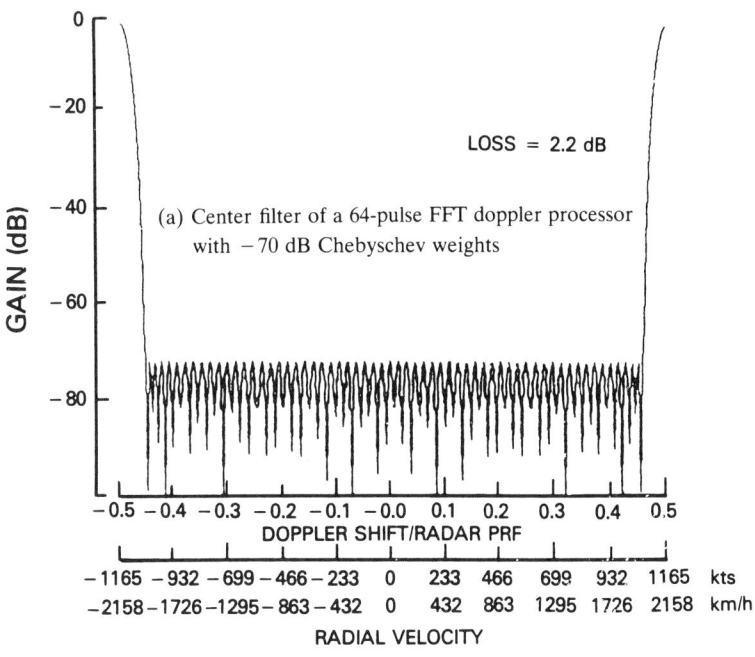
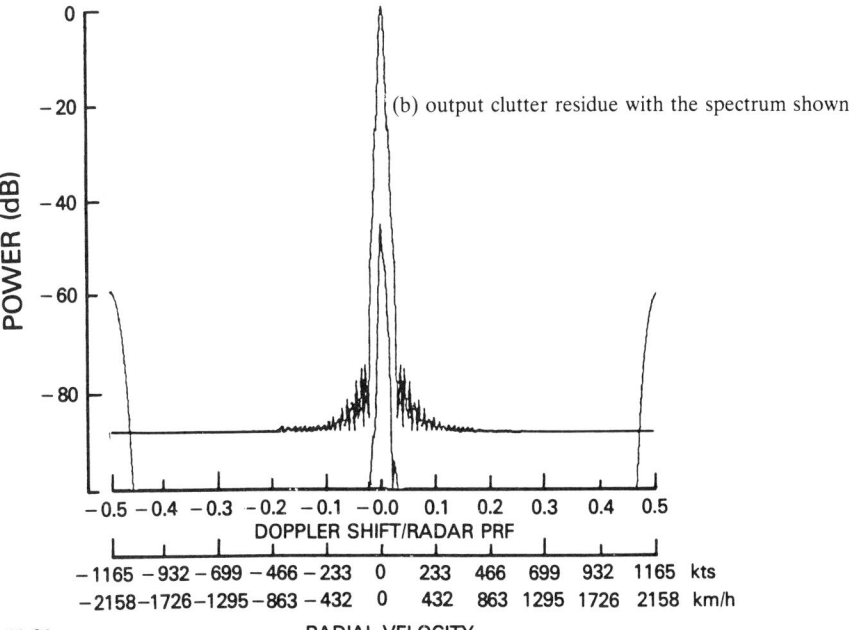

Figure 11.31

Figure 11.32 shows a cascaded MCF/FFT, where N_m pulses are processed with an MCF, a first level detector sets the desired P_d, for the N_a largest outputs of each range cell (which assumes at most N_a targets are expected in a range cell at one time; for a high resolution radar, N_a may be one), and the MCF output channels with detections are followed by an N_f-pulse FFT to reduce the clutter residue, and therefore reduce false alarms due to clutter. N_f would be made sufficiently large to reduce the clutter to the level to achieve the desired P_{fa}, where $I_{N_f} = 10 \log N_f$ is the improvement factor for the FFT. Figure 11.33 illustrates how an MCF channel output with a target detection is divided into N_f cells each having $1/N_f$ of the clutter if the MCF has prewhitened the clutter.

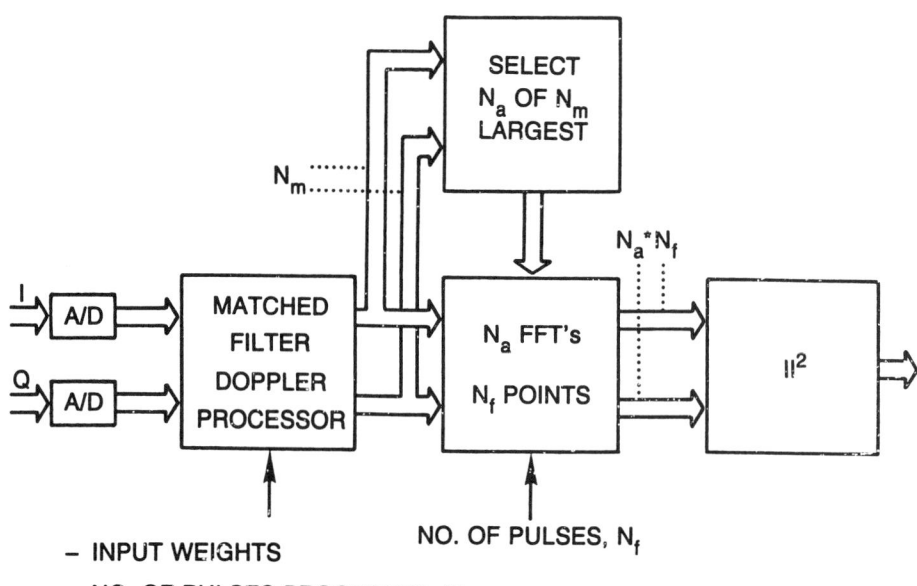

Figure 11.32 Cascaded matched clutter filter and FFT doppler processor.

Minimum Detectable Radial Velocity

An important factor in the design trade-offs in an SBR development is the ability of the radar to detect targets with low radial velocities. On one hand, the detection of low velocity targets may be undesirable if the targets of interest are airborne targets. In this case, ground vehicles would be considered false alarms, or even large stationary "discrete" clutter returns may cause false alarms due to system

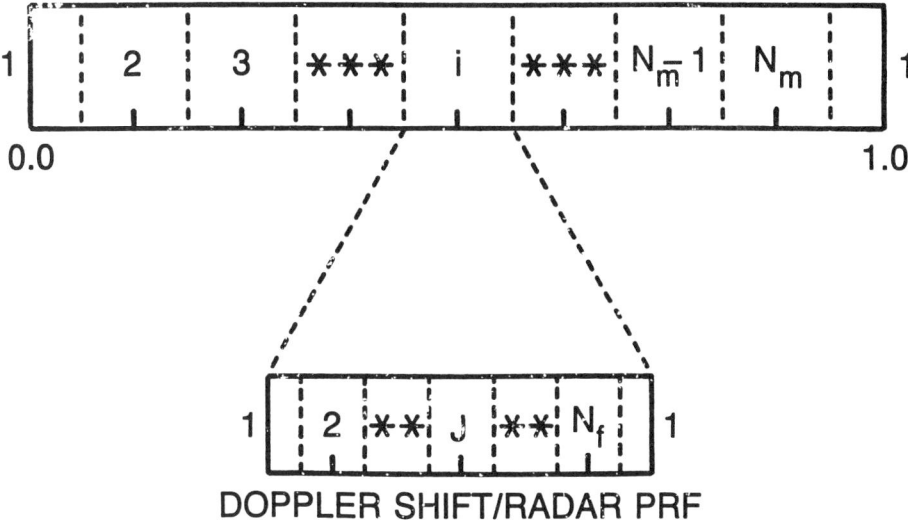

Figure 11.33 Illustration of the further division of the MTI detection region of one MCF output by an N_f-pulse FFT.

instabilities or platform motion. On the other hand, the detection of low velocity targets may be desirable because a high-speed target flying nearly tangential to the satellite position will have a low radial velocity. In addition, the detection of low-speed ground traffic may be desirable for some applications.

The performance of an SBR against low-speed targets is a function of not only the geometry, but also radar parameters, such as platform velocity (or indirectly the orbit altitude), LOS relative to the velocity, antenna aperture size, wavelength, and doppler processing techniques. An estimation of the MDV independent of the selection of doppler processors can be calculated by assuming that a target will not be detectable if its doppler is low enough to fall within the main-beam clutter. The MCF technique described above illustrates that a well designed doppler processor may be able to detect targets on the "skirts" of the main-beam clutter (Figure 11.30) so that this assumption predicts a somewhat higher MDV than will be realized.

Assuming a circular orbit and ignoring the earth's rotational velocity v_e, (11.32) can be simplified to $f_d = (2v_p/\lambda)(\sin EL \cos AZ)$. The angle to the first null on the antenna pattern can be approximated by $\theta_n = \lambda/L$ without aperture weighting to reduce sidelobes. L is the aperture length. These factors result in a somewhat lower MDV prediction than will be achieved with aperture weighting and considering v_e. Therefore, these compensating assumptions allow a good estimation of the expected MDV, calculated by

$$\text{MDV} = (\lambda/L)v_p \sin\text{EL} \tag{11.43}$$

Figure 11.34 illustrates the aperture requirements of (11.43) to achieve a desired MDV for three different wavelengths and two orbital altitudes. Weber and Haykin [38] showed that, for a nadir-pointing, phased array, rectangular antenna with the long dimension aligned with the ground-track velocity vector and aperture area held constant, the clutter spectral spread due to platform motion is minimized if the long dimension is maximized. This result is valid for all phase-scanned angles and limited only by other considerations affecting the length/width aspect ratio.

From (11.43), the MDV is proportional to sinEL with all other parameters held constant. From (11.21), sinEL is proportional to $\cos\psi$. We should not conclude that low velocity targets are more detectable at high grazing angles than at low grazing angles near the horizon. For a target in level flight, the radial component of its velocity vector in the direction of the SBR also is proportional to $\cos\psi$. Because the effects of these two factors cancel, the detectability of low velocity targets is essentially independent of grazing angle until the target returns and the doppler processing is overwhelmed by the increase in clutter level at high grazing angles.

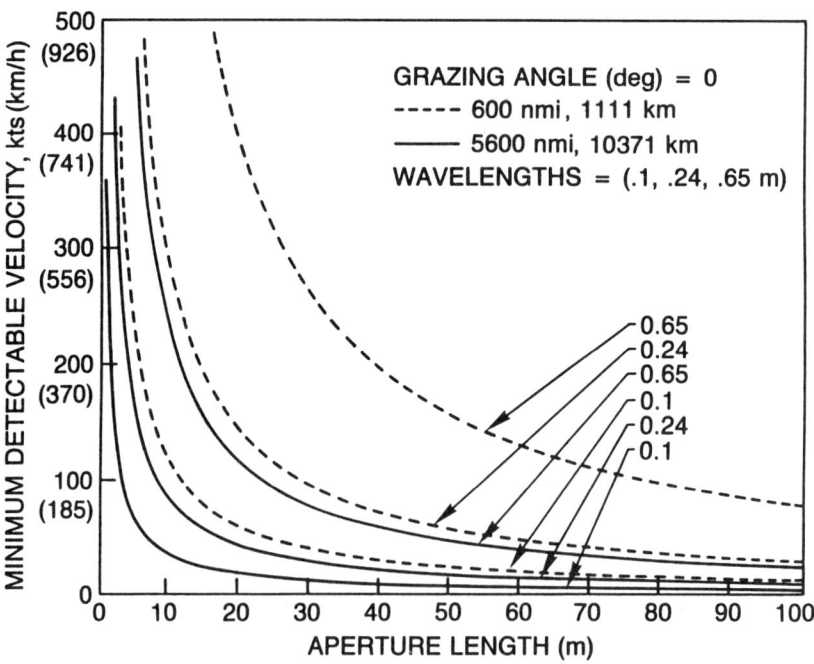

Figure 11.34 Minimum detectable velocity *versus* antenna aperture length at orbital altitudes of 600 nmi (1111 km) and 5600 nmi (10,371 km); UHF, L-band, and S-band are shown.

11.4.2 Motion Compensation Techniques

Clutter Locking

For clutter cancellation processors such as MTI and *displaced phase center antenna* (DPCA), the notch of the canceller response must be centered at the average value of the clutter spectrum. This is usually accomplished with *clutter-locking techniques* [11], such as *time-averaged clutter-coherent airborne radar* (TACCAR) and vector cancellers. The principal feature of these techniques is their ability to center the clutter-rejection notch automatically at the average doppler frequency of the clutter. Thus, these techniques are capable of rejecting clutter that has its own average velocity, such as weather, chaff, and sea clutter, as well as all clutter with an average doppler caused by platform motion.

The detailed design of these techniques takes many forms for particular applications, but Figure 11.35 illustrates the basic concept, which consists of phase-locking the radar receiver to the clutter returns at a selected range. To achieve this, the received signals are converted to an intermediate frequency and phase detected with a reference IF signal generated with a coherent local oscillator and the transmitted signal. A gate is used to select the output of the phase detector at the selected range. This output is integrated over a number of interpulse periods determined by the time constant of the phase-locked loop (PLL). The integrator is gated on at the selected range and its output is used to control the frequency of a *voltage-controlled oscillator* (VCO). The VCO and the LO are mixed to form the reference for converting the received signals to IF. This ensures that the received signals, after being converted to IF, and the transmitter are phase locked and that the clutter spectrum is centered at this IF, regardless of the average doppler shift of the clutter.

The time constant of the PLL must be larger than the reciprocal of the PRF to maintain enough stability so as not to add significant spreading of the clutter spectrum [39]. Therefore, these techniques cannot simultaneously cancel two types of clutter with different average dopplers, unless they occur at separate ranges and separate PLLs are used to provide a reference shifted at the different ranges to the proper loop [39].

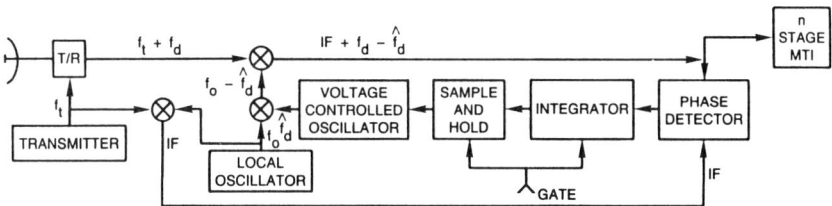

Figure 11.35 Simplified TACCAR clutter-locking system.

Instabilities or noise in the PLL affects clutter cancellation by spreading the clutter spectrum and can be evaluated along with other factors using (11.37). If there is an offset error, f_e, in the PLL estimation of the average clutter doppler, the improvement factor of an MTI of n cancellers is

$$I_n = \frac{\dfrac{2^n}{n!}\left(\dfrac{f_r}{2\pi\sigma_c}\right)^{2n}[1\cdot 3\cdot 4\ldots (2n-1)]}{\left(\dfrac{f_e}{\sigma_c}\right)^{2n} + \sum_{k=1}^{n}\binom{2n}{2k}\left(\dfrac{f_e}{\sigma_c}\right)^{2n-2k}[1\cdot 3\cdot 5\ldots (2k-1)]} \quad (11.44)$$

where f_r is the PRF and $\sigma_c^2 = \sigma_{im}^2 + \sigma_{pm}^2 + \sigma_{stab}^2$ [3] when the effects of internal motion, platform motion, and system stabilities are combined. Figure 11.36 illustrates the improvement factor for a single-canceller MTI versus σ_c/f_r for several values of f_e.

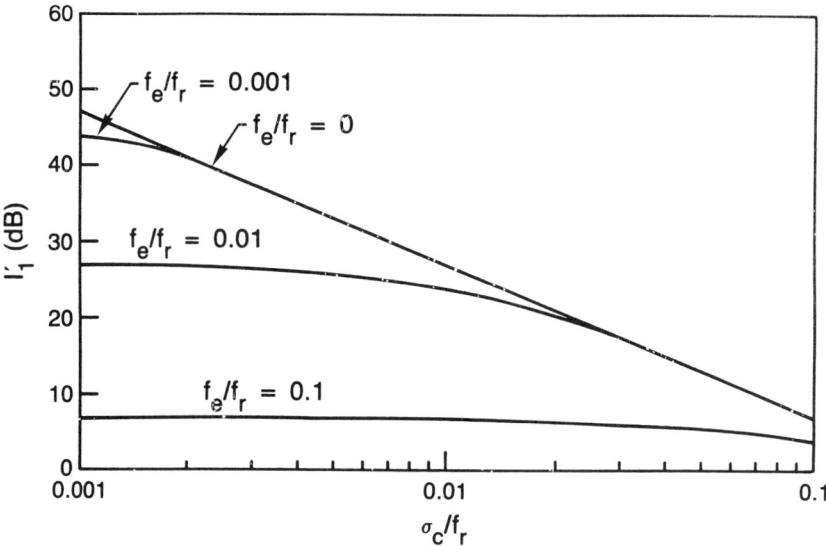

Figure 11.36 MTI improvement factor (single delay MTI).

Quadratic Phase Modulation Waveform (QPMW)

Figure 11.23 shows the isodops in a 10° window centered at 0° azimuth angle and 26° grazing angle (illustrated by the asterisk). As shown, the doppler is almost constant at a given range (horizontal cut on the figure), but variations of the doppler with range results in the clutter spectrum of Figure 11.24 when the antenna pattern of Figure 11.15 and a PRF of 10 kHz are used. The spectrum results from the range ambiguities for this PRF. The QPMW is designed to compensate for this effect [9, 40].

Figure 11.37 illustrates the variation in the spread of elevation angles between range ambiguities, R_a, near the satellite and near the horizon. The large variations in angle correspond to large variations in doppler. Curve 1 to Figure 11.38 shows this variation in doppler with range at 0° azimuth and 26° grazing angle. Curve 2 shows the result after the QPMW compensation. The location of the boundaries of the 3 dB beamwidth is shown for reference. As illustrated, the variation of the doppler over this beamwidth has been essentially eliminated, and the original spectrum of Figure 11.24 is transformed to that of Figure 11.39.

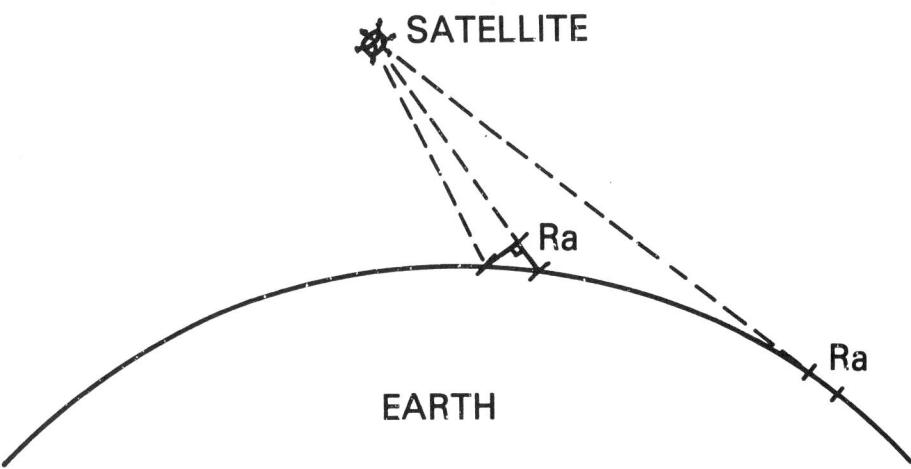

Figure 11.37 Equal slant range increments near the satellite and near the "edge" of the earth; R_a represents one slant-range ambiguity.

This is accomplished by transmitting a waveform with a linear frequency (quadratic phase) modulation from pulse to pulse as shown in Figure 11.40 and described by

$$f_T(t) = \alpha_T + f_{T_{\min}} \quad \text{and} \quad \alpha_T = \frac{f_{T_{\max}} - f_{T_{\min}}}{T_B} \tag{11.45}$$

Where $f_{T_{\max}}$ and $f_{T_{\min}}$ are the dopplers at the maximum and minimum ranges of the selected correction window (usually the 3 dB beamwidth). T_B is the burst length of the transmitted waveform. If the toe of the 3 dB beamwidth is above the horizon,

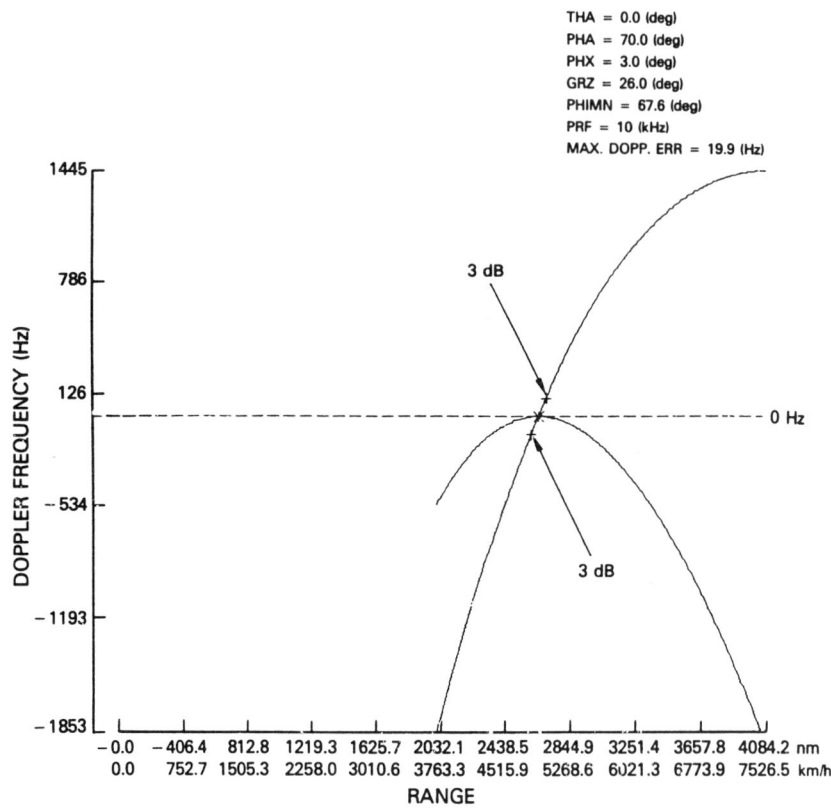

Figure 11.38 Doppler shift *versus* range for L-band, 5600 nmi (10,371 km) altitude, 0° azimuth angle, and 26° grazing angle, with and without the *quadrature phase modulation waveform* (QPMW).

this technique is probably disabled because the curvature of the dopplers with range (see Figure 11.38) nullifies the compensation. The spread in dopplers near the horizon is much less and the technique is unnecessary.

QPMW is most effective at higher grazing angles where the spread of dopplers in range is the greatest and also the most nearly linear. Figure 11.24 and 11.39 illustrate the effectiveness of the technique at a 26° grazing angle. In fact, the compensated spectrum of Figure 11.39 is so narrow that the weights for an MCF processor for this spectrum is difficult to compute because the resulting covariance matrix is nearly singular and difficult to invert. However, this very narrow spectrum could be very effectively rejected by a simple MTI or DPCA (as described in the following subsection) canceller and followed by an MCF or FFT processor.

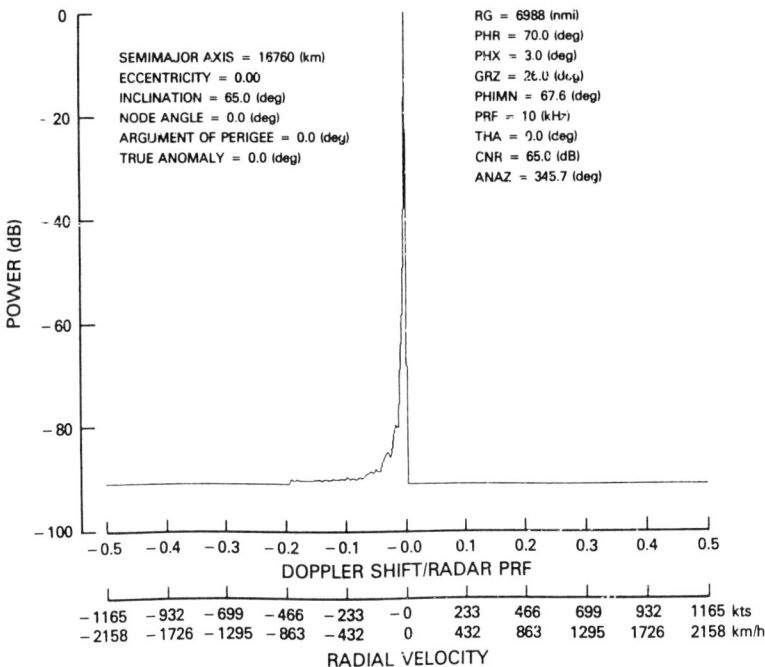

Figure 11.39 The corresponding spectrum of Figure 11.24 when the QPMW is employed.

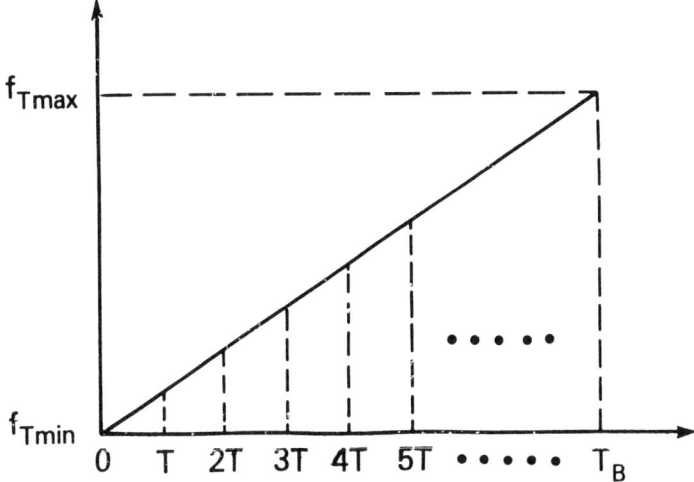

Figure 11.40 Transmitting frequency modulation function to compensate for the range-ambiguity platform motion spectrum.

The range ambiguity doppler spread decreases as approximately cosAZ as the antenna is scanned away from the ground track velocity vector, and the magnitude of the QPMW correction ought to decrease accordingly.

Displaced Phase Center Antenna (DPCA)

When the antenna of an SBR is scanned to AZ = 90°, Figure 11.21 shows that the dominant doppler spread occurs within a range cell. The DPCA technique compensates for this doppler spread due to platform motion. The doppler spread increases as approximately sinAZ, and the need of DPCA is most pronounced at AZ = 90°.

After a clutter-locking mode has been used to remove the average clutter doppler frequency, the remaining doppler shift due to platform motion for a scatterer at an angle θ with respect to the peak of the antenna beam is [41]:

$$f'_d = 2\frac{v_p}{\lambda} \sin EL \left[\cos AZ (\cos\theta - 1) - \sin AZ \sin\theta \right] \quad (11.46)$$

By assuming a high gain antenna pattern so that the range of angles for the main beam is small, the small angle approximation for (11.46) is

$$f_d \approx 2\frac{v_y}{\lambda}\theta$$

where $v_y = v_p \sin EL \sin AZ$.

This relation shows that there remains a doppler shift approximately proportional to the angle θ of the scatterer with respect to the peak of the main beam. Therefore, a platform motion spectrum results, which is weighted by the two-way antenna pattern. By assuming a Gaussian antenna pattern and homogeneous clutter [42], the limitation to the MTI improvement factor of an n-canceller MTI caused by platform motion is

$$(I_n)_{pm} = \frac{2^n}{n!}\left(\frac{1}{1.2\pi}\frac{L}{v_y T}\right)^{2n} \quad (11.47)$$

where L is the length of the aperture in the direction of the ground track velocity vector and $T = 1/f_r$ is the pulse repetition interval (PRI). The expression $x = v_y T/L$ represents the fraction of the antenna aperture by which the antenna is displaced during an interpulse period. When the platform motion and the internal motion effects are combined by using (11.37), the MTI improvement factor for a single canceller shown in Figure 11.41 results.

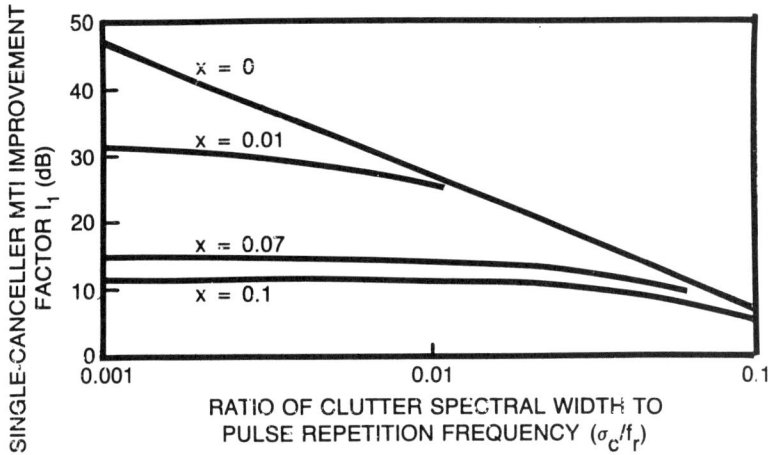

Figure 11.41 Effect of platform motion of the single-canceller ($n = 1$) MTI improvement factor I_1; x is the fraction of the antenna aperture that the antenna is displaced per interpulse period. The $x = 0$ curve represents no platform motion.

DPCA compensates for the platform velocity parallel to the plane of the aperture by physically or electronically displacing the phase center of the antenna in the opposite direction. To see how this is accomplished, consider the effect of the doppler shift on the interpulse phase advance of returns. Integrating (11.46) and making the small angle assumptions for a high gain antenna leads to $\eta = 4\pi(v_y T/\lambda)\theta$, which is illustrated in Figure 11.42. If the antenna pattern is $G(\theta)$, the correction vectors, e_1 and e_2, can be realized by an additional pattern defined by

$$\Delta(\theta) = jG(\theta) \tan\frac{\eta}{2} \tag{11.48}$$

and used as shown in Figure 11.43. The transfer function of an n-stage MTI implemented as shown in Figure 11.43 is

$$|H_\eta(f)|^2 = \left(1 + \tan^2\frac{\eta}{2}\right)\left[2\sin\left(\pi\frac{f - f_d'}{f_r}\right)\right]^2\left[2\sin\left(\pi\frac{f}{f_r}\right)\right]^{2n-2} \tag{11.49}$$

The three components on the right-hand side of this equation are (1) a clutter amplification factor caused by the quadrature correction signals, e_1 and e_2, (2) the transfer function of the DPCA canceller, and (3) the transfer function of the remaining $n - 1$ cancellers. Notice that the transfer function for the DPCA canceller has had the doppler shift caused by platform motion subtracted. This means

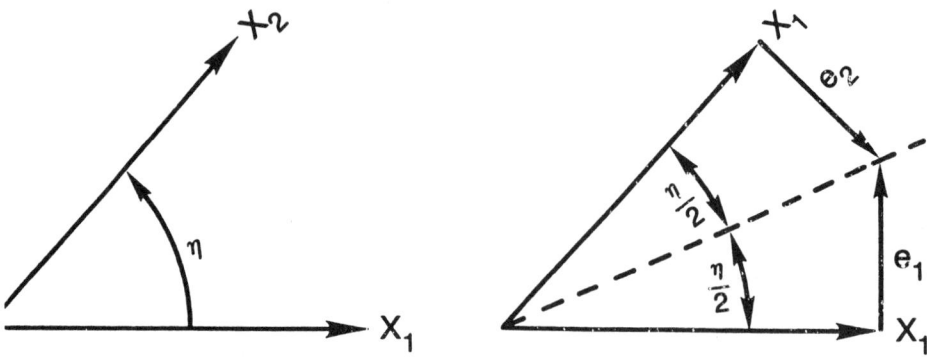

Figure 11.42 Vectorial representation of (a) pulse-to-pulse advance η and (b) platform motion compensation.

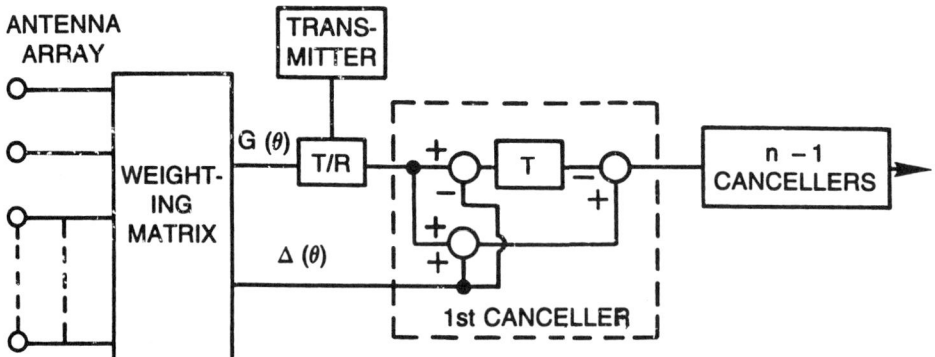

Figure 11.43 *Displaced phase center antenna* (DPCA) correction applied to the first canceller of an n-stage MTI.

that the clutter spectrum processed by the DPCA contains internal motion, but not platform motion. The clutter spectrum processed by the remaining cancellers, however, contains the platform motion as well as internal motion. This is a unique feature of DPCA which is true for any cascaded processor such as an MCF or FFT that follows the DPCA. The clutter spectrum processed by any doppler processor following a DPCA canceller will contain the platform motion. The result of this effect on the improvement factor of a double canceller, $n = 2$, is shown in Figure 11.44. If both cancellers had DPCA corrections, the dashed curves would all coincide with the $x = 0$ curve.

To realize an optimal DPCA implemented as shown in Figure 11.43, the clutter amplification factor of (11.49) must be eliminated. This amplification is caused by adding a quadrature correction instead of simply rotating the vectors of Figure 11.42 by $\eta/2$. Also, the correction patterns defined by (11.48) are not realizable. This can be explained by considering a phased array in which the weights

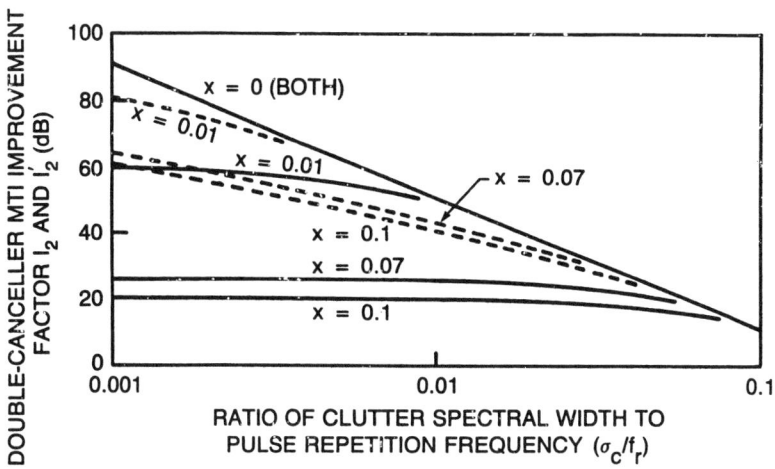

Figure 11.44 Comparison of MTI improvement factor for a double-canceller ($n = 2$) with DPCA I'_2, dashed curves) and without DPCA (I_2, solid curves); x is the fraction of the antenna aperture that the antenna is displaced per interpulse period. The $x = 0$ curve represents no platform motion.

on all the elements describe the antenna pattern, $G(\theta)$, Therefore, if the same elements are used to generate $\Delta(\theta)$, all of the degrees of freedom are needed to generate the $G(\theta)$ term in (11.48) and there are no more degrees of freedom to be used for the remaining factors.

A method using the LMS algorithm described earlier has been devised [43] to design an optimal correction pattern to address both of the above issues. The resulting weight vectors for the two optimal correction patterns are

$$B'_{\text{opt}} = (M_x^{-1} M'_{xx} - I)U \quad \text{and} \quad B''_{\text{opt}} = (M_x^{-1} M''_{xx} - I)U \tag{11.50}$$

The (n_1, n_2) element of the covariance matrix, M_x, is given by

$$m_x(n_1, n_2) = \int_{-\pi/2}^{\pi/2} |G_t(\theta)|^2 |E(\theta)|^2 e^{j(n_1 - n_2)\phi} d\theta$$

and the (n_1, n_2) elements of the cross-covariance matrices are given by

$$m'_{xx}(n_1, n_2) = \int_{-\pi/2}^{\pi/2} |G_t(\theta)|^2 |E(\theta)|^2 e^{j\omega'_d T/2} e^{j(n_1 - n_2)\phi} d\theta$$

and

$$m''_{xx}(n_1, n_2) = \int_{-\pi/2}^{\pi/2} |G_t(\theta)|^2 |E(\theta)|^2 e^{-j\omega'_d T/2} e^{j(n_1 - n_2)\phi} d\theta$$

where $G_t(\theta)$ is the transmitting antenna pattern, $E(\theta)$ is the element pattern of the array elements, $\omega'_d = 2\pi f'_d$, which is given by (11.46) or the high gain approximation, and $\phi = (2\pi d/\lambda) \sin\theta$; d is the distance between elements of the array, I is the identity matrix, and U is the weight vector of the receive pattern. When this procedure is applied to the antenna pattern of Figure 11.45(a), the correction pattern of Figure 11.45(b) results for AZ = 90°, $v_p \sin EL/\lambda = 0.7$ (PRF). The clutter cancellation for $\sigma_{im} = 0.01$ (PRF) shown in Figure 11.46 illustrates that the main-lobe clutter residue has the same spectrum as the input. Therefore, follow-on processors must process a clutter spectrum that includes platform motion effects. The cancellation ratio of 24 dB is limited by the value of σ_{im} chosen. When the procedure is followed with $\sigma_{im} = 0$, the main-lobe clutter residue is equal to the sidelobe clutter residue level. The sidelobe clutter residue is dominated by the correction-pattern sidelobe level, which results from the LMS algorithm, providing the solution that minimizes the total output clutter. Thus, the available degrees of freedom are used to match the ideal correction pattern over its main lobe, and the sidelobes that result are shown.

The correction patterns are multiplied by sin AZ because the component of platform motion for which DPCA compensates decreases as the antenna is scanned away from AZ = 90°. The DPCA-corrected canceller can be cascaded with additional doppler processors, such as other MTI cancellers, FFT, or MCF [44].

Another implementation of an N-pulse DPCA has been analyzed extensively by MIT Lincoln Laboratory for SBR. This implementation perfectly compensates for platform motion over the time interval of the N pulses if the antenna aperture physically faces perpendicular to the ground-track velocity vector. This implementation, illustrated in Figure 11.47, is ideally suited for a nadirpointing, rectangular-aperture, phased array SBR. The long dimension of the aperture must be aligned with the ground-track velocity vector.

The entire aperture is used for transmitting. On receive, the aperture is divided into as many subarrays as pulses to be compensated. A three-pulse, or double-canceller, DPCA is shown. The satellite moves a distance $v_p T$ between transmitted pulses. The use of subarrays on receive causes the phase center, identified by the plus sign in the figure, to be displaced in the opposite direction by $2v_p T$ between pulses. The "virtual" phase center of the two-way pattern remains fixed in space for the three pulses shown. The distance between the location of the aperture on transmit and receive corresponds to the round-trip propagation time to the earth and back. Except for the fact that the bistatic angle between the transmitting and receiving positions for each pulse changes, this is seen to be a perfect compensation.

The clutter processing for this concept consists of delaying received signals from the forward subarray by $2T$, delaying received signals from the middle subarray by T, and then multiplying the three signals with a selected weight vector (such as binomial or Emerson) and adding them.

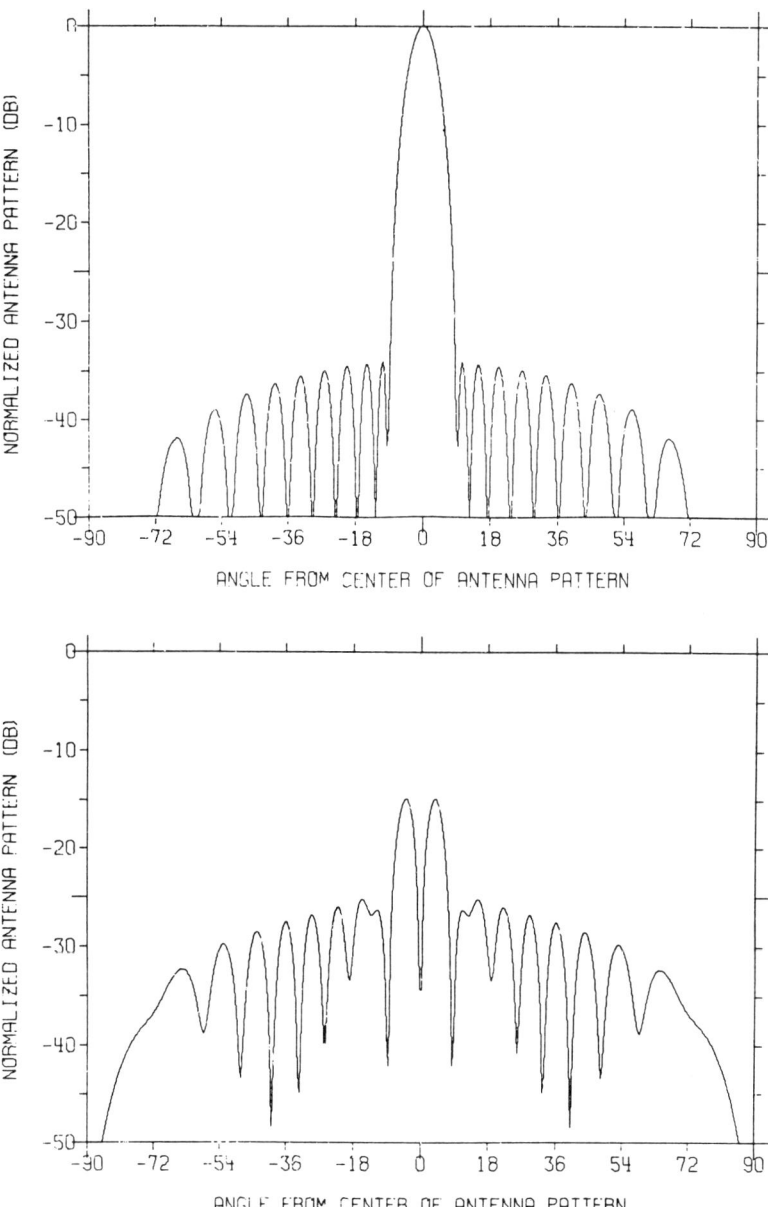

Figure 11.45 (a) Mean-squared antenna pattern for 12-element array with dipole elements at 0.9 wavelength spacing and 34 dB Chebyschev weights; (b) mean-squared amplitude of optimum platform motion correction pattern for antenna pattern of (a).

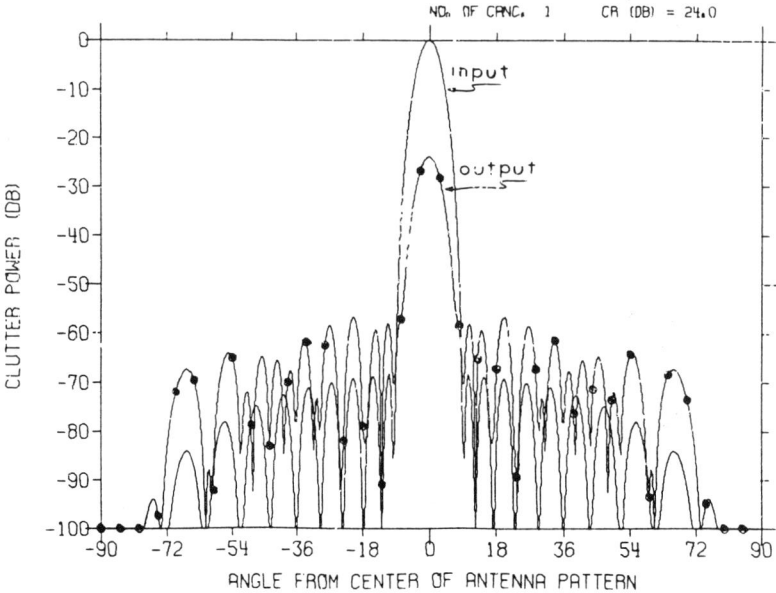

Figure 11.46 Clutter power at output and input of MTI canceller with platform motion compensation using correction pattern of Figure 11.45.

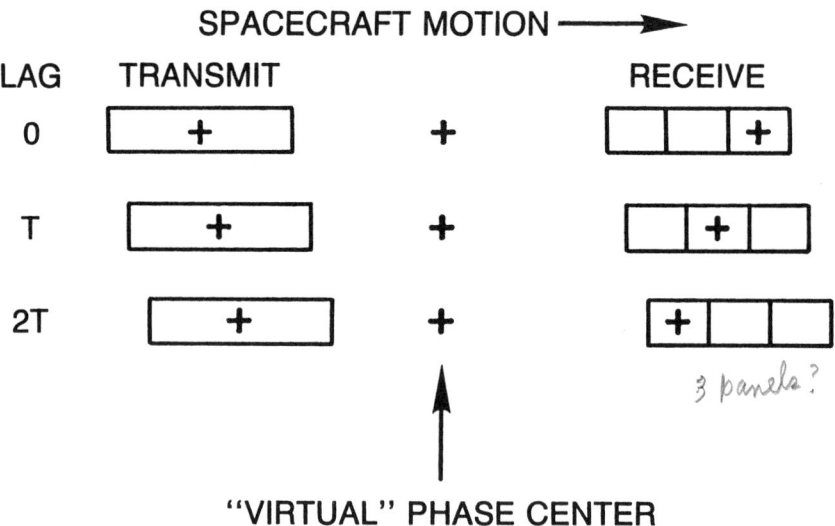

Figure 11.47 Illustration of a three-pulse DPCA method of maintaining the two-way "virtual" phase center fixed in space for those three pulses.

Error Analyses

R.W. Miller of MIT Lincoln Laboratory has analyzed the effect of errors on this latter DPCA implementation. These analytical equations can also be applied to the former implementation by recognizing that the two phase centers of the former consist of (1) the receiving pattern plus the first correction pattern and (2) the receiving pattern plus the second correction pattern. If the errors are equal but uncorrelated between phase centers, the improvement factor limit due to these errors is given by

$$I_r = \frac{1}{1 - \rho} \tag{11.51}$$

for a target with doppler frequency that has equal probability of any value within the PRF ambiguities; ρ is the clutter cross-correlation between phase centers. Miller derived engineering equations relating ρ to the following error mechanisms: phase-center offset errors, antenna deformation, receiver channel mismatch, T/R modules' amplitude and phase errors, analog-to-digital (A/D) conversion, and internal clutter motion.

(a) *Phase-center offset errors* result from (1) imperfect alignment of the axis of the antenna that contains the long dimension of the rectangular array (the longitudinal axis) with the ground-track velocity vector, and (2) an error in the DPCA delay time. If the longitudinal axis makes an angle ψ_e with the ground-track velocity vector and the delay error is τ_e, the cross-correlation between phase centers is

$$|\rho| = \exp\left\{-8\left[\frac{\pi\sigma_u}{\lambda}\left(\delta_x - \delta_z \frac{u_0}{w_0}\right)\right]^2\right\} \exp\left\{-8\left[\frac{\pi\sigma_v}{\lambda}\left(\delta_y - \delta_z \frac{v_0}{w_0}\right)\right]^2\right\} \tag{11.52}$$

where σ_u and σ_v are the respective two-way longitudinal and transverse beamwidths, $\delta_x = v_p \tau_e / 2$, $\delta_y = (L_{pc}/2) \cos\phi_e \sin\theta_e$, $\delta_z = (L_{pc}/2) \sin\phi_e$, L_{pc} is the distance between phase centers, θ_e and ϕ_e are the yaw errors and pitch errors respectively and related to ψ_e by $\cos\psi_e = \cos\theta_e \cos\phi_e$, $u_0 = \cos AZ$, $v_0 = \sin EL$, and $w_0 = \cos EL$. The receive antenna aperture is a fraction of the total aperture for DPCA, and so the two-way longitudinal beamwidth is given by $\sigma_u = (K\lambda/\sqrt{2})/\sqrt{1/L^2 + 1/L_p^2}$ and the two-way transversal beamwidth is given by $\sigma_v = K\lambda/2W$, where L is the length of the array, L_p is the length of the receive subarray, W is the width of the array, and K is 0.55 for no weighting and 0.78 for Hamming weights.

Miller pointed out that when the DPCA canceller is preceded by an FFT for doppler beam sharpening, the patterns need only be matched over the resulting narrower beamwidths, resulting in an easing of tolerance requirements. (This has been called *arrested synthetic aperture radar,* ASAR.) In this case, the effective transversal beamwidth becomes $\sigma_u = K\lambda/\sqrt{2}\, 2v_p T_I$. K is determined by the weighting function used for the FFT and T_I is the coherent processing interval (the number of pulses, N_I, integrated by the FFT times 1/PRF).

(b) *Antenna deformation errors* (or warping) analyzes errors including bowing and twisting. For these errors, the cross-correlation of the receive channels are

$$|\rho| = \exp\left[-\frac{\Delta\theta^2}{2\sigma_R^2}\left(\frac{\sigma_T^2}{\sigma_R^2 + \sigma_T^2}\right)\right] \tag{11.53}$$

where $\Delta\theta = (4d/L)(L_{pc}/L)$ for longitudinal bowing; d is the displaced distance of one end of the array relative to the center of the array. The transmitting longitudinal beamwidth is $\sigma_T = K\lambda/\sqrt{2}\, L_p$ and the receiving longitudinal beamwidth is $\sigma_R = K\lambda/\sqrt{2}\, L$. Transversal bowing has no first-order effects on beam decorrelation. Miller observed that if ASAR is used, this effect is extremely small, and both longitudinal and transversal bowing can be ignored. Twist is the simplest kind of deformation that affects ASAR processing. For this case, $\Delta\theta = (2d/W)(L_{pc}/L)$, where d is the deviation from flatness measured from the array center to a corner and W is the transverse dimension of the array. If ASAR is used, only the transverse deformation will affect beam decorrelation. Again, the transmitting transversal beamwidth is $\sigma_T = K\lambda/\sqrt{2}\, W$ and the receiving transversal beamwidth is $\sigma_R = K\lambda/\sqrt{2}\, W$.

(c) *Receiver mismatch errors* result from the receiver channels having different transfer functions. The cross-correlation of these channels is

$$|\rho| = \frac{1}{\sqrt{1 + \alpha^2 + \beta^2}} \tag{11.54}$$

where α^2 and β^2 are the mean-squared amplitude and phase errors, respectively.

(d) *T/R module amplitude and phase errors* were evaluated empirically by Miller, resulting in a model for these errors as given by

$$1 - |\rho| = \frac{0.31}{N}(\sigma_\alpha^2 + \sigma_\beta^2) \tag{11.55}$$

where N is the number of elements in each phase center subarray, the standard deviation of the phase errors $\sigma_\beta = \pi/180$ times the standard deviation in degrees, and the standard deviation of the amplitude errors $\sigma_\alpha = (\ln 10)/20$ times the standard deviation in decibels.

To make these random errors independent of the A/D quantization errors, the A/D converters must be designed to change states at different points. This is also necessary to eliminate *quantization lobes* in a low-sidelobe antenna.

(e) *Analog-to-digital conversion errors* result from the number of quantization levels between the most negative and the most positive signal levels. For clutter having an in-phase (I) and quadrature (Q) Gaussian amplitude pdf with standard deviation σ, a constant k is selected so that the span of the A/D quantization levels is $\pm k\sigma$. The channel decorrelation for an n bit A/D converter in both I and Q channels as each receiver is

$$1 - |\rho| = \frac{k^2}{12} 2^{-2(n-1)} \tag{11.56}$$

(f) *Internal clutter motion* affects a DPCA canceller in the same way as it does an MTI canceller, however these effects can be expressed in terms of DPCA errors and evaluated using Equation (11.51). The corresponding decorrelation is

$$\rho = \exp\left[-2\left(\frac{\pi \sigma_v L_{pc}}{v_p \lambda}\right)^2\right] \tag{11.57}$$

where σ_v is plotted in Figure 11.11 for several types of clutter.

Sidelobe Clutter

Miller has derived a unique expression for the sidelobe clutter radar cross section (RCS), which can be put into the radar equation to compute C/N or can be used in (11.1) to compute $(S/C)_{SL}$. The sidelobe clutter RCS is

$$\bar{\sigma}_{SL} = \frac{G_{SL}}{G_T G_R B T_I}\left(\frac{R_S}{h}\right)^4 \sigma_E \tag{11.58}$$

where G_{SL} is the two-way antenna sidelobe gain, G_T is the transmitting gain, G_R is the receiving gain, B is the pulse bandwidth, T_I is the coherent processing time, R_S is the radar slant range, and σ_E is the normalized RCS of the whole earth. In the above relation, all gains are relative to isotropic. The BT_I product appears because the sidelobe clutter is spread over many range-doppler ambiguity intervals, each of which contains BT_I resolution cells. The other factors in the relation simply allow for the differences in the relative range and antenna gains for the main lobe and sidelobes. Using the σ° models shown in Figure 11.48 for three types of clutter, σ_E is illustrated in Figure 11.49.

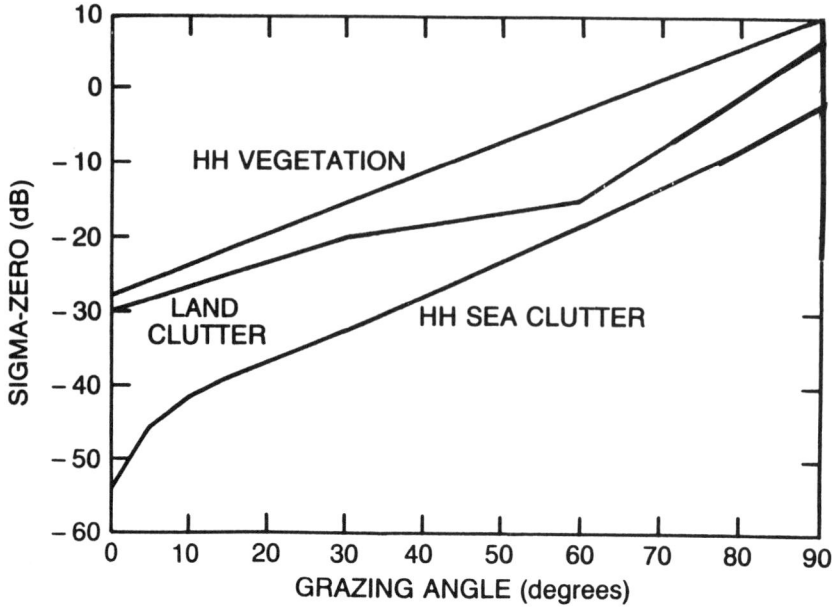

Figure 11.48 Typical values of $\sigma°$, the radar backscatter coefficient as a function of grazing angle.

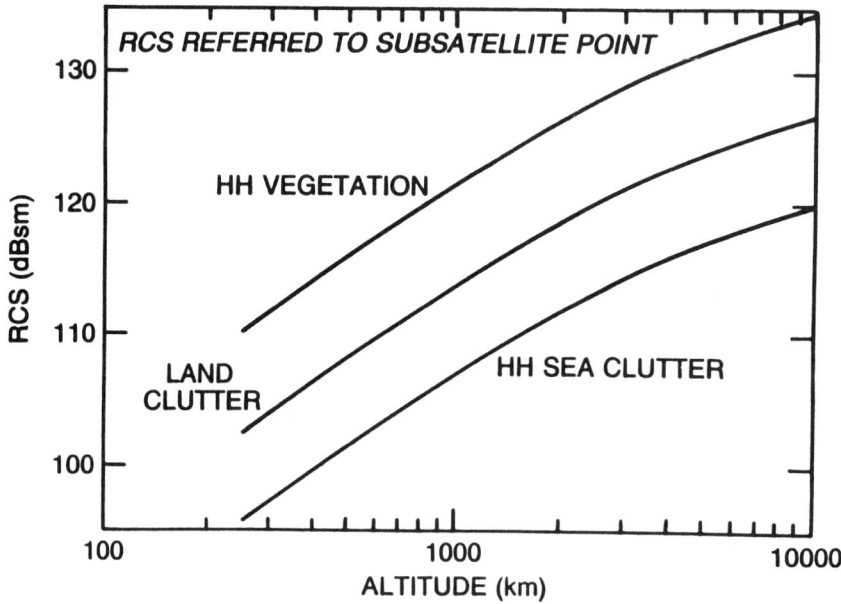

Figure 11.49 The integrated RCS of sidelobe clutter based on the clutter models of Figure 11.48. The integration is over the whole area of the earth in view of the SBR for each orbit altitude.

11.4.3 Doppler Beam-Sharpening

The clutter-rejection techniques considered thus far have been concerned with determining the signal-to-clutter ratio at the radar receiver and, using the doppler characteristics of the clutter, designing filters to reject the clutter and pass the targets. A different, but analogous, approach is to consider processing techniques to simply reduce the clutter competing with the target. Referring to (11.1) and (11.2), we see that the S/C can be increased by decreasing the effective area of the clutter, A_c (i.e., the resolution cell coinciding with the target).

A_c can be reduced by increasing the resolution of the radar (11.3). The range resolution, $c\tau/2$, can be reduced by decreasing the pulsewidth (the compressed pulsewidth of a pulse compression waveform), or, equivalently, by increasing the waveform bandwidth, $1/B$. The azimuth, or cross-range, resolution can be reduced by decreasing the azimuth beamwidth, which implies increasing the radar frequency or the length, L, of the antenna. The frequency or antenna size are usually determined by many factors, and to increase either or both to obtain the needed level of clutter rejection is impractical.

An alternative is to use a combination of a high resolution waveform for range resolution and an (SAR) technique (i.e., doppler beam-sharpening, DBS) for high cross-range resolution [3]. The difference between SAR and DBS as employed here is that the usual constraints on range and doppler ambiguities for SAR are removed for DBS as it is concerned with target detection rather than imaging (Chapter 4). The effects of the waveform and signal processing on the clutter level are the primary consideration here.

A synthetic aperture is formed by pointing the antenna at some angle α with respect to the platform ground-track velocity vector v_p and coherently processing N_I pulses over an interval T_I. The effective synthetic aperture length is $L_{\text{eff}} = v_p T_I \sin\alpha$ [45]. The angle α is related to AZ and EL of Figure 11.3 by $\cos\alpha = \cos\text{AZ} \sin\text{EL}$ and $\sin\alpha = \sqrt{1 - \cos^2\text{AZ} \sin^2\text{EL}}$. The beamwidth of this doppler-sharpened beam is $\theta_{\text{DBS}} = \lambda/2L_{\text{eff}}$ and the cross-range resolution at a slant range R_S is

$$\Delta_{\text{CR}} = \frac{R_S \lambda}{2v_p T_I \sin\alpha} = \frac{R_S \lambda}{2L_{\text{eff}}} \tag{11.59}$$

which leads an effective clutter area for N_a range ambiguities:

$$A_c = \left(\frac{c\tau}{2}\right) \Delta_{\text{CR}} N_a \tag{11.60}$$

and the S/C in this resolution cell is

$$\frac{S}{C} = \frac{\sigma_T}{\sigma° A_c} = \frac{2\sigma_T v_p T_I \sin\alpha}{R_S \lambda \left(\frac{c\tau}{2}\right) N_a \sigma°} \qquad (11.61)$$

where $\sigma°$ is the clutter backscatter coefficient discussed in Section 11.2.

There are limitations on the coherent integration time T_I, which limit the achievable cross-range resolution. Because the satellite ground track is essentially a straight line during T_I, the range to a scatterer at an angle α with this velocity vector changes during T_I. When the range changes by $\lambda/8$, the returns are considered decorrelated and the cross-range resolution is limited by this factor [46]. If there is no compensation, the SAR is said to be unfocused. The cross-range resolution of a conventional real-aperture radar is $\Delta_{\text{conv}} = \lambda R/L$, and for an unfocused SAR it is $\Delta_{\text{unf}} = \sqrt{R\lambda}/2$. If the SAR is focused, a correction for the range change is made, and the limitation to T_I is the time that the antenna footprint takes to sweep across the scatterer due to satellite motion. For a focused SAR, $\Delta_{\text{FOC}} = L/2$, which is independent of range. This surprising result is because the size of the footprint increases with range, which, in turn, allows for a longer T_I, resulting in a longer effective synthetic aperture, exactly compensating for the beam spread as the range increases. Obtaining an even smaller Δ_{FOC} is possible by scanning the antenna beam to keep the footprint on a spot for a longer time [47]. This is called *spotlighting,* and it is applicable when only a specific region is to be searched.

An SAR or DBS is a simple, low-power radar ideally suited for space, but the very long integration time and very high range resolution combine to require extensive signal processing. There are three additional limitations. First, the search rate or area coverage rate is low for a low-power, small-aperture radar. The search rate is limited by power-aperture product just like a real-beam radar. Second, because an SAR cannot look ahead, there is a "blind" swath along the ground track as the satellite passes over. Third, the location of a scatterer in azimuth is determined by the doppler of the scatterer in an imaging SAR. Of course, if the DBS radar is to detect moving targets, the target's doppler causes it to be "imaged" at the wrong azimuth. With all of these limitations, a DBS SBR is applicable to the detection of stationary ground targets, slow-moving ground targets or ships, and even high-speed air targets having sufficient power-aperture product combined with DPCA in an ASAR mode as described in Section 11.4.2.

11.5 SPACE-BASED INTERFERENCE CONSIDERATIONS

An SBR has the advantage over ground and air-based radars of having a long LOS. Hence, its target detection range is less LOS-limited. However, the SBR's long LOS has the disadvantage that earth-based interference sources are more likely to be within the field of view of the SBR. These interference sources may be man-made or natural, intentional or unintentional, and in the air, sea-based, or land-based. Nonetheless, there is a potential for many electromagnetic earth-based sources to interfere with the detection of desired targets.

The SBR visibility obviously increases as a function of the satellite altitude (in this chapter, all altitudes are measured from the earth's surface). To illustrate this, consider Figures 11.50 and 11.6 [48]. In Figure 11.50, we have

h = the SBR's altitude;
R_e = earth's radius (3440 nmi or 6371 km);
R_{MAX} = radius of field-of-view swath (also called the *swath width radius*, SWR).

The SWR *versus* the satellite altitude is plotted in Figure 11.6. The SWR for an airborne radar at 35,000 ft (10.7 km) is 185 nmi (343 km). From Figure 11.6 observe that even a low-altitude SBR's SWR is many times this value. Hence, the likelihood of being interfered (or jammed) increases accordingly.

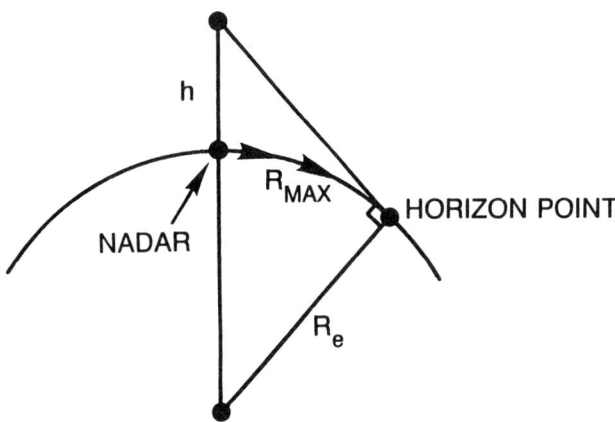

Figure 11.50 Swath width radius.

Consider the power of these interferers or jammers at the output of the SBR's receiver. A common system parameter used to measure this power is called the *signal-to-interference ratio* (SIR), given by the expression:

$$\text{SIR} = \frac{S}{N + J} = \frac{S}{N}\left(\frac{1}{1 + J/N}\right) \qquad (11.62)$$

where

S = signal power of the desired signal;
N = internal noise power (due to thermal noise);
J = jamming power of a given earth-based interference source;
S/N = signal-to-noise power ratio;
J/N = jamming-to-noise power ratio.

All of these quantities are measured at the output of the radar receiver's main channel before any signal processing occurs. The SIR is a function of the S/N and J/N. For jamming cancellation schemes, a cancellation ratio equal to or greater than the J/N is often cited as a basic system design goal. Related to this is the desired jammer cancellation ratio (CR_{dB}) which is given as

$$CR_{dB} = 10 \log_{10}(J/N) \text{ for } J/N \geq 1 \tag{11.63}$$

We must be careful when using this measure of canceller system effectiveness because it may be interpreted to imply that by merely raising the internal noise level, N, J/N decreases and hence the required cancellation ratio is decreased. In reality, the SIR will be a more meaningful canceller measure because it includes a factor, S/N, which decreases as the internal noise is increased. However, the usual figure of merit for a radar design is to detect the smallest possible target at the greatest possible range. This drives the radar engineer to design the most sensitive (lowest noise figure) receiver possible, which makes the radar even more susceptible to interference. This is especially true for a space-based radar that must detect targets at extremely long ranges. Thus, a useful figure of merit for a jammer cancellation technique is to reduce the jammer below the system noise level so that the detection range of the radar is not degraded by the jammer.

The jamming power can be found using the formula [8]:

$$\frac{J}{N} = \frac{\text{ERPD}_J}{(R_J/\lambda_0)^2} \cdot G_J \cdot L \cdot \frac{1}{4\pi k T_0 F_n} \tag{11.64}$$

where

ERPD_J = effective radiated power density of the jammer (W/MHz);
R_J = LOS distance from SBR to jammer (m);
kT_0 = 4×10^{-15} W/MHz (k is Boltzmann's constant);
T_0 = 290 K, standard reference temperature;
F_n = receiver noise factor [4].

Plots of $(J/N)_{dB}$ versus R_J/λ_0 and $ERPD_J$ are shown in Figure 11.51 for a system without losses, the gain in the direction of the jammer is 0 dB (isotropic), and the system temperature is 290 K, i.e., F_n is assumed to be unity. Note that if there are losses, the gain is nonisotropic, or the system temperature is not 290 K, the values given in this figure can be adjusted according. For example, a -10 dBi sidelobe gain results in the $(J/N)_{dB}$ being reduced by -10 dB or a 2 dB receiver noise figure results in the $(J/N)_{dB}$ being reduced by -2 dB. Observe from this figure that increasing R_J/λ_0 can significantly reduce the cancellation ratio requirement. This can be done by placing the SBR at a higher altitude, or merely increasing the transmitted frequency. However, we point out that the range to the target also increases and its detectability decreases so that this is not an acceptable solution.

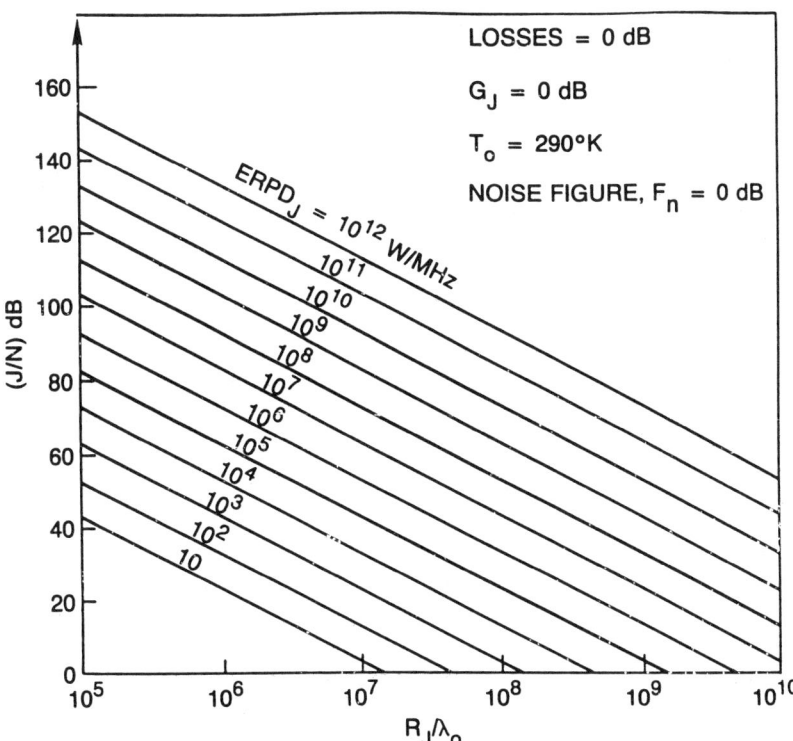

Figure 11.51 The received jammer-to-noise ratio (J/N) versus the ratio of the range to the jammer (R_J) and the wavelength (λ_0) of the radar (or jammer) frequency.

Besides considerable visibility to earth-based interferers, another consideration inherent to SBR jammer rejection is the fact that the SBR is moving with respect to the jammer, or, equivalently, the jammer is moving with respect to the SBR. A common technique for rejecting a jammer is to place a null in the main antenna's receiving radiation pattern in the direction of a jamming source. Any null on the receiving antenna pattern that is generated in the direction of the jamming source is not stable with respect to a moving jammer. For example, at $t = 0$, let a null be placed in the received antenna pattern at exactly the same angle as a jammer. A few seconds later, the jammer has moved with respect to the SBR due to its motion and the earth's rotation. Hence, there is a change of the jammer angle with respect to the SBR, the jammer is no longer at the null angle, and jammer power at the SBR's receiver output will increase. This increase (or degradation) can be measured or calculated as a function of time. A parameter called the *null stability time* is defined as the time that a desired cancellation ratio is maintained. Long null stability times are desirable so that the SBR's signal processor is not overly taxed by high update rates and the clutter rejection is not degraded by fluctuation in the antenna pattern during the coherent processing interval.

The null stability time, T_{null}, is a function of many parameters, including the antenna aperture's length and width, carrier frequency, jammer's angle off mechanical boresight, the jammer's bandwidth, satellite altitude, and nulling implementation. However, we can relate the null stability of an SBR system at one satellite altitude, h_1, with the same SBR system at another altitude, h_0, if the J/N are identical for both. This is because, for the case described, the null stability time is proportional to the sidereal period, T_{per} [48], of the satellite at a given altitude. Thus,

$$\frac{T_{\text{null}}(h_1)}{T_{\text{null}}(h_0)} = \frac{T_{\text{per}}(h_1)}{T_{\text{per}}(h_0)} = \text{null stability factor } (N_s) \tag{11.65}$$

The null stability factor as defined above is plotted in Figure 11.52 for $h_0 = 400$ nmi (741 km).

We briefly outline the procedure for finding $T_{\text{null}}(h_0)$:

1. Place a jammer on the earth at an angle, θ_J, off mechanical boresight, of an SBR located at altitude, h_0.
2. Compute the optimal SLC weights, $\mathbf{w}_{\text{opt}}$, needed to cancel the jammer (see Section 11.7 for this procedure).
3. Compute the cancellation ratio, $\text{CR}(\theta_J, \mathbf{w}_{\text{opt}})$ associated with these weights.
4. Compute $\text{CR}(\theta, \mathbf{w}_{\text{opt}})$ for $\theta \neq \theta_J$, but the same cancellation weights are used. Note $\theta = \theta_J + \Delta\theta$.

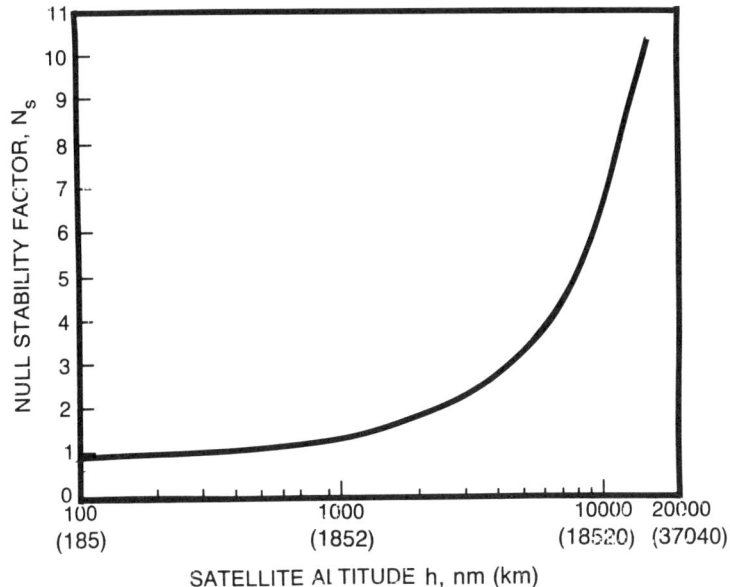

Figure 11.52 Null stability factor *versus* satellite altitude.

5. Let θ_L and θ_R be the angles on the left and right sides of θ_J, respectively, where the cancellation ratio drops 3 dB from $CR(\theta_J, \mathbf{w}_{opt})$.
6. Compute $|\theta_L - \theta_R|$ and the time, T, for the jammer's aspect angle to change by this amount; $|\cdot|$ denotes absolute value.
7. Find the maximum value of $|\theta_L - \theta_R|$ and the corresponding T over all θ_J.
8. Set $T_{null}(h_0)$ equal to this maximum time.
9. Use Equation (11.65) to compute $T_{null}(h_1)$.

11.6 INTERFERENCE/REJECTION TECHNIQUES

In this section, we describe a variety of intentional interference techniques or *electronic countermeasures* (ECM) and rejection techniques referred to as *electronic counter-countermeasures* (ECCM), which could be considered in the space-based radar environment. There are a substantial number of publications on these topics in the open literature [49–69], so we present only a brief introduction to this area and a description of the more important techniques for the SBR environment.

First, the DoD definitions of ECM and ECCM [69] are given as follows.

Electronic countermeasures (ECM) is that division of electronic warfare involving actions taken to prevent or reduce an enemy's effective use of the electromagnetic spectrum.

Electronic counter-countermeasures (ECCM) is that division of electronic warfare involving actions taken to insure friendly effective use of the electromagnetic spectrum despite the enemy's use of electronic warfare.

The effects of employing ECM and ECCM are listed in Table 11.7 [49]. For each ECM action, we attempt to negate it with an ECCM action, and *vice versa*. This "game" can be played back and forth until resource and cost limits become a factor. Basic ECM and ECCM techniques are listed in Table 11.8. Note that ECM techniques fall into the two basic categories: *denial* and *deception*. Denial techniques, such as barrage noise, overwhelm the SBR receiver whereby noise masks the desired targets. Deception techniques, such as repeaters, inject false targets into the SBR receiver-detector-tracker chain.

Table 11.7
Effects of ECM and ECCM [49]

ECM Effects
Denial of detection
Operator confusion-deception
Delay in detection-tracking information
Tracking of an invalid target
Overloading of computer (excessive number of targets)
Denial of measurement of target position, range rate
Target tracking loss
Errors in values of target position, range rate

ECCM Effects
Prevention of receiver saturation
Constant false-alarm rate (CFAR)
Enhancement of S/J ratio
Directional interference discrimination
Rejection of invalid targets
Maintenance of track

For the SBR environment, earth-based countermeasures installations could consist of wideband ESM intercept receivers, very large antennas, and very high average power sources. A major advantage of reflectors, as opposed to phased-array antennas, for this application is their lower cost and inherently wide bandwidth. Because the station should operate over a full hemisphere, a phased array system would require as many as four faces, resulting in a very high installation cost.

Table 11.8
Basic ECM and ECCM Techniques

ECM
CW
Long pulse
Spot noise
Barrage noise
Swept FM
Blinking
Impulse
Short pulse
Chaff
Radar absorbing materials (RAM)
Repeaters
False target generators

ECCM
Frequency agility/PRF agility
Pulse compression
High power
Coherent receiver
Low sidelobes
Sidelobe canceller (SLC)
Sidelobe blanker
High gain antenna
CFAR
Dicke-fix
MTI-doppler filtering
Large dynamic range
Sidelobe deception

Two parameters that strongly limit jammer antenna size are pointing accuracy and slew rates. For land-based antennas, the maximum required slew rate for earth orbiting satellites as low as 500 nmi (926 km) is 0.45°/s. Moreover, this rate is currently within the capability of the largest antennas, such as Goldstone. However, for sea-based antennas, the required slew rate is dictated by the ship's roll, pitch, and yaw rates, rather than by target movement rates. For land-based jammers, instrumentation accuracy and land surveying capabilities are sufficient to allow open-loop tracking using positional information from a space track network.

Of the ECCM techniques listed in Table 11.8, some that are specifically effective for the SBR environment are sidelobe blanking, frequency agility, sidelobe masking, and sidelobe cancellation. Figure 11.53 illustrates an ECCM response for countering each of the listed ECM techniques. The first three ECCM techniques are aimed at forcing the ECM jammer into the broadband mode, which

ECM	ECCM
REPEATER PULSE-JAMMERS	SIDELOBE BLANKING
SLOW TUNING NARROWBAND NOISE JAMMING	BURST-TO-BURST FREQ AGILITY
FAST TUNING NARROWBAND NOISE JAMMING	SIDELOBE DECEPTION RADIATION
WIDEBAND NOISE JAMMING	SIDELOBE CANCELLATION (ADAPTIVE NULLING)

Figure 11.53 ECM and ECCM responses.

would decrease the jammer's $ERPD_j$ and have a significant effect in reducing the required cancellation ratio. To illustrate this, if B_{op} is the operating bandwidth of the SBR (where the center frequency hops around this bandwidth) and B is the instantaneous SBR receiver (or information) bandwidth, the reduction of required CR is given by

$$\text{CR reduction} = 10 \log_{10} \frac{B_{op}}{B} \tag{11.66}$$

For example if B_{op} = 100 MHz and B = 1 MHz, the required CR is reduced by 20 dB. Each of these ECCM techniques is briefly described.

Sidelobe Blanking

This technique [53–56] detects repeater jammers or false target generators when their signals enter the sidelobes of the radar's main antenna. Figure 11.54 depicts the basic operating principles of this technique. An auxiliary antenna on the radar

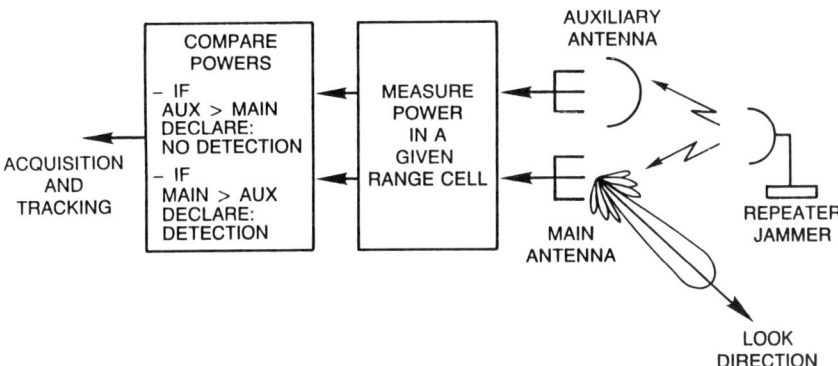

Figure 11.54 Sidelobe blanking.

is used and has more gain than the sidelobes of the radar's main antenna. This auxiliary antenna is followed by its own receiver chain and detector. Targets detected in identical range cells of the main and auxiliary receiver-detector are identified and their respective powers are compared. If the power of the target in the auxiliary channel is greater than that in the main channel (indicating that the false target entered through the sidelobes), the target is rejected as a false detection. We assume that main-lobe clutter has been cancelled in both the main and auxiliary channels.

Frequency Agility

This technique [53–56, 59] is designed to defeat slow-tuning narrowband noise jamming. As illustrated in Figure 11.55, the radar transmits a different frequency on each pulse burst. During the time needed for the jammer's ECM and support (ESM) system to establish the new radar frequency and retune its transmitter, the SBR receiver's frequency and the jammer's frequency are not the same. In fact, for the SBR environment, if the jammer is far enough away with respect to where the SBR is looking and the radar waveform is short, there may be no possibility of the jammer being on the same frequency at the same time as the SBR is receiving this frequency. However, the burst length of the random waveform necessary for clutter rejection may be so long that the jammer may always jam at the same frequency at some time during the waveform's reception.

Sidelobe Masking

Sidelobe deceptive radiation [53–56] is a technique whereby electromagnetic (EM) energy is directed at the enemy jammer. The characteristics of this EM energy,

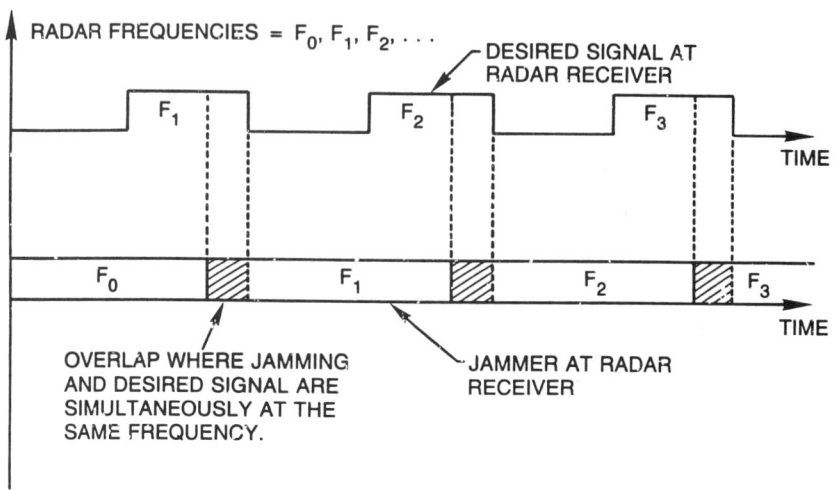

Figure 11.55 Frequency agility.

such as frequency and PRI are deceptive in nature, in that this signal cloaks or masks the characteristics of a primary signal. For example, if an enemy jammer is located in the sidelobes of the main antenna pattern (see Figure 11.1), the jammer receives a small fraction of the transmitted signal. If the jammer were to detect this energy, to locate the direction of arrival, and to determine the frequency of the signal, it could point a highly directive, high-power, narrowband jamming signal, which might have the same characteristics as the transmitted signal, at the radar and thus disrupt reception.

However, the jammer's knowledge of the actual transmitted frequency, F_1, can be degraded by also transmitting a low-power, multifrequency, deceptive signal (see Figure 11.56). The power needed for this deception is approximately equal to the expected sidelobe level of the main antenna pattern. As a result, the enemy jammer receives not only the real antenna frequency, F_1, but also energy at other frequencies. Hence, the enemy jammer is forced to jam over a band of frequencies because it is unsure of which frequency is being transmitted. Obviously, this decreases the jamming power spectral density about the actual received frequency, F_1, because the jammer power is dissipated over a wider frequency band.

The masking signal can be implemented by using a separate antenna, or the same antenna that generates the desired signal may be employed if the antenna has multibeam capability. However, in the latter case, a separate beam-forming network is necessary. Traditionally, the masking antenna tends to have a broad-beam, low-gain pattern. This is because, in most cases, the location of the jammer in angle is unknown. However, this technique is highly wasteful of power because much of the masking energy is not directed only at the jammer, but also at other angles.

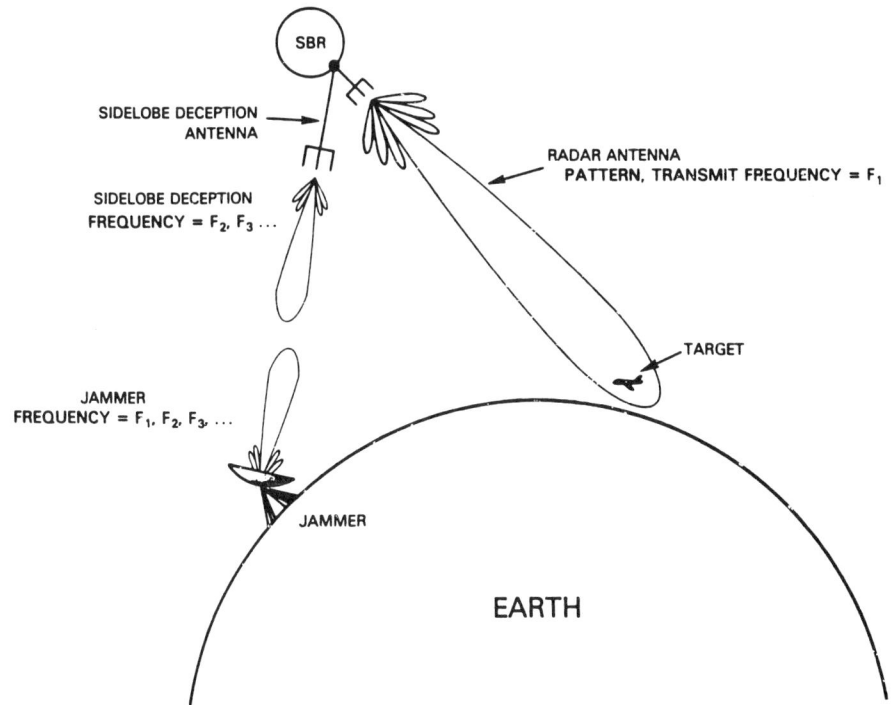

Figure 11.56 Sidelobe deception geometry.

Sidelobe Cancellation (SLC)

For this technique [70–81], a null is placed in the receiving antenna's sidelobes in the direction of an interfering source as illustrated in Figure 11.57. The basic SLC configurations (called an adaptive space-time canceller) is shown in Figure 11.58. (Note that previously, T was defined for clutter processing as the PRI; henceforth, in this chapter, it represents any arbitrary time delay.) The SLC bases its own design (its internal adjustments) settings on estimated statistical characteristics of the input and output signals. In particular, the SLC weights the coherent output from each sensor (antenna) and adds them to form a receiving beam. For an adaptive array, these weights may not be constant, but rather can change as a function of the spatial properties of the interference field. These weights are chosen so that when the signals from the auxiliary antennas are weighted and subtracted from the main antenna channel, the output interference residue of the main channel is minimized. In effect, an artificial null has been placed in the direction of the jammer in the main antenna pattern.

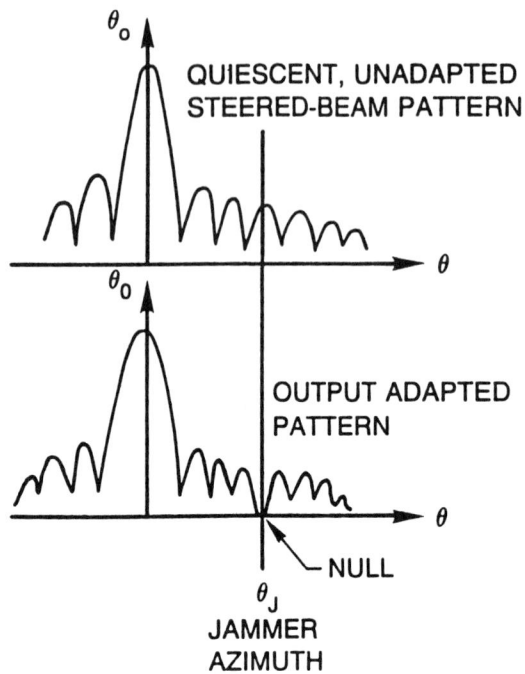

Figure 11.57 Sidelobe cancellation.

11.7 MAIN AND AUXILIARY ANTENNA CONSIDERATIONS

11.7.1 Main Antenna

The SBR adaptive ECCM is an important auxiliary feature that is fundamentally affected by the antenna design. Thus, the antenna design will determine realizable ECCM performance, and therefore the two cannot be separated. This interdependent situation and other considerations, such as cost, inaccessibility, reliability, large clutter return, and large ECM power levels, have led to the formulation of certain basic needs that shape the development of the combined antenna-ECCM system. These needs are as follows.

(a) *Low sidelobe antenna pattern:* This is necessary to reduce the adaptive dynamic range burden and improve detection. Implications include high precision, small error tolerance, and high cost.

(b) *Decouple clutter from jamming:* This is necessary to reduce precision requirements, convergence time, and the degrading interaction of the two different types of interference upon the cancellation of each.

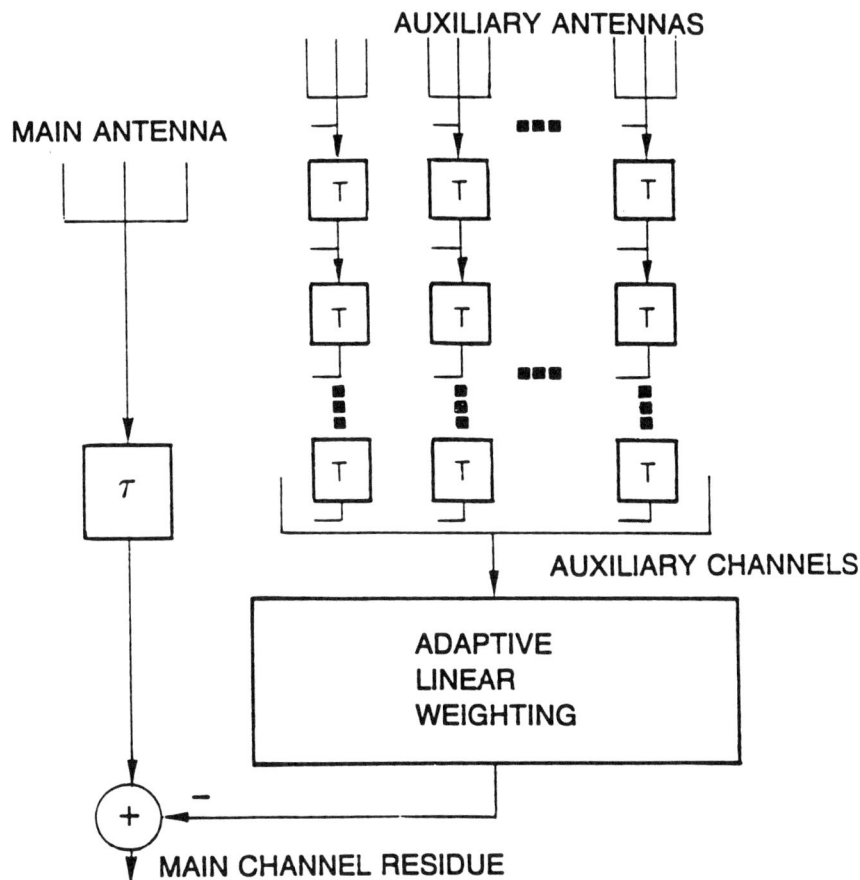

Figure 11.58 Generic sidelobe canceller (SLC).

(c) *Small number of adaptive degrees of freedom:* This is necessary because of the excessively high cost of a fully adaptive system for a very large phased array with thousands of elements. Also, for only a small number of jammers, adaptive complexity can be economically tailored to match the number of jammers.

(d) *Basic sidelobe canceller configuration:* This configuration has the same driving considerations as the preceding item and the factors of simplicity in implementation and minimum processing requirements.

Some basic candidates for the main SBR antenna are listed below:

1. a reflector space-fed from an active array;
2. an active lens space-fed from an active feed array;

3. dual reflectors with an active feed array;
4. an active phased array.

For interference rejection, a reflector configuration has the disadvantages of having inherently high sidelobe levels and the auxiliary antenna elements are restricted to be on the periphery of the reflector dish. A space-fed lens has the disadvantages of multipath from the lens to the feed, the potential of direct jamming of the feed array, complicated beam-forming for the auxiliary antenna channels, and if the auxiliary elements are on the lens, a communication link must be established between the lens and the feed. The dual reflectors have problems similar to both the reflector and space-fed lens. The phased array is the most versatile of the four configurations in that auxiliary beam-forming and very low sidelobes are possible. In addition, sidelobe masking is well suited for this antenna because it permits simultaneous transmissions in jammer directions while the main beam is directed elsewhere. However, the cost of mainbeam and auxiliary beam-forming is expensive.

11.7.2 Auxiliary Antennas

Because of the low sidelobe requirement for an SBR main antenna, it is important that the auxiliary antennas that are used for sidelobe cancellation and blanking do not perturb or raise this designed sidelobe level. For SLC purposes, the auxiliary antennas normally should be in close proximity or even integrated into the main antenna structure. Hence, electromagnetic coupling between these antennas should be minimized to maintain the low quiescent sidelobe level.

Two distinct ways of forming the auxiliary channels are called *element space* SLC and *beam space* SLC. For an element space SLC, individual elements or a cluster of tied auxiliary antenna elements are sent through an RF receiver. Thereafter, this RF output forms an auxiliary input (or channel) into whichever SLC scheme is implemented. For beam space SLC, the auxiliary elements or antennas form a beam or beams in the direction of the individual jammers. The beam-formed outputs are sent into individual RF receivers and their outputs form the SLC auxiliary inputs or channels.

The beam space SLC implementation perturbs the main antenna's quiescent sidelobes less than does element space SLC. However, its complexity and cost are high. To illustrate this point, consider two SLC options indicated in Figure 11.59, where we show cancellation of a single jammer via a single auxiliary element (dashed lines) and a single auxiliary beam (solid lines). Note that the jammer is nulled by either option, but the adapted pattern sidelobe effects are quite different. The auxiliary SLC beam will have very little effect on the remainder of the adapted pattern sidelobes because of its highly localized interaction beamwidth region, whereas the auxiliary SLC element interacts across the entire adapted pattern

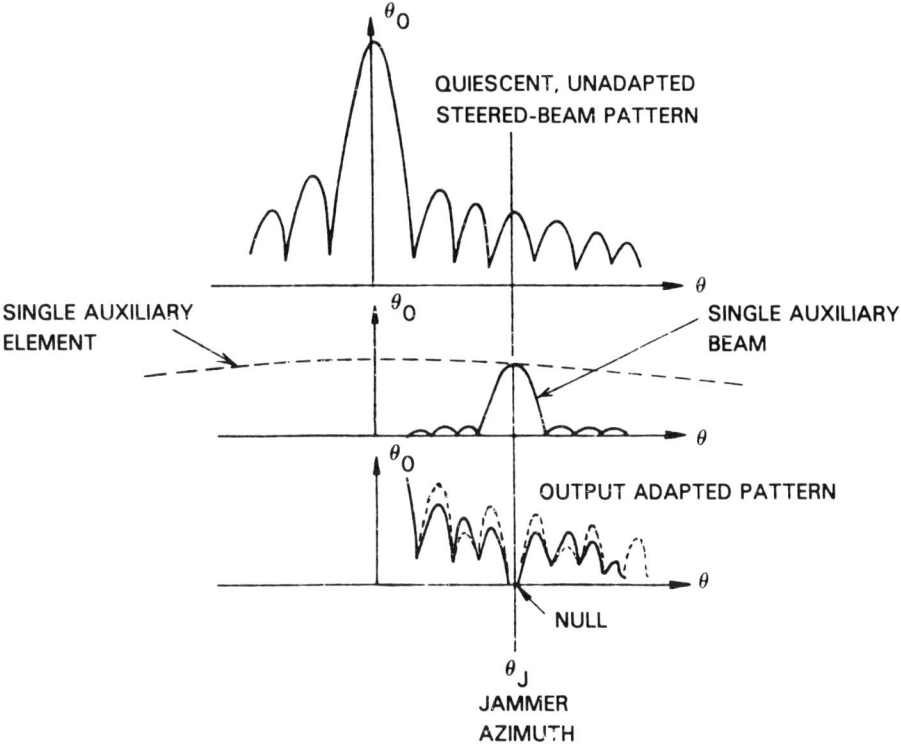

Figure 11.59 Conceptual cancellation comparison between a single auxiliary element and a single auxiliary beam.

sidelobe region because of its broad element pattern. In general, the use of auxiliary elements produces increasing adaptive sidelobe degradation as the number of jammers increases.

If a beam space SLC is employed, it is necessary to use techniques capable of providing the jammer's angular direction-of-arrival information that is sufficient to permit SLC beam assignment or pointing angle. Two solutions to finding the jammers in angle are (1) the main-beam array could be time-shared for searching and tracking of interference sources, or (2) an optimal spatial spectrum estimation technique [77] (such as super-resolution) could be implemented. The second approach needs a phased array for implementation. This array can either be part of the main antenna (if it is an array) or separate from the main antenna. The second approach potentially provides much higher angular resolution than the first method. The difference may be critical if jamming sources are closely spaced. Also, if the main antenna is a phased array, only a small fraction of the aperture elements need be used to obtain this superior resolution.

Now, we discuss the concept of subarray SLC beam-forming. The purpose of the subarray beamformer is to form the auxiliary channels of the SLC. Our discussion assumes that the main antenna is a phased array and the auxiliary channels are formed by using antenna elements on the main array. A fully adaptive array is one in which every element is individually adaptively controlled. In theory, a fully adaptive array provides the necessary degrees of freedom to lower all of the deterministic sidelobes to any arbitrary level. A partially adaptive array is one in which elements are controlled in groups (the *subarray* approach), or in which only certain elements, called *auxiliary elements,* are controllable.

The partially adaptive array has not been extensively studied as is evident by the lack of references available in the literature [82–87]. Chapman [84] initiated one of the earliest studies (*circa* 1974) exploring how to combine effectively the N elements of an entire planar antenna array into a collection of N_{aux} subarrays. He developed the *row-column precision array* (RCPA) in which each element signal is split into two paths: a row path and a column path. All of the elements in a given row or column are added together, and all of the row and column outputs are then adaptively combined. Morgan [83] studied the technique of using only auxiliary elements to perform adaption. He showed that correlation coefficient between the adaptive element spatial vectors plays a key role in characterizing the performance of a partially adaptive array. A significant difference in array performance can occur if the array elements selected for adaptive control are not properly chosen.

Obviously, the fully adaptive configuration is preferred because it offers the most control over the response of the array. However, the typical array may have many elements, posing several immediate problems. Implementing an adaptive array with many degrees of freedom can have considerable effect on the total cost of the array, which alone, however, is not sufficient grounds for rejecting the idea. Processor implementation, however, poses a more serious problem for a system of this size, as is discussed in more detail in the next section.

These considerations point toward the desirability of reducing the dimensionality of the processor, while maintaining as much control as possible over an aperture of a given size. Reduction of the dimensionality of the processor is achieved by use of subarrays as illustrated in Figure 11.60. Here, the dimensionality is reduced from N to N_{aux}, where $N_{aux} < N$, by using a transformation matrix H which is an $N_{aux} \times N$ matrix. It is not a trivial problem to specify an effective H matrix. If the H matrix is specified improperly, the overall signal-to-noise ratio can degrade significantly as a function of the direction of arrival angles of jamming. The procedure for specifying the H matrix is not simplistic as will become evident after reading the next subsection. However, some useful rules of thumb are given in [30] and [89].

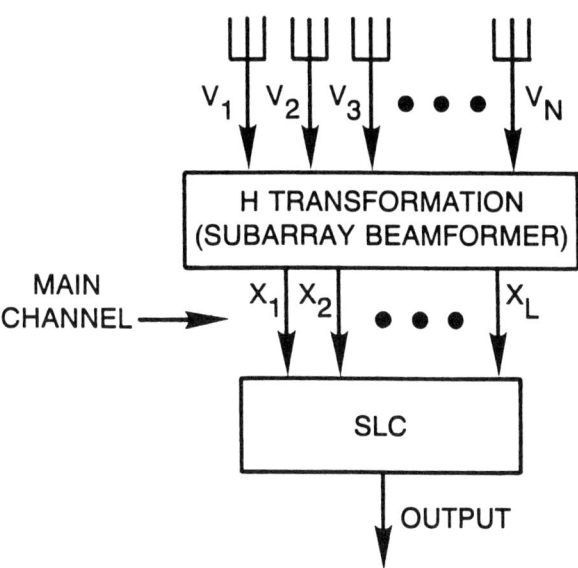

Figure 11.60 Adaptive subarray configuration.

11.7.3 Bandwidth-Aperture Dispersion

A wideband jammer in the sidelobes of an antenna is more difficult to null than a narrowband jammer because the inputs on the individual antenna elements become uncorrelated not only in phase but also in time. This is illustrated in Figure 11.61, where a jamming wavefront impinges upon a linear array at an angle θ off boresight. For antenna elements at half-wavelength spacings at some specified frequency, f_0, the jammer voltage on the nth element is

$$v_n(t) = x(t - \tau_n)e^{j(n-1)\pi\sin\theta} \qquad (11.67)$$

where

$$\tau_n = \frac{(n-1)d}{c}\sin\theta \qquad (11.68)$$

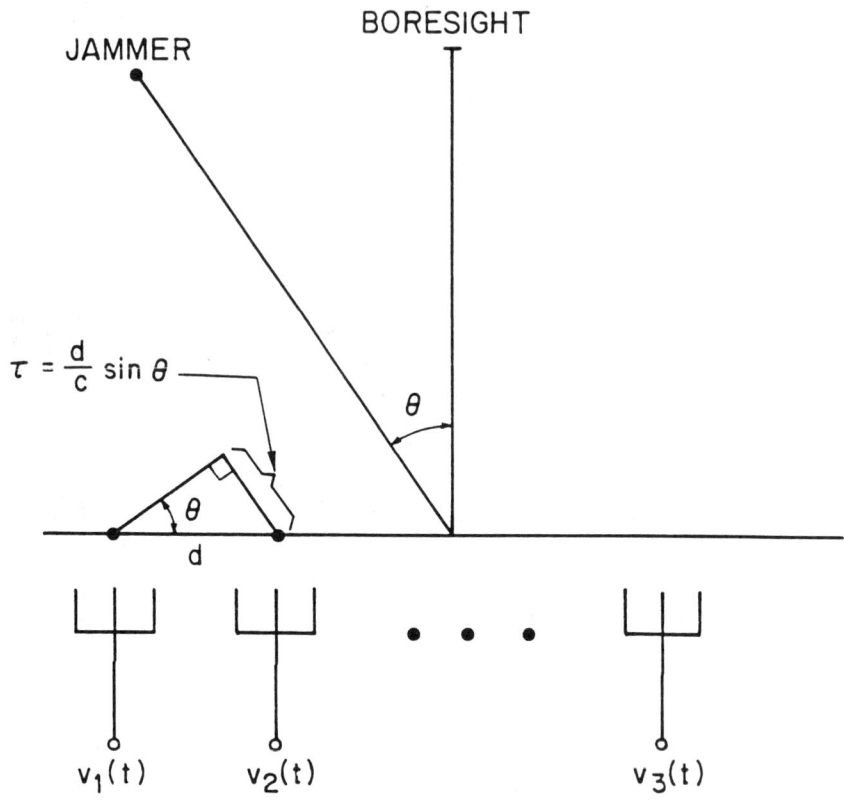

Figure 11.61 Linear array configuration and jammer.

In (11.67), $x(t)$ is the jammer noise modulation and $(n - 1)\pi \sin\theta$ is the relative phase of the jammer's carrier frequency at the nth element measured with respect to the first. In (11.68), c is the speed of light and τ_n is the time delay of the jammer modulation from the first antenna element to the nth.

Thus, the cross-correlation between the nth and mth antenna elements is

$$E\{v_n(t)v_m^*(t)\} = r(\tau_n - \tau_m)e^{j(n-m)\tau \sin\theta} \qquad (11.69)$$

where $r(\tau)$ is the cross-correlation function of $x_n(t - \tau_n)$ and $x_m(t - \tau_m)$. For a narrowband jammer $r(\tau) = 1$. However, for other than a narrowband jammer, $r(\tau)$ tends to decrease as τ increases from zero. Also, note from (11.68) that τ increases as the jammer off-boresight angle increases.

This decorrelation across the array makes it more difficult to null the jammer. Normally, a single degree of freedom (one auxiliary antenna element) is necessary to null a narrowband jammer. However, in general, a jammer may require more degrees of freedom, and hence more auxiliary elements or antennas to be properly nulled. The number of degrees of freedom necessary is a function of

- jammer power
- aperture size
- input frequency filters to the individual channels
- jammer angle off-boresight
- aperture weighting function
- jammer bandwidth-to-carrier frequency ratio
- position of the auxiliary antenna elements
- type of auxiliary beamformer

In general, the number of degrees of freedom necessary to null a wideband jammer increases in quantum fashion as the jammer power, aperture size, jammer off-boresight angle, and jammer bandwidth-to-carrier frequency ratio increase [88].

To specify a given number of degrees of freedom against a jammer, a careful analysis using the parameters listed above is necessary. We briefly outline this analysis procedure for adaptive arrays. Normally, the expected number of jammers, N_J and their power levels are specified. Let

$\theta_J(k)$ = angle off mechanical boresight of the kth jammer;
$\sigma_J^2(k)$ = power of the kth jammer at the array element level referenced to the internal power level;
$\omega_0 = 2\pi f_0$, angular center frequency of the array;
N = number of antenna elements;
N_{aux} = number of auxiliary channels;
r_{mm} = power in the main channel before cancellation;
$\mathbf{r}_{mm}^{(k)}$ = power in the main channel before cancellation due to the kth jammer;
$\mathbf{R}_{aa}$ = $N_{\text{aux}} \times N_{\text{aux}}$ covariance matrix of the auxiliary channels;
$\mathbf{R}_{aa}^{(k)}$ = $N_{\text{aux}} \times N_{\text{aux}}$ covariance matrix of the auxiliary channels due to the kth jammer;
$\mathbf{r}_{am}$ = cross-correlation vector of length N_{aux} between each auxiliary and the main channel;
$\mathbf{r}_{am}^{(k)}$ = cross-correlation vector of length N_{aux} between each auxiliary and the main channel due to the kth jammer;
CR = output cancellation ratio;
$\mathbf{w}$ = optimal weighting vector of length, N_{aux};
$S_k(\omega)$ = the baseband frequency spectrum of the kth jammer;
$g_n(m)$ = the gain of the mth antenna element used to form the nth channel.

Let $H_k(\theta_J(k), i, \omega, \omega_0)$ be the jammer off-boresight angle and baseband frequency transfer function of the kth jammer in the ith channel, where for the main channel i equals zero and ω is the baseband angular frequency. If the jammers are uncorrelated noise sources, then

$$r_{mm} = \sum_{k=1}^{N_J} r_{mm}^{(k)} + 1 \tag{11.70}$$

$$\mathbf{r}_{am} = \sum_{k=1}^{N_J} \mathbf{r}_{am}^{(k)} \tag{11.71}$$

$$\mathbf{R}_{aa} = \sum_{k=1}^{N} \mathbf{R}_{aa}^{(k)} + I \tag{11.72}$$

where I is the $N_{\text{aux}} \times N_{\text{aux}}$ identity matrix, which is due to the internal noise contribution to the covariance matrix.

The optimal weight vector is given by the expression [72]:

$$\mathbf{w} = \mathbf{R}_{aa}^{-1} \mathbf{r}_{am} \tag{11.73}$$

and the cancellation ratio is given by

$$\text{CR} = \frac{r_{mm}}{r_{mm} - \mathbf{r}_{am}^t \mathbf{R}_{aa}^{-1} \mathbf{r}_{am}} \tag{11.74}$$

where t denotes the conjugate transpose operation, and -1 indicates the matrix inverse. In addition, if $R_{l_1 l_2}^{(k)}$ is the l_1, l_2 element of $\mathbf{R}_{aa}^{(k)}$, and $r_l^{(k)}$ is the lth element of $\mathbf{r}_{am}^{(k)}$, then

$$R_{l_1 l_2}^{(k)} = \sigma_J^2(k) \int_{-\infty}^{\infty} H_k^*(\theta_J(k), l_1, \omega, \omega_0) H_k(\theta_J(k), l_2, \omega, \omega_0) S_k(\omega) \, d\omega \tag{11.75}$$

$$r_l^{(k)} = \sigma_J^2(k) \int_{-\infty}^{\infty} H_k^*(\theta_J(k), l, \omega, \omega_0) H_k(\theta_J(k), 0, \omega, \omega_0) S_k(\omega) \, d\omega \tag{11.76}$$

$$r_{mm}^{(k)} = \sigma_J^2(k) \int_{-\infty}^{\infty} |H_k(\theta_J(k), 0, \omega, \omega_0)|^2 S_k(\omega) \, d\omega \tag{11.77}$$

We note that, for a linear array with the antenna elements at half-wavelength spacings, we have

$$H_k(\theta_J(k), l, \omega, \omega_0) = \sum_{m=1}^{N} g_l(m) \, e^{-j(m-1)\pi \sin\theta_J(k)} \, e^{-j(m-1)\omega\pi \sin\theta_J(k)} \quad (11.78)$$

Tapped time delays can be placed at either the element or auxiliary level to form more auxiliary channels for sidelobe cancellation. However, the above $H(\cdot)$ transfer function must be modified accordingly to include the time-delay frequency-transfer function, $e^{-j\omega T}$, where T is the element or auxiliary tapped time delay.

The above methodology can be quite tedious for complicated jammer-main-auxiliary configurations. However, a simple rule of thumb [88] has been developed to relate the maximum achievable cancellation ratio for a linear array when only one degree of freedom or a single auxiliary antenna element is used, and if this element is at the array center, there is only one jammer, and the jammer has a rectangular spectrum. This rule is

$$\text{CR} = \left(\frac{\pi}{2c} B L_a \sin\theta_J\right)^2 \quad (11.79)$$

where

B = instantaneous bandwidth of SBR;
L_a = array length;
θ_J = jammer angle off mechanical boresight.

11.7.4 Phase Center Matching

If the phase centers of the main and auxiliary arrays are not nearly coincident, performance degradation occurs against closely spaced jammers [89]. The difference between the phase of a given jammer in the main array and that of the same jammer in an auxiliary antenna will be approximately identical for jammers separated by less than a beamwidth if the main and auxiliary antennas have the same phase center. Thus, only one complex voltage weight may cancel all of the jammers within a beamwidth. If the main and auxiliary phase centers are a significant distance apart, however, the phase differences are amplified, and multiple adaptive elements will be needed to cancel jammers located within the same sidelobe.

In addition to the loss in performance against closely spaced jammers, mismatching the phase centers for the same reason also causes loss in performance against well separated (more than a beamwidth apart) wide-bandwidth jammers having nonzero dispersion across the main antenna. This problem becomes more serious as the dispersion increases. This phenomenon can be explained by noting that displacement of the auxiliary antennas from the main array adds a phase term $\exp(j\omega\tau)$ to each auxiliary (11.67), where τ is the difference in arrival time of the jammer between the phase centers of the main and auxiliary arrays. A complex voltage weight applied to an auxiliary can remove this phase at only one frequency.

Displacement of the phase center of the auxiliary array from that of the main array also causes ripples in the adapted pattern (assuming the auxiliary antenna is separate from the main antenna). Because the combined apertures of the auxiliary and main arrays are larger than the main array aperture, there will be more sidelobes and randomness in the adapted pattern than in the quiescent main antenna pattern.

11.7.5 Other Issues

Some other important considerations for implementing a main-auxiliary SLC antenna system are the effects of random sidelobes of the main antenna, gain margin of the main antenna sidelobes to the auxiliary gain, multipath, and polarization. Without random sidelobes, the 3 dB beamwidth of the auxiliary array should be comparable to the null-to-null spacing of the sidelobes in the main array, thus prompting the placement of the auxiliary elements around the periphery of a planar array to match the sidelobe structure of the main antenna pattern [89]. The broadening of some sidelobes results from the presence of random element errors (or from applying a weighting function to the aperture to achieve low sidelobes) and indicates that it is often desirable to place a few elements within, as well as on, the periphery of the main aperture [89]. Auxiliary array configurations that utilized elements within or on the periphery of a planar array generally have produced a few more decibels of cancellation performance than configurations that used only elements on the periphery.

Gain margin is defined as

$$G_M = \frac{G_A}{G_{SL}}$$

$$= \frac{\text{gain of auxiliary antenna elements or beamformer}}{\text{sidelobe gain of main antenna in the direction of the jammer}}$$

(11.80)

The choice of G_M is a trade-off. In the steady state for an adaptive SLC, to have $G_M \gg 1$ would be desirable. If $G_M \gg 1$, the auxiliary weights will be small, which results in the internal noises in the auxiliary channels being attenuated. As a result, the steady-state output noise power residue is smaller. However, in the adaptive transient state (finite number of samples), the transient sidelobes are proportional to G_M so that it is desirable to have $G_M \ll 1$ because this attenuates the adaptive transient sidelobe levels [90–91]. A compromise is normally to choose $G_M \approx 10$ dB.

Multipath through the antenna beamformers can cause serious canceller degradations. This multipath can be caused by different voltage standing wave ratios (VSWRs) through the receiver chains, physical struts in the main or auxiliary beamforming, or possibly direct radiation of the jammer to the feed array for a space-fed lens antenna system. Compensation for this is normally provided by using adaptive transversal filters in the auxiliary channels [72].

If the transmitted signal has dual polarization (horizontal and vertical), the jammer may also have the dual-polarization capability. In any case, the cross-polarization component of both the radar and jammer antennas can require a dual-polarization capability. The main receiver channel actually consists of two channels, each with a given polarization. If there is jamming in both of these main channels, it is obviously necessary to have auxiliary channels at both polarizations for SLC purposes. However, a given auxiliary channel may not need to have both polarizations simultaneously.

11.8 SIDELOBE CANCELLERS

The SLC offers the most improvement of all the interference rejection techniques in the SBR environment. In this section, we discuss the SLC's algorithm, implementation, and limitations.

11.8.1 SLC Algorithm

For an adaptive processor with N degrees of freedom, each of the N input channels is multiplied by a complex weight and summed to form the output. The adaptive weights are determined according to a control law [74]:

$$\mathbf{Rw} = \mu \mathbf{s} \tag{11.81}$$

where $\mathbf{R}$ is an $N \times N$ matrix of the cross-covariances of the signals in the N channels, $\mathbf{w}$ is an N-element vector of the weights, μ is an arbitrary nonzero constant, and $\mathbf{s}$ is a vector representation of the desired signal in each channel. For a sidelobe cancellation configuration, $\mathbf{s} = (1\ 0\ 0\ \ldots\ 0)^T$, where T denotes vector transpose and the 1 in the first position indicates the contribution of the signal in the main channel. Solving for the weights is equivalent to solving a set of N simultaneous equations. A variety of algorithms can be utilized in solving for the optimal weights. Some of the more popular are listed:

- Howells-Applebaum control loop [72–74]
- Widrow's LMS algorithm [72, 75]
- Direct matrix inverse (DMI) [72, 76]
- Inverse matrix updating (IMU) [72]
- Gram-Schmidt orthogonalization [70, 92, 93]
- Kalman methods [72, 94]

The most straightforward way in which to implement a processor to solve for optimal weights is to use a form of the Howells control loop in analog hardware [72–74], the Widrow LMS algorithm [72, 75], or one of its derivatives in a digital machine. Both are realizations of the *steepest descent* method, an iterative technique employed in the minimization of a function. Such is a reliable method of solution in this application because it can be shown that the surface described by the output residue as a function of **w** is a concave quadratic hypersurface and, as such, contains only one minimum. The problem with this approach is that the convergence rate depends on the particular physical configuration of the interference sources and the array. Mathematically, the transient response of the processor is a function of the eigenvalues of the covariance matrix **R**. The spread in the amplitude of the eigenvalues can be considerable and, because the particular configuration of noise sources encountered is never known *a priori*, the resultant variance in settling time can be large, particularly if the order of N is large (i.e., a large number of degrees of freedom are used).

Reed *et al.* [72, 76] have proposed to solve directly for the weights by inverting an estimate of the true covariance matrix, thereby eliminating the uncertainty of the convergence rate of the iterative technique. This method is also subject to limits related to the order of the matrix **R**. The number of arithmetic steps required to obtain the solution by this method is proportional to the cube of the order N. This complexity coupled with the requirement for high numerical accuracy puts an upper bound on N, which is related to computation time available and cost that can be tolerated. However, the output S/N will be on the average within 3 dB of the optimum if $2N$ samples of each channel are used to estimate the covariance matrix.

The IMU [72], Gram-Schmidt orthogonalization [70, 92–93], and Kalman [72, 94] algorithms are recursive in nature, but converge to the optimal weights much faster than the Howells-Applebaum control loop or the Widrow LMS algorithm. The IMU and Kalman algorithms do not converge as fast as the DMI technique. However, the number of arithmetic steps required to obtain a solution by these methods is proportional to N^2 instead of N^3. The Gram-Schmidt algorithm converges as fast as the DMI technique and has many implementation advantages, such as using *systolic array* or *pipeline* processing [61, 70, 93].

11.8.2 Implementation

Three possible implementations of an SLC are all-analog, all-digital, and hybrid (part analog and digital). The all-digital SLC has the advantages of algorithm flexibility and high-speed convergence. The flexibility characteristic may be important in interference environments, where the jammers are blinking or changing polarizations. In this case, the SLC algorithm might be reconfigured in some optimal sense. The principal disadvantage of the all-digital SLC is that the RF-to-digital receiver chains, which are present in each channel (main and auxiliaries), must be highly matched. Hence, there will be tight specifications on the IF mixer, I and Q detector, and A/D converters.

The major disadvantages of the all-analog canceller is its lack of versatility and slow convergence. If the satellite motion causes null stability problems, the faster converging all-digital implementation is more desirable. The analog canceller's principal advantage is that the channels at RF can be relatively well matched so that they need not be equalized (as will be the case for the all-digital canceller).

A cascaded A/D canceller offers the advantages of both digital and analog systems. There are a variety of methods for implementing this hybrid SLC. Some of these are given here:

1. Compute and apply analog weights at RF followed by a digital SLC;
2. Compute analog weights digitally, apply at RF, followed by a digital SLC
3. Compute analog weights digitally and apply at RF. Do so recursively until the quiescent weights are achieved.

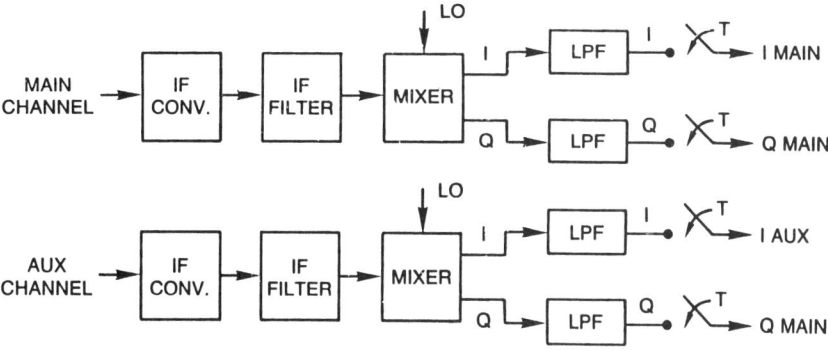

Figure 11.62 RF-to-IF-to-baseband-to-digital conversion chain for the main and auxiliary channels.

11.8.3 Limitations

An adaptive canceller combines auxiliary data channels with a main data channel so as to minimize the interference power residue of the main channel's output. Hence, adaptive cancellation is an effective way of eliminating unwanted data (or interference) from a main channel (the information channel) by use of correlated input data from auxiliary channels. Mismatch errors of any kind between channels of an adaptive canceller cause a reduction in the achievable cancellation ratio. These mismatch errors include small time delay differences, in-phase (I) and quadrature phase (Q) imbalances, A/D converter strobing errors, and frequency mismatch errors among the various channels. For a radar digital canceller, many of these errors occur in the typical receiver chain, shown in Figure 11.62, consisting of the RF-to-IF-to-baseband conversion, sample and hold (S + H) circuitry, and A/D conversion. If any link of this chain is not identical among the channels, mismatch errors cause the canceller's performance to degrade. Table 11.9 lists many of the mismatch errors that occur along the receiver chain. We have placed an asterisk next to the errors which are noncompensable in the SLC. (Note that if the strobing error is random, this is also noncompensable.)

Table 11.9
Listing of Receiver Mismatch Errors

RF Errors	filter
	nonlinearity*
	dc bias
	VSWR (time delay)
IF Errors	bandpass filter
	nonlinearity*
	dc bias
	harmonics*
	VSWR
I and Q Errors	low-pass filter
	phase
	amplitude
	dc bias
	harmonics*
	VSWR
	nonlinearity*
S + H, A/D Errors	strobing (inter, intra, fixed offset)
	quantization (no. of bits)
	linearity-dynamic range*
Digital Processor	quantization (no. of bits)

*Indicates error is not compensable.

To understand completely the effects of the receiver mismatch errors, the characteristics of the external interference environment must be known. However, in most instances, this is not possible. To specify canceller implementation limits, all of the inputs are tied together, exactly the same wideband noise is injected into each channel (main and auxiliary), and cancellation performance is measured at the output. The mode of calibration is called *self-cancellation* [95]. In a sense, the self-cancellation mode yields best-case (or an upper bound on) cancellation performance.

Because there are so many potential errors in the receiver channel, to give definitive formulas, which relate the achievable cancellation ratio as a function of all these errors, is difficult. However, if we assume that the errors are small and statistically independent, we can write [96] the inverse cancellation ratio as a sum of ratios computed with respect to a given error (i.e., all errors equal to zero except one when computing a given error term). Hence, each error through the receiver chain can be isolated and accounted in an *error budget*. This is a reasonable assumption for a high performance system because the errors must be small to achieve a high level of interference rejection.

For example, consider a single auxiliary channel that is a time delayed replica of the main channel. Let B be the bandwidth of the signals and τ the time delay in the auxiliary channel. For rectangular spectra and all other errors set equal to zero, the inverse cancellation ratio for a single weight canceller is given by

$$CR^{-1} = 1 - \left(\frac{\sin \pi B \tau}{\pi B \tau}\right)^2 \approx \frac{1}{3}(\pi B \tau)^2, \quad B\tau \ll 1. \tag{11.82}$$

We plot CR *versus* $B\tau$ in Figure 11.63. In addition, for a $B = 1$ MHz, CR is plotted *versus* τ. Note that for this example small time delays can severely reduce the cancellation ratio.

However, multipath, time delays, or frequency mismatch can be further compensated by using *adaptive transversal filters* (ATF) [72, 95–96]. The generic ATF is illustrated in Figure 11.64 where T is the sampling time. The auxiliary weights $w_1, w_2, \ldots, w_L$ are chosen adaptively so that frequency filter mismatch errors between the main and auxiliary channels are compensated. Another technique for frequency mismatch channel compensation is *band-partitioned canceller*) (sometimes called *subbanded canceller*) [97–98].

The design [95–96] of the ATFs is a function of the various filter errors (pole-zero perturbation, offset center frequency, bandwidth), filter type, sampling rate, and multipath errors. We have shown [96] that the ATF is effective in equalizing the I and Q low-pass filter errors only if the time samples of the I and Q channels are individually adaptively weighted. This technique has been called *real weighting*

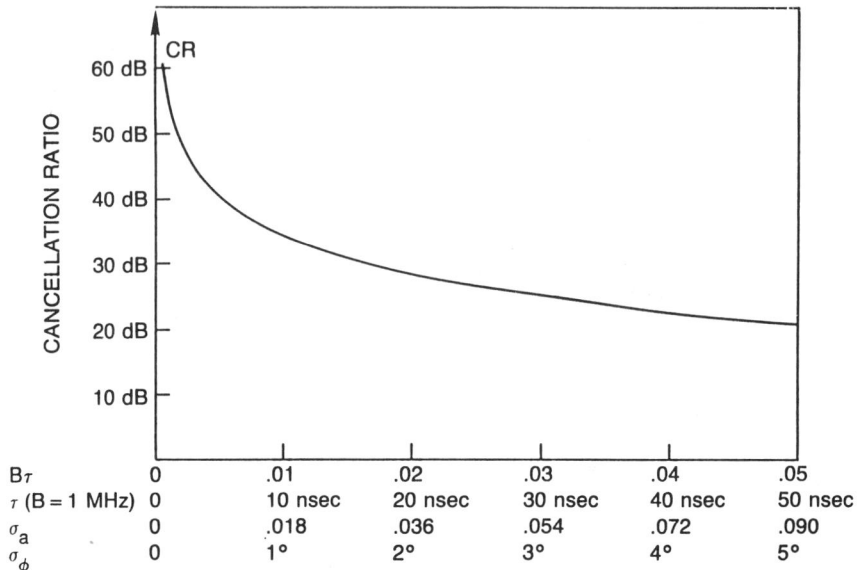

Figure 11.63 Cancellation ratio *versus* $B\tau$, time delay phase, or amplitude mismatch for a single auxiliary SLC.

or I and Q weighting, and it completely eliminates any constant I and Q mixer phase and amplitude mismatch [96]. If I and Q weighting is not employed, it can be shown that [96] the cancellation ratio (with all other errors set to zero) is related to the constant I and Q phase and amplitude errors by the approximate expression:

$$CR^{-1} \approx \sigma_a^2 + \sigma_\phi^2 \qquad (11.83)$$

where

σ_a^2 = variance of the I and Q amplitude mismatch (referenced to the ideal gain of one);
σ_ϕ^2 = variance of the I and Q phase mismatch (in radians).

The cancellation ratio *versus* σ_a and σ_ϕ is also plotted in Figure 11.63.

The A/D converter must operate at a speed high enough to preserve the information content of the radar signal, and the number of bits into which it quantizes the signal must be sufficient for the precision or dynamic range required.

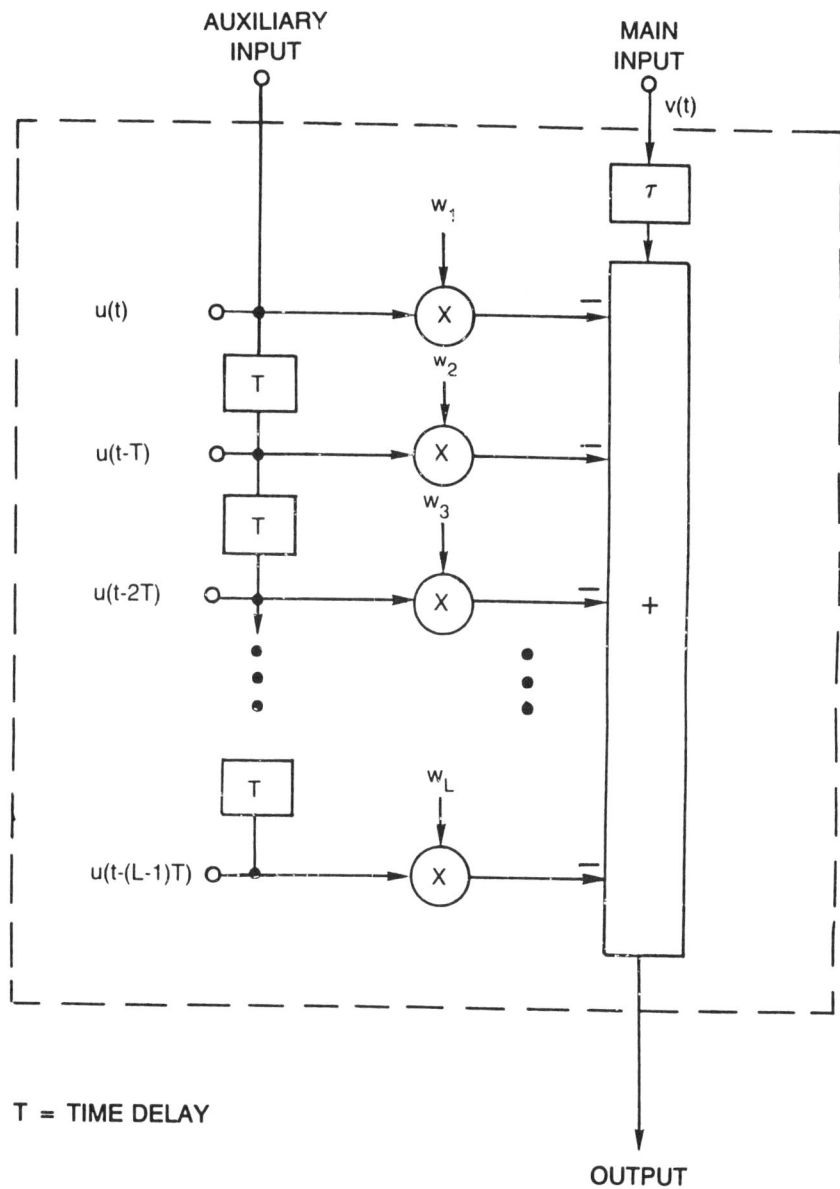

Figure 11.64 Adaptive transversal filter.

The number of bits in the A/D converter determines the maximum cancellation ratio that an SLC can achieve. In addition, the A/D converter must have the large dynamic range necessary to handle high levels of jamming as well as clutter. An n-bit converter divides the output of a phase detector into $2^n - 1$ discrete intervals. The number of bits, n, necessary for a given level of cancellation is approximately given by

$$CR_{dB} = 20 \log_{10} 2^n = 6n \tag{11.84}$$

Thus, 10 bits of A/D yield a maximum cancellation ratio of 60 dB.

The number of bits required in the digital signal processor for a required cancellation ratio is a function of the external noise environment, the internal noise level, and the type of SLC algorithm employed. Research in this area is found in the references [72, 79, 99].

11.9 INTERACTIONS OF CLUTTER AND INTERFERENCE PROCESSING

An interference-rejection system should not significantly degrade the detection of the desired signal energy in background clutter. Possible interactions of the two processing schemes are related to the sidelobe level of the main antenna, SLC convergence, main-beam distortion (particularly time-varying distortion), main-beam jamming, DPCA-jamming rejection.

Sidelobe Level

The quiescent main antenna pattern is normally designed to have sidelobes below a given level. For the SBR environment, this sidelobe level will probably be chosen so that the sidelobes clutter levels do not limit the output signal-to-clutter-noise ratio (i.e., the output sidelobe clutter level is below the receiver noise). The quiescent antenna pattern after adaptation can be significantly degraded due to the SLC. Hence auxiliary beam-forming was suggested as a means of localizing in angle the perturbations due to the retrodirective auxiliary pattern (see Section 11.7). If auxiliary beam-forming is not used, the quiescent adaptive antenna pattern is degraded due to the finite number of samples taken in each of the input data channels. The transient adaptive antenna pattern sidelobe level using element space auxiliaries has been shown [90–91] to be a function of the interference environment, the number of auxiliary channels, the number of independent samples per channel, and the gain margin of the main antenna's sidelobe level to the auxiliary level.

SLC Convergence

If the SLC is implemented before clutter processing, the main channel into the SLC contains the large clutter return, which is detrimental to the convergence of the auxiliary weights of the SLC. Hence, the jammer cancellation ratio is degraded accordingly. If we assume that the clutter power received in the main channel is much larger than the clutter entering the low-gain auxiliary channels, which will be illuminated through the low-gain transmitting antenna sidelobes, the clutter in the auxiliary channels can be neglected. We can thus show [72] that

$$\frac{K_{3dB}}{N_{aux}} = 2 + \left(\frac{C}{N}\right) \tag{11.85}$$

where

K_{3dB} = number of independent samples per channel needed so that the output residue is within 3 dB of the minimum;

N_{aux} = number of auxiliary channels;

$\frac{C}{N}$ = input clutter-to-noise power ratio in the main channel after beamforming.

We plot K_{3dB}/N_{aux} *versus* (C/N) in Figure 11.65. The independent samples are usually taken in succeeding range cells, so that the convergence time is the product of the number of samples needed and range resolution expressed in time. We observe, for the high levels of clutter expected in the SBR environment (greater than 30 dB), that the SLC requires at least 1000 independent samples per channel (which assumes that we use only one auxiliary). This number is excessive for most SBR ECCM considerations. Thus, we need to eliminate the clutter return before SLC processing. This can be accomplished in two ways:

1. Employ the sophisticated clutter processor before the SLC.
2. Employ a crude clutter processor (MTI) before the SLC to remove the clutter from the jamming signal, than compute the auxiliary weights in the SLC, apply these to the auxiliary data, and subtract this resultant from the original main channel data that contain the clutter and target returns. The sophisticated clutter processor can be employed afterward, without the MTI degrading the detection of low-velocity targets.

Method 1 is costly because, to maintain matched channels, the main and each auxiliary channel must be clutter processed identically and the jammer cancellation is applied to each of the doppler filters. Thus, there would be much duplication

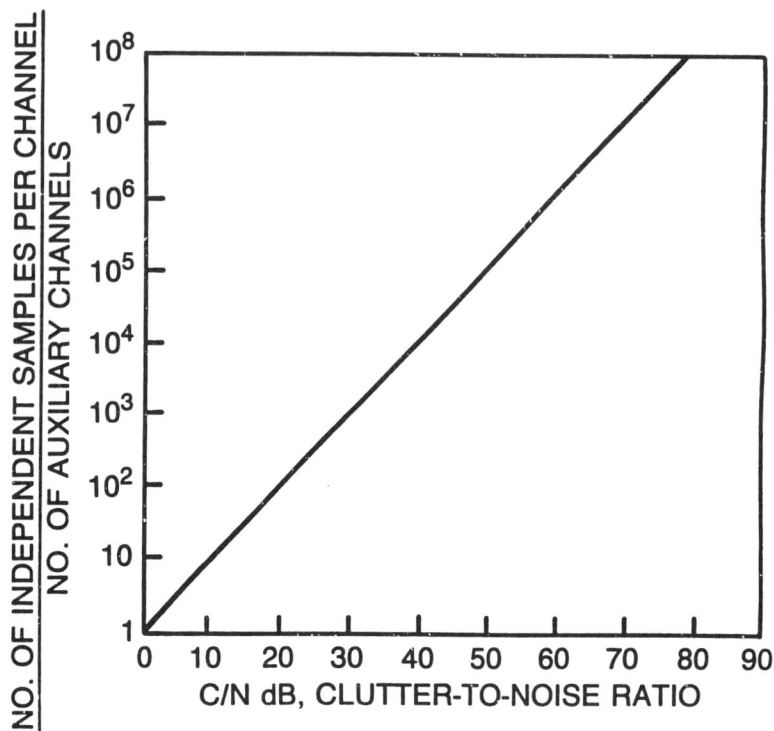

Figure 11.65 The ratio of the number of independent samples needed for convergence to the number of auxiliary channels *versus* the input main-beam clutter-to-noise ratio.

in the signal processor. There is also duplication in method 2, but the level is much smaller than with method 1. A small number of MTI cancellers in each channel should suffice to prevent the clutter from perturbing the convergence of the SLC. Note that in method 2 the computed weights are applied back to the auxiliary data which have not passed through the MTI processors.

Main-beam Distortion

The retrodirective auxiliary antenna pattern adds to the main antenna pattern so that the main beam will have a perturbation across it. This perturbation would have two effects: (1) the main beam could be broadened and (2) the main-beam's shape could change from one PRI to the next. If the main-beam is broadened, the clutter spectrum widens and hence clutter cancellation degrades. If the main beam's shape changes from the PRI to the next, all clutter-rejection schemes will have their performance degraded. Consider the effects on an ideal DPCA, where the forward and backward beams are identical. If we substract SLC-perturbed forward and backward beams, only the difference between two successive auxiliary antenna patterns remains. The level of this difference ultimately limits the clutter cancellation.

Main-Beam Jamming

Main-beam jamming can effectively shut down an SBR in terms of both clutter and ECCM. Even if a main-beam cancellation scheme could properly reject or null the jammer, the resultant antenna pattern would have a main beam that was greatly distorted and might either cancel or greatly attenuate the target. This distortion of the main beam would also affect the main-beam clutter spectrum, which, along with any null instabilities, might severely limit clutter cancellation. One way to reduce main-beam jamming is with a high-gain, narrow-beamwidth antenna so that the jammer may need to be on-board the target to ensure being in the main beam. Future research is needed into simultaneous rejection of main-lobe jamming and main-lobe clutter in an SBR environment.

DPCA-Jamming Rejection

For each DPCA main-beam channel, there is an associated set of auxiliary channels used for jammer cancellation. The number of auxiliary channels used for jammer cancellation is proportional to the number of DPCA main beams, N_{DPCA}. Hence, the hardware and processing necessary to generate these auxiliary channels is duplicated by a factor equal to N_{DPCA}.

REFERENCES

1. Blake, L.V., "Prediction of Radar Range," Chapter 2 in *Radar Handbook*, M.I. Skolnik (ed.), McGraw-Hill, New York, 1970, pp. 2-64.
2. Barton, D.K., *Radar System Analysis,* Artech House, Norwood, MA, 1976, pp. 95–108.
3. Schleher, D.C. (ed.), *MTI Radar,* Artech House, Norwood, MA., 1978, pp. 6–61.
4. Skolnik, M.I. (ed.), *Radar Handbook,* McGraw-Hill, New York, 1970, pp. 1.4–1.8.
5. Jordan, A.K., C.G. Purves, and J.F. Diggs, "Analysis of Skylab II S193 Scatterometer Data," Naval Research Laboratory Rep. 7877, May 1975.
6. Moore, R.K., K.A. Soofi, and S.M. Purduski, "A Radar Model: Average Scattering Coefficients of Land, Snow, and Ice," *IEEE Trans. Aerospace and Electronic Systems,* Vol. AES-16, No. 6, November 1980.
7. Long, M.W., *Radar Reflectivity of Land and Sea,* Artech House, Norwood, MA, 1983, pp. 143–312.
8. Nathanson, F.E., *Radar Design Principles,* McGraw-Hill, New York, 1969, pp. 228–275.

9. Lauer, C.J., "A Clutter Model for Spacebased Radar," Mission Research Corp. Rep. DNA-TR-83-35 (MRC-N-602), Santa Barbara, CA, November 1983.
10. Tomlinson, P.G., "A Model for Space Based Radar Clutter," Decision-Science Applications, Inc., RADC-TR-79-166, June 1979.
11. Shrader, W.W., "MTI Radar," Chapter 17 of Skolnik, *op. cit.* [4], pp. 17.9–17.38.
12. Ulaby, F.T., R.K. Moore, and A.K. Fung, *Microwave Remote Sensing, Active and Passive,* Vol. II, Artech House, Norwood, MA, 1982.
13. Lerch, Jr., C.S., "Satellite Surveillance Radar," Chapter 32 of Skolnik, *op. cit.* [4], pp. 32.12–32.13.
14. Brookner, E., "Aurora Clutter Versus Frequency for Space-Based Radar System," Raytheon EDL Memo. EB:79:6, March 1979, Wayland, MA.
15. Skolnik, M.I., *Introduction to Radar Systems,* 2nd Ed., McGraw-Hill, New York, 1980, pp. 223–512.
16. Valenzuela, G.R., and M.B. Laing, "On the Statistics of Sea Clutter," Naval Research Laboratory Rep. 7349, December 1971.
17. Valenzuela, G.R., and M.B. Laing, "Point-Scatter Formulation of Terrain Clutter Statistics," Naval Research Laboratory Rep. 7459, September 1972.
18. Trunk, G.V., "Small- and Large-Sample Behavior of Two Detectors Against Envelope-Detected Sea Clutter," *IEEE Trans. Information Theory,* Vol. IT-16, No. 1, January 1970, pp. 95–99.
19. George, S.F., "The Detection of Nonfluctuating Targets in Log-normal Clutter," Naval Research Laboratory Rep. 6796, AD677,926, October 1968.
20. Sheller, S.L., G.A. Andrews, and T.S. Weaver, "Modeling Clutter for a Space-Based Radar (SBR)," Naval Research Laboratory Rep. 8875, March 1985.
21. Barlow, E.J., "Doppler Radar," *Proc. IRE,* Vol. 37, No. 4, April 1949, p. 351.
22. Wright, J.W., "A New Model for Sea Clutter," *IEEE Trans. Antennas and Propagation,* Vol. AP-16, No. 2, March 1968.
23. Pidgeon, V.W., "Doppler Dependence of Radar Sea Return," *J. Geophysical Research,* Vol. 72, February 1968, pp. 1333–1341.
24. Valenzuela, G.R., and M.B. Laing, "Study of Doppler Spectra of Radar Sea-Echo," *J. Geophysical Research,* Vol. 75, No. 3, January 1970.
25. Fishbein, W., S.W. Graveline, and O.E. Rittenback, "Clutter Attenuation Analysis," US. Army Electronics Command, Tech. Rep. ECOM-2808, March 1967. Reprinted in Schleher, *op. cit.* [3], pp. 331–354.
26. Schrank, H., "Antenna Designer's Notebook," *IEEE Antennas and Propagation Society Newsletter,* April 1985.
27. Cheston, T.C., Chapter 11 of Skolnik, *op. cit.* [4], p. 11.1–11.36.
28. Farrell, J.L., and R.L. Taylor, "Doppler Radar Clutter," *IEEE Trans. Aerospace and Navigational Electronics,* September 1964.
29. O'Sullivan, M.R. and R.F. Pawula, "Clutter Spectra of Low PRF AMTI Pulse-Doppler Radar," *IEEE Trans. Aerospace and Electronic Systems,* Vol. AES-9, No. 1, January 1973.
30. White, W., and A. Ruvin, "Recent Advances in the Synthesis of Comb Filters," *IRE Nat. Conv. Rec.,* Vol. 5, Pt. 2, 1957, pp. 186–189. Reprinted in Schleher, *op. cit.* [3], pp. 143–156.
31. Schleher, D.C., and D. Schulkind, "Optimization of Nonrecursive MTI," *IEE Int. Radar Conf.,* London, October 1977, pp. 182–185. Reprinted in Schleher, *op. cit.* [3], pp. 277–280.
32. Andrews, G.A., "Optimal Radar Doppler Processors," Naval Research Laboratory Rep. 7727, May 1974.
33. Emerson, R.C., "Some Pulsed Doppler MTI and AMTI Techniques," Rand Corp., R-274, March 1954. Reprinted in Schleher, *op. cit.* [3], pp. 77–142.

34. Brennan, L., and I. Reed, "Theory of Adaptive Radar," *IEEE Trans. Aerospace and Electronic Systems*, Vol. AES-9, No. 2, March 1973, pp. 237–252. Reprinted in Schleher, *op. cit.* [3], pp. 473–488.
35. Andrews, G.A., "Comparison of Radar Doppler Filtering Techniques," Naval Research Laboratory Rep. 7811, October 1974.
36. Andrews, G.A., and S.L. Sheller, "A Matched Filter Doppler Processor for Airborne Radar," Naval Research Laboratory Rep. 8700, July 1983,
37. Andrews, G.A., "Performance of Cascaded MTI and Coherent Integration Filters in a Clutter Environment," Naval Research Laboratory Rep. 7533, March 1973. Reprinted in Lewis, B.L., F.F. Kretschmer, and W.W. Shelton, *Aspects of Radar Signal Processing*, Artech House, Norwood, MA, 1986, pp. 434–464.
38. Weber, P., and S. Haykin, "Space-Based Radar: Narrowing the Main-Beam Clutter Spectrum through the Use of a High-Aspect-Ratio Antenna," *IEE Proc.*, Vol. 135, Pt. F, No. 1, February 1988.
39. Andrews, G.A., "Airborne Radar Motion Compensation Techniques: Evaluation of TACCAR," Naval Research Laboratory Rep. 7407, April 1972.
40. Bird, J.S., and A.W. Bridgewater, "Performance of Space-Based Radar in the Presence of Earth Clutter," *IEE Proc.* Vol. 131, Pt. F, No. 5, pp. 491–500.
41. Andrews, G.A., "Airborne Radar Motion Compensation Techniques: Evaluation of DPCA," Naval Research Laboratory Rep. 7426, July 1972.
42. Staudaher, F.M., "Airborne MTI," Chapter 18 of Skolnik, *et al.* [4].
43. Andrews, G.A., "Radar Antenna Pattern Design for Platform Motion Compensation," *IEEE Trans. Antenna and Propagation*, Vol. AP-26, No. 4, July 1978.
44. Andrews, G.A., "An Airborne Radar Doppler Processing Philosophy," Naval Research Laboratory Rep. 8073, January 1977. Reprinted in Lewis, B.L., F.F. Kretschmer, and W.W. Shelton, *Aspects of Radar Signal Processing*, Artech House, Norwood, MA, 1986, pp. 465–499.
45. Cutrona, L.J., "Synthetic Aperture Radar," Chapter 23 of Skolnik, *op. cit.* [4].
46. Kovaly, J.J., "High Resolution Radar Fundamentals (Synthetic Aperture and Pulse Compression)," Chapter 17 of Brookner, E. (ed.), *Radar Technology*, Artech House, Dedham, MA, 1977, pp. 239–249.
47. Brookner, E., "Synthetic Aperture Radar Spotlight Mapper," Chapter 17 of Brookner, E. (ed.), *Radar Technolgy*, Artech House, Dedham, MA, 1977, pp. 251–258.
48. Eisele, J.A. and S.A. Nichols, "Orbital Mechanics of General-Convergence Satellites," NRL Report 7975, 50-43, April 1976.
49. Johnston, S.L., "Radar and ECCM," Chapter 4 in Scanlan, M.J. (ed.), *Modern Radar Techniques*, William Collins Sons, London, 1987.
50. Price, A., *History of US Electronic Warfare*, Vol. 1, Association of Old Crows, Alexandria, VA, 1984.
51. Ridenour, L.N. (ed.), *Massachusetts Institute of Technology Radiation Laboratory Series* (28 vols.) McGraw-Hill, New York, 1946.
52. Boyd, J.A., *et al.*, *Electronic Countermeasures*, Ann Arbor, MI, Peninsula Publishing, Los Altos, CA, 1965.
53. Schlesinger, R.J., *Principles of Electronic Warfare*, Prentice-Hall, Englewood Cliffs, 1961.
54. Schleher, D.C., *Introduction to Electronic Warfare*, Artech House, Norwood, MA, 1987.
55. Johnston, S.L. (ed.), *Radar Electronic Counter-Countermeasures*, Artech House, Norwood, MA, 1979.
56. Maksimov, M.V., *et al*, *Radar Anti-Jamming Techniques*, Radio i Svyaz, Moscow, 1976 (in Russian); English translation: Artech House, Norwood, MA, 1979.

57. Tispouras, D., et al., "ECM Technique Generation: Dual Coherent Source and Offboard Expendable Techniques," *Microwave J.*, Vol. 27, No. 9, September 1984, pp. 38–73.
58. "Special Issue on Electronic Warfare," *IEE Proc.*, Vol. **129**, Pt. F, No. 3, June 1982.
59. Van Brunt, L.B., *Applied ECM,* Vols. 1 and 2, EW Engineering, Dunn Loring, VA, 1978.
60. Hartman, R., "ESM Receiver Overview," *Proc. Military Microwaves Conf.* (MM-78), London (abstract only), October 1978.
61. *Proc. 1980 Adaptive Antenna Symp.* RADC-TR-80-378, December 1980.
62. Vakin, S.A., and L.N. Shustov, *Principles of Jamming and Electronic Reconnaissance,* Vols. 1 and 2, English translation: Foreign Technology Division, Wright Patterson Air Force Base, OH, May 1969.
63. Wiley, R.G., *Electronic Intelligence: The Analysis of Radar Signals,* Artech House, Norwood, MA, 1982.
64. Wiley, R.G., *Electronic Intelligence: The Interception of Radar Signals,* Artech House, Norwood, MA, 1985.
65. Fitts, R.E., *Fundamentals of Electronic Warfare,* US Air Force Academy, Colorado Springs, CO, March 1, 1972 edition.
66. Pett, M.C.D., "System Performance Trade-offs-Responsive and Repeater Jammers," *Proc. Military Microwaves Conf.* (MM-80), London, October 1980.
67. Grant, P.M., et al., "Introduction to Electronic Warfare," Special Issue on Electronic Warfare, *IEE Proc. F*, Vol. **129**, Pt F. No. 3, June 1982, pp. 113–132.
68. Eustace, H.F. (ed.), *International Countermeasures Handbook,* 3rd Ed., EW Communications, Watts Franklin, New York, 1977.
69. *Dictionary of Military and Associated Terms,* US Department of Defense, Joint Chiefs of Staff, JCS Pub-1, September 1974.
70. Lewis, B.L., F.F. Kretschmer, Jr., W.W. Wesley, *Aspects of Radar Signal Processing,* Artech House, Norwood, MA, 1986.
71. White, W.D., "Wideband Interference Cancellation in Adaptive Sidelobe Cancellers," *IEEE Trans. Aerospace and Electronic Systems,* Vol. AES-19, No. 6, November 1983, pp. 915–925.
72. Monzingo, R.A., and T.W. Miller, *Introduction to Adaptive Arrays,* John Wiley and Sons, New York, 1980.
73. Howells, P.W., Intermediate Frequency Side-Lobe Canceller, Unites States Patent 3,303,990, August 24, 1965 (filed May 4, 1959).
74. Applebaum, S.P., "Adaptive Arrays," *IEEE Trans. Antennas and Propagation,* Vol. 24, No. 5, September 1976, pp. 585–598; also Syracuse Univ. Research Corp., Rep. SPL TR 66-1, August 1966.
75. Widrow, B., P.E. Mantey, L.J. Griffiths, and B.B. Goode, "Adaptive Antenna Systems," *Proc. IEE,* Vol. **55**, No. 12, December 1967, pp. 2142–2159.
76. Reed, I.S., J.D. Mallett, and L.E. Brennan, "Rapid Convergence Rate in Adaptive Arrays," *IEEE Trans.*, Vol. **AES-10**, No. 6, November 1974, pp. 853–863.
77. Gabriel, W.F., "Spectral Analysis and Adaptive Array Superresolution Techniques," *Proc. IEEE* Vol. 68, 1980, pp. 654–666.
78. Riegler, R.L., and R.T. Compton, Jr., "An Adaptive Array for Interference Rejection," *Proc. IEEE,* Vol. **61**, No. 6, June 1973, pp. 748–758.
79. Rader, C.M., and A.O. Steinhardt, "Hyperbolic Householder Transformations," *IEEE Trans. Acoustics, Speech, and Signal Processing,* Vol. ASSP-34, No.6, December 1986, pp. 1589–1602.
80. Gabriel, W.F., "An Introduction to Adaptive Arrays," *Proc. IEEE,* Vol. **64**, No. 2, February 1976, pp. 239–272.
81. *IEEE Trans. Antennas and Propagation,* Special Issues on Adaptive Arrays, September 1976 and March 1986.

82. Nitzberg, R., "OTH Radar Aurora Clutter Rejection when Adapting a Fraction of the Array Elements," *EASCON 1976 Record,* IEEE Electronics and Aerospace Systems Convention, Washington, DC, September 1976, pp. 62.A–62.D.
83. Morgan, D.R., "Partially Adaptive Array Techniques," *IEEE Trans. Antennas and Propagation,* Vol. **AP-26,** No. 6, November 1978, pp. 823–833.
84. Chapman, D.J., "Partial Adaptivity for the Large Array," *IEEE Trans. Antennas and Propagations,* Vol. **AP-24,** No. 5, September 1976, pp. 685–696.
85. Vural, A.M., "A Comparative Performance Study of Adaptive Array Processors," 1977 IEEE Int. Conf. Acoustics, Speech, and Signal Processing, May 9–11, Hartford, CT, paper 20.6.
86. Howells, P.W., "High Quality Array Beamforming with a Combination of Precision and Adaptivity," prepared by Syracuse Univ. Research Corp., SURC TN74-150, June 1974.
87. Gerlach, K., and G.A. Andrews, "Adaptive Antenna Subarraying Using a Weighted Butler Matrix," Naval Research Laboratory Rep. 8676, September 1982.
88. Malloux, R.J., "Phase Array Theory and Technology," *Proc. IEEE,* Vol. 70, No. 3, March 1982, pp. 246–291.
89. Davis, R.M., and J.L. Gleich, "Element Placement in Adaptive Arrays and Sidelobe Cancellers," *Proc. 1980 Adaptive Antenna Symposium,* RADC-TR-80-378, Vol. II, December 1980.
90. Gerlach, K., "Sidelobe Level of an Adaptive Array Using the SMI Algorithm," Naval Research Laboratory Rep. 9079, February 1988.
91. Brennan, L.E., "Constraining Sidelobe Levels in Digital Sidelobe Cancellers," Adaptive Systems, Inc., Internal Report, 1984.
92. Alam, M.A., "Ortho-normal Lattice Filter-A Multistage, Multichannel Estimation Technique," *Geophys.,* Vol. 43, December 1978, pp. 1368–1383.
93. Gerlach, K., "Fast Orthogonalization Networks," *IEEE Trans. Antennas and Propagation,* Vol. AP-34, No. 2, March 1986, pp. 458–462.
94. Baird, C.A., "Recursive Processing for Adaptive Arrays," *Proc. Adaptive Antenna Systems Workshop,* March 11–13, 1974, Naval Research Laboratory Rep. 7803, pp. 163–182.
95. Gerlach, K., "Adaptive Canceller Limitations Due to Frequency Mismatch Errors," Naval Research Laboratory Rep. 8947, January 1986.
96. Gerlach, K., "Adaptive Canceller Limitations Due to I,Q Mismatch Errors," Naval Research Laboratory Rep. 9115, 1988.
97. Compton, Jr., R.T., "The Relationship Between Tapped Delay-Line and FFT Processing in Adaptive Arrays," *IEEE Trans. Antennas and Propagation,* Vol. AP-36, No. 1, January 1988, pp. 15–26.
98. Gabriel, W.F., "Adaptive Digital Processing Investigation of DFT Subbanding *versus* Transversal Filter Canceller," Naval Research Laboratory Rep. 8981, July 1986.
99. Nitzberg, R., "Effect of Computational Errors on Adaptive Processing When Using a Covariance Estimation Procedure," *Proc. 1980 Adaptive Antenna Symp.,* RADC-TR-80-378, December 1980.

Chapter 12
SPACE ANTENNA TECHNOLOGY
L.J. Cantafio
TRW

An important subsystem in a space-based radar (SBR) is the antenna. It becomes even more important in those radars where the transmitter, receiver, and phase shifters are integrated with the antenna to form active phased arrays. This chapter provides a review of the technology of antennas for SBR. Requirements, types, analysis, testing, and selected designs are discussed. The subject of space-based antennas is very extensive and can easily cover an entire book rather than only a chapter. Therefore, the reader should expect only limited treatments, and he or she should consult the references and bibliography for further details.

12.1 REQUIREMENTS

In Chapter 1 of this text [1], three types of space-based radar were identified: Type I is the small, short-range radar such as the rendezvous radars [2] used on the Space Shuttle [3, 4], Apollo [5], Gemini [6], and OMV programs [7]; Type II SBR includes the earth and planetary [14] resources radars used for mapping [8, 9], scatterometers [10], altimeters [11], weather [12, 13] and subsurface probing [9]; Type III SBR includes the large phased array surveillance radars proposed for multimission defense [15, 16], air traffic control [17], and disarmament treaty monitoring functions [18, 19]. The antenna requirements for the three types of radar differ significantly in most areas with the exception that all must be lightweight. Launch vehicle costs increase when payload weight increases.

12.1.1 Requirements for Type I SBR Antennas

The Type I antenna is usually small and must track with high precision. The Apollo rendezvous radar was required to track continuously in angle as the target flew overhead [2]. Therefore, an azimuth-over-elevation antenna assembly mount was designed, rather than the elevation-over-azimuth approach, to avoid the servo pole. The X-band 63.5-cm diameter antenna for the Apollo rendezvous radar was a four-horn monopulse Cassegrain-feed design that had a weight of 18 kg [5].

The Space Shuttle rendezvous radar antenna was required to search a $\pm 30°$ cone with a spiral scan. After acquisition, track errors were required to be less than 8 mr in angle (3σ) and less than 0.14 mr/s in angle rate [3, 4]. The radar antenna design was a 91.4-cm diameter parabola fabricated from graphite-epoxy composite material with a weight of 1.8 kg. The overall depth of the reflector was 31.8 cm. A focal-point monopulse feed design was used [4]. Figure 12.1 shows the radar antenna in its operational configuration as part of the Ku-band deployed assembly. A more detailed description of the shuttle rendezvous radar is given in Chapter 1 of this text [1].

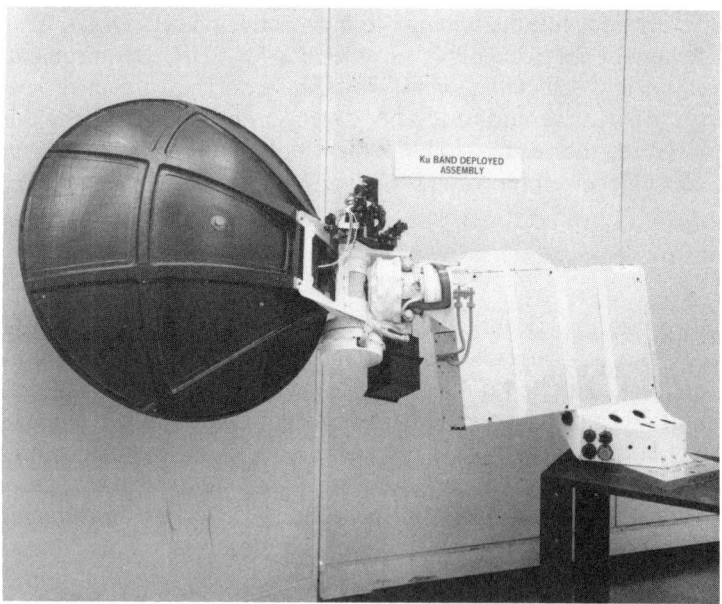

Figure 12.1 Shuttle rendezvous radar. (Courtesy of Hughes Aircraft Co.)

12.1.2 Requirements for Type II SBR Antennas

The antennas for Type II SBR are used in synthetic aperture functions, scatterometer functions to measure reflectivity, altimeter functions to measure satellite altitude above the surface (which is used to profile the earth or planet), weather radar measurements, and subsurface probing functions.

The requirements for a space-based SAR antenna depend on resolution, swath width, mode (side-looking or spotlight), launch vehicle, *et cetera* [20]. For a SEASAT class SAR [21, 22], the antenna beam must be held in an orientation that is normal to the spacecraft velocity vector. In advanced SEASAT class SAR [23–25], electronic scanning is planned, using distributed transmit and receive (T/R) modules throughout the array. The amplitude and phase errors in these modules must be held within certain limits to maintain low sidelobes and negligible antenna pattern distortion. For a large corporate-fed array, the module gain amplitude and phase error should be held to ± 0.75 dB and $\pm 10°$ when a module is used at each element [26]. The swath width is a function of several parameters, including orbital altitude, elevation beamwidth, and look angle. Figure 12.2 [27] shows a plot of swath width *versus* look angle for an antenna with an elevation beamwidth of 6° at an orbital altitude of 250 km. The synthetic aperture radar equation, as given

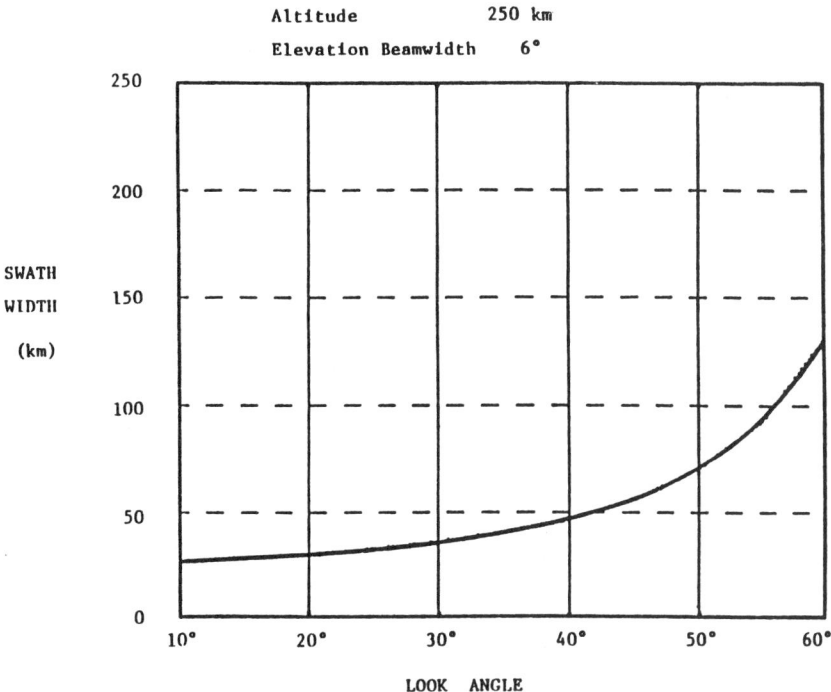

Figure 12.2 Swath width *versus* look angle [27].

in Chapters 1 and 4 of this book, shows that the average power required by the SAR transmitter is inversely proportional to the square of the effective area of the antenna. In a focused SAR that is side-looking, the limit of the cross-range resolution is one-half the real antenna size in the horizontal dimension. For a spotlight SAR, the cross-range resolution is independent of the real antenna size, and very large antennas may be used to reduce transmitter power requirements. The obvious conclusion from this discussion of SAR antenna requirements is that the space antenna is critically important in the design of a space-based SAR. A similar observation can be made for each of the antennas used in Type II SBR.

Requirements for altimeter-scatterometer antennas include available aperture area, peak gain, polarization, and minimum beamwidth. Spacecraft attitude error can impose a requirement for a minimum altimeter antenna beamwidth. The GEOS-C satellite had a spacecraft attitude error of one degree, and therefore the radar altimeter beamwidth was required to be not less than two degrees. The GEOS-C antenna [11, 28, 29] characteristics were as follows:

Type: paraboloid reflector
Available aperture: 59.06-cm diameter
Frequency: 13.9 GHz
Beamwidth (3dB): 2.6°
Peak Gain: 36 dB
Gain at 1° off boresight: 34 dB
Sidelobes: −20 dB

12.1.3 Requirements for Type III SBR Antennas

The antenna design requirements for the Type III antenna [20] include many factors, such as:

1. Coverage—electronic and mechanical
2. Beamwidth
3. Gain
4. RF power—peak and average, overall and per element
5. Sidelobes
6. Adaptive nulling
7. Temperature
8. Orbit
9. Distortion
10. Radiation
11. Weight
12. Outgassing
13. Meteoroid environment
14. Flexure loads—launch, deployment, orbital

15. Power distribution, condition, type level
16. Test
17. Shock
18. Vibration
19. Acoustic
20. Attitude control
21. Stowed configuration
22. Launch vehicle
23. Stiffness—stowed and deployed

This class of antenna is large and usually contains electronic scanning components that may be sensitive to radiation. Thus, shielding of components must be considered. Because the aperture area is large, the meteoroid environment ought to be considered. Detailed examples of the requirements imposed by all 23 factors are too extensive to be treated here. Selecting one factor, let us consider the meteoroid environment and its effect on a large (900 m²) phased array that uses SEASAT antenna technology [26]. What is the expected total area of meteoroid perforations for such a large opaque area? NASA has provided expressions for the meteorite mass and size for threshold penetration [30]. As suggested in NASA SP-8013 [30], a mean meteoroid velocity of 20 km/s and a mean density of 0.5 g/m³ will be assumed. The threshold penetration thickness, t, for thin ductile metal sheets is given by [31]:

$$t = K \rho^{1/6} m^{0.352} V^{0.875} \tag{12.1}$$

where

t = threshold penetration thickness (cm);
K = constant characteristic of the material;
ρ = density of the meteoroid (g/cm³);
m = mass of the meteoroid (g);
V = velocity of the meteoroid (km/s).

Typical K values are 0.3 for beryllium copper and 0.54 for 2024-T3 aluminum. Solving for m gives

$$m = \left(\frac{t}{K \rho^{1/6} V^{0.875}} \right)^{2.84} \tag{12.2}$$

An advanced SEASAT panel has a fiberglass face-sheet thickness of .051 mm (2 mils) and a copper layer thickness of .009 mm (0.35 mils). Combining the front and back layers gives a thickness, t, of 6.0×10^{-3} cm. This is an upper bound because two layers have more penetration resistance than a single layer that is twice as thick.

Using a K of 0.3 (to obtain a worst-case, i.e., minimum, threshold thickness), a density of 0.5 g/cm^3, and a velocity of 20 km/s, the resulting mass is 1.22×10^{-8}g. A corresponding diameter, assuming a spherical meteoroid, is 3.6×10^{-3} cm. An alternative relation for threshold penetration thickness (from [31], p. 68) is given by

$$D = \left(\frac{t}{(0.65) \ e^{-1/8} \ (\rho_m/\rho_t)^{1/2} V^{7/8}} \right)^{18/19} \tag{12.3}$$

where

D = penetration threshold diameter of the meteoroid (cm);
e = elongation of the sheet material (%);
ρ_m = density of the sheet material (g/cm^3);
ρ_t = density of the meteroid (g/cm^3);
V = velocity of the meteoroid (km/s);
t = thickness of the material (cm).

Values for V of 20 km/s, for t of 6.0×10^{-3}cm, and for ρ_m of 0.5 g/cm^3 will be used. To compare with the previous result, a value for e of 3% and for lt of 8.3 g/cm^3 (i.e., those of copper) will be used first. The resulting value of D is 4.3×10^{-3} cm, which is in reasonable agreement with the previous threshold diameter of 3.6×10^{-3} cm.

A more conservative approach is to combine the densities of copper (8.3 g/cm^3) and fiberglass (2.2 g/cm^3) with their respective thicknesses to obtain an overall density of the antenna sheet of $\rho_t = 3.1$ g/cm^3. Again, to be conservative, the elongation of fiberglass (about 0.2%) will also be used [32]. In this case, the resulting value of D is 1.9×10^{-3} cm, with a corresponding value for m of 1.9×10^{-9} g. These values will be used in the next steps.

The next question is how many particles having D greater than 1.9×10^{-3} cm will hit the antenna during a five-year mission. The corresponding flux is 4.8×10^{-6} particles/m^{2}s (from [31], Figure 7, reproduced as Figure 12.3). For a 500 nmi altitude, the defocusing factor, due to the earth's gravitational field is 0.97, and the body shielding factor is 0.74 (both from [30]). For a 900 m^2 area, the number of penetrations per mission is then (flux) $(1.02 \times 10'')$ or 4.9×10^5 penetrations/mission.

Figure 12.4 gives the meteroid diameter distribution of penetrating hits over a five-year mission on a 900 m^2 aperture. Thus, for example, a 1-mm diameter meteoroid hit should occur about ten times over a five-year mission, while a 0.1-mm diameter meteoroid hit should occur about 20,000 times.

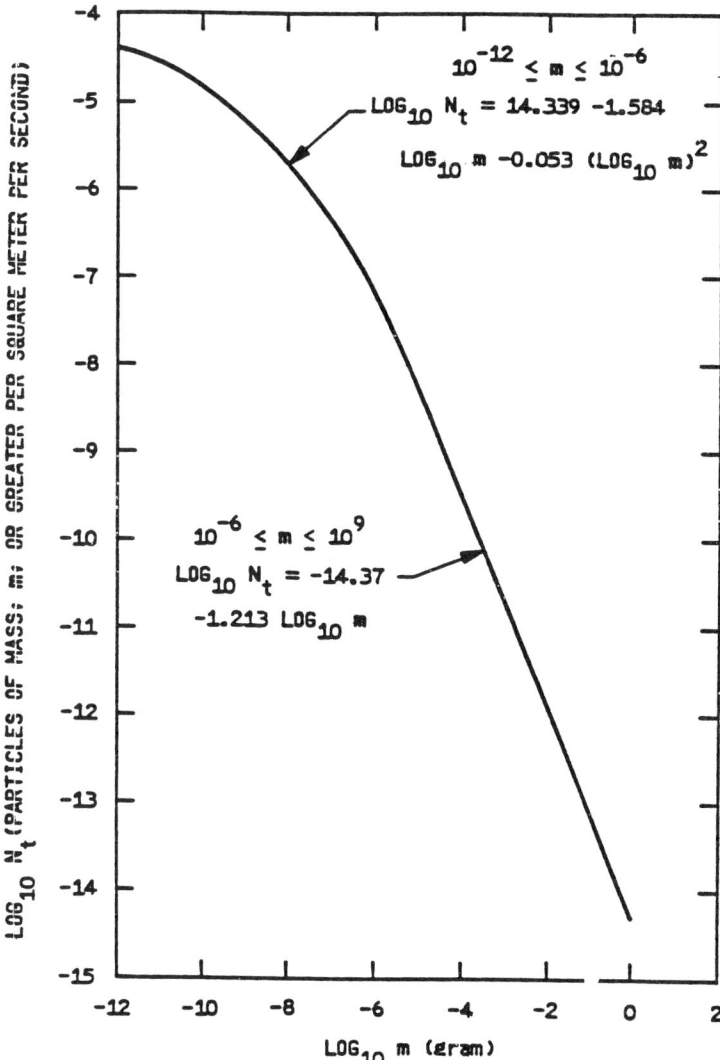

Figure 12.3 Average cumulative total meteoroid flux-mass model [31].

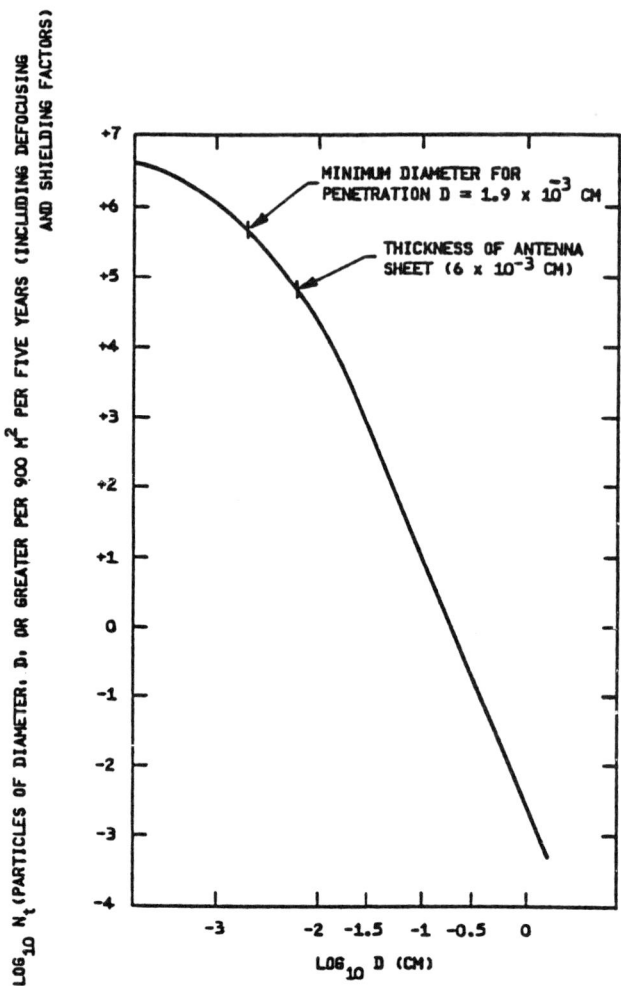

Figure 12.4 Average cumulative total meteoroid flux-mass model for 1 AU [34].

For meteoroids with diameters considerably smaller than the thickness of the target sheet, the penetration hole diameter should not be any larger than three times the meteoroid diameter [33]. For meteoroids with diameters equal to or larger than the target sheet thickness, the penetration holes will have diameters about ten times the meteoroid diameters.

We will now estimate the areas of the perforations produced by variously sized meteorids. This will be done in Table 12.1 by obtaining from Figure 12.3 the net fluxes for meteoroids between different diameters. These numbers will be multiplied by the maximum area that a perforation can have in that size interval [34]. (To be conservative, a perforation diameter that is ten times the largest meteoroid diameter within that size interval will be chosen.)

Table 12.1

Meteroid Diameter (cm)	Flux (Particles/ Mission)	Net Flux in Size Interval (Particles)	Maximum Area per Perforation/ Mission (cm^2)	Total Perforated Area per Size Interval (cm^2)
1.9×10^{-3}	4.9×10^5			
6.0×10^{-3}	6.9×10^4	4.2×10^5	2.8×10^{-3}	1.2×10^3
1×10^{-2}	2.3×10^4	4.6×10^4	7.9×10^{-3}	3.6×10^2
3.2×10^{-2}	6.3×10^2	2.2×10^4	8.0×10^{-2}	1.8×10^3
1×10^{-1}	9.6	6.2×10^2	7.9×10^{-1}	4.9×10^2
3.2×10^{-1}	0.15	9.5	8.0	76

Total Perforated Area = 3.9×10^3 cm^2

Therefore, the total area of meteoroid perforations is about 4,000 cm^2 or about 0.04% of the antenna area.

12.1.4 Lens Antenna Requirements

Lens antennas are frequently used in SBR. One of the reasons for selecting the lens configuration is the relatively relaxed surface tolerance requirements. The following analyses quantify the required tolerances for certain distortions, including distortions of the plane of the lens and element location errors within the plane of the lens. These requirements are functions of the focal length, the lens diameter, and allowable phase error.

Lens Antenna Surface Tolerance

The geometry of a lens antenna with a diameter D, is defined by a feed located at a distance F from the plane of the lens. The geometry is shown in Figure 12.5. The angle α is defined by

$$\tan \alpha = \frac{D}{2F} \tag{12.4}$$

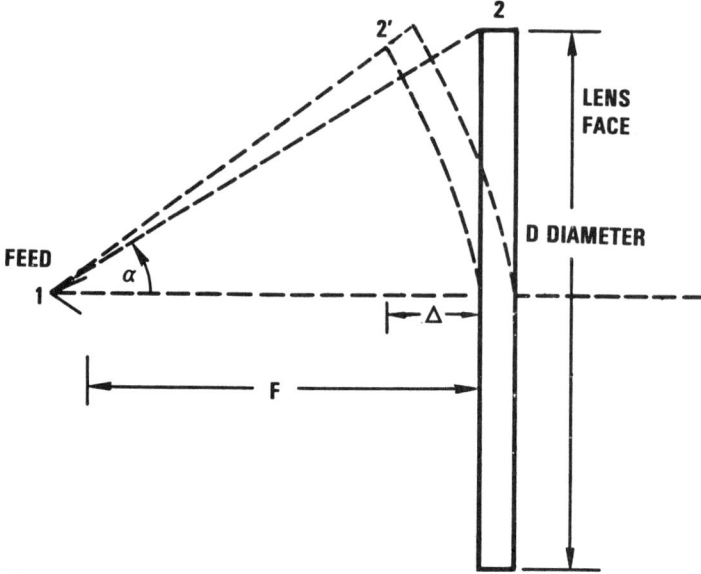

Figure 12.5 Lens antenna geometry.

Assume that the surface of the lens is distorted so that the edge of the lens defined by the point 2 is moved to the point 2'. The ray path to the point 2 is given by

$$\text{PL} = \frac{F}{\cos\alpha} \tag{12.5}$$

For the distorted position of the surface of the lens, there is an increase in the ray path length, Δ, and a decrease in the path length to the point 2', which is given by $\Delta/\cos\alpha$. The net change in path length is therefore ϵ, defined by

$$\epsilon = \Delta\left(\frac{1}{\cos\alpha} - 1\right) \tag{12.6}$$

If the allowed path length change is $\lambda/16$, we have

$$\epsilon = \frac{\lambda}{16} = \Delta\left(\frac{1}{\cos\alpha} - 1\right) \tag{12.7}$$

and the allowable surface distortion is

$$\Delta = \frac{\lambda}{16}\left(\frac{1}{\cos\alpha} - 1\right)^{-1} \tag{12.8}$$

For example, if the F/D ratio is 2.5, from (12.4), $\alpha = 11.31°$ and $\cos\alpha = 0.98$. Then, from (12.8), the allowable surface distortion is 3.16λ. Figure 12.6 shows the relationship between F/D ratio and tolerance for an allowable phase error (path length change) of $\lambda/8$ and $\lambda/16$. Typical values of maximum allowable irregularity in the wavefronts or equiphase surfaces formed by the lens are $\lambda/8$ and $\lambda/16$. The smaller the allowed phase error is, the lower are the sidelobes and the less is the gain loss. Actual values depend on the particular antenna design because the factors to be considered include F/D, scan angle, illumination function, and subarray design.

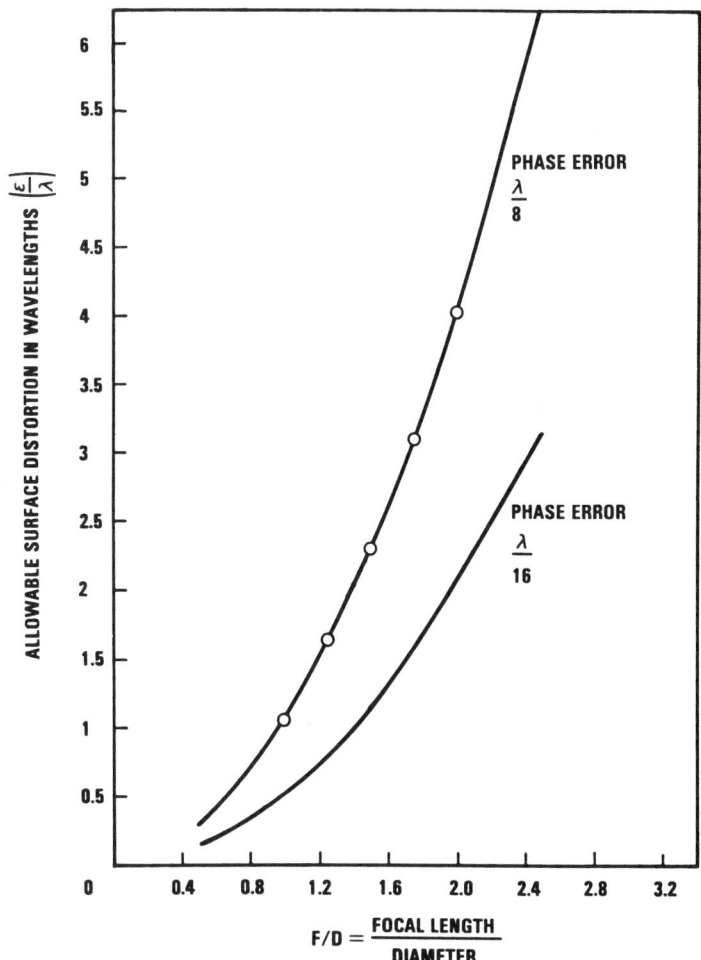

Figure 12.6 Allowable surface distortion for a lens antenna as a function of F/D and phase error.

Lens Antenna Element Location Error

This analysis relates the element location error and the focal length to lens diameter ratio, F/D, for a 762 λ diameter antenna.

The geometry of a lens antenna with a diameter D is defined by a feed located at a distance F from the plane of the lens. The geometry is shown in Figure 12.7. The angle α is defined by

$$\alpha = \tan^{-1} \frac{D}{2F} \tag{12.9}$$

The RF path length between the feed and an element near the rim of the lens located at the distance r from the center of the lens is

$$\text{PL} = [F^2 + r^2]^{1/2} \tag{12.10}$$

If the antenna lens is distorted by a length Δ, this element is incorrectly located at a distance $r + \Delta$ from the center of the lens, and the actual path length is given by

$$\text{PL}_a = [F^2 + (r + \Delta)^2]^{1/2} \tag{12.11}$$

The angle α' is then defined as

$$\alpha' = \tan^{-1} \frac{r + \Delta}{F} \tag{12.12}$$

The difference in the two path lengths is the path length error, or phase error ϵ, and is given by

$$|\epsilon| = \left| [F^2 + r^2]^{1/2} - [F^2 + (r + \Delta)^2]^{1/2} \right| \tag{12.13}$$

where $|\epsilon|$ is the absolute magnitude of the phase error. Consider two values of phase error, $\lambda/8$ and $\lambda/16$, where λ is the wavelength. Equation (12.13) may be written as

$$\frac{\epsilon}{F} = [1 + \tan^2\alpha]^{1/2} - [1 + \tan^2\alpha']^{1/2} \tag{12.14}$$

by using (12.9) and (12.12) and expanding (12.13). It is well known that

$$\sec\alpha = [1 + \tan^2\alpha]^{1/2} \tag{12.15}$$

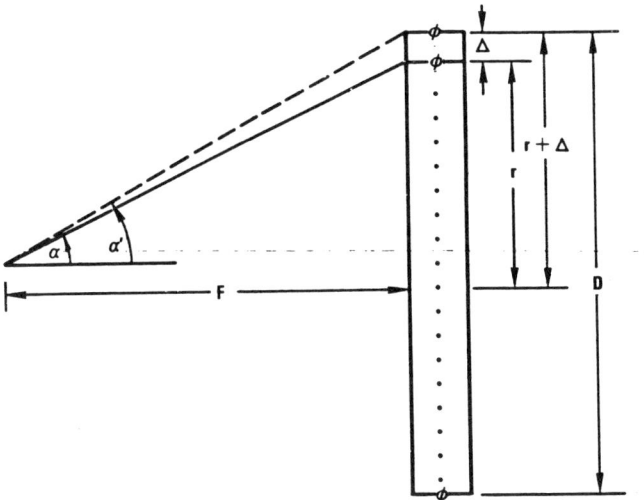

Figure 12.7 Lens geometry showing distortion.

Therefore, (12.14) may be written as

$$\sec\alpha' = \sec\alpha - \frac{\epsilon}{F} \tag{12.16}$$

Equation (12.16) has been solved for two values of phase error and a lens diameter of 762 λ. Table 12.2 shows the results for selected values of F/D. Figure 12.8 shows the plot of (12.16) for two values of phase error, and therefore gives the allowable element location error, and therefore the allowable distortion in the radial direction of elements located in the plane of the lens.

Table 12.2
F/D versus Tolerance

	Tolerance in Wavelengths	
F/D	$\epsilon = \lambda/8$	$\epsilon = \lambda/16$
0.1	0.1275	0.0637
0.25	0.1398	0.0699
0.5	0.1768	0.0884
1	0.2796	0.1398
1.5	0.3955	0.1977
2	0.5157	0.2578
2.5	0.6379	0.3188

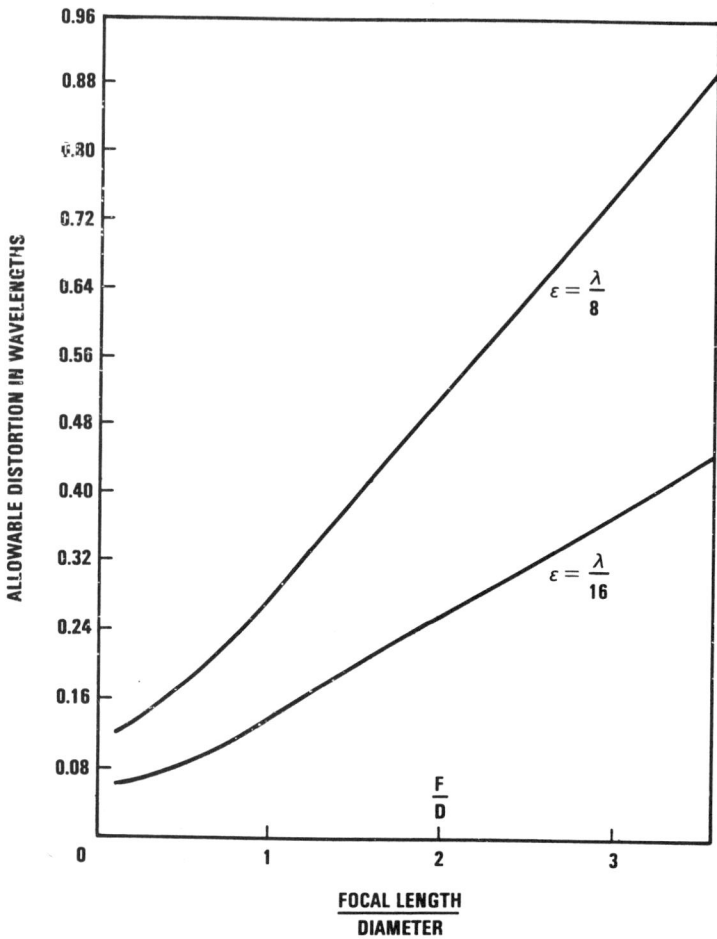

Figure 12.8 Allowable element location error for a 762λ-diameter lens antenna.

12.1.5 Other SBR Antenna Requirements

The thermal and natural radiation environments in space can have significant influence on the design requirements of a SBR antenna. Particular effects depend on the orbital altitude and the materials used in the antenna structure. (Chapter 16 of this text discusses the structural problem and solutions.) In general, the thermal environment will cause distortion. For a large corporate-fed phased array, this distortion will cause a decrease in antenna gain. Figure 12.9 shows the effect of random phase errors that are caused by the distortion, ϵ, when the error correlation interval is large with respect to a wavelength. Observe from Figure 12.9

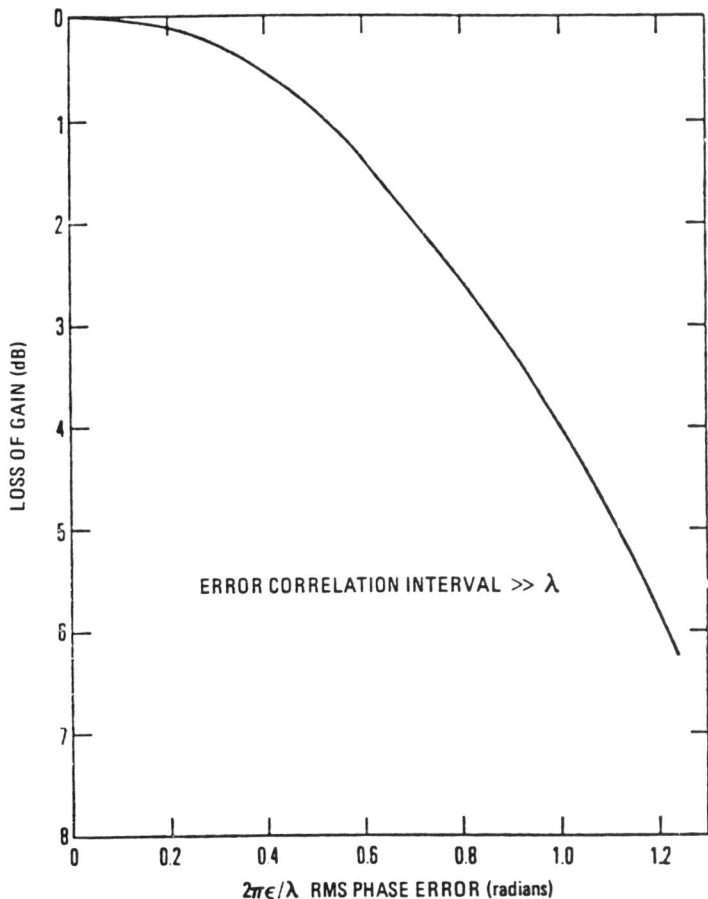

Figure 12.9 Antenna gain-loss due to random phase errors.

that a 2-dB loss in gain is obtained when the distortion is about one-tenth of a wavelength. Thus, for a planar corporate phased array antenna operating at a wavelength of 10 cm, the rms distortion of the plane of the array must be held to less than 1 cm if a 2-dB loss in antenna gain is to be maintained.

Thermal distortion in a 70-ft (21.34 m) diameter parabolic reflector was studied [35] at synchronous orbit. Reflector performance comparisons were made for titanium and graphite composite materials. Tolerances that must be held on reflector antennas are generally more severe than for corporate-fed phased arrays for the same performance. Figure 12.10 shows the results of the analysis. The graphite composite material performance is superior, giving an rms distortion of

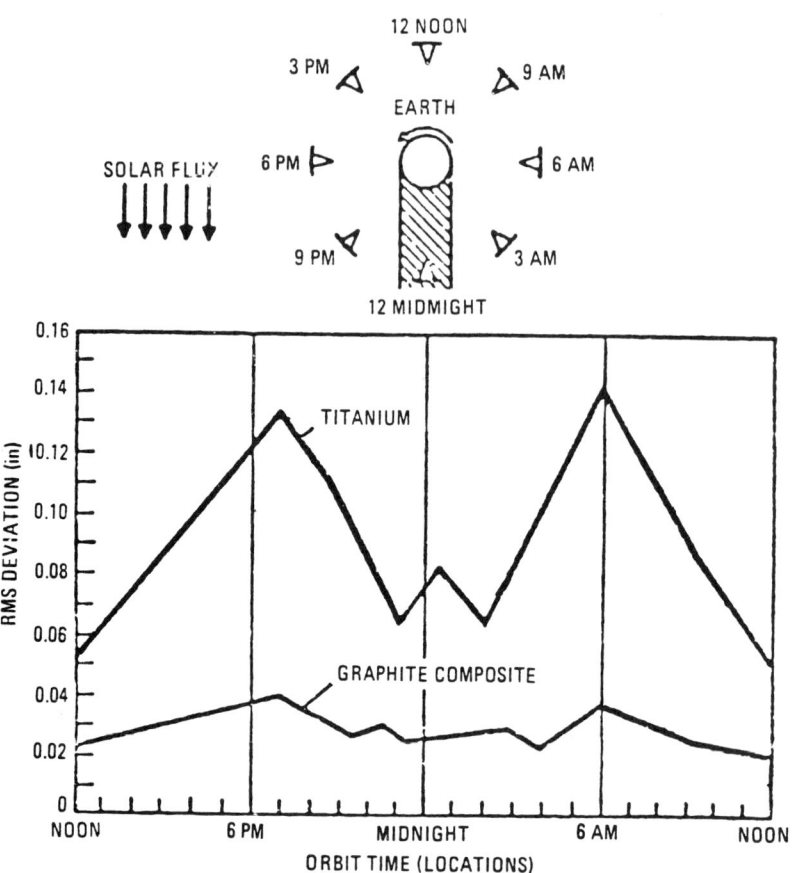

Figure 12.10 Thermal distortion [35].

about 0.076 cm. If this were one-fiftieth of the wavelength, the antenna could operate satisfactorily at a wavelength of 3.8 cm.

Consider a 70-m diameter lens phased array [36, 37] at an altitude of 5600 nmi as shown in Figure 12.11. The progress of the sun angle is also shown in Figure 12.11. Simulations have predicted the maximum and minimum temperatures given in Table 12.3 for selected parts of the space-fed lens antenna [38].

By choosing the proper materials, the design of this antenna class will experience lower distortions than those allowable. Figure 12.12 [38, 39] shows the loss in relative gain for a 71-m diameter space-fed antenna as a function of the deflection-distortion in wavelengths. We can see that the relative gain is down 1 dB when the distortion is about 5 λ at a 20° scan angle.

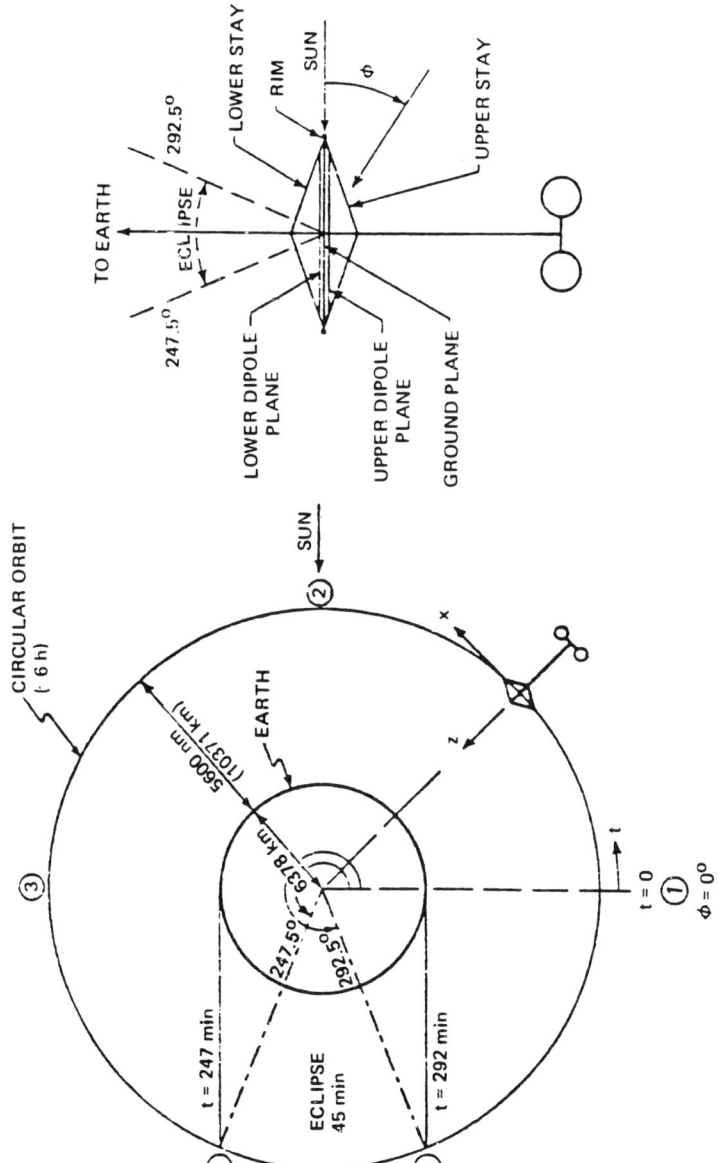

Figure 12.11 SBR at 5600 nmi orbital altitude and the sun angle progression [38].

Table 12.3

Location	Temperature(K)	
	Maximum	Minimum
Ground plane	264	224
Rim	182	160
Upper stays	231	186
Lower stays	217	201
Upper dipole plane	314	201
Lower dipole plane	274	220

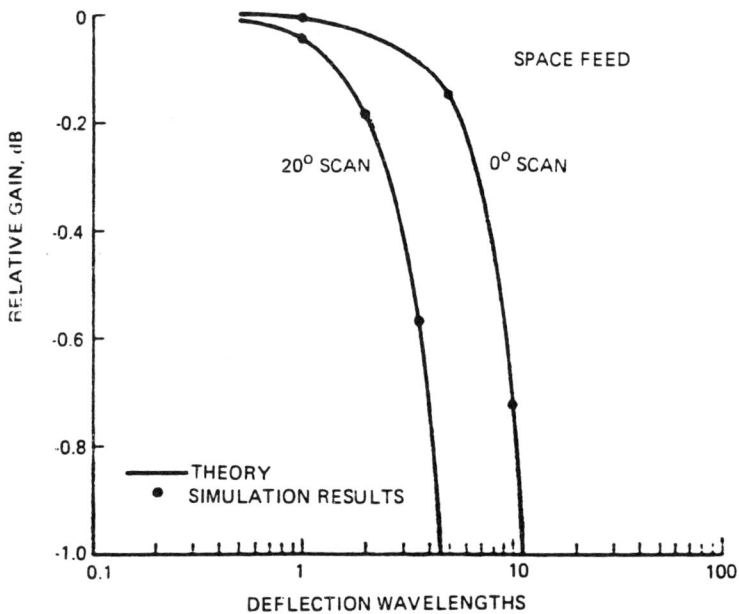

Figure 12.12 Gain-loss sensitivity to distortion for space-fed lens array [38].

12.2 SELECTED SPACE-BASED ANTENNA DESIGNS AND CONCEPTS

The development of SBR is strongly dependent on the technology of large space-deployable antennas. Large antennas must be used because the radar ranges are significantly greater than usual and the prime power in the radar is limited. The vacuum of space and the zero-gravity environment permit the deployment of antennas with low mass per unit antenna area. Antennas with large diameters, up to 1 km, have been discussed by United States developers including Grumman, Harris, General Dynamics, Rockwell, and Lockheed [40–48]. In the Union of Soviet Socialist Republics, antennas with diameters in the 1 to 10 km range have been discussed [49]. In addition to being large and deployable, the SBR antenna must maintain its desired shape, whether it be parabolic or planar. As shown earlier, small deviations can cause significant loss in antenna gain. Stable configurations are obtained by using materials with low *coefficient of thermal expansion* (CTE). Characteristics of selected materials for stable RF subsystems are shown in Table 12.4. Data include CTE, density, modulus, conductivity, and attenuation of WR-75 waveguide fabricated in each material. The following paragraphs describe selected antenna designs and concepts by United States and Union of Soviet Socialist Republics designers to illustrate the technology of large space-based antennas.

12.2.1 United States Space Deployable Antennas

In 1974 the United States deployed in geosynchronous orbit the Lockheed-NASA ATS-6 parabolic reflector shown in Figure 12.13. It was 9.1 m in diameter with a tolerance of 1.52 mm rms and a specific weight of 1.4 kg/m^2 [41, 50]. The ATS-6 antenna embodies the *"flex-rib"* technique. During the years subsequent to that launch, Lockhead has evolved the flex-rib deployment technology to additional reflector designs, the polyconic and the maypole. In a parametric study for NASA-JPL [41], Lockhead described a 945-m diameter reflector for a fully deployable communication satellite for synchronous altitude operation. The specific weight of the reflector was 0.03 kg/m^2 and the tolerance was predicted to be 0.18 mm. The directivity of the antenna would be 96.8 dB at a frequency of 7 GHz. Figure 12.14 shows the antenna concept in a gravity-gradient stabilized satellite configuration. A design criterion limited the satellite launch weight to the STS capability of 30,000 kg.

Harris Corporation developed the radial-rib double-mesh design, and in 1970 Harris built a 3.8-m diameter antenna [40]. Since then, Harris developed the TDRSS 4.88-m diameter antenna and three generic antenna designs including the radial rib, TRAC, and hoop-column concepts. The weight *versus* diameter capabilities of these three designs are shown in Figure 12.15 [47]. As part of the NASA deployable antenna flight experiment (DAFE) design study, Harris estimated that a 50-m diameter reflector assembly would have an overall weight of 819 kg (1805 lb). An estimate of the weight details is shown in Table 12.5 [47]. The specific mass of this design is 0.417 kg/m^2 and the estimated surface error was 4 mm rms. In a parallel DAFE competition, the Grumman Aerospace Corporation designed a 50-m diameter phased array lens antenna that would have a specific mass of 0.522 kg/m^2. The weight estimate for Grumman's shuttle-deployed test configuration is shown in Table 12.6. Also shown is the estimated weight for a parabolic reflector design with similar specific mass and performance [48].

Table 12.4
Potential Material Selection for Thermally Stable RF System [20]

Material	Expansion Coefficient (in./in°F × 10⁻⁶)	Density (lb/in.³)	Young's Modulus (× 10⁶ lbf/in)	Thermal Conductivity (Btu-in./hr-ft²°F)	WR75 Attenuation (dB/ft) at 11.95 GHz
Aluminum	13.1	0.10	10	1513	0.049
Titanium	5.1	0.16	16	444	0.274
Invar	1.1	0.29	20	93	0.370
Beryllium	6.8	0.07	40–44	1138	0.082
Graphite/Epoxy	0.03	0.06	17–25	75 (axial)	1.560 (bare)
				7.3 (transverse)	0.040 (coated)
Gold	6.8	0.70		2064	0.048
Copper	7.8	0.32		2944	0.040
Silver	11.0	0.38		3101	0.039
Rhodium	4.7	0.45		611	0.087
Kevlar[a]49	−1.1 longitudinal +33 radial	0.052	19	0.334 (axial) 0.285 (transverse)	

[a] DuPont trademark.

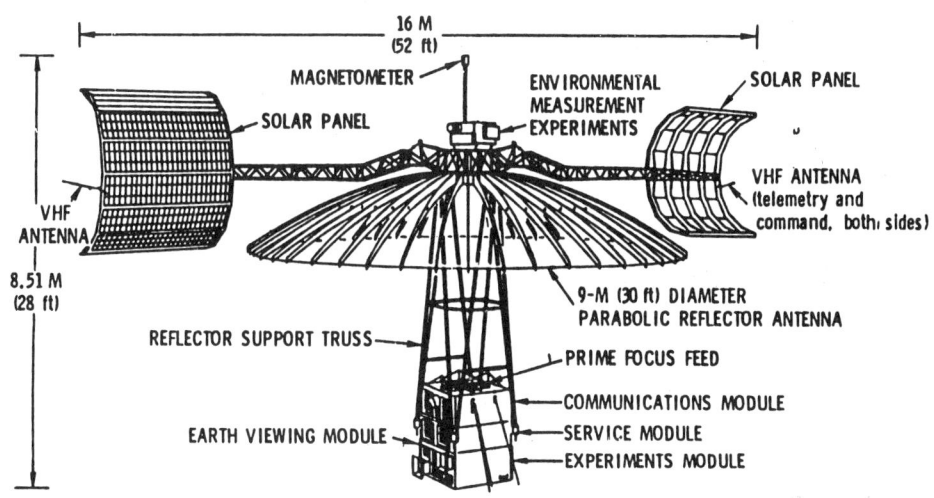

Figure 12.13 NASA ATS-6.

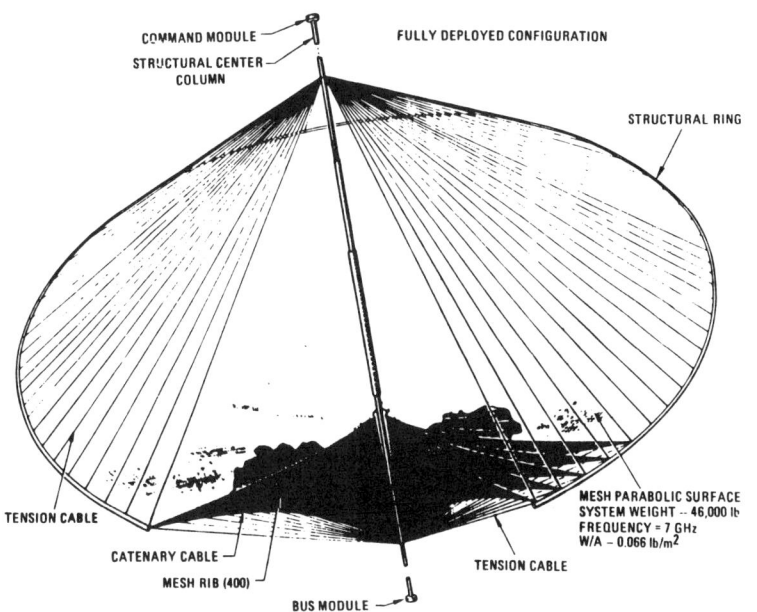

Figure 12.14 Lockheed Maypole 945-m antenna concept [41].

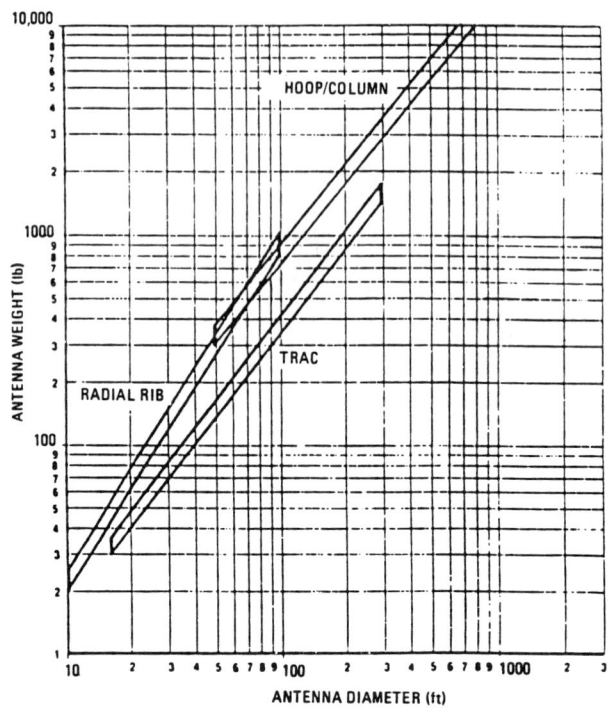

Figure 12.15 Weight *versus* diameter for three generic antenna designs [47].

Table 12.5
Harris/DAFE 50-m Diameter Antenna Weight Summary [47]

Description	Weight (pound)
Hub Upper Restraint Assembly	16.87
Hub-MDS-Shell Assembly	112.22
Hub Lower Restraint Assembly	18.07
Rib Assembly	640.15
Surface Assembly	100.00
Thermal Control Assembly Reflector	96.48
Feed and Cable Assembly	225.50
Feed Mast Assembly	378.00
Lower Support Assembly	70.00
Lower IF Fitting Assembly	25.00
Optical Alignment Assembly	3.00
Wiring-Cabling-Instrumentation Assembly	112.56
Stowage Assembly	7.50
TOTAL	1805.35

Table 12.6
Grumman NASA/DAFE Weights—50-m Diameter [48]

ITEM	50 M PHASED ARRAY (KG)		50 M REFLECTOR (KG)	
ANTENNA	1024		841	
– HUB	61		61	
– DRUM	89		89	
– STAY & PLATFORM	82		82	
– STAY PLATFORM SUPPORT MAST	–		10	
– RIM SYSTEM	179		179	
– MEMBRANE INCL. FIXED PHASE SHIFTERS	471		–	
– MESH, NODE FITTINGS AND TRUSS SUPPORTS	–		57	
– ANTENNA SUPPORT MAST	142		35	
– FEED SUPPORT MAST AND CANISTER	–		291	
– FEED MAST AND CANISTER	–		37	
ANTENNA INSTRUMENTATION	100		118	
– ANGLE MEAS, CORNER REFL, TEMP GAGES, SIGNAL COND, BATTERIES, TELEMETRY	100		100	
– FEEDS, ELECTR, RCV	–		18	
ORBITER BAY EQUIPMENT	2945		2927	
– FEED, ELECTR, XTR, RCV, PROC	46		28	
– JETTISON SUPPORT & ANTENNA MOUNT	16		16	
– FORWARD SUPPORT STRUCT	1045	(GFE)	1045	(GFE)
– AFT SUPPORT & ANTENNA PIVOT	1136	(GFE)	1136	(GFE)
– INSTRUMENTATION & WIRING	170		170	
– SECOND RMS	532	(GFE)	532	(GFE)
RF SATELLITE	66		66	
MISSION PAYLOAD SPECIALIST STATION	18		18	
25% CONTINGENCY (5% ON GFE)	496		450	
Δ SHUTTLE PAYLOAD	4649		4420	

A 15-m diameter deployable space antenna was built, analyzed, and tested by NASA and Harris Corporation. Hoop-column technology was used [51, 52]. Figure 12.16 shows the antenna in its deployed configuration. Table 12.7 provides the antenna parameters including measured gain, beamwidth, and rms surface error of 1.55 mm. The antenna design employs pretensioned cables and mesh to produce a deployable paraboloidal reflector surface. Fabrication errors and thermal distortions can significantly reduce surface accuracy and consequently degrade electromagnetic performance. Therefore, the ability to adjust the antenna surface shape is highly desirable. A shape adjustment algorithm was developed that consisted of finite-element and least-squares error analyses to minimize the surface distortions. Given a vertical surface error (e) to be corrected, a set of compensating control cable adjustments ($\Delta\mu$) may be computed by using least-squares error analysis of the following form [51]:

$$[S]^T[S][W][\Delta\mu] = [S]^T[W](e) \qquad (12.17)$$

where $[S]$ is the cable influence coefficients and $[W]$ is a diagonal weighting matrix.

Table 12.7
Hoop-Column Deployable Space Antenna [52]

Status: Tested 1985
MFR: Harris Corp. for NASA
Type: Parabola (24 gores) adjustable cables (96)
Diameter: 15 m

Stowed Dimensions: 1 m × 3 m
Deployed Dimensions: 15 m (diameter) × 14 m (height)
Mass: 410 kg

W/A: 2.32 kg/m^2
Surface rms Error: 1.55 mm after adjustment
D/ϵ: 9681
F/D: 0.6212
Frequency: 11.6 GHz
Beamwidth: 0.29° E-plane
Gain: 53.85 dB
Elements, Materials, Weight:
   Hoop-Graphite-Epoxy   136.6 kg
   Column-Graphite-Epoxy   141.6 kg
   Mesh-Gold Plate Molybdenum Wire   9.1 kg
   Cables-Graphite and Quartz   3.2 kg
   Feed Mast-Steel   11.5 kg
   Feed (sim.)   108 kg

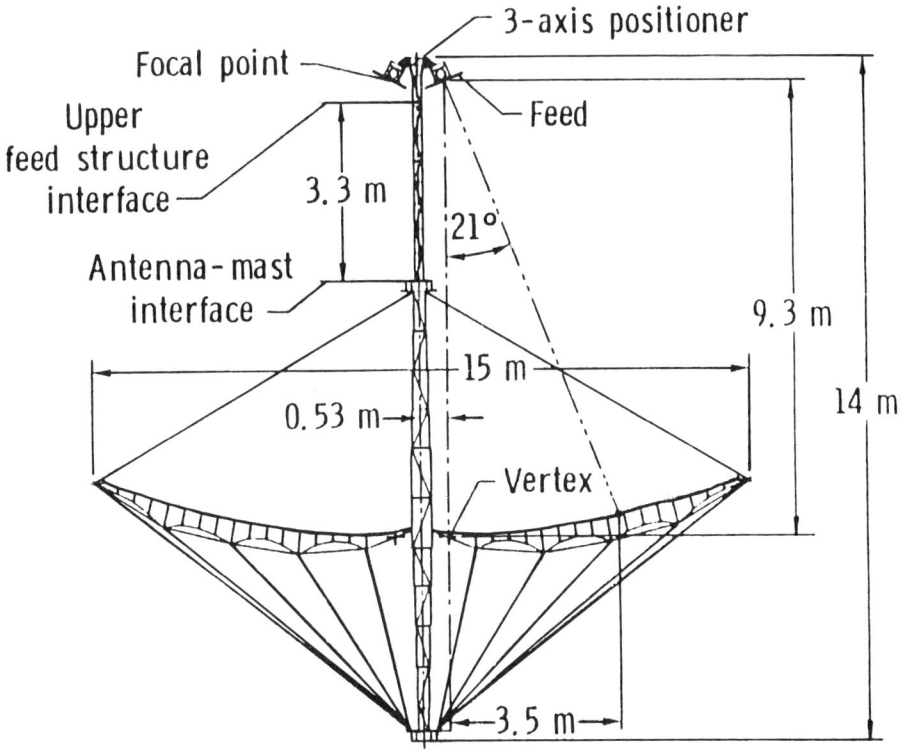

Figure 12.16 15-meter hoop-column antenna [51].

This algorithm represents a set of 96 simultaneous equations, which may be solved to obtain the best set of control adjustments to minimize a given surface error. Experimental results using 96 changes in control cable length verified the analyses in a procedure that reduced antenna surface error by 38% to 1.55 mm. This technique had the potential for on-orbit compensation for a variety of surface shape distortions. A candidate sensor for the required cable adjustments is the *surface accuracy measurements sensor* (SAMS) developed by TRW. The SAMS has demonstrated measurement accuracy of 30 μm over a distance of 60 m [20].

The DAFE studies were conducted for the NASA Marshall Space Flight Center (MSFC) by Harris and Grumman in a competition during the period from August 1980 to September 1981 [47, 48]. The two primary objectives of the study were to demonstrate by a flight experiment the capability to launch, deploy, retract, and return to earth a large (50 m) diameter space frame; and to verify by flight experiment the capability of the space frame to attain and to maintain the dimensional precision required to operate as a spaceborne antenna. Both contractors devised orbiter-attached experiments that would maximize program outputs while minimizing risks to both orbiter and the experiment. Although many flight configurations were designed, overall results were similar for both phased array and parabolic antennas. Grumman and Harris also devised measurement techniques that would provide a 1.27-mm rms accuracy required for the measurement of antenna deformation.

General Dynamics has designed space-erectable antennas and parabolic graphite-epoxy reflectors for space-based applications [43, 44, 46]. A 2.44-m diameter reflector was built and tested, and it had a surface tolerance of 0.0635 mm rms and a specific mass of 4.4 kg/m^2. Figure 12.17 shows the antenna during tests in which surface errors, focal length, and vertex displacement were measured. The space erectable designs have a specific mass of 0.49 kg/m^2, but the surface tolerance is on the order of 10 mm rms. Consequently, the space erectable antenna designs were primarily configured for relatively low frequency operation. Figure 12.18 shows the various design trades of weight *versus* diameter and packaging alternatives for the General Dynamics space-erectable antennas [43].

General Dynamics developed a single-layer membrane concept for the US Air Force RADC [53], which reduced the complexity of the three layer mesh used in large spaceborne lens antenna concepts. Microstrip elements were integrated with the T/R module subassembly as shown in Figure 12.19. Several critical issues such as weight, power distribution, producibility, stowability, and deployability of large SBR antennas were reduced by the advantages of the single-layer membrane. For example, there are concerns of entanglement and element alignment in the deployment of the three-layer mesh concept. The single-layer membrane has low-profile dielectrics with feathered edges, no element misalignment to create T/R module-to-dipole shear, and metallic mesh to provide uniform load distribution. A 127-element test array was built and showed excellent test results at 3 GHz (Table 12.8). During the study, a concept for a 70-m diameter SBR phased array at 3 GHz was designed, which indicated significant advantages over previous designs.

TRW developed an advanced antenna concept during work sponsored by the Jet Propulsion Laboratory (JPL) as part of the NASA Large Space Systems Technology (LSST) program [54]. The feasibility of stowing large solid antenna reflectors in the shuttle was examined. The antennas would be designed to operate in the 10 to 100 GHz range and maintain rms deviation on the order of 10^{-5} diameter fabrication error. Thermal deviation for a 30.5-m (100 ft) diameter antenna was estimated to be 0.0864 mm (0.0034 in) rms. The weight of antenna reflectors was estimated for diameters of 5 to 30.5 m (16 to 100 ft). Figure 12.20 shows the plot of reflector weight excluding the weight of feeds and subreflectors. The basic construction assumed a graphite epoxy aluminum honeycomb sandwich configuration. Figure 12.21 shows deployed and stowed configurations of a 10-m reflector designed by TRW.

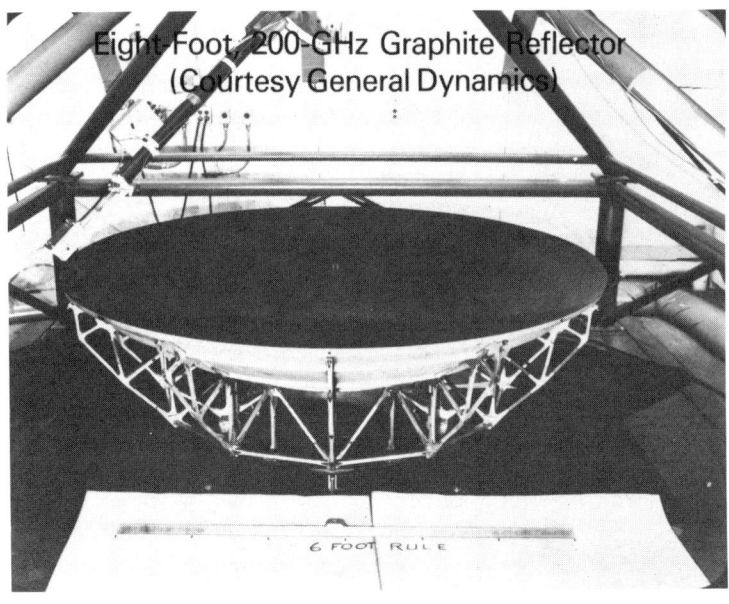

Figure 12.17 Eight-foot 200 GHz graphite reflector. (Courtesy of General Dynamics.)

Table 12.8
General Dynamics Single-Membrane Lens Antenna

Type: Microstrip active element phased array	
Frequency: 3 GHz	
Status: 1982 design and development of 127-element test array	
Bandwidth: 15%	
Configuration: See Figure 12.19	
Element Configuration: 0.7λ triangular lattice	
Concept:	
Diameter	70 m
Frequency	3 GHz
Weight	5017 kg
W/A	1.3 kg/m^2
Fundamental Frequency	0.129 Hz
MAX. Deflection	3.94 mm
Stowed Dimension	3.26 m (diameter)
T/R Module Weight	3.82 g
T/R Module Ave. Input Power	0.65 Wdc
Test Array Results:	
No. of Elements	127
Directivity (3 GHz)	29.5 dB
Feed Losses	7.5 dB
Measured Gain	18 dB
Calculated Gain	18.4 dB

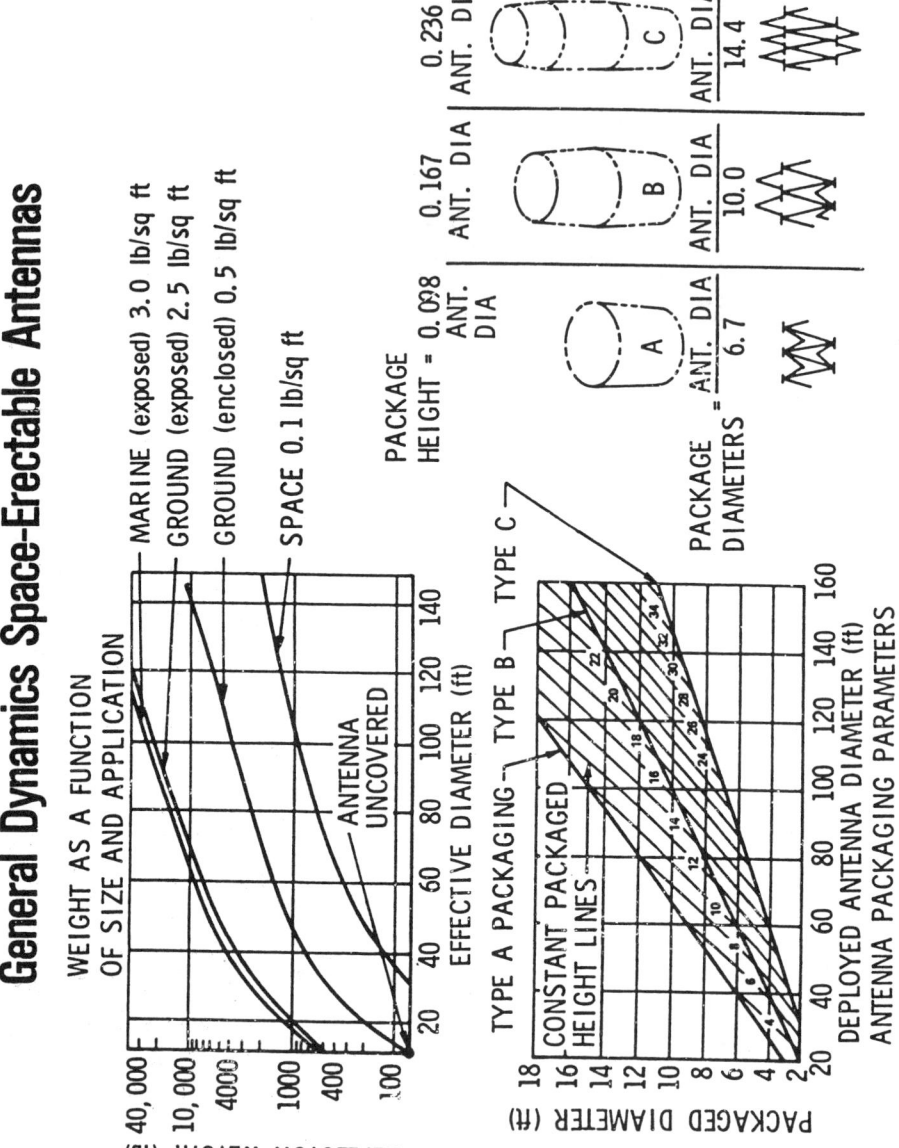

Figure 12.18 General Dynamics space erectable antennas.

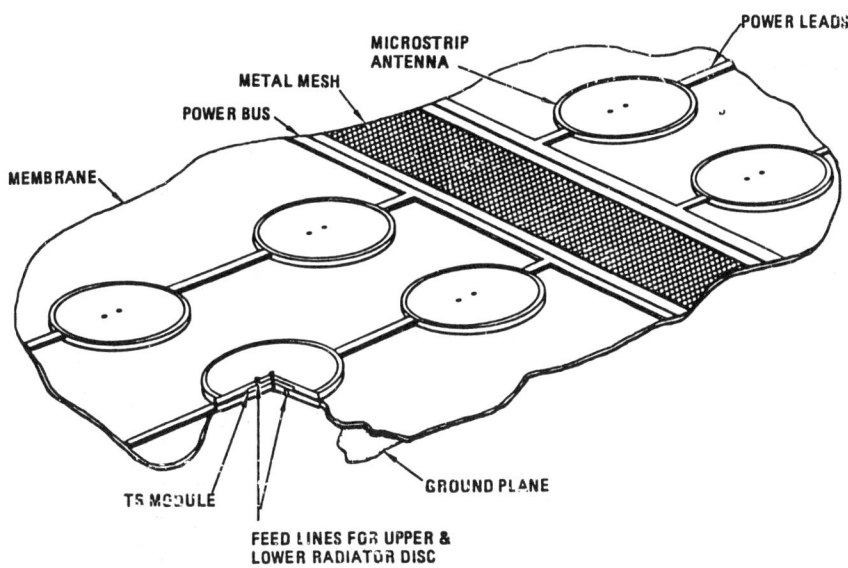

Figure 12.19 The microstrip antenna single-layer membrane concept.

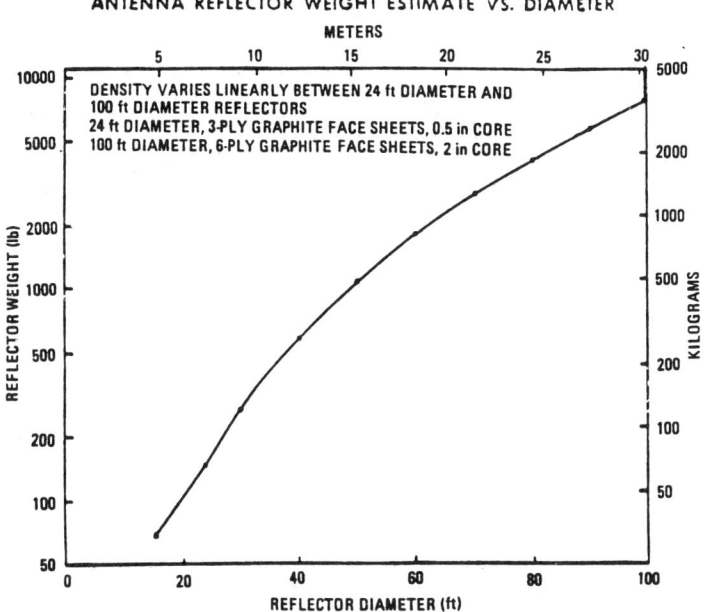

Figure 12.20 TRW antenna reflector weight estimate [54].

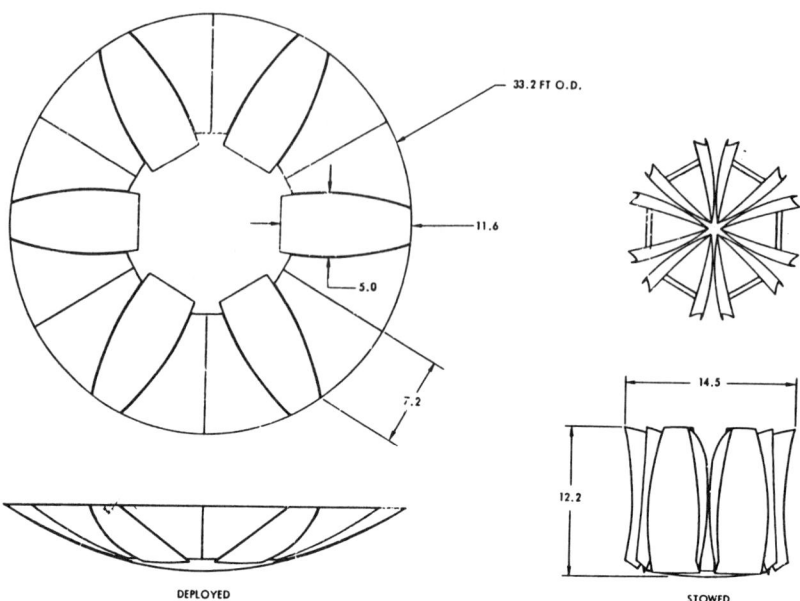

Figure 12.21 TRW Six-panel configuration.

Grumman has been developing large deployable phased-array antenna concepts since the 1970s. One configuration of a 70-m diameter lens is shown in Figure 12.22 [55] as a phased array SBR. Grumman began a program for NASA in April 1979 to provide a proof of concept for a key portion of the antenna (i.e., a 3.1-m membrane that comprises three planes shown in Figure 12.23). T/R modules were tested in the membrane antenna in 1987. The modules are mounted in the ground plane, and they are connected to the dipoles in the feed-side and target-side dipole planes. This antenna configuration is identical to that which Grumman studied in the DAFE program, discussed earlier. This antenna has also been independently simulated by Atlantic Research Corporation [56]. One result from the simulation showed an H-plane grating lobe singularity angle of 23.49° when the array elements were located on an equilateral triangular lattice. The altitude of the triangle was 0.413λ. The structure tolerance equations for the Grumman lens antenna are [56]:

$$E_a \approx \frac{\epsilon}{\cos\theta - \left(1 + \frac{\alpha^2}{4f^2}\right)^{-1/2}} \qquad (12.18)$$

$$E_r \approx \frac{\epsilon}{\cos\Psi \sin\theta - \left(1 + \frac{4f^2}{\alpha^2}\right)^{-1/2}} \qquad (12.19)$$

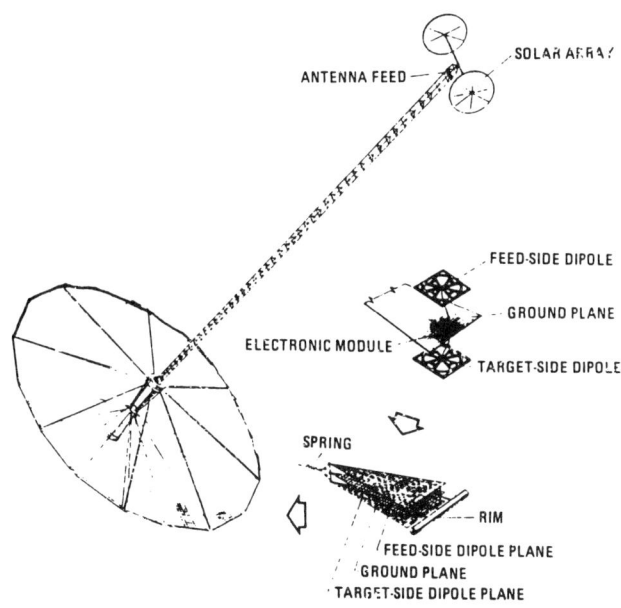

Figure 12.22 Space-fed phased array SBR antenna [55]

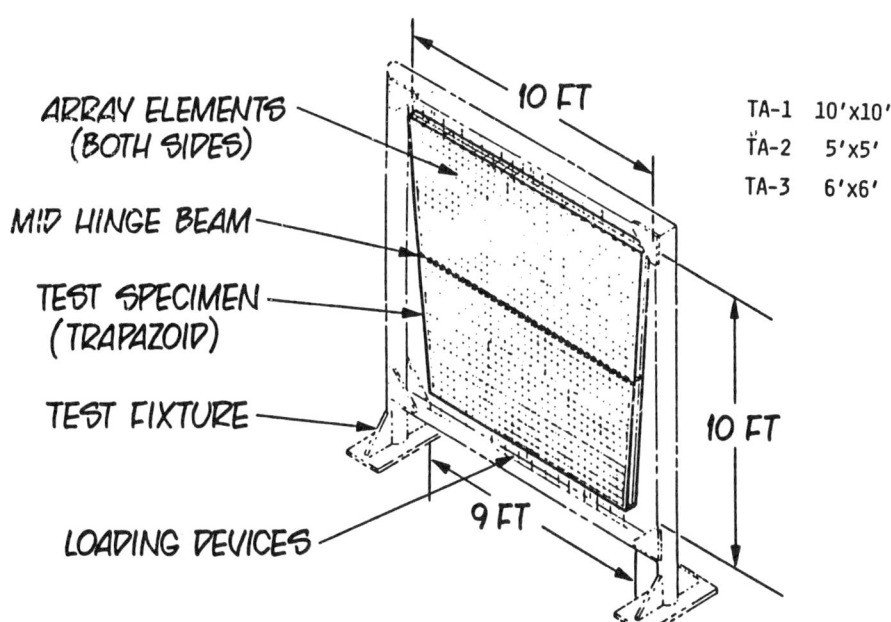

Figure 12.23 Phased array flexible membrane test specimen [55].

where

E_a = axial structural error;
E_r = radial structural error;
ϵ = electrical error allowable;
Ψ = location of the element in azimuth on the array;
α = fraction of radius at which the element is located;
θ = maximum scan angle;
f = F/D ratio.

Figure 12.24 shows typical antenna structure tolerances at the rim and at the drum for an allowable $\lambda/32$ phase error [56]. The allowable tolerances are relatively large compared with reflector antennas that are designed for the same phase error and level of sidelobes.

During a low-altitude SBR study, Grumman and Raytheon proposed a large, single-axis, rollout antenna concept that is shown in Figure 12.25. A 60 m × 25 m aperture is deployed by a dual-drum configuration, as shown in Figure 12.26, into a space-fed phased array with an F/D of 1.5. This deployment concept was considered to be a near-term solution with low risk, using techniques developed for solar array deployments. Table 12.9 gives details of the antenna concept, and we can see that the antenna mass per unit area is 2.5 kg/m^2 and the allowable deformation of the feed-array is ± 50.8 mm.

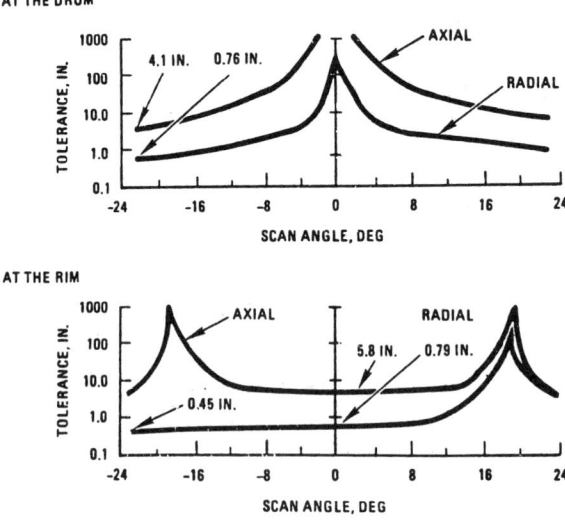

Figure 12.24 Grumman lens antenna structure tolerances [56].

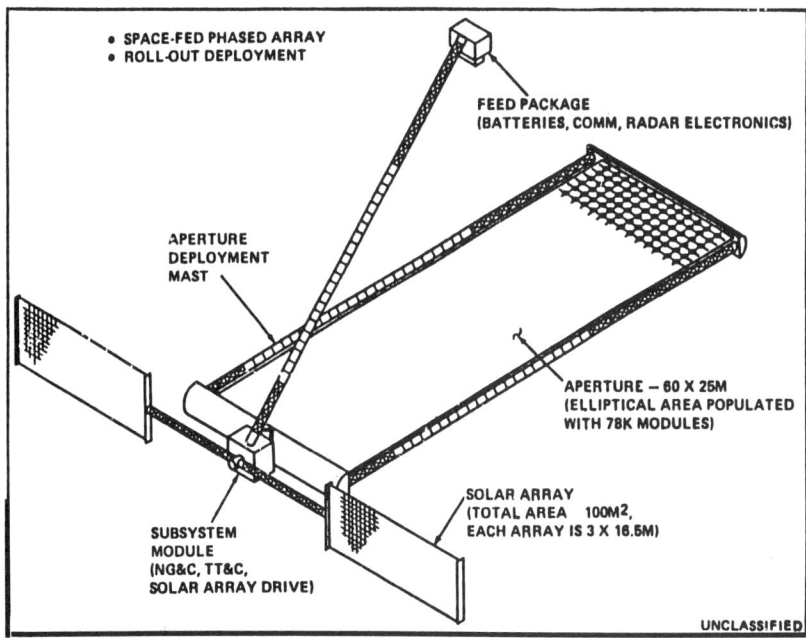

Figure 12.25 Grumman-Raytheon LASBR concept.

Table 12.9
Grumman Single-Axis Rollout Antenna Concept (Circa 1984)

Frequency:	L-band (1.31 GHz)
Bandwidth:	14% operating, ± 1 MHz instantaneous
Size:	60 × 25 m ellipse
Type:	Phased array lens ($F/D = 1.5$)
Elements:	78,510 on triangular lattice
Membrane:	3 planes
	AL ground plane
	Kapton antenna planes
Scan Angle:	± 60°
Power:	76 kW peak, 3.2 kW av
Feed:	Supported on astromast $L = 80$ m
	1.42 m wide × 0.59 m high × 0.46 m deep
	88 transmit modules
	Weight 342 kg
Mass:	3744 kg
Mass per Unit Area (W/A):	2.5 kg/m^2
Peak Deformation Estimates:	Membrane 12.95 mm
	Feed 0.16° tilt angle
	10.16 × 7.87 mm linear
Allowable Feed Deformation:	Linear ± 50.8 mm
	Angular ± 1°

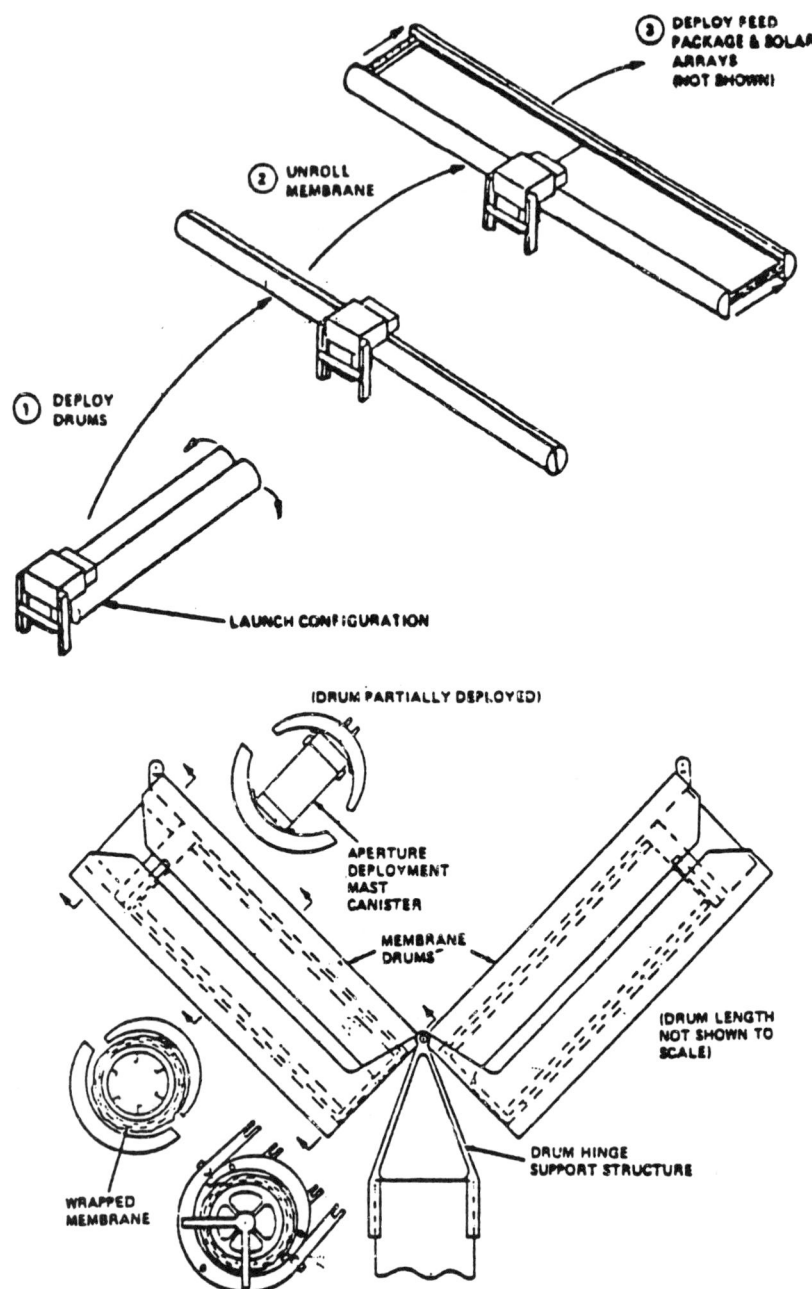

Figure 12.26 Dual-Drum deployment concept and structural details.

Antenna systems have been studied and fabricated under the LSST program. The work being performed is described each year at the annual technical reviews held at NASA-LARC [57, 58]. Each contractor is developing unique concepts for large deployable antennas in the 20 to 100 m diameter class to support frequencies up to 100 GHz. Antenna surface accuracy technology up to one-fiftieth of a wavelength is being developed. In early 1984, one contractor, Lockheed Missiles and Space Company (LMSC), demonstrated in a simulated zero-gravity environment the technology for a large space-deployable antenna [59]. LMSC fabricated a 22.5° sector of a 55-m diameter *"wrap-rib"* parabolic antenna and deployed it in a ground-based zero-gravity facility. The surface of the antenna is a knit mesh of 1.2-mil gold-plated molybdenum wire that is contoured by graphite-epoxy ribs. Each rib weighs 9.1 kg, is 27.5-m long, and is lenticular in shape, which allows the ribs to collapse as they are wrapped around a central hub for stowage prior to deployment. The ribs resume their required structural shape as they unwind (under constraint), thereby stretching the mesh into the proper parabolic shape (see Figure 16.13 in Chapter 16). Overall surface error is expected to be on the order of 0.3 mm rms. The Lockheed development program was initiated in March 1980 to demonstrate the readiness of large diameter offset-reflector technology through development of ground test capable, flight representative, full-size hardware.

The SEASAT-A antenna (designed by Ball Aerospace) is a 10.74 × 2.16 m microstrip array that is deployed after orbit insertion. Operating wavelength is 23.5 cm. The deployment sequence of the SEASAT antenna is shown in Figure 12.27. This antenna is very similar to the SIR-A. Both antennas are described in the literature [8, 60], and they are significant developments in large deployable antenna technology. The SIR-B antenna is similar, except that it was mechanically steered. The SIR-C antenna will be electronically steered and dual-frequency. Tables 12.10 and 12.11 provide a summary of the RF and mechanical characteristics of the Ball SEASAT class of antennas.

Ball (in 1981) designed an antenna for the low-altitude space-based radar (LASBR) mission [26]. The 13.8 m × 63.6 m array is a direct extension of space-proven SEASAT and SIR-A technology with stringent constraints of array two-way sidelobe and beam skirt performance. The design features a single-axis deployable truss, fabricated from graphite epoxy, microstrip honeycomb panels, and passive three-bit hybrid phase shifters at each of the 49,152 elements. The loss and weight penalty [$W/A = 4.02$ kg] of a corporate-feed network are compensated by using transmitting and receiving gain at each of 384 subpanels. Ball estimated that a 1987 launch date would have been feasible. Figure 12.28 shows the on-orbit configuration of the LASBR and Table 12.12 gives detailed characteristics. Figure 12.29 shows the antenna panel construction. Table 12.13 shows a summary of characteristics and status for selected Type III SBR antennas. Table 12.14 gives a summary of the parameters of four Type I SBR antennas.

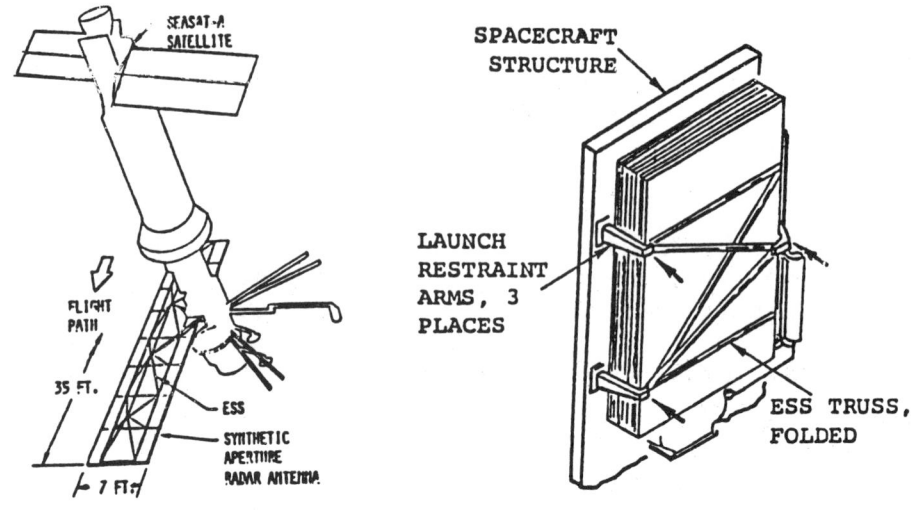

(A) **SEASAT SATELLITE CONFIGURATION**

(B) **STOWED TRUSS LOAD PATHS**

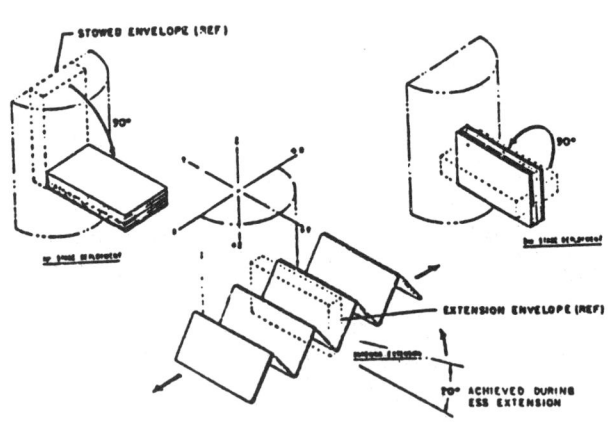

(C) **ENVELOPE DEPLOYMENT SEQUENCE**

Figure 12.27 SEASAT SAR antenna [60].

Table 12.10
SEASAT, SIR-A, SIR-B, and SIR-C Antennas' RF Characteristics

	SEASAT	SIR-A	SIR-B	SIR-C
Frequency	1275 MHz	1278 MHz	1282 MHz	1275 and 5300 MHz
Bandwidth (1.5:1 VSWR)	22 MHz	8 MHz	16 MHz	>20 MHz
Gain	34.9 dB	33.6 dB	33.0 dB	37.0 dB (L-Band)
				43.0 dB (C-Band)
Polarization	Horizontal Linear	Horizontal Linear	Horizontal Linear	Horizontal Linear and Vertical Linear
Beamwidths:				Adjustable through amplitude and phase
H-plane	6.2°	6.2°	6.2°	0.99° L-Band
E-Plane	1.1°	1.4°	1.1°	0.24° C-Band
Beam Pointing Angle	20.5°	47°	15° to 60° (Mechanical steering)	Tilted to 35° then ±25° electronically steered

Table 12.11
SEASAT, SIR-A, SIR-B, and SIR-C Antennas' Mechanical Characteristics

	SEASAT	SIR-A	SIR-B	SIR-C
Size (Deployed)	10.74 × 2.16 m	9.4 × 2.16 m	10.74 × 2.16 m	12.06 × 4.2 m
Size (Folded)	1.34 × 2.16 m	—	4.1 × 2.16 m	4.1 × 4.2 m
Weight	103 kg	181 kg	306 kg	900 kg
Support Structure	Graphite-Epoxy 3-D Truss	Rigid Aluminum 3-D Truss	Rigid Aluminum 2-D and 3-D Truss	Graphite-Epoxy 2-D Truss
Fold Mechanisms	Multifold (Spring Loaded)	Fixed	Two Folds (Motor Driven)	Two Folds (Motor Driven)
No. of Radiating Elements	1024	896	1024	864 (L-Band)
				5184 (C-Band)
No. of Panels	8	7	8	9
Feed System	Microstrip, Coax. and Suspended-Substrate	Microstrip, Coax.	Microstrip, Coax.	Microstrip, Coax., Waveguide
W/A (kg/m²)	4.44	8.9145	13.1906	17.7683

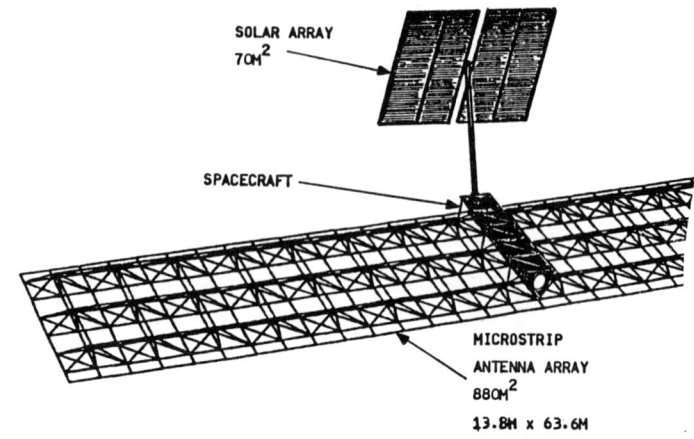

Figure 12.28 LASBR on-orbit configuration [26].

Table 12.12
LASBR Antenna System Performance [26]

Contractor:	Ball Aerospace Systems
Type:	Micro-Strip Active Corporate-Feed Phased Array
Center Frequency:	1.275 GHz
Bandwidth:	36 MHz tuned, 6 MHz instantaneous
Size:	13.8 m × 63.6 m
Directivity:	53 dB
Net Gain:	45.8 dB Tx, 45.6 dB Rx at 60° Scan
Beamwidth (at 50 dB, 2-Way):	0.58° AZ, 1.4° EL
Sidelobes:	−55 dB two-way, near-in
	−22 dB/octave, roll-off
	−6 dBi two-way, far-out
No. of Panels:	32
No. of Elements:	49.152 (on Equilateral Triangle)
No. of Phase Shifters:	49.152 (3 bits each)
No. of Subpanels:	384
No. of Transmit PA Modules:	248
No. of Receive LNA Modules:	384
Polarization:	Linear-Horizontal
Array Flatness:	±1 cm (correlated)
Structure:	Glass fiber honeycomb graphite epoxy tubes and fittings, astro-quartz cloth cover.
Weight:	3531 kg
W/A:	4.023 kg/m^2
Feasible Launch Date:	1987

Table 12.13
Satellite-Borne Deployable Antenna Parameters and Status

Nomenclature	Type	Status	Diameter (meters)	Tolerance (E_{rms} mm)	D/E	Max. Freq. (GHz)	W/A (kg/m^2)
HAC-P-22	Slotted WG	Space Qual '67	7.62 × 1.07	0.51	1,500	10	20.4
Harris DAT	Parabola	Devel. '73	15.24	1.27	12,000	10	1.4
Grumman Wire Wheel	Lens Phased Array	Study '74	300	10	30,000	0.79	0.03
LMSC ATS-6	Parabola	Deployed '74	9.1	1.52	6,000	8.1	1.4
GD-GRE	Parabola	Space Qual '75	2.44	0.0635	38,425	200	4.4
Ball-SEASAT	Planar Array	Deployed '78	10.74 × 2.16	3	3580	1.29	4.4
Harris Hoop/Column	Parabola	Prelim. Design '79	100	7.6	13,124	2	0.23
TRW Sunflower	Reflector	Study '79	30	0.404	74,280	100	4.8
JPL-EXP	Parabola	Study '80	30	1.0	30,000	15	?
Harris-TDRSS	Parabola	Space Qual '81	4.88	0.46	10,674	15.1	1.3
Grumman-DAFE	Lens Phased Array	Study '81	50	8	6,250	—	0.42
Ball LASBR	Active Corporate Phased Array	Study '81	63.6 × 13.8	10	6,360	1.275	4.02
GD	Single Membrane Phased Array Lens	Design '82	70	3.94	17,767	3	1.3
Grumman Rollout	Phased Array Lens	Study '84	60 × 25	12.95	4,633	1.31	2.5
Harris Hoop-Column	Parabola	Test '85	15	1.55	9,681	11.6	2.32

Table 12.14
Rendezvous Radar Antennas

Parameters			
Nomenclature:	Iracs	Apollo	Gemini	OMV
MFR:	Hughes	RCA	Westinghouse	Motorola
Frequency (GHz):	13.75–15.15	9.8328 Tx	L-band	9.5–9.8
		9.792 Tx		
Gain (dB):	38.4	31.5	15.3	30.5
Beamwidth (degrees):	1.68	4	50	5
Coverage (degrees):	±40 Cone	±70 × 225	70	±20 Cone
Polarization:	Linear	Circular	Circular	Linear
Sidelobe (dB):	−22	−13		
Angle Accuracy (3σ):	8 mrad	8 mrad (bias)	10 mrad	20 mrad
		4 mrad (random)		
Angle Rate Error:	0.14 mr/s	±0.4 mr/s		
Size (cm):	91.3 diameter	63.5 diameter	1.64λ × 1.64λ	35.6 × 38.1
Type:	Parabola	Parabola	Interferometer	Planar slotted array
	Focal point feed	4-horn	Array of 4 spirals	
	Monopulse	Monopulse	1 Tx, 3 Rx	
Weight (kg):	1.8 (Reflector)	18.8 (Total)	<1	5 (Array)
	45.4 (Total)			11.8 (Total)
Notes:	Weight includes reflector, feed gimbal, gyros encoders, WG switch.	Weight includes reflector, feed comparator, gimbal 4 gyros, resolvers, frequency multiplier chain, phase modulator, mixer, preamp.		Weight includes reflector, gimbal, comparator.

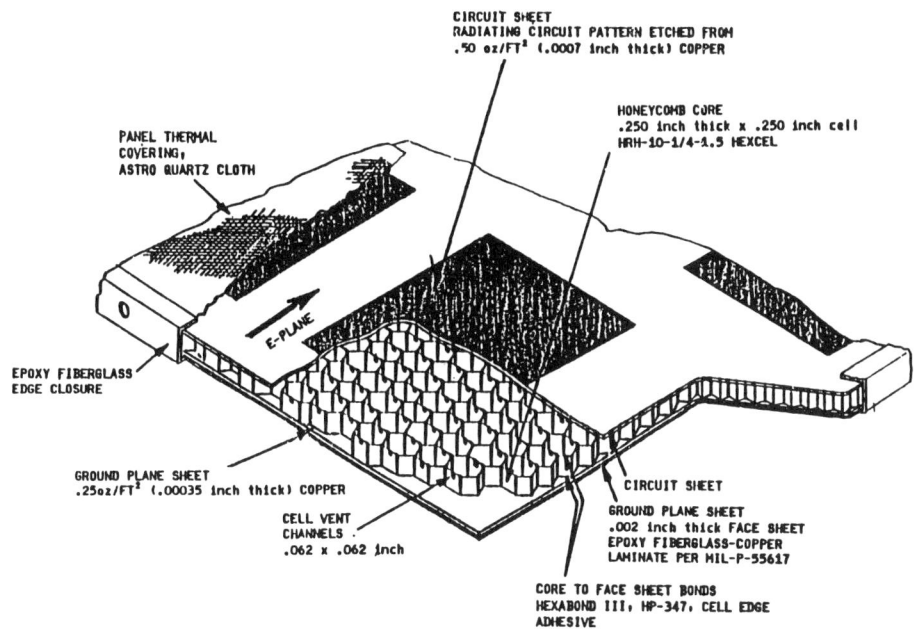

Figure 12.29 Improved SEASAT type of antenna panel construction [26].

12.2.2 Union of Soviet Socialist Republics Space Deployable Antennas

In July 1979, the Union of Soviet Socialist Republics deployed the KRT-10 antenna from the Salyut 6 manned space station [61]. This antenna is a paraboloid 10 m in diameter, having a weight of 200 kg (Figure 12.30). The antenna performed radiotelescope functions on pulsars and, in conjunction with a 70-m ground-based antenna in the Crimea, formed a very long baseline interferometer, with the baseline being larger than the earth's diameter [62]. The mesh size is 3 mm, and therefore can support propagation with short wavelengths. No operating frequencies are mentioned [62]; however, a beamwidth of 2° is given, allowing calculation of an operating frequency of 1048 MHz. The antenna was also used as a radiometer and observed an 8-km ground area from the 400-km orbit. The antenna was fixed to the spacecraft, and the entire spacecraft was moved to scan the antenna beam. When the antenna tests were completed, the KRT-10 was separated from the spacecraft. The Soviet prediction for future deployable antennas called for larger and technically superior antennas and radiotelescopes for use by communication system designers, geophysicists, and energy specialists.

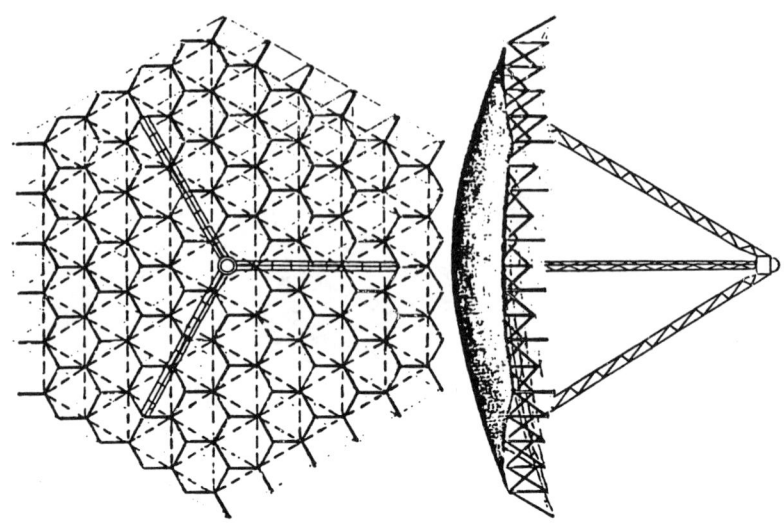

Figure 12.30 USSR KRT-10 [62].

On September 28, 1983, the Union of Soviet Socialist Republics launched the Cosmos 1500 satellite with a side-looking radar (SLR) for the purpose of all-weather probing of the surface and ice cover of the earth's seas and ocean [63]. The SLR is described in Chapter 1. The antenna is a slotted waveguide array of 480 slots with a length of 11.085 m and a height of 4 cm. The operational wavelength is 3.15 cm. Beamwidth of the antenna is $0.2° \times 42°$, providing a gain of 35 dB. The antenna is constructed of a copper waveguide that measures 23×10 mm in cross section. The slots are in the wide wall with variable spacing to provide a cosine on a pedestal amplitude distribution. Figure 12.31 shows the antenna during deployment [63]. The five sections of the antenna are mated and held in position by spring-loaded locks on the ends that are operated by release mechanisms at the end of the deployment cycle. Helical springs are provided on the flange faces along the wide wall for electrically tight joints. Relative leakage power between sections is down by 50 dB. After deployment, the antenna can be rotated through 35° from the nadir. Cosmos 1500 has been followed by Cosmos 1602, Cosmos 1766, and Cosmos 1869.

Bujakas *et al.*, from the Union of Soviet Socialist Republics Academy of Sciences in Moscow presented one of a series of papers at the 1977 IAF Congress that concerned large antennas in space for use in a very long baseline radio interferometer [49]. The radio telescope considered the use of antennas in space with diameters up to 10 km that operated in the 1 mm to 1 m range of wavelength.

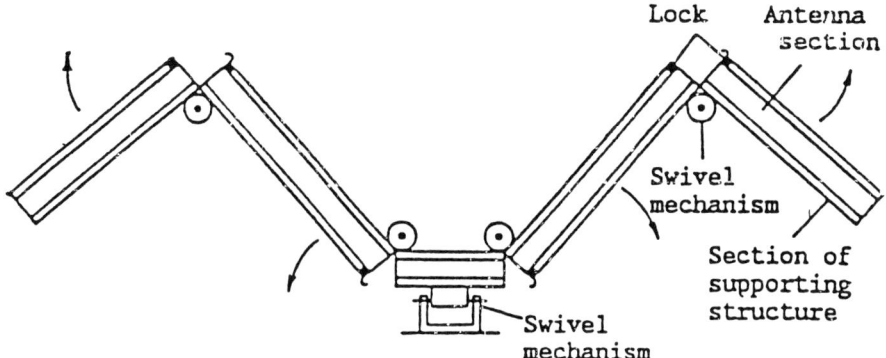

Figure 12.31 Schematic of antenna module for COSMOS 1500 [63].

The large diameter antenna reflector was assembled from 200-m modules that had a specific mass of 0.2 kg/m^2. These modules were composed of a spatial rod framework on which flat equilateral hexagons of about 4-m diameter were mounted. Figure 12.32 shows the Union of Socialist Republics concept for the infinitely built-up radio telescope. Calculations by the Soviets [49] predicted that the range of such radio telescopes would be between 1.5 and 1500 billion light years when the sensitivity was $3(10)^{-37}$ WM^{-2} Hz^{-1} and the resolution was $1.5(10)^{-10}$s of arc. The Soviets stated that the technology allowed development and construction of the 200-m module starting in 1977. Estimates for the total cost of developing a system that used reflectors 1 to 10 km in diameter were assessed at one to ten billion dollars.

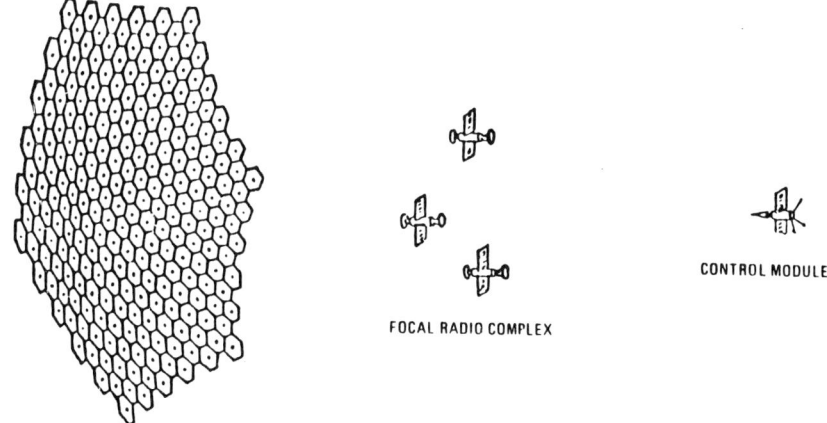

Figure 12.32 Infinitely built-up radio telescope [49].

12.3 TESTING SPACE ANTENNAS

The far-field distance for a 50-m antenna that operates at a 10-cm wavelength is 50 km. Consequently, large space-based antennas should be tested in near-field test facilities on the earth before deployment in space. The technology is well developed in near-field testing [64, 65]. We may reasonably expect the antenna far-field parameters below to be measured to the following accuracies, using a near-field test facility [67]:

On-Axis Gain (dB)	NLT 50 + 3.0
Null Depth (dB)	−45 + 3.0
Pattern Gain (dB)	10 + 1.5
Sidelobe (dB)	30 + 8
Beam Boresight (degrees)	10% of beamwidth
Beamwidth (degrees)	5%
Cross-Polarization (dB)	−25
Monopulse Error Slope (volts/degree)	0.75 (+1%)

When the SBR antenna is in orbit, the effective radiated power can be measured on earth to check the antenna gain. These tests can be combined with on-orbit measurements of antenna deployed dimensions using an on-board measurement system such as the TRW's SAMS [20]. SAMS is capable of measurement of antenna surface deformation with an error or 30 μm at ranges up to 60 m. Figure 12.33 shows the performance of SAMS [20].

12.4 FUTURE OF SPACE ANTENNA TECHNOLOGY

The deployment and operation of Type I and Type II SBR antennas have been demonstrated. Type III SBR antenna technology is in its early stages of development. Antennas with diameters of tens to hundreds of meters are being seriously considered for deployment in the 1990s. Figure 12.34 shows a JPL projection [66] of where the technology is headed in terms of antenna sizes and operating frequency. The author has added concepts to Figure 12.34 that have been studied since the JPL made the original projection. Along with these trends, large and complex phased arrays and feeds will also require development. The technology will require extension to the areas of higher frequency, broader bandwidth, multiple beam, multiple frequency, and adaptivity. There will be challenging work for spaceborne antenna engineers for decades to come.

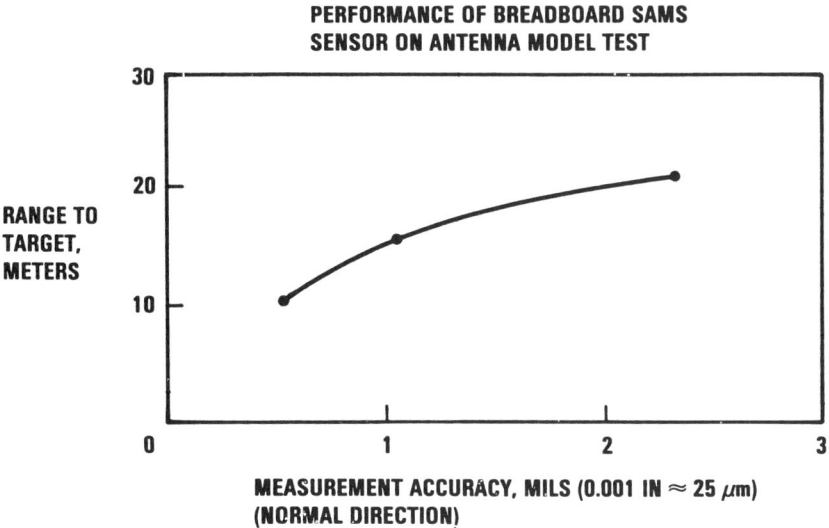

Figure 12.33 Performance of breadboard SAMS sensor on antenna model test.

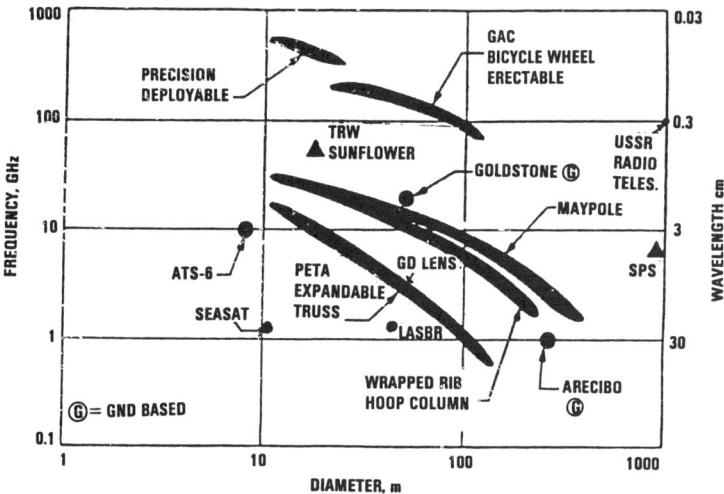

Figure 12.34 Projection of space antenna capability [66].

REFERENCES

1. Cantafio, L.J., "Space-Based Radar Systems," Chapter 1 of *Space-Based Radar Handbook*, L.J. Cantafio (ed.), Artech House, 1989.
2. Fenner, R.G., and R.F., Broderick, "Spaceborne-Radar Applications," Chapter 34 of *Radar Handbook*, M.I. Skolnik, (ed.), McGraw-Hill, 1970.
3. Hughes Aircraft Company, "Ku-band Integrated Radar and Communications Equipment for the Space Shuttle Orbiter Vehicle," Preliminary Design Review, Vol. I, March 14–24, 1978.
4. Griffin, J.W., et al., "Ku-band—The First Year of Operation," *Rec. IEEE 1985 Int. Radar Conf.*, May 6–9, 1985, Arlington, VA, IEEE Cat. No. 85CH2076-8, pp. 330–339.
5. RCA Government and Commercial Systems, Aerospace Systems Division, Burlington, MA, "The Apollo LM Rendezvous Radar and Transponder," Report LTM 3300-15D, February 1971.
6. Quigley, W.W., "Gemini Rendezvous Radar," *Microwave J.*, Vol. 8, No. 6, June 1965, pp. 39–45.
7. NASA George C. Marshall Space Flight Center, Alabama, "Orbital Maneuvering Vehicle," Request for Proposal, 1–6-pp-01438, November 1985.
8. Elachi, C., et al., Spaceborne Synthetic Aperture Imaging Radars: "Applications, Techniques and Technology," *Proc. IEEE*, Vol. 70, No. 10, October 1982, pp. 1174–1209.
9. Elachi, C., and J. Granger, "Spaceborne Imaging Radars Probe," *IEEE Spectrum*, Vol. 19, No. 11, November 1982, pp. 24–29.
10. Williams, F.C., et al., "The Pioneer Venus Orbiter Radar," 1976 WESCON Session 4, Los Angeles, September 1976, pp. 14–17.
11. Hofmeister, E.L., et al., "GEOS-C Radar Altimeter," Vol. 1, Data Users Handbook, General Electric Co., Utica, NY, May 1976.
12. NASA Announcement of Opportunity—The Earth Observing System (EOS), January 19, 1988, A.O. No. OSSA-1-88.
13. Malibu Research and TRW, "Lorra/Tramar Design—Feasibility Study," May 11, 1988, MRA P. 214-3.
14. "Soviet Radar Records Venus Surface Imager," *Aviation Week and Space Technology*, Vol. 119, No. 17, October 24, 1983, p. 18.
15. Ulsamer, E., "In Focus—Approach Set on Space Radars," *Air Force Magazine*, Vol. 67, No. 2, February 1984, pp. 17–18.
16. Brookner, E., and T.F. Mahoney, "Derivation of a Satellite Radar Architecture for Air Surveillance," *IEEE Eascon '83 Conf. Rec.*, Washington, DC, September 19–21, 1983.
17. Cantafio, L.J., and J.S. Avrin, "Satellite-Borne Radar for Global Air Traffic Surveillance," *IEEE ELECTRO '82 Professional Program Session Rec.*, Boston, MA, May 25–27, 1982.
18. Cantafio, L.J., "Space Based Radar Concept for the Proposed United Nations International Satellite Monitoring Agency," *Military Microwaves '84 Conf.*, London UK, October 24–26, 1984.
19. *The Implication of Establishing an International Satellite Monitoring Agency*, United Nations, Pub. No. E.83.IX. 3.
20. Cantafio, L.J., "Satellite-Borne Radar," Lecture IX in Advanced Radar Technology Short Course, Technology Service Corp., San Diego, CA, April 22, 1983.
21. NASA News Release, "New Satellite to Measure Ocean Surface Topography and Sea State," No. 75-88, NASA, Washington, DC, March 31, 1975.
22. "Functional Requirements for the SEASAT-A Synthetic Aperture Radar System," Jet Propulsion Laboratory, Pasadena, CA, FR No. FM51I774 Rev. August 2, 1976.

23. Davidson, S.E., "Antennas for Spacecraft Synthetic Aperture Radars," *Proc. IEEE 1986 National Radar Conf.*, Los Angeles, March 12–13, 1986, pp. 31–34.
24. Vant, M.R., and P. George, "The RADARSAT Prototype Synthetic-Aperture Radar Signal Processor," *Proc. IEEE 1986 National Radar Conf.*, Los Angeles, March 12–13, 1986, pp. 25–30.
25. Pike, T.K., "Analysis of ERS-1 SAR Performance Through Simulation," *Proc. IEEE 1986 National Radar Conf.* Los Angeles, March 12–13, 1986, pp. 13–18.
26. Larson, T.R., "A Microstrip Honeycomb Array for the Low Altitude Space Based Radar Mission," Ball Aerospace Systems Division, Report F81-06, August 1981.
27. Jedlicka, R.P., and P.A. Henry, "SAR Antenna Trade-off Study," New Mexico State Univ., Physical Science Laboratory, Rep. No. PS01085, Contract No. NAS9-95519, June 1985.
28. NASA Brochure, "Geodynamics Experimental Ocean Satellite Project of the Earth and Ocean Physics Applications Program," NASA, Wallops Flight Center, Wallops Island, VA, 1975.
29. NASA Brochure, *GEOS-C Mission Plan*, TK-6340-001 Rev. 3, NASA, Wallops Flight Center, Wallops Island, VA, December 18, 1974.
30. NASA SP-8013, "Meteoroid Environment Model," 1969.
31. NASA SP-3051, "Space Materials Handbook," 3rd Ed., Chapter 6.
32. DuPont publication A-54310 on Kapton films.
33. Naumann, R.J., "The Near-Earth Meteoroid Environment," NASA TN D 3717.
34. R.G. Thomson, "Analysis of Hypervelocity Perforation of a Visco-Plastic Solid Including the Effects of Target—Material Yield Strength," NASA TR R-221.
35. Fager, J.A., "Application of Graphite Composites to Future Spacecraft Antennas," AIAA Paper No. 76-238, *Sixth Communications Satellite Systems Conf.*, April 6–8, 1976.
36. Schultz, J.L., and P. Nosal, "Space-Based Radar," Horizons, Grumman Aerospace Corp., Vol. 15, No. 1, 1979, p. 10.
37. Fawcette, J., "Large Radar Satellite Proposed," *Microwave Systems News*, Vol. 8, No. 9, September 1978, pp. 17–20.
38. Mrstik, A.V., et al., "RF Systems in Space—Space-Based Radar Analysis," General Research Corp., RADC TR-83-91, Vol. II, Final Technical Report, April 1983.
39. Ludwig, A.C., et al., "RF Systems in Space—Space Antennas Frequency (SARF) Simulation," General Research Corp., RADC TR-38-91, Vol. I, Final Technical Report, April 1983.
40. Bearse, S.V., "Knitted Antenna Solving Knotty Problems," *Microwaves*, March 1974, p. 14.
41. Lockheed Missiles and Space Company, "Large Furlable Antenna Study," Report LMSC-D384797, January 20, 1975.
42. Cummings, Freeman, and Benz, "Deployable Parabolic Antenna," United States Patent No. 3,789,375, December 18, 1973, assigned to Rockwell, Inc.
43. Fager, J.A. and R. Garriott, "Large Aperture Expandable Truss Microwave Antenna," *IEEE Trans. Antennas and Prop.*, AP-17, No. 4, July 1969, pp. 452–458.
44. Das, A. and J.A. Delaney, "Spacecraft Phased Array Configurations," *IEEE Trans. Antennas and Prop.*, Vol. AP-17, No. 4, July 1969, pp. 522–524.
45. Grumman Aerospace Corporation, "Final Report—Spaceborne Radar Study," AFSC-ESD Contract F19628-74-R-0140, Rep. No. 74-21AF-I, June 28, 1974.
46. Hagler, T., "Building Large Structures in Space," *Astronautics and Aeronautics*, Vol. 14, No. 5, May 1976, pp. 56–61.
47. Harris Corporation, "Deployable Antenna Flight Experiment—Preliminary Definition Study," Third Quarterly Review, June 24, 1981.
48. Grumman Aerospace Corporation, "Deployable Antenna Flight Experiment Definition Study," Mid-Term Review, NAS-8-33932, March 20, 1981.

49. Bujakas, V.I., *et al.*, "Infinitely Built-Up Radio Telescope," Paper IAF-77-67, presented at IAF XXVIIIth Congress, September 25–October 1, 1977, Prague, Czechoslovakia.
50. Ulsamer, E., "ATS-6 NASA's Huge Transmitter in the Sky," *Air Force Magazine.* Vol. 57, No. 8, August 1974.
51. Belvin, W.K., *et al.*, "Quasi-Static Shape Adjustment of a 15 Meter Diameter Space Antenna," AIAA Paper No. 87-0869-CP, 28th Structures Dynamics Materials Conference, AIAA, Monterey, CA, April 1987.
52. Harris Corporation, "15-Meter Diameter Hoop Column Antenna," Final Report, NASA Contract No. NAS 1-15763, June 1986.
53. Henry, R.R., *et al.*, "Design and Development of a Microstrip Antenna Single-Layer Membrane Lens for Space Radar," General Dynamics Corp., RADC-TR-82178, June 1982.
54. Archer, J.S., "Advanced Sunflower Antenna Concept Development," LSST First Annual Technical Review, NASA LRC, November 7–8, 1979.
55. "Phased Array Development Program—Radiating Membrane," Grumman Aerospace Corporation Orientation Briefing, April 11, 1979.
56. Schuman, H.K., D.R. Pflug, and L.D. Thompson, "Space-Based Radar Array System Simulation and Validation," Atlantic Research Corporation, RADC-TR-81-215, August 1981.
57. Ward, J.C., Jr. "Large Space Systems Technology-1979," NASA Conference Publication 2118, First Annual Program Technical Review, NASA LRC. Hampton, VA, November 7–8, 1979.
58. Kopriver, F., III, "Large Space Systems Technology-1980," NASA Conference Publication 2168, Second Annual Technical Review, NASA LRC, Hampton Virginia, Vol. I, Systems Technology, Vol. II, Base Technology, November 18–20, 1980.
59. "Lockheed Tests Large Space Antenna," *Aviation Week and Space Technology,* Vol. 120, No. 18, April 30, 1984, p. 70.
60. Brejcha, A.G., L.H. Keeler, and G.G. Sanford, "The SEASAT-A Synthetic Aperture Radar Antenna," Synthetic Aperture Radar Technology Conference, Las Cruces, NM, March 8–10, 1978.
61. Covault, C., "Radio Telescope Erected on Salyut 6," *Aviation Week and Space Technology,* August 13, 1979, p. 54.
62. Danolov, Y., *et al.*, "The First Space Radiotelescope," *Nauka i Zhizn',* No. 11, 1979, pp. 2–6.
63. Kalmykov, A.I., *et al.*, "Side-Looking Radar of Kosmos-1500 Satellite," *Issledovanive Zemli Iz Kosmosa,* No. 3, May–June 1985.
64. Johnson, R.C., *et al.*, "Determination of Far-Field Antenna Patterns from Near-Field Measurements," *Proc. IEEE,* Vol. 61, No. 12, pp. 1668–1678 (December 1973).
65. Gillespie, E.S., ed., "Special Issue on Near-Field Scanning Techniques," *IEEE Trans. Antennas and Propagation,* Vol. APS-36, No. 6, June 1988.
66. JPL, "Programmatic Considerations for Large Space Antennas," Joint-Interagency Conf. on Applications of Large Space Antennas, June 15, 1977.
67. Kaheny, R.W., "Draft Purchase Description for the Antenna Near Field Measurement System," United States Department of the Army, Army White Sands Missile Range, NM, April 10, 1986.

Apologia

The original plan for this book called for 16 chapters, including one devoted to transmitting-receiving modules. However, due to circumstances beyond the control of the Editor and the Publisher, the proposed Chapter 13, titled T/R Modules, will not appear in this volume. For this omission, we wish to offer our sincere apologies.

Chapter 14
ON-BOARD RADAR SIGNAL PROCESSORS
E.E. Swartzlander, Jr.
TRW

Integrated circuit technology has evolved from primitive chips with only a few gates to the current *very large scale integration* (VLSI) circuits, which are complete systems with hundreds of thousands of gates. Significant further growth in chip complexity and performance is expected, which will permit more advanced radar processing to be implemented on-board, thereby reducing the telemetry loads, permitting encryption to improve data security, and providing other benefits.

Systems designed to exploit VLSI differ greatly from those employing earlier technologies because emphasis is on developing chips that are *generic* in the sense that they perform complete functions and can be used for multiple applications. This often implies that "extra" gates are included to make the chip more useful for other applications. This is in contrast to earlier technologies, where custom chips were developed to implement portions of a specific processor with little applicability to others. Effective VLSI designs result from coordination of the technological constraints, basic functional structure, and "user-friendly" external interfaces. The techniques described in this chapter have been followed to develop chips that are efficient, reproducible, and sufficiently general to apply for a variety of on-board signal processing systems. Other chapters in this book describe the space-based radar algorithms to be realized. Additional detail on VLSI signal processing beyond the scope of this chapter is provided in references [1–3].

14.1 INTRODUCTION

On-board radar processing algorithms are similar to those applied to ground, shipborne, or airborne radars. The longer ranges imply greater memory capacities, higher computational rates, and modest parametric changes (such as longer *fast*

Fourier transforms, FFTs), but the basic algorithmic flow is similar. On-board radar processing is becoming practical with the advent of VLSI. Radar processing is a major application of VLSI, and as such it provides justification and direction for continued development of VLSI technology.

The computational requirements seem to be increasing without bound. One explanation is that the historical growth in processing capability leads potential users to expect continually greater performance. Despite the five order of magnitude improvement in circuit performance, processor performance is still inadequate for the on-board implementation of many advanced radar systems at present.

Radar processing generally can be divided into high-throughput front-end processing, where data rates often exceed 100 *million samples per second* (MSPS) with algorithms like pulse compression and filtering that can be realized with fixed processors, and slower back-end processing, where evolving algorithms like *constant false-alarm rate* (CFAR) detection and track association require flexibility and reprogramming ability. Different classes of processors are used for these two regimes: *special-purpose processors* for the front-end and *generic signal processors* (GSP) for the back-end.

Section 14.2 examines the evolution of generic signal processors, which may be viewed as classical stored-program computers optimized for signal processing applications.

Section 14.3 covers the use of VLSI to implement a generic signal processor. This section includes brief descriptions of several chips and an architecture that exploits them to achieve throughputs of 100 *million operations per second* (MOPS).

Section 14.4 describes the development of a VLSI-based, special-purpose signal processing system. This specialized processor implements the fast Fourier transform; for this specific algorithm, the processor achieves a throughput of over 2000 MOPS! This illustrates the advantage of special-purpose processors, whereby sacrificing programmability and flexibility realizes great levels of performance.

Section 14.5 discusses the implementation of a digital beamformer for bistatic radar applications. Beyond the issues involved in developing a specialized signal processor, this example demonstrates the development of processor and chip architecture, including algorithm and technology selection.

14.2 GENERIC SIGNAL PROCESSORS

Generic signal processing is loosely defined as the implementation of signal processing algorithms on *reprogrammable* (i.e., generic) hardware. In contrast to conventional signal processing systems where the hardware and software are custom designed for a specific application, GSP reuses a hardware design (and major software modules) for many applications by reprogramming. The GSP approach incurs a penalty in recurring cost relative to the custom design that it replaces, but greatly reduces the nonrecurring cost. In most back-end, on-board applications, the performance requirements are modest and volumes are so low that GSP solutions are attractive. With customized arithmetic units executing several operations per cycle, generic signal processors can achieve throughputs of well over 100 MOPS with current technology for so-called "well behaved" applications that arise in radar signal processing [4].

During the four decades since the development of the first stored-program computers, there has been much evolution in processor architecture, in part as a response to the increasing complexity of the requirements. One of the major design decisions in developing a GSP-based signal processing system is the choice of processor architecture. In this section, we survey programmable processor architectures to identify the basic approaches available to the systems architect.

14.2.1 Computer Architectures

In contrast to many signal processing applications, where ultra-high-speed logic is required, generic signal processors are best implemented with technology that provides moderate speed with high levels of integration. Ultra-high-speed logic incurs severe problems in the areas of integration level, thermal management, packaging, and signal propagation between circuits. While there has been progress in all of these areas, much additional development appears to be necessary before production generic signal processors implemented with ultra-high-speed logic will be feasible. The low density of current high-speed technologies and the speed penalty of driving off-chip [5] necessitates employing architectural devices such as cache memories to minimize the amount of interchip communication, a process that increases the hardware and software complexity of the system.

Von Neumann Architecture

The first computers, developed in the late 1940s and early 1950s, implemented the Von Neumann architecture [6], shown on Figure 14.1. It comprises a central processing unit coupled to a read-write memory, currently implemented with *random access memory* (RAM), although some of the early computers used sequential access memories such as mercury delay lines or magnetic drums. The memory stores both the program (i.e., the instruction sequence) and data. During operation, the *central processing unit* (CPU) reads the first instruction from memory and executes it. Then, the next instruction is read from memory, executed, and the process continues until the program is completed.

The significant aspect of this architecture is that a single memory is used for both the program and the data; accordingly, a program may access the memory

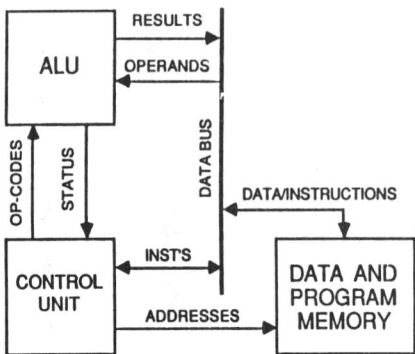

Figure 14.1 Von Neumann architecture.

and modify instructions to change the program. This stored-program concept represents the attraction as well as the disadvantage of the Von Neumann architecture. Part of the explanation for the preeminence of the Von Neumann architecture is that only a single memory unit is required. This consideration is much less significant now than it was in the early days of computing, when memory was very difficult (and expensive!) to implement, although radiation-hardened memory suitable for on-board applications is still problematical. A more significant attraction is the ease of program modification, while the disadvantage is that all operations performed within the machine (either arithmetic, control, or input-output) require memory access. Four memory access operations (i.e., an instruction fetch, two data reads, and one data write) are typically required in the execution of each instruction, and so the memory and the memory data bus must be four times faster than the other elements of the system to avoid memory bottlenecks. The requirement that the memory bandwidth be four times faster than other elements of the system contradicts one of the precepts of contemporary system design, which is that all chips operate similar clock rates.

Harvard Architecture

Another major class of architecture is the Harvard configuration shown in Figure 14.2. Here, separate memories are used for the program and for data storage. The program memory couples directly to the control unit. In implementing the program memory, RAM is often used early in the machine development cycle when the programs are subject to frequent change, and *read-only memory* (ROM) may be used for flight hardware when the design has been verified. The data memory is a RAM like the one in the Von Neumann architecture. In the Harvard architecture, the data memory is coupled directly to the *arithmetic logic unit* (ALU) without

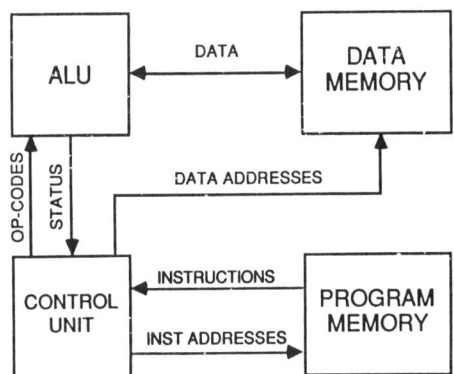

Figure 14.2 Harvard architecture.

use of a data bus. This architecture achieves a degree of concurrency in its operation: in a single cycle, an operand is accessed from the data memory for use by the ALU. Simultaneously, the control unit is accessing the next instruction from the program memory and computing the addresses for the next data access. Because the program memory and control unit operate concurrently, they can effectively use components with comparable speeds. Due to the number of operands per ALU operation, a communication bottleneck remains between the data memory and the ALU, unless the data memory is three times faster than the ALU. Thus, the memory bottleneck exhibited by the Von Neumann architecture is relieved somewhat (but certainly not eliminated!) by the Harvard architecture.

An advantage of the Harvard architecture over the Von Neumann architecture for space-based applications is that the program memory may be supported with special memory upset protection techniques. This is important because a single bit error, which may be induced by cosmic radiation, can be catastrophic if it changes the program. In contrast, such upset protection may not be necessary for the data memory because transient errors will simply cause one frame of data to be discarded.

Multiport Memory Architecture

The speed imbalance of the Von Neumann and Harvard architectures focuses on the need to provide multiple data to the ALU on each machine cycle. One solution [7] is to use multiport memory to implement the data memory. Such a multiport memory architecture can be implemented directly with a few chip types (a multiport memory, address generators, arithmetic elements, and a microprogrammed controller) as shown by Figure 14.3(a). Arithmetic elements can be implemented by using *registered arithmetic logic units* (RALU), *multiplier accumulators* (MAC), or other arithmetic elements. With this architecture, the data paths are not blocked or congested so long as the cycle times of all functions (i.e., the program and data memories, the arithmetic units, and the control unit) are comparable. In fact, with current technology, this architecture is capable of 25 MOPS. If the arithmetic elements are somewhat slower than the memories, multiple arithmetic elements can be paralleled as shown in Figure 14.3(b) to achieve a balanced architecture. Currently, multiport memories are severely limited in density compared to single-port memory, but if small data volumes are required, the performance benefits may outweigh the density limitations.

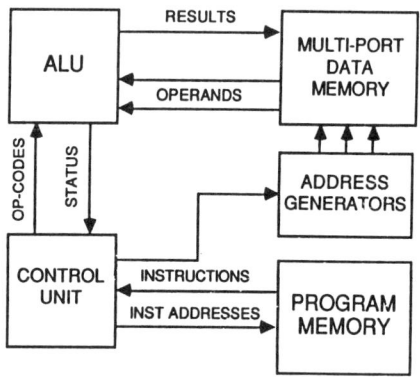

Figure 14.3(a) Multiport memory architecture, direct implementation.

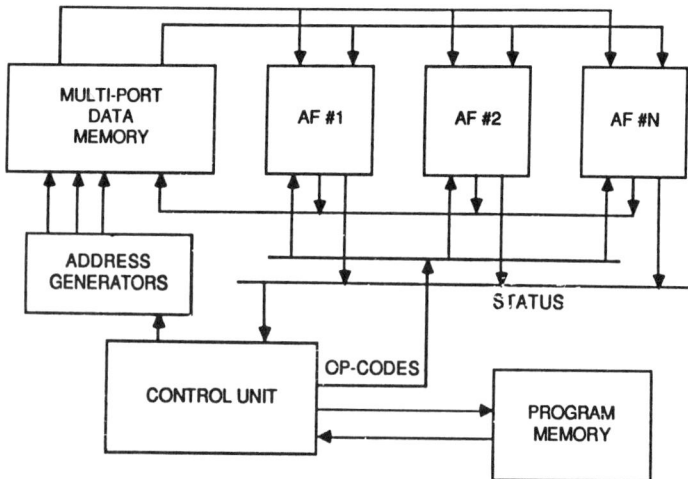

Figure 14.3(b) Multiport memory architecture, parallel elements.

In contrasting the Von Neumann, Harvard, and multiport memory architectures, the Von Neumann architecture is clearly the most flexible but achieves the lowest performance, while the multiport memory is the least flexible but achieves the highest performance. This is, in fact, a recurring theme in architecture design: flexibility and generality are obtained only by sacrificing performance.

14.2.2 Array Processors

The second category of GSP architecture is the *array processor* [8]. These are general-purpose (host) computers with special-purpose adjunct processors that are used to provide high performance for restricted classes of processing. Because the host computers are conventional single units, we will focus on the special-purpose adjuncts. The primary types of array processors in common use are parallel [8], systolic [9], pipeline [10], and programmable signal processors. Generally, parallel processors consist of a single specialized control unit that issues commands to an array of computational units that operate in parallel. Such processors often use a highly regular interconnection concept, allowing restricted communication between the computational units (e.g., with the INMOS transputer, a computational unit can be connected to up to four neighbors, typically located right, left, up, and down). Data routing between nonadjoining units (i.e., units that do not have a direct connection between them) requires a series of transfer steps during which no computation occurs.

Parallel Processors

As noted elsewhere [11], parallel processors can achieve substantial speed increases over single computers while using moderate-speed logic, but must be tailored to the problem. Commercial parallel processors are designed to be highly flexible so that they may be used for a variety of applications at a sacrifice of efficiency for any specific application. Pipeline processors also achieve high computational throughput with moderate device speeds. Like the parallel processors, they pose significant software development problems.

Systolic Architectures

The systolic computing concept extends performance by replacing the single ALU of a conventional processor (or a multiport memory GSP) with a serial cascade of several arithmetic elements, as shown in Figure 14.4. As noted by H.T. Kung, who popularized the systolic concept, "By replacing a single processing element [PE] with an array of PEs . . . a higher computation throughput can be achieved without increasing memory bandwidth" [12]. The systolic concept is most directly

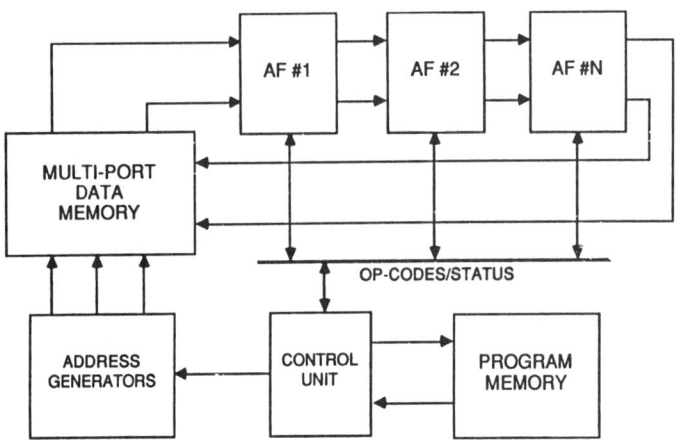

Figure 14.4 Systolic architecture.

applicable to computer-bound front-end radar signal processing applications [9], where the structure and interconnection of the arithmetic elements can be tailored for efficiently implementing important classes of algorithms. A closely related concept is pipelining, which has been applied to a variety of computer and signal processor applications [10].

Programmable Signal Processors

In the last two decades, the *programmable signal processor* (PSP) concept has evolved. Some number of arithmetic units and an assortment of input-output (I/O) channels are operated under stored-program control to provide high throughput. Although a single PSP design can be used with differing software to implement signal processors for a variety of applications, most PSP designs have special features to optimize their use within a single application area (e.g., SAR signal processing, radar beam forming). Programmability provides the flexibility to implement multiple operating modes within the single application area. Because of the desire to achieve efficiency with the PSP design, hardware is added to achieve a high level of data path flexibility. As a result, PSP hardware is more complex than that of a special-purpose processor for the same application. Also, the wide-word microcode used by early PSP designs is quite expensive to design, code, debug, modify, and maintain. The PSP is an attractive approach where a high degree of flexibility is required, such as, for example, in the early phase of system development when the software flexibility of the PSP allows correcting design errors and optimizing the algorithms.

Each of these array processor architectures is optimal for specific classes of applications. Generally, the PSP approach is best if a high degree of flexibility is required, as in the early development stages, where algorithms are not yet final or a processor must be reconfigured to accommodate evolving operational modes. Parallel, systolic, or pipeline processors are best for signal processing applications with well established algorithms that are unlikely to be changed.

14.3 VLSI IMPLEMENTATION OF A GENERIC SIGNAL PROCESSOR

We will describe some TRW chips, which are typical of the technology available for on-board systems, and show how they are used to implement a GSP for on-board radar signal processing.

14.3.1 VLSI Components

Here, we provide brief descriptions of the four-port memory, microcontroller, address generator, registered arithmetic logic unit, and multiplier accumulator. All of these are TRW chips designed for military and on-board space-based applications [13]. All operate at a 25 MHz clock rate and include a maintenance node that provides access through set-scan registers to allow an external processor to test the chips. The maintenance nodes may be chained into a maintenance network for system self-test.

Four-Port Memory

A block diagram of the four-port memory is shown on Figure 14.5. It consists of a four-bit wide "slice" of 1024 memory locations with two read ports and two write ports. Each read port comprises a ten-bit address input and an output enable control which activates the three-state output driver. Similarly the write ports consist of ten-bit address, four-bit data inputs, and write enable strobe signals. All addresses and data inputs and outputs are registered with controllable (transparent or nontransparent) registers to provide maximum application flexibility.

Microcontroller

The microcontroller, shown in Figure 14.6, generates program memory addresses based on program memory operation-codes, status signals, external inputs, or stored addresses. A "last-in, first-out" stack supports nested subroutine linking. A 16-bit address supports direct addressing of a 64 K word program memory, with a programmable offset to simplify expansion of the address space.

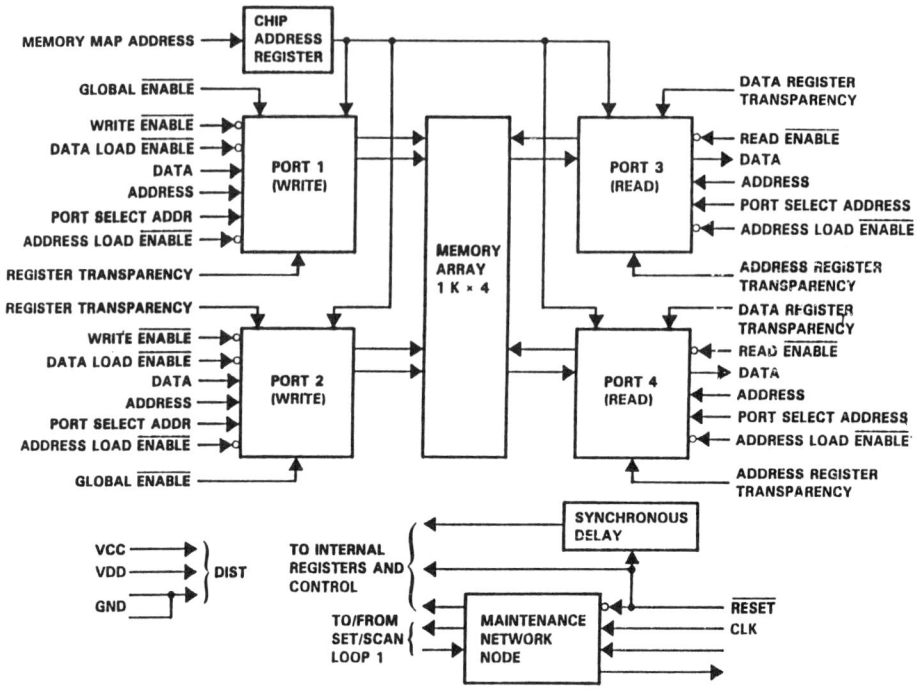

Figure 14.5 Four-port memory.

Address Generator

The address generator consists of four accumulators with independent base and step values to facilitate generation of address sequences. The block diagram is shown in Figure 14.7. The output address is selected from any of the four independently programmable accumulators, an immediate input, or addresses from the on-chip memory bank.

Registered Arithmetic Logic Unit (RALU)

The RALU, shown in Figure 14.8, consists of a cascadable 16-bit ALU, a 16-position barrel shifter, and a 32-word multiport RAM. The instruction set of the

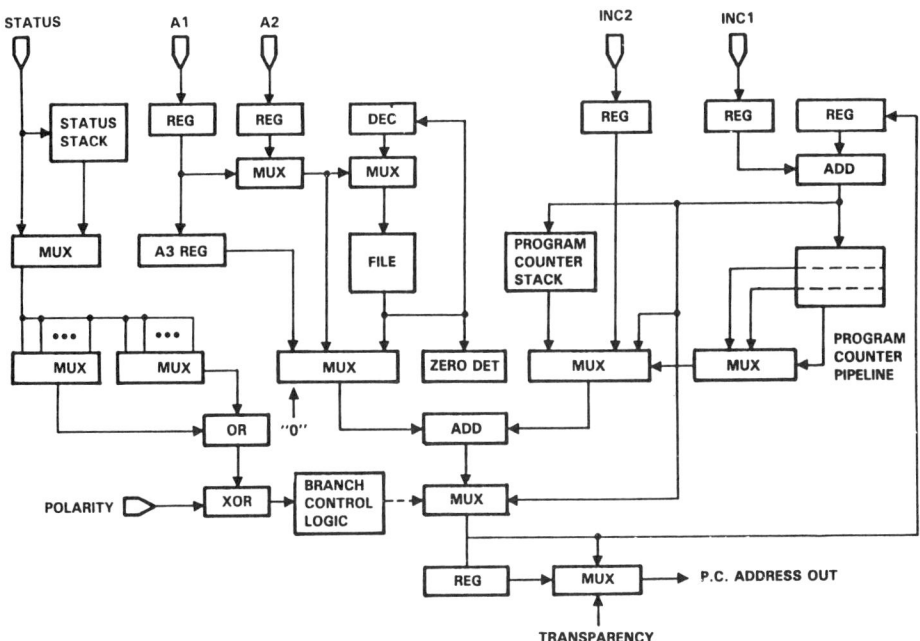

Figure 14.6 Microcontroller.

ALU supports all standard computer instruction set architectures, including floating-point arithmetic (facilitated by the barrel shifter). The RALU is cascaded as shown on Figure 14.9 to perform higher precision arithmetic.

Multiplier Accumulator (MAC)

Figure 14.10 shows the MAC, which consists of a 16 by 16 parallel multiplier, a 48-bit accumulator, and two 16-word RAMs (which operate together to emulate a 32-word multiport RAM). The MAC instruction set includes truncated and rounded 16-bit multiplication, full precision 32-bit multiplication, and 48-bit accumulation.

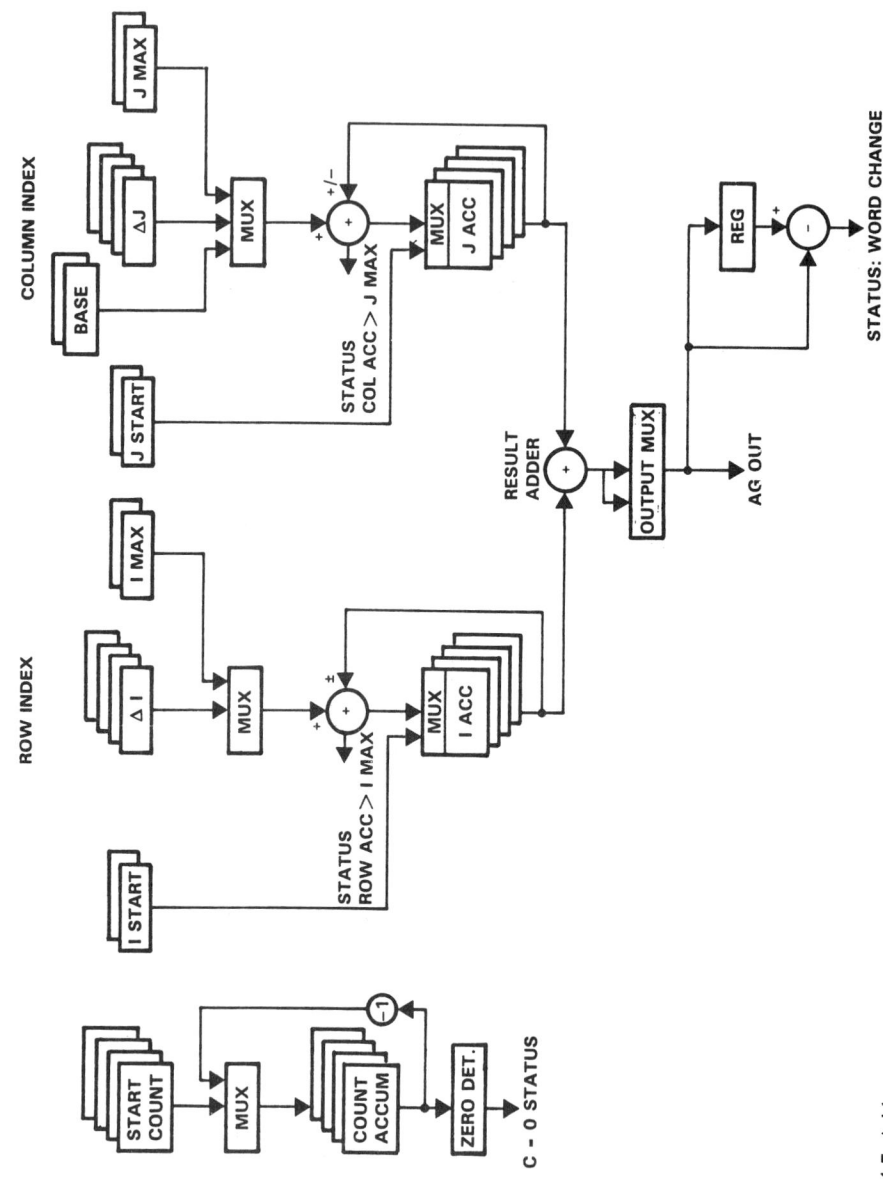

Figure 14.7 Address generator.

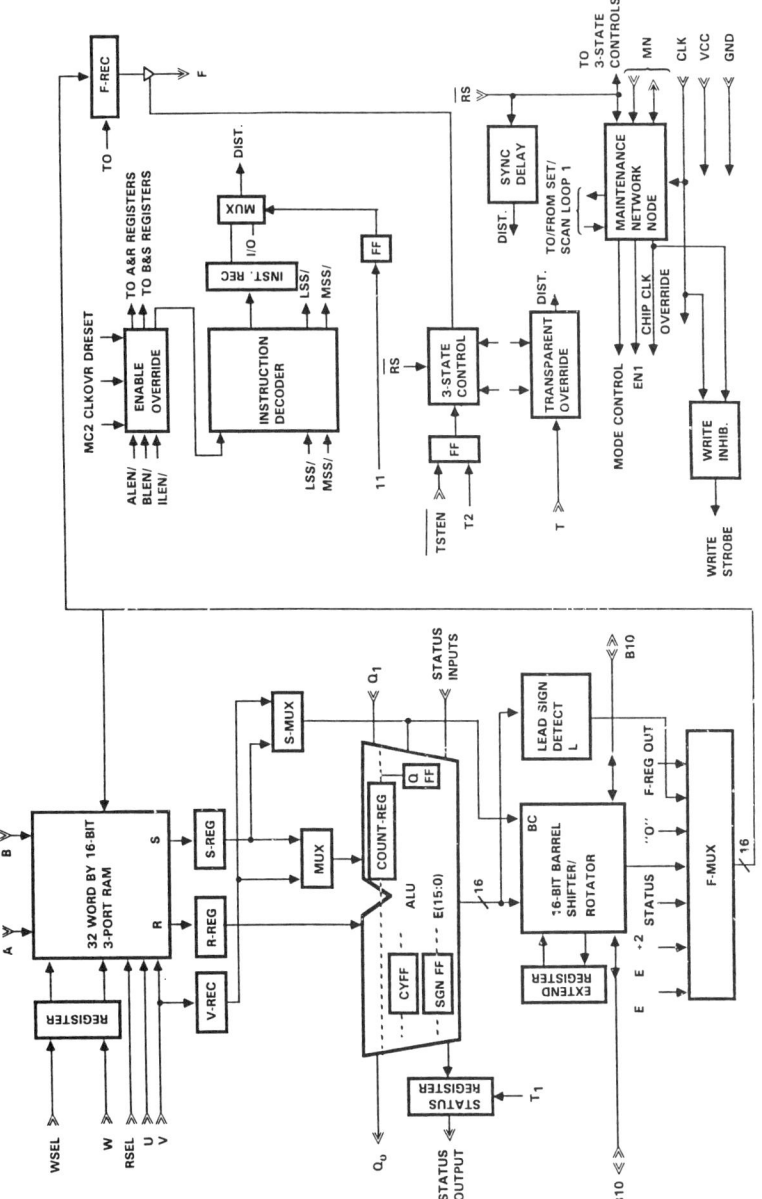

Figure 14.8 Registered arithmetic logic unit (RALU).

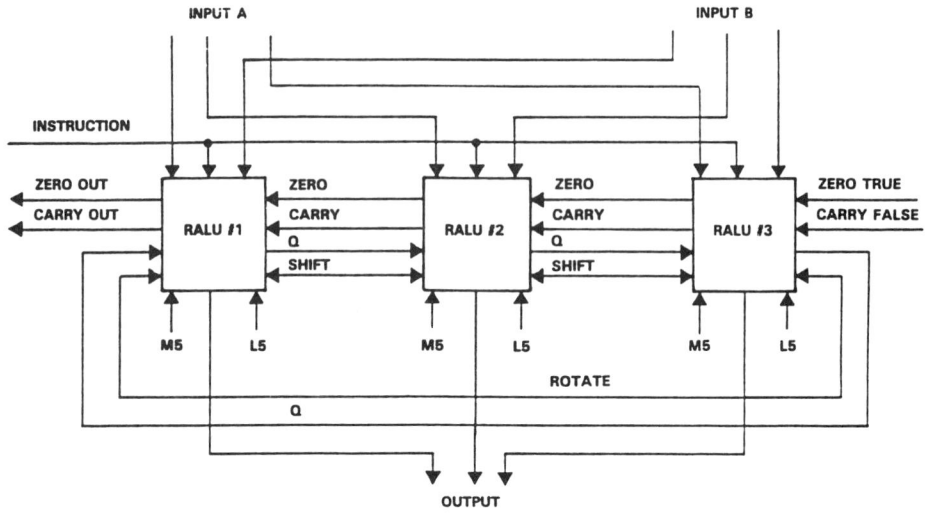

Figure 14.9 Cascading RALUs for greater precision.

14.3.2 Processor Implementation

We will describe the use of the VLSI chips of Section 14.3.1 to implement a GSP for on-board applications [14]. Although this processor was originally implemented with LSI technology, its architecture is well matched to insertion of the advanced VLSI technology. The realization described here retains compatibility with previous implementations, while extending the performance in speed and arithmetic capability and reducing the size, weight, and power requirements. Variants of this GSP, which involve tailoring the arithmetic unit to improve its performance for key algorithms, have been designed for other applications by using many of the same chips as this design. A total of six VLSI chips (with an average complexity of 17.5 K gates per chip) have been defined to realize this processor. Significant use of silicon compiler design methodologies achieves rapid (and cost-effective!) chip design.

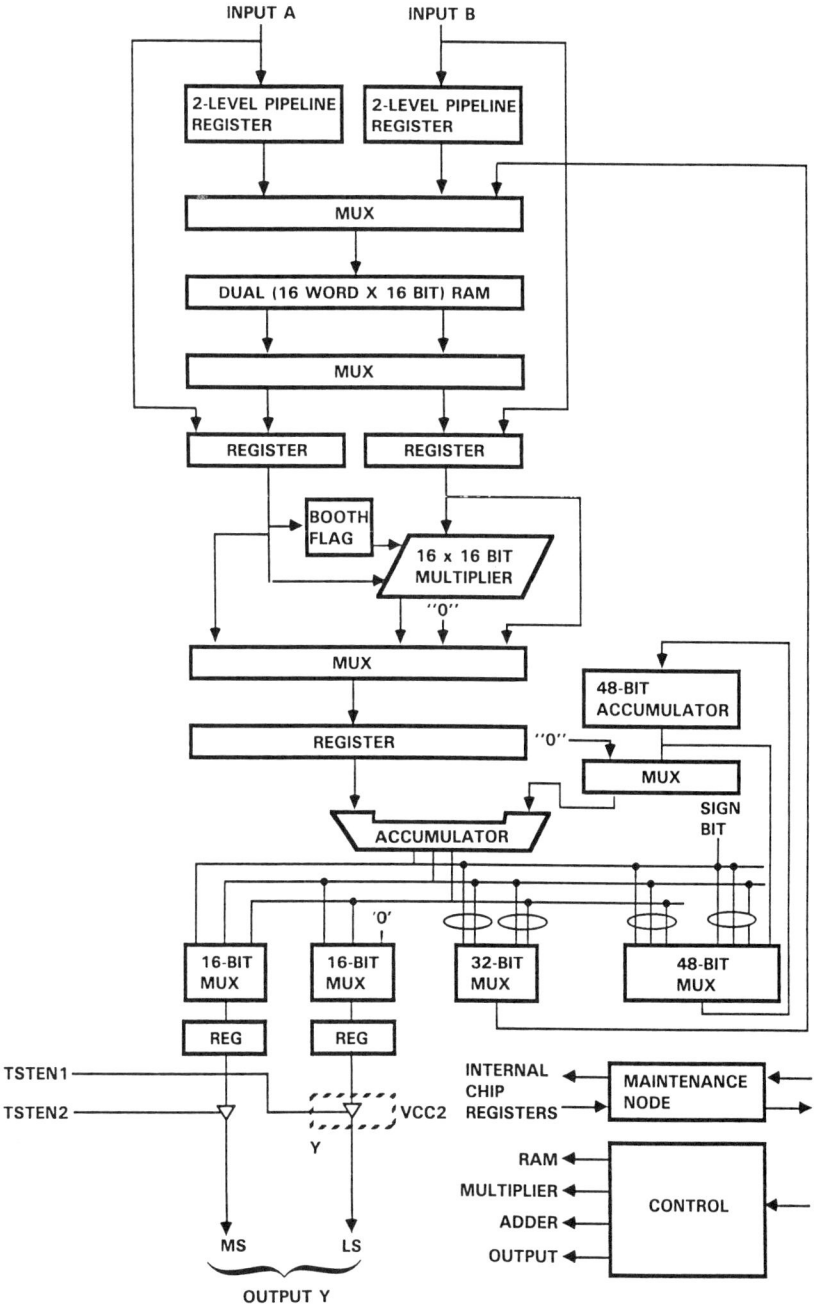

Figure 14.10 Multiplier-accumulator (MAC).

The GSP architecture, shown in Figure 14.11, consists of three major elements, a command interpreter, a system I/O unit, and an execution unit. The command interpreter performs the functions of the control unit and program memory of the multiport memory architecture shown previously in Figure 14.3(a). The command interpreter generates op-codes to control the execution unit, controls the address generators, and processes interrupts. The system I/O unit provides a data buffer and a *direct memory access* (DMA) interface between the "outside world" and the data memory.

The heart of the processor is the execution unit which implements the functions of the ALU, multiport memory, and address generators of the multiport memory architecture. The execution unit consists of a smart memory (a five-port memory with integral address generators), a complex arithmetic unit, and a microsequencer. The smart memory and complex arithmetic unit provide a tightly coupled processing loop, which operates at a 12.5 MHz cycle. Because each cycle accesses multiple data from memory, performs up to eight arithmetic operations, and writes results to memory, this configuration can achieve throughputs of up to 100 MOPS for specific radar algorithms. The implementation of the smart memory and the complex arithmetic unit are described below.

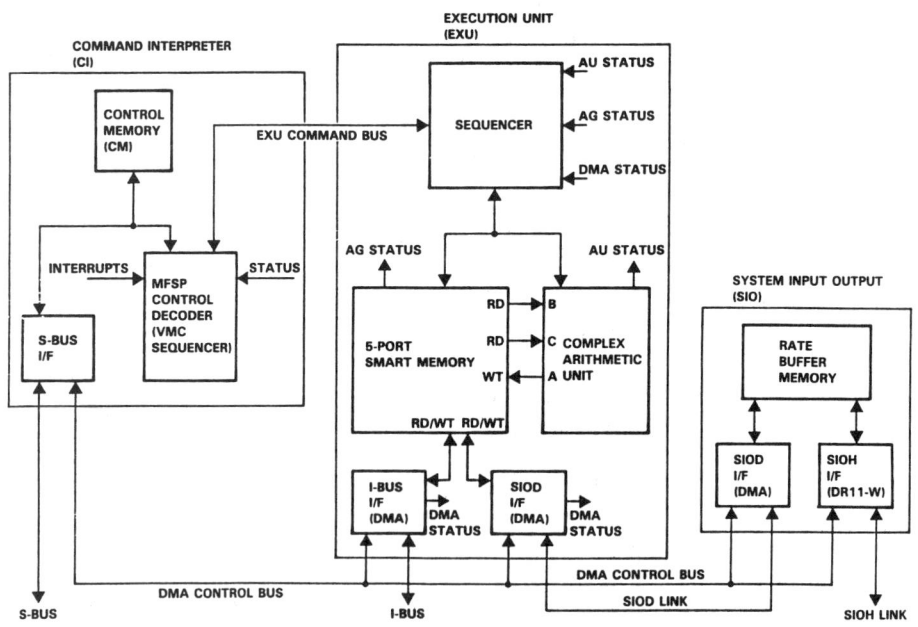

Figure 14.11 Macro function signal processor.

Smart Memory

The smart memory is shown in Figure 14.12. It consists of 5120 kilobits of memory and is realized by creating multiple virtual data ports, instead of using the four-port memory chip described in Section 14.3.1. The smart memory has a 64 K by 64-bit RAM and a 16 K by 64-bit ROM, with VLSI address generator and data formatter chips to create five virtual data ports, where three ports provide high speed interfaces between the smart memory and the complex arithmetic unit. The other two ports are for the input data bus and serial I/O ports, which are supplied with externally generated addresses. The ROM (which is relatively insensitive to single-event upset errors) is used to store most or all of the program, while the RAM is used for data storage.

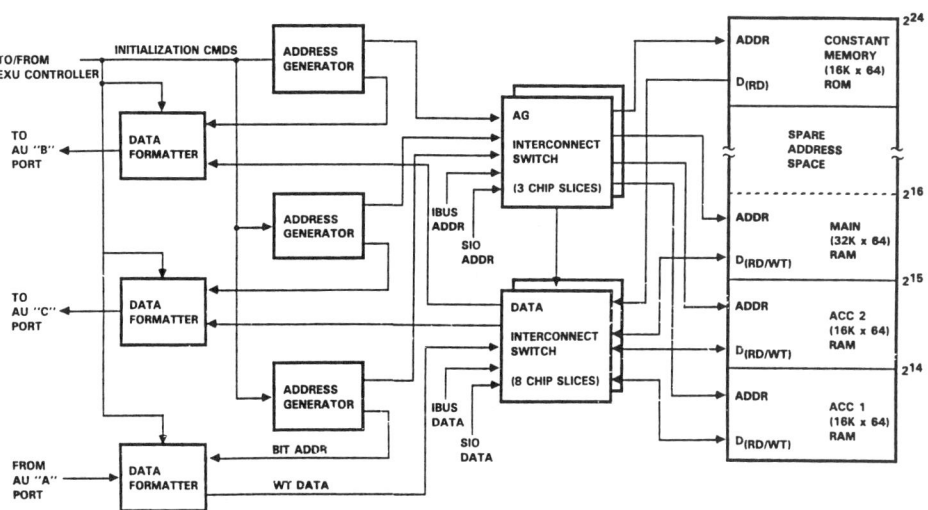

Figure 14.12 MFSP smart memory.

Complex Arithmetic Unit

The complex arithmetic unit is shown in Figure 14.13. It consists of four floating-point MACs and four floating-point RALUs, with interconnection switches to support realization of complex arithmetic operations as typically required to implement radar signal processing algorithms like the FFT. Each of the arithmetic chips includes an on-chip 32-word multiport register file to hold intermediate results and constants.

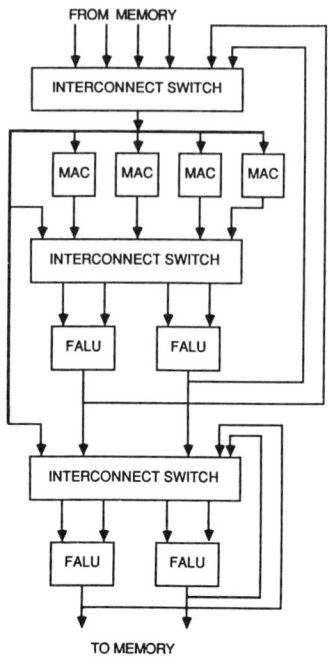

Figure 14.13 Complex arithmetic unit.

Other signal processing applications can use this processor by replacing the complex arithmetic unit with one tailored to the specific requirements. Such application-specific GSPs achieve high computational efficiency with modest nonrecurring cost because much of the design can be reused from one application to another.

14.4 SPECIAL-PURPOSE SIGNAL PROCESSOR DEVELOPMENT

We will describe the implementation of high-speed fast Fourier transform and inverse FFT signal processors with commercial and semicustom VLSI circuits.
Although this specific system is intended for a ground-based application, the approach, algorithm, and technology are directly applicable to on-board radar signal processors. The radix-4 pipeline FFT algorithm [15] is used to achieve data rates of 40 MSPS with modest 10 MHz clock rates. The interstage reordering is performed by delay commutators realized with semicustomized VLSI, while the arithmetic is performed by commercial floating-point adders and multipliers.

14.4.1 Processor Development Approach

The first step is determining the appropriate circuit technology by considering application environmental constraints (e.g., temperature, radiation), expected computational characteristics (e.g., throughput, arithmetic), and likely algorithm types (e.g., digital filtering, maximum entropy estimation, singular value decomposition). The characteristics of the selected circuit technology directly affect the algorithm design and determine appropriate architectures. For example, fast limited-complexity technologies, such as GaAs, are most effective in implementing simple serial architectures that perform recursive algorithms, while technologies that achieve higher complexity such as CMOS (i.e., complementary metal-oxide semiconductor) are most effective when implementing parallel or systolic array processor architectures.

The next step is to select an architecture and to perform an initial (high-level) design. This exercise identifies areas where a better understanding is required, such as the arithmetic rounding characteristics.

Functional simulation (where a program is developed to perform the functions of the algorithm as implemented by the high-level design) is used to resolve uncertainties about the operation of the algorithm.

Successive iterations of the technology selection, processor architecture design, and algorithm simulation serve to refine the design at ever increasing levels of detail. Upon completion of this process, a chip-level hardware design has been developed, software (if any) has been coded and debugged, and the algorithm execution has been extensively simulated (taking into account the arithmetic characteristics of the evolving hardware design). When the system is constructed, the simulation serves as a reference for component checkout, debugging, and system integration.

14.4.2 FFT Processor Implementation

On-board radar signal processors require many diverse functions: transformation, time and frequency domain processing, and general-purpose computation. For this specific frequency domain adaptive filter, the initial set of signal processing modules includes a data acquisition module, building block elements that are replicated to realize pipeline FFT and inverse FFT modules, a frequency domain filter module, a power spectral density computational module, and an output interface module [16].

All modules have separate data and control interfaces. The data interfaces satisfy a common protocol so that modules can be connected to form architectures that match the data flow of each specific application. The separation of the data and control is a contemporary realization of the Harvard architecture, which uses separate data and instruction memories to eliminate the Von Neumann memory bottleneck. In this context, the separation of data and control allows the (simple) data interfaces to operate at high speed, while the more flexible (and complex) control interfaces operate at a slower rate.

Although the Cooley-Tukey FFT algorithm developed in 1965 has made it possible to apply frequency domain digital processing to numerous on-board radar signal processing applications, many others (as described elsewhere in this book) require computational performance that exceeds present capabilities. Current data acquisition technology generates input data streams at rates of 25 to 250 MSPS, which can only be processed with special-purpose signal processors.

The FFT processor implemented here uses the radix-4 pipeline algorithm developed at Lincoln Laboratory [15]. With this algorithm, four complex data pass in parallel through a pipeline network comprised of computational elements and delay commutators as shown in Figure 14.14. Data rates of 40 MSPS are achieved by using 10 MHz clock rates because the radix-4 architecture concurrently processes four data streams. An important feature of this algorithm and architecture is that only two types of elements are used: computational elements and delay commutators. Only minor changes are required to implement forward and inverse transforms of lengths that are powers of 4. The changes involve varying the number of stages connected in series, changing the counter sequence and step size on the computational elements, and changing the length of the delays on the delay commutator. The computational element performs a four-point discrete Fourier transform. In this realization, 22-bit floating-point arithmetic is performed with single-chip adders and multipliers. The delay commutator reorders the data between computational stages as required for the FFT algorithm.

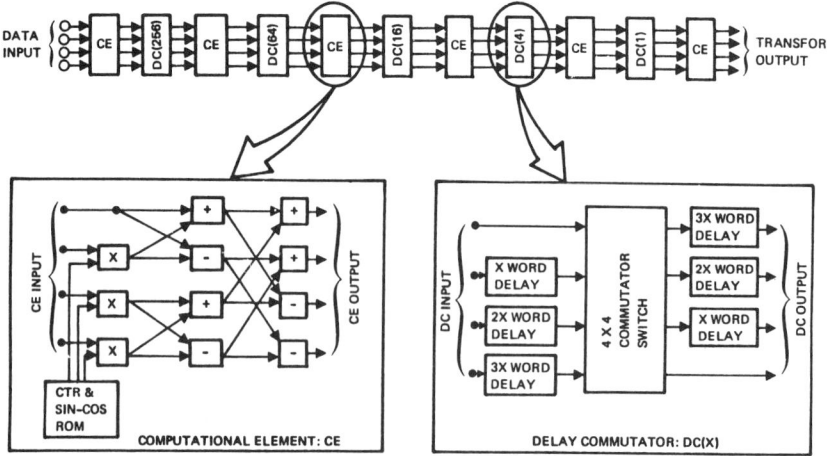

Figure 14.14 Pipeline network data flow.

14.4.3 The Delay Commutator Circuit

Careful examination of the FFT module design revealed that much of the complexity was due to the delay commutator element. Initial complexity estimates are 80 commercial integrated circuits for the computational element and 180 circuits for the delay commutator. Given the high complexity of the commercial implementation of the delay commutator, alternative approaches were examined. A B-bit wide data slice of a delay commutator that can be programmed for $X = 1, 4, 16, 64,$ and 256 requires approximately $400 \cdot (B + 1)$ logic gates and $3072 \cdot B$ shift register stages. Because a shift register stage is comparable in complexity to three random logic gates, this reduces to $400 + 9616 \cdot B$ gates. Table 14.1 compares a commercial implementation with three VLSI versions (based on gate arrays, standard cells, and custom technology). For the VLSI implementations, the maximum achievable bit slice widths are currently limited to 1, 4, and 10, respectively. With such width limitations, there would be 44, 11, or 5 delay commutator circuits required for the complex pairs of 22-bit data. For simultaneously minimizing system complexity and avoiding (expensive) customized VLSI development activity, the standard-cell approach was selected. The resulting delay commutator circuit is a 4-bit wide slice that uses programmable length shift registers and a 4 by 4 switch as shown in Figure 14.15. Data enter through shift registers with taps and multiplexers to set the delay at 1, 4, 16, 64, or 256 ($=X$) in the uppermost input register and at multiples of $2X$ and $3X$ in the middle and lower registers, respectively. Four 4:1 multiplexers implement the commutator function under the control of the programmable rate counter. The final 2-bit counter-decoder that controls the multiplexer settings can be reset and held to disable the commutator switch function for test purposes. Data from the 4:1 multiplexers are sent as output through programmable length shift registers, which are similar to the input shift registers.

Table 14.1
Comparison of Delay Commutator Implementations

Approach	Maximum Gates/Chip	Maximum Width	Chips/Stage	Relative Development Cost
Commercial	—	—	180	1
Gate Array	10000	1	44	2
Standard Cell	40000	4	11	3
Custom	100000	10	5	10

Operation of the delay commutator to reorder data is shown in Figure 14.16, where the data flow for a 64-point transform is shown [17]. The input data (with a spacing of 16) are applied to a radix-4 butterfly computational element producing output data with a spacing of 16. The reordering necessary to produce a data

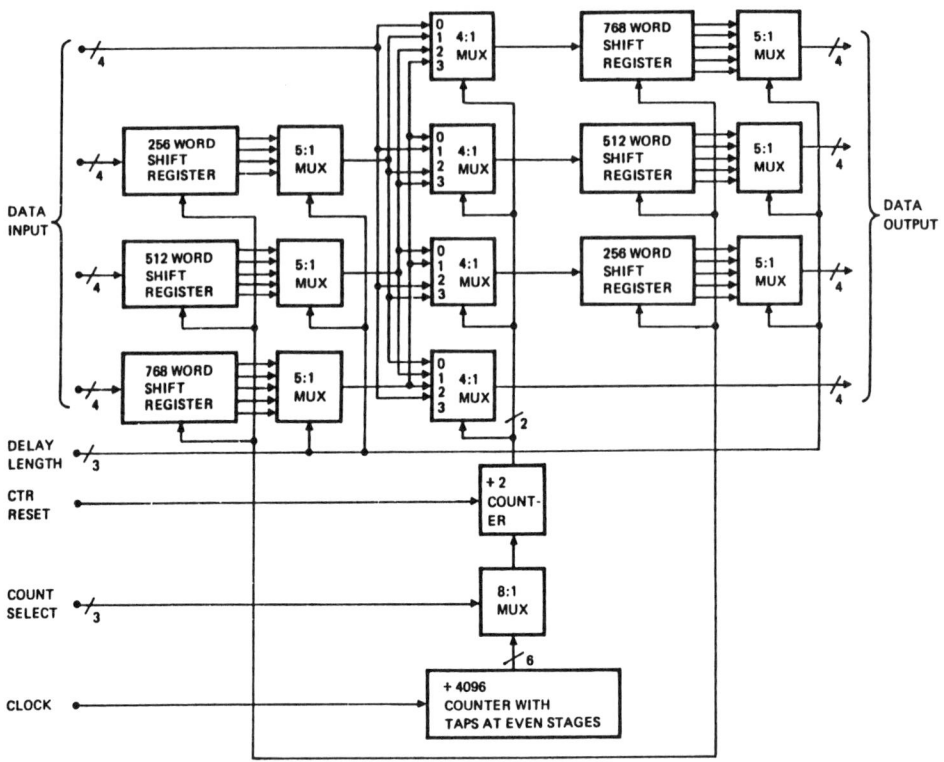

Figure 14.15 Delay commutator circuit.

spacing of 4 is accomplished with delays of 0, 4, 8 and 12; commutation at a rate of one-fourth the data rate; and delays of 12, 8, 4 and 0. The data (with a spacing of 4) are applied to a second radix-4 butterfly computational element. The resulting data are reordered by use of delays of 0, 1, 2, and 3; commutation at a rate equal to the data rate; and delays of 3, 2, 1, and 0. The final data have a spacing of 1 as required for the final radix-4 butterfly computational element.

This circuit was designed and implemented with standard-cell CMOS technology. The chip contains 12,288 shift register stages and about 2000 gates of random logic for a total complexity of 108,000 transistors [18]. At a clock rate of 10 MHz, the power dissipation is less than a half watt. The 8.6 mm by 9.5 mm chip (shown in Figure 14.17) is packaged in a 48-pin dual-in-line ceramic package. Each of the four bit slices is constructed with input registers in a column, switching logic in a second "random logic" column, and output register also in a column.

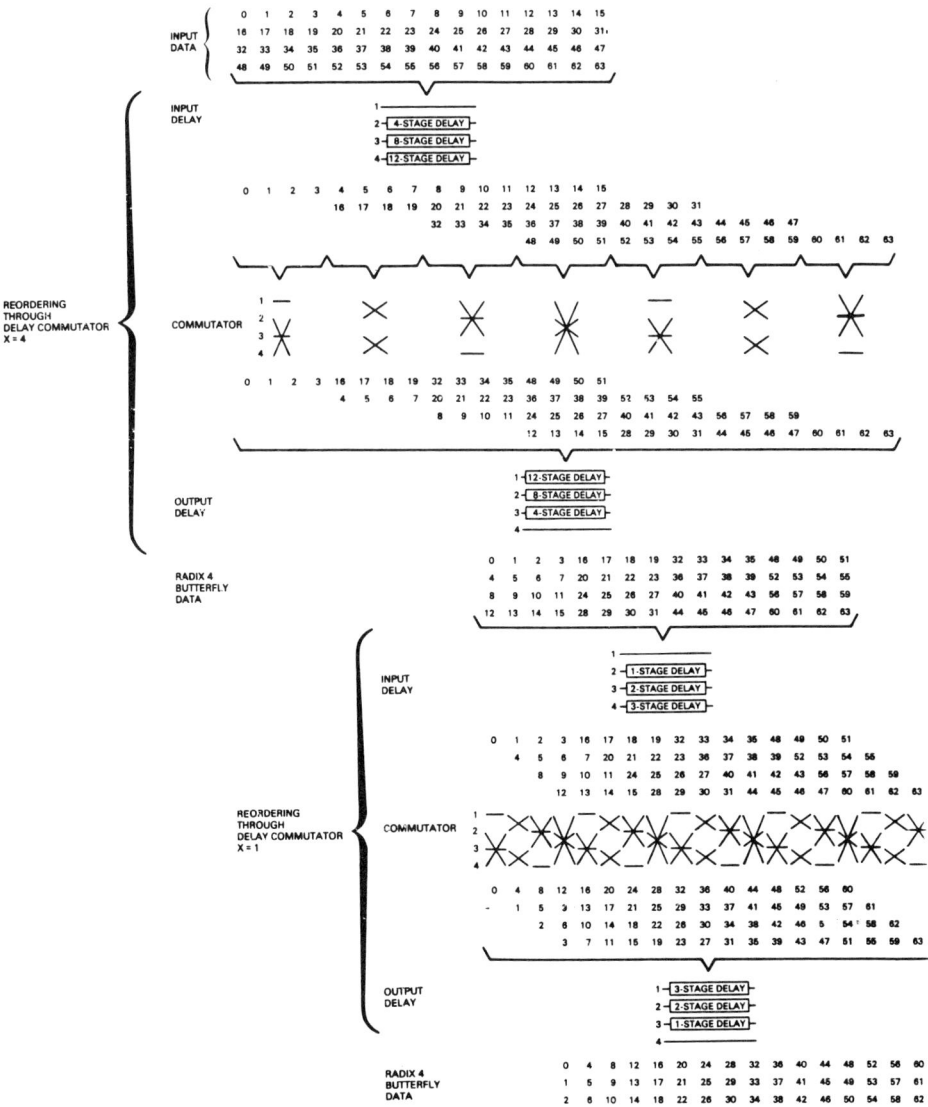

Figure 14.16 Delay commutator operation to reorder data for a 64-point transform.

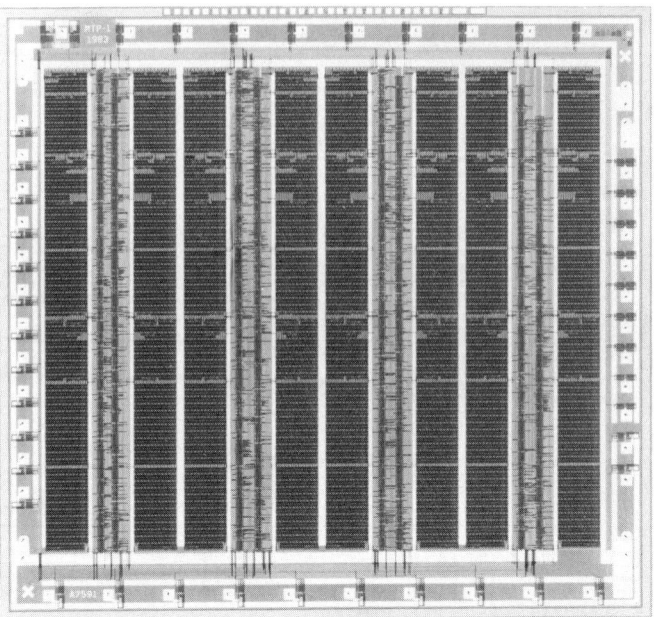

Figure 14.17 Chip for delay commutator chip in standard-cell CMOS technology.

The four nearly identical slices are about four times as tall as they are wide, producing a roughly square chip when properly stacked. There is minor variation in the random logic of each bit slice to account for sharing of the counters, decoders, and clock drivers.

Development of the delay commutator chip reduces the complexity of the 40 MHz 4096-point FFT from 1375 commercial chips to 546 chips (of which 66 are delay commutator chips). Such a reduction greatly improves system reliability as connections between circuits represent the dominant failure mechanism in modern systems [19]. With 60% fewer circuits (and a corresponding reduction in the number of interconnections), the reliability is greatly improved. For most on-board applications, 546 chips is still too complex, but recent progress with wafer scale integration promises to implement this processor on fewer wafers in the near future.

14.4.4 Arithmetic Realization

A critical design decision in radar signal processing concerns the arithmetic implementation. There are three somewhat contradictory requirements: (1) *high speed* to accommodate high signal bandwidths, (2) *high precision* to minimize computational error, and (3) *wide dynamic range* to avoid overflow. A wide variety of techniques are employed in on-board signal processors due to the differing relative importance of these requirements for specific applications. At the highest speeds, analog techniques, such as surface acoustic wave (SAW) devices, are often employed, although accuracy is marginal. At low speeds, where high accuracy is needed, minicomputers can perform high precision operations in firmware. In most radar applications, both high accuracy and high speed are important. Many users have compromised on 16-bit fixed-point arithmetic. Although it is fast, fixed-point arithmetic can produce overflow errors or loss of precision, unless data-dependent scaling is provided. Floating-point arithmetic has not been used due to size, cost, and speed limitations of available hardware. Recently, single-chip adders and multipliers using the 22-bit floating-point format (16-bit fraction, 6-bit exponent), have been developed. This format is a reasonable compromise among performance, speed, and size. Although 32-bit floating-point devices are available, for a given technology, the 22-bit format will always produce chips that are simpler and faster with adequate dynamic range and precision for most applications. Whereas 32-bit arithmetic is useful in scientific computation when inverting matrices or evaluating eigenvectors, such operations are usually performed at much lower rates than those required for radar signal processing, where input data are often limited in precision to 14 bits or less.

The chief advantage of floating-point arithmetic is increased dynamic range. As shown in Table 14.2, 22-bit floating point arithmetic provides 96 dB of precision (i.e., equivalent to 16-bit fixed-point arithmetic) over a dynamic range of 476 dB [20]. Although this dynamic range is much less than the 1686 dB provided by 32-bit floating-point arithmetic, the range is more than adequate for most space-based radar applications.

Device architecture for VLSI requires striking a delicate balance between extremely efficient custom chips optimized for use in a single system and generic integrated functions designed for broad applicability to many systems. The design issues require development of cellular arrays to permit design of very complex chips with reasonable amounts of time and personnel. We need to minimize the number of chip I/O signals (especially the number of outputs). Such I/O reduction minimizes the chip area required for bonding pads, reduces package size and cost, and decreases output driver power requirements. However, to ensure that the application flexibility of the chip is not compromised requires great care.

Table 14.2
Comparison of Arithmetic Implementations

Word Size	Number System	Dynamic Range	Precision
12 bit	Fixed Point	72 dB	72 dB
16 bit	Fixed Point	96 dB	96 dB
22 bit	Fixed Point	132 dB	132 dB
22 bit	Floating Point	476 dB	96 dB
32 bit	Floating Point	1686 dB	144 dB

14.5 DIGITAL RADAR BEAMFORMER CASE STUDY

Here we present a case study of customized VLSI chip development for a multibeam bistatic radar beamformer. The basic concept of digital beam-forming is briefly summarized to show how the beam-forming process operates, then the computational requirements are examined. Succeeding subsections summarize the algorithm selection process, and finally the ultimate VLSI chip is described to demonstrate why digital beam-forming is replacing analog techniques.

14.5.1 The Chip Development Process

The chip development process involves merging four disciplines: the application scenario, algorithms, architectures, and the current status of technology. Studies in each area are merged, as shown in Figure 14.18, to produce initial design studies, culminating in a feasibility assessment. In the case of this example processor, the initial assessment indicated that the technology was evolving and that foreseeable developments would be more cost-effective than analog phased array systems. Finally, refined application constructs (developed independently) are combined with the results of simulation and emulation studies to produce a finished design and specifications for the final chip development.

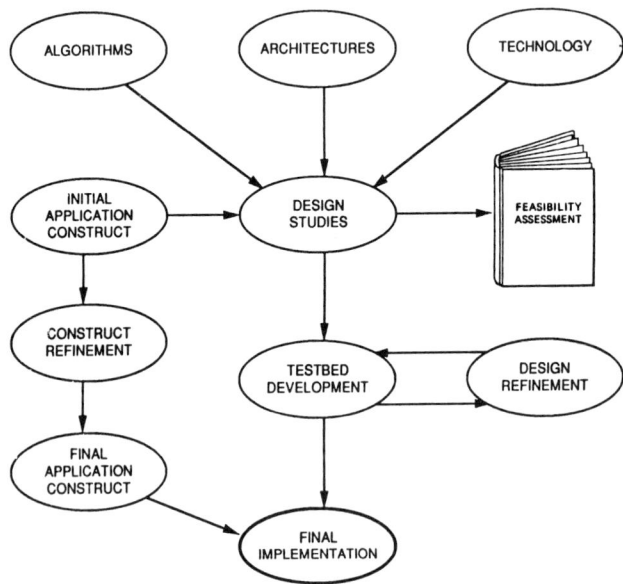

Figure 14.18 The development process for a customized VLSI chip.

14.5.2 Digital Beam-Forming

The basic idea of bistatic digital beam-forming is to down-convert, digitize, and sample data from a number of antennas. Based on the data samples, the arrival angles of the incoming radar returns are determined. Figure 14.19 shows a one-dimensional model of this process. Imagine that a transmitter has illuminated a large area and objects reflect energy to the receiver antenna array. The energy impinges upon the antenna array and is detected by receivers at each antenna element. The signal at the receivers forms a sinusoidal pattern in space with a frequency that depends on the angle of arrival.

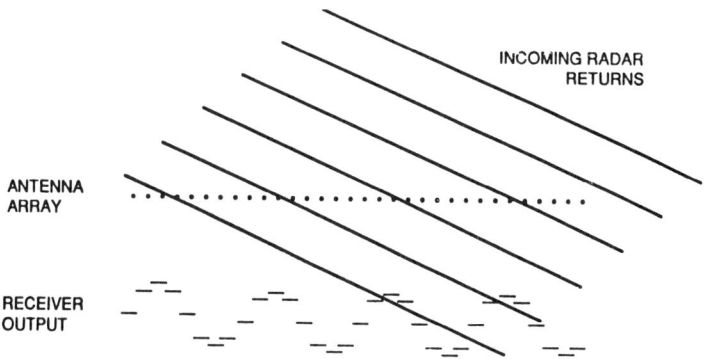

Figure 14.19 One-dimensional model of angles of arrival of incoming radar returns.

The basic beam-forming system comprises receivers and digitizers at each antenna element. The digital data are routed to a spectrum analyzer that determines whether there are components at specific frequencies. If only a single beam is to be formed, the receiver outputs are examined to determine whether the corresponding frequency is present. Alternatively, all frequencies can be examined for simultaneously forming all possible beams. The choice between the single beams, multiple beams, and all possible beams depends on the relevant radar applications.

Figure 14.20 shows the beamformer interfaces. The antenna and receiver array (with A/D converters at each of the receivers) produces the digital data that enter into the beamformer. The output of the beamformer goes to radar signal processors (generally one for each beam), or to a very fast processor that is shared among several beams. The radar signal processor outputs go to the user, which may be, for example, an airport approach control system. The radar data processor provides data back to the beamformer, giving the beam locations of interest for the next time interval.

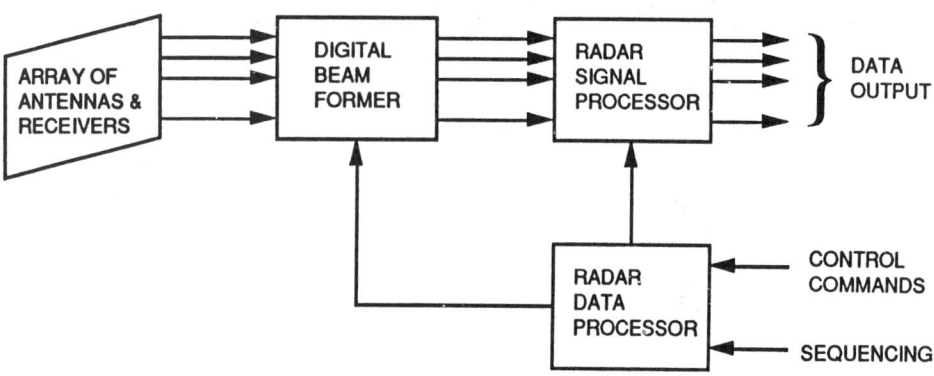

Figure 14.20 Basic beam-forming system.

The analog approach is shown on Figure 14.21. Analog (actually RF) signals at the antenna elements are coupled through phase shifters into a single receiver. The phase shifters are set to provide values of delay that electrically orient the antenna so that a maximum signal is obtained for a particular angular orientation. The advantage of the phased array system is that it uses a single receiver and a single A/D converter. That was a very big advantage when receivers cost $50,000 to $100,000 and A/D converters cost even more, but now it is practical to build high performance (but limited precision) single-chip receivers and single-chip A/D converters.

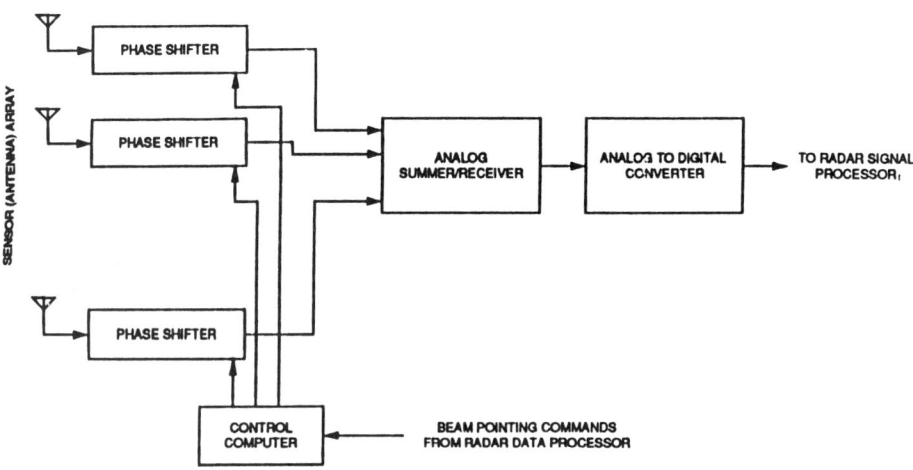

Figure 14.21 Analog beam-forming system.

For the digital approach of Figure 14.22, a receiver and an A/D converter are placed at each antenna element. The resulting digital data are then bused to the beamformer, which performs the spectral analysis. In contrast to the analog phased array approach, this scheme uses a receiver and an A/D converter for each antenna element, but with the current state of technology, replicating VLSI circuits is easier and cheaper than constructing precisely matched analog systems. This system forms any number of beams simultaneously by simply evaluating multiple components of the spectrum.

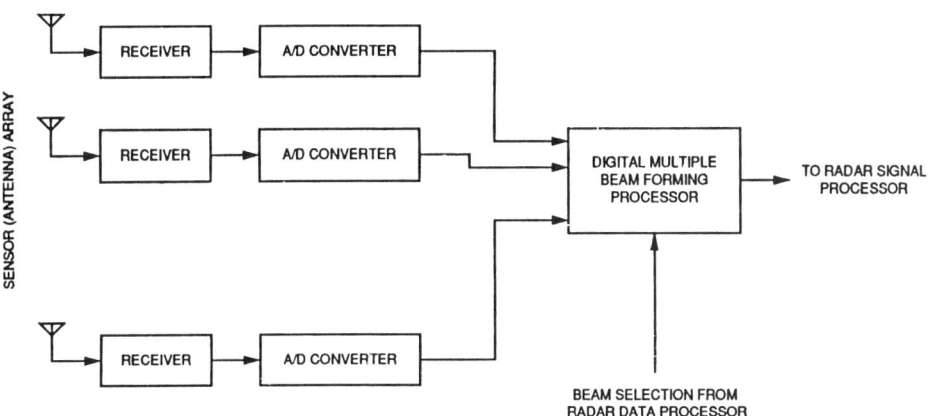

Figure 14.22 Digital beam-forming system.

14.5.3 Data Rates

In one specific scenario, the antenna is a 16 × 32 element array, with a total of 512 elements. To achieve the desired range resolution, each A/D converter is sampled at a rate of 10 MHz, and because there are 512 elements, the total input data rate is slightly over 5 GHz. This is the peak data rate. In fact, for most applications, the average data rate is quite low. A pulse of energy is transmitted, it travels through space up to the region of interest, and returns from that region are reflected back. Just before the return reaches the receiver, the receiver array is activated, the receivers are sampled at 10 MHz, and then, after the return has passed, the sampling is stopped. With duty cycles of 1 to 10%, the net data rate is in the 50 to 500 MSPS range. Rates in the 500 MSPS range can be routed using *emitter-coupled logic* (ECL) signal protocols. The system can use either a parallel interconnect structure with 8 to 16 channels at speeds in the tens of MSPS, or a single high-speed fiber optic data link.

14.5.4 Algorithm Selection

A critical aspect is the algorithm selection. Beamforming requires computing components of a two-dimensional Fourier transform, and so there are several candidate algorithms. They range from the *discrete Fourier transform* (DFT) and the fast Fourier transform to the *Winograd Fourier transform algorithm* (WFTA). In computing all of the frequency components of an N-sample data sequence with the DFT, on the order of N^2 multiplications are required. With the FFT, the multiplication count decreases to the order of $N \cdot \log N$. Finally, with the Winograd transform, the multiplications count decreases to the order of N. Comparison on the basis of multiplication count alone, however, is misleading for two reasons: it neglects other arithmetic operations that may be significant and it assumes that control and memory complexity is negligible, which is not valid for many special-purpose processors. This section focuses on two primary selection criteria: first, structural modularity so that more beams may be formed by adding computational elements; second, the total implementation complexity, not just the number of multipliers but the total number of integrated circuits, which is a good first-order indicator of the cost, size, power, and reliability.

The DFT beam-forming process is shown by Figure 14.23. There is a complex multiplier for each antenna element. Each receiver output is multiplied by the sine and cosine values necessary to produce the desired phase shift. The products are summed to generate the beam output. This process is repeated for each beam for each time sample. Thus, if the antenna array is M elements wide and N elements high, there will be $M \cdot N$ complex multiplications per beam. For K simultaneous beams, the total complex multiplication count is $K \cdot M \cdot N$. The total number of complex adds is $K \cdot [(M \cdot N) - 1]$. No memory is required. The only control requirement arises from the need to supply new sine and cosine values to the multipliers when it is necessary to change the beam direction.

Another approach is to use the FFT algorithm. The one-dimensional FFT is applied in multiple dimensions as shown in Figure 14.24. The data at the receivers are shown as an $N \cdot M$ matrix. It is transformed by columns to generate a matrix of intermediate results. Rows of the intermediate results are transformed to produce the beam patterns. If only K beams are to be formed, there need be only K row transforms. Each column transform requires an M-element FFT, which requires $(M/2) \cdot \log M$ butterfly computations. Because there are N column transforms, the total is $N(M/2) \cdot \log M$ butterflys. The row tranforms are N-element FFT computations for a total of $(N/2) \cdot \log N$ butterfly computations for each transform. With K row transforms the total is $(N/2) \cdot (M \cdot \log M + K \cdot \log N)$ butterflys. Because each butterfly requires one complex multiplication and two complex additions, the total arithmetic requirement is $(N/2) \cdot (M \cdot \log M + K \cdot \log N)$ complex multiplications and $N \cdot (M \cdot \log M + K \cdot \log N)$ complex additions. The FFT requires storage in an amount equal to the length of the transform.

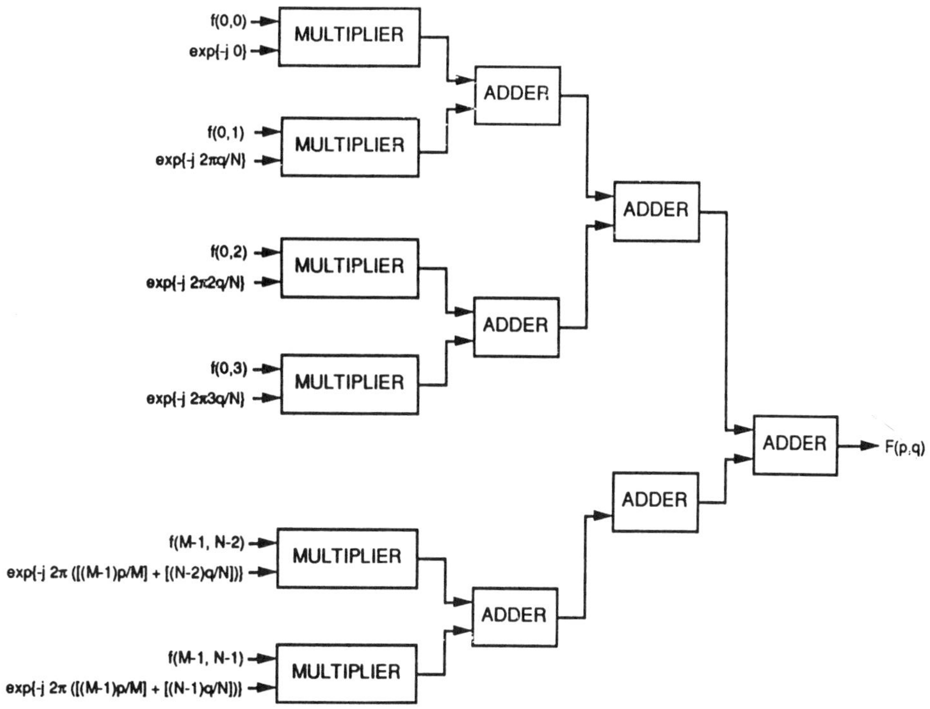

Figure 14.23 Beam-forming process using discrete Fourier transform.

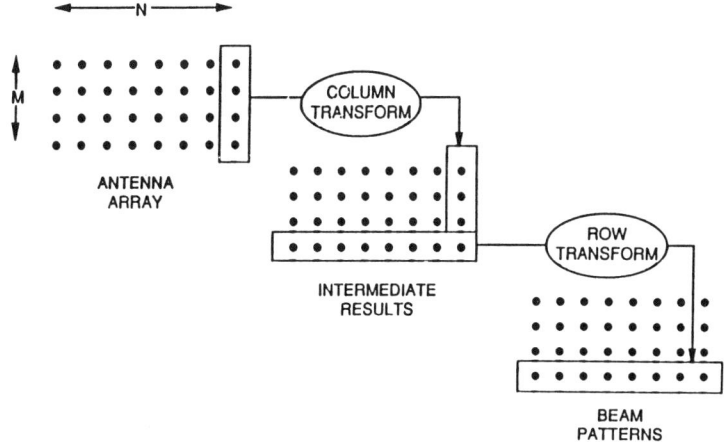

Figure 14.24 Beam-forming process using fast Fourier transform.

Storage must also be provided for the intermediate results so that a memory size of at least $N \cdot [(2 \cdot M) + K]$ complex words is required. Control requirements include generation of the address sequences for the FFT algorithm and of the sine and cosine values.

The WFTA approach is like the FFT approach in that the two-dimensional transform is performed via one-dimensional transforms. The major difference is that N-point one-dimensional WFTA computation requires approximately N complex multiplications and $N \cdot \log N$ complex additions. Applying this to the factorization of Figure 14.24, the N column transforms involve $N \cdot M$ complex multiplications and $N \cdot M \cdot \log M$ complex additions. The K row transforms require $K \cdot N$ complex multiplications and $K \cdot N \cdot \log N$ complex additions. The total operation count is $N \cdot (K + M)$ complex multiplications and $N \cdot (M \cdot \log M + K \cdot \log N)$ complex additions. The WFTA memory requirements are greater than for the FFT because the WFTA is not an "in place" algorithm. (Not in place means that data are accessed from memory and results are computed, but the results must be stored in locations other than where the data were accessed because the original data must be available for other computations.) Similarly, the control complexity is much greater for the WFTA than for the FFT. These approaches are compared on Table 14.3.

The DFT requires arithmetic in proportion to the number of simultaneous beams, but has low control and memory complexity. The FFT requires more arithmetic than a single-beam DFT, but far less than for the DFT as the number of beams approaches $M \cdot N$ (the maximum number of distinct beams). The comparison of total complexity is more involved as it requires examination of the relative difficulty of arithmetic, memory, and control. Detailed design studies suggest that, for a 512 element antenna, the arithmetic break-even occurs for three simultaneous beams. Consideration of the memory and control complexity raises the break-even point to the order of 25 beams (a small fraction of the 512 distinct beams!).

Table 14.3
Comparison of Implementation Complexity

Approach	Arithmetic		Memory	Control
	Complex Mults	Complex Adds		
DFT	KMN	$K(MN - 1)$	None	Low
FFT	$N/2(M \log_2 M + L \log_2 N)$	$N(M \log_2 M + L \log_2 N)$	MN	Moderate
WFT	$N(L + M)$	$N(M \log_2 M + L \log_2 N)$	$\gg MN$	High

Note: The array size is M by N, K is the number of simultaneous beams, and $L = \min(K, M)$

14.5.5 VLSI Implementation

The gate rate requirements can be calculated for the DFT beamformer. With an input of 100 MSPS, there will be 400 million multiplications per second. Because the MAC of Section 14.3.1 can perform 25 million multiplications per second, 16 MACs are required. For the FFT, the complexity is about a factor of 20 higher.

These chips are used to implement systems by coupling the data from the receiver array to the DFT beamformer elements in parallel with one element for each beam, as shown in Figure 14.25. Throughput requirements necessitate using 16 chips in parallel at each DFT element. Thus, there is a single source of data that is multiplexed from the receiver array and routed in a broadcast mode directly to the DFT elements.

The structure of a DFT circuit is shown on Figure 14.26. This is a multiplier-accumulator that operates on complex data. The multiplier-accumulator multiplies the digitized receiver outputs by sine and cosine values that are generated with a ROM. The ROM is addressed by a sequence generator (a simple accumulator) that governs which beam is formed.

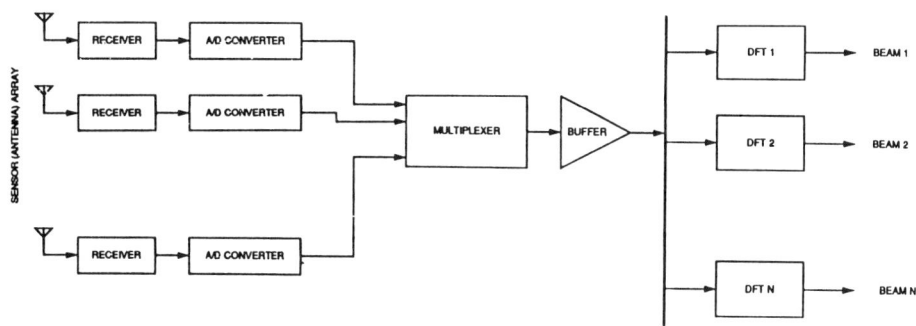

Figure 14.25 Beam-forming elements in parallel using discrete Fourier transforms (DFT1, DFT2, ..., DFTN).

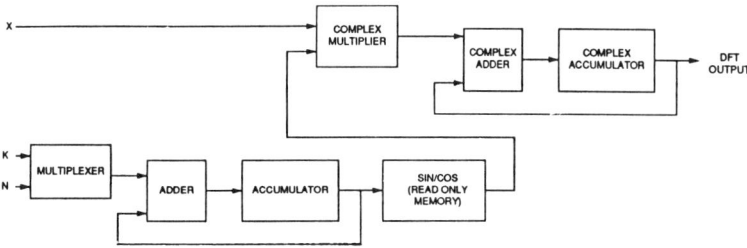

Figure 14.26 Multiplier-accumulator circuit for performing the discrete Fourier transform.

14.5.6 The Payoff

A critical concern in system development is the cost. If the VLSI-based design costs much less to implement than non-VLSI-based designs the benefits of the VLSI design will justify the required development effort. Figure 14.27 compares the cost of development of a ground-based bistatic digital beam-forming system with a comparable phased array system. In the 1978 time frame, the digital system is about 25% more expensive. The development of single-chip A/D converters caused the trade-off to reverse by 1980, where the phased array system is 300% more expansive. Improvements in digital technology have caused the digital costs to continue to decline (i.e., by a factor of 4 in the 1980–1985 period), while phased array costs remain relatively stable. This chart is based on production cost estimates for a specific system that forms a small number of simultaneous beams. Increasing the number of simultaneous beams would improve the relative cost advantage of the digital implementation because much of the digital cost is associated with the receiver and A/D converter array.

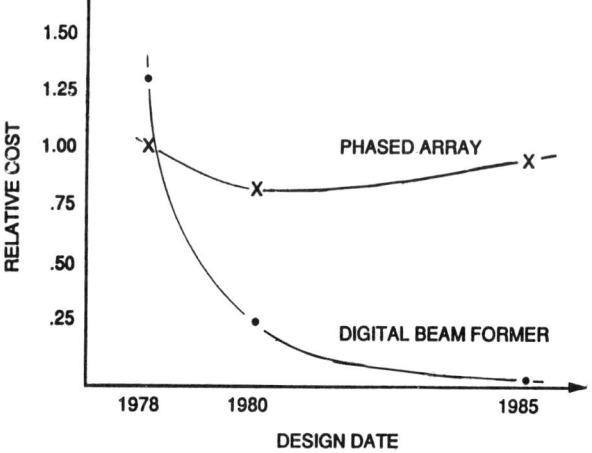

Figure 14.27 Comparative cost of development for digital and phased array beam-forming systems.

REFERENCES

1. Swartzlander, E.E., Jr., *VLSI Signal Processing Systems*, Boston, Kluwer, 1986.
2. Cappello, P.R., et al., eds., *VLSI Signal Processing*, Volume I, IEEE Press, New York, 1984.
3. Kung, S.Y., et al., eds., *VLSI Signal Processing*, Volume II, IEEE Press, New York, 1986.
4. Joint Director of Laboratories/Technology Panel on Signal Processing (JDL/TPSP) *Generic Signal Processing Workshop*, Griffiss AFB, NY: Rome Air Development Center Technical Report, RADC-TR-87-268, January 1988.
5. Bloch, E., and D.J. Galage, "Component Progress: Its Effect on High Speed Computer Architecture and Machine Organization," in D.J. Kuck, D.H. Lawrie, and A.H. Sameh, eds., *High Speed Computer and Algorithm Organization*, Academic Press, New York, 1977, pp. 13–39.
6. Burks, A.W., H.H. Goldstine, and J.Von Neumann, *Preliminary Discussion of the Logical Design of an Electronic Computing Instrument*, Institute for Advanced Study, Princeton, NJ, June 28, 1946.
7. Swartzlander, E.E., Jr., *High Speed Micro Signal Processor Study*, Wright Patterson AFB, OH: Air Force Avionics Laboratory Technical Report, AFAL-TR-77-63, April 1977.
8. Hwang, K., and F.A. Briggs, *Computer Architecture and Parallel Processing*, McGraw-Hill, New York, 1984.
9. Swartzlander, E.E., Jr., ed., *Systolic Signal Processing Systems*, Marcel Dekker, New York, 1987.
10. Kogge, P.M., *The Architecture of Pipelined Computers*, McGraw-Hill, New York, 1981.
11. Kuck, D.J. *The Structure of Computers and Computations*, Vol. 1, John Wiley and Sons, New York, 1978.
12. Kung, H.T. "Why Systolic Architectures?," *IEEE Computer Magazine*, Vol. 15, No. 1, January, 1982, pp. 37–46.
13. Swartzlander, E.E., Jr., "VHSIC-Based Innovative Architectures," *Johns Hopkins/APL Technical Review*, Vol. 1, 1988.
14. Swartzlander, E.E., Jr. and E.S. Yang, "AOSP Macro Function Signal Processor VHSIC Insertion," *Joint Director of Laboratories/Technology Panel on Signal Processing (JDL/TPSP) Generic Signal Processing Workshop*, Griffiss AFB, NY: Rome Air Development Center Technical Report, RADC-TR-87-268, January, 1988, pp. 99–103.
15. Gold, B., and T. Bially, "Parallelism in Fast Fourier Transform Hardware," *IEEE Trans. Audio and Electronics*, Vol. AU-21, 1973, pp. 5–16.
16. Swartzlander, E.E., Jr., L.S. Lome, and G. Hallnor, "Digital Signal Processing with VLSI Technology," *Proc. IEEE Int. Conf. Acoustics, Speech, and Signal Processing*, 1983, pp. 951–954.
17. Rabiner, L.R., and B. Gold, *Theory and Applications of Digital Signal Processing*, Prentice-Hall, Englewood Cliffs, NJ, 1975, p. 611.
18. Swartzlander, E.E., Jr., W.K.W. Young, and S.J. Joseph, "A Radix 4 Delay Commutator for Fast Fourier Transform Processor Implementation," *IEEE J. Solid-State Circuits*, Vol. SC-19, 1984, pp. 702–709.
19. Preston, G.W., "The Very Large Scale Integrated Circuit," *American Scientist*, Vol. 71, 1983, pp. 466–472.
20. Eldon, J.A., and C. Robertson, "A Floating Point Format for Signal Processing," *Proc. IEEE Int. Conf. Acoustics, Speech, and Signal Processing*, 1982, pp. 717–720.

Chapter 15
PRIME POWER SYSTEMS IN SPACE
J.E. Boretz
TRW

15.1 SPACE POWER SYSTEM CLASSIFICATIONS

15.1.1 Introduction

Space power systems can be broadly classified into several different general categories. The simplest classification is to identify systems as either *static* or *dynamic*, depending on whether electric power is created by a system with essentially no moving parts or involves some form of dynamic machinery. The next category differentiating these systems is the form in which the input energy is provided to the conversion cycle. Initially, systems are described as either *nonnuclear* or *nuclear* so that the spacecraft user knows whether to be aware of shielding and safety issues. The nonnuclear heat sources include solar and chemical energy devices, and nuclear heat sources are divided into radioisotope or reactor types. There are a wide variety of both static and dynamic energy conversion systems that can be combined with these various heat sources to form an overall space power system.

15.1.2 System Selection

The selection of a candidate space power system for a particular application is highly dependent on the overall prime power requirements of the space-based radar (SBR), its spacecraft and mission requirements, and the effects of the natural and perhaps hostile space environments. Hence, the selection of a candidate SBR prime power system involves a complex series of analytical and design trade-offs to arrive at an optimum configuration. These trade-offs are usually conducted by the space power system engineer in conjunction with the spacecraft systems engineer. The details of the myriad parameters which must be taken into consideration in this selection process are beyond the scope of this chapter. Suffice it to say that as an outcome of these trade-off studies, a candidate prime space power system is ultimately realized. However, for the purposes of this book, based on the SBR systems outlined in Chapter 1, we can describe a variety of typical space power configurations that have the potential for being considered as prime electrical power sources. These will be discussed primarily from the standpoint of their compatibility with required power levels, operating life, specific power and mass considerations, and their development status.

15.1.3 Space Power Configurations

The candidate space power systems described in this chapter are limited to typical generic configurations that represent the current technology. When considered appropriate, we allude to growth potential. Table 15.1 lists the various systems discussed here. In general, these systems will be described from the standpoint of power level, performance, mass, operating life, and their past and current applications. Technology in the space power systems arena is constantly evolving, and so a brief discussion of the future growth potential of these systems will also be provided.

Table 15.1
Candidate SBR Prime Power Systems

Section	Type of System
15.2	Solar Array-Battery (S/A-B)
15.3	Organic Rankine Cycle (ORC)
15.4	Closed Brayton Cycle (CBC)
15.5	Free Piston Stirling Engine (FPSE)
15.6	Supercritical Cycle (SC)
15.7	Potassium Rankine Cycle (PRC)
15.8	Alkaline Metal Thermoelectric Conversion (AMTEC)
15.9	Photochemical Halogen-Hydraulic Cycles (RAFT)
15.10	Thermoelectric Conversion Cycles (TEC)
15.11	Thermionic Conversion Systems (TDC)
15.12	Topping Conversion Systems (TOP/CS)
15.13	Flywheel Systems (FLY/W)
15.14	Advanced High Energy Density Rechargeable Batteries (HEDRB)
15.15	Regenerative Fuel Cell System (RFC)

15.2 SOLAR ARRAY-BATTERY SYSTEMS (S/A-B)

The solar array-battery system has been the most frequently utilized space power system since the 1970s. It has provided reliable and long-life electrical power to spacecraft and satellites in an output power range from a few hundred watts to 12.5 kW. Current plans are to use it on the Space Station in the 37.5 to 75 kW$_e$ power range with potential growth to as high as 150 kW$_e$. Some early work conducted on solar arrays is described in the references [1–4]. A recent and extremely comprehensive documentation of solar array system design is provided in Reference [5]. Battery technology and system design characteristics are provided in the references [6, 7].

15.2.1 Solar Array Performance

The performance analysis and design of a solar array is very complex and requires a broad database of the parameters governing output power, operating life, and reliability. We cannot adequately cover this complex subject here and the reader is referred to [5] for comprehensive coverage of this type of space power system.

The power of a solar array is established to meet end-of-life (EOL) requirements because the operating life of an array can be anywhere from three to fifteen years and various losses occur over so long a period of time. Solar arrays can be configured as cylindrical arrays (spinners) or oriented panel arrays (three-axis body stabilized). For a ten-year operating life, typical specific power levels at EOL are 11.3 W/kg and 25.5 W/kg, respectively. Because of the lower achievable specific power levels of the spin stabilized solar arrays, they have mostly been utilized for low-power-level applications (e.g.<1000 W). Body stabilized solar arrays have been utilized for power levels up to 12.5 kW (Skylab) and are under consideration for the space station for a power level of 37.5 kW. Use of multiple arrays for the space station will result in an output power level of 150 kW, with growth projected to values as high as 300 kW. These large solar arrays are launched into space in a stowed configuration. Hence, mechanisms must be provided for delatching and deployment. In addition for missions of very long duration (e.g., fifteen years), in-orbit replacement of the solar array wings and their associated battery and power conditioning units may be required. Replacement in space necessitates incorporating a retraction capability in the deployment mechanism as well as a relatching feature. Because of the need for the array wings to track the sun, a solar array drive mechanism is also required. Hence, a solar array system consists of many subsystems to facilitate all its required operational features. A typical breakdown of these subsystems is shown in Figure 15.1.

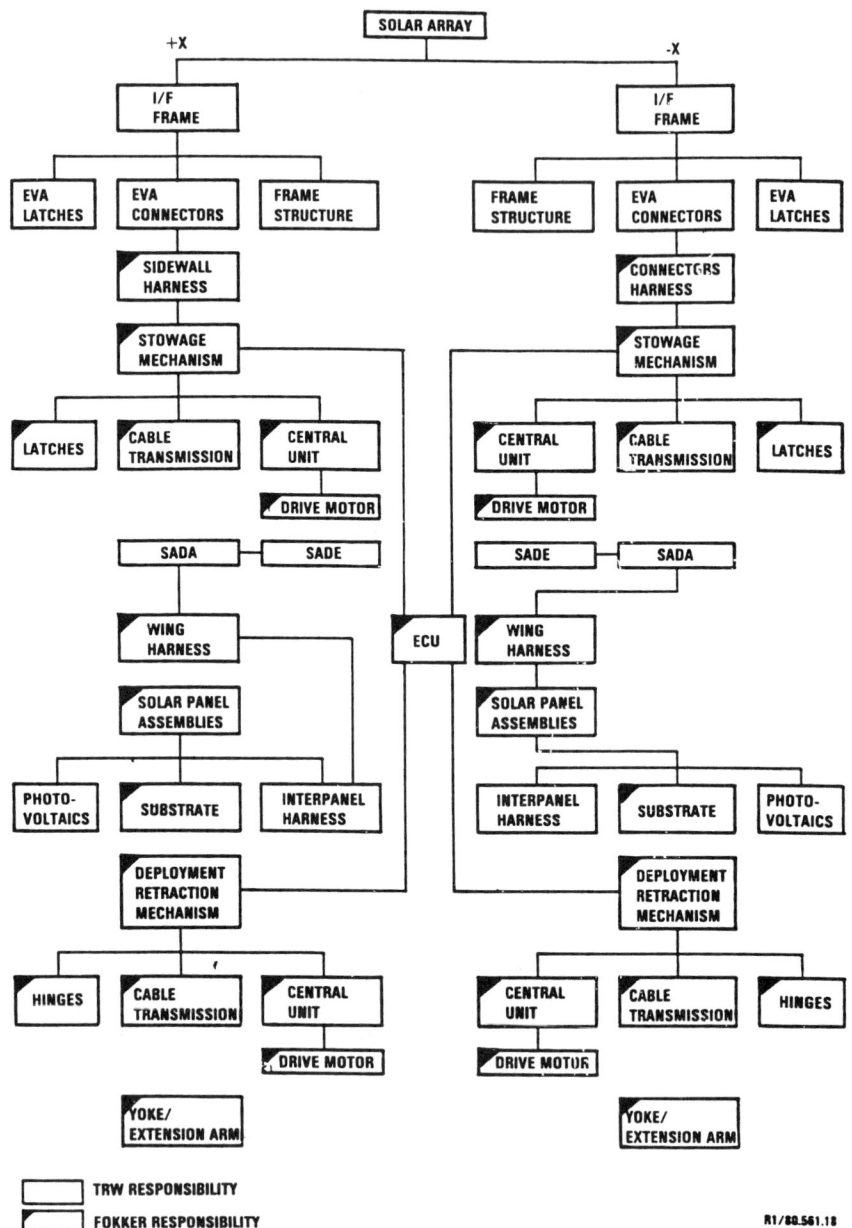

Figure 15.1 Solar array system breakdown.

A three-axis stabilized solar array consists of two wings composed of multiple panels. Depending on the application, these panels can be configured from so-called flexible Kapton blankets or rigid-honeycomb-facesheet covered substrate panels. A typical flexible-blanket solar array is depicted in Figures 15.2 and 15.3 (SPAR Corp.). For missions that impose higher maneuvering loads on the array, a rigid panel substrate array of the type shown in Figure 15.4 (Fokker Space Systems) is preferred.

In addition to the solar array, energy storage in the form of rechargeable batteries is required to provide power during the eclipse period.

Figure 15.2 Flexible blanket solar array.

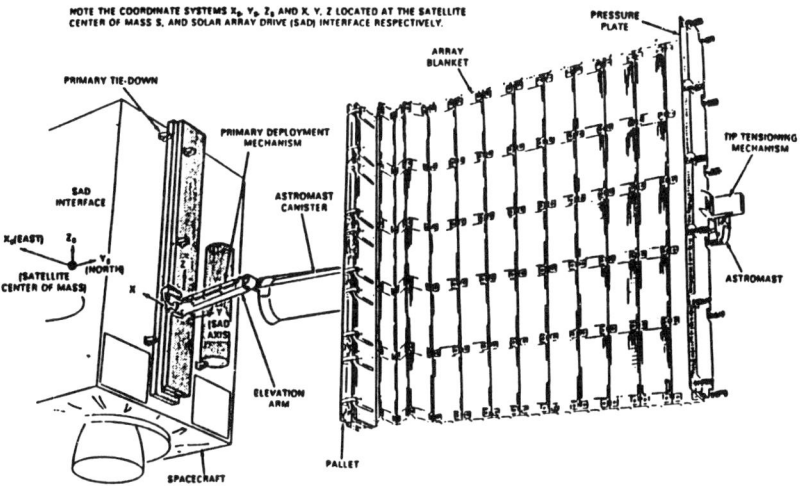

Figure 15.3 L-SAT array deployed configuration.

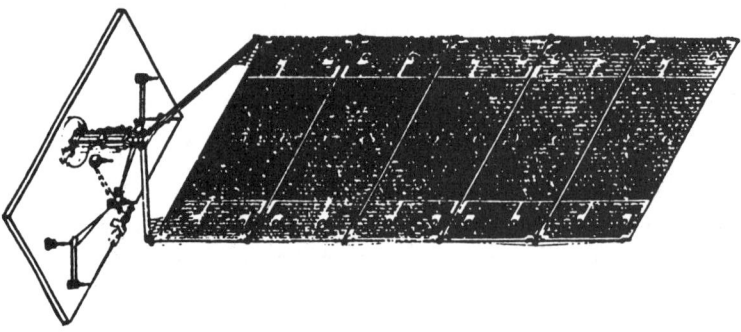

Figure 15.4 AXAF solar array wing deployed configuration.

15.2.2 Battery System Characteristics

Nickel-cadmium (NiCd) batteries have been the energy storage source of choice to date, but nickel-hydrogen (NiH$_2$) batteries have emerged as an attractive alternative owing to longer cycle life and higher energy density. Because of more stringent weight limitations of future military satellites, combined with their projected higher power level requirements, rechargeable batteries with higher energy densities than the NiCd and NiH$_2$ batteries will be required. Potential military space missions with power levels ranging from 5 to 30 kW of base load electrical power are identified in the Military Space Systems Technology Model (MSSTM), and studies of advanced satellite power systems have shown that the battery will remain the heaviest component of the power system. A battery with an energy density of 110 Wh/kg, for example, would still comprise as much as 20% of the total satellite weight.

In anticipation of future need for higher energy density batteries, the Air Force Wright Aeronautical Laboratories (AFWAL) initiated a study in 1979 to determine which batteries currently being developed had the potential to meet the energy density and cycle life goals for military satellites. The contractor, Hughes Aircraft Corp., identified two battery technologies that could possibly meet these goals (i.e., lithium alloy-metal sulfide (Li-alloy/MS) and sodium-sulfur (NaS)). Following these recommendations, a contract was initiated with Gould in March 1983 for Li-alloy/MS cell development. A second contract to design a NaS cell is underway at Eagle-Picher Corp.

These programs are designed to be the first task of a High Energy Density Rechargeable Battery (HEDRB) development program. The ultimate goal is a 110 Wh/kg battery with a ten-year life in midaltitude and geosynchronous orbits. Meeting the HEDRB energy density goal would reduce battery weight by 80% as compared to NiCd batteries and by 60% as compared to NiH$_2$ batteries. This translates to a savings of 1500 kg (*versus* NiCd) for a 30 kW satellite power system.

Figure 15.5 shows a comparison of battery weight *versus* power systems output for these batteries. The relationship between satellite battery weight and the system power level to be sustained by the battery is for a one-hour rate. The slope of each line is inversely proportional to the energy density of the particular battery. The programs are structured to deliver a space-flight-qualified battery by 1993, and includes cell development, module development, and life and flight testing. This is a very ambitious goal because the objective of achieving a 110 Wh/kg (50 Wh/lb) battery represents a threefold decrease in battery weight, as can be seen in Figure 15.6.

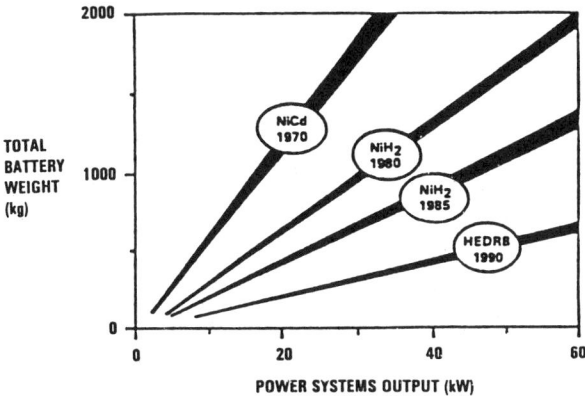

Figure 15.5 Battery weight *versus* output power.

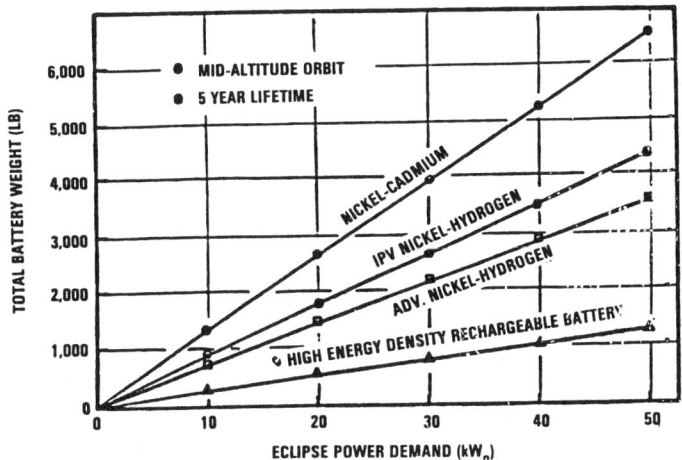

Figure 15.6 Total battery weight *versus* eclipse power demand.

To put the HEDRB development program in perspective, we will review current space satellite batteries. The more important parameters when discussing batteries are listed in Table 15.2 because their values are utilized in arriving at battery energy density and total system weight. Current battery configurations (NiCd and NiH_2) are depicted in Figures 15.7 and 15.8. The design and performance characteristics for these type batteries, as they were built for various space satellite programs, are summarized in Table 15.3. Notice that the energy density of these batteries varied from 12.9 to 16.6 Wh/lb. While the values shown in Table 15.2 do not reflect this, the NiH_2 batteries currently in use are actually 15% lighter than the NiCd batteries (e.g., 19.0 Wh/lb), as indicated by the values depicted in Figures 15.7 and 15.8. However, the NiH_2 batteries are approximately 2.5 times larger in volume than the NiCd units.

Table 15.2
Battery Terminology

Ampere-Hour
Defines Battery cell capacity, i.e., 40 AH means battery when new can deliver 40 amps for 1 hour.

Depth-of-Discharge (DOD)
Percent of battery capacity used during discharge. Most often measured starting at 100% charge, i.e., 75% DOD means 25% of battery capacity remains at end of discharge.

Watt-Hours
Defines battery stored energy and can be approximated by multiplying battery average discharge voltage by ampere-hour capacity.

Charge Rate
Typically $C/15$, where C = BATTERY CAPACITY (AH).

Battery Average Discharge Voltage = 1.25V × Number of Series Cells.

$$DOD = \frac{\text{Energy Delivered to Load in Eclipse}}{\text{Total Battery Stored Energy}} = \frac{\text{Load Power (watts)} \times \text{Discharge Time (hours)}}{\text{Capacity (ampere hours)} \times \text{Battery Average Discharge Voltage (volts)}} \times 100$$

As previously stated, because many military space missions are requiring much higher power levels (e.g., 5 to 30 kW) and considerable interest has been expressed in the use of batteries as chemical energy storage systems for very high power level burst-duration systems, a two- to threefold reduction in battery system weight is being sought. There are two candidate batteries under development to achieve these higher energy densities, sodium-sulfur (NaS) and lithium-iron disulfide ($LiFeS_2$). These are listed in Figure 15.9 with their projected performance goals. A brief description of these advanced batteries follows.

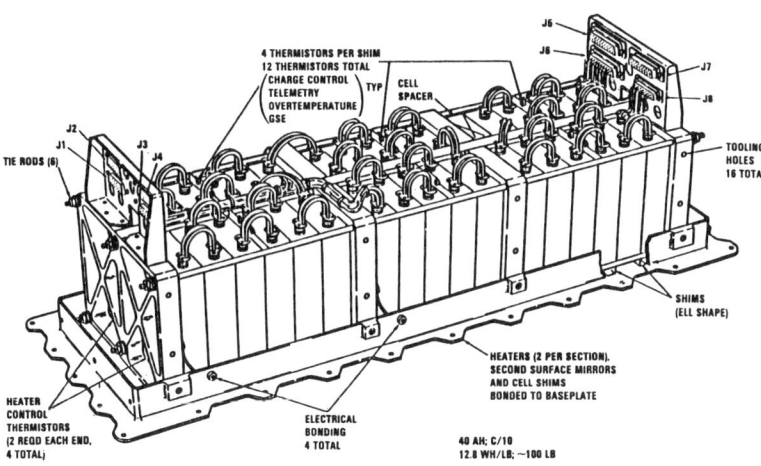

Figure 15.7 Typical nickel-cadmium battery

Figure 15.8 Nickel-hydrogen batteries.

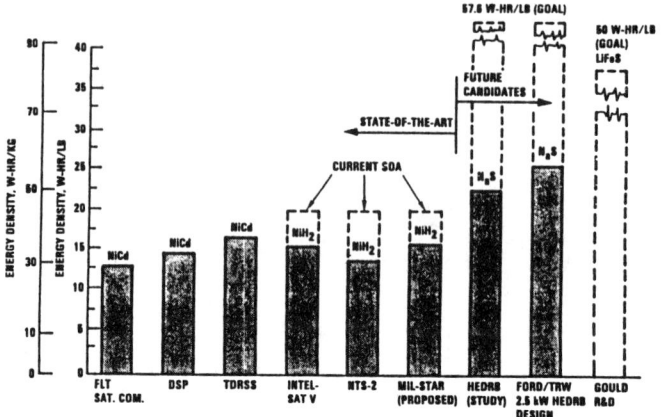

Figure 15.9 Energy density for satellite power system batteries.

Table 15.3
Battery Characteristics (as Built)

Program	Type	Cell Capacity (AH)	Series Cells	Weight (lbs)	Size L × W × H (in)	Energy Density (WH/lb)
DSP	NiCd	28	22	52.7	12 × 8.7 × 7.7	14.6
777	Nicd	18	22	38.0	12.1 × 8.7 × 5.5	13.0
FLTSATCOM	NiCd	28	24	65.0	15.5 × 11.5 × 8.5	12.9*
TDRS	NiCd	48 AH × 3	24	262.0	(2) 33.5 × 11.9 × 10	16.5**
3557	NiCd	40 AH × 3	22	213.6	(2) 32.4 × 11.1 × 8.5	15.4
Intelsat V	NiCd	34	28	71.5	18.5 × 13.4 × 5.1	16.6
Intelsat V	NiH_2	30	27	66.3	20.5 × 20.5 × 8.7	15.5
Lockheed/ USAF	NiH_2	50	21	110.00	17 × 25 × 10.5	12.1
NTS-2	NiH_2	35	14	45.0	(2) 16 × 8 × 9	13.8
MILSTAR (Proposal)	NiH_2	93	24	181.0	N/A	15.7

*Includes cell bypass circuits (approximately 9 lbs).
**Includes cell voltage scanners (approximately 7 lbs).

Note: TDRS and 3557 batteries are mounted in two assemblies, which together contain three batteries. Cell capacity is based on actual *versus* measured data, not nameplate ratings.

Sodium Sulfur Battery (NaS)

Sodium-sulfur cells (Figures 15.10 and 15.11) use a solid ceramic electrolyte to separate reactants and to provide a conductive path for sodium ions during operation. Because the electrolyte is neither electronically conductive nor permeable to the molten reactants (sodium anode and sulfur-sodium polysulfide cathode), it completely prevents internal self-discharge. The cells operate at temperatures between 300 and 350°C, a range where reactants and discharge products are liquid. During discharge, electrons are allowed to pass through an external circuit from anode to cathode (Figure 15.10), while sodium ions pass through the ceramic electrolyte to combine with sulfide ions to form sodium polysulfide. The cell's open-circuit voltage is essentially constant at 2.08 V until all sulfur has been reacted to form sodium pentasulfide (Na_2S_5). As additional sodium is reacted, the open-circuit voltage decreases approximately linearly to 1.75 V at a composition used to define 100% theoretical ampere-hour capacity. During recharge, the cell processes are reversed, and although not all sodium can be recovered from the cathode, cells typically deliver in excess of 85% of theoretical capacity at their design discharge-charge rates. Because the electrodes are liquid, shape change problems encountered by many high-energy batteries do not exist for the sodium-sulfur system.

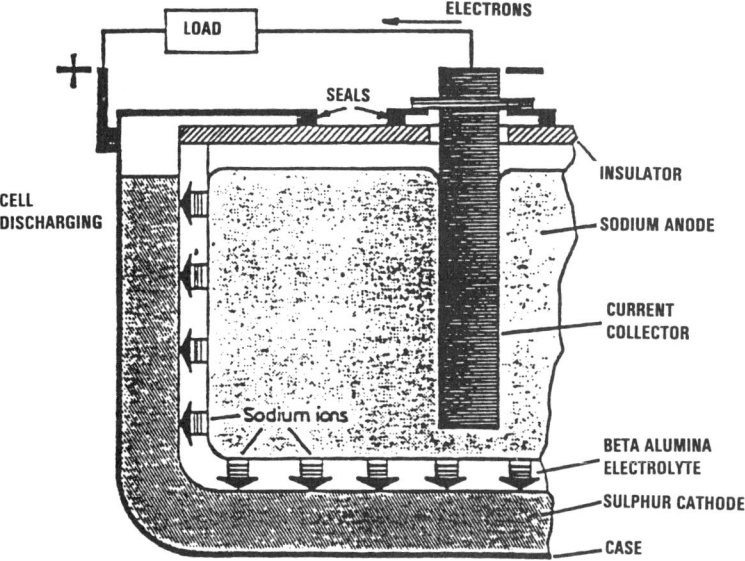

Figure 15.10 Sodium-sulfur battery concept

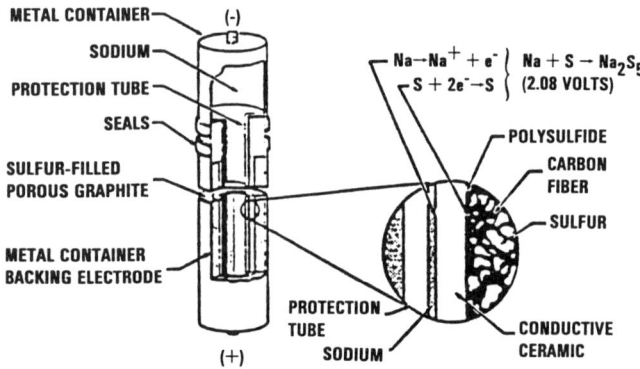

Figure 15.11 Schematic of sodium-sulfur cell.

In summary, tests of sodium-sulfur cells at high discharge rates typical of advanced satellite applications have successfully demonstrated:

1. rated specific energies in excess of 120 Wh/kg (54.6 Wh/lb);
2. specific power values up to 224 Wh/kg (102 W/lb) at a 75 A discharge current;
3. overall efficiencies of 87 to 89% over an operating temperature range of 325 to 375°C;
4. a projected life of 4000 cycles at 75% DOD, based on extrapolation from data over 200 cycles;
5. satisfactory freeze-thaw and failed-cell safety characteristics.

Lithium-Iron Disulfide (LiFeS$_2$)

A high-energy-density rechargeable battery made of lithium-metal sulfide is under development for advanced military satellites. Operating in conjunction with a photovoltaic array, the battery would provide system power during periods of earth eclipse. The higher system power requirements foreseen (e.g., 30 kW), dictate a battery energy density of at least 110 Wh/kg to keep battery weight manageable. Lithium-metal sulfide is a high-temperature (450°C), molten salt battery, and is optimistically expected to achieve the energy density goal by redesign of the cell and incorporation of higher energy electrodes. Automated cycle testing of 40 cells of present design simulates midaltitude and geosynchronous duty cycles, and examines the effects of *depth of discharge* (DOD) on performance and cycle life.

Owing to the extensive efforts, the Li-alloy/MS system, which the US Department of Energy had already sponsored at Argonne National Laboratories, was used as a starting point. The first step toward satellite cell development was to build a reasonable quantity (40) of cells to assess the current status of the technology, and a test program (designated Level One) was designed to evaluate the cells in simulated satellite orbit duty cycles. The orbits selected were midaltitude, a 5600 nmi orbit with a 6-hour charge-discharge cycle, and geosynchronous with 24-hour cycle. The cell tests were also designed to evaluate the effects of DOD on cycle life.

During Level One testing, cell development is to continue toward a second design (Level Two), aimed at advancing performance closer to that required for a satellite cell. After the Level Two design is completed, 60 cells will be built and evaluated in tests similar to those for Level One. Thereafter, cell development will continue toward achieving in a Level Three design the ultimate goals of the program (i.e., 165 Wh/kg and 1000 to 5000 cycles over a 3 to 10 year period).

Achieving a stable, long-life, iron disulfide electrode will require limiting cell operating temperatures below 420°C to avoid thermal decomposition of the disulfide to the monosulfide and free sulfur, leading to loss of positive electrode capacity.

In summary, lithium-alloy–iron sulfide cells, under development for advanced military satellite applications, are being tested for performance and cycle life in simulated midaltitude and geosynchronous orbit duty cycles, at DOD ranging from 40 to 80%. Advanced cell development is underway to improve substantially the energy density toward the ultimate mission goal of 165 Wh/kg (75 Wh/lb).

15.3 ORGANIC RANKINE CYCLE (SYSTEM NO. 1)

The *organic Rankine cycle* (ORC) system is similar to conventional steam Rankine cycle systems in use in central station plants. The ORC power conversion system can be coupled to most heat sources (e.g., isotope, chemical combustion, solar collector, reactor). In addition to the heat source, the major system components consist of a boiler, turbopump-generator, regenerator, and radiator. The selection of an organic working fluid instead of steam for space power systems is based on the desire to achieve high cycle efficiencies at moderate peak cycle temperatures. Because organic working fluids such as toluene and RC-1 have a positive sloping saturated vapor line, the turbine exhaust is in the highly superheated region. This permits use of efficiency enhancing regeneration as shown in Figure 15.12.

In the 1 to 10 kW_e output power range an isotope (Pu 238) powered ORC would be a good candidate to meet this power range. A typical system is depicted in Figure 15.13. A ground demonstration unit of this cycle operating at an output power level of 1.3 kW_e was developed by the Sundstrand Corp. under contract to the Department of Energy. The results are reported in [8, 9, 10]. A schematic of this system is shown in Figure 15.14. It could be upgraded to 2.0 kW_e with minor

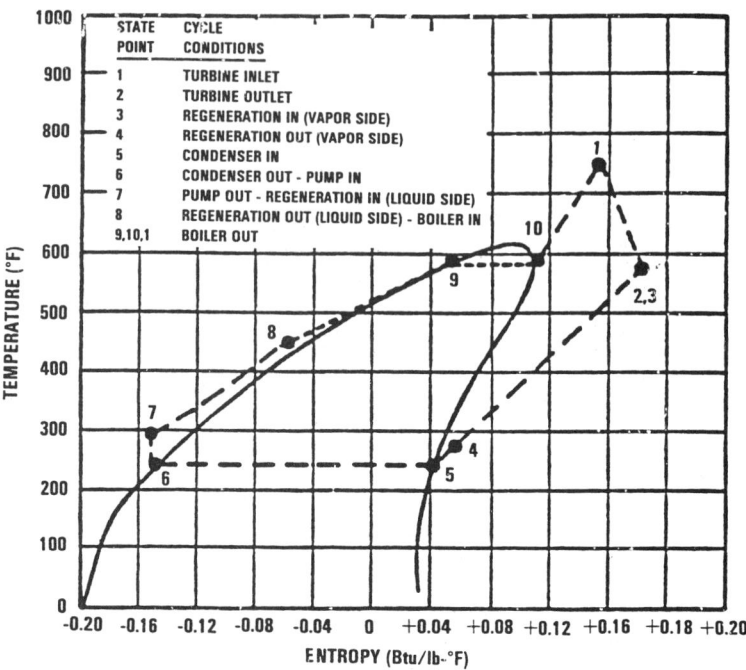

Figure 15.12 Typical ORCEPS cycle conditions.

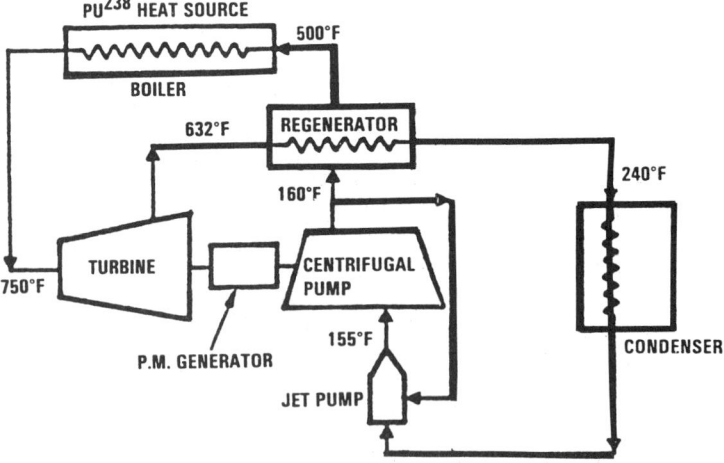

Figure 15.13 ORCEPS components and typical cycle conditions (working fluid-toluene).

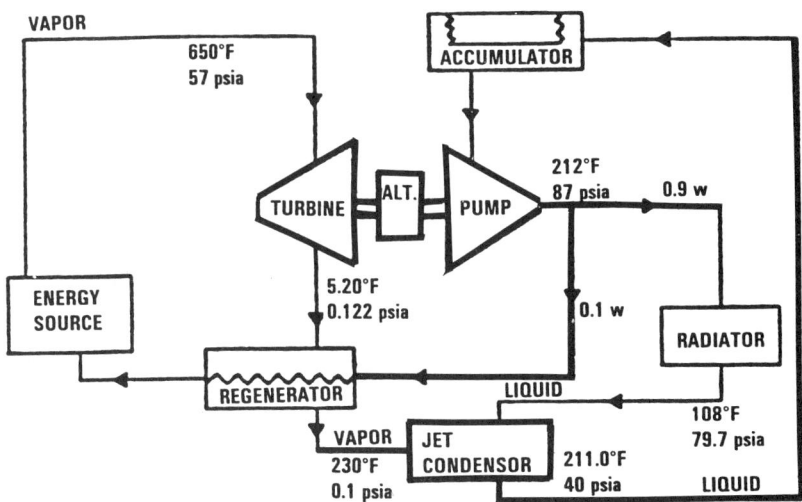

Figure 15.14 DIPS components and typical cycle conditions (working fluid—Dowtherm A).

changes to the components. In addition, after further evaluation, Sundstrand decided to change the working fluid from Dowtherm A to toluene (CP-25) and replace the jet condenser with a *rotating fluid management device* (RFMD). The changes were made to improve the system performance and cycle efficiency increase from 18 to 25% has been projected. Currently, a seven-year development program is planned by the Department of Energy for the Engineering Development Unit at a nominal power level of 6 kW_e. The goal of this *dynamic isotope power system* (DIPS) program is to select and demonstrate the necessary technology required to provide space power in the 1 to 10 kW_e range for use by emerging military space missions in the early 1990s and beyond. The main technology issues for the ORC were two-phase boiling and condensing in a zero-gravity environment, possible decomposition of the organic working fluid, isotope availability and cost, and safety. By operating in the supercritical region, the problem of two-phase boiling is eliminated.

A Grumman Corporation design for a condenser and heat pipe radiator developed for the Space Station's solar dynamic power system has eliminated the uncertainties associated with two-phase condensing. In addition, the Department of Energy advised the Air Force of the isotope availability and cost status, and they concluded that both issues were acceptable to future military space programs. Finally, many safety studies have been done, and by designing the isotope heat source for "total containment," the required levels of safety can be achieved.

In the 10 to 50 kW$_e$ power level range, the current emphasis has been on the development of a solar collector powered ORC in the 23 to 40 kW$_e$ range for the Space Station. The ORC cycle characteristics are similar to those outlined above for the DIPS with the additional requirement for an LiOH thermal energy storage subsystem to provide the heat input during the eclipse period of the orbit. Major technological issues that must be considered for a solar dynamic power system are collector concentration ratio and pointing accuracy, peak and part load operating characteristics, receiver aperture optimization, and structural dynamic interactions. In addition, deployable and erectable solar collector configurations were investigated. The database available for these power systems includes the *Sunflower* 3 kW$_e$ solar dynamic mercury Rankine cycle, previously developed by TRW, and the *ASTEC* 15 kW$_e$ solar dynamic rubidium Rankine cycle, which was developed by the Sundstrand Corp. for the Air Force. In addition, various terrestrial power plants were developed by JPL and Barber-Nichols Corp. for the Department of Energy at the 25 kW$_e$ level.

When operating in the 50 to 300 kW$_e$ power level, an alternative heat source for the ORC power system would be a nuclear reactor. For these higher power levels, the reactor has several advantages over the Pu 238 isotope heat source. First, it is considerably lower in cost and there is no problem with availability of materials. In addition, the reactor is not started until it is in orbit, which greatly simplifies launch operations and enhances safety. Both thermal and fast reactor types can be utilized. The thermal reactors consist of a U_{235}-ZrH core with beryllium control drums. A considerable technology base exists for these reactors from the earlier 3 kW$_e$ SNAP-2 program and 35–50 kW$_e$ SNAP-8 program. While previously developed to operate in conjunction with mercury Rankine cycles, thermal reactors are readily adaptable as heat sources for supercritical ORC power systems. Currently, a *compact nuclear power source* (CNPS) is being developed by the Los Alamos National Laboratory (LANL) to power a 25 kW$_e$ ORC. The design is oriented to a remote-site, arctic, terrestrial application, but it is readily adaptable for space applications. The CNPS combines a number of inherent safety features. Safeguard considerations are minimized by the use of 19.9% enriched U_{235} fuel. This coated-particle fuel retains virtually all of the fission products generated throughout a 20-year life. Transient effects are mitigated by the large graphite mass of the core. A strong negative temperature coefficient of reactivity is the salient inherent safety feature, constraining peak reactor temperatures to well below failure limits in any credible accident scenario. With mechanical stops on the control rods, core temperatures are limited to a maximum of 750°C (1382°F) under an uncontrolled reactivity insertion accident. In the lower power level range (e.g., 10–50 kW$_e$), it is feasible to consider a so-called *hybrid* OTC power system. This consists of topping a conventional ORC with a silicon-germanium (SiGe), static, *thermoelectric energy* (T/E) conversion system. By employing this concept, overall cycle efficiencies as high as 30% can be achieved. Because this concept utilizes a T/E hot junction temperature of 1000°C (1832°F), a fast reactor compatible with this very high peak cycle temperature is required. A fast-spectrum *lithium cooled reactor* (LCR) using uranium nitride (UN) as a fuel with a niobium-zirconium-0.1 carbon structure (Nb-Zr-0.1C) as a core would be a typical heat source. There are other potential candidate high-temperature reactors that can be considered (e.g., boiling potassium, gas-cooled pebble bed, or NERVA derivative reactors), but the fast-spectrum LCR is currently under development on the 100 to 300 kW$_e$ SP-100 program.

This reactor would have the highest probability of achieving technology readiness by 1995. A typical reactor-ORC power system is shown in Figures 15.15 and 15.16. A similar schematic employing the LANL compact nuclear reactor source is shown in Figure 15.17

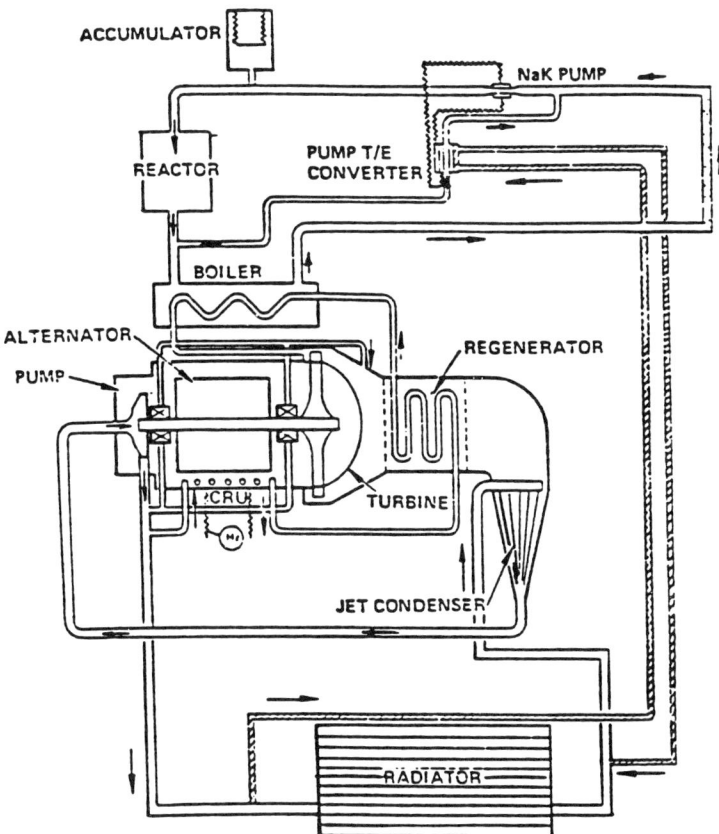

Figure 15.15 Typical reactor-organic Rankine cycle electrical power system (ORCEPS) schematic.

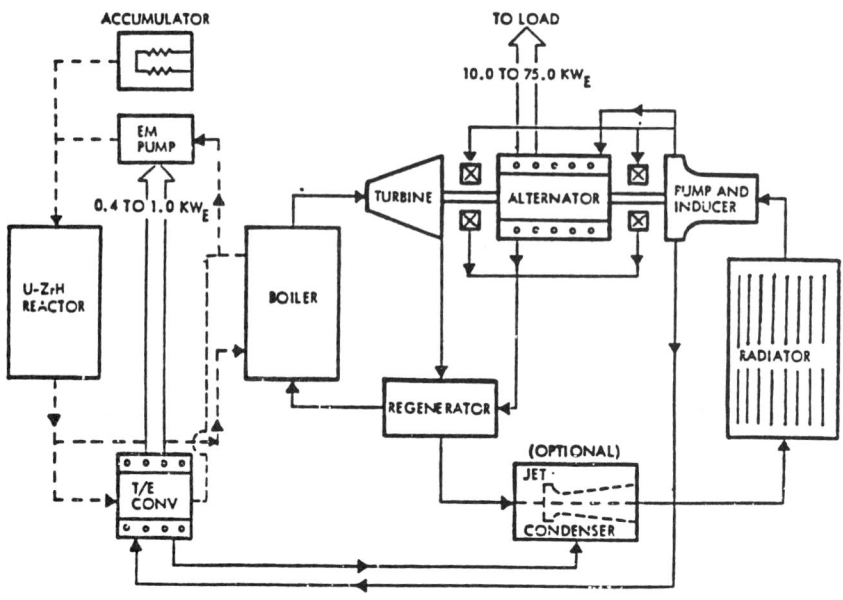

Figure 15.16 Candidate System No. 1—reactor-organic Rankine cycle electrical power system (ORCEPS) schematic.

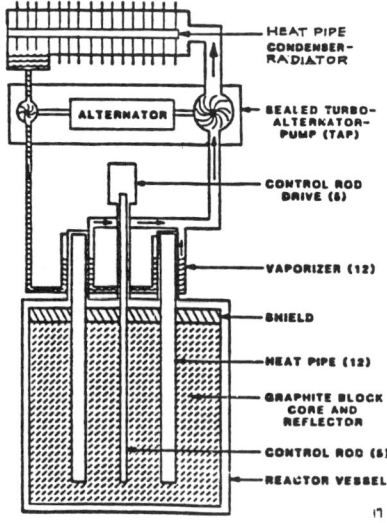

Figure 15.17 Compact reactor-ORC Power Source.

15.4 CLOSED BRAYTON CYCLE (SYSTEM NO. 2)

The *closed Brayton cycle* (CBC) system is similar to conventional gas turbine heat engines, except that the working fluid is recirculated in a closed-loop fashion, rather than being emitted as exhaust to the atmosphere. A typical CBC flow diagram is depicted in Figure 15.18. The major components composing a CBC space power system are a heat source (e.g., isotope, reactor, solar collector, or chemical reactor), combined rotating unit (turbine-compressor and alternator), recuperator, radiator, and gas management system (accumulator, sensors, valves, and controls). Despite the dynamic characterization of the CBC, it consists of only one moving part, the *combined rotating unit* (CRU). A general arrangement of the CRU is shown in Figure 15.19. With the use of working-fluid lubricated, aerodynamically loaded, foil bearings, the Garrett-AiResearch Corp. has achieved many millions of hours of satisfactory, long-life operation. Up to 40,000 hours of continuous operation have been demonstrated in testing and current projections are for 10 years (87,500 hours) of maintenance-free operation on the Space Station program.

The ideal CBC is composed of two *isentropic* (constant entropy) and two *isobaric* (constant pressure) processes. This is depicted by the solid lines in Figure 15.20. The cycle shown by the dashed lines is a *recuperated cycle,* wherein the heat rejected from the turbine is transferred through a heat exchanger (recuperator) to the working fluid at the compressor outlet. This reduces the amount of heat to be added externally from the heat source, and results in an increase in overall efficiency.

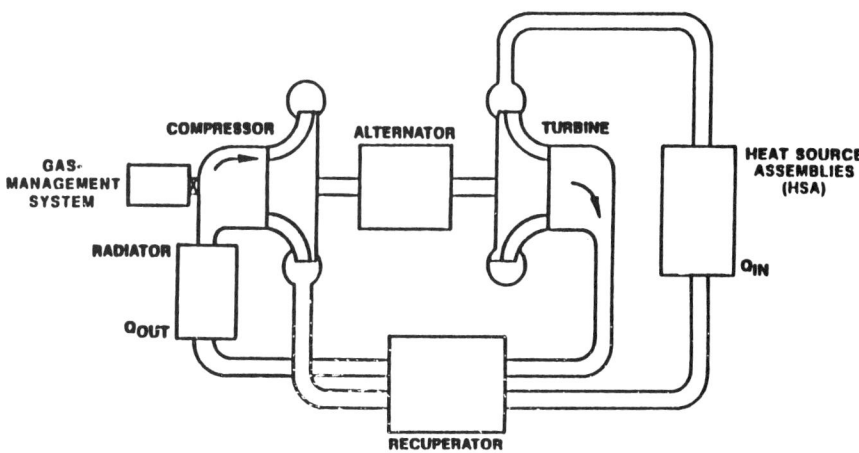

Figure 15.18 Typical closed Brayton cycle flow diagram.

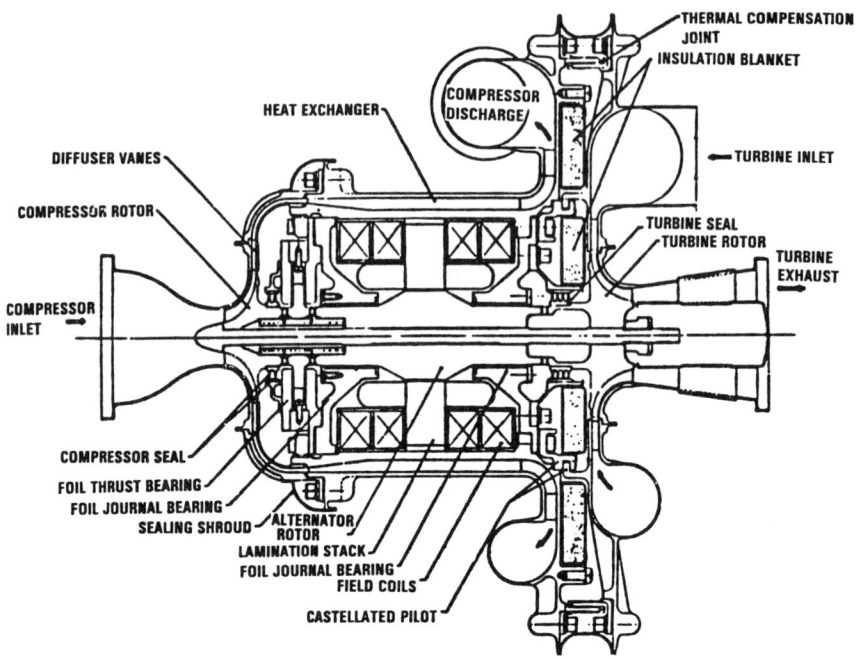

Figure 15.19 CBC combined rotating unit (CRU).

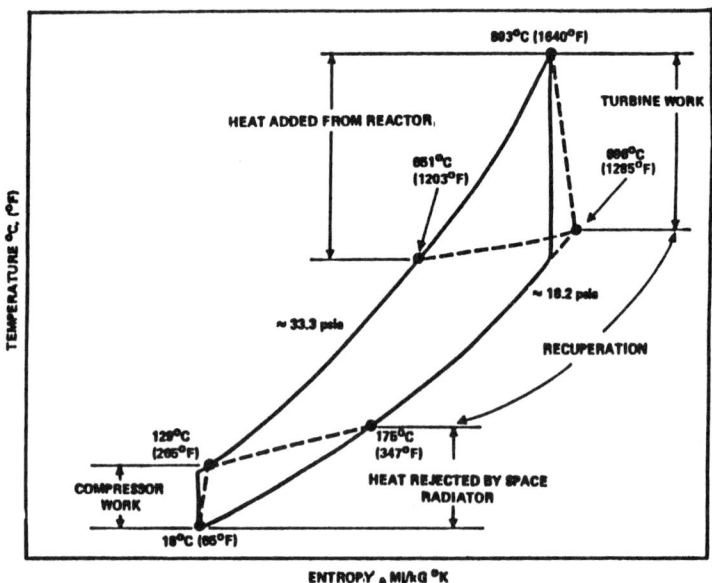

Figure 15.20 Temperature-entropy diagram of Brayton cycle power conversion system.

The real cycle is a gas cycle which in the closed form can use any of a number of gases as a working fluid. These include helium, nitrogen, argon, xenon, or a mixture of gases such as helium-xenon (He-Xe). The mixture attempts to approximate a gas of higher molecular weight than helium for performance, size, and weight optimization, while still retaining the good heat transfer characteristics of helium. For space power system applications, a He-Xe mixture with a molecular weight of 39.9 has been found to be a satisfactory compromise.

In the real cycle, the isentropic processes are replaced by polytropic processes, and the isobaric processes are modified by the pressure drops in the machinery and ducting components. These changes to the ideal cycle result in a decrease in cycle efficiency. In the design of a CBC energy conversion system, these modifications are minimized in an attempt to approach the ideal cycle performance.

Under Department of Energy sponsorship, the Garrett-AiResearch Corp. developed a Pu_{238} isotope-powered CBC in the 1.3 to 2.0 kW_e output power range. The acronym BIPS, standing for *Brayton isotope power system,* denotes this configuration. Currently, the Department of Energy applies the cycle generic designation DIPS to encompass all dynamic isotope power systems. This includes the ORC, CBC, and FPSE i.e., (free-piston Stirling engine).

The typical cycle conditions for the BIPS at an output power level of 1.35 kW_e is shown in Figure 15.21. For similar reasons as we discussed for the ORC-DIPS, the Pu_{238} isotope-powered CBC-DIPS is limited to an upper power level of 10 kW_e. With minor modifications to the components, the 1.35 kW_e CBC-BIPS can be upgraded to 2.08 kW_e. A schematic flow diagram at this higher power level is shown in Figure 15.22. The major changes required are an increase of two or three multihundred-watt (MHW) standard 2400 W thermal heat sources and somewhat larger cycle heat exchangers (e.g., heat source exchanger, recuperator, or radiator). Because of the improvement in the component performance with increasing power level, the CBC-BIPS overall cycle efficiency improves. Hence, the system-specific weight also increases from approximately 1.3 W/lb to 5.0 W/lb, shown in Figure 15.23.

The CBC is also capable of operation with a solar collector as the heat source. Because of the practical physical size limitations of solar collectors, the recommended power level range for these systems is 10 to 50 kW_e. A flow schematic of a typical 40 kW_e CBC solar dynamic system is given in Figure 15.24. Thermal energy storage for operation during eclipse periods is provided by a heat storage device that employs the latent heat of fusion of lithium fluoride to maintain the cycle turbine inlet temperature. The necessity for thermal energy storage for the eclipse portion of the orbits, combined with the losses in the receiver and solar collector results in a fairly low system-specific weight of 2.75 W/lb (for a 270 nmi orbit), despite the improved component performance at the higher power level range. For higher orbits (e.g., 5500 nmi), a specific weight of 3.5 W/lb. is possible. A typical thermodynamic cycle for a solar dynamic CBC is shown by Figure 15.25.

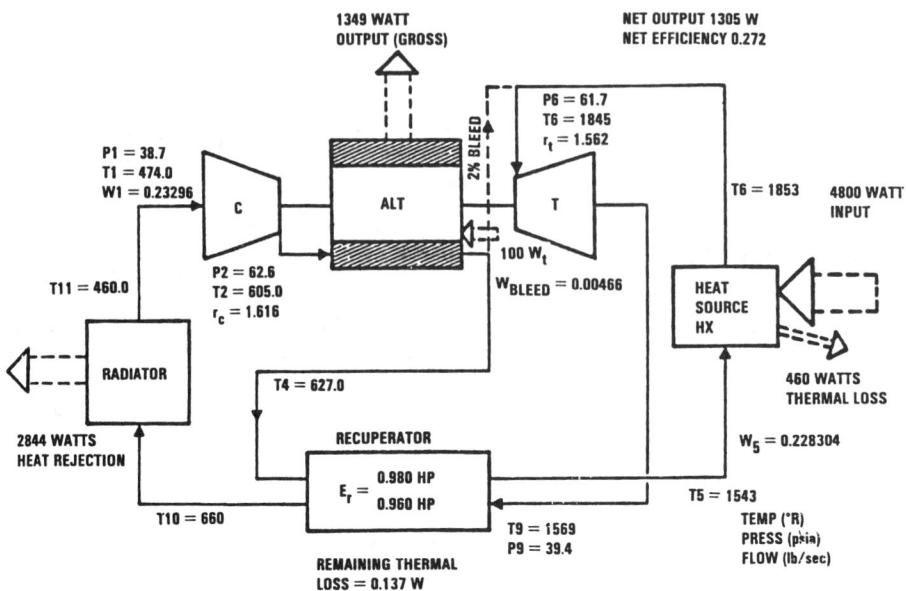

Figure 15.21 BIPS components and typical cycle conditions (working fluid—Helium-Xenon).

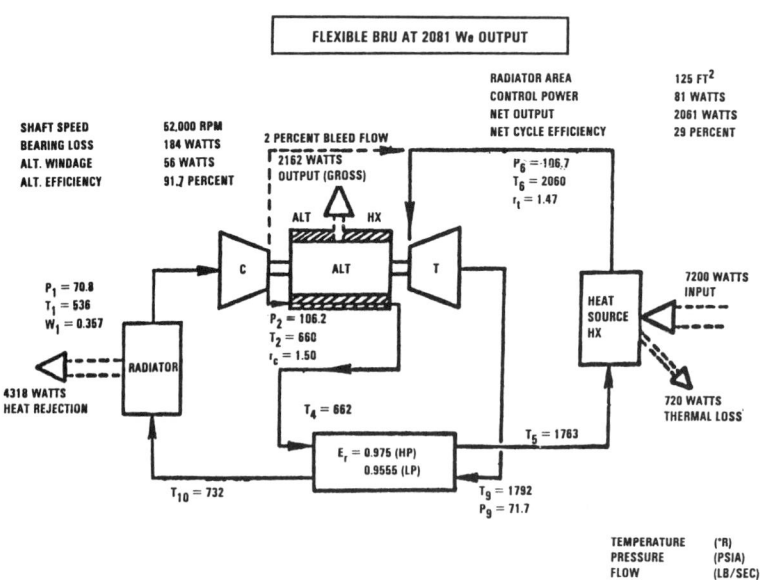

Figure 15.22 CBC-BIPS flow diagram for 2.08 kW$_e$ power level.

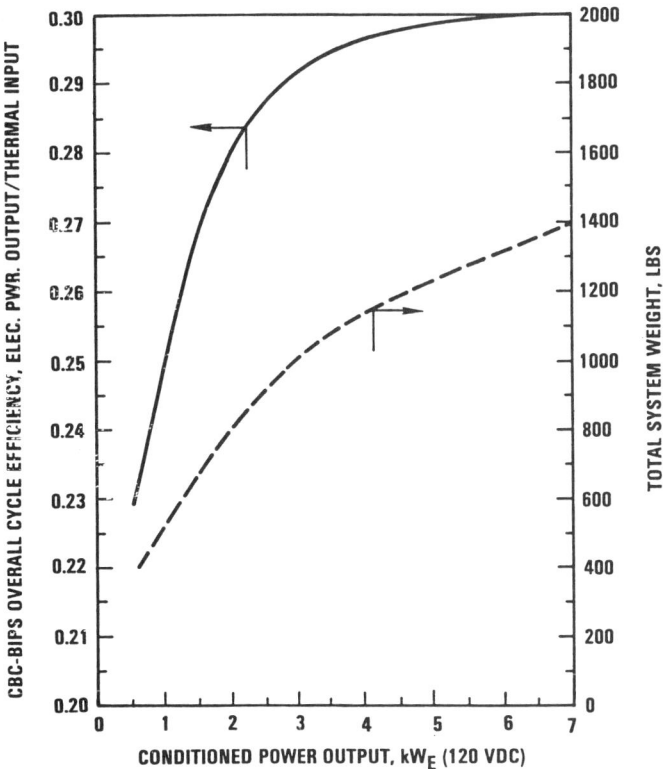

Figure 15.23 CBC-BIPS projected performance.

For continuous power (e.g., 7 to 20 years of life) in the 50 to 300 kW$_e$ power level range, the use of a fast, high-temperature reactor as a heat source with a CBC is recommended. As we previously discussed for the reactor powered ORC, the NERVA derivative reactor, gas and liquid metal cooled compact reactors, and gas cooled particle bed reactors are typical candidate concepts. These are depicted in Figures 15.26 and 15.27. Depending on the type of reactor employed, there could be many possible flow schematics that would comprise a CBC space power system. Generally, they would fall into the category of two-loop or three-loop systems as depicted in Figure 15.28. In the two-loop system, the reactor heat transport loop and the power conversion system loop are combined into one loop. A separate heat-rejection loop is used to reject the cycle waste heat and alternator waste heat to space. It can also cool the cold-junction leg of a thermoelectric converter which is used to power a liquid metal, electromagnetic pump, or the

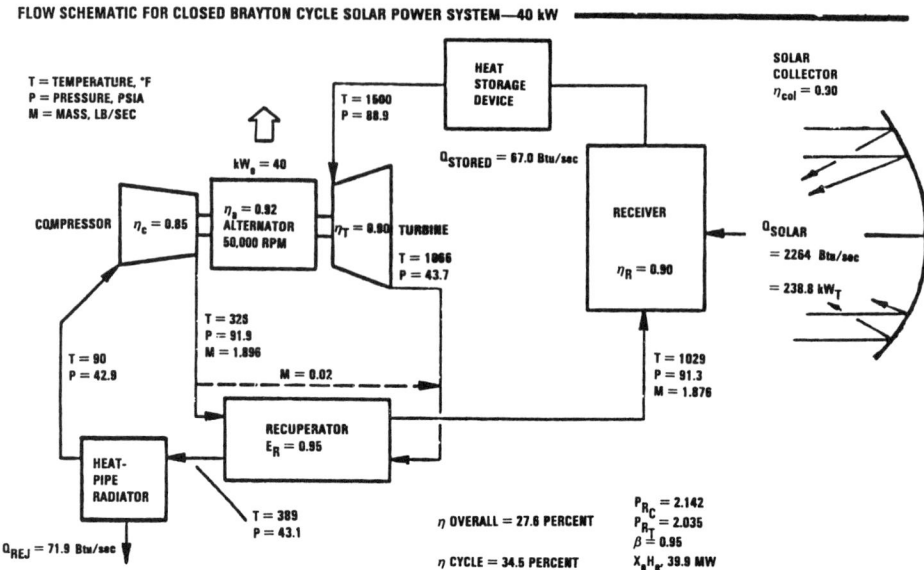

Figure 15.24 Typical CBC solar dynamic flow schematic.

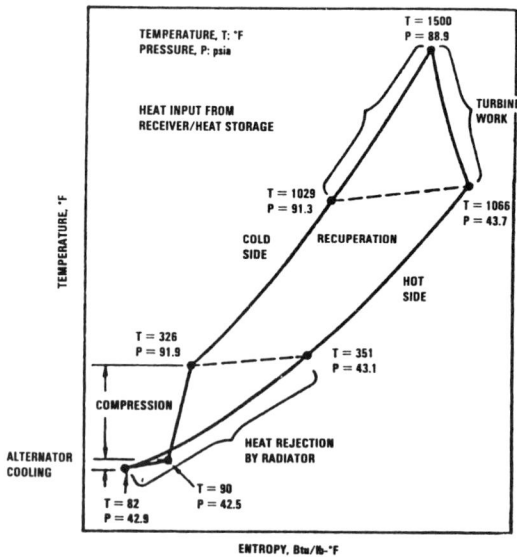

Figure 15.25 Temperature-entropy diagram for CBC solar dynamic power conversion system.

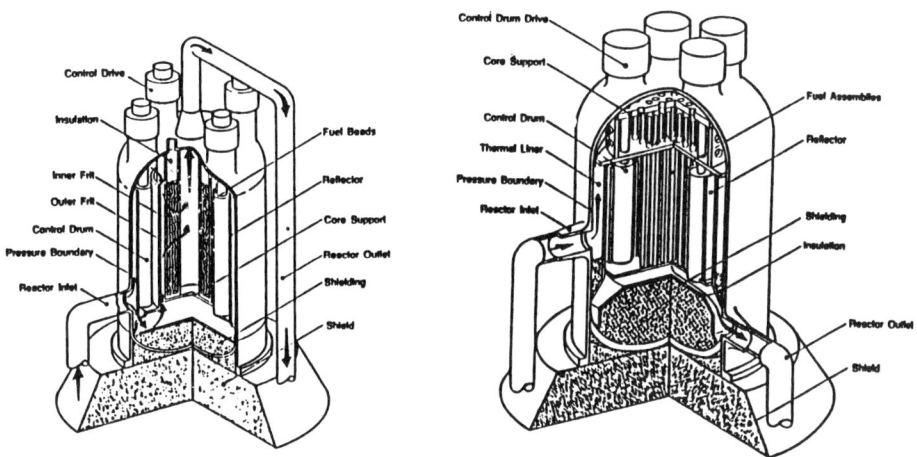

Figure 15.26 Liquid metal cooled space nuclear reactor concepts.

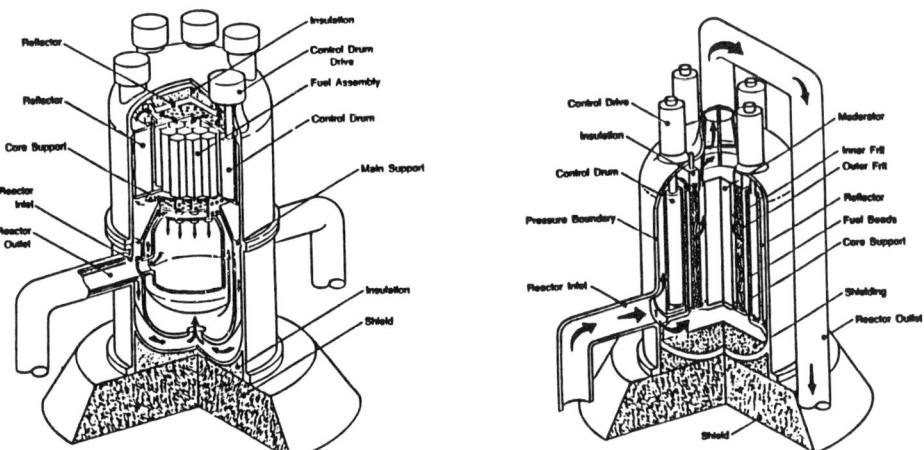

Figure 15.27 Gas cooled space nuclear reactor concepts.

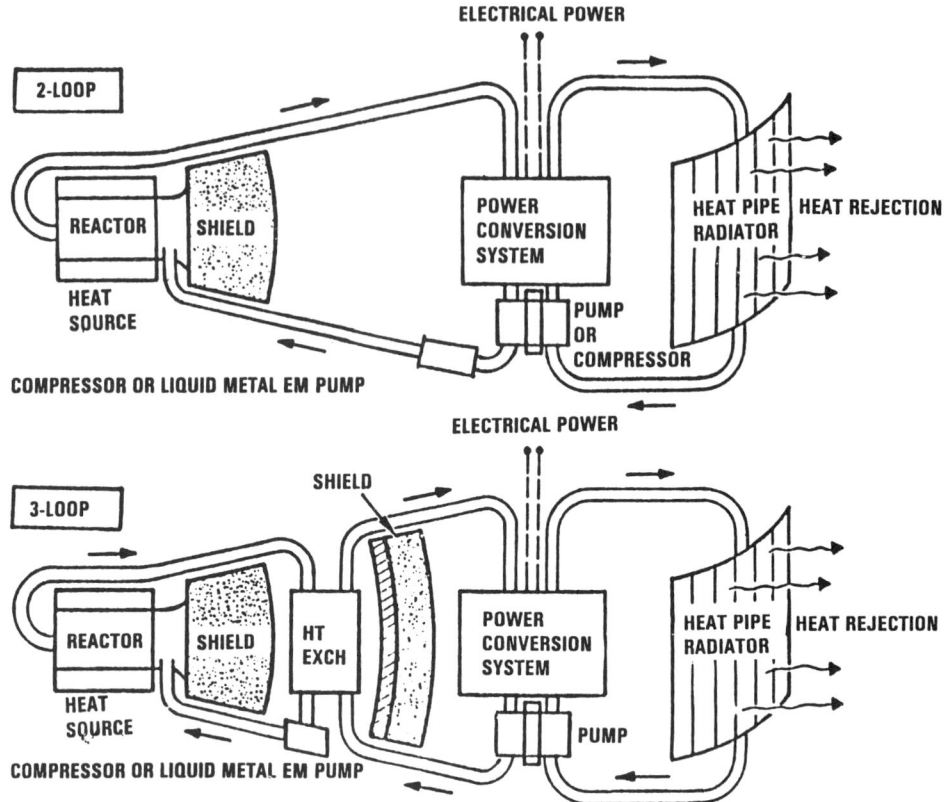

Figure 15.28 Two-loop and three-loop nuclear reactor space system schematic.

motor windings of an electrically driven compressor. An alternative is the three-loop system, which provides three separate loops to perform the heat transport, power conversion, and heat-rejection functions, as depicted for a lithium-cooled reactor on Figure 15.29. Other concepts are possible for use with the CBC to enhance their performance or increase their flexibility of operation, including the use of turbine reheating, compressor intercooling, and thermal or flywheel energy storage.

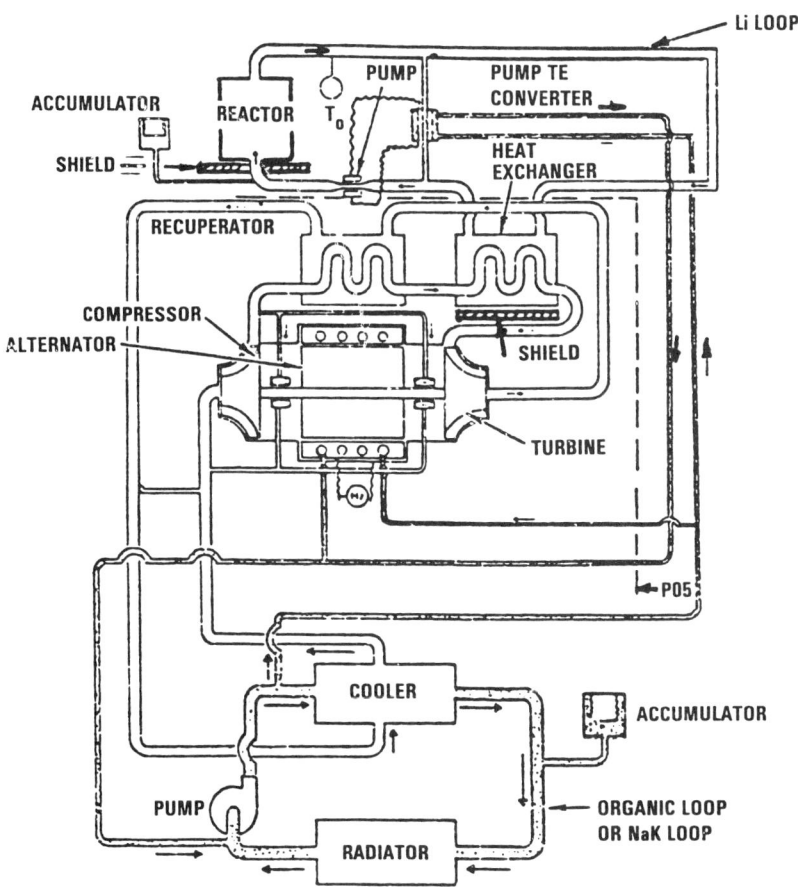

Figure 15.29 Typical reactor-Brayton energy conversion system.

15.5 FREE PISTON STIRLING ENGINE POWER SYSTEMS (SYSTEM NO. 3)

The Stirling engine can best be described as a closed-cycle piston engine with cyclic recirculation of the working fluid. Power is produced in the manner that is conventional for all gas cycle heat engines. The working gas (helium) is very nearly isothermally compressed at low temperature, heated at constant volume, expanded approximately isothermally at high temperature, and then cooled through a very nearly constant volume heat-rejection process. This type of operation is depicted in Figure 15.30. The major difference between the Stirling engine and conventional

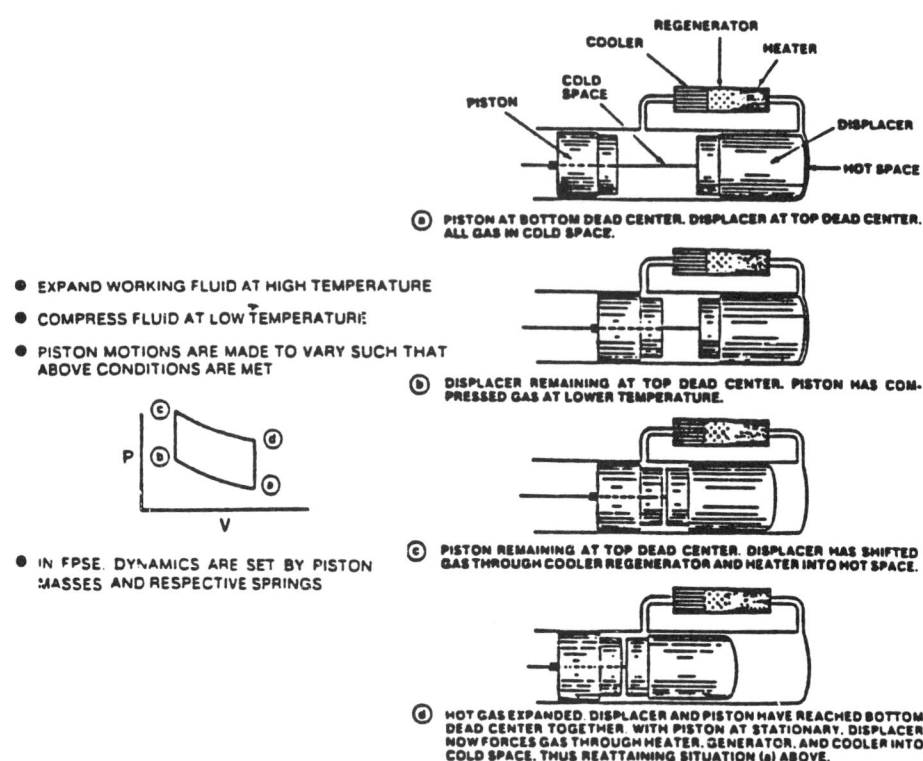

- EXPAND WORKING FLUID AT HIGH TEMPERATURE
- COMPRESS FLUID AT LOW TEMPERATURE
- PISTON MOTIONS ARE MADE TO VARY SUCH THAT ABOVE CONDITIONS ARE MET
- IN FPSE, DYNAMICS ARE SET BY PISTON MASSES AND RESPECTIVE SPRINGS

Figure 15.30 Stirling engine operation.

piston engines is the method of heat addition. In the Stirling engine, the required heat is added to the working gas inside the engine through a heat-exchanger wall. As this wall has a high heat capacity, rapidly heating and cooling the wall is not possible. Therefore, in one form of the Stirling engine concept, a second piston (called the *displacer*) is added to the machine to move the gas between two stationary chamber volumes (called the *hot space* and *cold space*). However, shuttling the gas between the hot and cold volumes would waste large quantities of heat, unless a matrix type regenerator were placed between the hot and cold sources.

When this is done, heat is stored in the regenerator as the gas moves toward the cold space (compression space) and released as the working fluid returns from the cold space to the hot space (expansion space). While the displacer piston motions act to heat and to cool the working gas, the power piston motions act on the working gas to compress and to expand it. The displacer and power pistons move in a phased relationship (see Figure 15.31(b) and (c)) so that the working gas is alternately compressed, heated, expanded, and cooled, thereby producing output power.

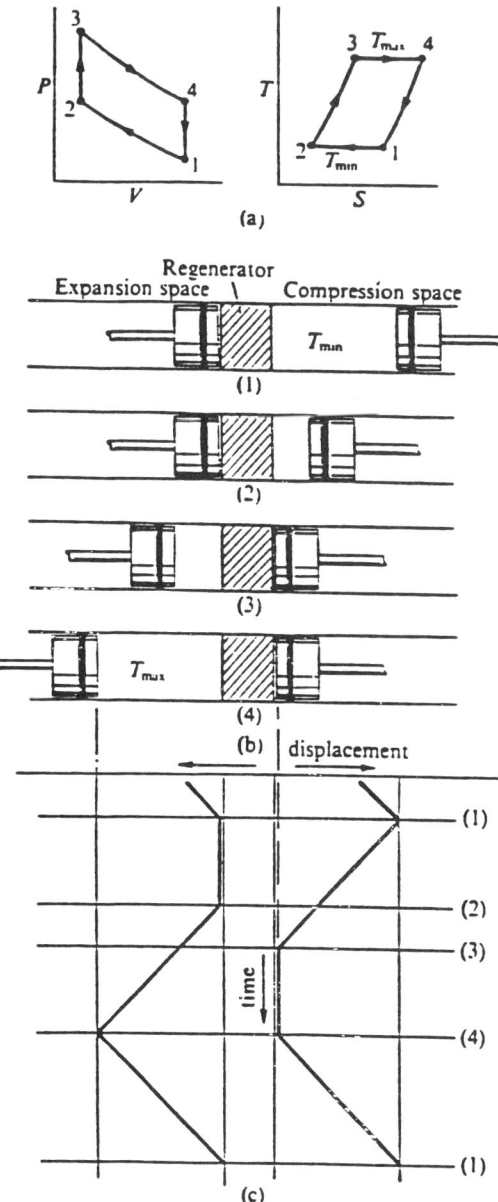

Figure 15.31 Stirling engine cycle diagrams.
  (a) *P-V* and *T-S* diagrams.
  (b) Piston arrangement at the terminal points of the cycle.
  (c) Time-displacement diagram.

Using these thermodynamic and cycle characteristics, there is a wide variation in general types of Stirling engine configurations. The original configurations utilized crankshaft drives to convert the engine's reciprocating motion into a rotary output. In addition to being fairly heavy, the engines also required a separate lubrication system. This quite often led to contamination of the helium working fluid and degradation of the engine performance. A flexible diaphragm (called a *rollsock*) was devised to eliminate lubricant leakage past mechanical seals. However, continued flexing of the rollsock proved to be a highly life-limiting feature. Two other Stirling engine concepts were devised in an attempt to reduce engine weight and extend operating life. These eliminated the crankshaft drive and substituted a swash plate drive or a Wankel type of rotary piston to achieve a rotary shaft output. However, these engines were still fairly complex and required separate lubricating systems. This made the engines unattractive for the long-life requirements of space applications. Finally, a so-called free-piston, single-cylinder engine employing a linear electromagnetic (EM) generator was conceived (see Figure 15.31(c)), which resolved the weight and lubrication technical issues and made the engine attractive for space applications. The *free-piston Stirling engine* (FPSE) systems operate without physical linkages, instead relying on internal gas pressures to impart the correct motions to the reciprocating elements. An FPSE operates, in effect, as a resonant mass-spring system driven by the engine thermodynamics (as a thermal oscillator), or a combination of the engine thermodynamics and an external electrical signal (as a thermal amplifier).

Similar to the ORC and CBC, the FPSE system can be powered by a Pu_{238} isotope heat source. Because of its potential for high overall conversion efficiency (e.g., 28 to 35%), this engine is practical to consider in the 1.0 to 12.5 kW_e output power range. A *Stirling isotope power system* (SIPS) was developed by General Electric for the Department of Energy to provide an output power of 1.112 kW_e. A schematic diagram of this system is shown in Figure 15.32, and performance characteristics are provided in [11]. During this period, the most technologically mature Stirling engine was the dual-piston, single-cylinder, rhombic drive type developed by North American Phillips. Hence, this engine was utilized for that application. However, since then, Mechanical Technology, Inc. has developed a 12.5 kW_e FPSE, and this engine in various modular forms is currently being considered for most space power applications. The engine has been designated the *space power demonstrator engine* (SPDE). A cross section of this configuration is shown in Figure 15.33.

The FPSE system can also be solar powered or fast-reactor powered to provide higher output power levels. The approach is to use two of the SPDEs in an opposed, back-to-back, piston arrangement to provide a total output power of 25 kW_e. This configuration is depicted in Figure 15.34. This arrangement not only produces increased output power, but also greatly reduces any dynamic unbalance.

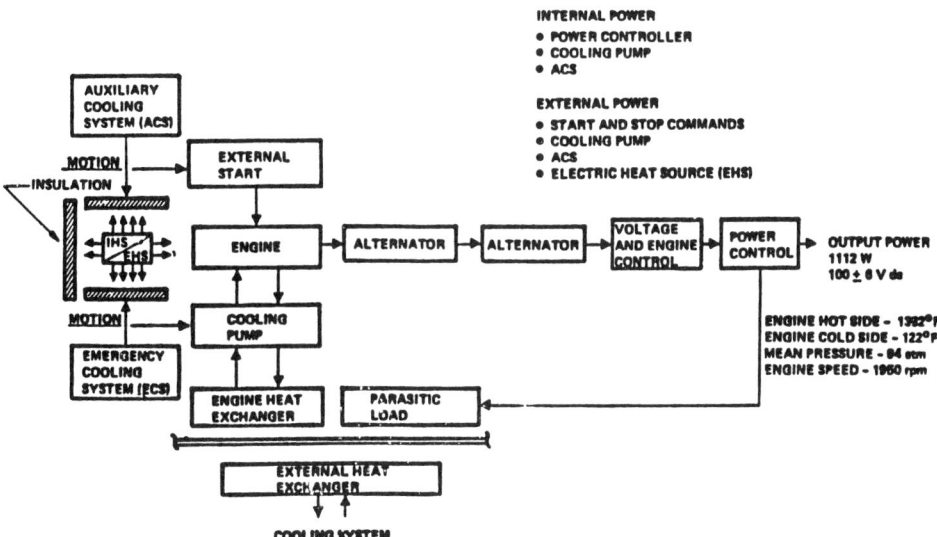

Figure 15.32 SIPS components and typical cycle condition (working fluid—helium).

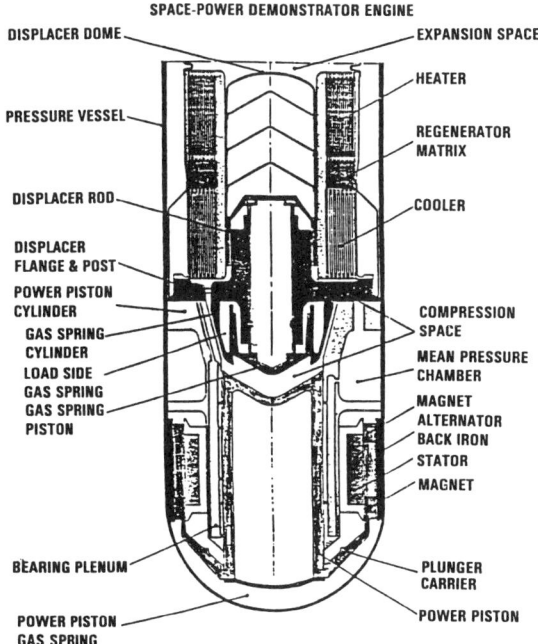

Figure 15.33 12.5 kW$_e$ free-piston Stirling engine.

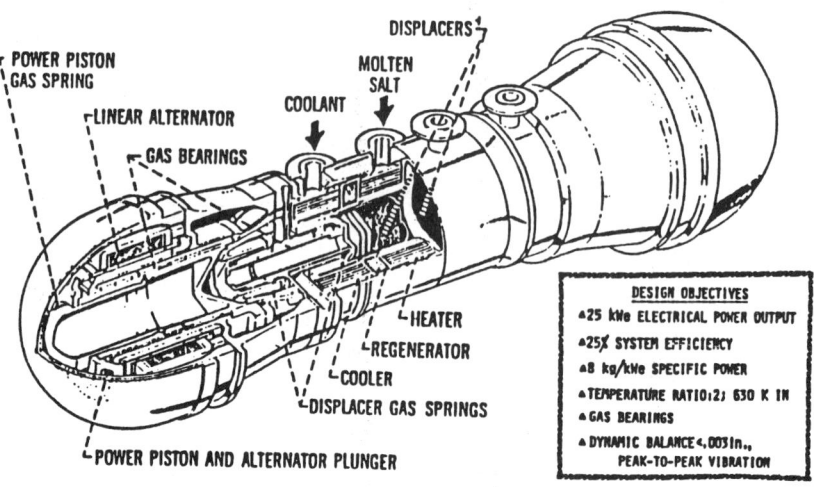

Figure 15.34 Candidate 25 kW$_e$ free-piston Stirling engine with linear alternators.

This candidate engine together with a lithium fluoride (LiF) thermal energy storage subsystem can be powered by a solar collector to provide an output power of 25 kW$_e$. For higher power levels, a limiting factor would be the maximum practical diameter of a deployable or erectable solar collector. A typical 25 kW$_e$ solar powered FPSE is depicted in Figure 15.35. To achieve higher power levels, the use of multiple systems is recommended.

For higher power level FPSE systems (e.g., 50 to 500 kW$_e$), a lighter weight approach would be to use a fast reactor as a heat source. This range of power systems was studied by Rockwell International under contract to JPL [12]. The overall arrangement for a baseline 100 kW$_e$ system is shown in Figure 15.36(a). The reactor is controlled by four sliding reflector segments with the actuators located below the neutron shield. This reactor uses UO$_2$ fuel encased in a refractory alloy cladding. The reactor is cooled by a eutectic sodium-potassium alloy (NaK). The coolant passes from the reactor through manifolds to five modular heat exchangers. These NaK-to-pipe heat exchangers surround the middle of each of the five 25 kW$_e$ FPSEs. Sodium pipes transport the heat from the *intermediate heat exchangers* (IHXs) to the Stirling engine heaters. The heaters are located in modular *heater-cooler-regenerator* (HCR) units, outside of the main pressure shell of

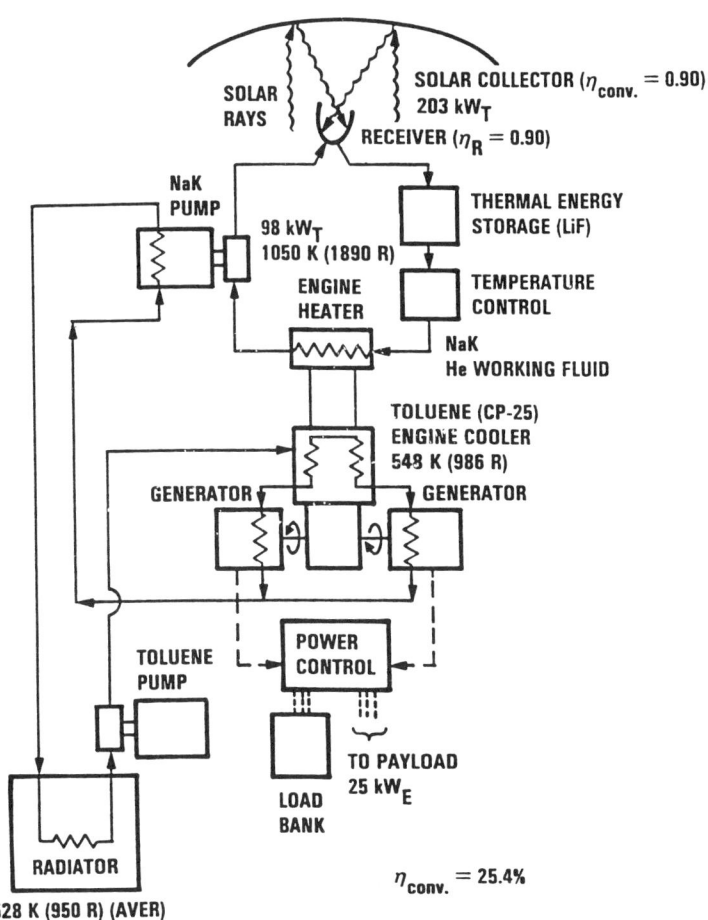

Figure 15.35 Solar powered FPSE with thermal energy storage system schematic.

the engine (see Figure 15.37). The specific weight of the high temperature (T_H = 1114 K), 100 kW$_e$ baseline system is 94.16 lb/kW$_e$. A backup concept operating at a lower hot-side engine temperature (T_H = 964 K) is shown in Figure 15.36(b). This more conservative design results in a reduced net cycle efficiency and an increase in specific weight (118 lb/kW$_e$). However, this backup concept represents a reduced development risk.

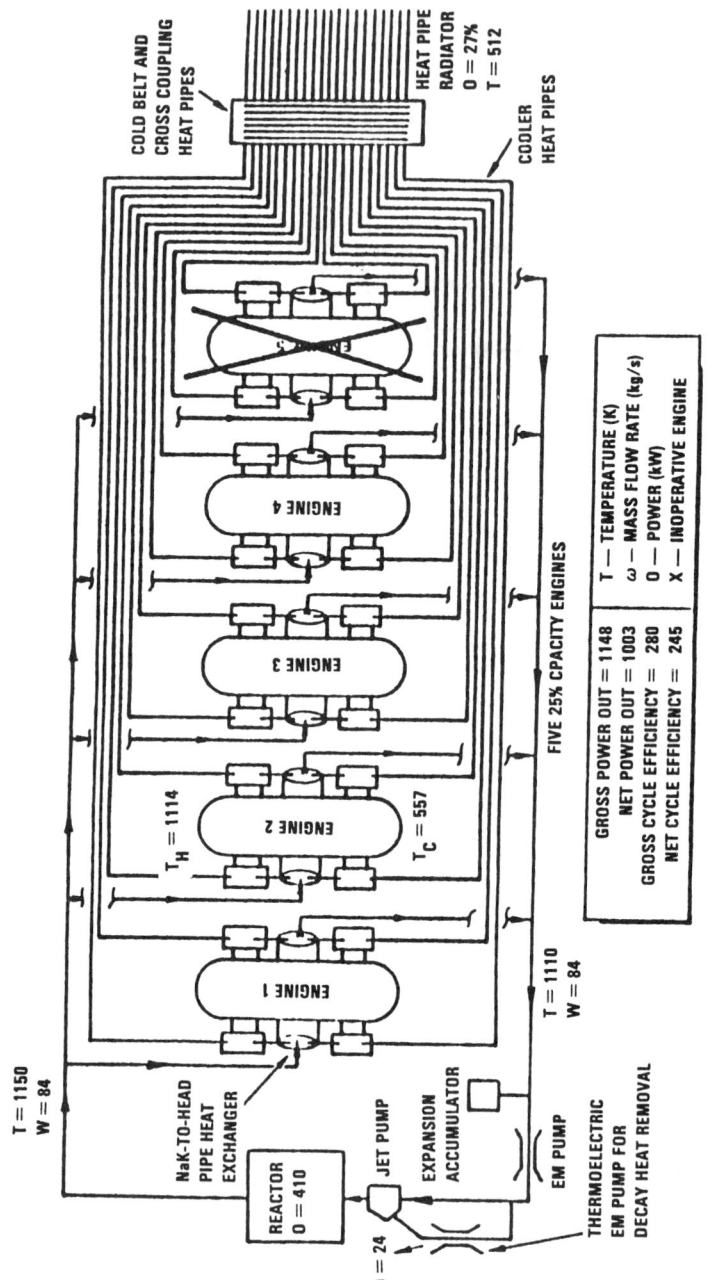

(a) OPERATING CONDITIONS FOR THE 100-kWe BASELINE SYSTEM (T_H = 1114 K)

Figure 15.36 Baseline 100 kW$_e$ reactor FPSE power systems.

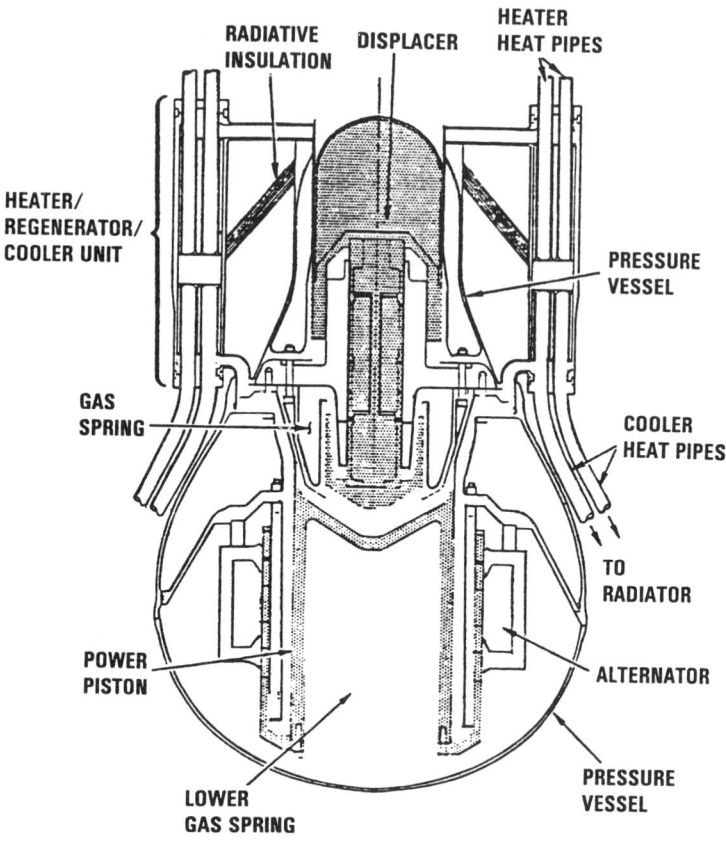

Figure 15.37 SP-100 Stirling engine module.

15.6 SUPERCRITICAL CYCLE POWER SYSTEMS (SYSTEM NO. 4)

The thermodynamic cycles most commonly used for closed cycle power systems are the Rankine, Brayton, and Stirling. By the use of simple regeneration, all three cycles can greatly increase their overall efficiency. For space applications, increased performance is extremely important because it can result in significant weight and size reductions in the overall power plant. The steam Rankine cycle cannot use simple regeneration for improving efficiency. Usually, the Rankine Cycles must resort to bleeding the working fluid during the expansion process (e.g., turbine reheating), but this requires a fairly complex and heavy multistage turbine, and is feasible only in large, stationary, terrestrial power plant installations. Rankine cycles using organic working fluids can employ regeneration, but are limited to fairly low peak cycle operating temperature to avoid pyrolytic decomposition.

The achievable performance of the Rankine cycles is thus limited. However, the advantage inherent in all Rankine cycles is that the required nonpower-producing pump work is only a small fraction (approximately 3%) of the total work output. This is offset to a certain extent for space applications by the complexity of designing heat exchangers for two-phase boiling and condensing in a zero-gravity environment.

The Brayton cycle employs a compressible gas (e.g., He-Xe mixture) and consumes a large portion of the total turbine work output (up to 60%) in the compression process, but compensates for this by the use of simple regeneration to improve cycle efficiency. The Stirling cycle uses a constant volume process, and therefore has the disadvantage of requiring a reciprocating engine with many complex sliding mechanical components. This cycle, however, can achieve high thermal efficiency by simple regeneration, but as a reciprocating machine it is not compatible with multimegawatt power levels.

A thermodynamic cycle that avoids all these problems, retains the various advantages, and operates in the supercritical pressure region of the particular working fluid selected is called the *supercritical cycle*. The supercritical region of a fluid is the thermodynamic regime above its critical pressure and temperature. A typical enthalpy-entropy diagram is shown in Figure 15.38. In this region, the density of the liquid and vapor at any given state point is the same. For space applications, this is an advantage as it eliminates the phenomenon of two-phase boiling and condensing in a zero-gravity environment. This greatly simplifies the cycle boiler and condenser designs. For closed cycles, the most common working fluids are CO_2, SO_2, or NH_3 (see Table 15.4). To date, most development work has been done with CO_2 (e.g., Feher cycle). A typical schematic of the closed form of the supercritical cycle is shown in Figure 15.39. Note that this is a high-pressure cycle, and because of this the performance characteristics of the turbomachinery are not suitable for power levels below 25 kW_e. Hence, isotope heat sources would not be utlized with supercritical cycles.

Table 15.4
Critical Temperatures and Pressures of Candidate Supercritical and Cycle Working Fluids

Compound	Formula	Critical Temperature		Critical Pressure	
		(K)	(°F)	(atm)	(psia)
Carbon Dioxide	CO_2	304.2	88.2	72.9	1072
Ammonia	NH_3	405.5	270.5	111.3	1636
Sulfur Dioxide	SO_2	430.7	315.6	77.8	1144

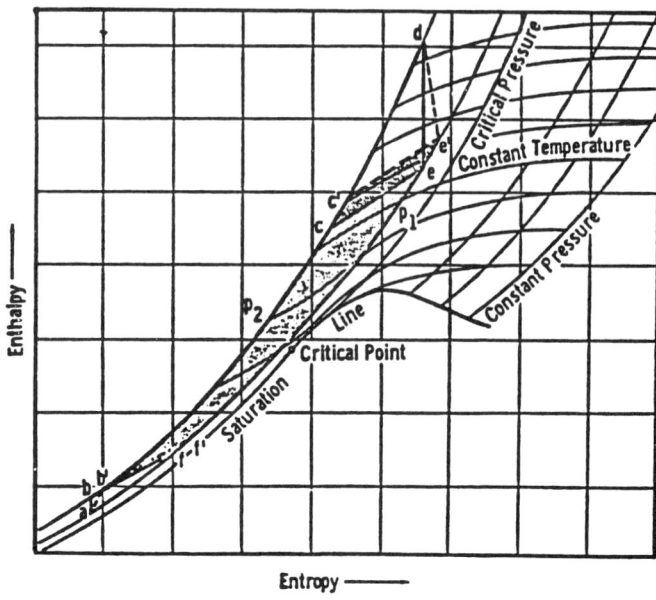

Figure 15.38 Enthalpy-entropy diagram for supercritical cycle.

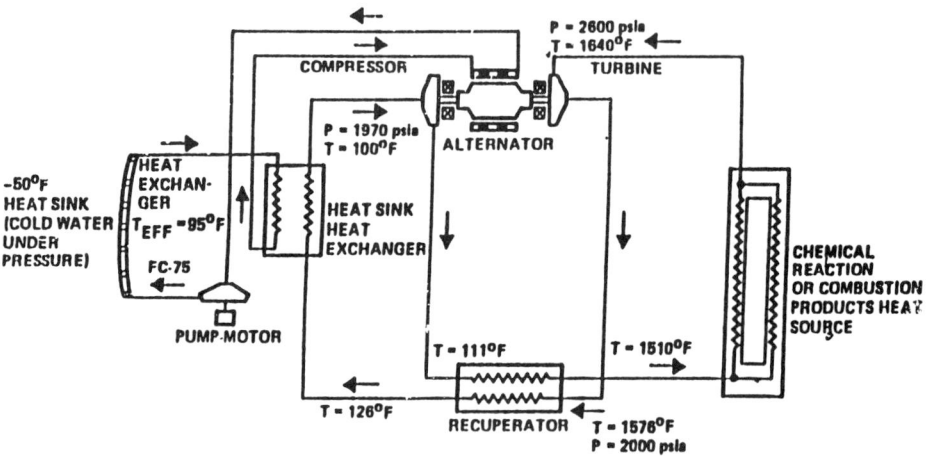

Figure 15.39 Typical schematic of closed form supercritical cycle.

Similar to the ORC, CBC, and FPSE power systems, the closed supercritical cycle can utilize solar collector or thermal energy storage and reactors to develop continuous power. As previously discussed, solar-collector heat sources limit the cycle output power to approximately 50 kW$_e$, and use of multiple units is recommended for higher power levels. Uranium oxide (UO_2) or uranium carbide (UC) fast-reactor heat sources are compatible with both continuous and burst power operations. The high-pressure turbomachinery of this cycle is unique in that it can provide high output power levels (up to hundreds of MW$_e$) with very lightweight and small-sized components. For the closed form supercritical cycles, chemical heat sources can be used to produce short-duration burst power. Typical chemical heat sources include the following:

- Be/LOX + LH_2
- Be/GOX + COl_2
- Li + SF_6 or Li + SOCl2

In the open form of the cycle, the heat source is inherent in the release of energy in the combustion process. With the exception of the Be/LOX + LH_2 combination, all other heat sources address effluent management by retaining the chemical reaction or combustion products within the cycle storage tanks. In this exceptional case, only gaseous H_2 is vented to space.

15.7 POTASSIUM RANKINE CYCLE (SYSTEM NO. 5)

The *liquid-metal potassium* (K) *Rankine cycle* (LMKRC) is similar to the steam and organic Rankine cycles. However, by using liquid potassium as a working fluid, it is capable of operating at very high turbine inlet temperatures. As a consequence, the LMKRC can also reject the cycle waste heat at fairly high temperatures without suffering any major degradation in overall cycle efficiency. High heat-rejection temperatures result in smaller radiator areas, and hence reduced mass. For high power levels, this cycle characteristic becomes quite important because radiator mass usually can constitute up to 50% of the total weight of a low-temperature heat-rejection cycle such as ORC or CBC.

During the early 1960s, both GE and Pratt and Whitney carried out preliminary research and development programs. These were designated SNAP-50, and they consisted of a 300 kW$_e$ liquid metal (lithium) cooled nuclear fast reactor as a heat source, combined with a LMKRC system. SNAP-50s were designed to produce continuous power for a 7 to 10 year operating life. Due to lack of a mission and high-temperature material corrosive problems, however, this program was terminated.

With the advent of the Strategic Defense Initiative (SDI) program and its requirements for very high level (hundreds of MW_e) burst and continuous power output space electrical systems, there has been renewed interest in LMKRC systems. Various companies and government laboratories (e.g., TRW, GE, Martin-Marietta, ORNL, ANL, and LANL) have been synthesizing various conceptual designs to provide both burst power and continuous power outputs at the multi-megawatt level.

Chemical reactors can provide an attractive heat source option for burst power applications. A typical example is depicted in Figure 15.40. This concept utilizes the exothermic reaction of titanium (Ti) and oxygen (O_2) to provide the thermal energy input to the LMKRC system. This concept is somewhat unique in that it employs helium-xenon (He-Xe) in a heat transport loop to transfer the chemical reactor thermal energy to the potassium loop. A temperature-entropy diagram for the LMKRC is shown in Figure 15.41. It consists of both two-phase boiling and condensing, and, as it must occur at zero-gravity in space, remains the technical issue regarding the technology readiness of the LMKRC. However, constant-temperature heat rejection at high temperatures remains an attractive feature of this cycle.

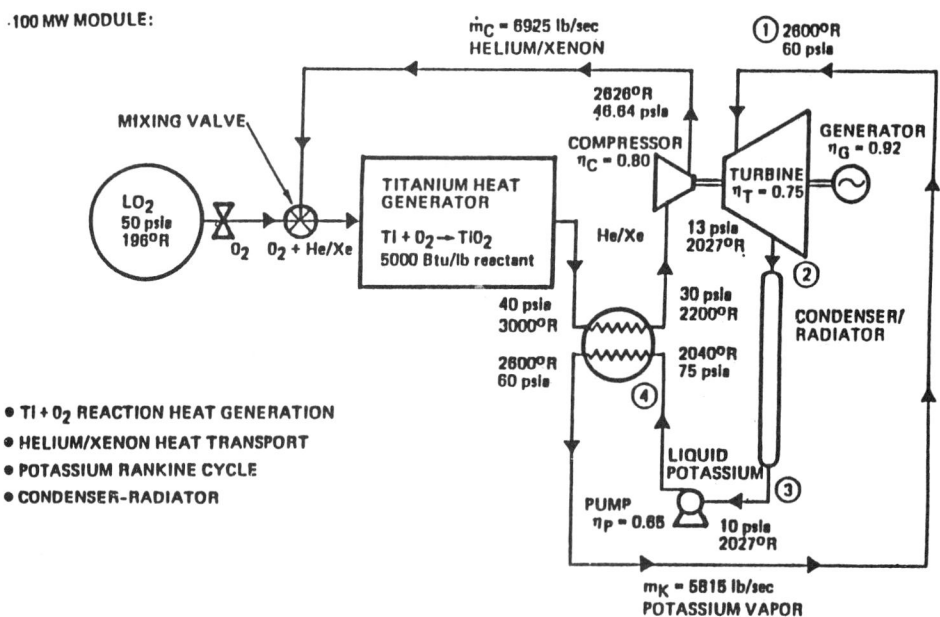

Figure 15.40 100 MW_e chemical reactor, burst power LMKRC system.

Titanium-Oxygen Reactor
Thermodynamic Cycle

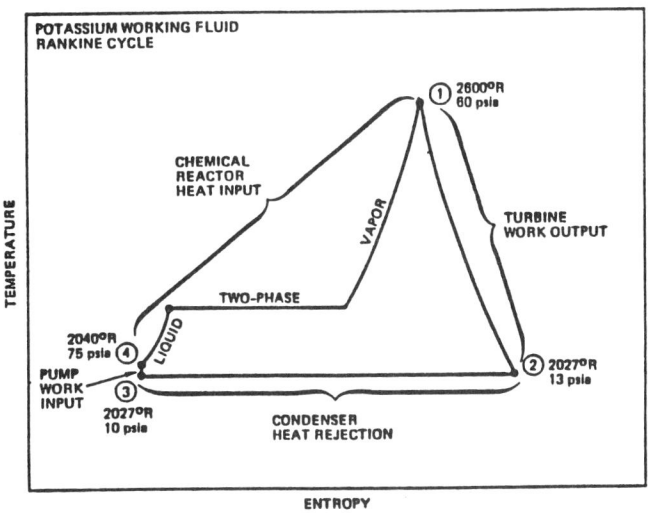

Figure 15.41 Typical LMKRC temperature-entropy diagram.

A nuclear reactor heat source is an attractive option for continuous power generation. A typical reactor-powered LMKRC system is shown in Figure 15.42. This system consists of a lithium (Li) cooled fast reactor utilizing a potassium Rankine cycle for energy conversion. The thermal energy from the reactor is transmitted to the potassium boiler through a heat exchanger in the lithium heat transport loop. The superheated potassium vapor generated goes through a set of turbines that drives a generator and liquid potassium cycle pump. The turbine exhausts are combined and condensed in a combination condenser and heat pipe radiator. The lithium is recirculated through the reactor by using an electromagnetic pump. Potassium is circulated by using a combination of a main pump and booster jet pumps. Paired, counter-rotating turbogenerators and turbopumps are utilized to eliminate unbalanced torques. While the overall efficiency of the LMKRC system is only 20%, this is more than offset by the high heat-rejection temperature which results in a small, lightweight radiator. Because of this, a specific weight of approximately 10.5 lb/kW$_e$ at the 1 MW$_e$ power level is possible.

Use of a reactor-powered LMKRC for burst power applications is also possible. This usually requires both fast response and peak power levels in excess of the steady-state output power, and so can usually be accomplished by some form

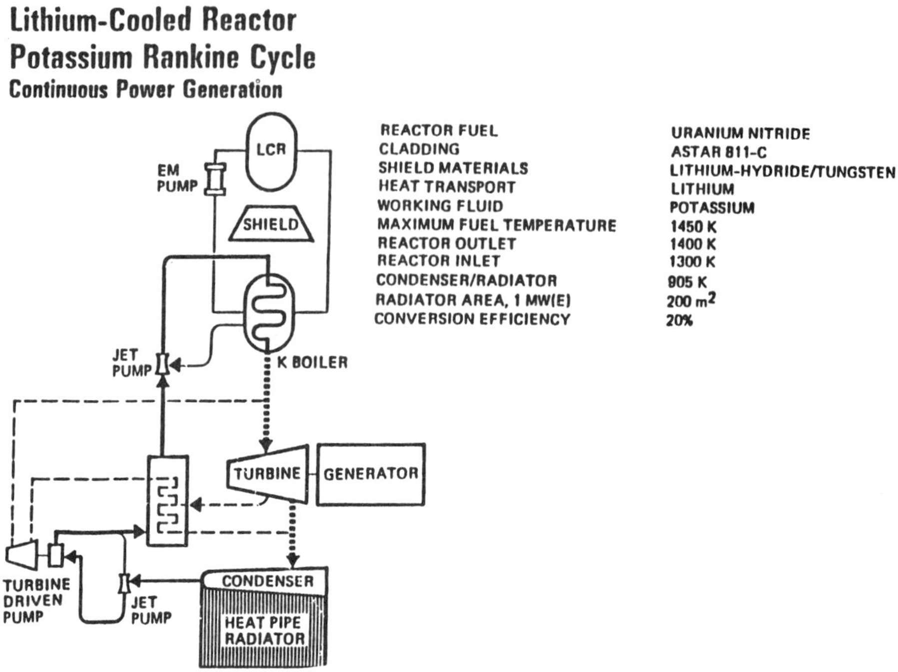

Lithium-Cooled Reactor
Potassium Rankine Cycle
Continuous Power Generation

REACTOR FUEL	URANIUM NITRIDE
CLADDING	ASTAR 811-C
SHIELD MATERIALS	LITHIUM-HYDRIDE/TUNGSTEN
HEAT TRANSPORT	LITHIUM
WORKING FLUID	POTASSIUM
MAXIMUM FUEL TEMPERATURE	1450 K
REACTOR OUTLET	1400 K
REACTOR INLET	1300 K
CONDENSER/RADIATOR	905 K
RADIATOR AREA, 1 MW(E)	200 m^2
CONVERSION EFFICIENCY	20%

Figure 15.42 Lithium-cooled nuclear reactor, LMKRC power system.

of energy storage. One approach is to use *thermal energy storage* (TES), available from the heat of fusion of a molten salt such as lithium fluoride (LiF) or lithium hydride (LiH). A typical reactor-LMKRC system with TES is depicted in Figure 15.43. An alternative approach is to utilize *mechanical energy storage* (MES) in the form of a hermetically sealed, motor-driven, flywheel generator with magnetic suspension bearings. A typical reactor-LMKRC with MES is shown in Figure 15.44. Finally, another option is to utilize *electrochemical energy storage* (ECES) in the form of a regenerative hydrogen-oxygen fuel cell, to provide the fast response burst power with recharging taking place between burst periods if the duty cycle allows. A schematic of this reactor-LMKRC with ECES is given in Figure 15.45. The selection of a particular optimum energy storage concept is greatly dependent on the mission load profile, and requires detailed trade-off analyses to arrive at a minimum-mass subsystem.

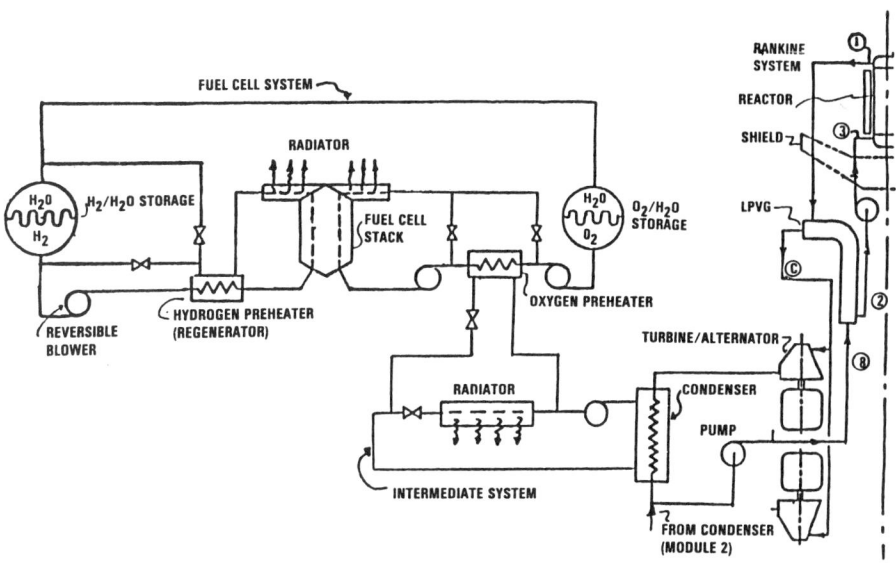

Figure 15.45 Reactor-LMKRC with electrochemical energy storage.

15.8 ALKALI METAL THERMOELECTRIC CONVERSION (AMTEC) SYSTEM (SYSTEM NO. 6)

The *alkali metal thermoelectric converter* (AMTEC), sometimes referred to as a *sodium heat engine* (SHE), has the potential for providing direct energy conversion for a space nuclear reactor power system. The AMTEC is a thermally regenerative electrochemical device for high efficiency, direct, thermal-to-electrical energy conversion. The AMTEC operation derives from the unique sodium ion conducting properties of *beta-alumina solid electrolyte* (BASE). Experimental results have shown that the AMTEC has conversion efficiencies that are much higher than those of other direct thermoelectric devices. An efficiency of 19% and a power density of 1 W/cm^2 have been demonstrated in laboratory devices, and calculations indicate that a 25–30% conversion efficiency may be achieved with an optimized design. Further, The AMTEC has no moving mechanical parts and its hot side temperature (900–1300 K) makes it compatible with either the *general-purpose heat source* (GPHS) or the projected SP-100 nuclear space power reactor. For these reasons, the AMTEC has been studied at JPL for several space power applications. The operating cycle of the AMTEC is illustrated in Figure 15.46.

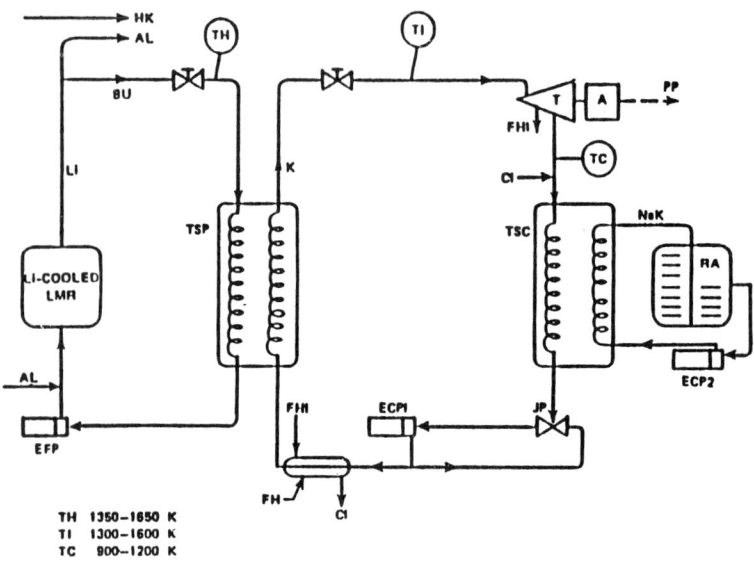

Figure 15.43 Lithium-cooled reactor, LMKRC with thermal energy storage.

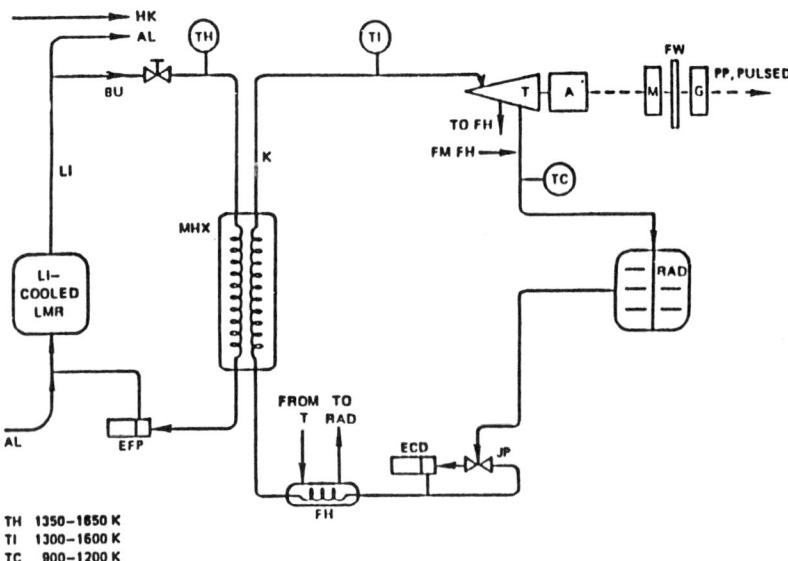

Figure 15.44 Lithium-cooled reactor, LMKRC with flywheel reactor storage.

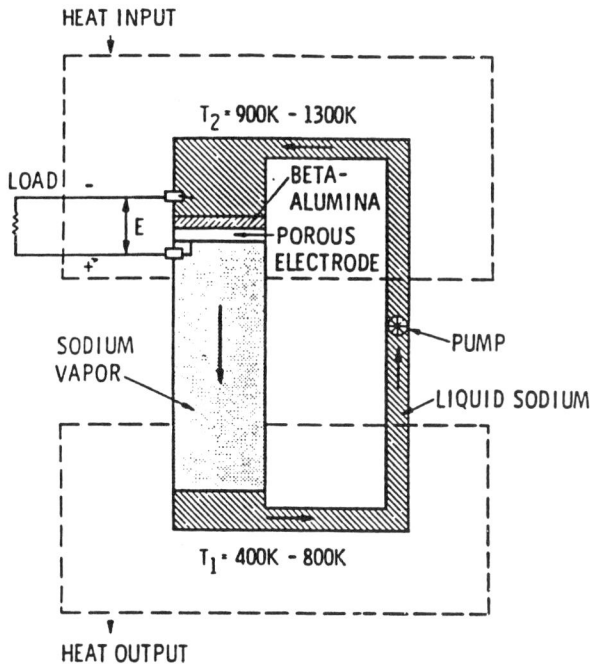

Figure 15.46 AMTEC device schematic.

15.9 THERMOELECTRIC CONVERSION CYCLES (SYSTEM NOS. 7, 8, 9)

The *thermoelectric conversion* (TEC) cycle has had the most extensive application of all other space power systems. This is a result of many factors. Perhaps of foremost importance is that the TEC cycle is compatible with all heat source forms (e.g., isotope, nuclear reactor, solar concentrator, and chemical combustor). In addition, as a static system with high-level redundancy provided by multiple thermoelectric couples, very high reliability levels could be projected. For planetary missions where the solar intensity is greatly diminished, an isotope powered TEC is the only practical candidate system. Now that much higher levels of survivability are required to meet various military space threats, an isotope or nuclear reactor powered TEC offers considerably higher levels of hardening capability. The earliest space use of TEC was in the low-power-level ranges (e.g., 5–150 W_e). A summary of these missions is shown in Table 15.5.

Table 15.5
Summary of Radioisotope Thermoelectric Generators Successfully Launched by the United States (1961–1982)

Power Source	Number of Power Sources	Initial Average Power per Power Source (W)	Spacecraft	Mission Type	Launch Date[a]	Initial Orbit	Status
SNAP-3B7	1	2.7	TRANSIT 4A	NAVIGATIONAL	6/29/61 (ETR)	≈ 890 × 1000 km 67.5°, 104 min	SATELLITE SHUT DOWN BUT OPERATIONAL
SNAP-3B8	1	2.7	TRANSIT 4B	NAVIGATIONAL	11/15/61 (ETR)	≈ 960 × 1130 km	SATELLITE CEASED TRANSMITTING
SNAP-9A	1	>25.2	TRANSIT 5BN-1	NAVIGATIONAL	9/28/63 (WTR)	32.4°, 106 min ≈ 1090 × 1150 km 89.9°, 107 min	SATELLITE CEASED TRANSMITTING
SNAP-9A	1	26.8	TRANSIT 5BN-2	NAVIGATIONAL	12/5/63 (WTR)	≈ 1080 × 1110 km 90.0°, 107 min	NAVIGATIONAL CAPACITY CEASED, BUT SNAP-9A TELEMETRY OPERATIONAL
SNAP-19B	2	28.2	NIMBUS III	METEOROLOGICAL	4/14/69 (WTR)	1070 × 1131 km 99.9°, 107 min	MONITORING CEASED
SNAP-27	1	73.6	APOLLO 12	LUNAR	11/14/69 (KSC)	LUNAR TRAJECTORY	STATION SHUT DOWN
SNAP-27	1	72.5	APOLLO 14	LUNAR	1/31/71 (KSC)	LUNAR TRAJECTORY	STATION SHUT DOWN
SNAP-27	1	74.7	APOLLO 15	LUNAR	7/26/71 (KSC)	LUNAR TRAJECTORY	STATION SHUT DOWN

Table 15.5
Summary of Radioisotope Thermoelectric Generators Successfully Launched by the United States (1961–1982)

Power Source	Number of Power Sources	Initial Average Power per Power Source (W)	Spacecraft	Mission Type	Launch Date[a]	Initial Orbit	Status
SNAP-19	4	40.7	PIONEER 10	PLANETARY	3/2/72 (ETR)	SOLAR SYSTEM ESCAPE TRAJECTORY	STILL OPERATING
SNAP-27	1	70.9	APOLLO 16	LUNAR	4/16/72 (ETR)	LUNAR TRAJECTORY	STATION SHUT DOWN
TRANSIT-RTG	1	35.6	TRIAD	NAVIGATIONAL	9/2/72 (WTR)	716 × 863 km 90.1°, 101 min	STILL OPERATING
SNAP-27	1	75.4	APOLLO 17	LUNAR	12/7/72 (KSC)	LUNAR TRAJECTORY	STATION SHUT DOWN
SNAP-19	4	39.9	PIONEER 11	PLANETARY	4/5/73 (ETR)	SOLAR SYSTEM ESCAPE TRAJECTORY	STILL OPERATING
SNAP-19	2	42.3	VIKING 1	MARS LANDER	8/20/75 (ETR)	TRANS-MARS TRAJECTORY	LANDER SHUT DOWN
SNAP-19	2	43.1	VIKING 2	MARS LANDER	9/9/75 (ETR)	TRANS-MARS TRAJECTORY	LANDER SHUT DOWN
MHW-RTG	2	153.7	LES-8	COMMUNICATIONS	3/14/76 (ETR)	35,787 km 25.0°, 1436 min	STILL OPERATING
MHW-RTG	2	154.2	LES-9	COMMUNICATIONS	3/14/76 (ETR)	35,787 km 25.0°, 1436 min	STILL OPERATING
MHW-RTG	3	159.2	VOYAGER 2	PLANETARY	8/20/77 (ETR)	SOLAR SYSTEM ESCAPE TRAJECTORY	STILL OPERATING
MHW-RTG	3	156.7	VOYAGER 1	PLANETARY	9/5/77 (ETR)	SOLAR SYSTEM ESCAPE TRAJECTORY	STILL OPERATING

[a]Key to launching stations: ETR—Eastern Test Range, WTR—Western Test Range, KSC—Kennedy Space Center.

Thermoelectric systems utilize the *Seebeck effect* in a circuit for power generation. The manner in which this is done is shown in Figure 15.47. In some applications, a secondary heat-rejection loop is employed. The best type of material for TEC has been determined to be those categorized as semiconductors because the TEC efficiency is directly related to a parameter called *figure of merit (Z)*. Semiconductor materials have the best potential for achieving high values for Z. Semiconductor materials for thermoelectric power generation perform in accordance with their basic parameters which relate to the electromotive force (EMF) generated, thermal conductivity, and electrical resistance. These variables are listed below:

- Seebeck coefficient, S, V/°C;
- Thermal conductivity, K, W/°C-cm;
- Electrical resistivity, p, Ω-cm.

It is obviously desirable for S to be as high as possible and K and p to be low. This leads to the well known figure of merit for thermoelectric materials:

$$Z = \frac{S^2}{pk}, \quad V^{-1}$$

The relationship of Z to the thermoelectric efficiency is depicted in Table 15.6. Note that the efficiency is directly related to the Carnot efficiency and a so-called material efficiency. Hence, to achieve high TEC efficiency, much research has been carried out by many companies (e.g., GE, Westinghouse, TEECO, GA Technologies, Teledyne) and government agencies (e.g., JPL). Effort has been directed toward developing materials which yield high values of Z and can withstand high temperatures without melting or other deterioration such as the migration of doping substances.

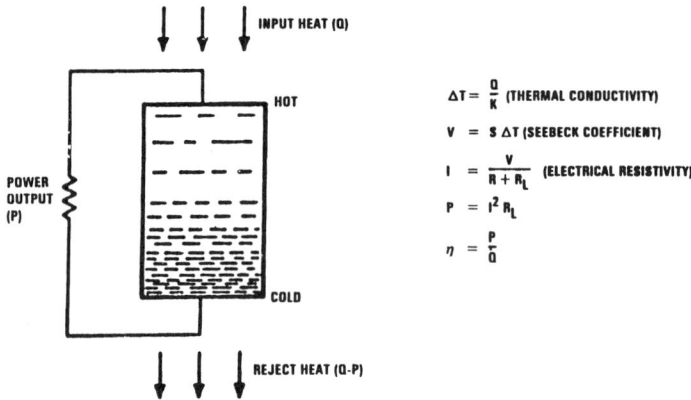

Figure 15.47 Thermoelectric converter performance parameters.

Table 15.6
Thermoelectric efficiency

$$\text{Thermoelectric Efficiency} = \begin{pmatrix} \text{Carnot} \\ \text{Efficiency} \end{pmatrix} \times \begin{pmatrix} \text{Material} \\ \text{Efficiency} \end{pmatrix}$$

$$\eta = \frac{T_H - T_C}{T_H} \times \frac{\sqrt{1 + Z\overline{T}} - 1}{\sqrt{1 + Z\overline{T}} + \frac{T_C}{T_H}} \approx \frac{1}{4} Z \Delta T$$

where

T_H = hot junction temperature,
T_C = cold junction temperature,
$\overline{T}$ = average temperature,
Z = figure of merit.

A TEC conversion system that is in the very early phases of development is the SP-100 nuclear reactor power systems. The Department of Energy after conducting tradeoff between reactor powered thermoelectric, Brayton, Stirling, and thermionic conversion cycles at GE, LANL, Westinghouse, GA Technologies, and Rockwell International, finally selected a 100 kW$_e$ power-level reference design. The baseline configuration is shown in Figure 15.48, which employs a 25-meter separation distance between the reactor and the payload. The total integrated radiation dose at the payload dose plane at the end of 10 years is 1×10^{13} neutrons/cm^2 and 5×10^5 rads. These levels are set by the acceptable values for hardened electronics. The performance is shown for three different thermoelectric materials. However, development problems were encountered with LaSx and SiGe-GaP, so that the current design is based upon SiGe (e.g., fallback case). With a total mass of 3292 kg (7242 lb), the specific weight of the SP-100 becomes approximately 33 kg/kW$_e$ (73 lb/kW$_e$). The question often arises as to why the TEC was selected in preference to the higher conversion efficiency thermionic system. Both are static systems, and the previously mentioned trade-off studies resulted in equal masses. The decision was based on the Department of Energy's assessment of the lower development risks associated with lower peak cycle temperatures of the TEC system, the much greater technological base, and the extensive space flight experience on previous programs.

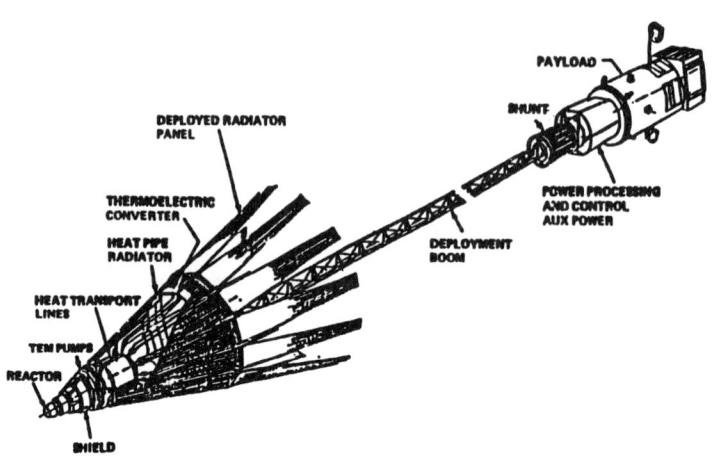

Figure 15.48 SP-100 baseline system configuration.

15.10 THERMIONIC CONVERSION SYSTEMS (SYSTEM NO. 10)

Thermionic power conversion is based upon the use of a vacuum diode utilizing a temperature difference, and employs electrons as the working fluid. Similar to thermoelectric converters, a thermionic converter is subject to Carnot cycle efficiency limitations, the maximum efficiency being limited by the absolute temperature of the source and sink.

The principle of the thermionic converter is essentially the same as that of the rectifier tube. Electrons are given off at a heated emitter and collector (anode), and then pass through the external load back to the emitter (Figure 15.49).

At increased temperatures, the energy level of some of the electrons is raised to the point that they overcome the potential barrier for electrons within the emitter surface. As the number of electrons in the interelectrode space increases, a negative space-charge barrier, a cloud of electrons is formed. As this barrier builds up, the number of electrons that can leave the emitter (this number being the *surface work function*) is sharply reduced. This occurs despite the fact that the electrode spacing may be only 0.002 to 0.003 inch. To alleviate this problem, the plasma diode is used so that the space-charge barrier is neutralized by injecting positively charged particles into the interelectrode space. A small pressure of cesium ions is introduced between the electrodes to form a neutral plasma with the electrons of the space charge. This greatly reduces it and permits some relaxation of the extremely small

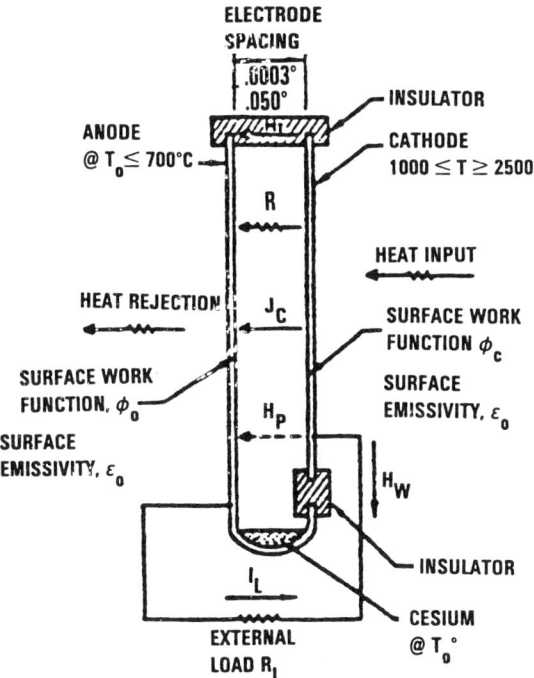

Figure 15.49 Schematic diagrams of thermionic converter.

separation tolerances of the electrodes. This relaxation is very important because, to be practical energy sources, the emitter and collector must have comparatively large areas as the power output is a direct function of these areas. Maintaining microscopic separation between these two large, flat surfaces at temperatures above 2000°F for any protracted period becomes very difficult. Material sublimation and surface distortion under these conditions due to temperature cycling usually cause premature shorting. The source of the cesium vapor is normally a cesium reservoir, which is maintained at the saturation temperature corresponding to the required vapor pressure. Figure 15.50 indicates the behavior of the space charge under the influence of the number of cesium ions striking a square centimeter of surface per second, according to the temperature of a cesium-coated tungsten wire, by showing the cathode emission rate as a function of temperature. The thermionic conversion efficiency is plotted in Figure 15.51, which shows the effects on conversion efficiency of anode and cathode work functions at various temperatures. The principal barriers to efficient, lightweight systems are low current densities, material temperature limitations, and plasma energy losses, or very close electrode spacing. A typical thermionic conversion system (without energy source) is shown on Figure 15.52.

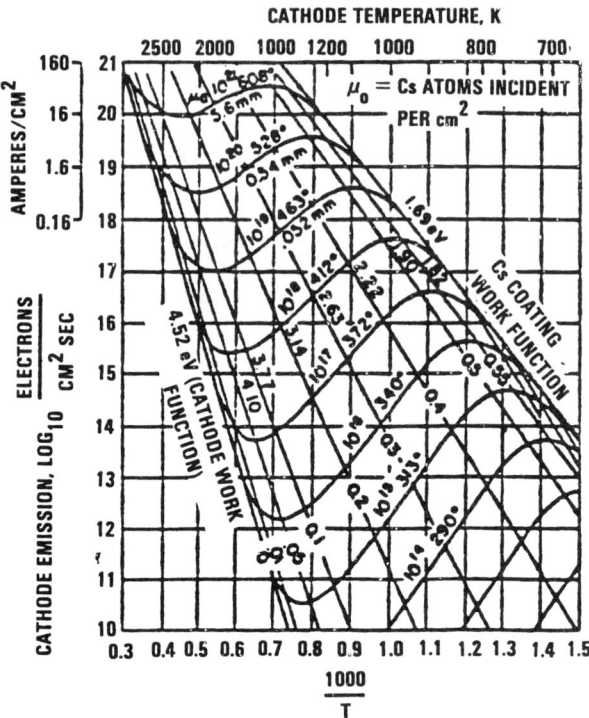

Figure 15.50 Influences of cesium ion concentrations on electric space charge.

Two basic system approaches have been pursued for thermionic reactors, *in-core* and *out-of-core*. The former approach, which incorporates the converters inside the core, is illustrated in Figure 15.53. This design concept has been utilized for the TOPAZ reactor and in-pile tests conducted in the US in the 1960s and early 1970s.

The out-of-core system shown in Figure 15.54 transfers the thermal energy outside the reactor to a bank of thermionic converters. For a given power level, the reactor and shield have a lower mass and the conversion system is decoupled to a large extent from the reactor. In addition, mechanical coolant pumping is eliminated and heat pipes lend themselves to redundant systems.

To realize the system advantages of thermionic conversion (e.g., small area radiators), the emitters of the converters must operate at temperatures of at least 1750 K. The selection of the maximum emitter operating temperature is a compromise between system performance, mission life, and material's development status. Typical *thermionic fuel element* (TFE) materials are tungsten emitters, niobium collectors, and tantalum transition pieces.

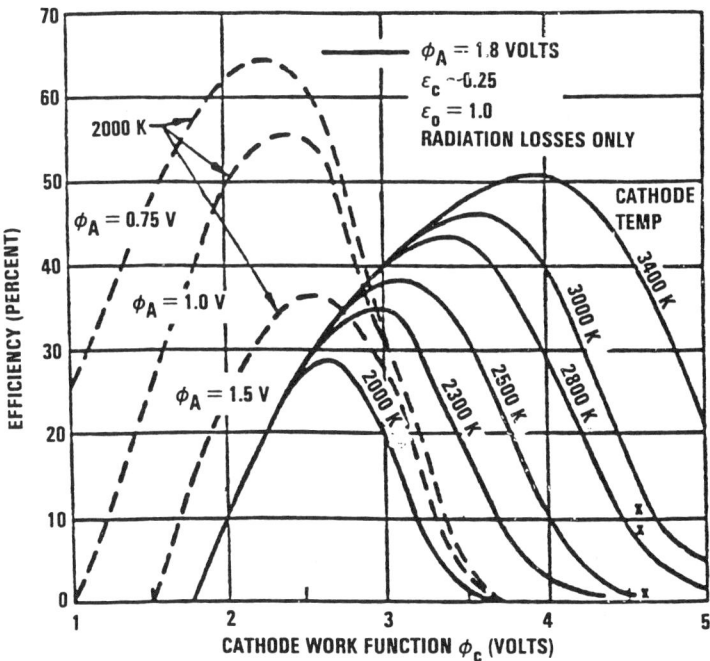

Figure 15.51 Thermionic converter efficiencies *versus* electrode work function and temperature.

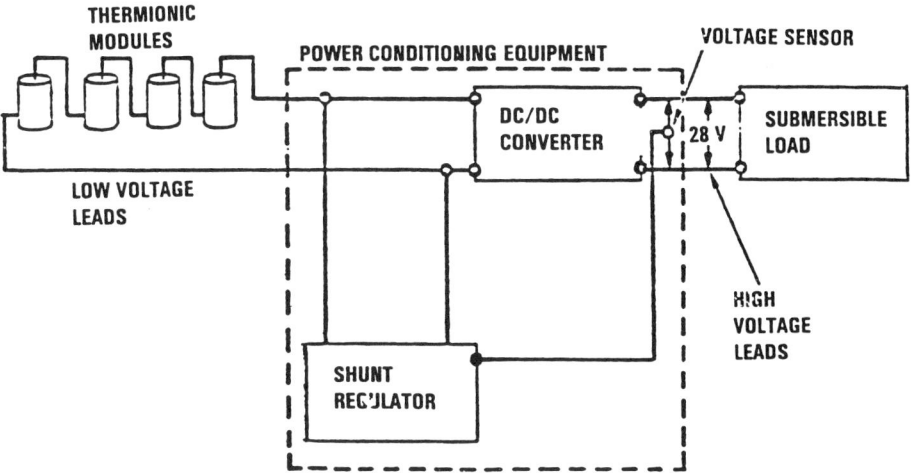

Figure 15.52 Thermionic system schematic.

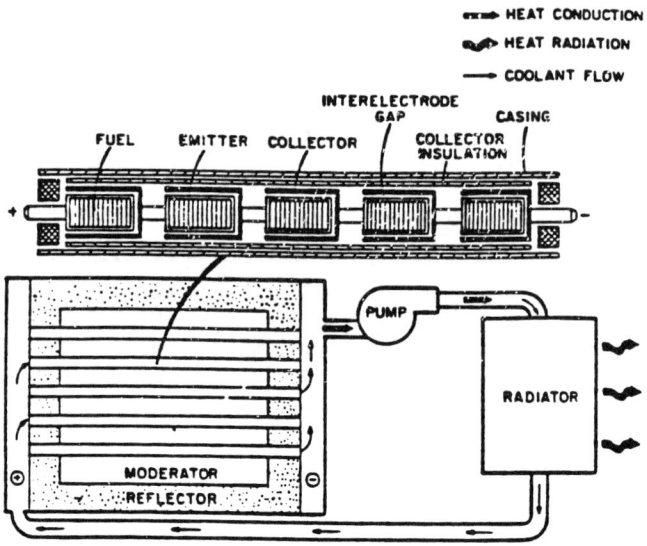

Figure 15.53 In-core thermionic reactor concept.

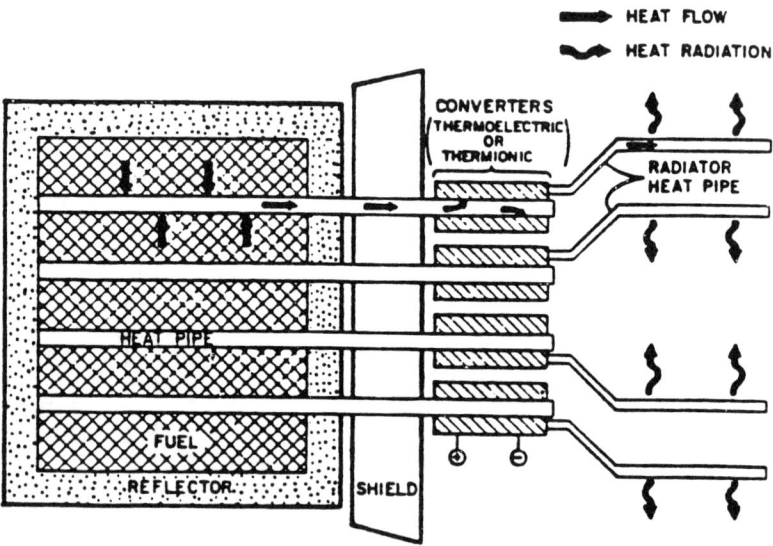

Figure 15.54 Out-of-core thermionic reactor concept.

15.10.1 Candidate Thermionic Conversion Systems

The reactor-TFE power system was initially under development from 1958 to 1962 under the SNAP-70 program. These systems were given the acronym STAR for *space thermionic auxiliary reactor*. Various in-core and out-of-core configurations were investigated over a power range from 70 to 2000 kW$_e$. The average specific weights were determined to be approximately 4 to 15 lbs/kW$_e$. Due to lack of any missions and the very early development status of these STAR systems, the SNAP-70 program was terminated in 1962. However, interest in thermionic conversion subsystems continued, and companies like GA Technologies, Rasor Associates, Thermo-Electro Corp., GE, and others as well as JPL conducted performance and life testing on a variety of TFEs.

With the initiation of the SP-100 program, the Department of Energy provided funding to GA Technologies to study the STAR concept for this 100 kW$_e$ reactor-powered application. The department also initiated limited life testing at GA Technologies and LANL. The overall arrangement of the baseline reactor-TFE SP-100 power system is shown in Figure 15.55, and its design concept is based on in-core energy conversion by using the thermionic conversion process. The reactor core is cooled by a eutectic alloy of sodium-potassium (NaK) that is circulated by an induction EM pump. The entire primary coolant system is constructed of niobium-1% zirconium alloy. The cycle reject heat is dissipated by a conical heat pipe radiator structure. A nuclear shield is located between the radiator and reactor to reduce the nuclear radiation to the spacecraft. The reactor-radiator is separated from the spacecraft by an extendable mast (not shown) deployed prior to start-up of the reactor-TFE system. A flow diagram of the NaK coolant loop is shown in Figure 15.56. Note that whereas the use of EM pumps maintains the static status of this concept, multiple dynamic electric drive motors are utilized to operate the reflector control drums. An energy balance for the cycle is shown in Figure 15.57. With 1233 kW$_t$ of heat input power and a net electrical output power (after conditioning) of 100 kW$_e$, the overall system efficiency of this cycle becomes 8.1%.

The main subsystems comprising the reactor-TFE system are the UO$_2$ fast reactor, the TFE subassemblies, and the radiator. The thermionic reactor is shown in Figure 15.58. With the exception of the rotating reflector control drums actuated by electric motor drives, this reactor is primarily a static device with considerable potential for achieving high reliability. The main reactor characteristics are summarized in Table 15.7. A total of 176 TFE with 44 connected in series in parallel strings achieves additional increased reliability. The individual thermionic cell assembly (Figures 15.59 and 15.60) is very simple in construction. This simplicity is retained when cells are combined into six-cell TFE assemblies as shown in Figures

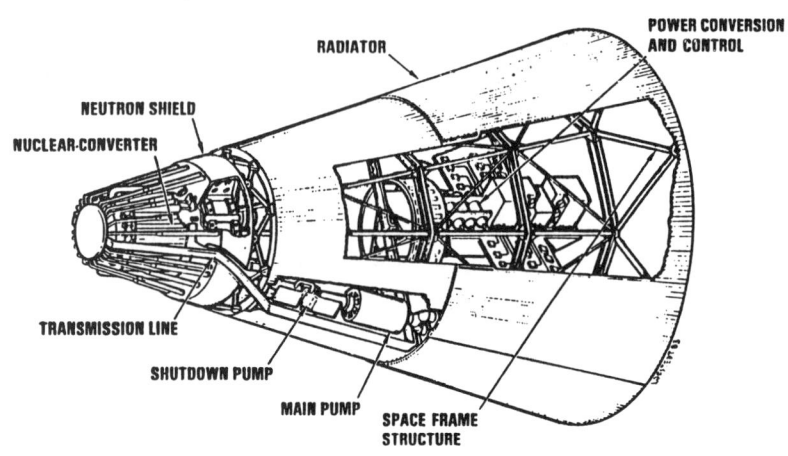

Figure 15.55 SP-100 space thermionic power system (100 kW$_e$).

Table 15.7
Reactor Parameters

Core-Reactor Diameter, mm	504
Core-Reactor Length, mm	913
No. of TFEs	176
No. in Series Connection	44
Power Output at Reactor Terminals, kW$_e$	108.6
Output Voltage, V	106.3
Conversion Efficiency	0.088
Thermal Power, kW$_t$	1233
Coolant	NaK
Coolant Inlet-Outlet Temperature, K	970/1100
Reactor Vessel Material	Nb-1Zr
Reflector Material	BeO (inside vessel)
	Be (outside vessel)
No. of Control Drums	16
Control Drum Diameter, mm	94
No. of In-core Shutdown Rods	7

15.61 and 15.62. The NaK coolant loop rejects the cycle waste heat to a great number (338) of sodium (Na) heat pipes. Because these are all individually connected to the coolant-loop NaK heat exchanger (see Figure 15.63), additional system design redundancy is achieved.

The general arrangement of the original baseline SP-100 design (*circa* 1984) is shown in Figure 15.64. This design employed a 12.5 cone angle, which resulted

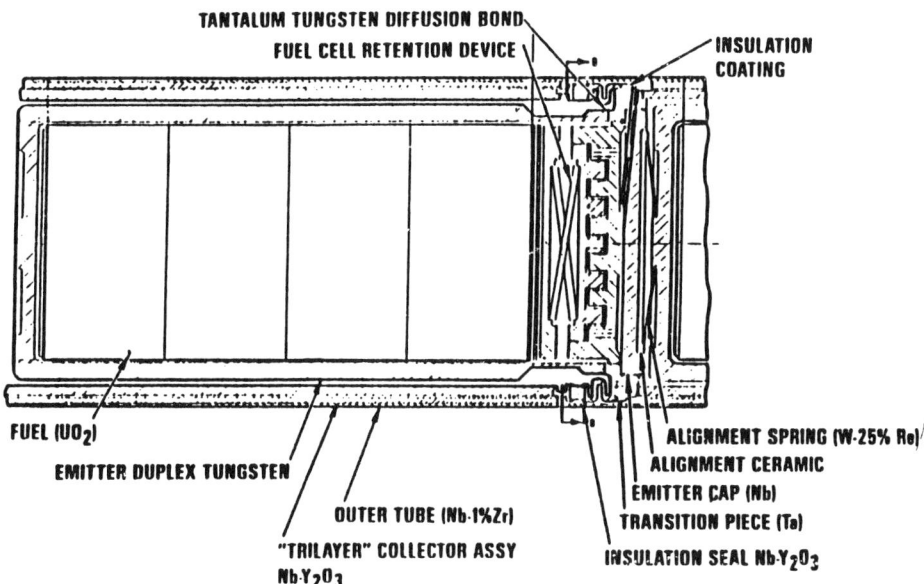

Figure 15.59 Thermionic cell assembly.

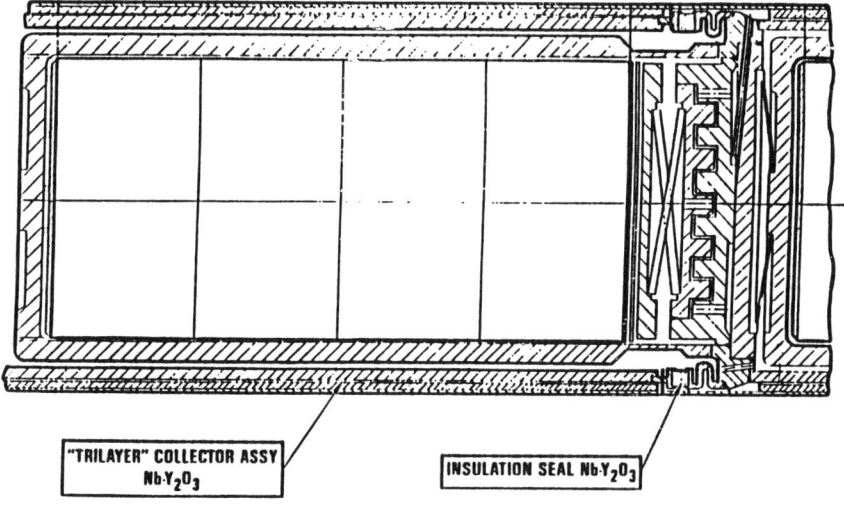

Figure 15.60 Insulator locations.

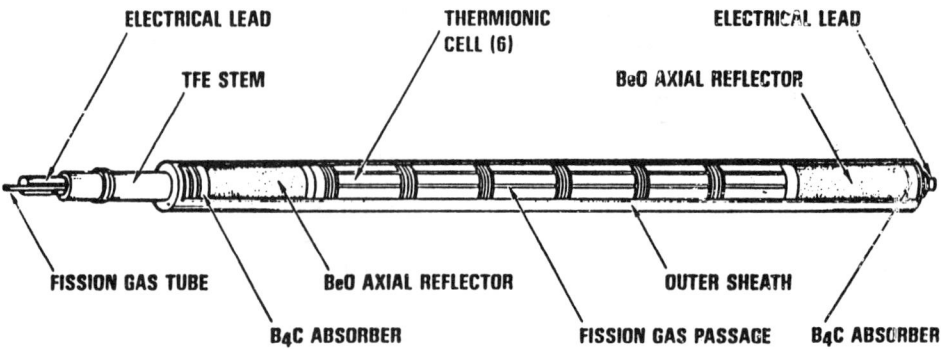

Figure 15.61 SP-100 TFE assembly (6 cells).

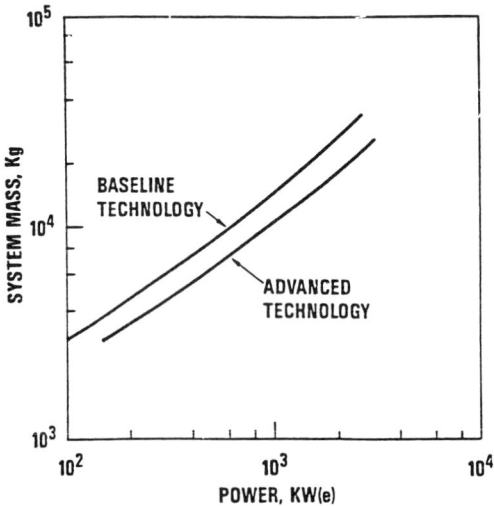

Figure 15.62 SP-100 thermionic fuel element assembly.

are summarized in Tables 15.8 and 15.9, respectively. In addition, by increasing the average emitter temperature from 1700 to 1760 K and lowering the radiator average surface temperature from 994 to 969 K, the overall cycle efficiency increased from 7.7 to 8.1% (see Table 15.10). This resulted in a reduction in core thermal power from 1328 kW_t to 1233 kW_t, a saving of approximately 7%.

In arriving at the 1985 baseline design, several trade-off studies were conducted. Typical of these was the influence of cone angle and separation distance on total system mass. These results are shown in Figure 15.66. The 1985 baseline system parameters and mass are summarized in Tables 15.11 and 15.12. Typical

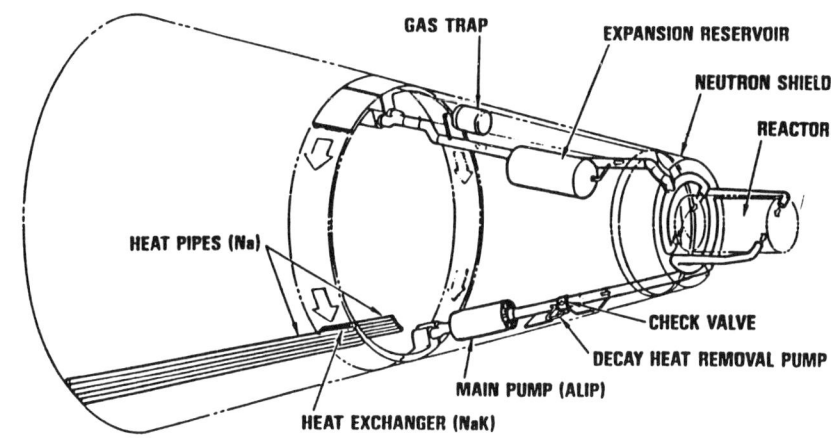

Figure 15.63 SP-100 Reactor-TFE coolant loop.

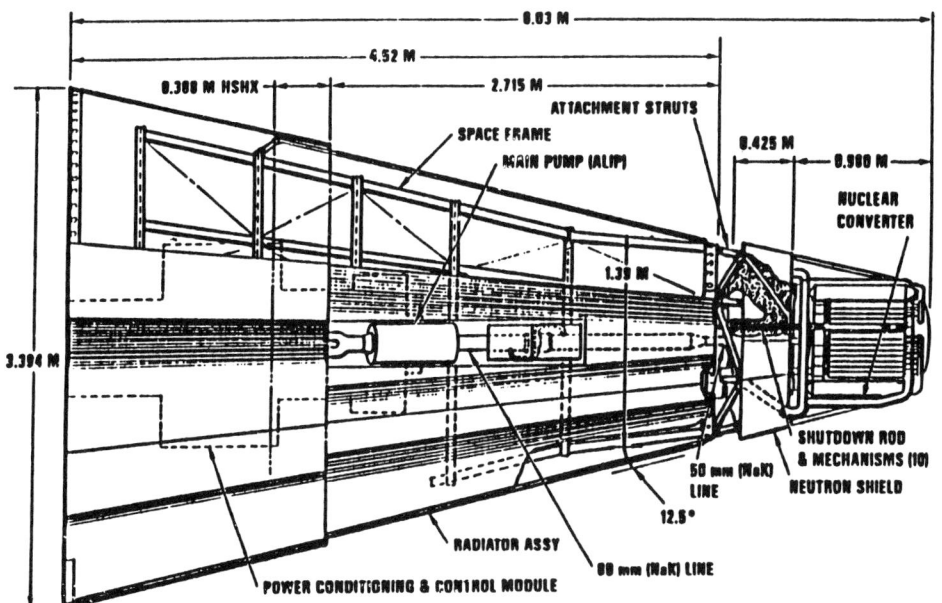

Figure 15.64 SP-100 1984 baseline design general arrangement.

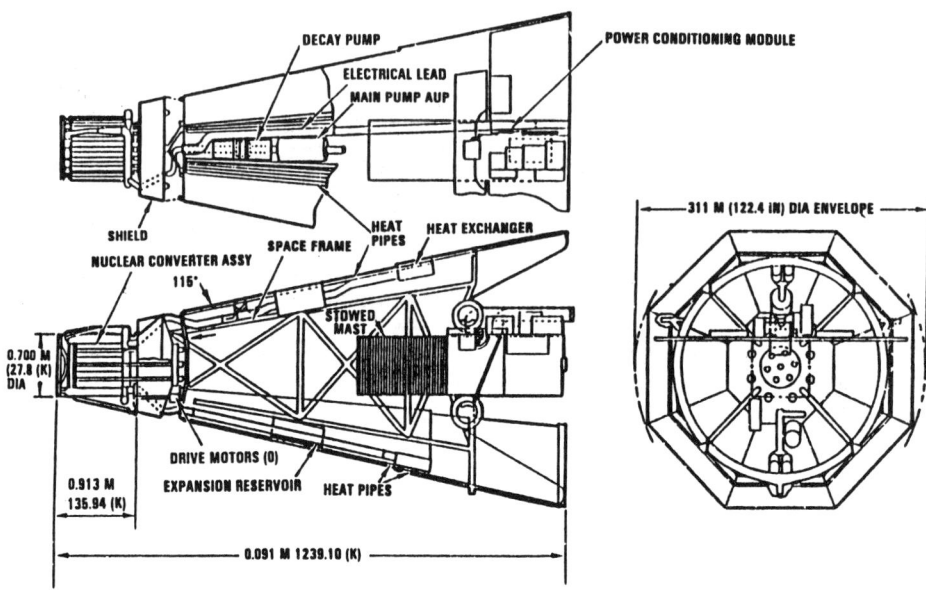

Figure 15.65 SP-100 1985 baseline system general arrangement (stowed configuration).

Table 15.8
TFE-Cell Parameters

Outside Diameter, mm	33.38
Total TFE Sheath Length, mm	513.5
No. of Cells	6
Cell Length, mm:	
−1	85.32
−2	61.30
−3	56.69
Emitter Diameter, mm	27.94
Emitter Clad Thickness, mm	1.78
Fuel Diameter, mm	24.1
Fuel Length, mm:	
−1	69.32
−2	45.30
−3	40.69
Interelectrode Spacing, mm	0.51
Collector Thickness, mm	0.94

Table 15.9
Baseline Radiator Design Summary

Heat Rejection (EOL), kW$_t$	1124
Surface Temperatures (EOL) (max, avg, min), K	1021, 969, 919
Installed Area (outside surface), m^2	24.1
Total Radiating Area, m^2	36.1
Total No. of Heat Pipes	338
Material	Nb-1 Zr (throughout)
Emissive Coating Material	TiC (ϵ = 0.9)
Emissive Coating Thickness, mm	0.1
Meteoroid Armor Thickness, mm	0.3
No. of Heat Pipe Penetrations at EOL (10 yr)	23

Table 15.10
Baseline System Parameter Evolution

	1983 Design	1984 Design	1985 Design
Cone Half-Angle, degrees	13.5	12.5	11.5
Maximum NaK Temperature (EOL), K	1100	1100	1100
Core Thermal Power (EOL), kW$_t$	1306	1328	1233
Reactor Output Power (EOL), kW$_e$	108.4	108.3	108.6
Reactor Output Voltage (EOL), Vdc	105	107	106.3
Core ΔT (EOL), K	100	120	130
Heat Rejected (EOL), kW$_t$	1198	1220	1124
Electric Output (EOL), kW$_e$	100	100	100
Overall Efficiency (EOL)	0.077	0.075	0.081
Piping Diameter, mm	100	80	80
Heat Pipe Radial Heat Flux, W/cm^2	25	40	40
Radiator Average Surface Temperature (EOL), K	994	979	969
System Length, m	5.76	6.03	6.09
System Maximum Diameter, m	3.59	3.4	3.11
System Mass, kg	2996	2911	3000
Average Emitter Temperature (EOL), K	1700	1700	1760
Coolant Flow Rate, kg/s	13.4	11.4	9.7
Total Pressure Drop, kPa	34.4	44.8	30.0
Pumping Power, kW(e)	2.55	3.3	2.12

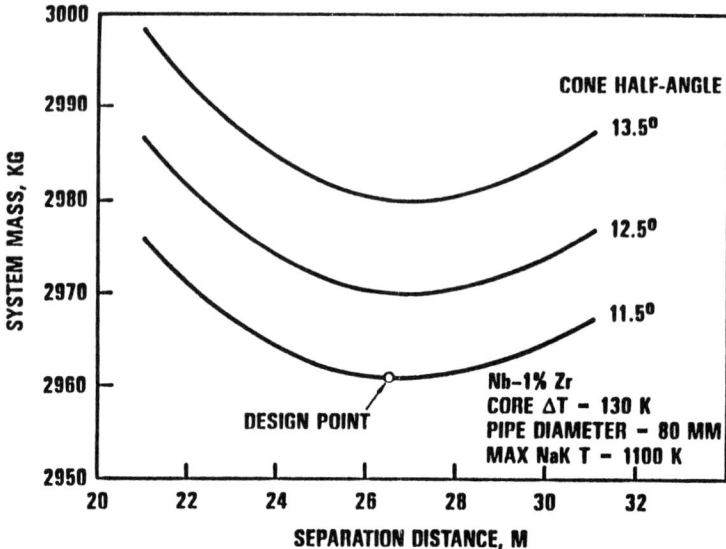

Figure 15.66 Influence of payload separation distance on system mass.

Table 15.11
System Parameters

Cone Half-Angle, degrees	11.5
Maximum NaK Temperature (EOL), K	1100
Core ΔT (EOL), K	130
Heat Rejected (EOL), kW_t	1124
Electric Output (net, EOL), kW_e	100
Core Thermal Power (EOL), kW_t	1233
Piping Diameter, mm	80
Heat Pipe Radial Heat Flux, W/cm^2	40
System Length, m	6.09
System Maximum Diameter, m	3.11
System Mass, kg	3000
Average Emitter Temperature (EOL), K	1740
Lifetime, yr	>7

temperatures throughout this system when operating at nominal full power conditions are depicted in Figure 15.67, and a schematic of a typical power distribution system is shown in Figure 15.68. The growth potential as projected by GA Technologies for this SP-100 reactor-TFE system is summarized in Table 15.13. A 2 MW_e version is shown in Figure 15.69. Deployed radiators are utilized to meet the heat-rejection requirements and still remain within the confines of the STS

Table 15.12
1985 Baseline Mass Summary

Item	Mass (kg)
Reactor	976
Shield	659
Heat Rejection	661
Boom	56
Cables	121
Power Conditioning	156
Reactor Control	260
Structures	80
SP-100 System	2969
STS Mount	31
TOTAL	3000

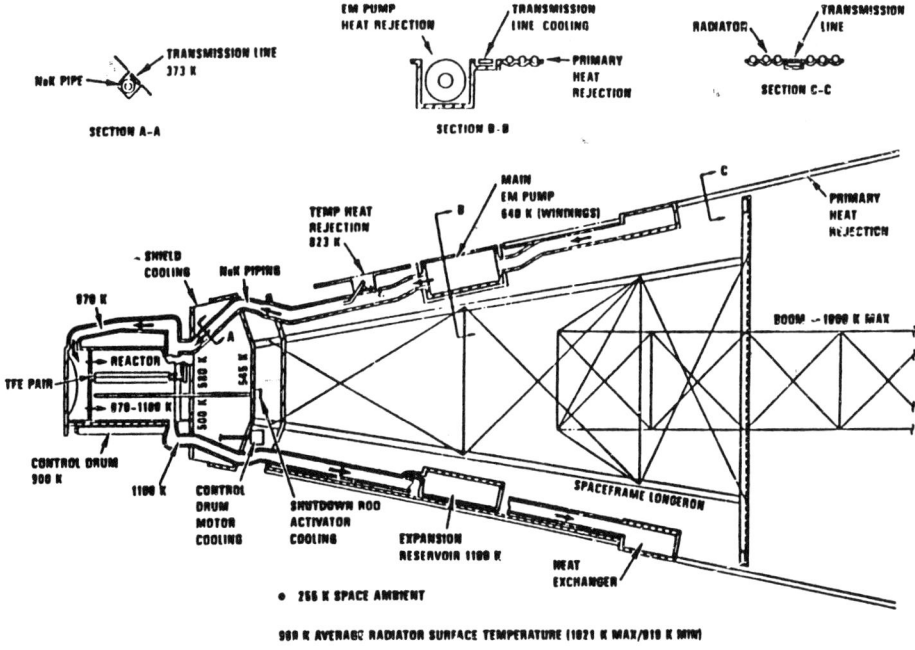

Figure 15.67 System thermal management schematic.

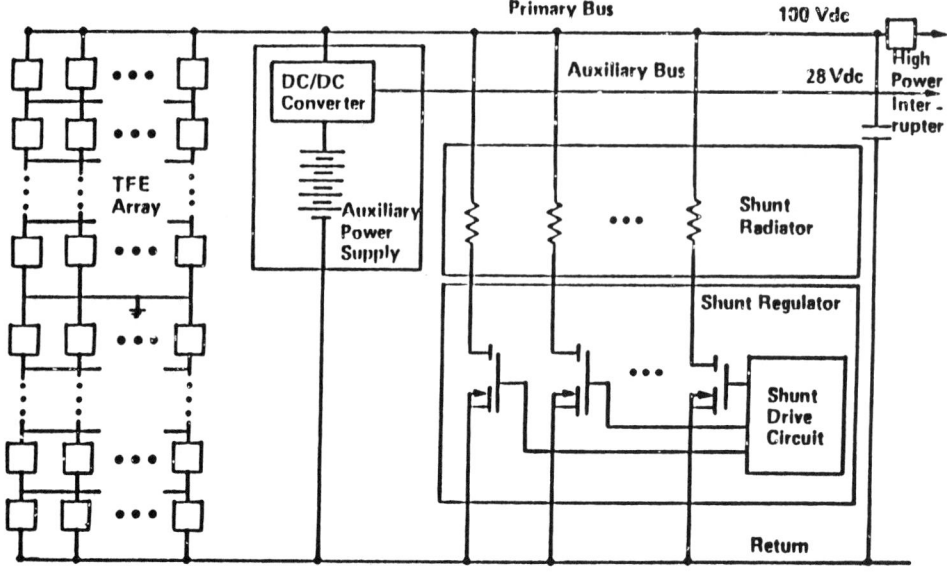

Figure 15.68 SP-100 power distribution schematic.

Table 15.13
Summary of Technology Growth Potentials

Technology (Radiator)	Power (kW$_e$)	Mass (kg)	L/L_{STS}
Fallback (fixed)	300	11,200	1.0
Fallback (deployed)	850	26,000	1.0
Baseline (fixed)	300	6,000	0.6
Baseline (fixed)	600	10,000	1.0
Baseline (deployed)	1,000	14,000	0.75
Baseline (deployed)	2,000	26,000	1.0
Advanced (deployed)	1,000	9,600	0.6
Advanced (deployed)	3,000	26,000	1.0

cargo bay. Finally, overall system mass reductions are projected (see Figure 15.70) as the design progresses from current baseline to advanced technology.

The in-core thermionic-TFE system is expandable to the multimegawatt power-level range (e.g., 1.0 to 20 MW$_e$). However, to reject the cycle waste heat, the fixed radiator must be replaced with a multipanel, deployable radiator, as depicted in Figure 15.71. These radiators are two-sided, flat-plate configurations, employing sodium heat pipes. A schematic of this conceptual design is shown in Figure 15.72 together with a list of its main features. The baseline reactor coolant

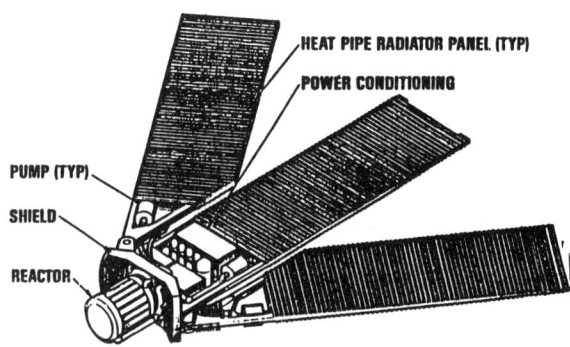

Figure 15.69 2-MW$_e$ thermionic power system concept.

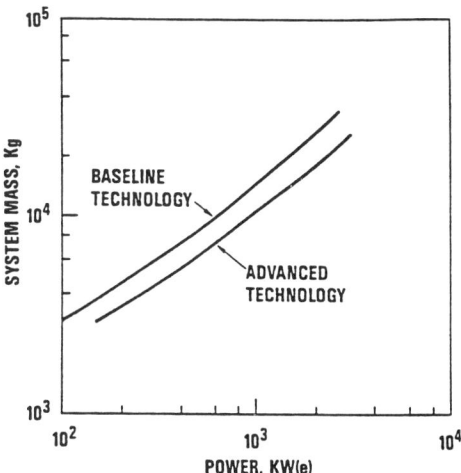

Figure 15.70 Reactor-TFE growth mass.

for this system is NaK, similar to that of the SP-100 version. However, for these higher output levels, a significant saving in power source mass can be achieved by substituting lithium (Li) as a coolant. The magnitude of this weight saving as a function of generator output is shown in Figure 15.73. The mass of the major subsystems as well as the total mass of a 1.0 MW$_e$ version of the reactor-TFE system is summarized on Table 15.14. Note that a specific weight of 8.9 kg/kW$_e$ (19.6 lb/kW$_e$) is achievable at an overall conversion cycle efficiency of 9%. For a projected improvement to 20% overall cycle efficiency and with Li as a coolant, a reduction in specific weight to 7.3 kg/kW$_e$ (16 lb/kW$_e$) appears feasible. Finally, the more significant attributes of this high-power reactor-TFE system are listed in Table 15.15.

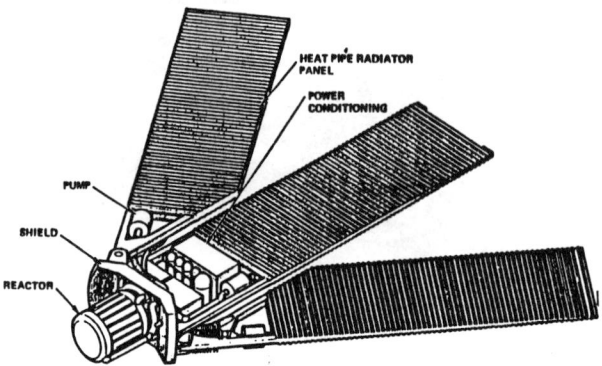

Figure 15.71 1.0 MW$_e$ reactor-TFE space power system.

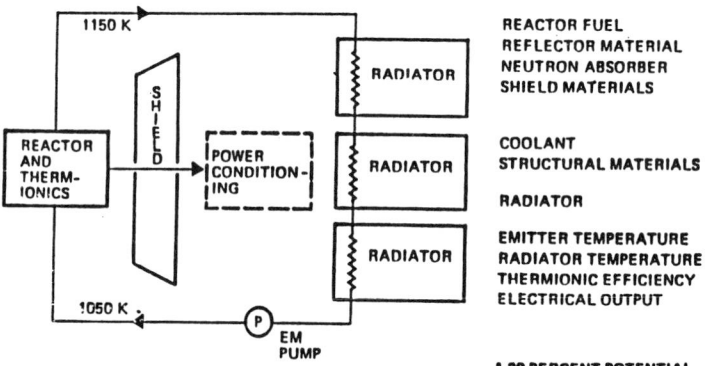

Figure 15.72 Schematic diagram of in-core reactor-TFE system.

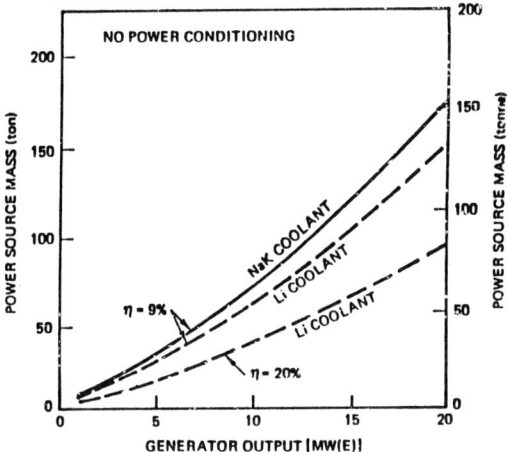

Figure 15.73 Power source mass *versus* generator output power.

Table 15.14
In-Core Thermionics (STAR-M)—Major Components Summary

Component	Performance	Mass (lb)	Mass (kg)
Reactor-Conversion	9% thermionics	10,560	4,800
Shield	LiH + ZrH/B	4,460	2,000
Heat Rejection	Sodium heat pipes Two-sided flat plate radiator 1100 K (1980°R) NaK coolant	4,620	2,100
	Total	19,580	8,900

Note: 1 MW(E), Continuous Power Generation

Table 15.15
In-Core Thermionics (STAR-M)—Significant Attributes

Static Energy Conversion
9% Conversion Efficiency
1900 K (3420°R) Emitter
1100 K (1980°R) Radiator
Sodium Heat Pipe Technology
NaK Heat Transport Fluid (Lithium Possible)
Direct Current Generation: 30–1000 V
Graceful Degradation
Potential for 20% Conversion Efficiency

REFERENCES

1. Boretz, J.E., "Large Solar Arrays—The Emerging Space Power Workhorse," *NASA/MSFC Symposium*, 1968.
2. Boretz, J.E., "Technology Problems Associated with Large Solar Arrays for Long Duration Space Missions," Florida Section of AIAA, Cape Kennedy, FL, 1969.
3. Boretz, J.E., *et al.*, "Study to Establish Criteria for a Solar Cell Array for use as a Primary Power Source for a Lunar Based Water Electrolysis System," Phase I, Final Technical Report, June 1968 (TRW Rep. No. 09681-6002-R000; NASA/MSFC Contract No. NAS8-21189).
4. Boretz, J.E. and J.L. Miller, "Advanced Design Modules for Lunar Surface Solar Array Power Systems," NASA/MSFC Contract No. NAS8-21189, January 1968.
5. Rauschenbach, H.S., "Solar Cell Array Design Handbook," Van Nostrand Reinhold, New York, 1980.
6. Bauer, P., "Batteries for Space Power Systems," NASA Office of Technology Utilization, Washington, DC, 1968.
7. Scott, W.R. and D.W. Rusta. "Sealed-Cell Nickel-Cadmium Battery Applications Manual," NASA Ref. Pub. No. 1052, 1979 (NASA/GSFD Contract No. NAS5-23514).
8. "Organic Rankine Kilowatt Isotope Power System," Final Phase I Report (COO-4299-032), Rockford, IL, Sundstrand Energy Systems, July 1978, Contract No. EN-78-C-4229.
9. "Dynamic Isotope Power System Technology Verification Phase," Final Report (DOE/ET/33001-T41; DE 83001134), Sundstrand Corp., Advanced Technology Group, April 1982, Contract No. DE-AC02-77ET-33001.
10. "Dynamic Isotope Power Subsystem (DIPS) Applications Study," Volume I—Summary, Final Report (DOE/ET/3267-1 [Vol. I]), Philadelphia, PA, General Electric Space Division, November 1979, Contract No. ET-78-C-01-3267.
11. Boretz, J.E., "Radioisotope Dynamic Power Systems," (Paper No. 849318), *19th IECEC Meeting,* San Francisco, CA, August 1984.
12. "SP-100 Reactor-Stirling Power System GES Baseline System Definition and Characterization Study," (Rockwell International Report No. RI/RD85-217-1), JPL Contract No. 956935, August 1986.

Chapter 16
SPACE-BASED RADAR STRUCTURES
E. Kovalcik
TRW

16.1 GENERAL REQUIREMENTS FOR SBR STRUCTURES

Space-based radar (SBR) antennas require structures designed for the space environment and to be launched into orbit by using available launch systems. Generally, these structures must be lightweight, geometrically accurate, and dimensionally stable. SBR antenna applications require structures that range in size up to several hundred meters in diameter (areal structures) to provide the necessary radiating apertures. Also, linear structures measuring hundreds of meters in length are needed to support feed systems at the focus or feed point.

Figure 16.1 [1] depicts the dimensional requirements and current technical capabilities for space-based reflector systems. Several regimes of size and surface accuracy are shown. Below the 4.5 m shuttle diameter limit, single-piece reflector structures can be used, and high surface-precision and high frequency applications therefore can be accommodated. Above the shuttle limit, deployable or erectable structures must be used. The solid-surface precision deployable regime can accommodate high frequencies, but current precision deployable structures are limited to approximately 30 m in diameter. Erectable solid-surface reflectors have the potential for large diameters and high frequencies, but considerable evaluation and development of astronaut assembly techniques or robotics are required to implement erectable precision structures in this regime. For larger apertures, the mesh deployable regime is up to several hundred meters of diameter, but surface accuracy is limited to lower frequency applications. Depending on diameter and precision, specific weight for areal space structures range from 1.5 to 7.5 kg/m^2 (0.3 to 1.5 lb/ft^2) for solid surface reflectors, and from 0.5 to 1.5 kg/m^2 (0.1 to 0.3 lb f/t^2) for mesh reflectors.

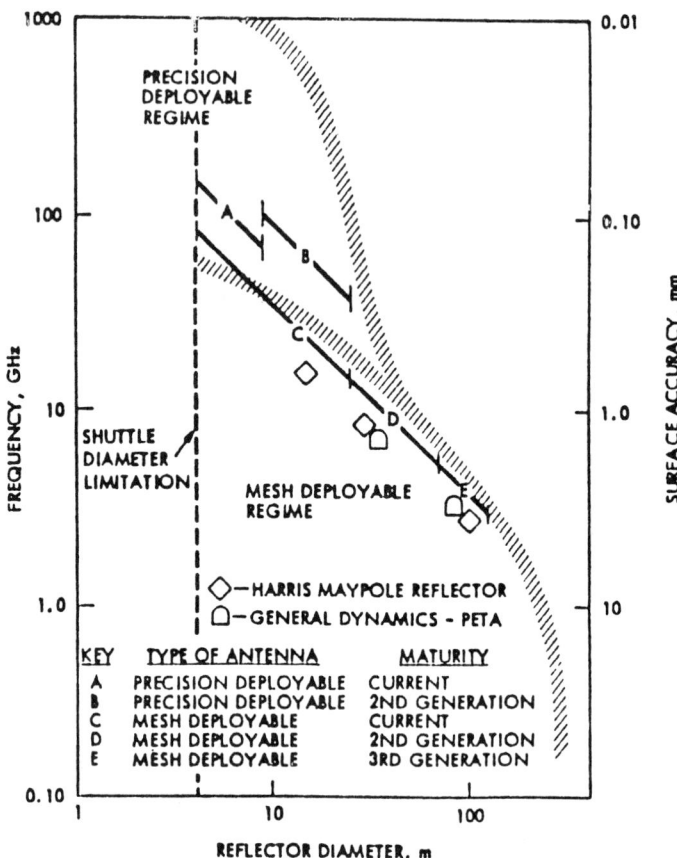

Figure 16.1 Projected performance for space-based reflectors.

Structures for large linear antenna configurations or feed-support applications must be stowed compactly for transport to space and deployed to meet accurate geometric location requirements as well as large operational dimensions. Figure 16.2 [2] summarizes the range of large truss beams available for space-based antennas. The figure is a plot of beam-bending stiffness as a function of beam depth. The points shown are requirements obtained from various mission studies, and the dashes outline these points to show the general stiffness trend. Studies have shown that the coiled longeron beam is limited to beam diameters of less than one meter due to difficulties in packaging the stored energy in the beam. Applications that require larger diameters use deployable beams with hinged longerons or erectable beams. Also indicated in Figure 16.2 are several of the more recent truss structure

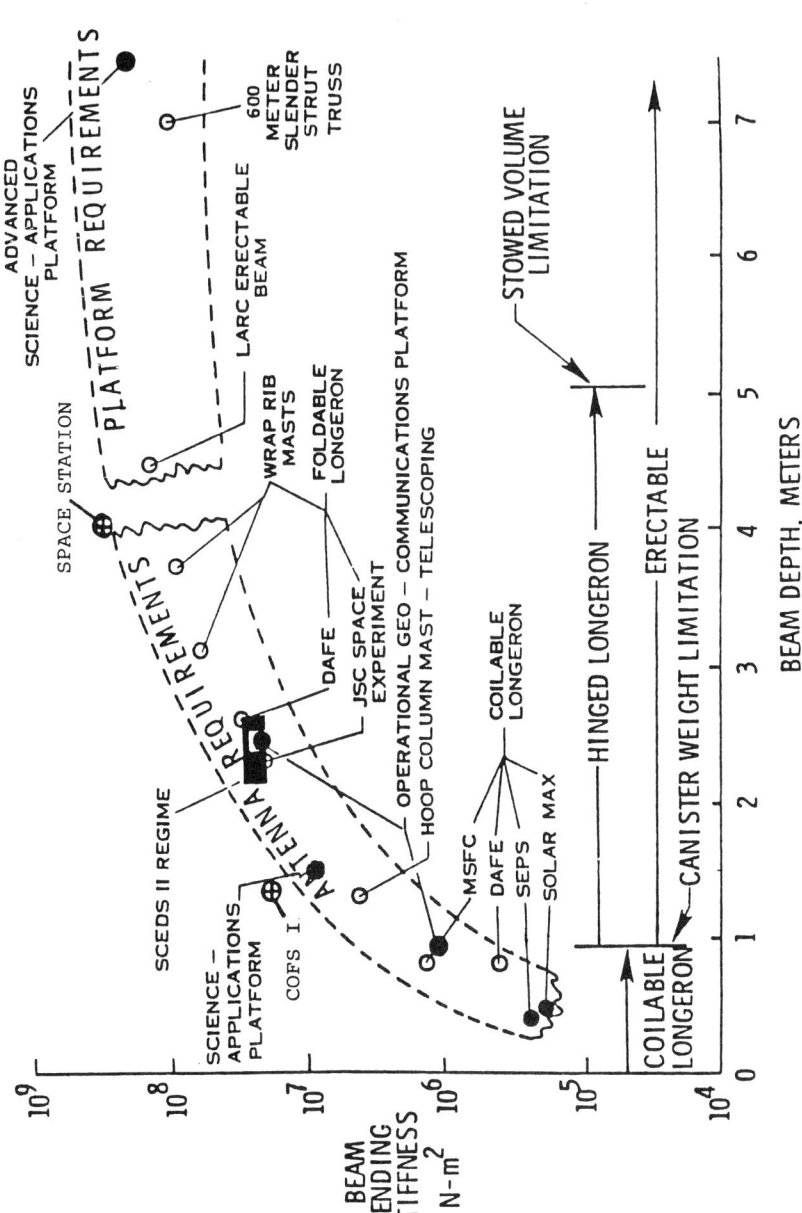

Figure 16.2 Large space truss beam characteristics.

development programs that typify the current state of space truss technology. These include COFS I [3] and Space Station [4]. Depending upon beam depth and stiffness, specific weight of linear structures can range from 0.14 to 1.4 kg/m (1.0 to 10 lb/ft) of length for structures designed to deploy in space.

There are many structural concepts available to meet space-based radar applications. Reflecting or radiating surfaces can be solid surfaces or mesh, and they require lightweight and dimensionally stable structural configurations. Linear structures are also highly developed, and the available structural concepts for both areal and linear requirements will be described in the following sections.

16.2 AREAL STRUCTURAL CONCEPTS

16.2.1 Solid-Surface Construction

Typical solid-surface reflector constructions are shown in Figure 16.3. These include waffle or egg-crate core sandwich, foam core sandwich, honeycomb core sandwich, and single thickness membrane. The sandwich type of construction provides good structural efficiency, but complicates producibility, especially for doubly curved reflector surfaces. The membrane construction provides lower stiffness, and it must be supported by a separate structure for large apertures.

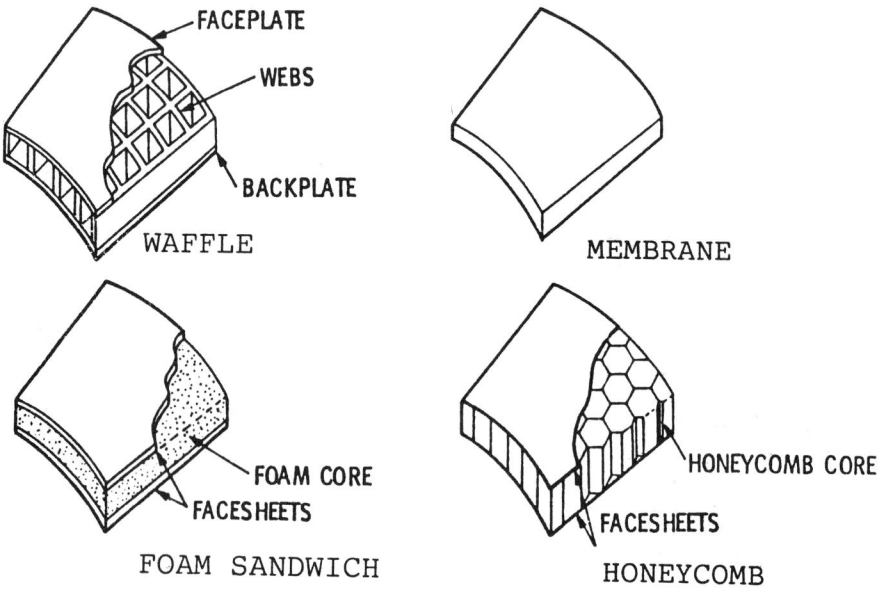

Figure 16.3 Solid-surface reflector construction.

16.2.2 Mesh Surface Construction

Several types of mesh surface are in general use for space applications. Tricot knitted mesh is shown in Figure 16.4. The mesh is knitted by using gold-plated molybdenum monofilament wire. This type of mesh has been used for TDRSS, ATS-6 [5], and several other space applications. Another type of mesh construction is shown in Figure 16.5. This is a welded grid mesh used aboard US Navy FLTSAT-COM spacecraft for UHF communication [6]. The mesh is constructed by using stranded wire, composed of seven strands of stainless steel and silver drawn wire. Spot welds at each grid junction form the wire strands into a continuous mesh surface. Strands of the mesh are specially formed to provide elasticity and mesh surface control tension under the extremes of the orbital thermal environment.

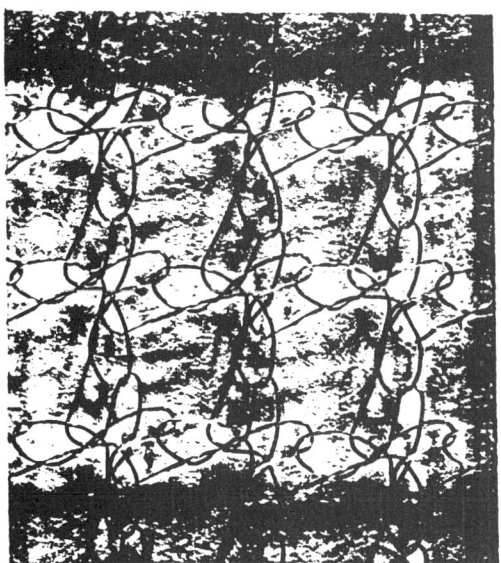

Figure 16.4 Tricot knit mesh.

16.2.3 Single-Piece Solid-Surface Reflector Concepts

For small-aperture requirements that are less than the launch vehicle diameter, single-piece solid-surface reflectors have utilized several structural concepts as shown in Figure 16.6 [24]. These include truss rib-stiffened membrane, thick sandwich, and rib-stiffened sandwich. These solid reflectors have been developed, and they are currently available for many antenna applications. The surface accuracy

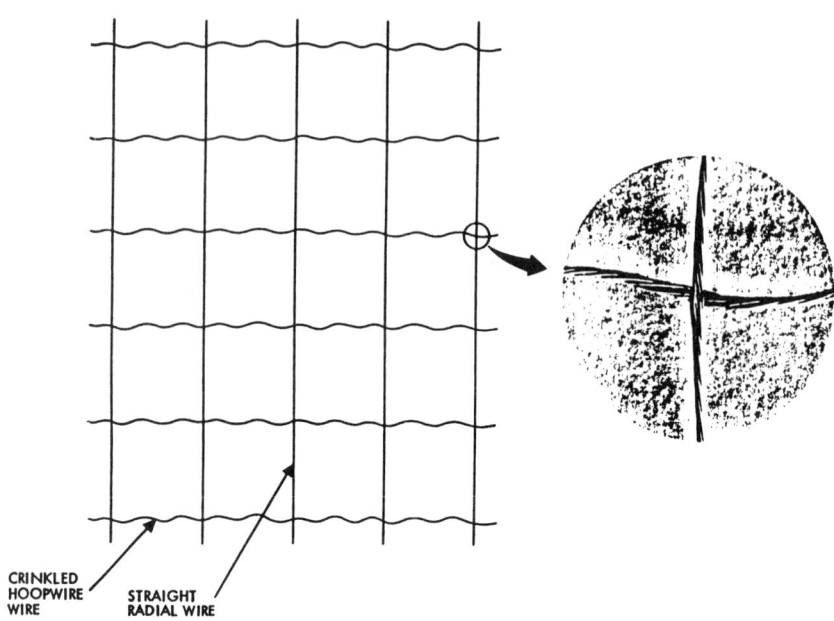

Figure 16.5 Welded grid mesh.

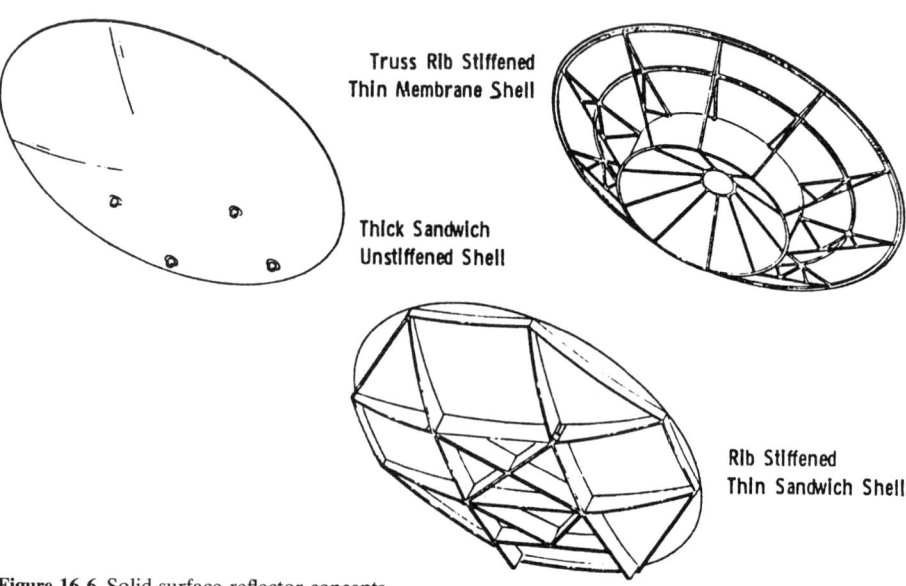

Figure 16.6 Solid-surface reflector concepts.

of the thick sandwich shell must be obtained by direct replication of the master fabrication tooling. The truss or rib-stiffened shell approach allows postfabrication adjustments to be made between the stiffening system and the shell at multiple adjustment points. Single-piece solid-surface reflectors have been widely used for scientific instruments, communication, and surveillance in space missions.

16.2.4 Deployable Solid-Surface Reflector Concepts

Space-based radar applications, which require large apertures and high surface precision, must be designed as deployable or erectable solid-surface reflectors. Due to the limited volume available on launch vehicles, large reflectors must be designed for launching in a packaged or stowed form to be deployed in orbit. Because space-based missions have widely differing requirements of configuration and size, there is no straightforward design approach available for deployable systems. However, many concepts have been studied and developed. Some of the most prominent solid-surface deployable concepts are listed in Table 16.1 and discussed in the following paragraphs.

Planar Array with Extendible Support Structure

This flight proven, three-dimensional mast was designed by Astro Aerospace, and the rigid array panels were developed by Ball Aerospace in conjunction with the Jet Propulsion Laboratory (JPL) [7]. The *extendible support structure* (ESS) was used on the SEASAT spacecraft in 1978, when it successfully deployed and supported the 10.74 m long, multiple-panel, synthetic aperture radar (SAR) antenna. Figure 16.7 depicts the ESS mast mated with rigid panels in a flat-pack arrangement. Stowed, partially deployed, and fully deployed geometries are illustrated in the figure. In other SBR applications, the length and cross section of the mast can be varied to match the array geometry, ensuring adequate strength and stiffness for any given size. The remote longeron is a multimember pantograph that is used to actuate mast deployment. The pantograph members are driven by an electric motor. When stowed, mast members intermesh with the panels to yield a stowed array envelope which is only slightly larger than that of the panels alone.

Wire-Wheel Phased Array

The wire-wheel phased array technology is being developed and designed by Grumman Aerospace Corp. This SBR concept uses an electrically scanned radar beam generated by the phased array elements. A large circular membrane plastic plane supports the array modules, and the central extended mast supports the feed system that radiates signals through space to the array and acts as a lens. Deployment of the antenna aperture plane is depicted in Figure 16.8 [8].

Table 16.1
Solid-Surface Deployable Antenna Concepts

NUMBER AND NAME	ILLUSTRATION	DESCRIPTION AND OPERATION	SOURCE	FLIGHT EXPERIENCE
1. Planer array with truss beam extendable support structure		o Rigid honeycomb sandwich panels containing array radiating elements o Truss beam supports the panels o Truss nodes are hinged and some members folded to allow stowage as a flatpack o Outer longeron is a multimember pantagraph that is used to actuate the beam extension o Extension is controlled by a motor driven actuator and springs at the hinge joints	Astro Aerospace Ball Aerospace	Flown on SEASAT Spacecraft
2. Wire-wheel phased array lens antenna		o Flexible plastic membrane dipole lens o Membrane gores wrap around central deployment drum o Rigid rim members deploy the gores using motor drives at the rim hinges o Upper and lower stay lines from the central mast to the rim are used to stabilize the deployed structure o Linear deployable mast supports the feed system	Grumman	Developmental

Table 16.1 (cont'd)

NUMBER AND NAME	ILLUSTRATION	DESCRIPTION AND OPERATION	SOURCE	FLIGHT EXPERIENCE
3. Sunflower Precision Deployable		o Rigid honeycomb sandwich petalous segments of a paraboloid o Individual segments are supported by stiff backup frame which allows post-fab surface adjustment o Frames/segments are hinged to central section and each other o Panels are folded and restrained for launch and automatically deployed on-orbit o Deployment is controlled by a motor driven actuator and springs at hinge joints o Hinges lock segments into continous stiff parabolic structure	TRW	3 meter microwave antenna qualification for space developmental work on 10 m to 15 m diameters
4. Hex Panel Precision Deployable		o Rigid honeycomb sandwich or membrane hex shaped segments o Deployable truss members are stowed behind each segment o Hex segments and folded truss are packaged in a canister arrangement for launch o Canister deploys each segment by rotating around deployed segments edges and locking truss hex peripheries o Errectable and robotic assembly techniques have also been studied	JPL Harris Astro Aerospace	Developmental work at component level

Table 16.1 (cont'd)

NUMBER AND NAME	ILLUSTRATION	DESCRIPTION AND OPERATION	SOURCE	FLIGHT EXPERIENCE
5. Inflatable Parabolic Reflector		o Metalized plastic film used as reflector o Plastic film cone and torus o Deployment using gas pressure inflation for self-rigidizing	L'Garde	Passive sphere flown in ECHO series Parabolic reflector under development
6. Eleactrostatically Figured Membrane		o Metalized plastic film used as flexible reflector with rigid rim o Rigidized command surface controls the reflector shape electrostatically o Shape measurement system provides signal for active surface control	MIT Lockheed	Developmental

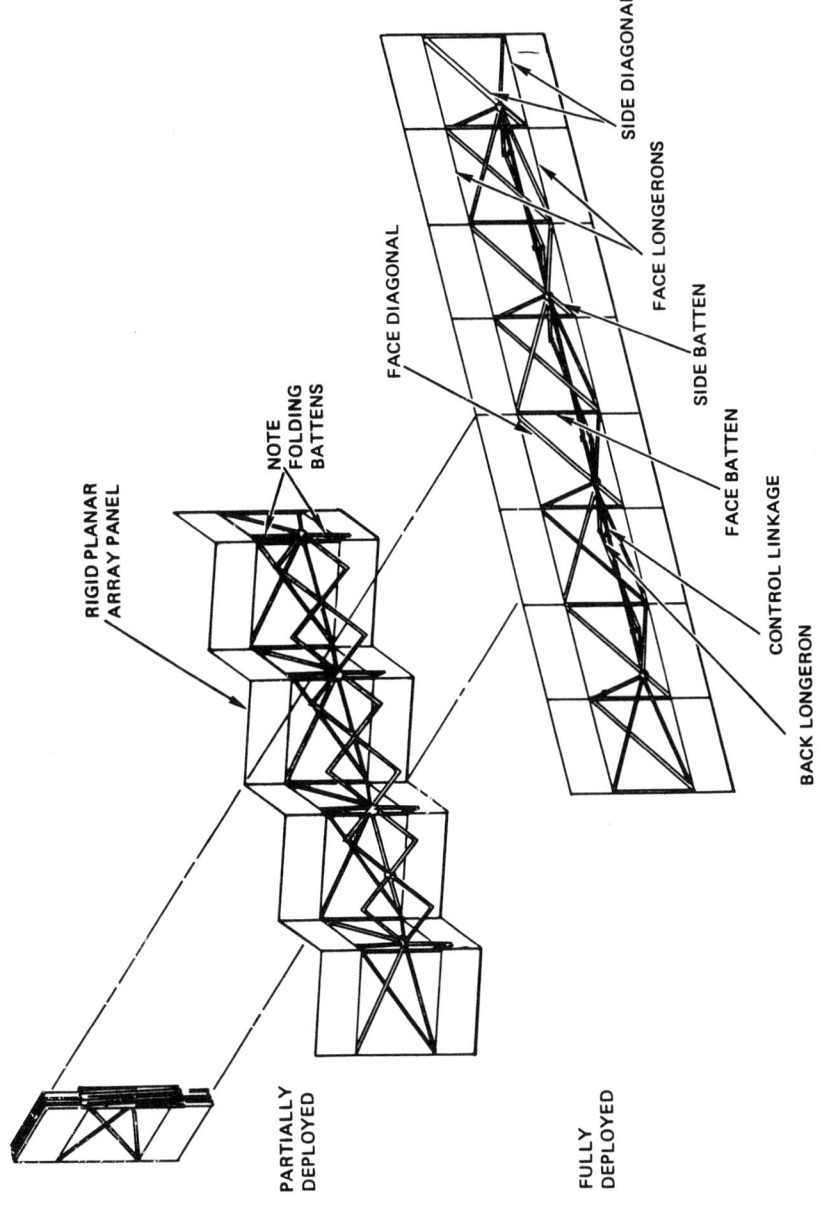

Figure 16.7 Planar array antenna with extendible truss beam.

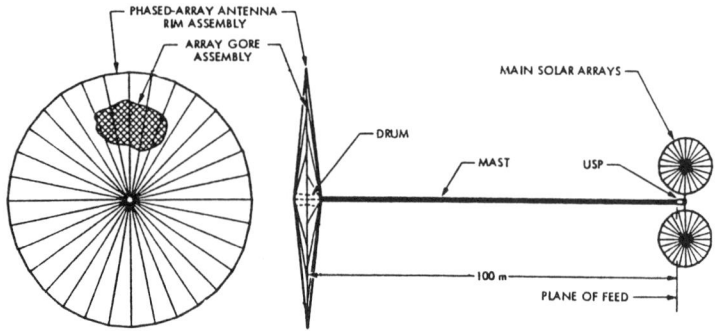

(a) Phased array antenna.

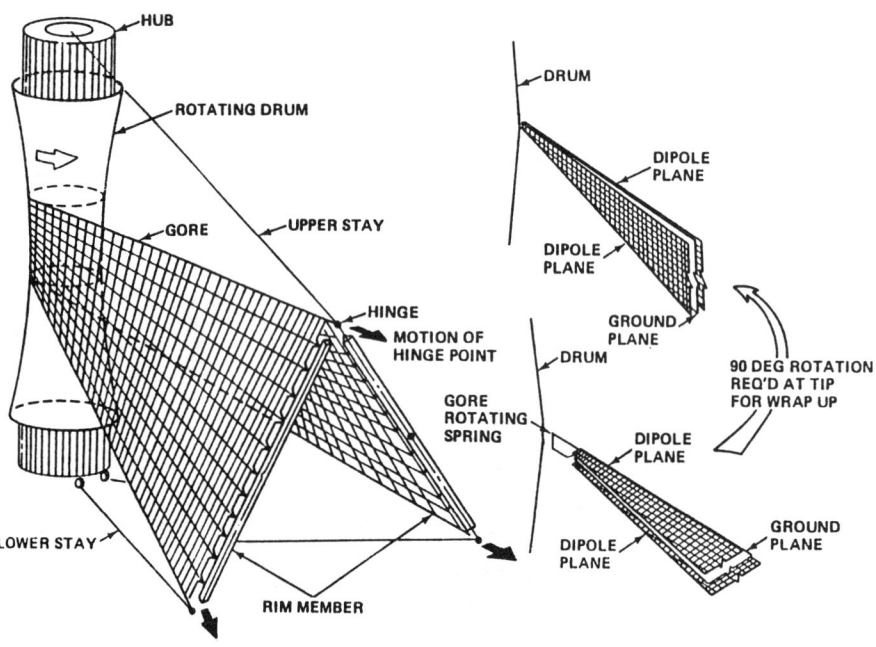

(b) Gore deployment concept.

Figure 16.8 Wire wheel phased array deployment.

Another lens array concept which has been studied [25] also uses the flexible membrane plastic array material stowed on a cylindrical drum. The membrane is deployed into a flat rectangular lens configuration by the extension of booms on either side of the drum.

Sunflower Precision Deployable Reflector

Figure 16.9 shows the Sunflower stowage and deployment concept, which has been developed by TRW [9]. Stowage configurations can be varied to meet envelope requirements by increasing or decreasing the number of panels to fit within the allowable space for the stowed envelope. Deployed diameter to stowed diameter ratio of 3:1 can be accommodated. Packaging configurations for diameters up to 15 m have been designed for the shuttle envelope by adding a second row of panels hinged to the periphery of the inner set of panels.

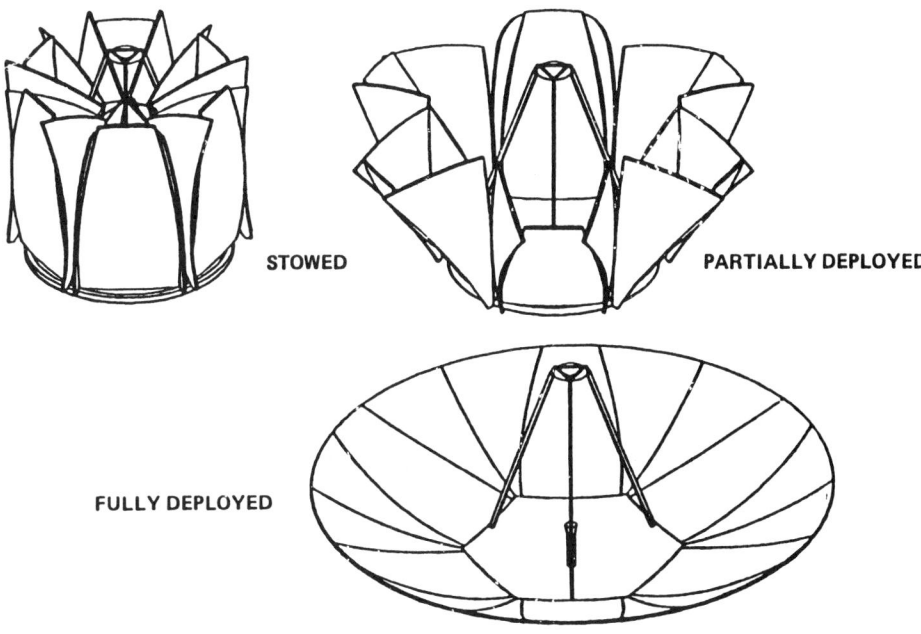

Figure 16.9 Sunflower antenna concept.

The most recent development work related to the Sunflower concept has introduced a separate support frame for each reflector segment, as shown in Figure 16.10. This allows postfabrication adjustment of the reflector panels, or active on-orbit control, to optimize surface shape.

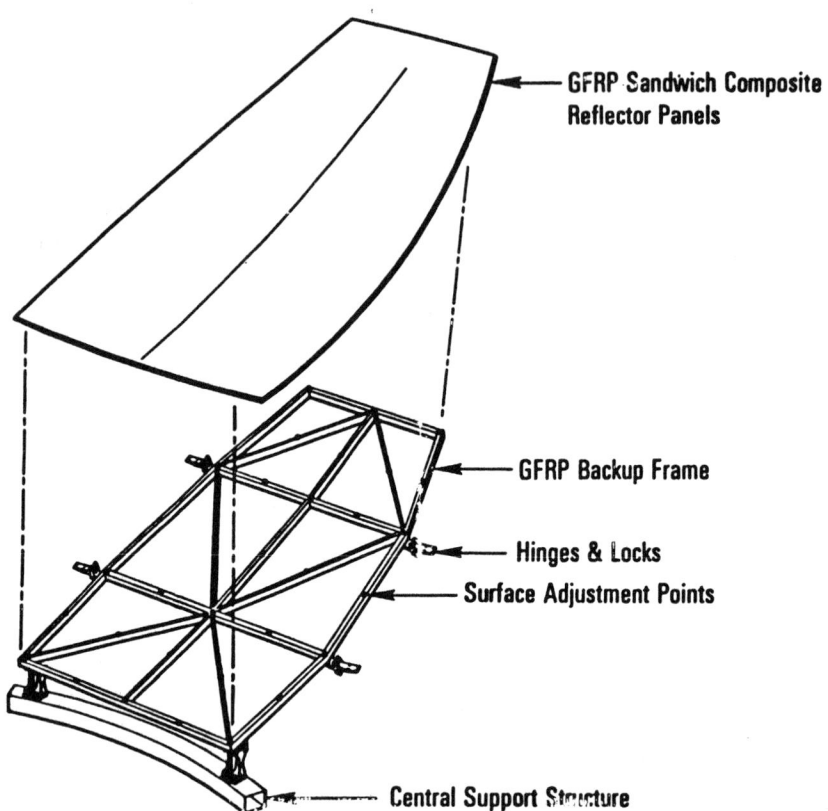

Figure 16.10 Sunflower postfabrication adjustment concept.

Hex Panel Precision Deployable Reflector

Several developmental activities have addressed the hexagonally shaped module approach for large space structures [10], [11]. The individual panels are sized to fit into the shuttle bay such that in the stowed condition the panels are stacked into a compact bundle, as shown in Figure 16.11. The required deployed diameter for various space missions can be accommodated by increasing the number of panel modules to form the desired reflector aperture.

Sequential deployment is accomplished by using an auxiliary support frame or canister, which supports the stacked modules and walks itself around the structure as adjacent panel frames become interlocked. Other concepts utilize space-erectable truss and panel modules, or some combination of automatic deployment with astronaut or robotic erection of reflector elements in space.

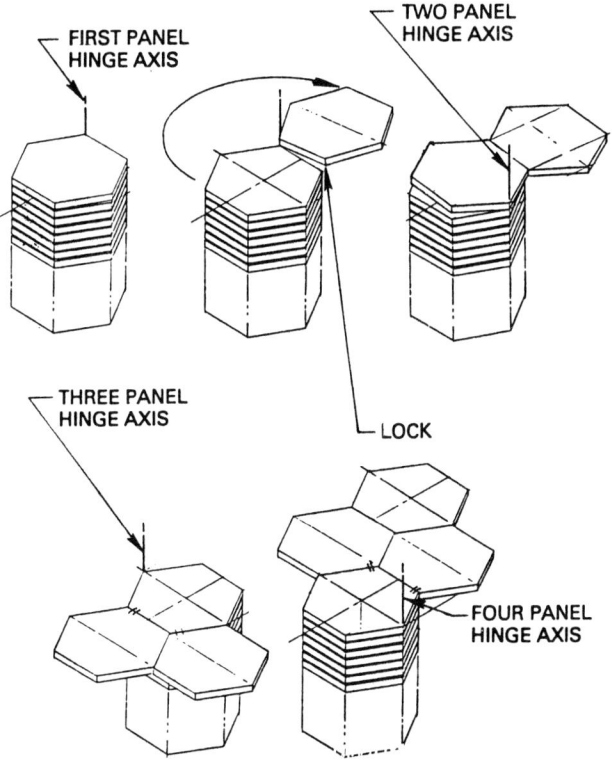

Figure 16.11 Hex panel reflector deployment concept.

Inflatable Parabolic Reflector [12]

The inflatable reflector concept uses a thin plastic film to combine a rigidizing torus with a parabolically shaped and conically shaped membrane envelope. The membranes are folded compactly for launching and inflated by release of internal gas pressure. The concept has low weight and excellent mechanical packaging efficiency. Therefore, this concept is feasible for very large aperture requirements. However, achievable surface accuracy is limited to lower frequency applications.

Electrostatically Figured Membrane [13]

The figured membrane concept uses a metalized plastic film as the reflecting surface. Because the reflector and command surfaces are constructed of thin plastic material, the concept promises low mass and compact stowage for launching. Development efforts have progressed in the control-related technology. Detailed materials, structures, and deployment system development is required to implement this unique concept.

16.2.5 Deployable Mesh Concepts

Large-aperture deployable mesh reflector concepts have received considerable developmental attention, and the most prominent structural concepts will be discussed in the following sections. Because the mesh reflective surface is flexible, it can be packaged for launch into a compact stowed envelope. When deployed on orbit the flexible mesh reflector is supported and maintained to approximate the desired reflector shape by a framework of ribs, cables, or trusses.

Both single-mesh and double-mesh concepts have been developed for space-based reflectors. The single-mesh approach approximates the desired reflector shape by stretching the mesh between stiff ribs formed to the required shape. Thus, the mesh gores deviate from the perfect reflector shape due to the mesh being flat between ribs; or, for tricot meshes, the mesh tension results in bulges or pillowing of the reflector mesh surface. Figure 16.12 demonstrates the accuracy that can be obtained with a parabolic radial rib antenna. The flat mesh gores between ribs result in an inherent rms approximation of the desired perfect shape, and precision is dependent on the number of ribs and the curvature of the paraboloid.

The double-mesh approach utilizes a second mesh, or other cordal strands, behind the reflective mesh such that short ties between the two pull the reflective mesh closer to the perfect paraboloidal shape. With many tie points, the rms approximation can be improved over the single-mesh capability. The TDRSS 15 GHz, 4.88 m antennas use a double-mesh design, and the resulting surface accuracy is plotted in Figure 16.12 for comparison with the single-mesh capability.

Table 16.2 summarizes the current technology of deployable mesh reflector concepts, and a brief structural-mechanical description is presented in the following paragraphs.

Wrap-Rib [14]

The wrap-rib reflector consists of a series of radial ribs shaped to form a parabola. A lightweight reflective mesh is stretched between these ribs to form the parabolodial reflecting surface. A sketch of the deployed wrap-rib antenna is shown in Figure 16.13. The flexible ribs are wrapped around a power-driven rotating spool that constrains the stored energy of the wrapped ribs and deploys the reflector surface at a controlled rate. The stowed configuration may be as small as one-fortieth of the deployed diameter.

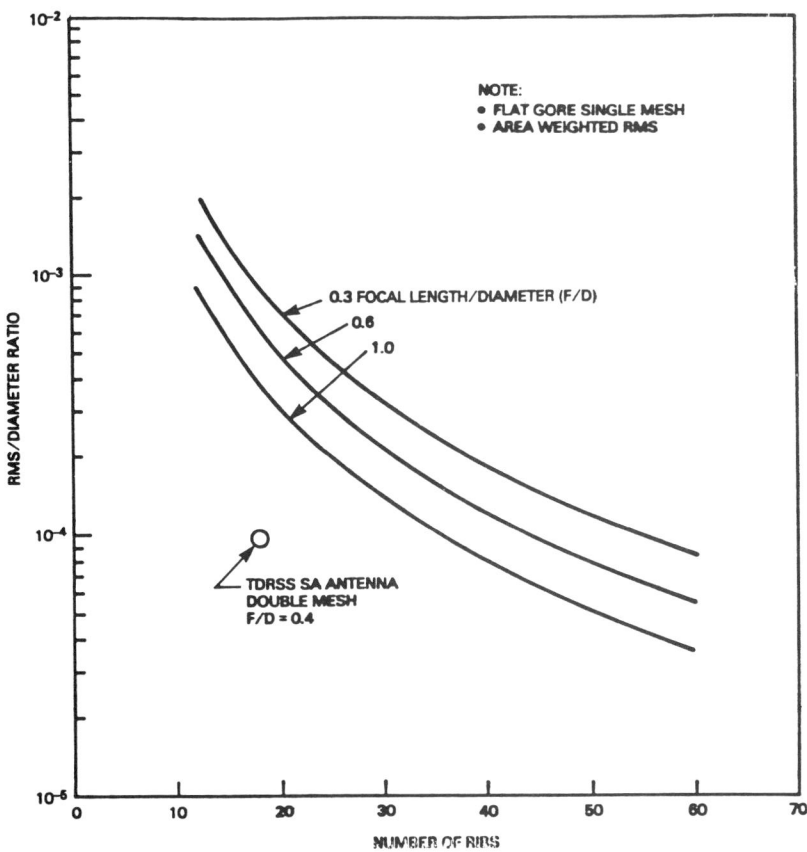

Figure 16.12 Paraboloid approximation accuracy for rib-gore antennas.

Radial-Rib [5, 6]

Figure 16.14 depicts the radial-rib deployment concept. The antenna is folded and it opens like an umbrella. The flexible mesh is attached to rigid ribs formed to the required deployed shape. We can see that the deployed diameter is limited by the stowage envelope length, and for shuttle deployment the diameter is limited to approximately 50 m. For larger aperture requirements, the ribs must be hinged and folded back on themselves, complicating the mesh stowage and deployment kinematics.

Table 16.2
Mesh Deployable Antenna Concepts

NUMBER AND NAME	ILLUSTRATION	DESCRIPTION AND OPERATION	SOURCE	FLIGHT EXPERIENCE
1. Wrap Rib		o Mesh attached to flexible radial ribs o Ribs wrap circumferentially around at central hub for stowage o Ribs and mesh deploy using stored energy in wrapped ribs o Deployment can be controlled by motor driven rotating spools	Lockheed	Flown on ATS-6
2. Radial Rib		o Mesh attached to rigid radial ribs o Ribs are hinged to a central hub and folded toward the central axis o Ribs and mesh deploy radially using stored spring energy and/or motor drive	Harris TRW	Flown on TDRSS and FLTSATCOM
3. Hoop and Column		o Mesh attached to inner and outer rings o Outer ring (hoop) and mesh are folded against central column for stowage o Hoop and mesh deployment is controlled by motor actuators o Central column telescopes to deployed position o Support cables from column preloads the hoop and mesh in the deployed configuration	Harris	Developmental

Table 16.2 (cont'd)

NUMBER AND NAME	ILLUSTRATION	DESCRIPTION AND OPERATION	SOURCE	FLIGHT EXPERIENCE
4. Tetrahedral Truss		o Mesh reflective surface supported by deployable tetrahedron truss system o Truss members are hinged together with spider elements at each joint o Truss members and mesh fold into a compact package o Deployment is accomplished by energy o Stored in leaf springs at truss member joints o Deployment proceeds as a continual expansion of the total truss system o Tension ties across each truss bay are used to shape the mesh which spans the bay	General Dynamics NASA Langley Lockheed	Developmental
5. Box Truss		o Mesh reflective surface supported by deployable box truss system o Springs actuate the deployment o Deployment proceeds sequentially, bay-by-bay o Over center latching in deployed position	Martin Marietta	Developmental

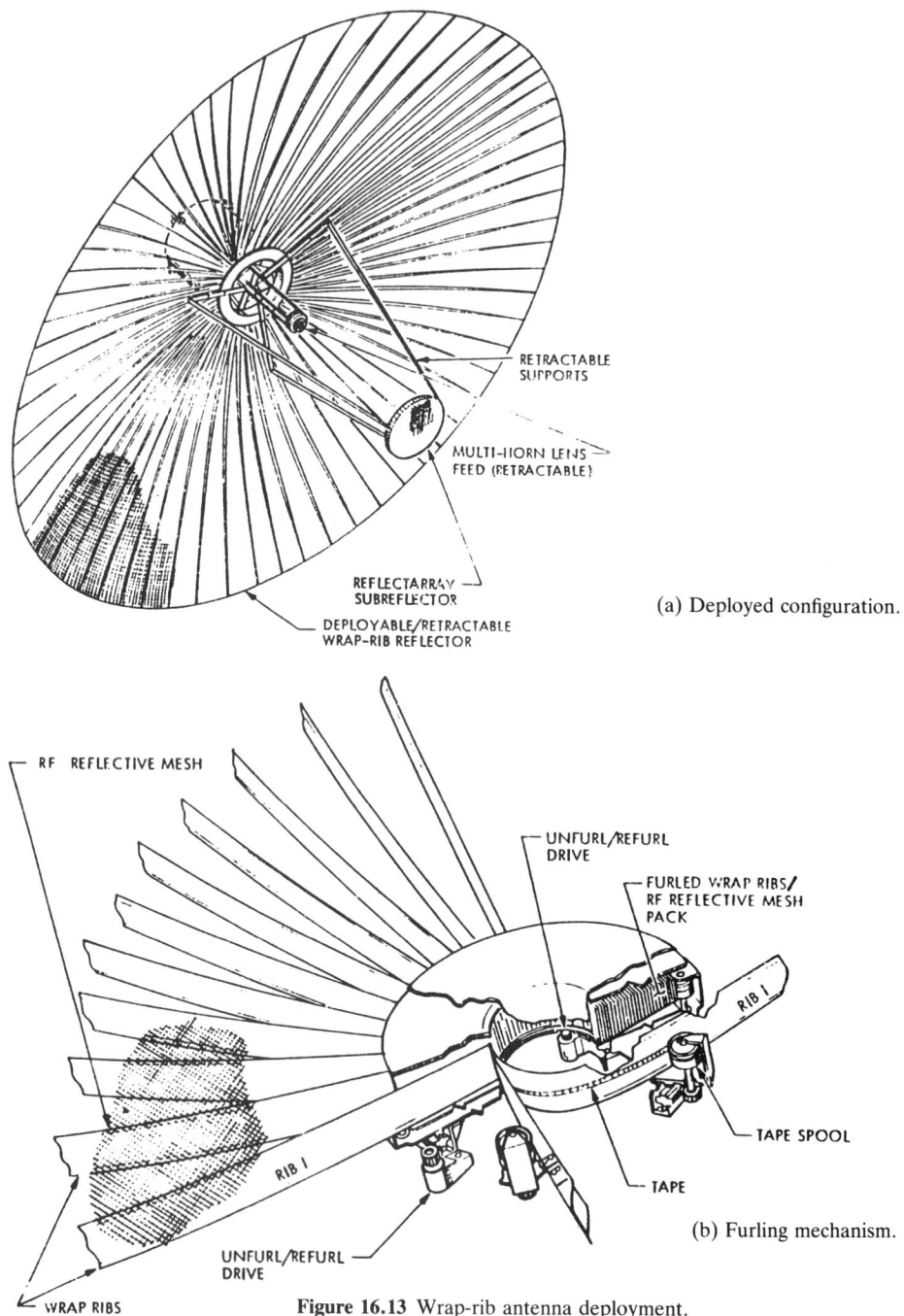

(a) Deployed configuration.

(b) Furling mechanism.

Figure 16.13 Wrap-rib antenna deployment.

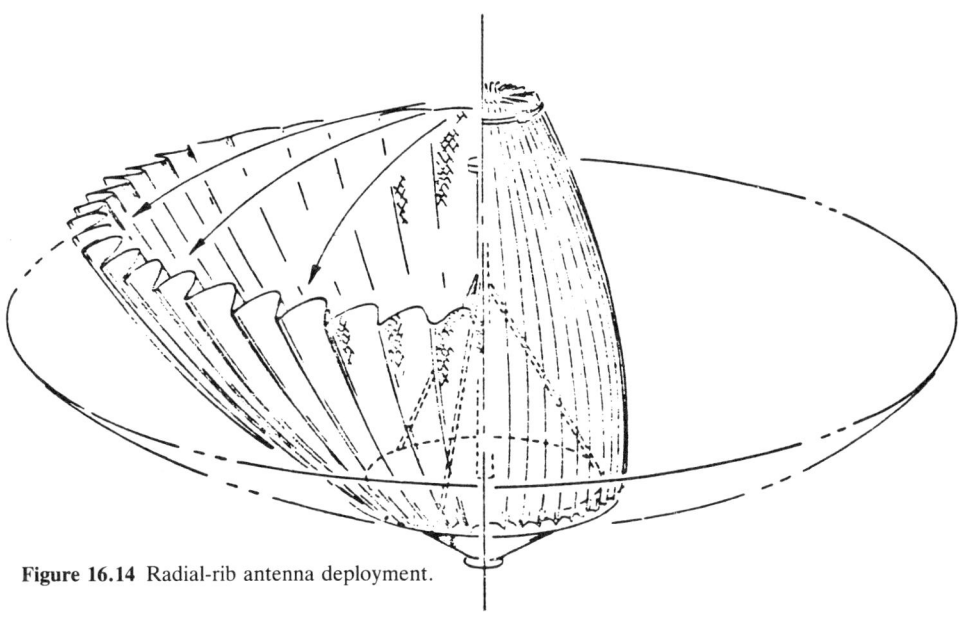
Figure 16.14 Radial-rib antenna deployment.

Hoop and Column [15, 16]

Figure 16.15 [16] depicts the hoop and column deployment sequence. The hoop provides a rigid structure to which the reflective mesh surface attaches, and consists of rigid segments that articulate at hinges joining adjacent segments. Torsion springs at each hinge provide the energy to deploy the hoop and attached mesh.

Tetrahedral Truss [17, 18]

Figure 16.16 shows the tetrahedral truss deployment concept for mesh reflectors. Each basic element of the structure is a deployable tetrahedral truss system hinged at each node by a spider link. Deployment is provided by springs used to hinge the bay struts at their centers or by springs at each node. The stowed spring energy deploys the truss and mesh, and then locks the reflector in the deployed position. The reflective mesh is supported across each truss bay by a series of tension ties and a webbing attachment system that creates an interface of the ties with the mesh.

The truss provides an inherently deep and stiff structure, and can be sized for very large aperture requirements with efficient stiffness-to-weight characteristics. Stowed packaging efficiency is also very good. A 150 m diameter reflector can be stowed in the shuttle payload compartment along with a typical spacecraft bus.

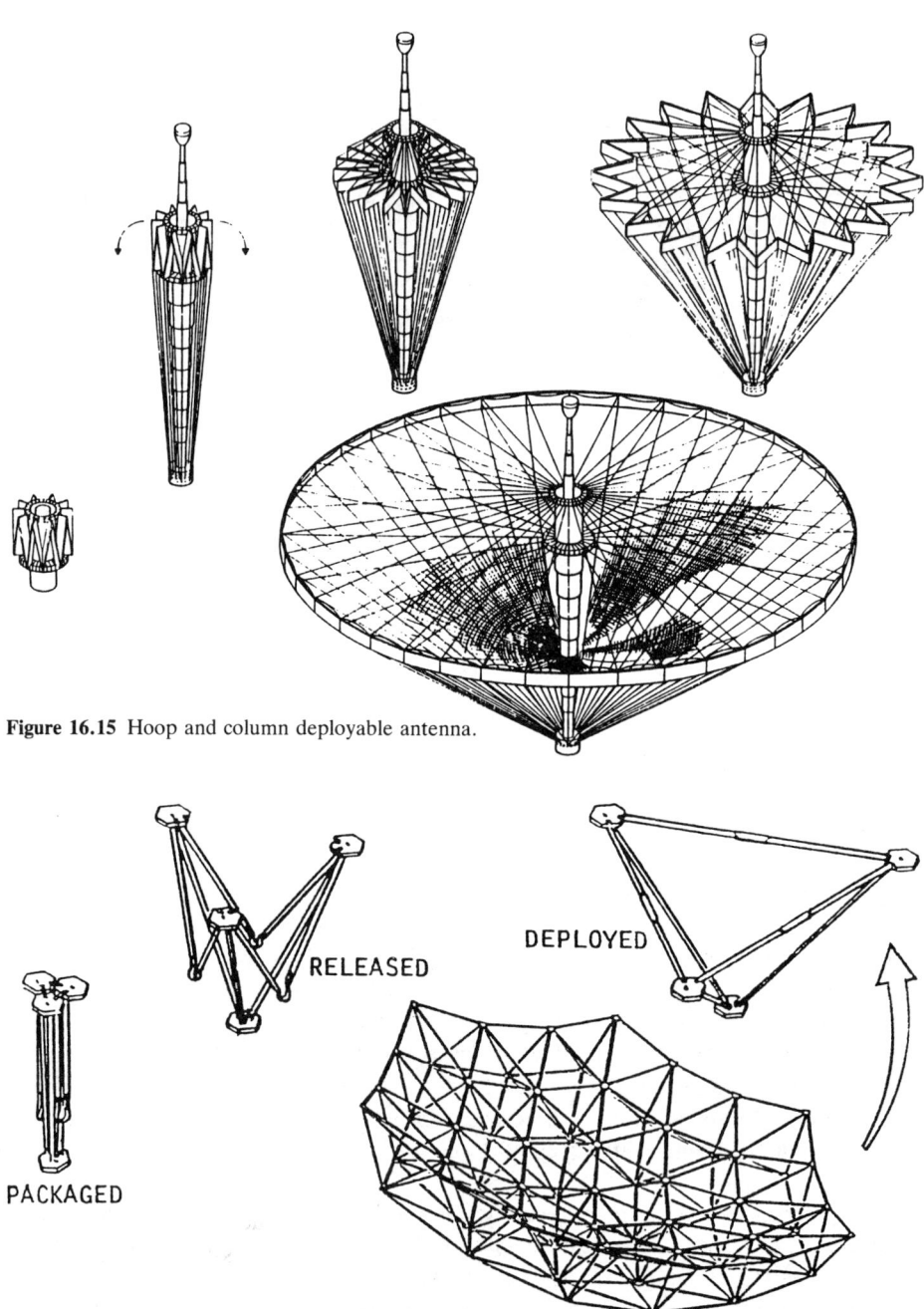

Figure 16.15 Hoop and column deployable antenna.

Figure 16.16 Tetrahedral truss deployable structure.

Box Truss Reflector [19]

The mesh reflective surface is suspended above the box truss structure by a series of stand-offs and tie cords to minimize the mesh-pillowing effect. As shown in Figure 16.17, the truss structure is a deployable frame consisting of two equal-length vertical members, two members hinged in the middle, and telescoping diagonal braces. The hinged members connect the ends of the verticals and fold inward to stow between the adjoining verticals. The braces control the shape of the deployed frame. Deployment power is provided by springs located at the hinges of the two middle-hinged members. Although the deployable box truss structure has been developed primarily for mesh reflective surfaces, very recent studies [26] have proposed a completely deployable box truss system which includes flat or curved solid surface plate elements in the folding concept. A 3.8 m diameter hexagonal truss, similar to the box concept, was recently deployed in space by a French-Soviet team to demonstrate the structural feasibility of the concept [27].

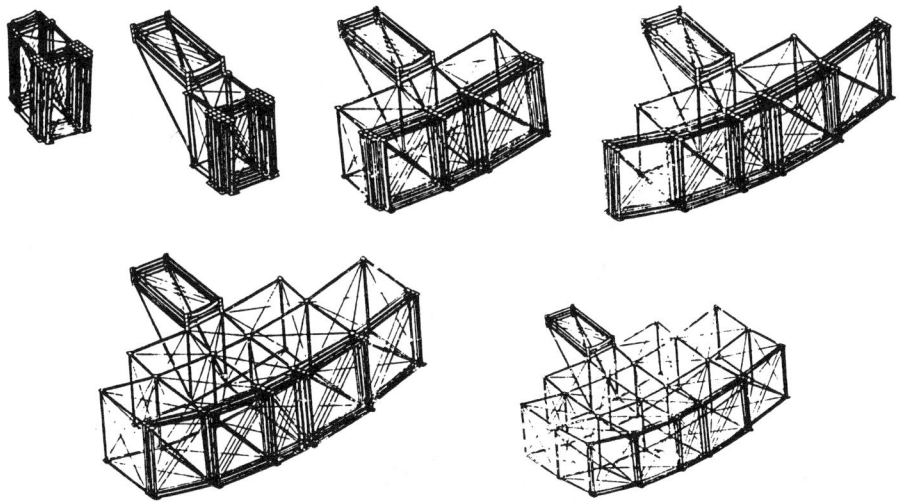

Figure 16.17 Box truss deployable structure.

16.3 LINEAR DEPLOYABLE STRUCTURE CONCEPTS

Space-based radar systems require large structures (hundreds of meters in length) to meet certain mission performance requirements. These structures must be stowed compactly for transport to space and deployed to meet precise geometric position requirements and large operational dimensions.

As we discussed previously (see Figure 16.2), long, linearly deployable booms have been studied and developed for space applications. Table 16.3 presents a compilation of linearly deployable boom concepts, and we give mechanical-structural descriptions of the more prominent concepts in the following paragraphs.

Table 16.3
Linear Deployable Structure Concepts

NUMBER AND NAME	ILLUSTRATION	DESCRIPTION AND OPERATION	SOURCE	FLIGHT EXPERIENCE
1. CONTINUOUS LONGERON COILABLE LATTICE MAST (LANYARD DEPLOYED)		• MAST STRUCTURE CONSISTS OF THREE CONTINUOUS LONGERONS, BATTENS AND THREE DIAGONALS PER BAY • LONGERON AND BATTEN MATERIAL IS FIBERGLASS EPOXY • DIAGONAL MATERIAL IS STRANDED WIRE OR FIBERGLASS • MAST SELF-DEPLOYS DUE TO STRAIN ENERGY IN COILED LONGERONS AT A RATE CONTROLLED BY PAYOUT OF A MOTORIZED RESTRAINING LANYARD	ASTRO AEROSPACE ABLE ENGINEERING	USED ON NUMEROUS SPACECRAFT FOR DEPLOYMENT OF SCIENTIFIC INSTRUMENTS
2. CONTINUOUS LONGERON COILABLE LATTICE MAST (CANISTER DEPLOYED)		• MAST STRUCTURE SIMILAR TO LANYARD DEPLOYED MAST • DEPLOYMENT IS ACTUATED BY LARGE MOTORIZED, THREE-THREADED, ROTATING NUT MECHANISM WITH STOWAGE CANISTER	ASTRO AEROSPACE ABLE ENGINEERING	OAST SOLAR ARRAY FLIGHT EXPERIMENT (SAFE I) ON STS-41 IN 1984, TO BE USED ON SPACE STATION AND OLYMPUS SOLAR ARRAYS
3. LATCHING, ARTICULATED LATTICE MAST (CANISTER DEPLOYED)		• MAST STRUCTURE CONSISTS OF THREE LONGERONS, BATTENS AND SIX DIAGONALS PER BAY • THE LONGERONS ARE SEGMENTS OF METALLIC, FIBERGLASS OR GRAPHITE TUBES/RODS, WHICH ARE ARTICULATED AT THE BATTEN FRAMES WITH UNIVERSAL HINGE FITTINGS • DIAGONAL MEMBERS, TYPICALLY METALLIC CABLES • DEPLOYMENT REQUIRES LATCHING OF THREE DIAGONALS PER BAY • DEPLOYMENT IS ACTUATED BY LARGE MOTORIZED, THREE-THREADED ROTATING NUT MECHANISM WITH STOWAGE CANISTER	ASTRO AEROSPACE ABLE ENGINEERING	PROTOTYPES DEVELOPED FOR VARIOUS GROUND APPLICATIONS AND HIGH POWER ARRAYS. NO FLIGHT EXPERIENCE
4. FOLDABLE ARTICULATED SQUARE TRUSS (LANYARD OR CANISTER DEPLOYED)		• MAST STRUCTURE CONSISTS OF THREE OR FOUR LONGERONS, BATTENS AND SIX DIAGONALS PER BAY • LONGERONS ARE SEGMENTS OF METALLIC, FIBERGLASS OR GRAPHITE TUBES/RODS, WHICH ARE ARTICULATED AT THE BATTEN FRAMES WITH SPECIAL HINGE JOINTS • DIAGONAL MEMBERS TYPICALLY METALLIC CABLES WITHOUT ANY LATCH MECHANISM • BATTENS COILABLE FIBERGLASS • MAST SELF-DEPLOYS DUE TO STRAIN ENERGY IN COILED BATTENS AT A RATE CONTROLLED BY PAYOUT OF A MOTORIZED RESTRAINING LANYARD OR CANISTER	ABLE ENGINEERING	TO BE APPLIED ON TETHERED SATELLITE EXPERIMENT
5. STACK BEAM (STACKING TRIANGULAR ARTICULATED COMPACT BEAM)		• MAST STRUCTURE CONSISTS OF THREE LONGERONS, BATTENS AND THREE DIAGONALS PER BAY • ELEMENTS MADE FROM METAL, FIBERGLASS OR GRAPHITE EPOXY • THE LONGERON AND DIAGONALS ARE SEGMENTS OF TUBES/RODS WHICH HAVE HINGES AT THEIR MIDPOINTS AND AT THE BATTEN FRAMES • HINGES HAVE TORQUE SPRINGS TO OBTAIN HINGE RESTRAINING MOMENT CAPABILITY	ASTRO AEROSPACE	DEVELOPMENTAL

Table 16.3 (cont'd)

NUMBER AND NAME	ILLUSTRATION	DESCRIPTION AND OPERATION	SOURCE	FLIGHT EXPERIENCE
6. METALLIC STRIP BOOM, EXTENDABLE REEL STORED (STEM, BI-STEM, EDGELOCK)		• TUBES FORMED BY ONE OR MORE METALLIC CYLINDRICAL, THIN STRIPS • STRIPS ARE STOWED BY ELASTICALLY FLATTENING THE SECTION AND REELING THEM ON SPOOLS • TUBE IS FORMED BY MOTORIZED ROTATION OF THE SPOOLS • MATERIAL TYPICALLY STAINLESS STEEL OR BERYLLIUM COPPER • SOME VERISIONS PERMIT INTERLOCKING OF THE STRIP EDGES TO IMPROVE TORSIONAL STIFFNESS	ASTRO AEROSPACE FAIRCHILD	IN ONE FORM OR ANOTHER, USED IN NUMEROUS FLIGHT PROGRAMS FOR ANTENNAS, GRAVITY GRADIENT BOOMS, ETC. USED ON CTS, FRUSA AND SPACE TELESCOPE SOLAR ARRAYS
7. METALLIC STRIP BOOM LENTICULAR WELDED BEAM		• TUBE FORMED BY TWO METALLIC OR GRAPHITE HALF-LENTICULAR, METALLIC STRIPS • STRIPS ARE WELDED (BONDED) AT THEIR EDGES TO FORM LENTICULAR CROSS SECTION TUBE • TUBE IS STOWED BY ELASTICALLY FLATTENING THE SECTION AND ROLLING IT UP ON A MOTORIZED REEL	LMSC BOEING ASTRO AEROSPACE	USED ON MARS VIKING BIOLOGICAL EXPERIMENT PACKAGE
8. TELESCOPING CYLINDERS		• CONCENTRIC METALLIC OR GRAPHITE TUBES IN GRADUATED DIAMETERS • SECTIONS ARE EXTENDED AND LATCHED IN THE FULL EXTENDED POSITION USING GAS ACTUATION	BRITISH AEROSPACE	SMALL VERSION FLOWN ON BRITISH X4 SATELLITE
9. 1D AND 2D CROSS-SECTION PANTOGRAPH		• FOLDED MULTI-LINK ARMS ATTACHED TOGETHER TO CREATE ONE-DIMENSIONAL OR TWO-DIMENSIONAL CROSS-SECTION TRUSS BEAM • LINKS ARE METALLIC • ACTUATION BY SPRINGS AT THE HINGE POINTS	ASTRO AEROSPACE COMSAT (SNIAS) LMSC TRW	ASTRO ESS VERSION FLOWN ON SEASAT
10. FOLDING BEAM		• GEOMETRY MAY BE TRUSS TUBULAR OR SOLID • HINGES ON EITHER END • LATCHES AFTER DEPLOYMENT • USUALLY DEPLOYED BY CABLE/PULLEY SYSTEM	TRW BOEING FAIRCHILD	USED FREQUENTLY IN SPACE FLIGHT; USUALLY AS RELATIVELY SHORT MEMBERS (LESS THAN 30 ft)
11. LAZY TONG		• SOLAR ARRAY PANELS MUST BE HINGED TOGETHER WITH ROTARY ATTACHMENT TO BEAMS AT PANEL EDGES • PANELS ALIGN AND LATCH TO ACCOMMODATE COLUMN LOADS • CAN USE VARIOUS PRIME MOVERS TO ACTUATE DEPLOYMENT	LMSC FAIRCHILD	THE PEGASUS SPACECRAFT DEPLOYED FLAT-PANELS 14 BY 48 ft (EACH WING) AS METEROID DETECTORS USING THIS SYSTEM

16.3.1 Lattice Mast Booms

Both Able Engineering and Astro Aerospace design a line of linear deployable booms that uses a truss structure consisting of three longerons stabilized by battens and diagonal elements, as illustrated in Figure 16.18. There are two general types: continuous longeron systems that coil when stowed, and articulated longeron systems that fold up when stowed. Both types retract into very compact cylindrical stowage volumes, with the stowage height typically being 2–10% of the deployed length, depending on longeron diameter requirements. Fiberglas is used as the material in the continuous longeron version to allow elastic coiling when the boom

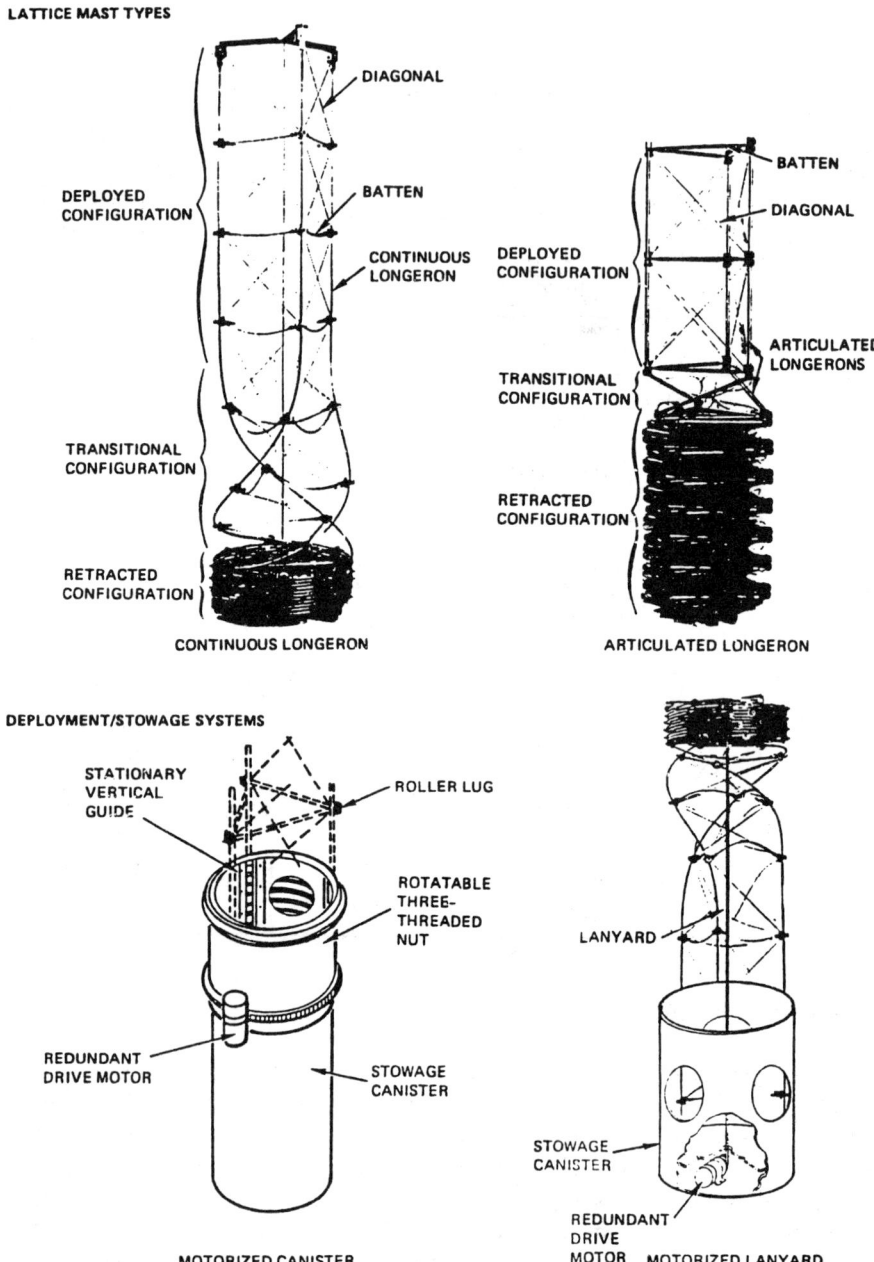

Figure 16.18 Lattice mast systems.

is retracted. Metal or composite material can be used for the longerons in the articulated version.

Both the continuous and articulated longeron systems can be deployed and retracted by a motorized canister. The continuous longeron system also can be deployed and retracted by using a motorized lanyard system.

The continuous longeron system, either lanyard-deployed or canister-deployed, is space-qualified, having flown as instrument booms or in solar array applications. The continuous longeron version will be used in the Space Station and the LSAT-Olympus solar arrays.

16.3.2 Folding Articulated Square Truss Mast (FASTMAST)

The FASTMAST system was designed by Able Engineering of Goleta, California. It is qualified for flight on the Tethered Satellite Experiment, and capable of self-deployment and retraction operations.

Figure 16.19 illustrates the mast in various phases of deployment. The mast consists of four multihinged longerons stabilized at alternate hinge nodes by rigid square batten frames. The longerons are located at the batten frame corners. The frames are spaced along the longerons so that each mast cell is cubic. Each longeron also is hinged midway between frames so that when stowed the longeron folds along one of the cube faces. Therefore, each mast bay stows in a length equal to twice the local diameter of a longeron. Preloaded fiberglas leaf springs interconnect the midpoint hinges, resulting in a system that develops forces that keep the longerons straight when the boom is deployed and can cause or aid boom deployment. When deployed, the exterior face of each bay is equipped with redundant tension cables across the corners to stabilize the extended boom. There are no mechanical latches required to achieve deployed stiffness and strength. The longeron diameters can be modified along their lengths to match varying applied loads, thereby minimizing weight.

An important feature of the FASTMAST is that only one bay is in transition at a time. This provides a very stable system in that the mast develops full strength and stiffness, even when only partially deployed. Another important feature is that the mast stacks flatly without rotating, thereby easing the integration of electrical cables and simplifying alignment.

The number of longerons is variable. Increasing the number of longerons increases the mast load capability and provides structural redundancy. The mast load capacity also can be varied by increasing the overall cross-sectional geometry.

The mast can be deployed or retracted by two methods. One method uses a motorized canister system, containing an elevating mechanism that is a large drum nut with four helical threads machined on the inside. A small roller engages the thread of the nut at the corner of each square batten frame of the mast. When the

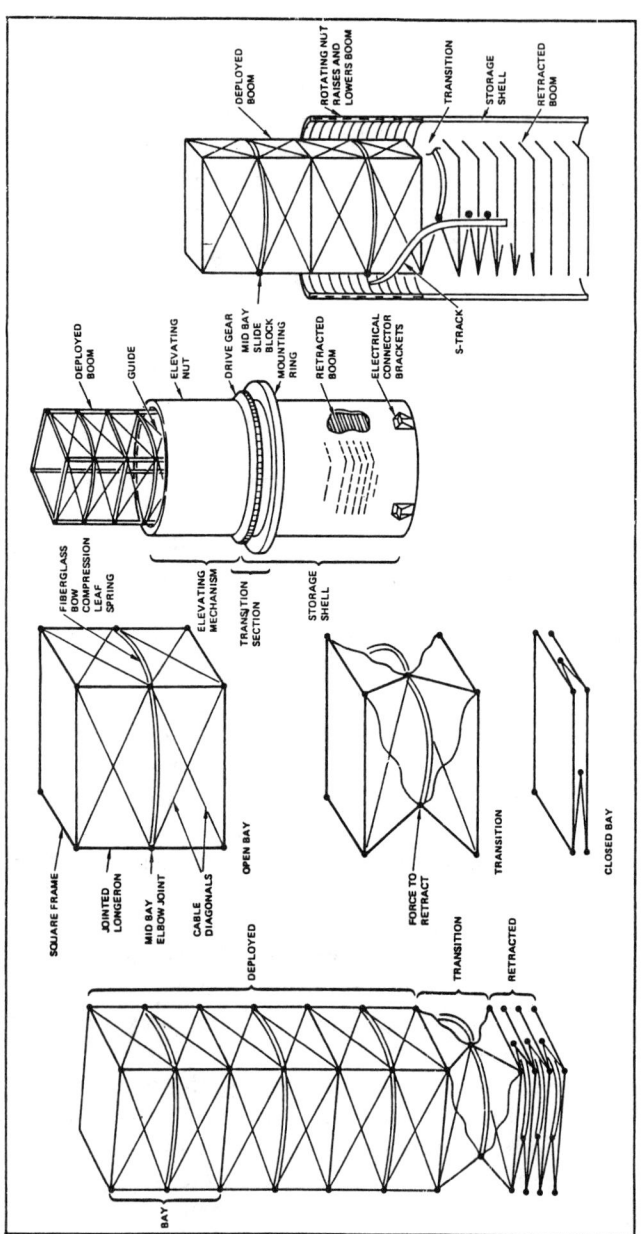

Figure 16.19 Foldable articulating square truss mast (FASTMAST).

drum nut is rotated by a motor, the mast deploys or retracts, depending on the direction of rotation. An S-shaped guide track in the transition section of the canister assists in opening or closing each mast bay in a sequential manner.

Another deployment method uses a lanyard integrated into the structure instead of the elevating drum nut mechanism. A motor-driven lanyard at the root of the mast is used to control the rate of deployment or to retract the mast. This method results in a lighter weight mast system. However, full strength and stiffness are not obtained until the mast is fully deployed.

16.3.3 Metallic Strip Booms

Astro Aerospace designs a line of actuator boom mechanisms that uses the metallic strip principle, as illustrated in Figure 16.20. *Storable tubular extendible member* (STEM) booms are thin-walled, cylindrical cross section devices that incorporate prestressed metallic tapes (beryllium copper or stainless steel). The prestressed-preshaped tapes are stowed on spools in a flattened condition, in a manner similar to a carpenter's metal tape measure. The boom can be formed from a single strip of metal tape that overlaps itself to form a nearly circular cross section (the STEM configuration), or can be formed from two strips of metal tape that nest together (BI-STEM). Another version has two metal strips interlocking along their edges to create greater torsional stiffness and strength. These devices are packaged in an extremely efficient manner, but are limited in strength and stiffness.

Telescoping concentric tubes can also be used for deployable tubular boom applications. However, these uses are generally limited to short lengths and low stiffness requirements.

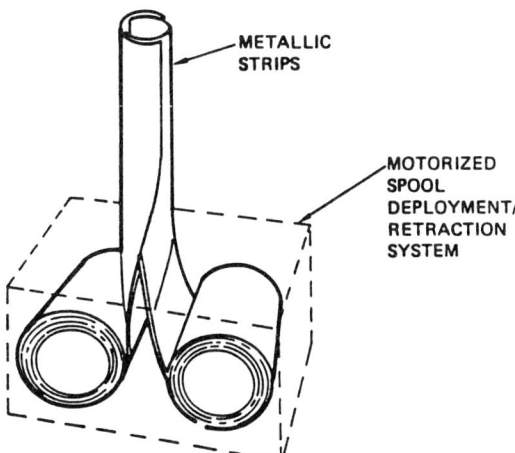

Figure 16.20 Storable tubular extendible member (STEM) boom concept.

16.3.4 Folding Beam Concepts

Pantographs, lazy tongs, and other folding beam structural systems have been used frequently in space-based applications. Folding beams generally are configured for specific functions, such as solar array deployment or to position scientific instruments away from the spacecraft body. These applications have been for deployed lengths of less than 10 m.

16.4 ERECTABLE STRUCTURAL CONCEPTS

The automatically deployable structures previously described are generally highly complex kinematic systems, which require extensive development and testing, especially to demonstrate the deployed geometric precision required for space-based radar requirements. Techniques for manual erection or assembly by astronauts could simplify the structural design, and also provide increased geometric precision for large structures as compared with automatically deployed structures. However, the system's reliability might be improved by a completely automatic assembly, which would avoid the human error parameter. Ultimately, the assembly of large, precise space structures will address the cost factor, and the assembly cost must be compared for manual, automatic, or robotic assembly methods.

For near-term space missions, two general categories can be identified for assembly of space structures. These categories are distinguished by missions where astronauts can be employed in the assembly operations, such as Space Station or low earth orbit missions, as compared to missions at higher orbits (including synchronous) where astronauts' capabilities cannot be utilized. These categories and associated assembly methods are listed in Table 16.4. The nonastronaut-supported mission is much more restricted in the available assembly methods. However, extravehicular assembly has severe time limitations and potentially high operational costs. Automatically deployable assemblies are highly dependent on the specific mission function and structure geometry. Hybrid combinations of various assembly techniques may ultimately result in the least costly.

Development of manual and robotic assemblies for space structures has been related primarily to Space Station requirements. NASA has developed the assembly of truss structures by astronauts working in space suits. The effort involves design, fabrication, and testing of truss struts and joints, and includes extensive experimentation in underwater zero-gravity simulation tanks as well as the shuttle ACCESS and EASE flight experiments [20, 21].

The developmental work demonstrates that struts with specially designed joints can be assembled—on time—by astronauts trained in underwater zero-gravity simulators. However, other assembly operations, such as mounting of modules and joining electrical cables to the structure, are in a low state of technology. Robotic assembly techniques have received little attention and will require signif-

Table 16.4
Large Precision Structure Assembly Techniques

Astronaut-Supported Missions
EVA—Manual Assembly
EVA—Robotic Asssembly
IVA—Robotic Assembly
Automatic
Hybrid—Auto-EVA-IVA

No Astronaut Support
Automatic
Robotic
Hybrid—Auto-Robotic

Definitions:
EVA—Extra vehicular activity.
IVA—Inter vehicular activity using remote manipulator system.

icant development before implementation into the design and assembly of large space-based structures.

Based on the technological status of manual and robotic assembly techniques, near-term space-based radar structures must be designed for automatic deployment or utilization of astronaut assembly techniques being developed for Space Station. Future erectable concepts will require extensive development of astronaut-compatible joints, modules, and techniques, including extensive water tank training and precursor shuttle flights to validate the assembly structure.

16.5 STRUCTURAL DESIGN CONSIDERATIONS FOR SBR ANTENNAS

Structures for space-based radar applications are generally large and require geometric precision to meet mission performance requirements. The large aperture size and feed distances require that the structure be deployable. The space-based radar antenna, the spacecraft, and any upper stage must fit within the volumetric and structural constraints of the available launch vehicle. Thus, compact stowage concepts must be selected for structural components and equipment.

The large size also means that the structure is outside the thermally controlled spacecraft bus. Thus, the structure generally must use passive thermal control and sees large temperature extremes in the space environment ($\pm 100°C$). To accommodate these large temperature extremes, the selected structural materials should have a low *coefficient of thermal expansion* (CTE). *Graphite fiber-reinforced plastic* (GFRP) composite is the most commonly selected material for antenna structures because it can be designed to have a CTE close to zero. For applications which may require that the structure be RF transparent, Kevlar material is selected because it has dielectric properties and its thermal expansion is very small. (Kevlar is the trade name of aramid, polyphenylene, manufactured by DuPont.) However, for low RF applications where geometric accuracy requirements are not as critical, more conventional, lower cost structural materials can be utilized.

Geometric precision is not only controlled by the proper selection of materials, but also by careful control of the structure's manufacturing operations. Tooling and manufacturing processes must be developed to obtain the desired dimensional precision. In some cases, postfabrication adjustment of the geometry may be designed into the structure to guarantee required precision. For very stringent geometric accuracy requirements, active on-orbit shape adjustment may be warranted.

The large size of SBR antennas also requires efficient structural concepts to minimize launch weight. Thus, mesh or honeycomb sandwich and truss structures are efficient structural concepts that minimize weight.

Large space-based structures must also consider deployed stiffness and the combined behavior of the structure and spacecraft attitude control system. Present practice for spacecraft systems is the control of rigid body motions and the avoidance of flexible structure and control interaction. However, for very large structures, the structure and control systems should be designed interactively, and the two systems might include the ability to suppress flexible body responses to acceptable levels.

Table 16.5 summarizes the key structural design considerations discussed in this section, which are inherent to the design of large, precise structures for space-based radar systems.

Table 16.5
SBR Antenna Structural Design Considerations

Structure Requirement	Design Approach
Large Dimensions	Deployable
	Compact Stowage for Launch
	Passive Thermal Control
Geometric Precision	Low CTE Materials
	Low Creep Materials
	Tooling and Fabrication Control
	Postfabrication Adjustment
	Active On-Orbit Adjustment
Low Mass	Efficient Structural Configurations
	Efficient Structural Materials
Deployed Stiffness and Pointing	Stiff Structural Concepts
	Stiff Structural Materials
	Structure and Control System Interaction
	Active Suppression of Response

16.6 MATERIAL SELECTION FOR SBR STRUCTURES

The selection of materials for space-based structures is dependent on factors such as light weight, high stiffness, structural load range, and the cost and availability of material. As discussed in the previous section, SBR structures also require low thermal expansion and dimensional stability, and advanced composite materials can be designed to meet these special requirements.

Mechanical and thermal design properties for typical spacecraft materials are presented in Table 16.6 [22]. We can see that graphite epoxy composites offer excellent strength, good stiffness characteristics, and low CTE. Composite materials can be tailored by design to achieve near-zero CTE and a wide range of strength and stiffness.

Figure 16.21 presents strength and stiffness characteristics for various graphite epoxy composite materials in comparison to several metals and matrix materials. Composite materials consist of high strength, high modulus fibers embedded in a homogeneous matrix material. Boron, silicon carbide, or graphite fibers are embedded in an epoxy resin matrix, aluminum, or other metals for metal matrix composites. Structural members generally are constructed by using multiple layers of thin-ply unidirectional fiber tapes. If all the layers and fibers are unidirectional, the structural member is *orthotropic*, and it is termed *zero-degree* (0°) fiber orientation. When layers are cross-plied at balanced, uniform orientations, the structural member becomes *isotropic*. Thus, a variety of strength, stiffness, and CTE properties can be tailored in the design of composite materials.

Figure 16.22 [23] shows the near-zero CTE range that can be obtained with graphite epoxy resin composites. Invar, quartz, and ULE (titanium silicate glass) are shown for comparison, and they demonstrate low CTE, but also low specific stiffness characteristics. Material properties for aluminum honeycomb sandwich structures with graphite composite face sheets are shown in Table 16.7. Again, near-zero CTE can be tailored with the proper combination of sandwich face sheets and core material.

As we mentioned, Kevlar (aramid) can be used in radar applications that may require RF electrical transparency or dielectric characteristics. Kevlar 49 has a relatively low CTE and tensile strength comparable to high strength graphites. Because composite materials are expensive, lower cost spacecraft materials ought to be used in structural design if mission requirements can be met with their less favorable properties.

Table 16.6
Spacecraft Material Properties [22]

Material	Material Type	Density, f (kg/m³ × 10³)	Longitudinal Ultimate Tensile Strength (N/m² × 10⁶)	Transverse Ultimate Tensile Strength (N/m² × 10⁶)	Longitudinal Tensile Yield Strength (N/m² × 10⁶)	Young's Modulus (N/m² × 10⁹)	Shear Modulus (N/m² × 10⁹)	Specific Longitudinal Ultimate Strength (N · m/kg × 10³)	Specific Stiffness, E/f (N · m/kg × 10⁶)	Specific Heat, C (J/kg · K)	Thermal Expansion α (10⁻⁶/K)	Thermal Conductivity K (W/m · K)
Aluminum, sheet	2014-T6	2.80	441	—	386	72	27.6	157.6	25.9	962	22.5	155
	2024-T36	2.77	482	—	413	72	27.6	174.2	26.1	879	22.5	121
	6061-T6	2.71	289	—	241	67	26.2	106.8	24.9	962	23.4	166
	7075-T6	2.80	523	—	448	71	26.9	187.1	25.4	837	28.9	134
Beryllium												
Extrusion	Be-38% Al	1.85	620	—	413	293	138	335.4	158.4	1862	11.5	179
Lockalloy	Cross rolled	2.10	426	—	431	186	138	203.2	88.6		17.0	212
Sheet		1.85	448	—	289	293	138	242.2	158.4	1862	11.5	179
Wrought	Hot pressed	1.83	275	—	179	293		150.6	160.1	1862	11.5	179
Boron epoxy												
[0]		2.01	1337	71		206	4.8	665.4	102.9	920	4.2	1.9
[0₂/±45]		2.01	717	107		115		356.7	57.6		4.6	0.4
Graphite/epoxy												
[0]	V_f 55%	1.49	1337	66		151	5.9	897.6	101.7		−0.36	
[0]	HTS	1.49	641	289		82		430.3	55.5			
[0₂/±45]	HTS	1.49	641	29		186	5.9	419.6	115.6			
[0]	HM	1.61	675			186	4.1	419.6	115.6			
[0]	UHM	1.69	620	20		289		367.1	171.3		−1.0	
[0₂/±45]	UHM											
Invar 36	Annealed	8.08	489		257	144	55	60.6	17.9	514	1.26	13.5
						44	16.5	124.9	25.3	1046	25.2	43.6
Magnesium extrusion tubes	AZ31B	1.77	221		110	44	16.5	152.0	25.3	1046	25.2	43.6
Magnesium sheet	AZ31B-H24	1.77	269	275	199	200	75.8	172.3	26.3	477	11.0	15.4
Steel PH15-7 MO	RH1050	7.6	1309		1171	200	75.8	110.0	25.5		11.3	38.1
4130 Chr Mdy	1350°F temp	7.83	861		710	200	75.8	110.0	25.5			
Ti6 Al-4 V sheet		4.43	1103		999	110	42.7	249	24.9	502	8.8	7.4
forgings and bar		4.43	1034		965	110	42.7	233.4	24.9	502	8.8	7.4
Kevlar 49 [0]		1.38	1378	29		75	2.1	999.06	54.9	1000	−4.0	1.7
Boron/Al [0]	50% Monolayer	2.60	1491	137	—	214	—	573	82		4	—

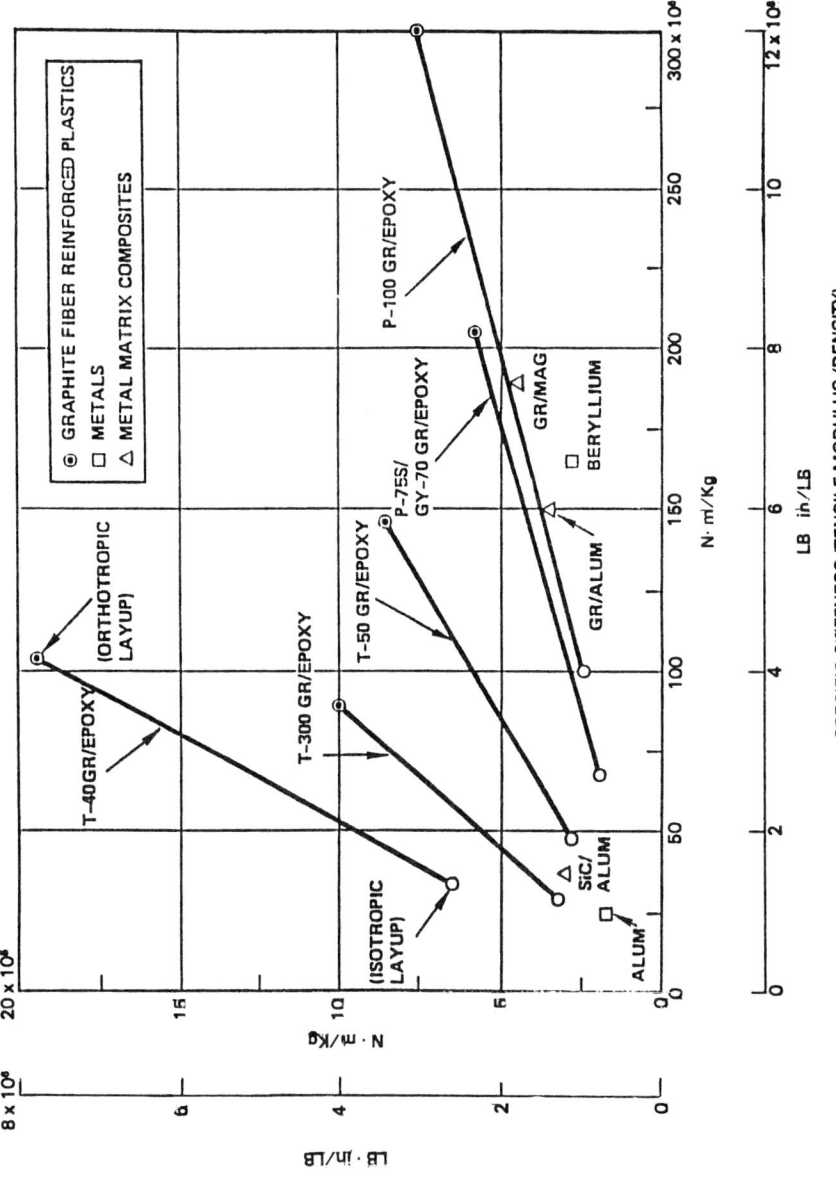

Figure 16.21 Advanced materials performance characteristics.

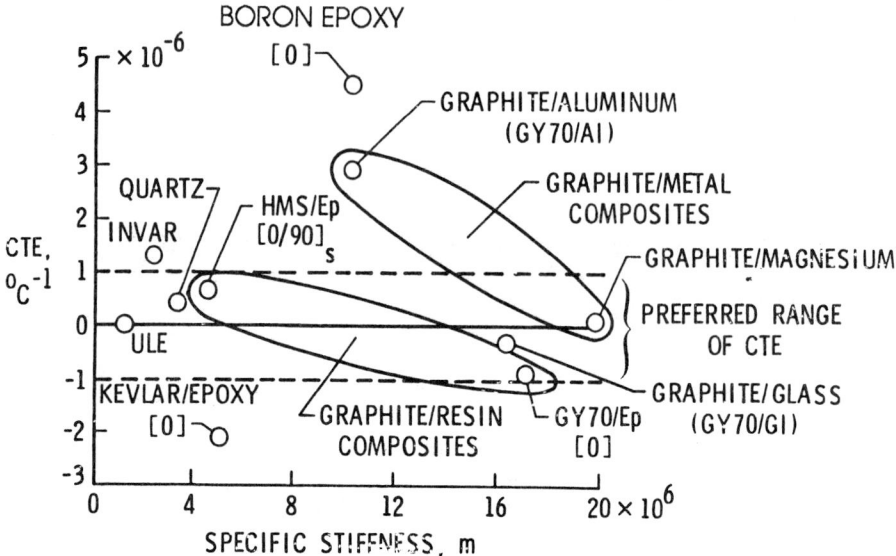

Figure 16.22 Low-CTE materials characteristics.

Table 16.7
Honeycomb Sandwich Properties

Fiber	Resin	Face Sheet			Core			Cure Process*	Sandwich Properties			Properties	
		Ply Thick (in)	No. of Plies	Orientation	Cell Size (in)	Density lb/ft³	Thickness (in)		Weight lb/ft² of Surface	Test Dir. (Core Ribbon Direction)	CTE in/in/°F × 10⁻⁶	E_t MSI	G MSI
GY70	934	0.0052	6	0, ±60, 90, ∓30	1/8	3.1	1/4	1	0.705	0	−0.06	15	2.9
		0.0054						2	0.715	90	−0.17	13	2.9
		0.0054	3	0, ±60	1/8	3.1	1/4	1	0.445	0 / 90	0.05 / −0.03	15	—
		0.0054	5	36, −72, 0, 72, −36	1/8	3.1	1/4	2	0.615	0 / 90	0.66 / −0.06	—	—
		0.0054	4	0, 90, 45, 135	1/8	3.1	1/4	2	0.525	0	0.3	—	—
		0.0054	4	0, 90, 90, 0	1/8	3.1	1/4	2	0.525	90	0		
GY70 (37 ends/in)	934	0.003	4	0, 90, 90, 0	1/4	1.6	1/4	1	0.271	0 / 90	0.21 / 0.09	21	
		0.003	3	0, ±60	1/4	1.6	1/4	1	0.21	0 / 90	−0.13 / −0.26		
GY70 (43 ends/in)	934	0.003	6	0, ±60, 90, ∓30	1/8	3.1	1/4	2	0.47	0 / 90	0.21 / 0.16		
		0.003	6	0, ±60, 90, ∓30	1/4	1.6	1/4		0.44	0 / 90	0.28 / 0.30		
										0	0.57		
										90 / 0	0.42 / 0.32		
										90	0.32		
GY70 (56 ends/in)	934	0.003	3	0, ±60	1/4	1.6	1/4	1	0.23	0 / 90	0.29 / 0.20		
		0.003	4	0, 90, 90, 0	1/4	1.6	1/4	1	0.28	0 / 90	0.24		

*1—Autoclaved faces, secondary bonded to core.
2—Single-stage layup and low-pressure autoclave cure.

REFERENCES

1. Freeland, R.E., and T.G. Campbell, "Deployable Antenna Technology for the Large Space Systems Technology Program," *Proc. AIAA/NASA Conf. Advanced Technology for Future Space Systems*, Hampton, VA, May 1979.
2. Mikulas, M.M., Jr., and H.G. Bush, "Advances in Structural Concepts," *Proc. Large Space Antenna Systems Technology*, NASA Conf. Pub. 2269 (Part 1), Hampton, VA, November–December 1982, p. 270.
3. Lenzi, D.C., and J.W. Shipley, "Mast Flight System Beam Structure and Beam Structural Performance," *Proc. NASA/DoD Control-Structures Interaction Technology*, NASA Conf. Pub. 2447 (Part 1), Norfolk, VA, November 1986, pp. 265–279.
4. Housner, J.M., "Structural Dynamics Model and Response of the Deployable Reference Configuration Space Station", NASA Technical Memorandum 86386, NASA Langley Research Center, May 1985.
5. Freeland, R.E., "Survey of Deployable Antenna Concepts," *Proc. Large Space Antenna Systems Technology*, NASA Conf. Pub. 2269 (Part 1), Hampton, VA, November–December 1982, pp. 414–417.
6. "Deployable UHF Spacecraft Antennas," TRW report prepared for Electronics Systems Division, US Air Force Systems Command, Contract F19628-74-C-0189, June 1974, pp. 6–8.
7. Brejcha, A.G., L.H. Keeler, and G.G. Sanford, "The Seasat-A Synthetic Aperture Radar Antenna," *Proc. Synthetic Aperture Radar Technology Conf.*, Las Cruces, NM, March 1978.
8. Freeland, R.E., "Survey of Deployable Antenna Concepts", *Proc. Large Space Antenna Systems Technology*, NASA Conf. Pub. 2269 (Part 1), Hampton, VA, November–December 1982, pp. 405–407.
9. Palmer, W.B., and M.M. Giebler, "Study of Advanced Sunflower Precision Deployable Antenna," TRW Rep. No. MEL-79-B-126, Jet Propulsion Laboratory, Contract No. 955340, November 1979.
10. Hedgepeth, J.M., and W.H. Greene, "Structures for Large Precision Reflectors", *Proc. Large Space Antenna Systems Technology*, NASA Conf. Pub. 2269 (Part 1), Hampton, VA, November–December 1982, pp. 361–380.
11. "Solar Concentrator Advanced Development Program," Task 1 Final Report, Harris Corp. Report prepared for NASA-Lewis Space Center, Contract NAS 3-24670, June 1986.
12. Friese, G.J., M. Thomas, and W.F. Hinson, "Inflated Antennas," *Proc. Large Space Antenna Systems Technology*, NASA Conf. Pub. 2269 (Part 1), Hampton, VA, November–December 1982, pp. 545–574.
13. Lang, J.H., "Electrostatically Figured Membrane Reflectors: An Overview," *Proc. Large Space Antenna Systems Technology*, NASA Conf. Pub. 2269 (Part 1), Hampton, VA, November–December 1982, pp. 575–582.
14. Woods, A.A., Jr., and N.F. Garcia, "Wrap-Rib Antenna Concept Development Overview," *Proc. Large Space Antenna Systems Technology*, NASA Conf. Pub. 2269 (Part 1), Hampton, VA, November–December 1982, pp. 423–468.
15. Sullivan, M.R., "Hoop-Column Antenna Development Program," *Proc. Large Space Antenna Systems Technology*, NASA Conf. Pub. 2269 (Part 1), Hampton, VA, November–December 1982, pp. 469–512.
16. Russell, R.A., T.G. Campbell, and R.E. Freeland, "NASA Technology for Large Space Antennas," AGARD Rep. No. 676, *NATO 49th Structures and Materials Panel Meeting*, Porz-Wahn, Federal Republic of Germany, October 1979, pp. 2–17.
17. Fager, J.A., "Status of Deployable Geo-Truss Development," *Proc. Large Space Antenna Systems Technology*, NASA Conf. Pub. 2269 (Part 1), Hampton, VA, November–December 1982, pp. 513–543.

18. Freeland, R.E., "Survey of Deployable Antenna Concepts," *Proc. Large Space Antenna Systems Technology*, NASA Conf. Pub. 2269 (Part 1), Hampton, VA, November–December 1982, pp. 387–389.
19. Coyner, J.V., Jr., "15-Meter Deployable Aperture Microwave Radiometer," *Proc. Large Space Antenna Systems Technology*, NASA Conf. Pub. 2269 (Part 1), Hampton, VA, November–December 1982, pp. 131–155.
20. Heard, W.L., Jr., J.J. Watson, J.L. Ross, S.C. Spring, and M.L. Cleave, "Results of the ACCESS Space Construction Shuttle Flight Experiment", *AIAA Space Systems Technology Conf.*, Paper No. 86-1186 CP, San Diego, CA, June 1986.
21. Akin, D.L., "EASE, Experimental Assembly of Structures in EVA, Overview of Selected Results," NASA CP2490, October 1987, pp. 199–218.
22. Agrawal, B.N., *Design of Geosynchronous Spacecraft*, Prentice-Hall, Englewood Cliffs, NJ, 1986, p. 245.
23. Blankenship, C.P., and J.C. Yu, "Structures and Materials Technology for Space Station," AIAA SDM Issues of the International Space Station, Paper No. 88-2446, Williamsburg, VA, April 1988.
24. Archer, J.S., "High Performance Parabolic Antenna Reflectors," *AIAA 7th Communications Satellite Systems Conf.*, Paper No. 78-593, San Diego, CA, April 1978.
25. Hill, H., D. Johnson, and H. Frauenberger, "Development of the Lens Antenna Deployment Demonstration (LADD) Shuttle-Attached Flight Experiment," *Proc. NASA/DoD Control-Structures Interaction Technology*, NASA Conf. Pub. 2447 (Part 1), Norfolk, VA, November 1986, pp. 125–144.
26. Takamatsu, K.A. and J. Onoda, "The New Deployable Truss Concepts for Large Antenna Structures or Solar Collectors," *AIAA/ASME/ASCE/AHS/ASC 30th SDM Conference*, Paper No. 89-1346-CP, Mobile, AL, April 1989.
27. "France Delivers Space Structure," *Flight International*, August 1988, pp. 24–25.

INDEX

Absolute calibration, 290
Acceleration, 3
Accuracy, 3, 13, 27, 505, 506
    of measurement, 290
    of near field test, 524
Adaptive nulling, 373–377, 442–451, 465–475
Address generator, 539–544
Air traffic control, 2, 26, 32, 42
Airborne moving target indicator (AMTI), 373–377, 413–442
Airborne pulsed doppler, 373–377, 413–442
Aircraft 14, 17–18, 21, 30
Algorithm, 548–555, 560–562
    DFT, 560–562, 563
    FFT, 560–562, 563
    WFTA, 560, 562
Algorithm simulation, 531–532, 549
Algorithms, wind vector, 295–299
Alignment, 506
ALSE, 23–26
ALTAIR 1, 83–90
Altimeter, 17, 21, 23, 28, 29, 229–280
Altimeters, 29, 484
Altitude, 57
Aluminum, characteristics, 500
Ambiguity diagram of radar altimeters, 230–232
Amplitude calibration, 290
Amplitude measurement, 289
AMTEC system, thermoelectric, 610
Analog-to-digital converter, 557–560, 564
Analog-to-digital conversion, 146, 152, 153
Angle resolution, 284
Angle speed discrimination, 286
Angular momentum, 48
Annular coverage, 72, 74
Anomoly
    eccentric, 51–52, 57

    mean, 51
    true, 48, 54
Antarctic, 19
Antenna, 7, 9, 14, 15, 20, 21, 24, 25, 30–33, 36, 37–39, 42–43, 44–46, 481–500, 502, 518
    requirements, 481
    types I, 482
    types II, 483–484
    types III, 484–489
    testing, 524
    designs, 498–499
    lens, 489–494
    future, 524–525
Antenna dimension, 122, 132, 151
    noise, 136, 138
    radar, 131, 139
Antenna structure tolerance, 512
Antenna technology projection, 524–525
Antennas for radar altimeters, 233–236
Apocenter, 21
Apogee, 4, 19, 49, 50
Apollo Lunar Module Radar Altimeter, 262–264
Apollo Lunar Sounder Radar, 2, 23–26
Apollo Rendezvous Radar, 481–482
Apsidal rotation, 60, 62
ARC, 292
Architecture
    computer, 533–537
    Harvard, 534–535
    multiport memory, 535–537
    processor, 549
    Von Neumann, 533–534
Arctic, 19
Argument of perigee, 54, 67, 71
Arithmetic realization, 554–555
Array antenna, 14, 27, 35

Array processors, 537–539
Arrays, 35, 499
Astronaut, 637, 650, 666–667
ATC radar parameters, 32–33
Atlantic Research Corp., 499–520
Atmospheric effects, 10
Atmospheric absorption, 349
ATS 499
ATS-6 reflector, 501
Auroral scintillation, 83–85
Auxiliary antennas, 373–377, 451–464
Auxiliary circle, 52
Azimuth, 53, 55, 63
Azimuth-over-elevation, 482

Backscattering cross section per unit surface area, 233–236
Ball Aerospace, 515, 643
Ballard, 4
Ballistic missile detection, 7
Bandwidth, 156, 518
Bandwidth-aperture dispersion, 451–464
Barton, D.K., 42, 372–395
Batteries, system characteristics, 572–573, 574
Beams
   bending stiffness, 638
   coilable longeron, 639
   erectable, 637
   folding, 666
   hinged longeron, 639
   linear deployable, 659–666
   truss, 638
Beam space SLC, 451–464
Beamwidth, 127, 284, 483–484
Beryllium characteristics, 500
Beste, 4
Binary integration, 113
Binomial distribution, 100
BIPAR, 173, 181, 188, 189
BISAR, 91
BMEWS, 7
Boltzmann's constant, 377–395, 442–447
Bombers, 7
Booms
   articulated longeron, 661
   canister-deployed, 663
   continuous Longeron, 661
   FAST Mast, 663
   folding beam, 665
   lanyard-deployed, 663, 665
   lattice mast, 660
   linear deployed, 661–665
   retraction, 663
   STEM, 665
   telescoping, 665
Box truss, 659
Bragg resonance, 294
Bragg scattering, 293–294
Brayton cycle (Closed), 585
   combined rotating unit, 585
Brookner, E., 34
Bujakas, 522
Burst combining, 109–117
Butterfly operation, 560–562

Calibration, 44–46
   external, 291
   internal, 291
   known-cross-section target, 292
Cancellation ratio (CR), 442–447, 477–451, 465–472
Capillary pumped loop (CPL), 344
Cascaded analog to digital canceller, 373–377, 465–472
Cassegrain feed, 482
Cassini mission, 29
CCD, 156–157
Celestial sphere, 32
Challenger, 2
Channel coherence function, 94–95
Chemical reactor, burst power, 607–608
Chi-square distribution, 92–93
Chi-squared, 290
Circular orbit, 7, 66, 69
Clutter cancellation, 4, 9, 42–43
Clutter cross section per unit area (σ°), 377–395
Clutter cross section per unit volume, 377–395
Clutter doppler spectrum, 377–395
Clutter locking, 413–442 425
Clutter power, received, 377–395
Clutter rejection, 373–377
Clutter-to-noise (C/N) ratio, 134, 143, 377–395, 472–475
Clutter velocity standard deviation, 377–395
CMOS technology, 157, 551–554
Coefficient of thermal expansion (CTE), 667, 669, 672
Coherence bandwidth, 88, 95
Cold plate (see Heat acquisition)
Command interpreter, 544–548

Complex arithmetic unit, 544–548
Component thermal requirements
   radar system components, 320–325
   electrical power components, 334, 335
Composite materials, 670
Compression & coding techniques, Huffman, 153
Computer architecture, 533–537
Constant propagation channel, 92–95
Constellation, 4, 5, 72, 74–77
Constants, 61
Continuous coverage, 7, 74, 78–79
Contractors, SBR, 26
CONUS, 34
Conversion factors, 61
Cooled reactor, 41–42
Cooley-Tukey fast Fourier transform (FFT) algorithm, 532–539
Coordinate systems, 51
   bistatic, 178
Cooper, characteristics, 493, 500
Corner reflector, 145
Corporate-fed antenna, 14, 32
Correlation function of received complex voltage, 94
COSMOS-1500 SBR, 18, 19–21, 521–523
Cross section, bistatic, 187
CTE, 498–499
Cumulative distribution of received power, 92–93

DAFE, 499, 502–503, 505
Data calibration, 162
   collection, 149
   rate, 130–131, 146, 153
   recorders, 139, 152–155
Delay commutator circuit, 551–554
Deployable antenna flight experiment (DAFE)
   Harris Corp., 449–520
   Grumman, 499–520
Deployable antenna parameters, 519
Detection, 19, 27
Detection sensitivity, 115–117
Digital beam forming, 557–564
Dipole antenna, 23–26
Dipole plane, 498
Direct matrix inverse (DMI), 465–472
Direct memory access (DMA), 544–548
Discontinuities, subsurface, 23
Discrete Fourier transform (DFT), 560–564

Discrimination radar, 35
Displaced phase center antenna (DPCA), 377–395, 413–442, 465–475
Displacement, 506
Distortion, 484, 491, 493, 495–498, 504–505
DNA wideband satellite, 86–87
Doppler, 130, 141–142, 162
   ambiguities, 395–413
   beam sharpening (DBS), 413–442
   derived range rate, 204
   frequency, 282
   frequency contours, 282
   shift, bistatic, 177, 182, 184
   wind radar, 312–315
Double coverage, 74–76, 78
Double threshold detection, 96–97
DPCA, 43, 430
   analog-to-digital conversion errors, 413–442
   antenna deformation errors, 413–442
   phase center offset errors, 413–442
   receiver mismatch errors, 413–442
Dual drum, 499–520
Duty cycle
   effect on thermal control, 334
Dwell time, 143
Dynamic range, 554–555

Earth-centered inertial, 51–52
Earth clutter
   effects on rendezvous radar, 211–214
Eccentric anomoly, 51
Eccentricity, 69–70
Effective radar cross section, target, 377–395
Effective radar cross section, clutter, 377–395
Eggbeater, 66–67
EIA, 35, 37
Electronic counter-countermeasures (ECCM), 447–451
Electronic countermeasures (ECM), 447–451
Electrostatically figured, 651
Element Space (SLC), 451–464
Elevation, 53, 54
Elevation-over-azimuth, 482
Ellipse, 49
Elliptical orbit, 60
Emerson weights, 413–442
Energy, 48
Enthalpy-entropy diagram (supercritical cycle), 603, 604
Environment, 135

Environment, space radiation, 42
EOS, 28-29
Epoxy, graphite, composite, 495
Equator, 77
Equatorial orbits, 7
ERS-1, 28, 138
    scatterometer, 29
Equatorial radius, 58, 61
Equatorial spread, F, 89
Equatorial plane, 52
Equations (orbit), 56-58
Equinox, 52
Erectable antennas, 499-521
Error,
    amplitude, 483
    phase, 483, 491-493
    module, 483-484
European Space Agency (ESA), 29
Escape velocity, 50
ETR, 40
EVA, 28
Evaporator, 344
Execution unit, 544-548
Exponential probability density function, 373-377
Extendable support structure (ESS), 643
Extended interaction amplifier, 35, 37
Fast Fourier transform (FFT), 413-442, 560-562
FEM (free electron maser), 36
FFT processor implementation, 549-550
Fixed point arithmetirc, 554-555
Flex-rib, 499-521
    deployment, 499-521
Flight path angle, 49, 55
Floating point arithmetic, 554-555
Fluid loops, electromechanically pumped two-phase, 344
Fluid loops, single phase, 347
FMCW radar altimeter, 230-232, 245-252
Footprint velocity, 184
Four-port memory, 539, 539-544
Free piston stirling engine (FPSE)
    solar powered, 596
    reactor powered, 596
Frequency agility, 413-442
Frequency hopping, 96-97
Frequency for radar altimeters, 233-237, 277-279
Future antenna technology, 524-525

Future SBR systems, 37
Fuze, 40

GaAs, 157
Gain, 484, 495, 522, 524
Gain loss, 495, 498
Gain margin, 451-464
Gaussian probability density function, 377-395
Gemini, 2, 481
General assembly of the United Nations (UN), 37
General correlation function, 102-104
General Dynamics, 506
    graphite epoxy reflector, 400-521
    space erectable antennas, 508
Generic signal processors, 532-548
Geocentric latitude, 52-53, 55
Geographic distribution of ionospheric scintillation, 85-90
GEOID, 14
Geometric theory of diffraction (GTD), 352
Geometry, bistatic, 176
GEOS, 15-18
GEOS-C, 2, 18, 483-484
    antenna, 483-484
(GEOS-3) radar altimeter, 268-272
GEOSTAT radar altimeter, 276-277
Ginsberg, 4
Global coverage by polar orbit, 10
Gold, characteristics, 493
GPS (global positioning system), 162
Gram-Schmidt orthogonalization, 465-472
Graphite, 495, 500, 504, 507
Graphite-epoxy, characteeristics, 493, 500
Graphite fiber reinforced plastic, 667, 670
Gravitational constant, 61
Grazing angle, 126, 377-395
    limitations, 5
Ground trace, 66, 69, 70, 80
Ground-plane, 498
Ground resolution, 38
Grumman Aerospace Corp., 505-506, 510, 512-513, 643
    70 m Diameter Lens, 499-521

Hardening, 10
Harris, 504-506
Harvard architecture, 533-537
Heat acquisition devices, 342
Heat acquistion and transport, 342-343
Heat exchanger, 342

Heat pipe, 41
Heat rejection, 340, 347
Heat transport capacity, 344
Heat transport length, 347–348
Hex panels, 650
Hoop-column, 504, 657
   technology, 499–521
Howells-Applebaum control loop, 465–472
Hughes Aircraft Co., 10

Imaging system, coherent, 23, 25
Inclination, 54
Independent bursts, 100–101
Independent samples, 281
Inflatable antenna, 651
Integrated circuits, hardening, 10
Integrated main lobe-to-sidelobe ratio, 377–395
Integrated radar and communication subsystem (IRACS), 200–204
Interference, 42, 43
Interference rejection, 373–377, 465–472, 472–475
International satellite monitoring agency, 36, 37
Invar, characteristics, 493, 500
Inverse matrix update (IMU), 465–472
Inverse synthetic aperture (ISAR), 121, 141
IRACS, 2, 10, 13, 200–205
Irregularity production, 85–90
ISMA radar, 37, 38
Isodops, 282–283, 395–413
Issues, SBR, 38–39

Jamming effective radiated power density, 442–447
Jamming power, received, 442–447
Jamming-to-noise ratio (J/N), 442–447
Japanese, H2, 29–30
Jointly normal variates, 14–15
JPL, 14, 29, 499

Kalman methods, 465–472
Keplerian motion, 47
Kevlar, 668, 672
   characteristics, 493
Kosmos 1500, 39
KRT-10, 521–523

L-band, 138
LASBR, 515, 518
   Ball, 499–521, 518
   Grumman, 499–521, 513
Latitude, 52–53

   geocentric, 52, 55
Launch azimuth, 63
Launch costs, 39–40
Launch vehicle capabilities, 43
Launch window, 65–66
Leakage of COSMOS 1500 waveguide, 521–523
Least-mean-squares (LMS) algorithm, 413–442
Lens antenna
   requirements, 489–494
   surface tolerance, 489–494
   geometry, 490
   element location error, 489–494
   F/D *versus* tolerance, 493
Lens array, 643, 644
Lens, tolerance, 489–491
Leondes, 4
Lincoln laboratory, 34, 549–550
Lithium-iron disulfide, battery, 578–579
LMSC antenna for LSST, 499–521
Lockheed, 34, 499, 501
Log-normal probability density function, 377–395
LORRA, 30, 31
Low altitude SBR, 502, 518, 573
Lower dipole plane, 498
LSST, 506
Luders, 4
Lunar material loss tangent, 25

M out of N detection, 83–85, 100
Mahoney, 34
Main beam clutter, 377–395
   distortion, 472–475
   footprint, 373–377
   jamming, 472–475
Major axis, 60
Marcum-Q function, 99
Mars, 29
Maser, 36
Mass of the earth, 34
Matched clutter filter (MCF), 413–442
Materials, stable RF, 500
Maximum range, 371–395
Maxwell Montes, 23
Mean anomaly, 57
Mean motion, 51
Measurement ranges, 370
Mechanical characteristics, antennas, 517
Membrane, 640, 641, 648, 651
Mesh, 504

deployable, 652–659
  double, 652
  single, 652
  tricot, 641, 652
  welded, 641
Metal matrix composite, 670
Meteorite, mass, size, 485
  environment, 485–489
  velocity, 485–489
  penetration thickness, 485–489
Method of moments (MOM), 351
Microcontroller, 539–543
Microstrip, 517
Military SBR, 33–35
Miller, R.W., 413–442
Millimeter wave, 35
Minimum detectable velocity (MDV), 413–442
Module-errors, 483–484
Molniya, 68, 70
Molybdenium, characteristics, 493
Monitoring satellite SBR, 36–38
Monostatic radar, 92–93
Monte Carlo simulation techniques, 102–103
Motorola, 40
MPP, 60
MTI improvement factor, 377–395, 413–442
Multiple missions, 7
Multiplier accumulator (MAC), 539–544, 541, 563
Multi port memory architecture, 533–537

NADIR hole, 5, 7, 14, 37
Nakagami-m distribution, 83–85, 90–91, 92–93, 102–103
National Space Agency of Japan, 28
NASA, 2, 14, 23, 26, 29, 30, 153, 158, 160, 485, 499, 503–506
NASA-JPL, 499
Nautical mile, 4
Near field testing, 524
Neyman-Pearson hypothesis test, 377–395
Node, 55
Noise
  solar system sources, 225–226
Noncoherent integration, 96–97, 101–102
Nonfluctuating target, 98–99
NSCAT, 29
Nuclear explosions, 84
Nuclear prime power, 41–42
Null stability errors, jamming, 442–447

Oblateness, 58
Ocean waves, 292
OMS, 43
OMV (orbital maneuvering vehicle), 481
On-axis, 524
On-board processor, 524
One-phase fluid loop, 347
One-way propagation, 92–93
Orbit, bistatic, 167, 180
Orbit selection, 3–5
Orbital elements, 53
Orbital maneuvering vehicle (see OMV)
Orbital plane, 65
Organic rankine cycle, 579–584
Orientation, spacecraft, 306

Pantograph, 643, 666
Parallel processors, 537–539
Parasitic radar, 173, 188, 189
Partially adaptive array, 451–464
Particles, meteoroid, 489
Patera, Cleopatra, 23
Path, direct, 176
Path length, bistatic, 177
Path scatter, bistatic, 176
Payload separation distance, 176
Pencil-beam scatterometers, 615
Penetration, thickness, 485, 486
Penetrations, meteoroid, 484–489
Perforations, meteoroid, 484–489
Perforations, area, 484–489
Perigee, 4, 49, 50
Period, 57
  of satellite, 4
Phase center matching, 451–464
Phase change material (PCM), 348
Phase shift keyed radar altimeter, 230, 252–257
Phased array, 2, 9, 34
Physical theory of diffraction (PTD), 352
Pioneer Venus Orbiter radar altimeter, 274–276
Planar array, 643
Platform, polar, 30
Polar orbits, 19
Polyconic, 499
  deployment, 499–521
Polyus-V, 21–23
Potassium rankine cycle, 605–609
Power, 136, 152
Power spectrum of radar altimeters, 230–232
Precision of measurement, 290

Pretensioned cables, 504
PRF Maximum,
 no range ambiguities, 395-413
 no doppler ambiguities, 395-413
Prime power, 41-42
 systems, 568
Probability of false alarm, 99
Processor architecture design, 549
Processor development, 549
Processor, digital, 160
 ground, 155, 156, 160
 on-board, 160
Processor implementation, 544-548
Programmable signal processors, 537-539
Propagation time, bistatic, 176
Protons, 10
Pulse compression radar altimeter, 242
Pulse-compression ration, 138
Pulse
 energy, 133
 length, 127, 129, 131
 ranging, 122, 124
 repetition frequency (PRF), 127, 129, 132, 133, 143

Quadrature detector, 97-98

Radar, 146, 481-483
 antenna, 123, 131, 138, 139
 beam, 124, 127, 131
 dimension, 122
 tracking performance, 96-97
Radar burst, 96-97
Radar cross section (RCS)
 aircraft, 362
 bomber, 364
 cars, 369
 cylinder, 354
 dihedral reflectors, 366
 E-band horn, 354
 fighter, 362-363
 flat plate, 354
 parabolic dish, 361
 satellites, 354, 363
 ships, 366, 368
 solar panels, 355
 tanks, 369
 telescope, 359
 thermal blanket, 356
 trihedral reflectors, 366-367
 trucks, 369

X-band horn, 354, 357
Radar Look, 96-97
RADARSAT, 29, 138
RADC, 506
Radial Rib, 499-521, 653
Radiation environment effects, 10
Radiators (see Thermal radiators)
Radio telescope, 521
Radiometer, scatterometer, 304
Random phase errors, 489-494, 495
Range ambiguities, 402-405
Range-angle discrimination, 286
Range on earth from nadir, 395-413
Range resolution, 395-413
Range-speed discrimination, 288-289
Range tracker for short pulse radar altimeter, 237-242
Rankine cycle, organic, 579
Rankine cycle, potassium, 605
Rationale for SBR, 7-8
Rayleigh distribution, 83-85
Rayleigh probability density function, 377-395
Rayleigh region, 350
Raytheon, 512
RCS statistics, 370
Real antenna, 484
Real aperture radar, 309-312
Real beam, 123
Receiver noise factor, 444
Received signal power, 91
Receiver, 136
 mismatch errors, 413-442, 465-472
 model, 97-98
 noise, 136
Recorders, ground, 155
Reflectivity, 483
Reflector, 484, 495, 499, 507
 electrostatic, 651
 erectable, 637, 650, 666-667
 hex deployable, 650
 inflatable, 651
 mesh deployable, 637, 652-659
 mesh surface, 641
 single piece, 637, 641-643
 solid deployable, 637, 643
 solid surface, 637, 640
 sunflower, 649
Registered arithmetic unit (RALU), 539-544
Relative calibration, 29
Remote sensing, 2, 13, 28, 29

Rendezvous
    maneuver, 197–200
    missions, 198, 204–206
    radar systems, 196–226
Rendezvous radar, 1, 2, 13, 26, 27
    antenna noise temperature, 225–226
    background noise sources, 225–226
    effects of earth clutter, 211–225
    frequency agility, 211–225
    radar-earth geometryy, 214–216
Rendezvous radar antennas, 20
Requirements, 2, 3, 5, 15
Requirements, antenna, 481–498
Resolution, 284–286
    along-track, 124, 131, 134, 139
        azimuth, 123
        ground range, 151
        range, 122, 131, 151
        slant range, 151
        spatial, 124, 126
    doppler, bistatic, 183
    radiometric, bistatic, 187
    requirements, 38
    spatial, bistatic, 180, 182
RF losses, 138
Rhodium, characteristics, 493
Rician probability distribution, 97–98
Rim, temperature, 498
rms amplitude errors, antenna, 395–413
rms phase errors, antenna, 395–413, 495
Robotics, 637, 650, 666–667
Rockwell, 499
Rollout, antenna, 512
Rorsat, 33
Rosette constellation, 32

S-209 radar, 23, 25
Salyut, 521
SAMS, 499–521
Sandwich construction, 640, 672
SAR, 1, 10
    spotlight, 483–484
    unfocussed, 289
Saturn, 29
Saturn rocket radar altimeter, 258–260
SBR (space based radar), 481–498, 510–513, 515
Scanning scatterometer, 304
Scattering coefficient, ocean, 289
    aximuthal variation, 294
    windspread variation, 293–294
Scattering measurement, 281
Scatterometer, definition, 289
    systems, 299
Scatterometry, 289
Scintillation, 83–85
    index, 92–93
SEASAT, 121, 123, 125, 131, 134, 138, 152, 156, 160, 643
    advanced panel, 485
    objective, 14
    parameters, 17
    subsystems, 14
    transmitter, 14
SEASAT-A, 515
SEASAT-A radar altimeter, 272–274
SEASAT antenna mechanical characteristics, 499–521
SEASAT SASS scatterometer, 305–307
SEASAT wind vector results, 297–299
Second-order fading statistics, 94–95
Semimajor axis, 49, 50, 57, 68
Semilatus rectum, 42
Shape adjustment
    active, 644, 668
    electrostatic, 651–652
    post fabrication, 643, 649, 668
Shielding, 485
Short pulse radar altimeter, 230–232, 237–242
Shuttle, 2, 9–10, 13, 481, 482, 637
    bay, 650
    compartment, 657
    envelope, 649
    truss experiments, 666
Shuttle, deployed, 499
Sidelobe, 483–484
    blanking, 442–447, 447–451
    cancellation (SLC), 442–447, 447–451, 465–472, 472–475
    clutter, 373–377, 377–395, 413–442
    deception, 447–451
    jamming, 373–377, 447–451
    masking, 447–451, 451–464
Side-Looking, 20, 483–484
Sigma Zero ($\sigma°$), 289
Signal-to-interference ratio (SIR), 442–447
Signal processing, 135, 144, 146, 157
Silver, characteristics, 493
Single burst detection, 99
Single membrane lens, 499–521
Single-phase fluid loop, 347

SIR, 10, 15, 28, 134, 138, 152, 155
SIR-A, 2, 15
  antenna, 499–521
SIR-B, 2, 15
  antenna, 499–521
SIR-C, 28
  antenna, 499–521
SIR-D, 28
Skylab radar altimeter, 18, 264–265
Skylab radscat, 302–305
Slant range, 377–395
SLC convergence, 472–475
Slotted waveguide antenna, 521–523
  array, 19
Slow fading, 94–95, 96–97
SLR, 19
Smart memory, 544–548
Solar dynamic system, CBC, 585
Soil moisture, 309
Sounder, 24–25
SP-100 baseline system, 615
Space, 481–498, 510–513, 515
Space environment, 325–326
  earth infrared heating, 326
  albedo heating, 327
  solar heating, 327, 328
  eclipse, 333
Space-deployable, erectable, 506
Space-fed lens, 512
  antenna temperatures, 498
Space Shuttle rendezvous radar, 482–483
Space Station tracking system, 28
Space transportation system (see STS), 197, 200
  rendezvous radar, 197–200, 200–204
Spacecraft
  Apollo, 197
  Challenger, 197
  Gemini, 197
  man maneuvering unit (MMU), 198
  orbital manuevering vehicle, 205, 206
  space transportation system (STS), 197, 200
Special purpose signal processor development, 548–555
Stable RF system, 500
Standard cell CMOS technology, 551–554
Stereo, radar, 172
Stirling engine, 593
Stretch pulse compression radar altimeter, 242–244
Strip mapping, 139

Structures
  areal, 637, 640–659
  deployable, 643–659
  design considerations, 667–668
  erectable, 637, 666–667
  ESS, 643
  linear, 637, 638, 659–663
  materials, 669
  mesh, 637, 652–659
  solid surface, 637
  thermal control, 668
STS, 10–13, 197, 200
Subsurface reflection depths, 25
Superconductor, 157
Supercritical cycle, 602–605
Surface accuracy measurements system (SAMS), 505
Surface tolerance, 489–491, 495–496
Surface precision, 637, 652
Surveyor lunar lander radar altimeter, 260–262
Survivability, 41
Swathwidth, 122, 129, 131
  bistatic, 181
  radius (SWR), 442–447
Swerling II model, 91, 104–108, 115–117
Synthetic aperture radar, 1, 2, 28, 29, 373–377, 132-165
  413–442
  bistatic, 191
System input output unit, 544–548
System noise temperature, 377–395
Systolic processors, 537–538

Target
  clutter, 135
  cross-section, 134
  visibility, 135
TDRS, 158
TDRSS, 13
TDRSS antenna, 499–521
Technology, antenna, 481
Technology selection, 549
Temperature
  entropy diagram-CBC, 585–593
  entropy diagram-ORC, 579–584
  on lens antenna, 498
  space fed lens, 498
Thermal bus
  definition, 348
Thermal control system components
  capability-pumped loops, 344

heat acquisition (cold plates), 323, 342, 343
heat pipes, 342
pumped one-phase fluid loops, 347
two-phase fluid loops, 344
thermal radiators, 332
thermal storage, 347–348
Thermal distortion, 495–496
Thermal management systems (TMS), 341
Thermal radiators, 342
Thermal storage, 348
Thermoelectric conversion, alkali metal, 610
Thermoelectric conversion cycles, 611–616
Thermionic conversion systems, 616–620
Thickness shielding, 42
Three-wavelength SBR, 26
Time averaged clutter, coherent airborne radar (TACCAR), 413–442
Time of flight, 50–51
Titanium, characteristics, 493
TIROS, 2
Titan, 29
Titan IV, 30
Tolerance, 495
  lens antenna surface, 489–491
Tooling, 643, 668
TOPEX, 29
T/R module characteristics, 33
TRAC, 499
Tracking and data relay satellite system, 13
TRAMAR, 30
Transmitter, 136
  geostationary, 172, 193
Truss
  beams, 638, 643
  booms, 661
  box, 659
  rib stiffend, 641
  tetrahedral, 657
TRW, 2, 487, 499–521
  antenna, 499–521
Two-phase fluid loop, 344, 347
Types of SBR, 481
  Type I, 2, 482
  Type II, 2, 483
  Type III, 3, 484

Ulsamer, 34
United Nations (UN), 26, 36–38
US space deployable antennas, 499–521
USSR space deployable antennas, 521–523

Van Allen Belt, 10
Varian, 35
Velocity of satellite, 3, 4
Venera, 21, 29
Venus, 21, 29
VLSI chip development, 556
VLSI components, 539–554
Von Neumann architecture, 533–537
Walker orbit, 4, 5
Walker's notation, 77
Wavelength, 121, 151
Waves, ocean, 292–293
WBMOD computer code, 115–117
Widrow's LMS algorithm, 465–472
Wind-vector algorithms, ocean surfaces, 295–297
Wind-vector measurement, ocean, 292–293
Wind speed variation, scattering coefficient, 294–295
Wind vector alias, 297–298
Winograd Fourier transform algorithm (WFTA), 560–562
Wire-wheel array, 648
WR-75, 499
Wrap rib, 652
WTR, 40

X-Band, 138, 153

Yvpatoriva, 21

900180